Near-Infrared Technology
in the Agricultural and Food Industries

Edited by
Phil Williams
Karl Norris

Published by the
American Association of Cereal Chemists, Inc.
St. Paul, Minnesota, USA

Cover designs illustrate the influences of particle size (front) and moisture (back) on the near-infrared spectra of wheat.

Library of Congress Catalog Card Number: 87-072083
International Standard Book Number: 0-913250-49-X

Second Printing, 1990

Reference in this publication to a company or product name by personnel of the U.S. Department of Agriculture is intended for explicit description only and does not imply approval or recommendation of the product by the U.S. Department of Agriculture to the exclusion of others that may be suitable.

Printed in the United States of America

American Association of Cereal Chemists, Inc.
3340 Pilot Knob Road
St. Paul, Minnesota 55121, USA

This monograph is dedicated
to Maxine and Corrie.

CONTRIBUTORS

F. E. BARTON II, U.S. Department of Agriculture-Agricultural Research Service, Russell Research Center, Athens, Georgia, U.S.A.

G. S. BIRTH, U.S. Department of Agriculture-Agricultural Research Service, Russell Research Center, Athens, Georgia, U.S.A.

T. FEARN, Flour Milling and Baking Research Association, Chorleywood, Herts, England

H. HECHT, Chemistry Department, South Dakota University, Brookings, South Dakota, U.S.A.

W. HRUSCHKA, U.S. Department of Agriculture-Agricultural Research Service, Instrumentation Research Laboratory, Beltsville, Maryland, U.S.A.

M. IWAMOTO, National Food Research Institute, Ministry of Agriculture, Forestry and Fisheries, Yatabe, Ibaraki, Japan

V. P. KRISCHENKO, Central Institute for Agrochemical Services, Moscow, U.S.S.R.

H. MARTENS, Norwegian Computing Center, Oslo, Norway

W. F. McCLURE, Department of Biological and Agricultural Engineering, North Carolina State University, Raleigh, North Carolina, U.S.A.

I. MURRAY, North Scotland College of Agriculture, Aberdeen, Scotland

T. NAES, Norwegian Food Research Institute, Aas-NLH, Norway

K. H. NORRIS, U.S. Department of Agriculture-Agricultural Research Service, Instrumentation Research Laboratory, Beltsville, Maryland, U.S.A.

B. OSBORNE, Flour Milling and Baking Research Association, Chorleywood, Herts, England

J. A. PANFORD, Grain Research Laboratory, Canadian Grain Commission, Winnipeg, Manitoba, Canada

S. G. STEVENSON, Grain Research Laboratory, Canadian Grain Commission, Winnipeg, Manitoba, Canada

R. TKACHUK, Grain Research Laboratory, Canadian Grain Commission, Winnipeg, Manitoba, Canada

P. C. WILLIAMS, Grain Research Laboratory, Canadian Grain Commission, Winnipeg, Manitoba, Canada

PREFACE

Near-infrared reflectance technology is the most practicable and exciting analytical technique to hit the agricultural and food industries since Johann Kjeldahl introduced the Kjeldahl test. Furthermore, it's quicker and cheaper than other tests and doesn't use sulfuric acid or sodium hydroxide. Some things are easier to measure with near-infrared than others. Water is easy and methionine is harder.

Imagine yourself walking through a forest. You see lots of trees, big ones and small ones, lots of shrubs and flowers and some grass. You've seen them all before and know what they are. By and by, you see a flower you've never seen before. So you take a book from the library and find out just what that flower is. Then you go back to the forest and, after walking a long time, you see another flower that is the same. You think you know what it is, but you are not quite sure. So you go back to the library and check the book out, then back to the forest. After a while, you get to know that rare flower. Meanwhile, you have seen lots of the trees and shrubs and other flowers that you recognized the first time. It has taken more energy and reading (self-programming) to get to know that rare flower, but then you know it. The observations you have made can be regarded as optical data used for calibration and the reference book as the analytical (wet chemistry) data. Despite the large number of optical data collected during your search, your recognition of the flower will only be as reliable as your reference book.

Near-infrared reflectance spectroscopy also involves optical data, the amount of radiation reflected from a sample relative to the radiation striking it. If you try to look into a room from the outside on a sunny day, it is not easy to see the objects inside the room. Most of the light is reflected back into your eyes, so all that your brain records is the shape of the window. This reflected light is *specular reflectance* or *Fresnel reflectance*. If you shade your eyes so that the sunlight is not reflected directly back at you, you will be able to see into the room and your eyes (the detectors) will receive light from the same light source (the sun) reflected back from objects inside the room. This is analogous to *diffuse reflectance*, since the light reaching your eyes carries information reflected back from the objects inside the room. Your brain (the computer) records what is present in the room because years of observation (calibration) have programmed your brain to recognize familiar signals in terms of colors, shades, furniture, pictures, appliances, people, pets, and other things—even brand names of appliances, names of artists, breeds of dogs, etc. Now suppose the room contains a unique object, such as a new model of a radio or VCR that you haven't seen before; your brain forms an opinion as to what it is but is not really sure, and you are not able to relate (display) exactly what the object is. These objects are analogous to *outliers*, and you will not be able to describe what you are seeing accurately until you have been told or have read about the object and this new knowledge has become part of the enormous bank of data already stored in your memory. The recall mechanism cannot recall knowledge that has not been added to the memory bank. Like the flower in the forest, the object belongs in the room, and you shouldn't remove it.

Near-infrared technology is like a "supermicroscope"—it sees past the surface of the sample and penetrates into the molecules themselves, so that theoretically we can measure everything that's there. All we have to do is teach the instrument to recognize what it is seeing (calibration) and make sure it is telling us the truth about what it sees (verification). There are billions of molecules of protein, water, oil, starch, etc., in the sample that the instrument views, and many of these absorb radiation at the same or closely adjacent wavelengths. We have to enable our instruments to differentiate between them. The best way to achieve this is to ensure that we have told our instrument all it needs to know about the material we expect it to analyze for us. We thus have to know all the features about the material before we can teach the instrument, which amounts to comprehensive sample accumulation and reference analysis by reliable standard methods in equally reliable laboratories. This process is the heart and soul of success in near-infrared technology. The instrument companies have done their part—all modern near-infrared instruments are good, generally more precise in their specifications than the laboratories they can and will replace. So it's up to us to enable these instruments to realize their tremendous potential for saving time, saving money, improving accuracy and precision, saving space, preventing pollution, simplifying operation, demonstrating versatility, and all their other features. The field and its potential have been concisely and comprehensively summarized by D. L. Wetzel (Near-infrared reflectance analysis: Sleeper in analytical techniques, Anal. Chem. 55:1165A-1176A, 1983), who also points out the potential of near-infrared technology in the agricultural and food industries.

We have assembled this book to help in the use of near-infrared instruments and technology. The first five chapters deal with fundamentals: the physics and chemistry without which the biology and its appreciation would not be possible; the mathematics and statistics to reduce the physical and chemical information to its simplest terms (believe me); and the building stones of the computerized spectrophotometer, the "supermicroscope." Chapters 6 and 7 attempt to explain how the system works, what can foul it up, and what we can do to prevent foul-ups and get the best out of the instruments. Chapter 8 deals with the complex material loosely referred to as fiber. It is a transition chapter, leading from the chemistry of the fibrous substances through the instrumental methods available for the study of these and other constituents of biological materials, into the contribution of near-infrared technology.

Chapters 9 through 13 summarize what has been done in terms of published or talked-about work in the near-infrared field. We have not included every published item in these chapters, and those of you who don't see your work referred to, please forgive us: the objective of these reviews is to illustrate the diversity of applications. We have no knowledge of what are probably some of the most ingenious applications, which are specific applications in industry. Also, we have no information from Latin America or from the continent of Africa beyond knowing that a number of instruments have been located there. Chapter 14, on whole grain applications, represents what is in many ways the ultimate application—testing with no sample preparation. We do not refer in detail to applications in industries such as plastics, textiles, pharmaceuticals, or petrochemicals, which are legion. This book deals with applications in the agricultural and food regimes. Principles described apply to any use of near-infrared technology, but industrial requirements involve more on-line installations and different techniques for sampling and sample preparation. Chapter 15 deals with qualitative applications, which are likely to become much more widely exploited in the future, for example, for the identification of materials and physical effects such as weather damage and staling.

Following the chapters are several types of supplementary information. We have included a set of spectra to illustrate what many materials and their constituents look like. The Appendix presents AACC Approved Method 39-10 for near-infrared testing. This method, which is concisely written, contains the basic principles for application to any commodity or constituent. The bibliography contains nearly 1,000 references, including supplementary references that deal mainly with subjects that are fairly closely related to near-infrared technology. The spectra and the bibiliography are indexed, to aid the reader in using them. Finally, the index to the text will serve partly as a dictionary because it leads the reader to many definitions.

Some repetition will be apparent, but repetition is the foundation stone of remembrance, and we believe that this book will serve as a starting point for future, more comprehensive works on the subject. We have waited until now to assemble it because a great deal of knowledge has been realized and clarified in the last 12 years. The success of near-infrared technology has been attributed to the development of low-cost amplification, detection, and computing components. In terms of instrumentation, this explanation is true, but the success of the method to us, the users, results from the simplicity of sample preparation and instrument operation and the avoidance of chemicals, polluting or otherwise. Most chemical and instrumental techniques require specialized sample preparation such as drying (moisture reduction), digestion, extractions, and other procedures. Not so with near-infrared technology, since moisture does not prevent accurate testing—it only changes the conditions required, such as the wavelength of measurement. We have attempted to explain all of this in the text. It's been fun writing it and we hope you will enjoy reading it.

Our deepest thanks are accorded to our secretaries, Susan Wiebe and Kathi Suderman, who have prepared the entire manuscript. Debbie Sobering and Tricia Starkey have spent many hours in proofreading and general faultfinding. John Antoniszyn has been involved with both proofreading and discussions on the text. Finally, Sue Stevenson, in addition to her authorship, assembled the bibliography.

Phil Williams
Karl Norris

CONTENTS

CHAPTER PAGE

1. **The Physics of Near-Infrared Reflectance.** Gerald S. Birth and Harry G. Hecht 1
   - I. Introduction 1
   - II. Physical Principles 2
   - III. Illustrations of Diffuse Reflectance 4
   - IV. Quantitative Representation of Absorption and Scatter in a Diffusing Medium 8
   - V. The Measurement of Diffuse Reflectance 10
   - VI. Kubelka-Munk Coefficients for Small Grains 13
   - VII. Summary 14

2. **Chemical Principles of Near-Infrared Technology.** I. Murray and P. C. Williams 17
   - I. Introduction 17
   - II. Energy and Molecules 17
   - III. Vibrational and Rotational Energy 19
   - IV. Dipoles and Absorbance 20
   - V. Absorbance and Reflectance and the Origins of Specta 21
   - VI. Chemical Structure and Spectra of Simple Substances 22
   - VII. Molecular Vibration and Spectral Bands 22
   - VIII. Illustration and Discussion of Some Simple Characteristic Spectra 24
   - IX. Assignment of Wavelengths in Agricultural Products 31

3. **Data Analysis: Wavelength Selection Methods.** William R. Hruschka 35
   - I. Introduction 35
   - II. Calibration, Measurement, and Validation 36
     A. Calibration—B. Measurement and Validation—C. Developing a Calibration Model
   - III. Sources of Error 40
     A. Sampling Error—B. Reference Method Error—C. NIR Method Error and Smoothing
   - IV. Single-Term Linear Regression and the Correlation Plot 43

V. Multiterm Linear Regression 45
A. Basic Properties—B. Calculation

VI. The Derivative 47
A. Basic Properties—B. Calculation

VII. The Fourier Transform 49
A. Basic Properties—B. Applications

VIII. Other Methods 51
A. Component Spectrum Reconstruction—B. Fast Correlation Transform—C. Normalizing Spectra—D. Discriminant Analysis

IX. Conclusion 52

Appendix 53

4. Multivariate Calibration by Data Compression. H. Martens and T. Naes 57

I. Introduction 57
A. Multivariate Calibration and Predicition—B. Calibration—C. Prediction

II. Linear Prediction and Ways to Find the Calibration Coefficients 59
A. Linear Prediction Equation—B. Multiple Linear Regression as a Calibration Method—C. Different Classes of Calibration Methods

III. Statistical Calibration Methods for Multicollinear Data 60
A. The General Model Framework—B. Conventional Calibration Methods: Selecting the "Best" Wavelengths—C. Hruschka Regression: Selecting the "Best" Calibration Samples—D. Fourier Transform Regression: Concentrating the Data to the Main Spectral Features—E. Principal Components Regression: Concentrating the Data to Their Most Dominant Dimensions—F. Partial Least-Squares Regression: Concentrating the Data to Their Most Relevant Dimensions—G. Calibration Based on Beer's Model for Mixtures

IV. Prediction Ability and Outlier Detection 69
A. Evaluating Prediction Ability—B. The Importance of Outlier Detection—C. Analysis of Near-Infrared Residuals—D. Leverage: Position Relative to Rest of Calibration Sample Set—E. Analysis of the Chemical Residuals—F. Combined Criteria

V. Data Pretreatment 72
A. Response Linearization—B. Multiplicative Scatter Correction

VI. Illustration by Artificial Data 73
A. Artificial Input Data—B. Graphical Study of the Input Data—C. Effect of Using Insufficient Range of Calibration Samples—D. Using a Complete Calibration Data Set—E. Paratial Least-Squares Regression—F. Outliers—G. Conclusions

VII. Results for Real Data 77
A. The Real Data Sets—B. The Effect of Overfitting—C. Comparison of Calibration Methods—D. Transformations of Near-Infrared Data—E. Improvements in the Partial Least-Squares Calibration Method

VIII. Discussion 82
A. Statistical Calibration Methods—B. Factors Affecting Choice of Method—C. Data Pretreatment—D. Error Detection—E. Updating

IX. Conclusions 84

Appendix A: Symbols and Abbreviations 84

Appendix B: Matrix Operations Illustrated for Multicomponent Analysis 84

5. Near-Infrared Instrumentation. W. F. McClure 89

I. Introduction 89

II. Components of Near-Infrared Systems 89
A. Lenses and Mirrors: Collecting Radiation—B. Radiation Sources—C. Monochromators—D. Filters—E. Detectors

III. Computerized Spectrophotometry: The COMP/SPEC 97
A. General Design—B. Optomechanical—C. Optoelectronic—D. Digital Interface

IV. Performance of the COMP/SPEC 101
A. Photometric Noise—B. Wavelength Precision—C. Fourier Analysis of Instrument Performance

V. Software for COMP/SPEC 103
A. COMP/SPEC File Structure—B. Scanning/Analysis—C. Analytical Software Program—D. Computerized Spectrophotometric Analytical System

6. Commercial Near-Infrared Reflectance Analyzers. P. C. Williams 107

I. Introduction 107

II. Technical Principles as They Apply to Instrumentation 107
A. Reflectance: Specular, Fresnel, and Diffuse—B. Band Overlap—C. Transmittance and Absorbance—D. Optical Density—E. Instrument Geometry—F. Reflectance Standards—G. Wavelength Selection—H. Mathematical Treatment—I. Optical Systems—J. The Basic Near-Infrared Analytical System

III. Commercial Bench-Type Instruments: First Generation 114
A. DICKEY-john Grain Analysis Computer Model 1—B. Neotec Grain Quality Analyzer—C. Anacon NIR Analyzer—D. Pacific Scientific/Neotec Automated Digital Analyzer—E. Technicon/DICKEY-john InfraAlyzer 2.5

IV. Commercial Bench-Type Instruments: Second Generation 119
A. Other Neotec GQA Models—B. Neotec Feed Quality Analyzer—C. DICKEY-john GAC III Series—D. Neotec GQA 101 Series—E. Neotec GQA Model 102—F. Labormim Infrapid 31—G. Neotec GQA Ultraspeed—H. Technicon InfraAlyzer Model 400—I. Pacific Scientific Feed Quality Analyzer Model 51A—J. Percon Inframatic 8100—K. Pacific Scientific Model 7000

V. Commercial Bench-Type Instruments: Third Generation 124
A. Percon Inframatic 8120—B. Trebor GT-90 Whole Grain Analyzer

VI. Simple Computerized Spectrophotometers 126
A. Pacific Scientific/Neotec Research Composition Analyzer, Model 6350—B. Labormim Infrapid 61—C. Technicon InfraAlyzer Model 500—D. LT Analyzer Model 7000—E. Trebor Models GT-70 and 7700—F. LT Quantum 1200—G. Applications of Computerized Spectrophotometers

VII. Operation of Commercial Bench-Type Instruments 130
A. Calibration—B. Operation—C. Monitoring

VIII. Operation of Computerized Spectrophotometers 136
A. Preparatory Steps—B. Exploratory Steps—C. Application Steps

7. Variables Affecting Near-Infrared Reflectance Spectroscopic Analysis. P. C. Williams 143

I. The Philosophy of Error 143
A. What is Error?—B. Definitions

II. Statistical Terms Necessary to the Evaluation of Accuracy 143

III. Precision 144

IV. Interpretation of Statistical Evaluation of NIR Analysis 146

V. Sources of Error in NIR Testing 148
A. Factors Associated with the Instrument—B. Factors Associated with the Sample—C. Operational Factors

8. Analytical Application to Fibrous Foods and Commodities. F. E. Barton II 169

I. Introduction 169

II. Structure and Composition of Forages 169

III. The Analysis of Forages 170

IV. NIR as an Analytical Method 175

V. Advantages of the Chemometric Method 180

9. Near-Infrared Research in Europe. B. G. Osborne, T. Fearn, and S. G. Stevenson 185

I. Introduction 185
A. History and Extent of NIR Usage in Europe—B. Some Early Problems and Their Solutions

II. Wheat-Breeding Programs and New Variety Trials 188
A. Criteria of Milling and Baking Quality of Wheat—B. Role of NIR

III. Milling Industry Applications 190
A. Grain Trading—B. Control of the Milling Process

IV. Baking Industry Applications 192
A. Flour Intake Quality Control—B. Monitoring of Product Composition

V. Additional Cereals and Oilseeds 193

VI. Mixed Feeds and Forages 195

VII. Agricultural Products for Human Consumption 196

10. Application of Near-Infrared Reflectance Spectroscopy in North America. Janet A. Panford 201

I. Introduction 201

II. Application to Plant and Animal Products 201
A. Wheat—B. Barley—C. Maize (Corn)—D. Other Cereal Grains—E. Oilseeds—F. Forages—G. Fruits and Vegetables—H. Tobacco—I. Pulses—J. Meat and Meat Products—K. Dairy Products—L. Ready-to-Eat Foods—M. Other Products

III. Research Applications 208

IV. Breeding Programs 208

11. Use of Near-Infrared Reflectance Spectrophotometry in Investigations in the USSR
V. P. Krischenko 213

I. Introduction 213

II. Instrument Evaluation 213

III. Analysis of Forages 215

IV. Universal Constants 216

V. Determination of Lysine in Maize by NIR Spectroscopy 216

VI. Determination of Precision and Analytical Error 216

VII. Determination of Fiber 218

VIII. Rapid Moisture Analysis 218

IX. Wheat Gluten Determination 219

12. Near-Infrared Research in Asia. Mutsuo Iwamoto 223

I. Introduction 223

II. Distribution of NIR Instruments in Asia 223

III. Trends of NIR Research in Japan 224
A. Rice—B. Wheat—C. Buckwheat—D. Soybean—E. Red Pepper—F. Laver—G. Feed—H. Green Tea—I. Other Foods

13. Near-Infrared Research in Australia and New Zealand. S. G. Stevenson 227

I. Introduction 227

II. Wool and Synthetics 227

III. The Dairy Industry 228

IV. Forage Applications 229

V. 1984 NIR International Symposium 230

14. Analysis of Whole Grains by Near-Infrared Reflectance. R. Tkachuk 233

I. Introduction 233

II. Definitions 233

III. Instrumentation 233

IV. Computer Programs 234

V. Data Acquisition 234

VI. Data Storage and Transformations 236

VII. Correlation Analysis 236

VIII. Regression Analysis and Optimum Wavelength Selection 236

IX. Reflectance and Transflectance Properties of Wheat 236

X. Analysis for Wheat Protein 238

XI. Wheat and Barley Moisture 239

XII. Rapeseed, Protein, Oil, and Chlorophyll 239

15. Qualitative Applications of Near-Infrared Reflectance Spectroscopy.
P. C. Williams and K. H. Norris ... 241

I. Introduction ... 241

II. Determination of Gross Composition ... 241

III. Detection of Impurities ... 241

IV. Detection of Areas of Maximum Differences in Concentration ... 242

V. Classification of Unknown Substances ... 243

Near-Infrared Spectra ... 247

Key to Near-Infrared Spectra ... 289

Appendix—Near-Infrared Method for Protein Determination: AACC Method 30-20 ... 291

Bibliography ... 293

Index of Topics in the Bibliography ... 315

Index to Text ... 321

CHAPTER 1

The Physics of Near-Infrared Reflectance

GERALD S. BIRTH
Russell Research Center
U.S. Department of Agriculture
Athens, Georgia, U.S.A.

HARRY G. HECHT
Chemistry Department
South Dakota University
Brookings, South Dakota, U.S.A.

I. INTRODUCTION

Physics is the science of matter and of interactions between matter and energy. It is divided into several fields, including electricity, optics, dynamics, particle science, and others. Near-infrared reflectance technology includes the phenomena associated with near-infrared reflectance itself, the instrumentation, sample preparation, laboratory analysis, and data-processing. Of these, physics is involved mainly with the instrumentation and the surface phenomena, including diffuse reflectance. The line between applied physics and engineering is a fine one where electro-optical instrumentation is concerned, and in this book instrument design and the manipulation of light and wavelength are accorded separate treatments. Accordingly this chapter will address the physics involved with light scatter and diffuse reflectance, subjects that in themselves are worthy of specialized attention.

As white light passes through a material it may be selectively absorbed, rendering a wavelength dependence to the emergent radiation that is perceived as color. This phenomenon is well known, providing a basis for structural study and quantitative analysis using transmission spectroscopic techniques. Color may also be imparted by the selective scatter of radiation. A familiar example is the blue color of the sky, which is a result of the fact that the scatter of sunlight by atmospheric particles is more efficient at shorter wavelengths.

Both of these phenomena occur within the dense scattering media. If the concentration of scattering centers is sufficiently large, a significant portion of the incident light will be returned to the surface, and the material may be described in terms of its reflecting properties. The reflectance may range from that of an ideal mirror-type surface, which reflects light only at the single angle dictated by geometric optics, to that of an ideal matte surface, which reflects uniformly at all angles.

The reflectance of a matte surface is often described as *diffuse*. It appears that the first serious attempt to explain the nature of diffuse reflectance was made by Bouguer in 1760. He assumed that a matte surface such as a powder, for example, could be viewed as an assembly of microcrystalline faces statistically distributed over all possible angles. Diffuse reflectance was thought to be a result of the mirror-type reflections from the various crystallites. In other words, Bouguer viewed diffuse reflectance as strictly a surface phenomenon.

Seeliger (1888) perceived that diffuse reflectance occurred not only at the surface, but from the interior as well. He considered radiation to penetrate the surface and be partially absorbed, with a portion being returned as a result of reflection, refraction, and diffraction at the various interfaces. An attempt was made by Lommel (1887) to generalize the Seeliger theory by including the effect of mutual irradiation of each volume element by scatter from every other volume element. The result was too complicated for practical use, but it represented an important conceptual advance in the understanding of the diffuse reflectance process.

It was shown by Pokrowski (1924) that both the Bouguer and the Lommel-Seeliger processes are operative in many cases, giving rise to radiation that may be viewed as a superposition of two components. One component is diffuse and the other is a mirror-type reflection. Various workers (Grabowski, 1914; Berry, 1923; Schulz, 1924; Middleton and Mungall, 1952) have discussed the distribution of microcrystalline interfaces necessary to reconcile this theory with the observations. The agreement is generally not perfect but it is sufficient to substantiate this basic concept, upon which much of the subsequent discussion is based.

The application of reflectance techniques to natural products involves some special considerations. The interaction between light and plant leaves, for example, has been a subject of great interest to scientists for many years. To gain an improved understanding of this interaction, Willstätter and Stoll (1918) applied ray tracing to a drawing of a leaf cross section. They proposed a theory to explain reflectance from a leaf on the basis of critical reflection of light at the spongy mesophyll cell wall-air interfaces. Gates et al (1965) measured the reflectance, transmittance, and absorbance of foliage for several plant species in the 0.4–10.0 μm spectral region. This was a major advance in understanding the interaction between light and plant material because previously published spectra were limited primarily to the visible. A ray tracing included in their report was based largely on critical reflection at cell walls within the leaf, similar to the Willstätter and Stoll approach. Gates et al postulated that the smaller bodies in the plant cells were of a suitable size to produce considerable light scattering in the leaves. Later, when the remote-sensing technology was developed, additional investigations were carried out to gain a more complete understanding of the reflection process from leaves. Kumar and Silva (1973) developed a ray-tracing procedure for a soybean leaf cross section. They ignored the smaller bodies in the cells and considered only air, cell wall, cell sap, and chloroplasts. Using the best estimates of index of refraction for those components of the leaf, they computed the reflectance and transmittance of a leaf to be 45.6% and 54.4%, respectively, for the nonabsorbing near-infrared spectral range, i.e., 0.7–1.3 μm. These authors compared this to a measured value of 47% reflectance. On the basis of that work and experimental results by others, they concluded that reflection and refraction at the larger interfaces in the leaf were a more significant factor in creating diffuse reflectance than the redistribution of radiation due to Rayleigh and/or Mie scattering by the smaller bodies in the leaves.

Gausman et al (1969) used physical and mathematical models to analyze the reflectance of cotton leaves. In their analysis of leaf reflectance, they concluded that variations in internal structure of upper leaves of cotton plants caused by different salinity stresses resulted in variations of reflectance within the 0.75–1.3 μm range. The differences in reflectance were attributed to differences in light scattering in the leaves, rather than to differences in absorption.

II. PHYSICAL PRINCIPLES

Diffuse reflectance spectroscopy is an extension, via instrumentation, of human vision. Instrumentation provides output in numerical form, so that mathematical treatments can be used; precise wavelength and radiometric measurements; and the potential for using radiation other than the visible. As wavelength increases, the photon energy decreases and the energy of a photon determines the nature of the absorption process. The relation between wavelength of the radiation and dimensions of the object with which the radiation interacts is a factor in the redistribution of the radiation, or light scattering. Probably the primary interest in diffuse reflectance spectroscopy is in the rapid and nondestructive determination of the concentration of certain constituents in a material. The instrumentation and procedures are similar to those used in absorption spectroscopy. Ideally, the absorption spectrum should be the output of the instrumentation, but most instruments measure the radiation that is not absorbed, rather than the radiation that is absorbed. Only if these measurements are made properly can valid absorption spectra be obtained. Fortunately, in many cases an absolute measure of absorption is not required, but the effect of variables that are dependent on time and wavelength must be eliminated. The objective is to record accurately the wavelength-dependent nature of absorption relative to a stable standard. According to the Beer-Lambert law, the concentration of an absorber is directly proportional to A, the sample absorbance.

$$A = \log \frac{I_o}{I_t} \tag{1}$$

where I_o = intensity of the incident radiation and I_t = intensity of the transmitted radiation.

This relationship is fundamental to spectroscopy, and it may be applied not only to transmission measurements, but also to the diffuse reflectance of light-scattering materials. It assumes that the path length of the radiation through the medium containing the absorber is constant. In absorption spectroscopy, the path length is maintained constant by the design of the cuvette that contains the absorber. On the other hand, the path length of the radiation for diffuse reflectance spectroscopy is a function of the microscopic structure of the material containing the absorber. This difference in *effective path length* constitutes the chief difference between diffuse reflectance spectroscopy and absorption spectroscopy. The action of radiation propagating through a diffusely reflecting material is termed *light scattering* and is the primary focus of this chapter.

The refractive index of a material is an important factor in the application of geometric optics. For a nonabsorbing medium the refractive index, n, is given simply as the ratio of the velocity of light in a vacuum, c, to that in the sample, v, so:

$$n = \frac{c}{v} \tag{2}$$

For specific values of n, a handbook (Weast, 1984) should be consulted.

An optical interface is the boundary between two media having different indices of refraction. When a radiant beam is incident on an interface, the beam is divided into two parts as shown in Fig. 1: the transmitted ray, T, and the reflected ray, R. The direction of the transmitted ray is defined by Snell's law:

$$(\sin \theta_i)n_1 = (\sin \theta_t)n_2 \ . \tag{3}$$

The angle of the reflected ray, θ_r, is equal to the angle of incidence θ_i. The fraction of the incident ray that is reflected is a function of the two indices of refraction and can be computed with the Fresnel equations:

$$R_\perp = \left[\frac{\cos\theta - \sqrt{(n_2/n_1)^2 - \sin^2\theta}}{\cos\theta + \sqrt{(n_2/n_1)^2 - \sin^2\theta}}\right]^2 \tag{4}$$

$$R_\parallel = \left[\frac{(n_2/n_1)^2\cos\theta - \sqrt{(n_2/n_1)^2 - \sin^2\theta}}{(n_2/n_1)^2\cos\theta + \sqrt{(n_2/n_1)^2 - \sin^2\theta}}\right]^2 \tag{5}$$

where $R_\parallel$ is radiation polarized parallel to the plane of incidence, and $R_\perp$ is radiation polarized perpendicular to the plane of incidence.

The plane of incidence is defined by the incident ray and the normal to the interface. The reflectance as computed with these equations is illustrated in Fig. 2 for angles of incidence from 0° to 90°. The significant points are identified by arrows in the figure: (1) At normal incidence there is no polarization effect and the total reflectance, $R_t = (R_\parallel + R_\perp)/2$ is a minimum. (2) The radiation polarized parallel to the plane of incidence is totally transmitted for an angle of incidence of 56°. Although this angle, Brewster's angle, is a function of the refractive index, it is between 50° and 60° for many common materials. (3) The reflection increases to 100% for high angles of incidence. Except for normal incidence, no radiation propagates away from the interface in a direction coinciding with the direction of the incident ray when n_1 and n_2 differ. The Fresnel equations yield $R = 0$ when $n_1 = n_2$. There is no change in the direction of the transmitted ray in that case. Thus, when a transparent solid is immersed in a fluid having a refractive index identical to the solid, there is no change in the direction of a radiant beam propagating from the fluid into the solid or vice versa. This analysis of change in direction of a radiant beam is an important factor in light scattering associated with diffusely reflecting materials. The regular reflectance (R_R) described here produces the shine or gloss associated with polished surfaces. In literature on color and appearance, this is called the *specular component* of reflectance. Other terms used are *Fresnel reflectance* and *surface reflectance*.

A primary concern of reflectance spectroscopy is the relation with absorption. The effect of absorption can be introduced through use of a complex refractive index:

$$n_2(1 - i\kappa) \tag{6}$$

in the Fresnel equations, where κ is known as the absorption index. As absorption increases, the regular reflection remains constant until the absorption index becomes very high. For example, when the absorption index approaches a value large enough to influence the regular reflection, then the absorbance is nearly equal to 1.0 for radiation propagating a distance equal to one wavelength into a material. The corresponding absorptivity (absorbance normalized to 1 cm), $a = A/\lambda$, would be of the order 10,000 cm^{-1} for radiation of wavelength $\lambda = 1.0$ μm. This level of absorption is much greater than that normally encountered with organic materials in the visible and near-infrared spectral regions.

This can be illustrated by comparing the transmittance spectra of a thin layer of water and the reflectance spectra (R_R) of a volume of water (Fig. 3). Consider the absorption maxima (transmittance minima) that occur at 3.0, 4.7, and 6.0 μm. With a 20-μm sample thickness the transmittance is near zero at 3.0 and 6.0 μm, but at 4.7 μm the transmittance is more than an order of magnitude greater. By noting how the reflectance changes in these three spectral regions, we see evidence that when the transmittance is so low that it is almost unmeasurable even for a very short path length, then data on absorption can be obtained from reflectance measurements. However, when the transmittance is relatively high compared with that at 3.0 or 6.0 μm, then reflectance does not appear to be a function of absorption, so transmittance measurements must be used. That would be the case for water in the 4.7-μm wavelength region.

As stated in the introduction, the light reflected by scattering media generally consists of both diffuse and specular components. Either could be used for a determination of the optical characteristics of the medium, but when both are present, the absorption features tend to become obscured. The reason for this can easily be seen by a simple qualitative argument. The diffuse component arises through multiple scatter by particles near the surface but inside the material. If the radiation travels a mean path length, $\bar{l}$, in the sample before it returns to the surface, it would be expected that the diffuse reflected radiation would be attenuated by a factor:

$$R_D \propto 10^{-\bar{\kappa}\bar{l}} \qquad (7)$$

Here $\bar{\kappa}$ is the volume-averaged absorption constant, which is related to the absorption index through $\kappa = 4\pi\kappa/\lambda$. On the other hand, the regular reflectance for normal incidence, corresponding to equations (4) and (5) for $\theta_i = 0°$ and taking absorption in medium 2 into account, may be written:

$$R_R = \frac{(n_2 - n_1)^2 + n_2^2\kappa^2}{(n_2 + n_1)^2 + n_2^2\kappa^2} \qquad (8)$$

We see that $R_D \rightarrow 0$ and $R_R \rightarrow 1$ as κ becomes large. Hence, these two components are complementary, and as a result, the reflected radiation is less distinct than the true absorption spectrum if both are present. The diffusion of light is an inherent characteristic of the materials concerned, and so methods are considered that may be used to eliminate the regular component.

The scatter of radiation is a result of reflection, refraction, and random diffraction at the surfaces of various particles. They are assumed to have a random distribution so that no coherence is maintained in the process and any polarization there may be of the incident beam is also lost. As the size of the particles is reduced, the number of scattering interfaces is increased, and a greater proportion of the light is returned to the surface without significant penetration of the scattering medium. (This effect is illustrated in the following section.) If

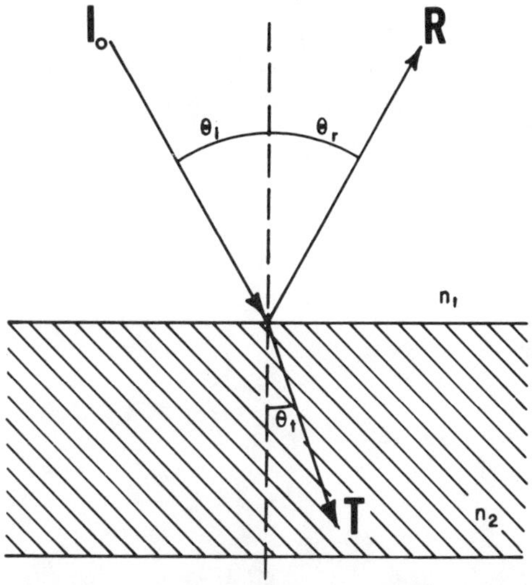

Fig. 1. Reflectance and transmittance for a ray of optical radiation incident on plane interface. Media have no absorption but differ in refractive indexes where $n_2 > n_1$. I_o = incident ray, R = reflected ray, T = transmitted ray.

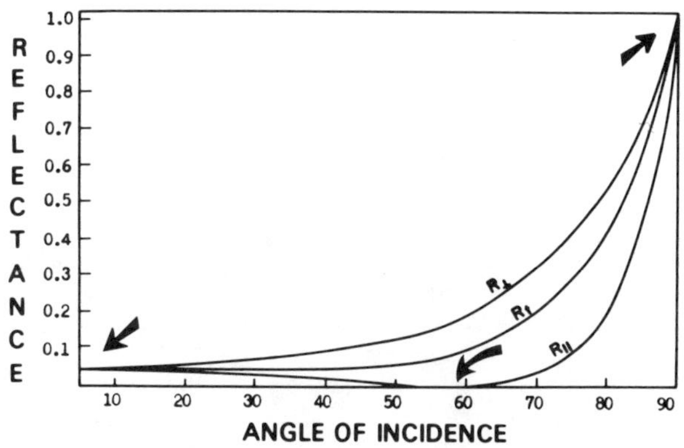

Fig. 2. Intensity of the reflected ray (regular reflection) shown in Fig. 1 as related to angle of incidence. Refractive indices are $n_1 = 1.0$ and $n_2 = 1.5$ for these curves. $R_\perp$ indicates perpendicular to the plane of incidence, $R_\parallel$ indicates parallel to the plane of incidence, and $R_t = (R_\parallel + R_\perp)/2$. Plane of incidence is defined by incident ray and normal to the interface. Radiation not reflected is transmitted through the interface.

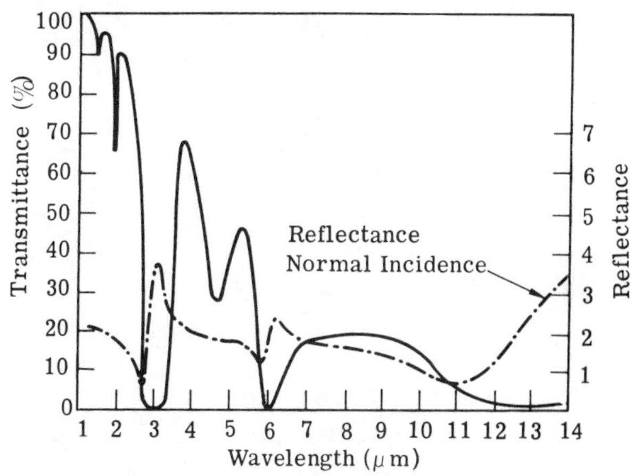

Fig. 3. Transmittance of seawater with 20-μm path length and reflectance of free seawater surface. (Reprinted, with permission, from Suits, 1985)

the mean path length, $\bar{l}$, is reduced sufficiently, the diffuse reflectance, R_D, may become sufficiently large that the regular component R_R may be negligible compared with it.

In the process of diffusion the radiation becomes entirely depolarized, but any radiation regularly reflected maintains its state of polarization. Thus by crossing an analyzer to the polarizer of the incident beam, the regular component can be entirely eliminated. A comparison of reflectance spectra with and without polarizer and analyzer can be used to assess the effectiveness of various efforts to eliminate the regular component. Using such techniques, Kortüm and Vogel (1958) showed that for weakly absorbing materials, such as $CuSO_4 \cdot 5H_2O$, the reflected light is purely diffuse if the material is ground to a fine powder.

For strong absorbers, a large proportion of regular reflection remains even when the particles are reduced to the smallest sizes attainable through grinding (Kortüm and Vogel, 1958). In such cases it may be necessary to resort to other means. For example, the absorber can be dispersed in a nonabsorbing medium, or on the surface of nonabsorbing particles. In extreme cases it may even be necessary to use polarized irradiation and a crossed analyzer, although this is not usually necessary with natural products whose absorbances tend to be rather low.

III. ILLUSTRATIONS OF DIFFUSE REFLECTANCE

Overall scientific literature is not consistent in the definition of diffuse reflectance. For example, many textbooks define diffuse reflectance as a matter of surface roughness. Because of this, some authors have used the term *body reflection* to describe that radiation which is diffused in the body of the material. For this discussion, diffuse reflectance refers only to that radiation which becomes diffuse due to the internal structure of the sample.

The potential error in measuring absorption when both diffuse and regular components are included in the radiometric measurements is illustrated in Figs. 4a and 4b. A test pattern consisting of a series of straight lines that intersect at a point was

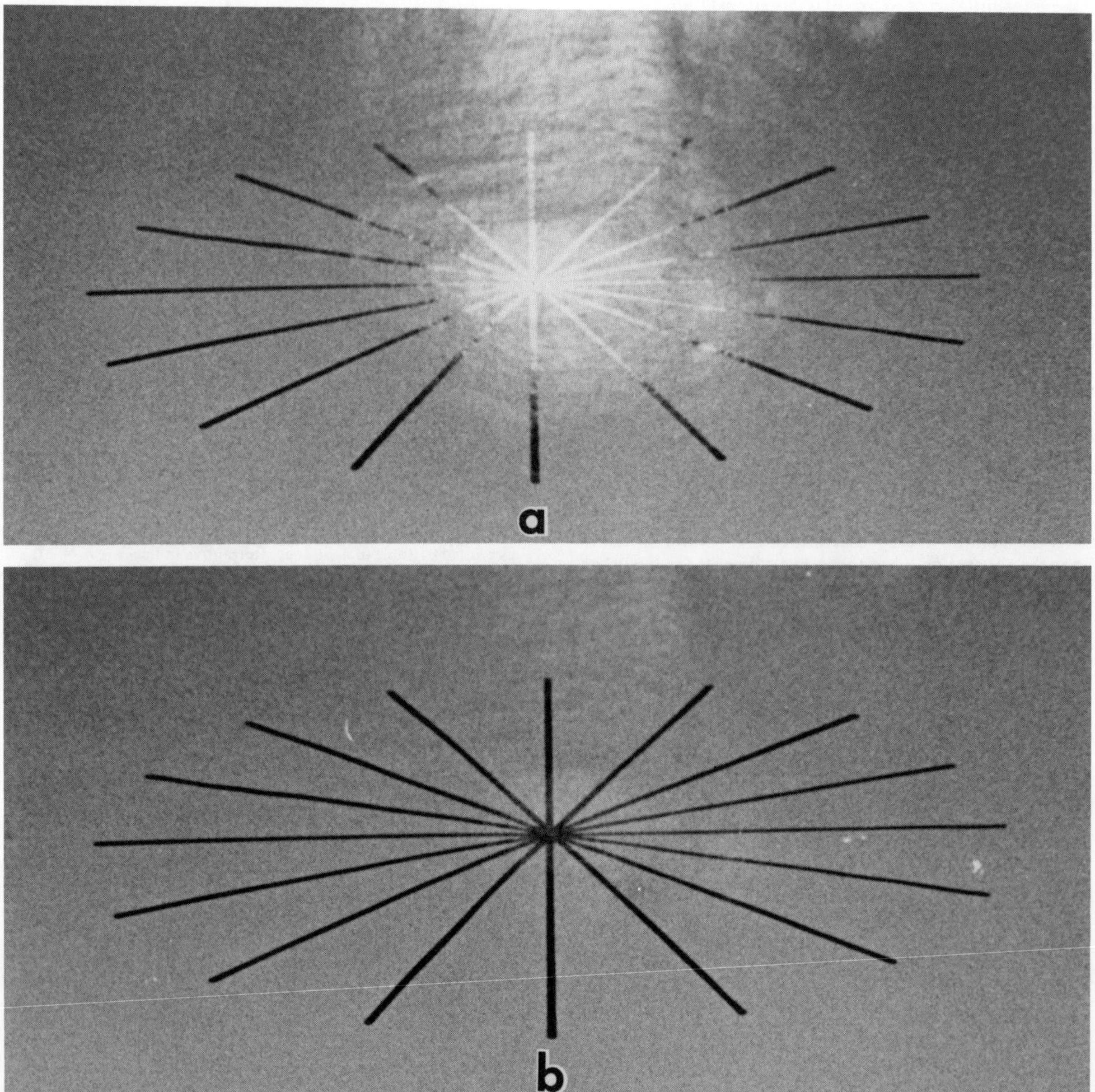

Fig. 4. Photograph (a) of test pattern consisting of black lines on white paper and using geometric conditions required to record regular reflection (center of pattern). With regular reflection, more light is reflected by black lines than by white paper. In (b), with angle of incidence near 56°, regular reflection was blocked with light-polarizing material. Result is that essentially only radiation reaching camera is due to diffuse reflectance, so lines appear completely black.

photographed using geometric conditions such that part of the regular reflectance was recorded on the film. Since the regular reflection is very directional, only that radiation directed toward the camera lens would be recorded on film as regular reflection.

The transmitted radiation entered the paper and became diffuse as a result of light scattering—that is, random reflection and refraction inside the paper. For areas having no ink, the radiation left the paper without attenuation but was distributed over the entire hemisphere surrounding the illuminated area. This spatial distribution resulted in less than 0.5% of the radiation transmitted into the paper actually reaching the camera lens even though the radiation encountered little or no absorption in the paper. Since the regular reflection was not diffused upon reflection from a smooth surface with high absorption (the black ink), a large fraction of that radiation reached the camera lens and the lines appeared bright. The combination of regular reflection and the transmitted radiation, which results in diffuse reflectance, can cause a condition where the total reflection is independent of absorption. Where the regular reflection was not recorded on film, the lines appeared black because the transmitted ray was almost totally absorbed by the black ink. The regular reflectance can be eliminated by polarizing the radiation in the vertical plane provided the angle of incidence is equal to Brewster's angle. This is illustrated in Fig. 4b.

The influence of the index of refraction of the surrounding medium is illustrated in Fig. 5a, where a polished apple has the shine expected to be associated with a polished surface. Figure 5b is the same apple but photographed under water, where the regular reflectance is reduced by a factor of about 10. Since the regular reflection is directional for smooth surfaces, it can frequently be avoided by proper design of the source-sample-detector geometry.

In the absence of regular reflection, the information obtained visually is primarily a response to diffuse reflectance. That is the process by which the colors of materials such as plant foliage, flowers, textiles, painted surfaces, and the printed word are seen.

In many cases line drawings have been used to illustrate how radiation becomes diffuse, but such an approach can lead to confusion when lines are used to indicate both the structure of the material and the radiation. Another approach was used to create the illustration shown in Fig. 6, which is a series of photographs with water in a glass tank. A laser is used, with the beam passing vertically through the water. In Fig. 6a, there are no particles in the water and the only place where any radiation was directed toward the camera lens was at the air-water interface and at the bottom of the tank at the water-glass interface. In Fig. 6b, milk was added to the water in the proportion 1:10,000. This created single-particle scattering, in which the radiation was propagated directly from the interaction with particles in the path of the laser beam to the camera lens without any interaction with other particles in the water. In Fig. 6c (milk 1:1,000), there was some multiple scattering because radiation was distributed over a larger volume of the water than was occupied by the laser beam; that is, the radiation reflected from other particles after the initial reflection from particles in the path of the laser beam. In Fig. 6d (milk 1:100), multiple scattering predominated and all directional characteristics of the laser beam in the water have been lost. This level of scattering involved a particle density that was much less than encountered in most diffuse reflectance measurements; however, the general semicircular distribution of the radiation was characteristic of the distribution for many materials when photographed in this manner. In Fig. 6e (milk 1:10), the distribution of radiation was similar to the previous figure, but the radius of the distribution was reduced. Figure 6f (whole milk, no dilution) showed that the radius of the distribution continued to decrease as the particle density increased. It appeared that all of the radiation left the milk close to the point of incidence. Conversely, the radiation transmitted entirely through the volume of milk was very low. The procedures for measuring light scattering with diffuse reflectance give a quantity that is proportional to particle density in the sample, so that the scattering increased from a value of zero in Fig. 6a to the highest value for the condition shown in Fig. 6f; hence, the scattering was inversely proportional to the radius of the distribution of radiation shown in these illustrations.

Figures 7 and 8 illustrate similar distributions in solid materials. The laser beam was conveyed through a metal tube to

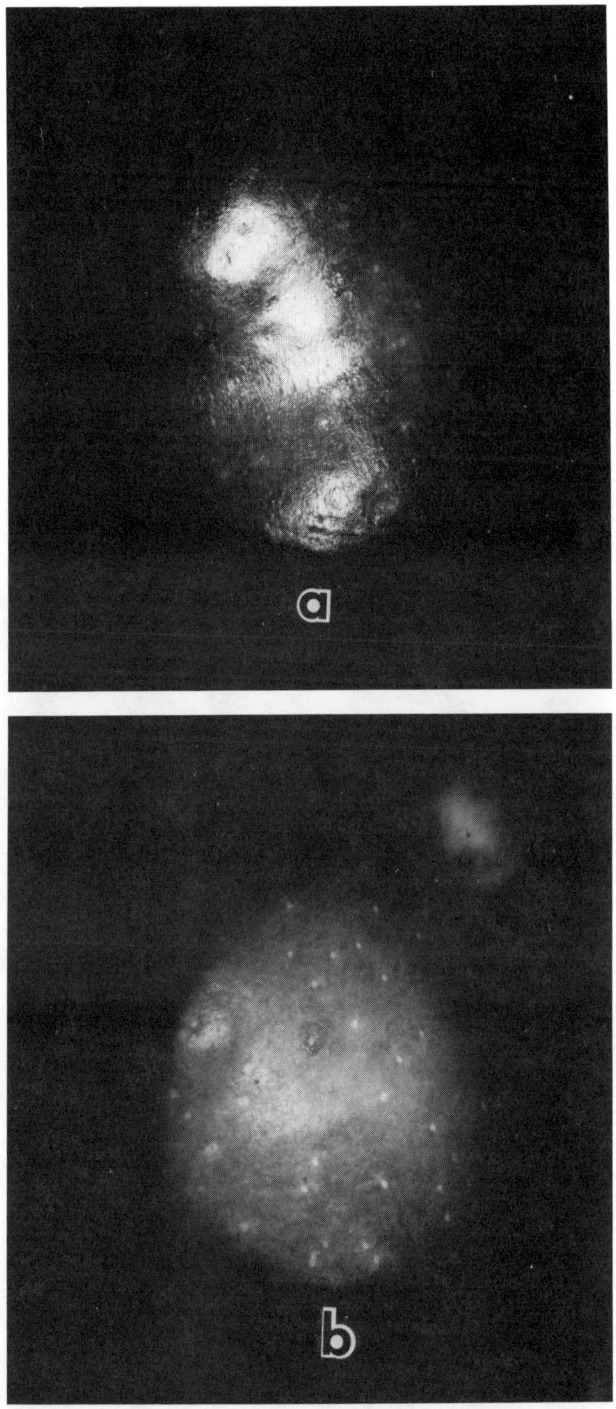

Fig. 5. (a) Gloss or shine due to regular reflection gives highlights to polished apple. (b) Regular reflection is reduced considerably by photographing apple when it is in water. Shape of image is combination of diameter of light beam illuminating apple and shape of fruit. Geometry for source, apple, and camera is identical for both photographs.

avoid overexposure at the point of incidence. Figure 7 shows a potato with the laser beam incident on a cut surface at the top. The distribution of radiation in the flesh was similar to that shown in Figs. 6d, e, and f; however, the radius of the distribution should not be compared with the milk, because some environmental light was used to show the outline of the potato, whereas the milk was photographed in a darker environment. Actually, the milk (Fig. 6f) had higher light-scattering properties and the radius of the distribution was less for the milk than the potato. Figure 8 is a similar illustration of wood. Figures 7 and 8 can be compared to observe the distribution of radiation in a high-moisture material (potato) as compared with a dry material (wood). In general, dry materials have higher light-scattering properties than materials with high water content. The cellular structure can influence how the radiation is redistributed in a material. This is illustrated by comparing Figs. 8a and 8b. The radiation propagated along the wood fibers more readily than across the grain of the wood fibers.

In contrast to the change in the distribution of radiation as particle density increases, many materials (Figs. 7 and 8) do not consist of particles but have a cellular structure. The radiation becomes diffuse after reflection at a series of optical interfaces; however, a certain amount of randomness in the position of orientation of the interfaces is essential to bring about complete diffusion. A light ray in an optical glass fiber may undergo millions of reflections and not become diffuse. Figure 9 is a photograph of a laser beam transmitted through an onion skin. The dome-shaped cells act like miniature lenses to refract some of the radiation in a direction away from the centerline of the laser beam. There is a general pattern to the cellular structure, but with small deviations. With additional layers of this cellular structure there would be additional randomization and a net result of complete diffusion of the radiation.

The same distribution of radiation illustrated in Fig. 6 can be obtained with solid materials by varying the thickness of the material rather than changing the particle density. The light-scattering properties of a solid are considered to be constant, but the transmitted ray must pass through a certain amount of the material to become diffuse. This is illustrated in Fig. 10, where potato flesh was used for the scattering medium and thin slices of tissue were supported on fine wires. The laser beam was conveyed through a metal tube to the tissue as was done for Figs. 7 and 8. These photographs were taken with the sample in water with sufficient particles added to create single-particle scattering so that the distribution of the reflected and transmitted radiation could be observed. In Fig. 10a the slice was 0.25 mm thick. For that thickness there was almost no diffuse reflectance, and the transmitted radiation had predominantly the same directional characteristics as the laser beam. With a thicker slice, Fig. 10b (0.75 mm), the transmitted radiation was partially diffuse but the distribution had a bias in the direction of the laser beam. Figure 10c (1.60 mm) shows that the transmitted radiation was nearly diffuse or spread out in all directions, and in addition, the reflectance was becoming significant. In Fig. 10d (2.50 mm), the transmitted radiation was essentially diffuse. Figures 10e and 10f illustrate that as thickness increased the reflectance increased and transmittance

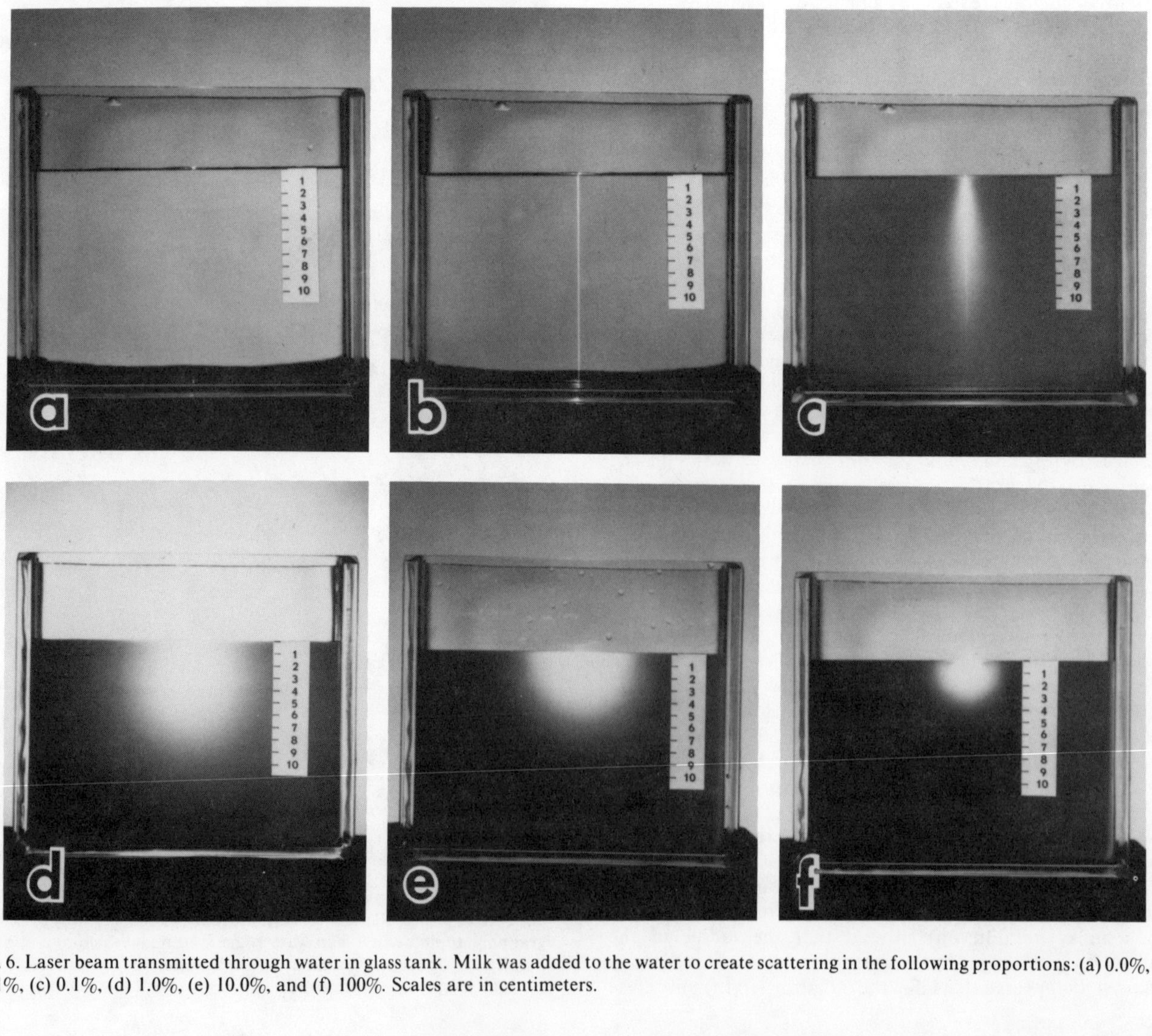

Fig. 6. Laser beam transmitted through water in glass tank. Milk was added to the water to create scattering in the following proportions: (a) 0.0%, (b) 0.01%, (c) 0.1%, (d) 1.0%, (e) 10.0%, and (f) 100%. Scales are in centimeters.

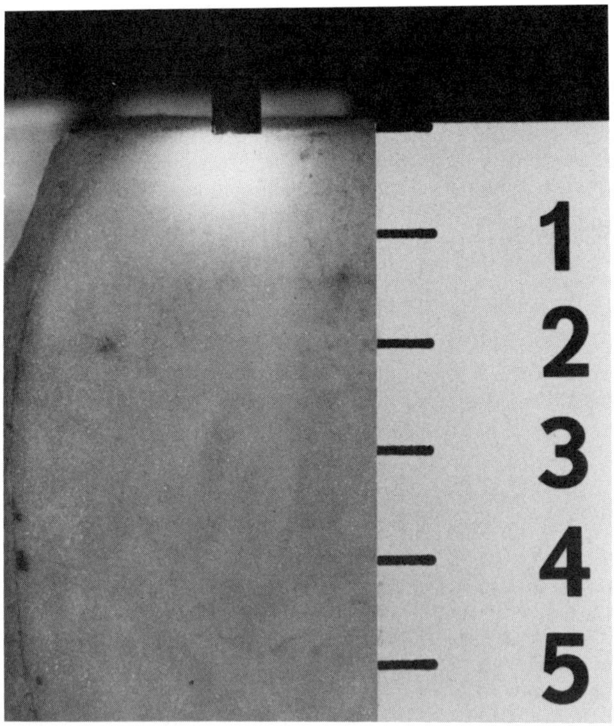

Fig. 7. Semicircular distribution around point of incidence that is evident in Fig. 6 also takes place in solids, as shown here for a potato. Scale is in centimeters.

decreased. Except for a small amount of radiation that was absorbed, the radiation that was not reflected was transmitted. The material thickness required for complete diffusion is a measurable quantity and has been termed *diffuse thickness*. For potato flesh the diffuse thickness is 3.70 mm (Birth, 1978).

The combination of light scattering and absorption is of primary interest (Plates 1 and 2). Plate 1 shows the relation between particle size and absorption. Green bottle glass was pulverized and sieved to obtain a range of particle sizes, and the effective path length increased as particle size increased. As the path length of the radiation into the material increased, the absorption due to the green pigment increased, so for large particles the material appears dark as a result of the relatively large interaction between the green pigment and the radiation. For smaller particles the mean path length is less and the material appears light, becoming almost white for the smallest particles. The scattering, which is inversely proportional to the mean path length, is a result of reflection and refraction of radiation (as shown in Fig. 1) at each optical interface. The irregular surface of the particles combined with their random orientation results in the radiation becoming diffuse.

Both the intensity of the reflected ray and the change in direction of the transmitted ray are functions of the change in the index of refraction as the incident ray propagates into the second medium (Fig. 1). If there is no change in the refractive index as the radiation propagates from one medium into another, there will be no reflected ray and no change in the direction for the transmitted ray, so the light scattering in these samples of pulverized glass is a function of the refractive index for the medium surrounding the particles. If this medium were changed to one having a refractive index closer to the refractive index of the glass particles, scattering would decrease, the mean

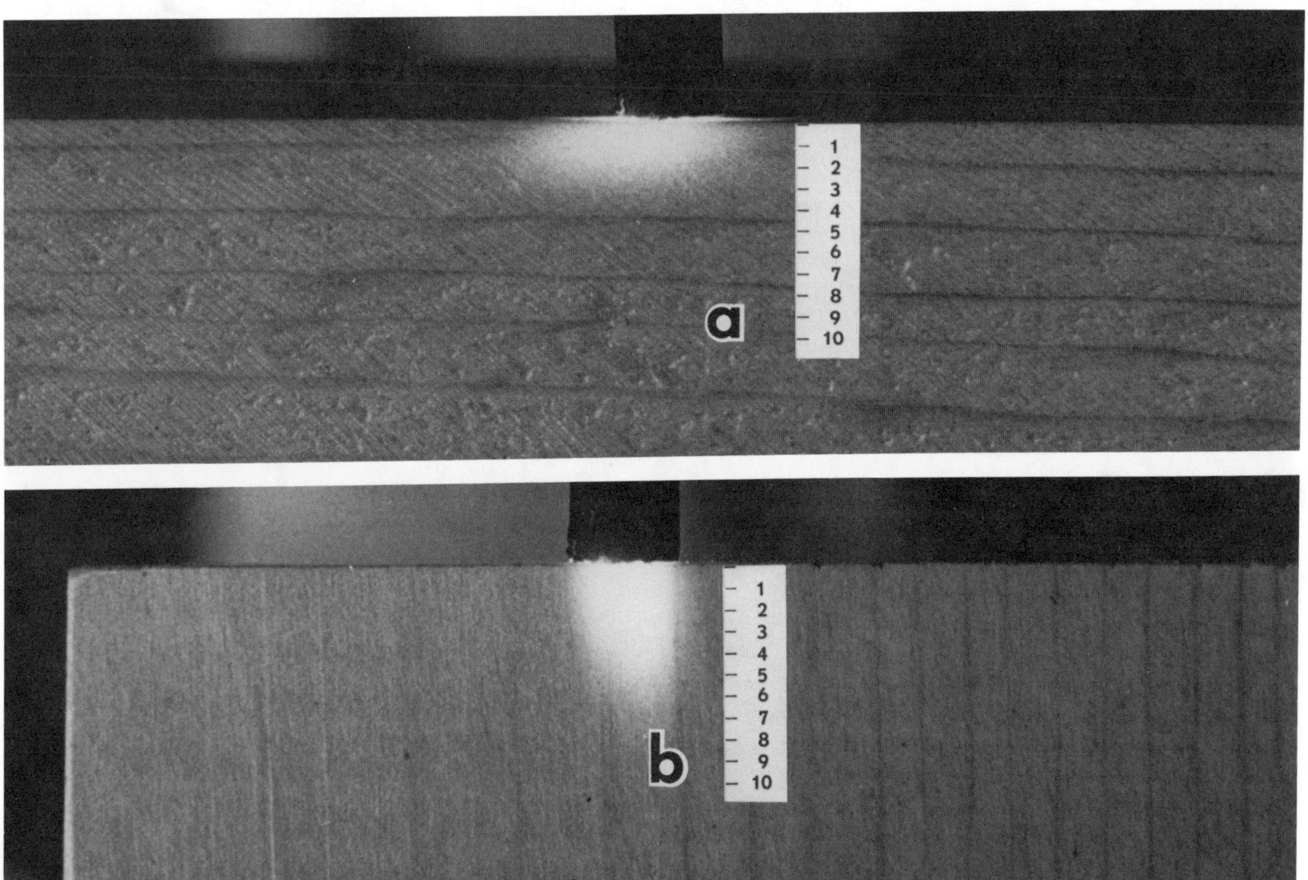

Fig. 8. (a) Distribution of radiation in wood, a relatively dry material, is elliptical rather than circular, with long axis parallel to wood fibers. Scale is in millimeters, so radius of distribution is small compared with that shown in Figs. 6 and 7. (b) Illumination on end grain. Elliptical distribution is still present and parallel to wood fibers.

path length would increase, and the sample would appear darker. Plate 2 shows three samples all having the same particle size but with different media surrounding the particles: (a) air, (b) water, and (c) mineral oil. In each case, the surrounding medium is colorless and free of any particles that could influence the scattering. If the refractive index of the surrounding fluid exactly matched that of the glass particles, the samples would be transparent. When these samples are viewed with transmitted light, samples (a) and (b) appear dark, whereas sample (c) appears translucent. The physical constants for the glass and the surrounding media are given in Table I. The reflectance of the glass was computed with equation (8). The green dye in the glass does not absorb sufficiently to influence the regular reflection. This illustration shows why high-moisture materials have lower scattering than the low-moisture materials that are compared in Figs. 7 and 8.

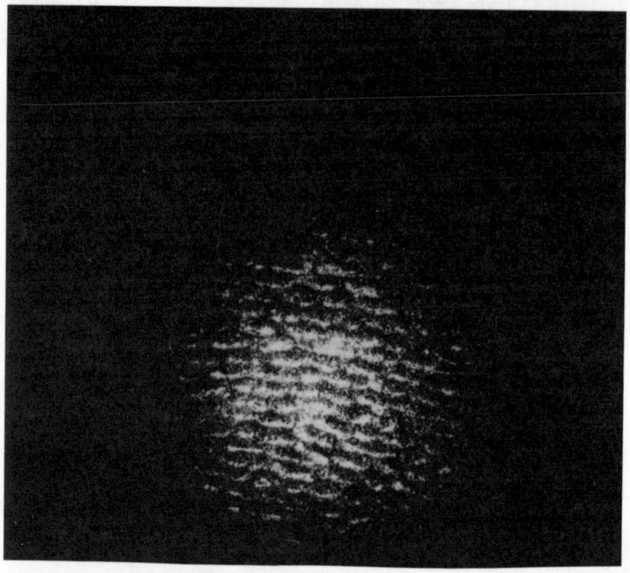

Fig. 9. Laser beam transmitted through onion skin, 0.05 mm thick.

IV. QUANTITATIVE REPRESENTATION OF ABSORPTION AND SCATTER IN A DIFFUSING MEDIUM

The transmittance, T, of an absorbing medium is defined through the Beer-Lambert law given in equation (1), which may be written in the alternative form:

$$T = \frac{I_t}{I_0} = 10^{-A} \quad . \tag{9}$$

The absorbance, A, as noted above, is linearly related to the concentration of the absorber. Quantitatively $A = \epsilon c l$ where ϵ is the molar absorptivity, c is the molar concentration, and l is the path length. Cells of fixed path length are generally used, so l may be regarded as a constant. A plot of the measured absorbance A vs. c is constructed using solutions of known concentration, and this "Beer's law plot" may then be used to determine the concentration of an unknown solution in terms of its measured absorbance. The method may be readily extended to multicomponent analysis by measurement of the absorbance

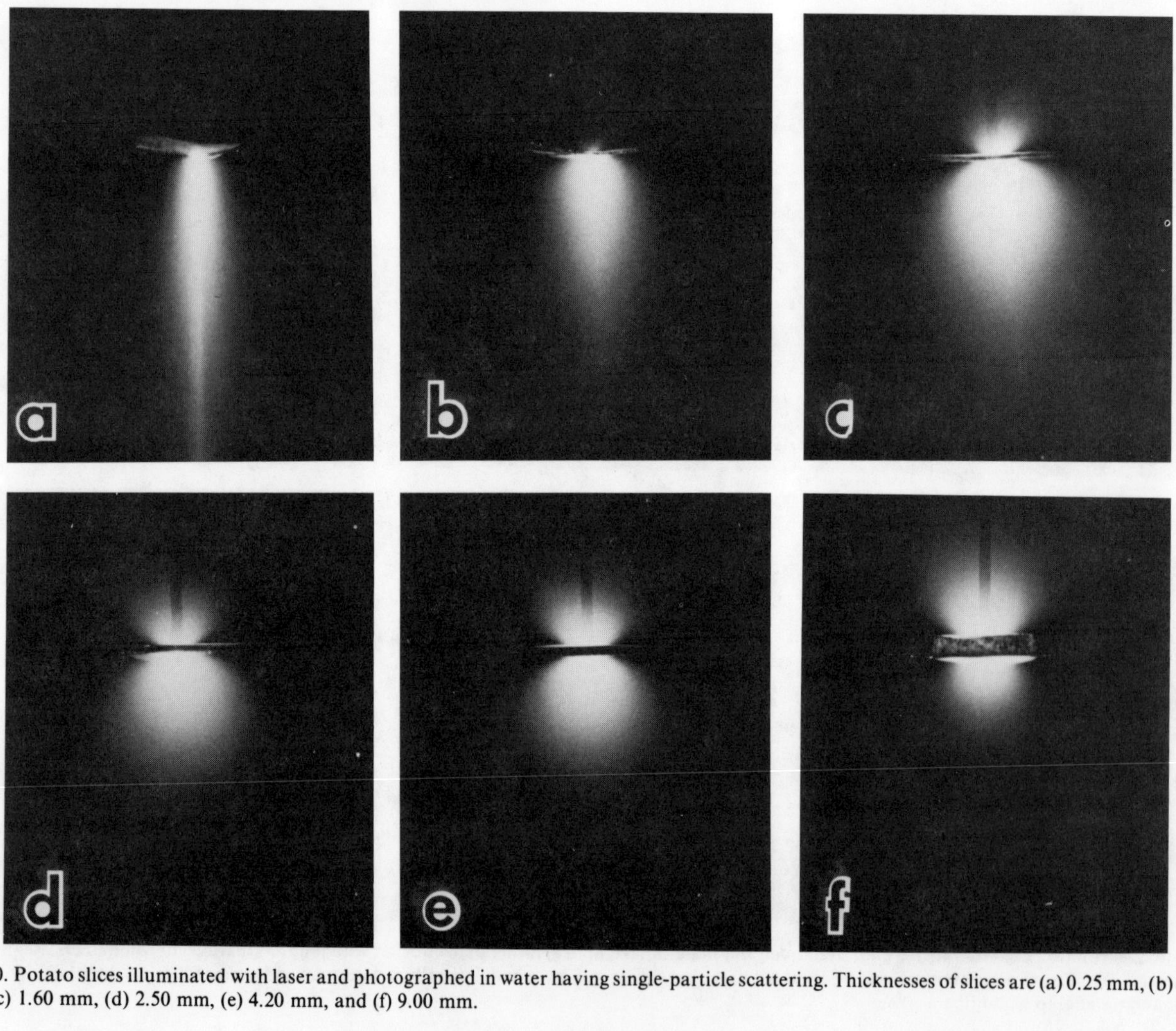

Fig. 10. Potato slices illuminated with laser and photographed in water having single-particle scattering. Thicknesses of slices are (a) 0.25 mm, (b) 0.75 mm, (c) 1.60 mm, (d) 2.50 mm, (e) 4.20 mm, and (f) 9.00 mm.

at as many wavelengths as there are components in the solution, assuming the wavelengths are chosen such that the components have widely differing absorptivities at each of the wavelengths used for measurement. The concentration of a material is related to the transmittance by:

$$c = \frac{1}{\epsilon l} \log \frac{1}{T} \quad . \tag{10}$$

Now if the medium both scatters and absorbs radiation, it may not be possible to measure the transmitted radiation in a reproducible manner. In dense scattering media the path length may be difficult to control; in fact, many systems of practical interest scatter so intensely that very little light is transmitted. Then the reflected radiation must be relied on for an optical analysis.

The mean depth of penetration of radiation into the sample, x, depends on the absorption and scattering characteristics of the medium. It is a quantity that is generally not known. If this is assumed to be determined primarily by the physical characteristics (particle size, shape, density, etc.), then it may be roughly constant within a series of similar samples. With the detector of the spectrophotometer repositioned in any of the various ways discussed in the next section, the instrument will read T, but what is in fact determined is the reflectance R_D. R_D therefore plays the same role as T in the above, and it would be expected that the concentration of the absorber should be related to it through:

$$c \propto \log 1/R \quad . \tag{11}$$

The above argument leads to a result that is very useful, but it lacks rigor in the sense that no functional relationship between c and $\log 1/R$ has been developed; however, empirically derived equations have been used extensively to implement this relation for practical applications. A treatment that is much more satisfying theoretically has been advanced by Kubelka and Munk (1931). The K-M theory assumes that the radiation that passes through a scattering medium may be divided into two fluxes. One (I) represents the radiation passed forward in the direction of the incident beam, and the other (J) represents radiation scattered back toward the illuminated surface (Fig. 11).

An equation can easily be written to describe the change dI in the intensity of the beam, I, as it passes through a small sample layer of thickness, dx. The amount of radiation absorbed is taken to be proportional to an absorption constant, K, and to the intensity of the beam itself. Since this represents a decrease, we include a minus sign and write $dI = -KIdx$. There will also be a loss due to scatter, which is represented through a second constant, S, and the total decrease in intensity is thus $dI = -KIdx - SIdx$. Now the radiation that is scattered from the I beam augments the J beam, and vice versa. Therefore the complete equation that describes the change in intensity of the I beam is:

$$dI = -(K + S)\, Idx + SJdx \quad . \tag{12}$$

By exactly the same argument, the equation for the J component is:

$$dJ = (K + S)\, Jdx - SIdx \quad . \tag{13}$$

These are a set of coupled differential equations. They may be solved using standard techniques to obtain expressions for the intensities of the beams I and J anywhere within the sample as a function of certain boundary conditions. The interested reader is referred to a paper by Kubelka (1948) where equations are given for the reflectance and the transmittance of scattering layers for various assumed conditions, such as on an ideal black background, on an ideal white background, without backing, etc. The boundary conditions of principal interest are $I = I_o$ at $x = 0$, and $J = 0$ at $x = d$. Then the solutions for T and R_D are of the form:

$$T = \frac{I_{x=d}}{I_o} = \frac{4\beta}{(1+\beta)^2 e^{\kappa d} - (1-\beta)^2 e^{-\kappa d}} \tag{14}$$

$$R_D = \frac{J_{x=o}}{I_o} = \frac{(1-\beta^2)(e^{\kappa d} - e^{-\kappa d})}{(1+\beta)^2 e^{\kappa d} - (1-\beta)^2 e^{-\kappa d}} \tag{15}$$

with:

$$\kappa = \Big[K(K + 2S)\Big]^{1/2} \text{ and } \beta = \Big[K/(K + 2S)\Big]^{1/2} \quad . \tag{16}$$

If $S = 0$, it follows that $\beta = 1$ and the above equations reduce to $R_D = 0$ and $T = e^{-\kappa d}$. This is just the Beer's law-type behavior that can be expected in the absence of scatter.

A question of considerable importance is how the two constants K and S introduced in the K-M theory should be interpreted. One would expect a relationship to exist between K and the ϵc of Beer's law, and the fact that $T \rightarrow e^{-\kappa d}$ as $S \rightarrow 0$ implies that they are in fact equal. The relationship is not quite so simple, however. The K-M theory treats all beams passing through the layer dx in the general direction of $x = d$ as part of I, and all beams passing through dx in the general direction of $x = 0$ as a part of J, but their actual path length through the layer dx

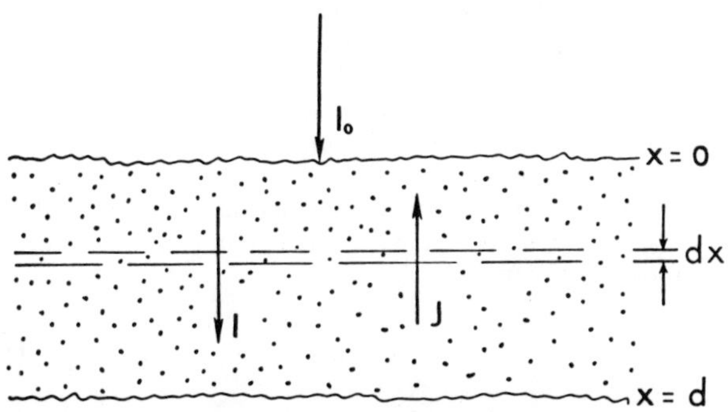

Fig. 11. Kubelka-Munk model for diffuse reflectance of scattering medium. It is assumed that the radiation field can be divided into two components, one traveling toward illuminated surface (J) and one traveling from illuminated surface (I).

TABLE I
Data for Pulverized Glass Immersed in Three Different Fluids

Material	Refractive Index	Reflectance of Glass[a]	Sample Reflectance[b]	Scatter Coefficient[c]
Glass	1.52	···	···	···
Air	1.00	0.043	0.74	46
Water	1.33	0.0044	0.44	6
Mineral oil	1.48	0.00018	0.20	1

[a]Computed values for glass in corresponding medium.
[b]Sample is pulverized glass with particles <0.125 mm. Reflectance was measured with 1.0-cm sample thickness at 5,500 A.
[c]Computed with K-M equations.

will be longer unless they happen to be traveling exactly perpendicular to the surface. It can be shown that if there are many beams equally distributed over all possible directions (i.e., an isotropic distribution within the medium), then the mean path length is twice the actual perpendicular length, dx, and $K = 2\epsilon c$. Careful measurements by various workers have shown that the proportionality factor is usually 2 or greater, values of 2.5–3.0 being typical. The larger values are probably due to the effects of total internal reflection, which tend to increase the path length even more for large angles of incidence; but the problem is a complex one involving many factors, such as the relative refractive index, particle size and shape, etc. (Hecht, 1976). Fortunately, it is generally not required that this factor be known precisely for quantitative analytical applications.

The scattering coefficient S is likewise due to the complex interplay of many factors (Hecht, 1976). It may be related to the elementary scattering process (Mudgett and Richards, 1973), but this is not usually done. As an order of magnitude estimate, S may be taken as equal to δ^{-1}, where δ is the diameter of the particles in a densely packed scattering medium (Simmons, 1972).

The K-M theory leads to a particularly simple solution for the reflectance if the sample thickness is sufficiently large so that no light passes entirely through. The reflectance in such a case is given by:

$$R = \frac{1-\beta}{1+\beta} = \frac{1-[K/(K+2S)]^{1/2}}{1+[K/(K+2S)]^{1/2}} \quad . \tag{17}$$

With some algebraic manipulation, this relationship may be written in the alternative form:

$$\frac{K}{S} = \frac{(1-R)^2}{2R} = f(R) \ . \tag{18}$$

The quantity $f(R)$ defined in this way is known as the K-M function.

In general, S depends on the wavelength of the light used for measurement. Fortunately, for many systems the wavelength dependence is small at least over a fairly limited wavelength range. Then a plot of $f(R)$ vs. wavelength should show the same spectral features as the transmission spectrum, since the two are the same except for the presence of a nearly constant scaling factor. The curve obtained in this way is often referred to as the characteristic color curve, and it can be used for identification purposes in just the same way as the transmission spectrum.

The K-M analysis requires a reference standard that is essentially nonabsorbing—i.e., one that has reflectance approaching 100%. Failure to meet this requirement may result in K having negative values for samples that have nearly zero absorption in certain spectral regions. On the other hand, the log $1/R$ expression of reflectance does not require the standard to be 100% reflecting because each application involves the use of a unique empirically derived equation. In either case, the standard must be stable with time and not be affected by normal environmental factors.

The above discussion has shown that the K-M constant K is directly proportional to the concentration, c, of the absorber, and may be written:

$$c \propto f(R) \ . \tag{19}$$

This represents a more rigorous relationship than $c \propto \log 1/R$, because log $1/R$ has a nonlinear relation with the concentration of absorbers that has not been defined mathematically. Despite that nonlinear relationship, many correlations with the concentration of an absorber show good linearity with computations based on log $1/R$ because the actual change in R for the range of measurement is, in many cases, relatively small. In addition, the computations are usually based on the shape of the reflectance spectrum rather than the specific value of R, so that equations using first or second derivative of log $1/R$ have been used extensively. In the case of samples where there is an appreciable range in reflectance due to differences in absorption or scattering, there could be sufficient change in R that a linear regression analysis based on a function of log $1/R$ would not give satisfactory results. Both relationships (equations [11] and [18]) are in wide use for quantitative analysis, particularly for spectra in the near-infrared region. This should not be surprising, in view of the fact that the two are functionally similar over narrow reflectance ranges (Fig. 12). Analysis is usually based on measurements relative to a reflectance standard (see the following section), and this also tends to normalize out any inherent differences there may be in the two functions. Rather sophisticated models have been developed by various other workers (Hecht, 1976), but the two relationships mentioned here are used almost exclusively for near-infrared reflectance analysis because of their simplicity and broad applicability.

V. THE MEASUREMENT OF DIFFUSE REFLECTANCE

It is possible to measure absolute diffuse reflectance, but the methods involve specialized equipment and a great deal of attention to detail. Although this is done in a few laboratories where the primary concern is the standardization of methods

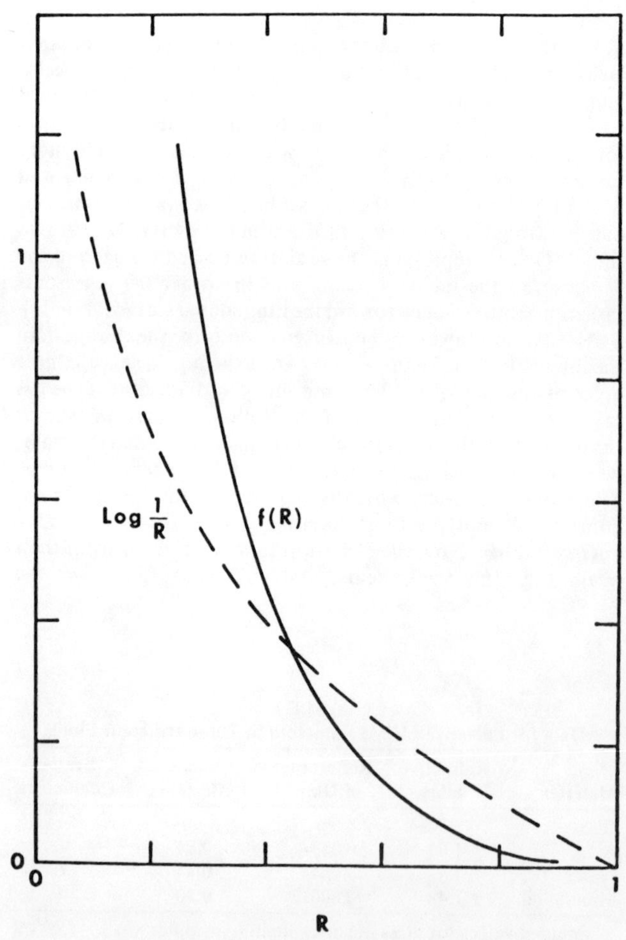

Fig. 12. Comparison of functions log $1/R$ and $f(R)$ commonly used to relate near-infrared reflectance to concentration of absorbing material.

Plate 1. Pulverized green bottle glass. Particle sizes are: (A) less than 0.125 mm, (B) 0.25–0.50 mm, and (C) 1.0–2.0 mm.

Plate 2. All three containers have pulverized glass with particles measuring less than 0.125 mm. Media surrounding the particles are: (A) air $n_1 = 1.0$, (B) water $n_1 = 1.33$, and (C) mineral oil $n_1 = 1.48$.

and materials, such procedures are not the normal concern of those involved with applied work. The common laboratory measurement is made relative to a reflectance standard of known characteristics.

For many years the primary standard of diffuse reflectance has been magnesium oxide, whose surface has been freshly smoked using a burning magnesium ribbon. Such surfaces tend to be rather fragile, and they deteriorate in time due to nitride formation, particularly if exposed to ultraviolet radiation. Thus the use of secondary standards has become common, even though they may have somewhat lower absolute reflectance than freshly smoked MgO. The reflectance relative to fresh MgO of quite a number of materials that may be used as secondary standards, such as Li_2CO_3, NaF, NaCl, $MgSO_4$, $BaSO_4$, SiO_2, Al_2O_3, and glucose, have been determined by Kortüm et al (1963). Among the more commonly used secondary standards at the present time are special preparations of $BaSO_4$, such as that marketed by Eastman for use as a white reflectance standard. The calibration of various barium sulfates, magnesium oxides, and opal glasses for use as working standards has been reviewed by Budde (1976). Powdered Teflon is becoming a popular reflectance standard for the near-infrared region (Norris et al, 1976).

Ideal diffuse reflectance follows a cosine law, as noted many years ago by Lambert. To the extent that the Lambert cosine law is obeyed, the reflectance is independent of the angle of illumination or the angle of viewing. The law is seldom obeyed perfectly in practice, however, and the measurement conditions should be specified for careful work. This is usually done by giving the angle of incidence followed by the angles of observation, either in parentheses or as subscripts to the symbol R. Some typical examples of measurement geometries are shown schematically in Fig. 13. The first two involve angular illumination and viewing, which is typical of instruments of older design. The two measurement geometries shown should yield identical results, in view of the well-known reciprocity law of Helmholtz, $R_{\theta,\phi} = R_{\phi,\theta}$. The $R_{45,0}$ design has been used by the Zeiss Company in the RA2 reflectance attachment, and the $R_{0,45}$ by Beckman and Unicam in attachments for the DU and SP.500 spectrophotometers, respectively. In the latter case, an annular mirror is used, which actually collects all radiation within the range of 35° to 55° and focuses it on the photomultiplier detector.

Naturally a great deal of light is lost in any angular measurement, and the sensitivity is greatly enhanced using an integrating sphere (Fig. 13). The sphere is coated with a highly reflective material, so that the radiation diffusely reflected in all directions by the sample (represented by the letter d) is reflected back into the sphere. This leads to a uniform illumination of the sphere wall, and thus the placement of the detector for this $R_{0,d}$ arrangement is not critical. Most modern instruments make use of an integrating sphere with either the $R_{0,d}$ or $R_{d,0}$ design.

As with other spectrophotometric measurements, either a single-beam or a double-beam mode may be used (Fig. 14). In the first case, the spectrophotometer is balanced to read 100% $R(T)$ with the reflectance standard covering the sample port. Then the percentage of reflectance is determined with the sample moved into this position. Double-beam operation makes use of phase-sensitive detection with a chopper, which alternately directs the beam at the reflectance standard and the sample. In this case, the sample and the standard remain in position at the two different ports. This introduces negligible error to the measurement as long as the areas of the ports are

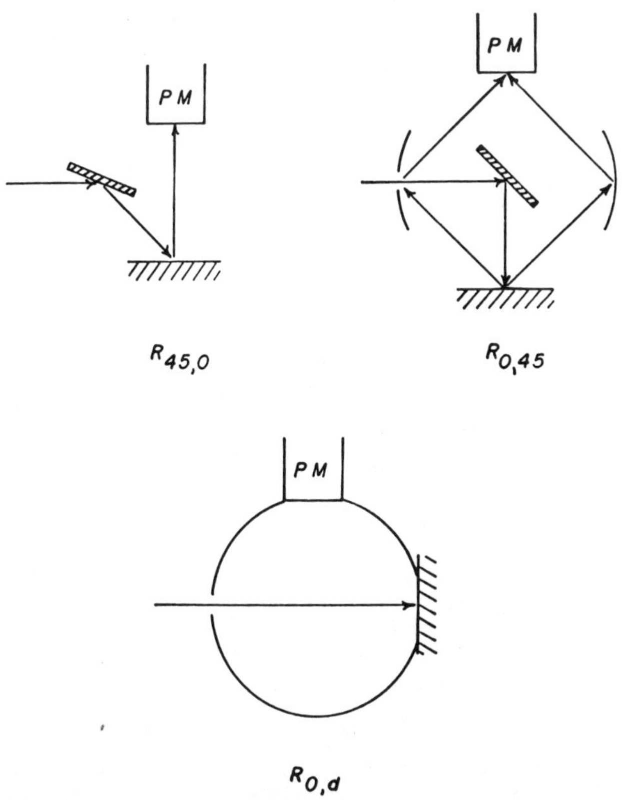

Fig. 13. Schematic representation of common measurement geometries for determination of diffuse reflectance. First subscript gives angle of incidence, second gives angle of observation. PM represents the photomultiplier tube.

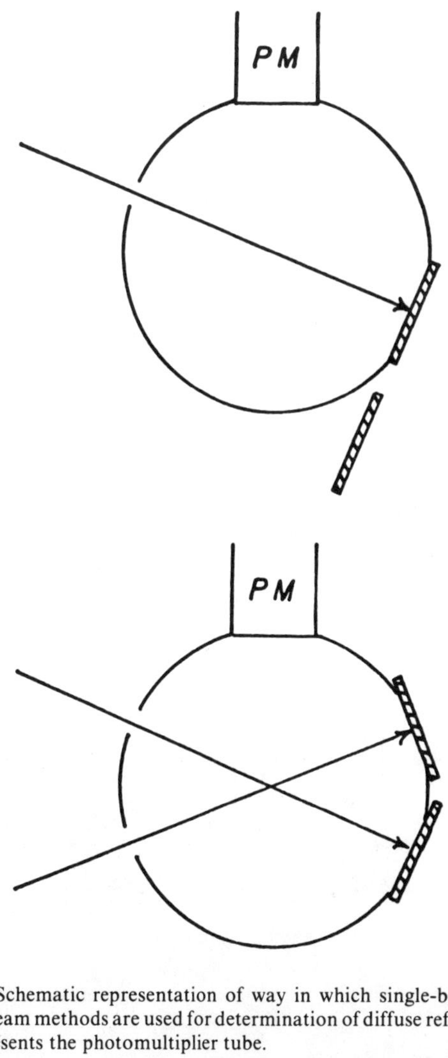

Fig. 14. Schematic representation of way in which single-beam and double-beam methods are used for determination of diffuse reflectance. PM represents the photomultiplier tube.

much smaller than the total surface area of the sphere. In an instrument design by Technicon, the light beam is directed toward the sphere wall for standardization.

The straightforward application of these measurement principles presents no particular difficulties within the visible spectral range, but some problems are encountered when one attempts to extend them to the infrared. Placing the sample before the monochromator simplifies the optics, but then it is difficult to prevent heating of the sample. Infrared detectors are of necessity rather small in the surface area, and this makes it difficult to focus the reflected radiation on the detector if the beam has been monochromated before striking the sample. Since the absorption is generally rather weak, an integrating sphere is of limited utility because its coating is apt to absorb more than the sample and the detector combined.

These difficulties have been discussed by Kortüm and Delfs (1964). They can be minimized in a reflectometer design that places the sample and detector at the two focal points of a half-ellipsoid with mirrored surfaces. High-quality diffuse reflectance spectra have been obtained by Kortüm and coworkers (1964 and 1973) within the range of 1,300–5,000 cm^{-1} using this design. These same principles have been used in the design of an attachment to fit many spectrophotometers that are commercially available (Harrick, 1967).

The problems mentioned above are not so severe as long as the measurements are limited to the near infrared; i.e., from the visible to approximately 3 μm or 3,000 cm^{-1}. A variety of geometries may be used within this region, including directional detection and the integrating sphere. The large solid-angle detector described by Norris et al (1976) and shown in Fig. 15 combines good collector efficiency with simplicity of construction. For this design, the test material is held in a glass-covered sample container. Thus, automatic devices can be used to load the samples into the reflectance measuring system. In addition, the samples can also be stored in their respective containers. However, the measurement system must be designed so that the radiation reflected from the glass cover does not contribute to the measured radiation. The detectors are positioned as close as possible to the illuminated area to intercept a large solid angle of the diffusely reflected radiation. This design is extensively used in near-infrared reflectance instruments.

Although reflectance is the most common nondestructive optical measurement for diffusely reflecting materials, it is practical to use transmittance geometry as well. In general, the transmitted radiation will be much less than for reflectance, so it is necessary to design a system that is very efficient to obtain an adequate low-noise signal from the detector. The first consideration is the wavelength range. Generally, absorption decreases with decreasing wavelength, and the 700-nm to 1,000-nm wavelength range includes weak absorption bands that are associated with most of the substances in biological materials. The most common detector for this wavelength range is silicon, and with a suitable source a dynamic range of six to seven optical density units can be obtained. The low absorption due to constituents in the sample, combined with the wide optical density range of silicon detectors, makes it possible to use transmittance geometry. The effective path length for transmittance measurements is usually longer than for reflectance, so that compensates for using the weaker absorption bands. The output from the detector is a response to transmitted radiation, but must be interpreted in the context of radiation absorbed. In addition, substances absorb at specific wavelengths, so wavelength analysis is an important part of the data processing. The radiation available at the detector can be changed by varying the sample thickness.

Figure 16 (Rosenthal, 1984) illustrates an instrument design for using transmittance geometry to measure the major constituents of wheat. The source of radiation is a 3 by 4 array (L) of infrared-emitting diodes. These diodes are selected for emission at wavelengths required for the data analysis. Filters or dispersion systems normally used with incandescent sources are not required in this design, but the resolution and precision of the measurement may be improved by incorporation of interference filters to correspond with each diode. The transmission instrument offered by Trebor Industries utilizes such filters. A Fresnel lens is used to collect the radiation from the infrared-emitting diodes so that the same area on the diffuser (D) is illuminated regardless of which diode is turned on. The diffuser is used so that the geometry of the illuminating radiation is the same for each diode. The sample thickness (*d*) can be modified as required to obtain suitable detector output. The detector (S) is located as close to the sample as possible to intercept as large a solid angle of the transmitted radiation as practical. The fluted wheel at the bottom discharges a portion of the grain so that another measurement can be made on a

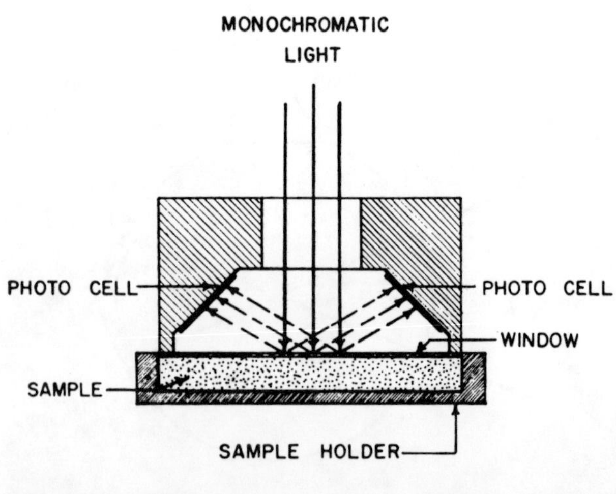

Fig. 15. Large solid-angle detector geometry for measurement of diffuse reflectance in near infrared. (Reprinted, with permission, from Norris, 1984)

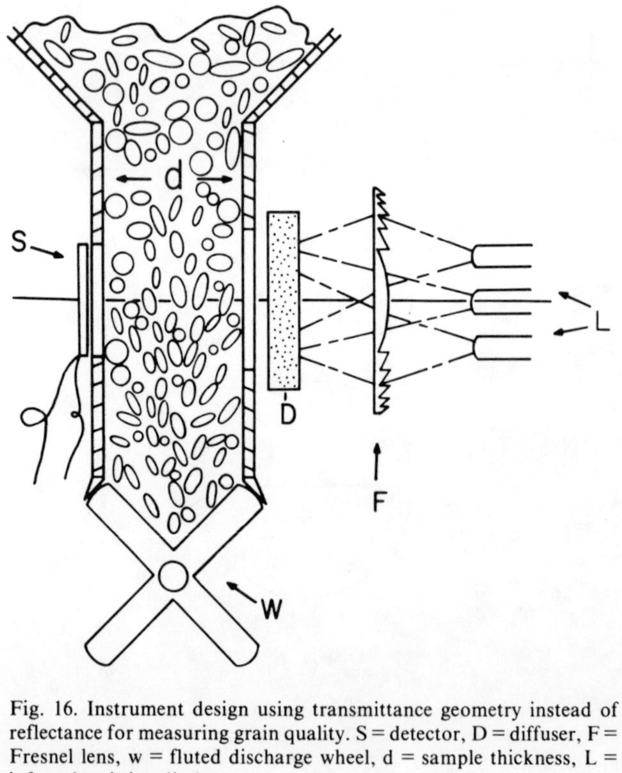

Fig. 16. Instrument design using transmittance geometry instead of reflectance for measuring grain quality. S = detector, D = diffuser, F = Fresnel lens, w = fluted discharge wheel, d = sample thickness, L = infrared emitting diodes.

different grain sample automatically. This procedure provides for using relatively large samples to reduce sampling errors. Since the measurement is not destructive, the grain sample can be returned to the source. The primary advantage of this design is that it uses relatively low-cost components and does not require the sample to be ground.

VI. KUBELKA-MUNK COEFFICIENTS FOR SMALL GRAINS

Most data processing of reflectance measurements on small grains uses the relationship given in equation (11); i.e., log $1/R$. However, some experimental results have been obtained to compute the K-M coefficients (Birth, 1985). Data for computing these coefficients were obtained using a single-beam spectrophotometer with an integrating sphere similar to that shown in Fig. 14.

For the reflectance measurements, the background of the sample was a black cavity that prevented the transmitted radiation from being reflected back into the sample. That procedure yielded data that are labeled R_0 in the K-M notation. Other measurements required for the computations are transmittance (T) and sample thickness (d).

The equations used to process the data are:

$$a = \frac{1 + R_0^2 - T^2}{2R_0} \tag{20}$$

$$b = (a^2 - 1)^{1/2} \tag{21}$$

$$K/S = a - 1 \tag{22}$$

$$Sd = 1/b\,(\sinh^{-1}(b/T) - \sinh^{-1} b) \tag{23}$$

$$S = Sd/d \tag{24}$$

$$K = (K/S)\,S\ . \tag{25}$$

Data for a sample of pulverized green bottle glass were recorded to evaluate the procedure. The sample consisted of all particles that passed through a sieve having 0.125-mm-diameter holes. Reflectance and transmittance measurements were recorded for five preparations of the sample. Each set of data was used to compute spectral curves of absorption and scattering coefficients. The five data sets were used to compute the average curve and the standard deviation. An important aspect of the results (see Fig. 17) is the comparison between the absorption coefficient and the transmittance of the intact colored glass expressed on an absorptivity scale (see Fig. 18). The agreement is good; however, the ratio between the two scales is about 5.0, whereas a value of 2.0 is predicted theoretically. Kortüm (1969) has shown results where this ratio was 2.8. The difficulty of attaining the theoretical value was discussed in Section IV of this chapter. The failure to predict the actual value of the absorptivity results in increased reliance on empirical procedures.

The scatter coefficient is not expected to have much wavelength dependence for a sample having particles several times larger than the wavelength of the radiation. However, the index of refraction does decrease with increasing wavelength and could result in the slight decrease in the scatter coefficient with increasing wavelength. The spectral scatter coefficients for four small grains (after grinding) are shown in Fig. 19. Since the coefficients exhibit very little wavelength dependence, the mean values and standard deviations were computed. Those quantities are given in Table II, and the particle analysis is available in Table III.

The absorption coefficients for the ground grain are shown in Fig. 20. The weak absorption near 1,000 nm is probably due to carbohydrates. The values for barley and oats are probably too high, but the value for wheat is reasonable. Since the absorption coefficients vary with wavelength, the standard deviation is expressed as a percentage of the mean; i.e., the coefficient of variation.

The K-M coefficients for intact grain are shown in Figs. 21 and 22. These data indicate that the light scattering is less than 1/10 of that for the ground samples. This difference is probably associated with the relatively large voids between the kernels as compared with the voids in the ground grain.

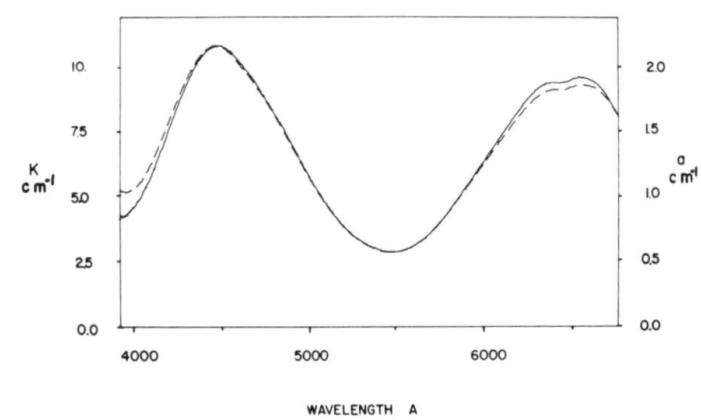

Fig. 18. Spectral absorption (K) characteristics of pulverized green bottle glass (dashed line) and absorptivity data (a) for intact clear glass (solid line). Coefficient of variation for pulverized glass was 4.3%.

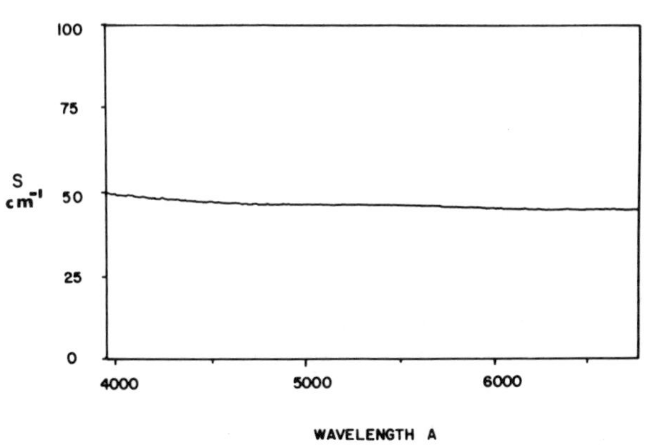

Fig. 17. Light-scattering characteristics of pulverized glass having particle size less than 0.125 mm. Data were computed with K-M equations using reflectance and transmittance measurements. Standard deviation for these data is 2.43.

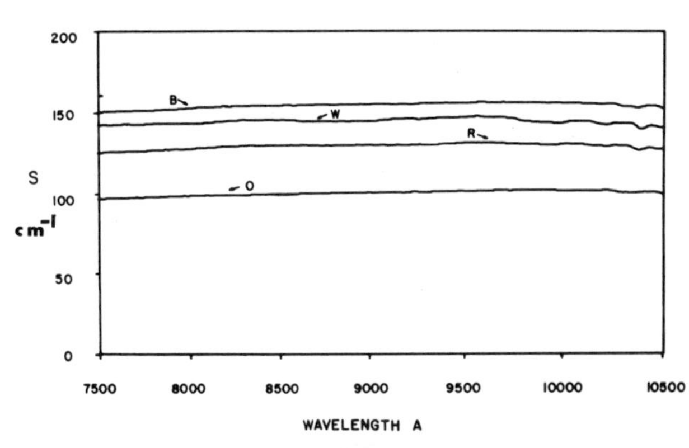

Fig. 19. K-M scatter coefficient vs. wavelength for four ground grains. Since there is relatively small wavelength dependence, mean and standard deviation were computed for each grain (see Table II). Letters identify the specific grains (barley, wheat, rye, and oats).

There is a large difference in the scatter coefficient between the ground grain and the intact grain, whereas the absorption for wheat is nearly the same for both conditions, being slightly higher for intact grain. The absorption curves should be nearly

TABLE II
Average K-M Scattering Coefficients (*S*) of Four Grains

	Ground Grain			Intact Grain		
	Mean *S* (cm^{-1})	SD (cm^{-1})	d^a (cm)	Mean *S* (cm^{-1})	SD (cm^{-1})	d^a (cm)
Oats	100	6.5	0.15	8.0	0.44	1.0
Barley	154	16.9	0.10	8.9	0.33	0.5
Rye	129	12.7	0.20	7.7	0.22	1.0
Wheat	145	15.9	0.20	6.9	0.13	1.0

[a]Sample thickness used for recording data.

TABLE III
Particle Size Distribution of Ground Grain

	Sieve Size (mm)				
	1.0	0.5	0.25	0.125	>0.125
Oats	0.01[a]	0.11	0.38	0.31	0.19
Barley	0.02	0.14	0.31	0.19	0.33
Rye	0.01	0.06	0.22	0.24	0.47
Wheat	0.01	0.05	0.16	0.20	0.58

[a]Weight fraction of sample remaining on respective sieve.

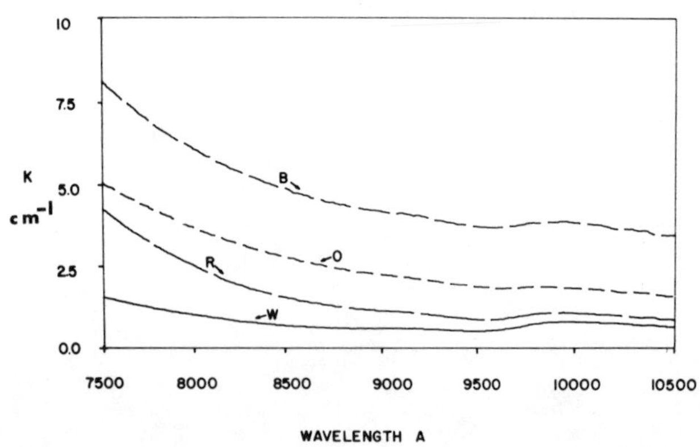

Fig. 20. Spectral absorption coefficient curves for ground grain. Coefficients of variation for these data were: barley, 5.1%; oats, 5.8%; rye, 8.5%; and wheat, 11.2%.

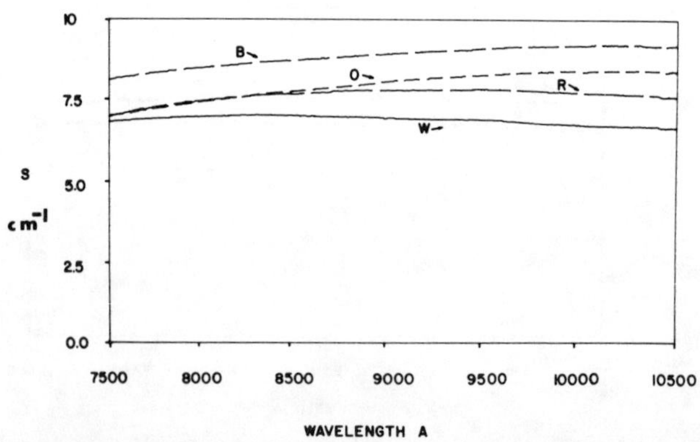

Fig. 21. Spectral scatter coefficient data for four grains (barley, oats, rye, and wheat), using intact grain. Averages and standard deviations are available in Table II.

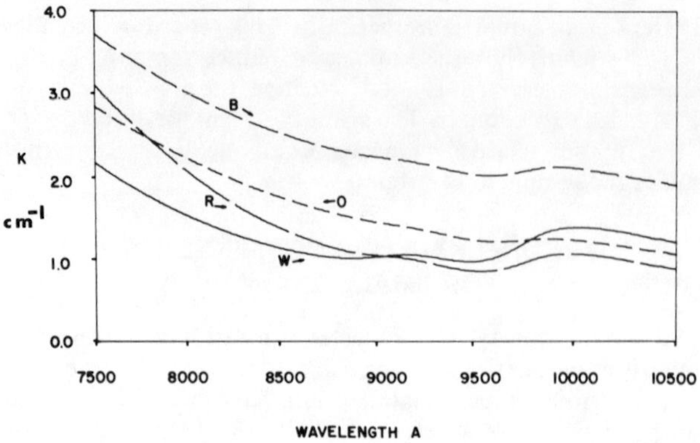

Fig. 22. Spectral absorption data for four grains, using intact grain. Coefficients of variation for these samples were: barley, 3.6%; oats, 5.7%; rye, 7.3%; wheat, 2.8%.

the same since the same material was used for both samples; however, the quantity of material was less for the ground grain, so that sufficient radiation would be transmitted to be measurable. On the other hand, the value of *K* for ground barley was nearly twice the value obtained using intact grain. In addition to the differences in the size of the voids in the two sample preparations, the distribution of constituents is also much different, so although the two sets of curves are expected to be similar, one should not expect them to be the same.

VII. SUMMARY

Only one theoretical analysis of diffuse reflectance has been described in this chapter. Although several attempts at developing a theoretical treatment have been made, the approach that gives usable results is empirical. One of the most significant values of the theoretical analyses is that they help to bring about a better understanding of the phenomena.

When the problem under consideration is a function of light scattering, empirical procedures would still be applicable, but the data-processing techniques may be different from those used for compositional analysis. Wavelength analysis is a crucial part of data treatment for determining composition, but light scattering may have very little wavelength dependence. If there is a portion of the reflectance spectra with no absorption, then the reflectance is only a function of light scattering and sample thickness.

LITERATURE CITED

BERRY, E. M. 1923. Diffuse reflection of light from a matt surface. J. Opt. Soc. Am. 7:627-633.

BIRTH, G. S. 1978. The light scattering properties of foods. J. Food Sci. 43:916-925.

BIRTH, G. S. 1986. The light scattering characteristics of ground grains. Int. Agrophys. 2:59-67.

BOUGUER, P. 1760. Traite d'optique sur la gradation de la lumiere. Paris.

BUDDE, W. 1976. Calibration of reflectance standards. J. Res. Natl. Bur. Stand. 80A:585-595.

GATES, D. M. et al. 1965. Spectral properties of plants. Appl. Opt. 4:11-20.

GAUSMAN, H. W. et al. 1969. Reflectance of cotton leaves and their structure. Remote Sensing Environ. 1:19-22.

GRABOWSKI, L. 1914. On the theoretical photometry of diffuse reflection. Astrophys. J. 39:299-306.

HARRICK, N. J. 1967. Internal Reflection Spectroscopy. Interscience, New York. 327 pp.

HECHT, H. G. 1976. The interpretation of diffuse reflectance spectra. J. Res. Natl. Bur. Stand. 80A:567-583.

KORTÜM, G. 1969. Reflectance Spectroscopy: Principles, Methods, Applications. Springer-Verlag, New York. 366 pp.

KORTÜM, G., BRAUN, W., and HERZOG, G. 1963. Principles and techniques of diffuse reflectance spectroscopy. Angew. Chem. Int. Ed. 2:333-341.

KORTÜM, G., and DELFS, H. 1964. Reflexionsspektrometrische messungen im infraroten spektralgebiet. Spectrochim. Acta 20:405-413.

KORTÜM, G., and KNEHR, H. 1973. Infrared reflection spectroscopic studies on the absorption of nitrogen oxide by zinc oxide. Ber. Bunsenges. Phys. Chem. 77:85.

KORTÜM, G., and VOGEL, J. 1958. Über reguläre und diffuse reflexion an pulvern und ihre abhängigkeit von der korngröbe. Z. Phys. Chem. N.F. (Frankfurt) 18:230-241.

KUBELKA, P. 1948. New contributions to the optics of intensely light scattering materials. J. Opt. Soc. Am. 38:448-457; errata 38:1067.

KUBELKA, P., and MUNK, F. 1931. Ein beitrag zur optik der farbanstriche. Z. Tech. Phys. 12:593-601.

KUMAR, R., and SILVA, L. 1973. Light ray tracing through a leaf cross section. Appl. Opt. 12:2950-2954.

LOMMEL, E. 1887. Die photometrie der diffusen Zurückwerfung. Sitzungsber. Munch. Akad. II. Kl., 17:95.

MIDDLETON, W. E. K., and MUNGALL, A. G. 1952. The luminous directional reflectance of snow. J. Opt. Soc. Am. 42:572-579.

MUDGETT, P. S., and RICHARDS, L. W. 1973. Kubelka-Munk scattering and absorption coefficients for use with glossy, opaque objects. J. Paint Techol. 45:43-53.

NORRIS, K. H. 1984. Reflectance Spectroscopy in Modern Methods of Food Analysis. AVI, Westport, Conn. 421 pp.

NORRIS, K. H. et al. 1976. Predicting forage quality by infrared reflectance spectroscopy. J. Anim. Sci. 43:889-897.

POKROWSKI, G. I. 1924. Zur theorie der diffusen lichtreflexion. I. Z. Phys. 35:34-37.

ROSENTHAL, R. D. 1984. Apparatus for near infrared quantitative measurement with temperature variation correction. U.S. patent 4,466,076, August 14.

SCHULTZ, H. 1924. Uber glanz und glanzmessung. Z. Tech. Phys. 5:135-141.

SEELIGER, R. V. 1888. Zur Photometrie zerstreut reflectirender Substanzen. Sitzungsber. Munch. Akad. II. Kl., 18:201-248.

SIMMONS, E. L. 1972. Relation of the diffuse reflectance remission function to the fundamental optical parameters. Opt. Acta 19:845-851.

SUITS, G. H. 1985. Natural sources. Chap. 3 in: The Infrared Handbook, revised ed. W. L. Wolfe and G. J. Zissis, eds. The Infrared Information and Analysis (IRIA) Center, Environmental Research Institute of Michigan, Ann Arbor.

WEAST, R. C., ed. 1984. Handbook of Chemistry and Physics. CRC Press, Cleveland.

WILLSTATTER, R., and STOLL, A. 1918. Untersuchungen uber die Assimilation der Kohlensause. Springer-Verlag, Berlin. 448 pp.

CHAPTER 2

Chemical Principles of Near-Infrared Technology

I. MURRAY
North Scotland College of Agriculture
Aberdeen AB9 1UD
Scotland, United Kingdom

P. C. WILLIAMS
Grain Research Laboratory
Canadian Grain Commission
Winnipeg, Manitoba, Canada

I. INTRODUCTION

The composition and behavior of all plant and animal materials are direct consequences of their chemical makeup. The chief components of all natural substances are proteins, nonfibrous carbohydrates such as starch and sugars, fats (oils), fibrous carbohydrates such as cellulose, moisture, minerals, and vitamins. Proteins, carbohydrates, and fats (or oils) are complicated molecules that are themselves composed of simpler chemical compounds such as amino acids, monosaccharide and disaccharide sugars, fatty acids, and glycerol. Their spatial arrangement, intercombination, degree of hydration, and electrostatic and covalent bonding capacity interact to create a wide range of chemical and physicochemical properties. These properties are utilized in the furnishing and preparation of foods and food substances for animals including humans.

The determination of the chemical constitution and physicochemical behavior of foods and their raw materials is essential to their efficient utilization in nutrition and marketing. For over 100 years, agricultural and food chemists have developed and applied chemical methods for the determination of composition and behavior. Many of the older methods still practiced represented the best efforts of earlier generations of chemists to assign a numerical value to a constituent that was believed to affect the processing quality of a commodity. These tests were empirical, being operationally defined and thus difficult to standardize, and very susceptible to operator and interlaboratory error. Near-infrared reflectance (NIR) spectroscopy is a technology that is for the most part based on the actual number of molecules of individual constituents present, and theoretically it presents a highly practicable analytical tool. Improvements in the development of modern techniques for standard analyses have facilitated the necessary calibration.

All organic matter consists of atoms, mainly carbon, oxygen, hydrogen, nitrogen, phosphorus, and sulphur, with minor amounts of other elements. These atoms combine by covalent and electrovalent bonds to form molecules. Because of the nature of the bonds, and the electrostatic charges on the atoms and molecules themselves, the molecules are constantly in motion; this is known as the *ground state*. The molecules vibrate at frequencies corresponding to wavelengths in the infrared region of the electromagnetic spectrum. NIR spectroscopic analysis affords a method for the harnessing and translation of these vibrations into simple, very rapid, and nonpolluting analytical procedures.

II. ENERGY AND MOLECULES

When molecules are irradiated with an external source of energy they acquire the potential for energy changes. The electromagnetic spectrum consists of energy vibrations ranging from wavelengths several meters in length (electron spin resonance and nuclear magnetic resonance) to less than 10^{-2} nanometers (cosmic rays). A micrometer (micron) is 1×10^{-6} meter (one thousandth of a millimeter), and a nanometer (nm) is one thousandth of a micrometer. These are the units most commonly encountered in NIR technology. The electromagnetic spectrum is summarized in Fig. 1. The visible area covers from about 300 to 750 nm, the near-infrared region lies between 750 and 2,600 nm, and the infrared region between 2,600 and 25,000 nm.

Electromagnetic radiation may be expressed in terms of frequency, ν; wavelength, λ; or wave number, $\bar{\nu}$. Frequency, which is expressed as reciprocal seconds (sec^{-1}), refers to the number of complete "waves" that occur in unit time (usually 1 sec). Wavelength (λ) is the distance in micrometers or nanometers between equivalent points on successive waves, and the wave number is the number of waves in a 1-cm wave train, expressed in reciprocal centimeters. The three parameters are related by:

$$\lambda\nu = c \qquad (1)$$

$$\bar{\nu} = 1/\lambda \qquad (2)$$

where ν is the frequency in cycles per second (sometimes called hertz, Hz), λ is the wavelength, and c is the velocity of light, 2.998×10^{10} cm/sec. Molecules and radiation have two things in common, namely energy and frequency. The units of radiation are referred to as photons, and the energy of a photon is given by:

$$Ep = h\nu \text{ or } E = hc\bar{\nu}\ , \qquad (3)$$

where h is Planck's constant, 6.626×10^{-27} erg-sec, so that the energy of a photon is related to the wavelength at which the irradiation is emitted. Photon energy can be absorbed by molecules, in which case the rotational, vibrational, or electronic energy of the molecule will be elevated by a discrete amount ΔEm that is exactly equivalent to the energy of the photon. Molecules can only absorb photons that are of an energy coincident with the characteristic vibrations of the molecule and that can result in the molecule being excited to a higher energy level. Figure 2 illustrates the absorption of a photon by a molecule. Photons that possess twice or three times the amount of energy necessary to elevate a molecule to a different level will bring about changes to the second or third energy levels, thereby creating first and second or higher overtones.

When energy in the form of infrared radiation is absorbed by a molecule, the energy status of the molecule changes. The vibrational and rotational quantum numbers are representative of these changes. The magnitude of rotational energy is lower than that of vibrational energy, but the energy level of a molecule is the sum of the vibrational and rotational energies. Most of the molecules in a substance are normally in the lowest energy level, or ground state. For this state, the vibrational energy can be represented as ν and the rotational energy by j. When an external source of energy is supplied, for example, by irradiation with a tungsten/halogen (white light) lamp, some of the molecules will absorb photon energy and jump from the ground state to the next highest energy level, represented by ν_1 for vibrational and j_1 for rotational energy. The changes in energy are represented respectively by $\Delta\nu$ and Δj, and the combined changes in vibrational and rotational energies constitute the absorption process. The fundamental absorption band is usually designated $\nu_1 j_1$. Fundamental absorptions usually occur in the mid-infrared region between 2,500 and 15,000 nm. Other molecules will absorb enough energy to reach the second energy level, represented by $\nu_2 j_2$. Those changes in vibrational and rotational energies constitute the absorbance represented by the first overtone band. Since fewer molecules reach this level, the first overtone band is generally much weaker in intensity than the fundamental absorption. Still fewer molecules reach the third level ($\nu_3 j_3$), which is the second overtone. First and second overtones occur at approximately one half and one third of the wavelength of the fundamental. For example, a fundamental at 3,100 nm may be expected to have a first overtone near 1,640 nm, and a second overtone near 1,100 nm. Higher energy levels are available, but fewer and fewer molecules attain these levels, so that third and higher overtones appear as weaker bands relative to the fundamental.

For convenience in most discussions concerning near-infrared spectroscopy, frequency is referred to in terms of wavelength. The formula for computing wavelength from frequency is:

$$\lambda\ (\text{nanometers}) = \frac{10{,}000{,}000}{\nu_1\ (\text{cm}^{-1})}\,. \tag{4}$$

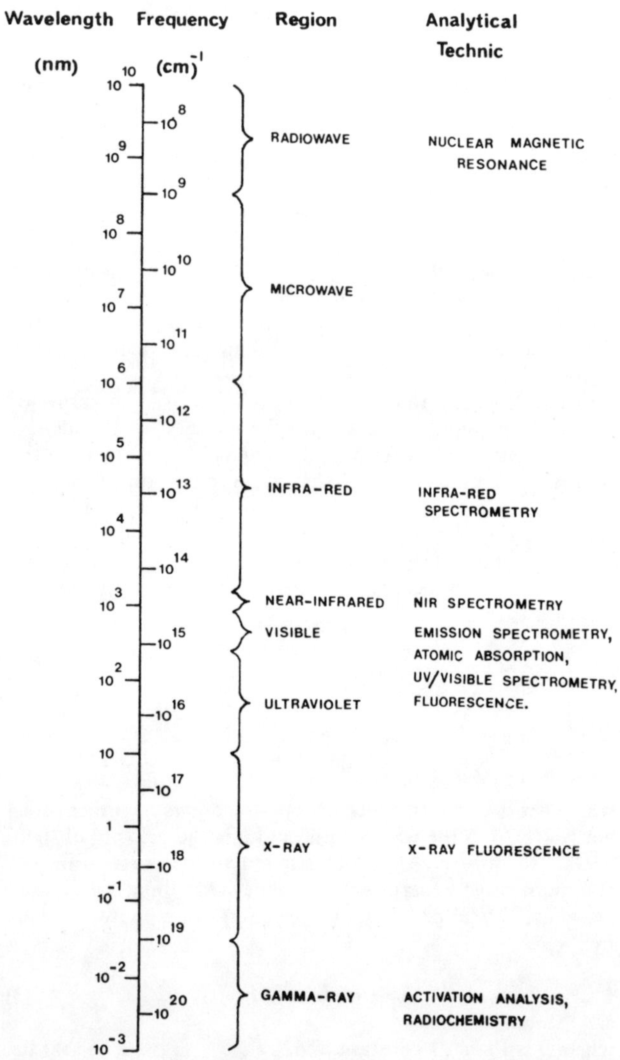

Fig. 1. Electromagnetic spectrum.

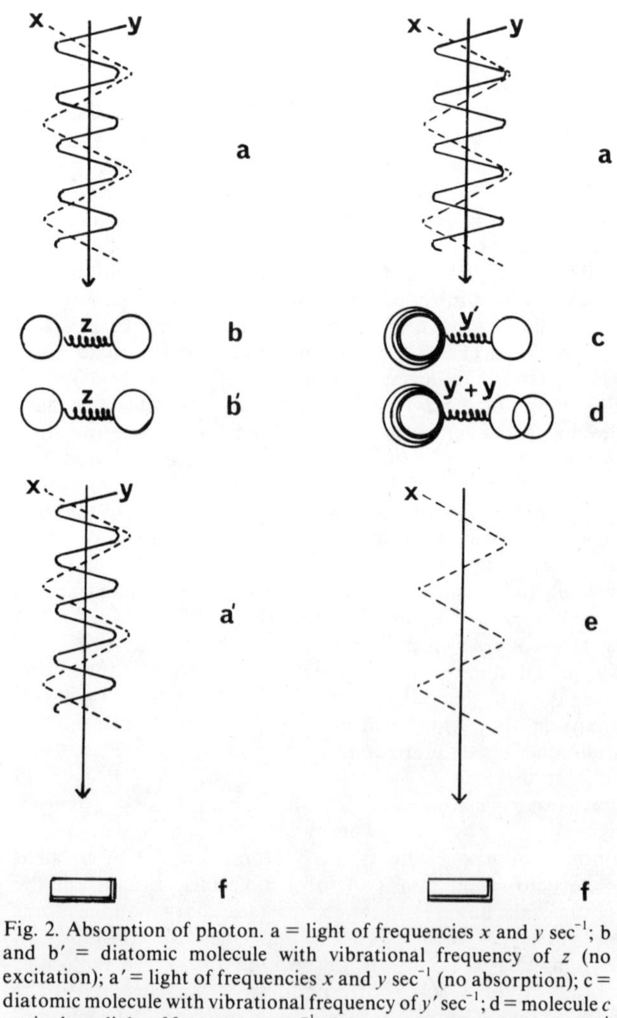

Fig. 2. Absorption of photon. a = light of frequencies x and y sec^{-1}; b and b′ = diatomic molecule with vibrational frequency of z (no excitation); a′ = light of frequencies x and y sec^{-1} (no absorption); c = diatomic molecule with vibrational frequency of y' sec^{-1}; d = molecule c excited; e = light of frequency x sec^{-1} (light photons of frequency y sec^{-1} have been absorbed); and f = detector.

For example, an absorption band occurring at a frequency of 2,300 cm^{-1} is equivalent to one with a wavelength given by:

$$\lambda nm = \frac{10{,}000{,}000}{2{,}300} = 4{,}348 \text{ nm} .$$

Similarly, the approximate wavelength of the first overtone (ν_2) of such a vibration is given by:

$$\lambda nm = \frac{10{,}000{,}000}{2\nu_1} = \frac{10{,}000{,}000}{4{,}600} = 2{,}174 \text{ nm} .$$

and the approximate wavelength of the second overtone by:

$$\lambda nm = \frac{10{,}000{,}000}{3\nu_1} = 1{,}449 \text{ nm} .$$

Because the vibration of covalent bonds in a molecule is not strictly simple harmonic motion obeying Hooke's law, the quantized vibrational energy levels are not equally spaced within the potential energy well. This phenomenon is known as *anharmonicity* and results in the frequency of overtones being slightly less than whole-number multiples of the fundamental frequency. As a result, first, second, and higher overtones, or harmonics, occur at slightly longer wavelengths than expected.

For reasons of sensitivity and signal-to-noise ratio, lead sulfide detectors are usually used between 1,000 and 2,600 nm, and silicon detectors are used at shorter wavelengths. Only the overtones from certain fundamental vibrations attain practical significance in quantitative NIR analysis. Others are less effective because of low intensity, occurrence at wavelengths outside of the optimal range of the detectors, and interference from other absorbers.

III. VIBRATIONAL AND ROTATIONAL ENERGY

The two main types of molecular motion are caused by vibrational and rotational energy transitions. Vibrational motion is created by movements of the atoms toward and away from each other in a manner analogous to a continuously oscillating spring. Rotational motion is created by rotation about the molecular axes. Vibrational energy changes are usually accompanied by a larger number of rotational energy changes appearing as rotational fine structure in the gaseous phase spectra. As a result, vibrational spectra appear as bands. All molecules and parts of molecules have vibrations at characteristic wavelengths. In each normal mode of vibration, all of the functional groups of atoms of the same type in a molecule vibrate with the same frequency. Similarly, all molecules of the same substance vibrate at the same frequency in the ground state, although this may not be strictly true in the case of complex molecules such as proteins. For example, these may become hydrated to different degrees in different samples, due to variations in the active water molecules available. This may cause differences in the vibration pattern of these molecules, which in turn may result in small wavelength shifts between samples.

There are two main modes of molecular vibrations, stretching and bending. Stretching is movement along the axes of bonds, so that the distance between atoms changes rhythmically. Bending vibrations may involve changes in bond angles between atoms, or movements of groups of atoms with respect to the rest of the molecule without movement of the atoms in the group with respect to one another. Twisting, wagging, scissoring, and rocking vibrations involve changes in bond angles. Only vibrations that result in rhythmic changes in the dipole moment of a molecule can cause absorbances in the infrared.

Polyatomic molecules possess three internal degrees of freedom, corresponding to the x, y, and z Cartesian axes. A molecule with n atoms possesses $3n$ degrees of freedom. For nonlinear molecules, three of these describe rotation and three describe translation of the molecule as a rigid unit. The remaining ($3n - 6$) degrees of freedom describe motions of the nuclei relative to each other, with the system fixed in space. In other words, they describe vibrational motions, all of which are theoretically capable of absorbing photon energy, and so there are $3n - 6$ possible fundamental absorbances for a nonlinear molecule. Of these, $n - 1$ are stretching motions and $2n - 5$ are bending motions. There are more deformation or bending than stretching vibrations, but stretching vibrations are usually the more intense. For linear molecules, only two degrees of freedom are required to describe rotation because rotation about the molecular axis is not measurable, so that there are ($3n - 5$) degrees of freedom available for vibrational motions. For molecules with large numbers of atoms, many fundamental vibrations are possible. For example, the alanine molecule, with 13 atoms, has 33 possible fundamental vibrations and nearly 100 absorbances are possible, assuming only fundamentals and first and second overtones and excluding combination bands, difference tones, and higher overtones.

For practical purposes, the absorbance of overtones higher than the third is of little significance in quantitative NIR spectroscopy. The intensities are very low and likely to be masked by higher intensity overtones of different molecular types occurring in the same frequency range. Information is repeated at intervals toward shorter wavelengths, and in an attenuated form, which can perform the function of the analyst's dilution series. Intensity of absorption falls off very rapidly at higher overtones. Table I illustrates the rate of fall of the intensity of absorbance for a typical polyatomic molecule.

Figure 3 illustrates the types of vibrations possible in a triatomic linear molecule. Of these vibrations, ν_1 represents the symmetrical stretching motion, ν_2 the asymmetrical stretching, and ν_3 the bending motion. Figure 4 illustrates these same vibrational modes in a nonlinear triatomic molecule. If the electrical symmetry of the molecule is unaffected by the sequential displacements of the atoms, the vibration does not couple with the energy of the photon. While such a vibration may occur when molecules collide, no absorption of photons accompanies the vibration. For example, there is no net change in dipole moment for the molecule as a whole accompanying the vibration ν_1 for a linear triatomic molecule, but there is a net change in dipole moment for this vibration in a nonlinear triatomic molecule. Consequently, there are only two fundamental bands detectable or active in the vibrational spectrum of the linear triatomic molecule, whereas three are seen for a nonlinear triatomic molecule. Symmetry conditions existing for molecules composed of more than three atoms can also result in overlap of absorption bands accompanying fundamental vibrations, and fewer than $3n - 6$ fundamental bands are observed.

The above $3n - 6$ rule for the number of absorption bands applies only to the number of *fundamental* bands. It does not

TABLE I
Typical Changes in Intensity of Absorbance at Different Overtone Levels

Transition (ν_0 to ν_n)	Overtone No.	Relative Absorbance for 1-cm Cell
1	Fundamental	100
2	First	9
3	Second	0.3
4	Third	0.01

apply to the number of overtone and combination bands. Combination bands occur when the absorbed photon excites two or more vibrations simultaneously. For this to happen, the energy of the photon has to equal the sum of the energies of the coupling vibrations. Combination bands are also influenced by the anharmonicity of the vibrations and, for this reason, the important bands involve motions of hydrogen atoms and occur in the near infrared. To a first approximation, the absorption intensities decrease with the number of different vibrations that couple. The strongest combination bands are found between the wavelengths of the C-H stretching vibration and its first overtone, i.e., between approximately 1,500 and 3,000 nm.

It should also be noted that an "inactive" vibration such as ν_1 for a linear triatomic molecule may couple with an "active" vibration such as ν_3 to produce an "active" combination band.

Yet another source of absorbance is Fermi resonance. Sometimes the frequency of a fundamental occurs close to that of an overtone or combination band involving a different fundamental. When this happens, two bands may appear instead of the single absorbance anticipated for the fundamental. For example, the $\nu = 1$ level of the fundamental may be close to the $\nu = 2$ level of the other overtone. The two absorbances appear with similar intensity at frequencies higher and lower than the expected positions of the fundamental and overtone (Fig. 5). Both bands involve both the fundamental and the overtone. The intensification of the overtone is caused by the influence of the fundamental. The fundamental and the overtone should be of the same symmetry type for Fermi resonance to occur.

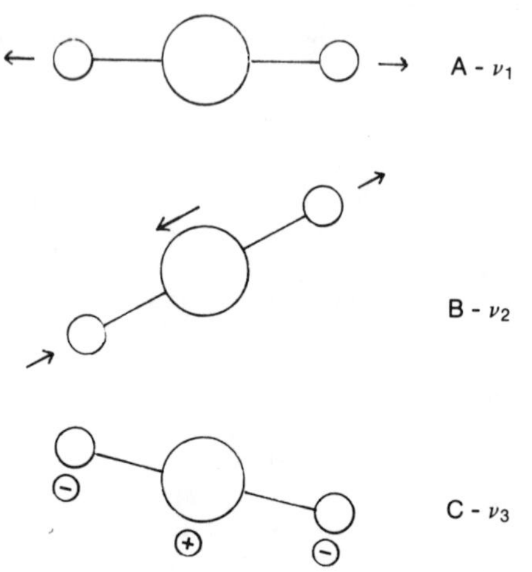

Fig. 3. Typical motions of linear triatomic molecule: (A) symmetrical stretching; (B) asymmetrical stretching; and (C) bending (scissoring), showing orientation above (circled +) and below (circled −) plane of page.

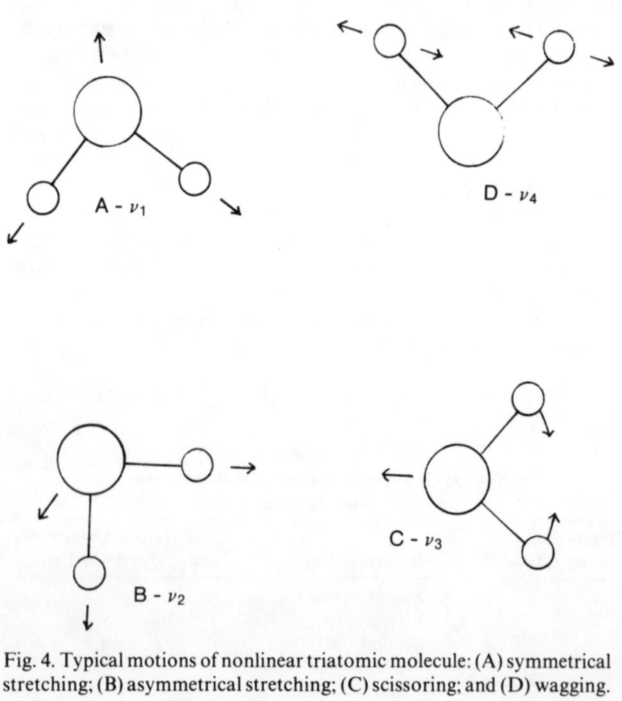

Fig. 4. Typical motions of nonlinear triatomic molecule: (A) symmetrical stretching; (B) asymmetrical stretching; (C) scissoring; and (D) wagging.

The popularity of NIR spectroscopy for the analysis of foods and related materials is associated with the intensities of absorption (reflectance) of solids in this region. The absorption of fundamental bands is so high that most materials appear essentially "black" at wavelengths longer than 2,500 nm and also almost "white" at wavelengths shorter than about 1,000 nm (ignoring electronic absorption, which can introduce absorption in the visible and ultraviolet regions). Thus, combination and overtone bands are of most interest to the analyst. However, this raises a question about the applicability of the above rules for vibration and absorption in solids, particularly crystalline solids (and most natural substances are crystalline over short ranges). In crystals, the molecules are "frozen" into exact spatial relationships. The force fields of adjacent molecules then interact in complicated ways that can cause vibration bands to split and greatly increase the number of absorption bands. These bands are seldom resolvable. While it is advisable to understand the basics of the absorption process in the near infrared, it should be recognized that NIR techniques are always likely to be empirical.

In practice, the number of fundamental bands observed is rarely exactly $3n - 6$ for any molecule. The number can be decreased by several factors, including a high degree of molecular symmetry, which causes certain pairs or triads of frequencies to be exactly identical. These vibrations are observed only as a single band. Some fundamentals occur so close together that they are not separated by the spectrometer, and others are too weak to be observed. Finally, some bands occur outside of the range of the spectrometer.

IV. DIPOLES AND ABSORBANCE

Any pair of bound atoms in a molecule will exhibit a natural oscillation and will vibrate at a specific frequency given by the equation:

$$\nu = \sqrt{\frac{k}{\mu}} \cdot \frac{1}{2\pi} \tag{5}$$

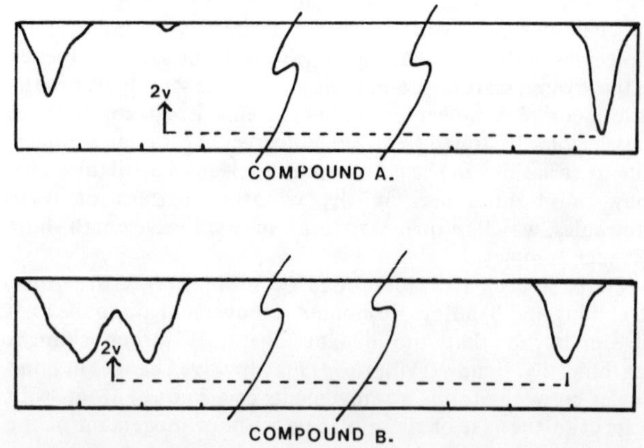

Fig. 5. Fermi resonance. Band expected to occur in spectrum of compound because of similarity to another compound actually appears as two bands.

where k is the force constant and μ is the reduced mass of the weights:

$$\mu = m_1 m_2 / (m_1 + m_2) . \tag{6}$$

Here m_1 and m_2 are the masses of the atoms. The masses of atoms and the binding forces holding atoms together in a chemical bond combine to form harmonic oscillators vibrating at frequencies corresponding to infrared electromagnetic radiation.

If a pair of atoms has an unequal distribution of electric charge it is said to form a dipole and hence to possess a local electric field. This field will strongly couple to the field of any passing light beam. Diatomic molecules such as pure hydrogen have a symmetrical distribution of charge, thus they form no dipole and do not absorb at infrared frequencies, whereas molecules such as hydrogen chloride form a dipole and absorb strongly. Only molecules that possess a permanent displacement of electric charge (dipole) are capable of absorbances in the infrared region. The magnitude of a dipole is a function of either charge in the dipole multiplied by the charge spacing or molecular distance. Dipole magnitude is usually expressed in Debye units and ranges from 0 in the case of molecules such as CH_4 with complete symmetry to highly polarized molecules such as orthochloronitrobenzene with a dipole of 4.7 units. If two different molecules vibrate at the same frequency, they will absorb photons at the same frequency, but the molecule with the greater dipole will absorb the radiation with more intensity.

A permanent dipole means that the centers of positive and negative electricity of the molecules, or parts of molecules, do not coincide. Practically every substance occurring in plants and animals and their derived products consists of molecules that individually possess dipoles of various magnitudes. The complexity of the molecules themselves together with other factors such as hydrogen bonding complicate the absorption spectra of natural materials, including all commodities of plant or animal origin.

Hydrogen bonding can occur whenever the *s*-orbital of the hydrogen atom can overlap the *p*-orbital of an acceptor atom such as oxygen, nitrogen, or a halogen. Hydrogen bonding alters the force constant, and as a result causes changes in the frequencies of stretching and bending vibrations. In general, *x*-H stretching bands move to longer frequencies, and H-*x* bending vibrations move to shorter frequencies.

V. ABSORBANCE AND REFLECTANCE AND THE ORIGINS OF SPECTRA

The intensity of absorption can be described in terms of transmittance:

$$T = I / I_o \tag{7}$$

where I is the intensity of energy emerging from and I_o the energy incident on the sample. For absorption spectra, the intensity I can be expressed more precisely by the Beer/Lambert law:

$$\log_{10} (I_o / I) = \log_{10} (I / T) = kcl = A . \tag{8}$$

A is a parameter called absorbance, sometimes called optical density; k is a proportionality constant called the molecular absorption (sometimes called molecular extinction) coefficient that is a characteristic of each molecular species; c is the concentration of absorbing molecules and l is the path length of the irradiating energy through the sample. It follows that, for monochromatic light traversing a fixed path length, the absorbance is directly proportional to concentration of the absorbing species. A linear relationship exists between absorbance and concentration, provided Beer's law is upheld, and no association occurs between absorbing molecules.

In the field of NIR spectroscopy, reflectance (R) is analogous to transmittance; equation (8) can thus be expressed in terms of reflectance as:

$$\log_{10} (1 / R) = kcl = A \tag{9}$$

or, more familiarly:

$$A = \log_{10} (1 / R) . \tag{10}$$

In reflectance mode, the sample pack has to be sufficiently thick to be effectively opaque so that no light reaches the back of the cuvette. Thus, all the incident light flux I_o is either absorbed or reflected. Only part of the reflected light can be measured using detectors mounted on a 45° cone or integrating sphere. In reflectance mode, both specular and diffuse components are superimposed so that the path length cannot be kept constant and will vary in every sample packing. Incident light normal to the sample surface will interact with particles lying at all possible orientations. Reflectance, refraction, transmittance, and absorbance will take place depending on both chemical and physical properties of the sample pack.

The net consequence of this scattering is to alter the proportion of absorbed and reflected radiation so that path length becomes an extra unknown along with concentration in the Beer/Lambert law. Thus, even if only one component were known to absorb at just one wavelength, it would be necessary to have another measurement wavelength to solve a pair of simultaneous equations relating path length and concentration to optical density at the two wavelengths.

Features that affect the mean path length include particle size, shape, crystalline form, bulk density, the refractive index of the pore space when filled with air, water, or oil, and absorption itself. Thus, finer grinding leads to paler color and lower optical density, whereas moist samples appear darker with higher optical density. It is the shapes of NIR spectra that convey chemical information, so that scanning instruments offer advantages in allowing shape analysis using derivatives or "quotientized" spectra (Norris and Williams, 1984).

In practice, most measurements require information from more than one region of the NIR spectrum, so that a matrix solution to a set of simultaneous equations forms part of the calibration process. Calibration is a regression modeling procedure that identifies the minimum subset of wavelength terms that best explains the chemical property across a population of similar commodities showing multivariate changes in composition. The particle size or scattering becomes the first principal component in reflectance and is satisfied as part of the calibration model.

Very few substances fail to absorb in the NIR. Sulfur, halon, and ceramic are nonabsorbers and accordingly are very useful as diffuse reflectance reference standards.

If the sample being irradiated was totally reflective and contained molecules all of which were 100% nonpolar and symmetrical, the "spectrum" would theoretically appear as a straight line across the wavelength range. In practice, at each wavelength point molecules occur that are capable of absorbing energy. By Avogadro's number, there are 6.02×10^{23} molecules in a gram molecule of any substance. Assuming the average sample scanned by an NIR instrument to be about 0.5–1.0 g, there will be roughly from 10^{15} to 10^{21} molecules of each individual constituent present in any sample, depending on their molecular weight and concentration, all vibrating at different frequencies. These will absorb energy according to the Beer/Lambert law to a greater or lesser degree at different wavelength points, depending upon their concentration. Accordingly, the amount of absorbance/reflectance that takes place will differ at different wavelength points, thereby creating the undulating pattern known as the spectrum. The shape of the spectral trace in terms of "peaks and valleys" of absorbance is characteristic of all of the absorbing molecules present in the sample and is affected by the composition of the molecules, by the presence and magnitude of dipoles, by interactions between molecules, by the resolution capability of the spectrometer, by the efficiency of the detector at different wavelengths, and by other factors.

Several factors affect the linearity of the Beer/Lambert law across the region of 1,000–2,600 nm in NIR diffuse reflectance analysis, the most important being differences in the linearity of response of the lead sulfide detectors that are responsible for making the measurement and differences in path length caused by particle characteristics. The influence of temperature also differs in different wavelength regions, and can have both bathochromic (shifting of bands to longer wavelength) or hypsochromic (shifting of bands to shorter wavelength) effects, depending on the temperature and position of the band. These factors are compensated for as far as possible by instrument design, and residual deviations from linearity are corrected by computerization of the log $1/R$ data recorded from the sample.

Absorption of near-infrared radiation can be attributed largely to the low mass of the hydrogen atom. Not only does this low mass cause resonant molecular vibration at high frequency, but the resultant oscillations are often anharmonic and as a consequence exhibit strong overtone absorption at near multiples of the fundamental frequency. Most of the fundamental oscillations of the bound hydrogen atom fall in the frequency range 3,700–2,800 cm^{-1}, equivalent to a wavelength range of 2,700–3,600 nm. Most overtones are found in the area from 25,000 to 4,500 cm^{-1} or from 400 to 2,200 nm. The area between 1,800 and 2,500 nm is referred to as the combination area, since many of its residents are combination bands. The hydrogenic fundamental and overtone vibrations couple with other modes of molecular motion to produce combination bands throughout the near infrared. This leads to an astronomical number of possible molecular motions and absorption bands even in organic molecules of relatively low molecular weight. The net absorption spectra represent the data that are employed in qualitative and quantitative analysis.

Coblenz (1905) found that the positions at which vibrations occur (in terms of wavelength) are characteristic of individual molecules and of groups of atoms (substituents, radicals, parts of chains and other groupings). In general, a fundamental vibration caused by a given grouping such as a -CH_3 group, belonging to a given molecule, will always occur in the same area of the spectrum, although the fine-tuning of its actual position can be affected by substituents to the grouping by temperature and other factors, including minute differences between instruments. The frequency or wavelength of the vibrations depends on the atoms present, their relative masses, the force constants of the bonds, and the arrangement of the atoms in space, and also possibly by the degree of hydration of the molecule. Information obtained with regard to characteristic absorbances of known atomic groupings, influence of other substituents on the intensity and positions of absorbances of known atomic groupings, and characteristic absorbances of simple compounds can be used to ascertain the structure of unknown organic substances in qualitative investigations.

In the mid-infrared, the vibrational frequencies of characteristic functional groups are relatively insensitive to the nature of the rest of the molecule. However, in the NIR spectral signature, these fundamentals display their overtone and combination tones, which provide not only concentration of such groups but their interaction one with another depending on local polar, steric, and hydrogen bonding effects. These factors, coupled with certain instrumental advantages of enhanced signal to noise and repetitive, attenuated absorbance, confer analytical advantages on NIR for quantitative analysis with minimal sample preparation.

VI. CHEMICAL STRUCTURE AND SPECTRA OF SIMPLE SUBSTANCES

All food and feed substances of plant and animal origin are composed of constituents possessing functional groups of atoms that absorb in the near-infrared region. These groupings include -CH-, -OH, -NH-, and other groupings. Their characteristic absorbance patterns can be traced down the spectra as overtones, and they can also be studied by preparing spectra of pure compounds in which the number of functional groupings increases in simple increments. An example is the aliphatic series of fatty acids, where the number of -CH_2- groups increases from 0 to 16. Figure 6 shows the spectra of the main components of plant- and animal-derived foods, namely water, oil (fat), starch, cellulose, and protein. The figure illustrates the atomic groups responsible for principal absorbers. Some of these absorbers—for example, CH stretching absorbers—occur in protein, starch, and cellulose as well as oil, and additional absorbers such as the CO and NH groupings also occur as overtones or combinations in oil and proteins. Figure 7 illustrates stacked spectra of the homologous series of aliphatic fatty acids (Murray, 1986). The principal change in the molecules is the addition of CH_2 groupings, increasing chain length, and progression from liquid to solid state at room temperatures. Other differences between compounds include changes in refractive indices, changes that are reflected in increased intensity of absorbance. The absorbers at 2,150 and 2,190 nm indicate the presence of *cis*-double bonds. The positions of the main CH absorbers near 1,200, 1,400, 1,750, 2,310, and 2,340 nm are clear. Changes in shape over the 2,300–2,500-nm region arise from increasing chain length. The CH absorbers are also apparent in the hydrocarbons cyclohexane, polythene, and benzene (Fig. 8). In the benzene spectrum, the absorbers near 1,670 and 2,150 nm are characteristic of aromatic rings. Figure 9 illustrates the positions of OH absorbers in simple sugarlike substances, but many of the same bands appear in starch and cellulose. Figure 10 portrays the spectra of the homologous series of aliphatic straight-chain alcohols. The CH stretching bands present in the fatty acids are clearly visible near 1,200, 1,400, 1,725, 1,750, 2,310, and 2,340 nm. The hydroxyl group shows two very broad absorptions centered about 1,600 and 2,100 nm. The broader band near 2,100 nm is characteristic of alcohols (especially primary alcohols), and appears in all compounds possessing alcoholic OH groups, such as sugars, starch, and cellulose.

Powdered whole grains and seeds, other food and feed ingredients, forages, dairy products, meat products, and other materials of plant and animal origin have spectra that are more undulating, with less sharp bands than pure compounds. The principal bands are still apparent in many cases. For example, wheat bran (Fig. 11) has a strong water band at 1,940 nm, the characteristic hydroxyl "hump" peaking at about 2,100 nm, which is present in all starch- and cellulose-containing substances, and also the two bands near 2,310 and 2,340 nm familiar to those who work with oil and oilseeds. Similarly, the spectrum of a sugar-coated breakfast cereal (Fig. 12) has a sharp band near 1,420 nm that is characteristic of crystalline sucrose and most crystalline monosaccharides. The 2,100-nm peak is also present.

VII. MOLECULAR VIBRATION AND SPECTRAL BANDS

The wavelength of an absorption band often reveals the nature of the chemical bonds responsible for the absorption. This part of the discussion is concerned with absorption bands occurring at the upper end of the vibrational spectrum. Such bands involve oscillations of dipoles either held together with high force or containing at least one atom with a low mass. Hydrogen, the lightest atom, with a mass of one, is an obvious contender. Equation (6) then becomes:

$$\mu = m/(1 + m)\,, \tag{11}$$

where m is the atomic weight of any atom combining with hydrogen. In organic molecules hydrogen is most often bound to carbon, nitrogen, or oxygen and the resulting reduced masses are 0.92, 0.93, and 0.94, respectively. However, the frequencies characteristic of C-H, N-H, and O-H bands are also influenced by the strengths of the chemical bonds and these are influenced by remaining portions of the molecule.

A rotating dipole also presents an oscillating field to any passing electromagnetic radiation, and the two fields will interact at appropriate frequencies. Absorption arising from this interaction typically occurs in the far-infrared region. In condensed states, molecular collisions effectively dampen out any resonant rotation so that rotational spectra are usually seen only in the vapor state. In this state these rotational modes can couple to vibrational modes causing the absorption band to split into a series of bands occurring at middle and near-infrared frequencies. The rotation responsible for absorption is that of the entire molecule, not just the X-H portion, and as the molecular weight increases, the characteristic rotational frequencies decrease. When coupled to a molecular vibration the split bands appear closer together.

Most fundamental resonant molecular vibrations, even those involving hydrogen atoms, occur at frequencies outside the near-infrared region. Those bands seen in the near infrared are overtones or combinations of the fundamentals, which are usually of higher intensity than overtones. The fact that any overtones are seen is evidence that the X-H vibration is not strictly harmonic. The deviation from harmonicity, which is called anharmonicity, increases with the amplitude of atomic oscillation. Vibrations of the hydrogen atom appear to be particularly susceptible to anharmonicity. Two consequences are of interest in this discussion. The frequency of overtone bands does not correspond to exact multiples of the fundamental vibrational frequencies and the intensity of absorption is related to the anharmonicity.

A simple equation can be written to describe the anharmonicity of a particular vibration. This equation relates

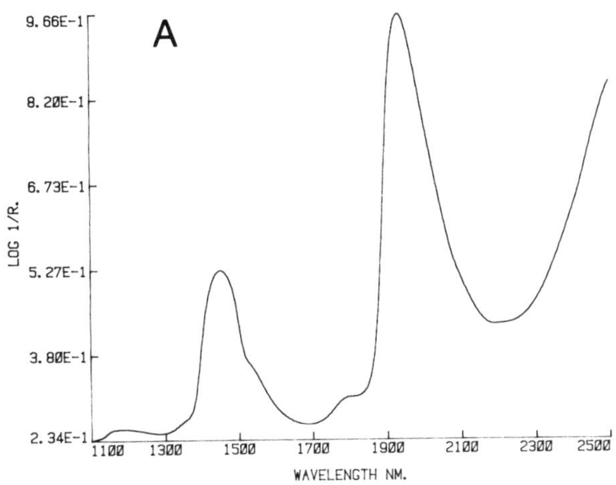

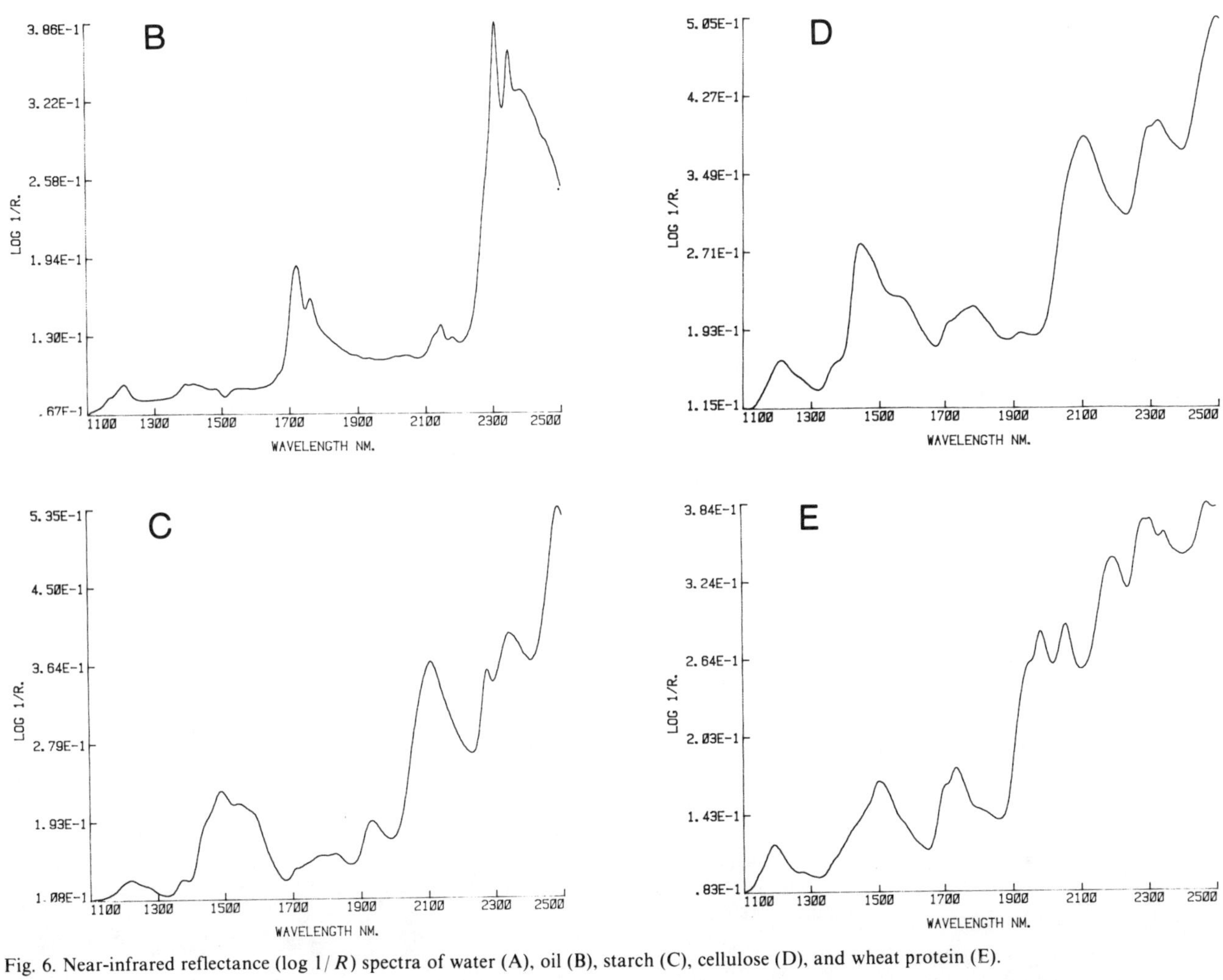

Fig. 6. Near-infrared reflectance (log $1/R$) spectra of water (A), oil (B), starch (C), cellulose (D), and wheat protein (E).

the observed frequency ν_j of the jth harmonic to the frequency ν_o of the fundamental:

$$\nu_j = j\nu_o\,(1 - jx)\,, \qquad (12)$$

where x is a positive number indicative of the anharmonicity. For CH vibrations, x is of the order of 0.005–0.05. It should be noted that the first overtone is the second harmonic ($j = 2$) and the second overtone is the third harmonic ($j = 3$), etc. The intensity of an overtone is dependent on the degree of anharmonicity in the vibrations. Typically, first overtones of vibrations involving hydrogen atoms have intensities one tenth to one hundredth that of the corresponding fundamental.

VIII. ILLUSTRATION AND DISCUSSION OF SOME SIMPLE CHARACTERISTIC SPECTRA

The principles discussed so far are illustrated by the absorption spectrum of HCl. All spectra shown here were obtained on a Beckman model 5270 spectrophotometer. At normal temperature and pressure, HCl is a gas; hence, it should exhibit rotational splitting of the characteristic vibration bands. The center of these bands occurs at about 2,980 cm^{-1} or 3,460 nm, which is outside the region of interest here; however, the center of the first overtone series can be expected at about twice the fundamental frequency and is indeed found at 5,670 cm^{-1} (1,764 nm), as seen in Fig. 13. From equation (12), the anharmonicity constant x is 0.0095. This suggests that the second overtone (third harmonic) would appear at about 8,423 cm^{-1} (1,190 nm). Intensity should be about an order of magnitude less than the first overtone.

Each of the vibration-rotation bands is seen to be split into unequal pairs of bands. The origin of this condition is found in the natural abundance of chlorine isotopes of atomic weights 35 and 37. The force constants binding these atoms to hydrogen are almost identical, so the expected difference in frequency between pairs of bands will be determined by mass only and can

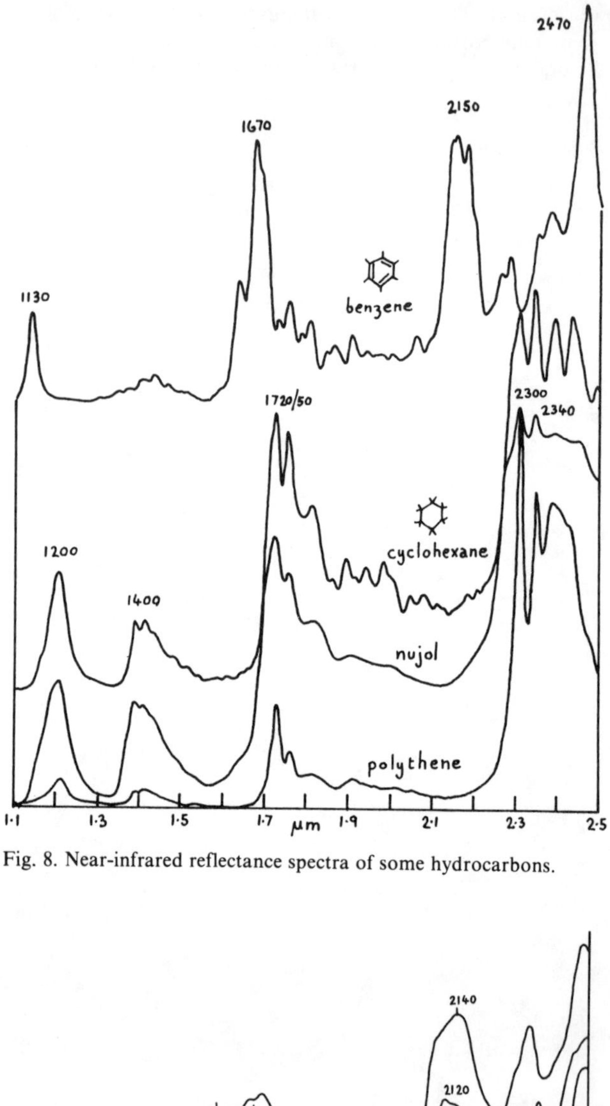

Fig. 8. Near-infrared reflectance spectra of some hydrocarbons.

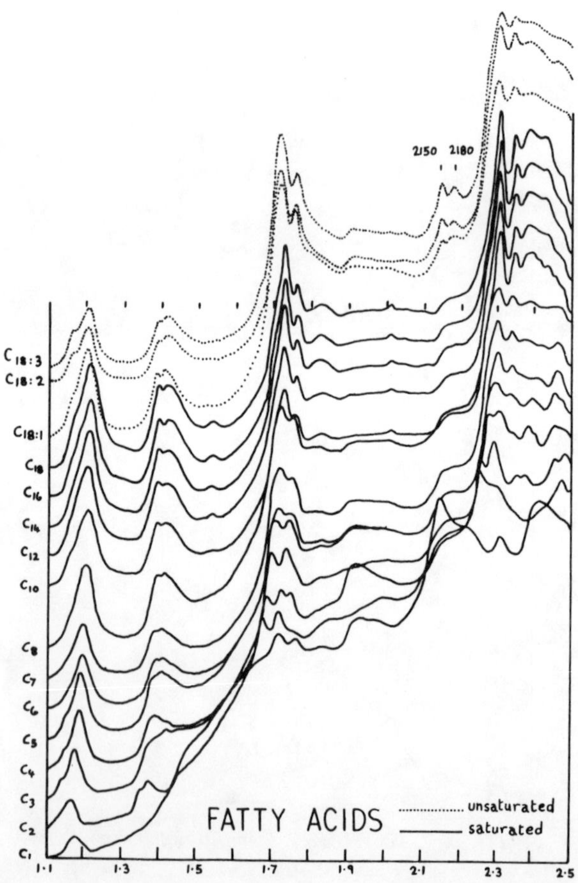

Fig. 7. Near-infrared reflectance spectra of aliphatic fatty acids.

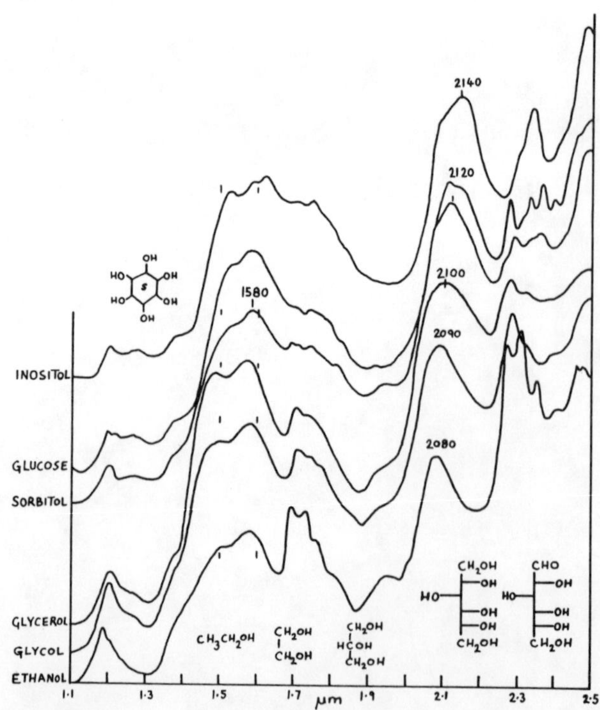

Fig. 9. Near-infrared reflectance spectra of OH-containing organic compounds.

be calculated from equations (5) and (11). The ratio of frequencies should be 1.00075, and this is what is observed in Fig. 13. The natural abundance ratio of chlorine isotopes is 75/25—very nearly equal to the ratio of absorbances of the band pairs. In addition, the series of paired bands can be grouped into two branches identified by the symbols *P* and *R*.

The other important isotope is deuterium. The DCl force constant is nearly equal to that for HCl, hence the ratio of frequencies of equivalent bands should be 0.717. This is a large shift readily detectable even under broad-band conditions, and it has been widely used to elucidate molecular structure.

As the number (n) of atoms in an absorbing molecule increases, the number of fundamental vibrations increases at the rate of $3n - 6$. Furthermore, the fundamentals may interact with overtones and overtones with each other to produce combination bands. As n increases, a prodigious number of possible absorption bands can result. Fortunately, only a small number of these vibrational modes are likely to involve a large amplitude motion of a hydrogen atom; hence, spectra in the near-infrared region usually exhibit useful structure.

These factors are illustrated in the spectrum of chloroform. In the liquid phase, the rotational bands are suppressed

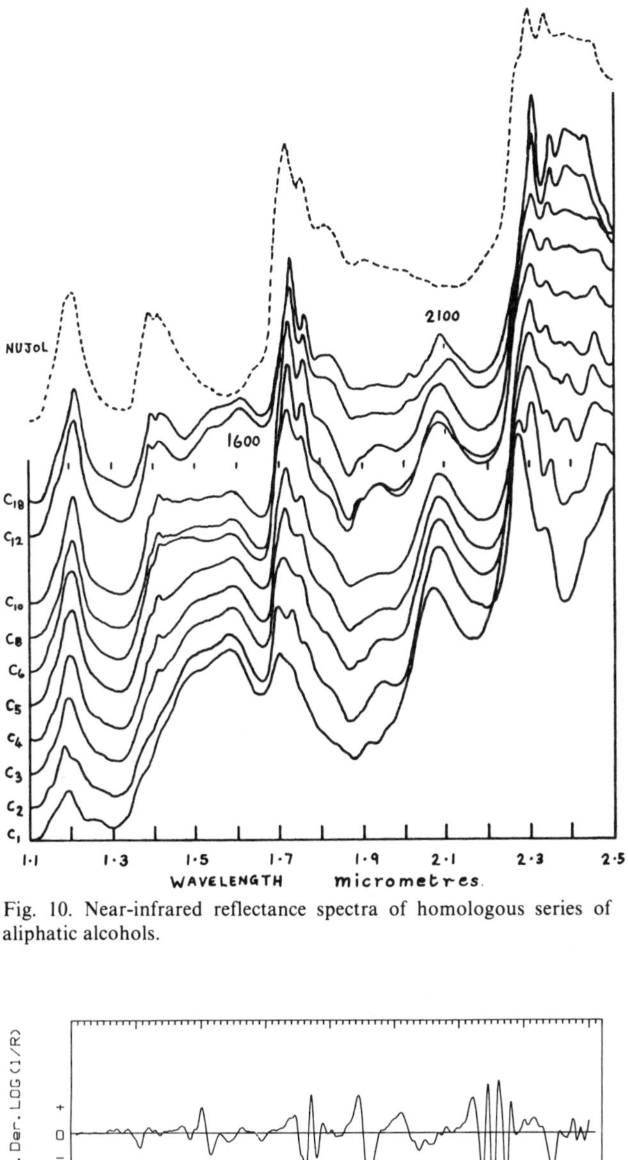

Fig. 10. Near-infrared reflectance spectra of homologous series of aliphatic alcohols.

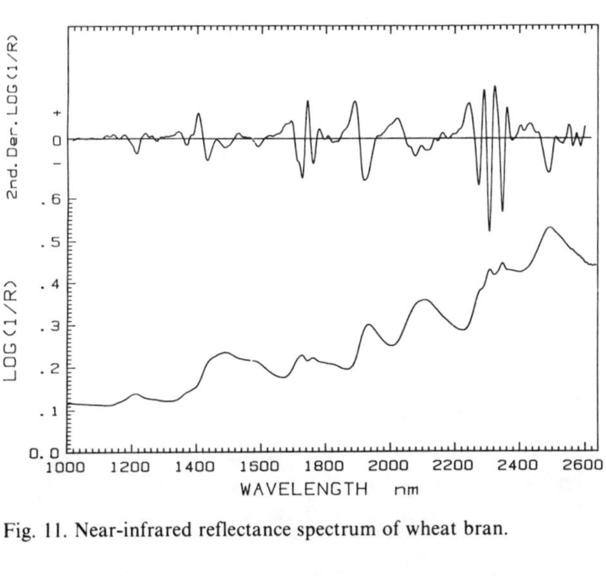

Fig. 11. Near-infrared reflectance spectrum of wheat bran.

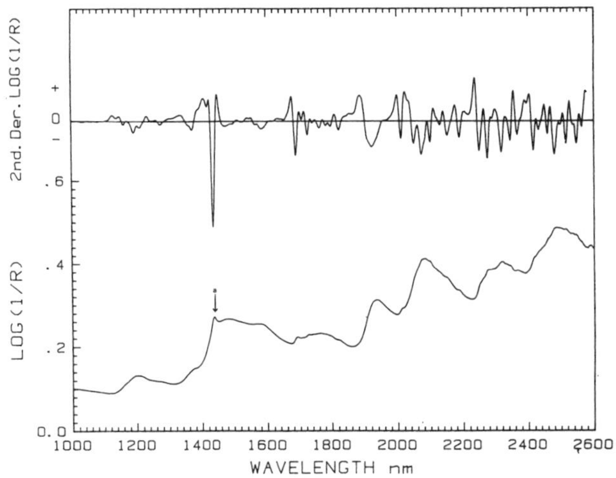

Fig. 12. Near-infrared reflectance spectrum of sugared rice cereal. Sharp band indicated by *a* is characteristic of crystalline sucrose and most monosaccharides.

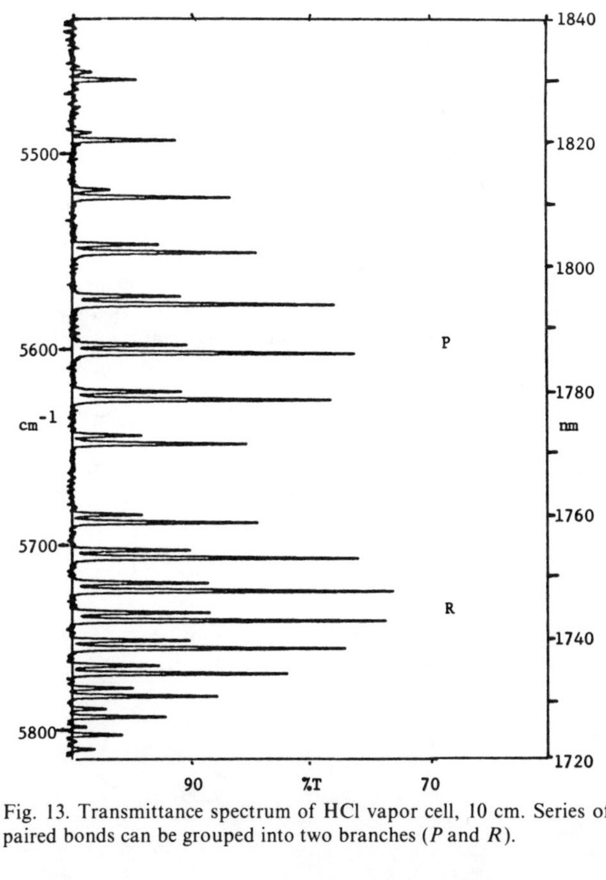

Fig. 13. Transmittance spectrum of HCl vapor cell, 10 cm. Series of paired bonds can be grouped into two branches (*P* and *R*).

and the band widths are too large to permit a resolution of isotope structure (except when deuterated). Because there is only one hydrogen atom in the molecule, the number of overtone and combination bands is manageable. However, the total number of atoms in the molecule is five, hence nine fundamental vibrations can be expected. For reasons of symmetry, three of these fundamentals are degenerate; Table II lists the frequencies of the six active fundamentals. The two bands labeled ν_1 and ν_4 are of particular interest because they involve, largely, C-H stretching and C-H bending modes, respectively. The other fundamental modes largely involve the C-Cl dipole motions. However, all the fundamental and the overtone vibrations also involve the carbon atom. As a result, all active vibrations interact to produce combination bands.

Figure 14 shows the near-infrared spectrum of liquid chloroform dried with molecular sieve pellets. There are over 40 resolved bands in this spectrum, and their absorptivities vary over a factor of 400,000. Interpretation of the chloroform spectrum is relatively simple once one knows the frequencies of the fundamentals (Table II). The first step involves identification of the overtones. The first overtones can be expected at slightly less than twice the frequencies of the fundamentals; they will most likely be the most intense bands in the vicinity of the estimated frequencies. The only first overtone found in the spectral interval of Fig. 14 is $2\nu_2$; however, a mid-infrared spectrum reveals the overtone $2\nu_4$. Overtones of the remaining fundamentals are difficult to identify, probably indicating little anharmonicity for these vibrations. For this reason it is unlikely that overtones of these bands will produce detectable combination bands.

Once the first overtones are identified, equation (12) can be used to estimate the frequencies of the higher ones. However, this equation provides only a first approximation. The strongest band in the vicinity of the calculated frequency is likely to be the harmonic. Table III lists the observed frequencies and the differences, $\Delta\nu$, between the integral multiple of the fundamental and the observed frequencies.

The combination bands may now be identified. Table IV lists the expected frequencies (and wavelengths) and the observed wavelengths. The agreement between calculations and observations is remarkably good. The bands are grouped for ease of calculation. All members of the first group involve the fundamental C-H stretching mode ν_1. The second group involves the combination mode $\nu_1 + \nu_4$, etc. The ν_{calcd} of the first group are sums of the respective fundamental frequencies without any harmonicity correction. The numbers appearing in the λ_{calcd} column are reciprocals of those in the ν_{calcd} column. Differences between the λ_{calcd} and λ_{obs} values can be attributed to anharmonicity. (Wavelength rather than frequency comparisons are made, simply because the instrument records linearly in wavelength.) A significant anharmonicity is seen for the first group, but it is not enough to confuse the band assignments. Such is not the case for the higher-term combination bands. To avoid this problem, the second group uses the observed frequency ($1/\lambda_{obs}$) of the 1+4 band in deriving ν_{calcd}. This process is repeated for each group.

All but three of the observed bands are accounted for in the above analysis. The 1+4 and 1+2*4 bands appear as barely resolved doublets. These are probably produced by Fermi resonance.

All of the bands in Fig. 14 appear to have about the same half-band width. This is not dictated by instrument resolution but by Doppler shifts and molecular collisions. In the vapor phase, molecular collisions are reduced (dependent on pressure) and techniques exist to reduce Doppler broadening so that half-band widths are greatly reduced. However, this may not help in band identification because of the complicating rotational structure. For example, Figs. 15 and 16 show the

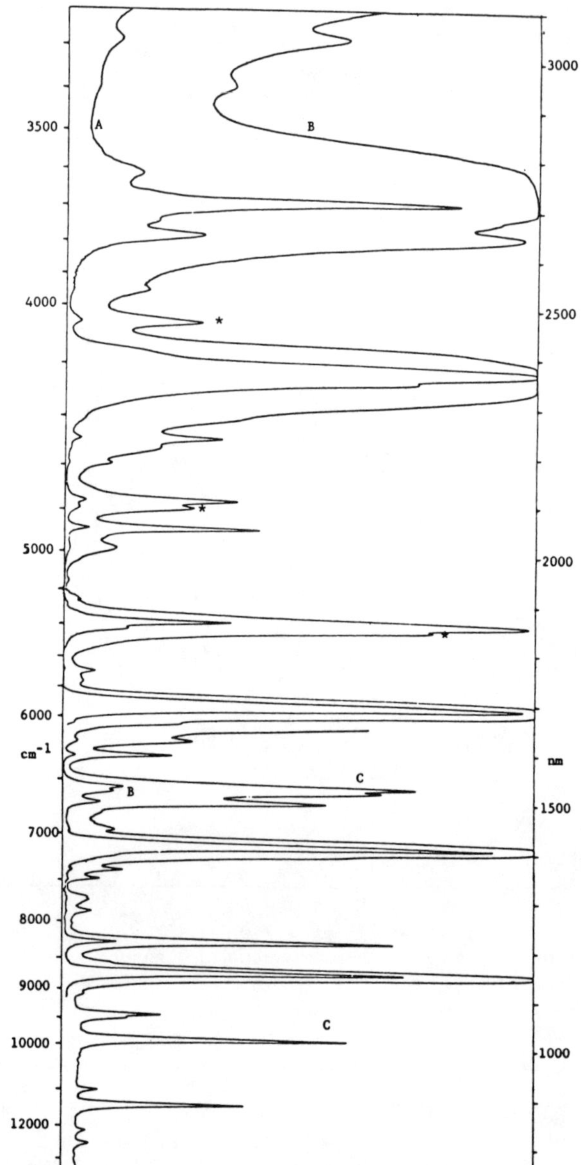

Fig. 14. Transmittance spectrum of liquid $CHCl_3$ cells: (A) 0.1 cm, (B) 1.0 cm, and (C) 10.0 cm. * = unidentified bands.

TABLE II
Frequencies (sec $^{-1}$) of Chloroform Fundamentals

ν_1	3,040
ν_2	667
ν_3	370
ν_4	1,214
ν_5	755
ν_6	261

TABLE III
Harmonic Frequencies (sec $^{-1}$)

Harmonic	ν_1	$\Delta\nu_1$	ν_4	$\Delta\nu_4$
1	3,040	0	1,214	0
2	5,910	170	2,400	28
3	8,680	440	3,592	55
4	11,320	840	4,744	112
5	13,870	1,330		

vapor spectrum in the vicinity of the two strongest bands, $\nu_1 + \nu_4$ and $2\nu_1$. The high mass introduced by the chlorine atoms causes the rotation bands to be so close together that they cannot be resolved by this spectrophotometer, and one sees only the envelope of these bands. The *P* and *R* branches can be discerned in the case of $2\nu_1$. The presence or absence of a *Q* branch, which is a function of vibrational symmetry, can be useful in band identification. The *Q* branch appears in Fig. 16 as a sharp peak between the *R* (lower) and *P* (higher) branches. (Note that the λ and ν axes are reversed between Figs. 13 and 16.)

The third molecule to be discussed here is water. Because water is probably the most important single constituent of the materials likely to be encountered by the readers of this book, it will be discussed in some depth. The near-infrared region is widely used for the analysis of water. In addition, the water spectrum introduces the important field of hydrogen bonding.

From the $3n - 6$ rule, one can expect three fundamental vibrations: symmetrical and antisymmetrical stretching of the O-H bonds and the bending or scissors motion of the molecule. The symmetrical stretching fundamental appears at 3,652 cm^{-1} and the antisymmetrical stretching fundamental appears at 3,756 cm^{-1}. The fundamental bending mode absorbs at 1,595 cm^{-1}. In spite of this apparent molecular simplicity, the vapor-phase spectrum is exceedingly complex because of the many rotational modes. Figure 17 shows this vapor spectrum in the vicinity of 1,400 nm, with all of these bands associated with the first overtone of the O-H stretching vibrations. The low

TABLE IV
Combination Bands of Chloroform

Term	ν_{calcd}[a]	λ_{calcd}	λ_{obs}[b]
1+6	3,301 (sec^{-1})	3,029 (nm)	3,042 (nm)
1+3	3,410	2,937	2,940
1+2	3,707	2,698	2,713
1+5	3,795	2,635	2,652
1+4	4,254	2,351	2,370
(1+4)+6	4,480	2,232	2,234
(1+4)+3	4,589	2,179	2,181
(1+4)+2	4,886	2,047	2,049
(1+4)+5	4,974	2,010	2,008
1+2*4	5,433	1,841	1,843
(1+2*4)+6	5,687	1,758	1,757
(1+2*4)+3	5,796	1,725	1,735
(1+2*4)+2	6,093	1,641	1,655
(1+2*4)+5	6,181	1,618	1,618
1+3*4	6,640	1,506	1,517
(2*1)+6	6,171	1,620	1,619
(2*1)+3	6,280	1,592	1,590
(2*1)+2	6,577	1,520	1,525
(2*1)+5	6,665	1,500	1,494
2*1+4	7,124	1,404	1,405
(2*1+4)+6	7,378	1,355	1,357
(2*1+4)+3	7,487	1,336	1,341
(2*1+4)+2	7,784	1,285	1,290
(2*1+4)+5	7,872	1,270	1,270
2*1+2*4	8,331	1,200	1,210
(3*1)+6	8,941	1,118	?
(3*1)+3	9,050	1,105	1,109
(3*1)+2	9,347	1,070	1,068
(3*1)+5	9,435	1,060	1,059
3*1+4	9,894	1,011	1,018
(3*1)+(2*4)	11,080	903	908
(4*1)+2	11,990	834	830
(4*1)+4	12,534	798	800

[a]Calculated.
[b]Observed.

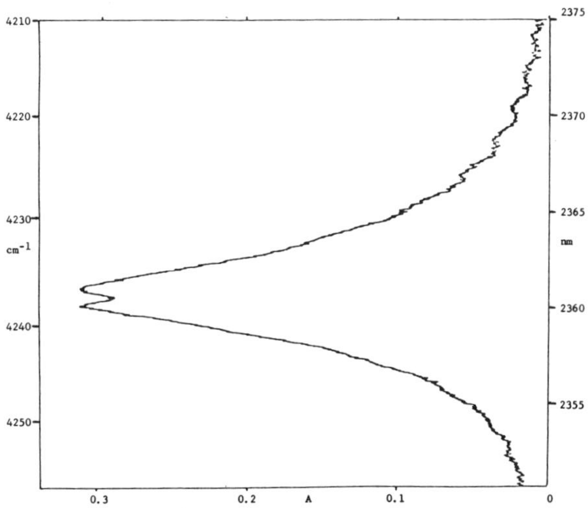

Fig. 15. Absorbance spectrum of $\nu_1 + \nu_4$ band of $CHCl_3$ vapor cell, 10.0 cm.

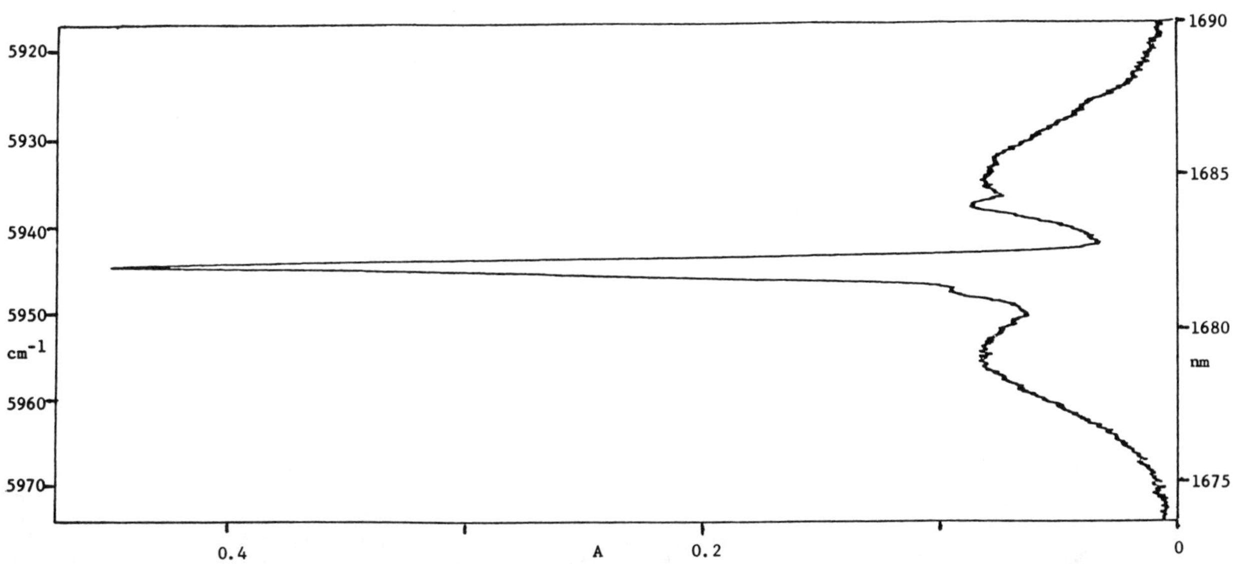

Fig. 16. Absorbance spectrum of $2\nu_1$ band of $CHCl_3$ vapor cell, 10.0 cm.

symmetry of this molecule contributes to the rich rotational spectrum.

In the liquid phase, water is poorly described by the simple formula H_2O. Instead, water molecules are associated with each other to produce something resembling a polymer that is better described by the formula $(H_2O)_x$. The forces causing the molecular association are described as hydrogen bonds.

The potential for a hydrogen bond exists when a hydrogen atom is bound to a highly electronegative atom. The binding electrons will then reside mostly around the electronegative atom, leaving the proton deficient in electrons. The proton will then be attracted to any atom capable of sharing (donating) an electron. Oxygen is an example of a highly electronegative atom, and any hydrogen bound to an oxygen atom will form a weak bond to any available electron donor atom. When such a bond forms, the force constant for the covalent O-H bond will decrease and the frequency of the O-H absorption band will decrease. The hydrogen bond, being usually weak, is readily disrupted by thermal collisions and the O-H absorption band is broadened. Even though the oxygen atom is considered highly electronegative, its nonbonding electrons are available for donation to any electron-deficient hydrogen atom. The structure of water may then be represented by Fig. 18, where the hydrogen bonds are indicated by dotted lines.

Figure 19 is a spectrum of pure water in the near infrared. Note that the absorption bands are significantly broader and more intense than the absorption bands of chloroform (Fig. 14). Broadening is a consequence of thermal motion, which distorts the weak O-H···O linkage. Unfortunately, the thinnest (0.1 mm) cell available was too thick to show the shape of the fundamental band, but it is very broad, extending from 2,700 nm to beyond 3,100 nm.

It is possible to observe the absorption spectrum of water in the unassociated state by dissolving the water in a nonpolar solvent such as carbon tetrachloride. The high transparency of CCl_4 is a side benefit. Water is only slightly soluble in CCl_4, so relatively long path cells must be used. Traces of impurities, particularly chloroform, are then likely to be seen.

Figure 20 shows the absorption spectrum of 10 cm of CCl_4. There is enough dissolved water in this sample to show the fundamental O-H symmetrical and antisymmetrical stretching vibrations at 2,760 nm (3,623 cm^{-1}) and 2,695 nm (3,710 cm^{-1}),

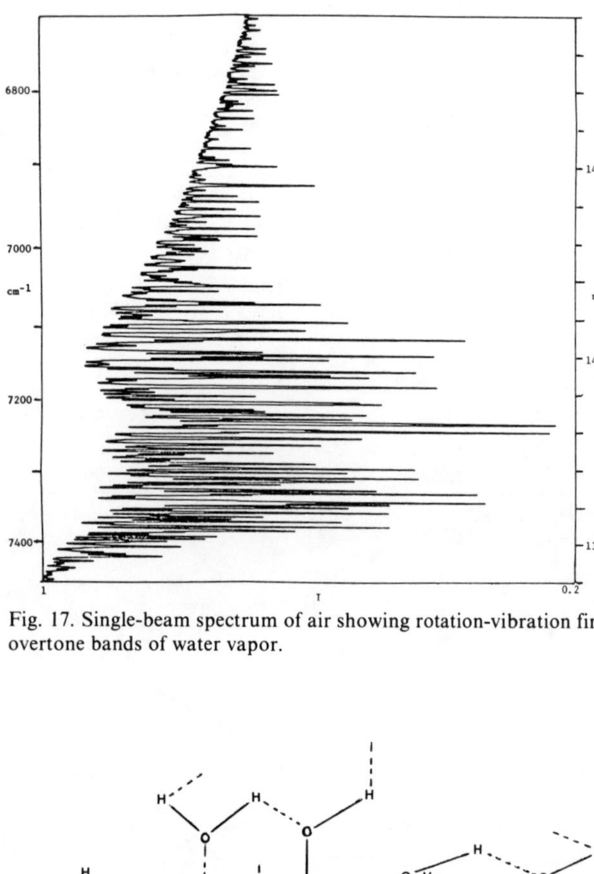

Fig. 17. Single-beam spectrum of air showing rotation-vibration first-overtone bands of water vapor.

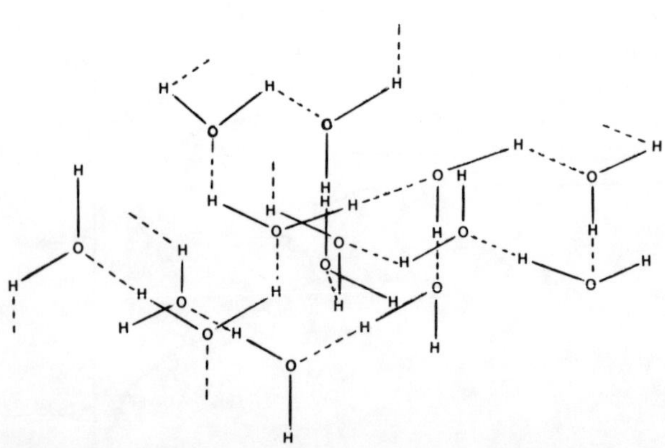

Fig. 18. Structure of solid water. Molecules are dispersed at random in three dimensions, but all molecules are linked together through hydrogen bonds (dotted lines).

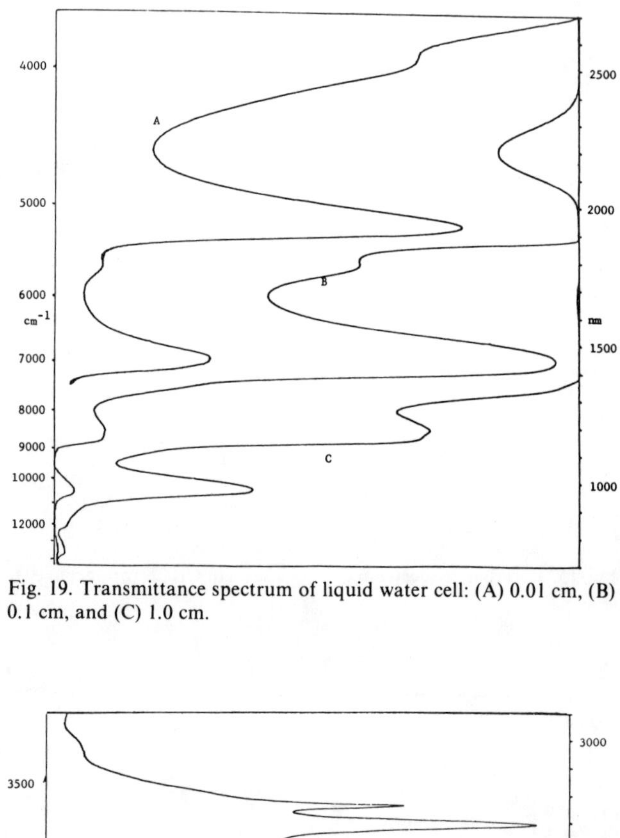

Fig. 19. Transmittance spectrum of liquid water cell: (A) 0.01 cm, (B) 0.1 cm, and (C) 1.0 cm.

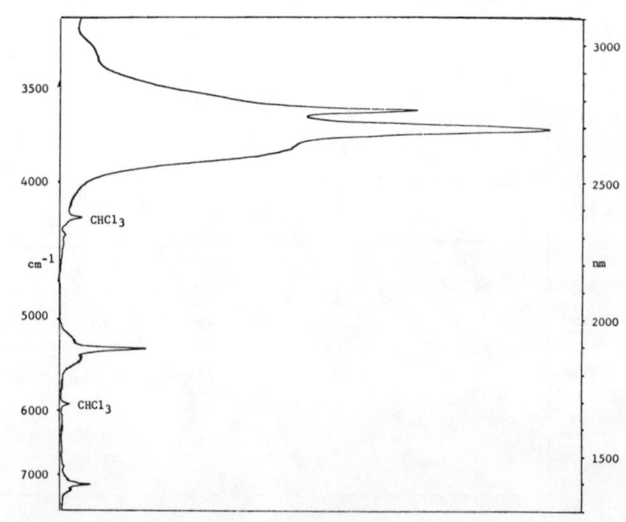

Fig. 20. Transmittance spectrum of CCl_4 saturated with H_2O cell, 10 cm.

respectively. The first overtone bands are seen at about 1,390 nm (7,190 cm^{-1}). The combination band of the O-H bend and stretch (probably antisymmetrical) is seen at 1,895 nm (5,279 cm^{-1}. The exact wavelengths of these bands vary with the polarity of the solvent. After taking concentration into account, it is found that association increases the intensity of the fundamental O-H···O band relative to the unassociated fundamental O-H band. The reverse is true of the first overtone. It appears likely that association lowers the anharmonicity of the vibration.

Returning to Fig. 19, we see that the water band whose maximum occurs at 1,925 nm must be the combination band in the associated state, whereas the band at 1,435 nm must be the first overtone. Higher-term overtone and combination bands are readily seen. In still longer paths, these bands extend into the visible region and are responsible for the blue color of water and ice.

Almost all compounds containing the O-H group are subject to hydrogen bonding. Exceptions occur where the O-H group is sterically hidden. Some of these compounds are soluble in CCl_4. At intermediate concentrations it is usually possible to see absorption bands of molecules in both associated and unassociated states. Quantitative measurements enable the calculation of the fraction of molecules in the "free" or unassociated state. The smaller the fraction of molecules in the free state for a given solute concentration, the stronger the hydrogen bond.

TABLE V
Simple Bond Combinations Likely to Cause Absorbances in Near-Infrared Region

-C-C-	H----O	N-H	-O-H	=P-P	-S-H
C=C	Intramolecular	-NH_2	-O-O-	=P-C-	-S-S-
-C-H	and	-NH_3	-O-S-	=P-H	-S-C-
CH_2	intermolecular	NH_2	S=O		S=C
-CH_3	hydrogen	-NH	-O-N=		-S-P=
-CH_2-phenyl	bonds	-N=N-	O=N-		
-CH=CH_2 vinyl		N-N	-O-C-		
-C-N=		N-P	O=C		
=C=N-		-N=P-	-O-P		
-C=N-		=N-S-	O=P-		
		-N=S-			

Hydrogen bonds are an essential ingredient of many biological functions. They also influence solubility and acidity properties. For these and other reasons, the study of hydrogen bonding assumes importance and near-infrared spectra provide a prime technique for pursuing such studies. The strength of hydrogen bonds depends upon both the electron-withdrawing power of the atom to which the hydrogen is covalently linked and the electron-donating properties of the atom to which the hydrogen is bridged. For example, the nitrogen atom does not withdraw electrons from a covalently bound hydrogen as strongly as does oxygen, but nitrogen is a stronger electron donor. Coordinating an O-H-containing compound to an atom such as tin greatly increases the electron-withdrawing power of the oxygen atom, and such compounds form very strong hydrogen bonds to good electron donors.

As the number of atoms (particularly hydrogen) in a molecule increases, the number of possible overtone and combination bands increases. They then overlap to produce an apparent absorption continuum. Still, the overtones and certain combination bands involving motion of the hydrogen atoms will be the most intense in the near infrared and will stand out from the continuum. The frequencies of the C-H, N-H, O-H, and S-H vibrations are sufficiently different to allow their use for structural analysis. To a limited extent, a differentiation among C-H groups linked to aromatic, aliphatic, olefinic, and acetylenic bonds can be made. The most definitive region for such differentiation is often between 2,000 and 2,500 nm where the combination bands of X-H stretching and bending modes are found.

The near-infrared region must be considered an adjunct to the mid-infrared when band assignment is attempted. However, for quantitative analytical use the near-infrared region can be used alone. The advantages of the near infrared are usually associated with sample absorptivity and instrumentation. Absorptivities of the fundamental bands are usually so large that very thin cells must be used for transmittance measurements. Quantitative measurements may then be compromised by a lack of knowledge of cell thickness or confused by interference fringes. Solids and viscous liquids are handled with difficulty in thin cells. Thicker and more rugged cells can be used in the near infrared. Signal-to-noise ratios are usually higher in the near infrared than in the mid-infrared, and the instrumentation is less expensive.

TABLE VI
Approximate Location of Absorbers Involving Nitrogen and Hydrogen

Wavelength (nm)			
1st Overtone	**2nd Overtone**	**3rd Overtone**	**Tentative Assignment**
	2,540–2,600	1,910–2,080	Amide III: Combination N-H stretch with C-O stretch, secondary amides
		1,950–1,980	Amide III: N-H stretch, *trans*-secondary amides
	2,330–2,390	1,760–1,800	Amide IV: N-H bend, primary amides
	2,270–2,320	1,710–1,730	N-H bend, *cis*-secondary amides
	2,200–2,250	1,640–1,680	NH_3+ NH deformation; "amino acid II"
	2,140–2,180	1,600–1,630	N-H bend, *trans*-secondary amides
	2,050–2,140	1,540–1,600	Ring deformation, pyrimidines, quinolines
	2,080–2,220	1,560–1,670	Amide II: N-H deformation coupled with C-H stretching, secondary amides, especially peptides
	2,040–2,110	1,530–1,580	N=N stretching, unsaturated nitrogen compounds
	2,030–2,100	1,520–1,570	NH_3 deformation; "amino acid I"
	2,020–2,150	1,510–1,610	N-H deformation, primary and secondary amines
1,810–1,970	1,200–1,310		Unknown absorber in most amino acids
1,660–2,500	1,110–1,670	1,000–1,250	N-H stretch, symmetrical, all amino acids and hydrochlorides
1,590–1,650	1,060–1,100		Asymmetrical N-H stretch, all amino acids and hydrochlorides
1,600–1,630	1,070–1,090		N-H stretch, secondary amides, *cis* and *trans*
1,570–1,600	1,040–1,060		N-H stretch, secondary amides, *cis*-bonded NH
1,520–1,620	1,010–1,080		Hydrogen bonding, peptide links, protein helices
1,530–1,580	1,040–1,070		N-H stretch, primary amides, bonded NH
1,500–1,530	1,000–1,020		N-H stretch *trans*-bonded NH, primary amides
1,510–1,530	1,000–1,020		Unidentified absorber, all proteins
1,490–1,510	~1,000		N-H stretch, imines (e.g., histidine)
1,490–1,510	1,000–1,020		N-H stretch, alkylated primary amines

The discussion above is largely based on earlier studies by Kaye (1954, 1955) and Murray (1986). For further information on band assignments and instrumentation, the texts by Bauman (1962), Alpert et al (1970), Stewart (1970), and Colthup et al (1975) are recommended. The literature on NIR is very extensive but widely dispersed. One of the best guides to this literature is the series of annual reviews on infrared spectroscopy appearing in *Analytical Chemistry*, and Wetzel (1983) has recently summarized the salient points of NIR technology.

TABLE VII
Approximate Location of C-N Bands in Near-Infrared Region

Wavelength (nm)			
1st Overtone	**2nd Overtone**	**3rd Overtone**	**Tentative Assignment**
		2,420–2,440	C-N stretch, primary-tertiary amines
		2,400–2,420	C-N stretch, primary amines, primary alpha-carbon atoms
		2,310–2,340	C-N stretch, primary amines, primary alpha-carbon atoms
		2,110–2,140	C-N stretch, secondary amines, secondary carbon atoms
	2,480–2,600	1,850–2,000	C-N stretch, acrylamines, alkyl amines, primary-tertiary
	2,450–2,550	1,840–1,870	C-N stretch, *cis*-secondary amides
	2,340–2,380	1,770–1,800	C-N stretch, amides with no N substitution
	1,970–2,100	1,470–1,570	C-N stretch, unsaturated nitrogen compounds
2,310–2,350	1,540–1,570	1,150–1,170	C-N stretch, -N=C=N-
1,490–1,510	900–1,000	...	C-N stretch, amines

TABLE VIII
Approximate Location of C-O Bands in Near-Infrared Region

Wavelength (nm)			
1st Overtone	**2nd Overtone**	**3rd Overtone**	**Tentative Assignment**
		2,220–2,380	C-O stretch, primary alcohols
		2,150–2,180	C-O stretch, tertiary alcohols
		2,050–2,100	C-O stretch, phenols
		2,100–2,180	C-O stretch, long-chain fatty acids
	2,590–2,640	1,920–2,080	C-O stretch, amide III combination, secondary amides
	2,330–2,540	1,780–2,080	Coupled C-O and O-H stretch, carboxylic acids
	2,380–2,500	1,780–1,920	C-O symmetrical vibrations, zwitterions
	2,140–2,360	1,600–1,770	C-O stretch, amino acid ionized carbonyls
	1,980–2,220	1,480–1,670	C=O stretch, urea, amide I, especially lower frequencies
	2,070–2,150	1,550–1,620	C-O bending, COŌ zwitterions
	2,080–2,140	1,560–1,610	C-O stretch, COOH, amino acids
	2,000–2,050	1,510–1,530	C=O stretch, solid primary amines, amide I
	1,990–2,030	1,490–1,520	C=O stretch, internally bonded, saturated aliphatic carboxylic acids
	1,950–1,990	1,460–1,490	C=O stretch, α-β unsaturated aldehydes
	1,970–2,080	1,470–1,570	C=O stretch, α-β unsaturated ketones
	1,920–1,960	1,440–1,470	C=O stretch, ketones
	1,930–1,970	1,440–1,470	C=O stretch, saturated aliphatic carboxylic acids
	1,910–1,930	1,430–1,450	C=O stretch, saturated aliphatic acids and esters
	1,800–1,920	1,350–1,440	C=O vibrations, open-chain acid anhydrides
2,330–2,420	1,550–1,610	1,160–1,210	COŌ stretch, or combination band, most amino acids
2,330–2,360	1,530–1,570	1,160–1,180	COŌ stretch, or combination band ionized amino acids

TABLE IX
Approximate Location of O-H Bands in Near-Infrared Region

Wavelength (nm)			
1st Overtone	**2nd Overtone**	**3rd Overtone**	**Tentative Assignment**
		1,950–2,020	O-H deformation, secondary alcohols
	2,330–2,540	1,780–2,080	C-O/O-H stretch coupled, carboxylic acids
	2,510–2,600	1,860–1,900	O-H deformation, primary alcohols
	2,440–2,500	1,820–1,870	O-H deformation, secondary alcohols
	2,000–2,090	1,510–1,540	O-H deformation, hydroxyl
2,060–2,150			C-O, O-H stretching combination, primary alcohols
1,920–1,950			O-H stretch/O-H deformation combination hydroxyl
1,620–1,700	1,110–1,140		O-H stretch, carboxylic acid dimers
1,560–2,000	1,030–1,330		O-H stretch, intramolecular OH bonds, polymers
1,470–1,560	950–1,040		O-H stretch, internal OH bonds, single bridge, polymeric
1,400–1,450			O-H stretch, internal OH bonds, single bridge
1,400–1,430			O-H stretch, COOH groups
1,390–1,420			O-H stretch, intramolecular OH bonds, single bridge
1,380–1,400			O-H stretch, phenols
1,360–1,390			O-H stretch, tertiary alcohols
1,370–1,390			O-H stretch, primary alcohols
1,360–1,380			O-H stretch, secondary alcohols

IX. ASSIGNMENT OF WAVELENGTHS IN AGRICULTURAL PRODUCTS

There are a large number of possible molecular band groups that can occur in the agricultural and food materials to which NIR spectroscopy is routinely applied. The simplest of these are listed in Table V. By far the most important are those involving hydrogen.

The assignment of overtones and other bands occurring in the near infrared involves identifying atomic associations likely to occur in the molecules of the substance under study (such as $-CH_2-$, $-CH_3$, -NH-, -OH, etc.). The next step is to search the literature for information concerning the assignment of fundamental absorbers in the infrared region, then calculating the position where overtones and combinations of the fundamental absorbances are likely to occur in the near-infrared region. Relatively little information has been documented with regard to the assignment of wavelengths to absorbers from plant materials and food constituents in the near infrared. Law and Tkachuk (1977), Osborne (1981), and more recently Coblentz (1982) have made some assignments. Those for individual constituents and commodities may not be absolutely precise for several reasons: (i) the anharmonicity referred to above; (ii) interactions between constituents within a commodity (particularly those involved in changes in hydration and hydrogen-bonding status); (iii) interactions between sample and instrument; (iv) minute differences in the internal geometry and optical components of the individual instrument;

TABLE X
Approximate Location of C-H Bands in Near-Infrared Region

Wavelength (nm)			
1st Overtone	**2nd Overtone**	**3rd Overtone**	**Tentative Assignment**
		2,120–2,150	$(CH_3)_2$-C; skeletal vibrations
	2,460–2,540	1,850–2,120	C-H bend, CH in long-chain fatty acids
		1,980–2,040	$(CH_3)_2$-C; skeletal vibrations
	2,540–2,580	1,900–1,940	-CH=CH-:-CH=CH_2; CH in phase deformation
	2,460–2,500	1,850–1,880	-CH-; CH deformation
	2,410–2,460	1,820–1,850	-$(CH_3)_3$; CH deformations
	2,390–2,440	1,810–1,840	-$C(CH_3)_3$; CH deformation
	2,370–2,400	1,770–1,790	C-H in-phase deformation, CHO groups
	2,340–2,370	1,750–1,780	-CH=CH_2 in-phase CH_2 deformation
2,300–2,350			C-H stretching, methylene groups, combination
	2,320–2,520	1,740–1,890	C-H stretch aliphatic compounds
	2,270–2,300	1,710–1,740	C-CH_3; CH asymmetrical deformation
	2,260–2,300	1,700–1,730	-CH_2-; CH asymmetrical deformation
	2,190–2,360	1,640–1,770	Pyrimidines and quinolines, ring deformation
	2,200–2,250	1,650–1,680	Benzene rings deformation
2,140–2,190			C-H stretching, *cis* unsaturation, combination
2,100–2,200			C-H stretching, skeletal in-plane deformation, combination
	2,060–2,150	1,550–1,620	Benzene rings deformation
	2,070–2,100	1,550–1,580	-C=C- stretch, conjugated chains
	2,020–2,060	1,510–1,550	-C=C- stretch, nonconjugated chains
1,720–1,860	1,150–1,240		C-H stretch, carbonyl compounds
1,680–1,740	1,120–1,170		C-H stretch, CH_3 groups (A_1)
1,700–1,760	1,130–1,170		C-H stretch, CH_2 groups (A_2)
1,700–1,740	1,140–1,160		C-H stretch, -CH= (A_3)
1,640–1,670	1,090–1,120		-CH=CH-; CH stretch, *cis* and *trans*
1,610–1,670	1,090–1,120		C-H stretch, quinolines
1,610–1,660	1,070–1,110		C-H stretch, pyrimidines
1,620–1,650	1,080–1,100		C-H; C-H stretch, aromatics
1,610–1,640	1,070–1,090		C-H stretch, pyridines
1,620–1,640	1,070–1,090		-CH=CH_2 vinyl, C-H stretch

TABLE XI
Assignment of Inorganic Absorbers in Near-Infrared Region

Wavelength (nm)			
1st Overtone	**2nd Overtone**	**3rd Overtone**	**Tentative Assignment**
	2,460–2,670	1,480–1,600	P=O free
	2,410–2,500	1,440–1,500	NO_3-
	2,330–2,400	1,390–1,440	NH_4+
	2,290–2,340		P-(phenyl ring)
2,270–2,500	1,510–1,670		C=N, SCN=
	2,230–2,360	1,340–1,420	Carbonates
2,030–2,200	1,350–1,470		P-H stretching
1,850–1,950	1,230–1,300		P-OH stretching
		1,818–2,200	Silicates
		1,810–2,000	Phosphates, PO_4, all
		1,760–1,860	Sulphates, SO_4=
		1,750–1,900	Phosphates, PO_3=
1,690–1,910	1,130–1,307		-SH stretch (very weak)
		1,600–1,800	P=O hydrogen bonded
		1,600–1,630	NO_2-
		1,510–1,830	Phosphates, PO_2-
1,510–1,650	1,010–1,100		NH_4+

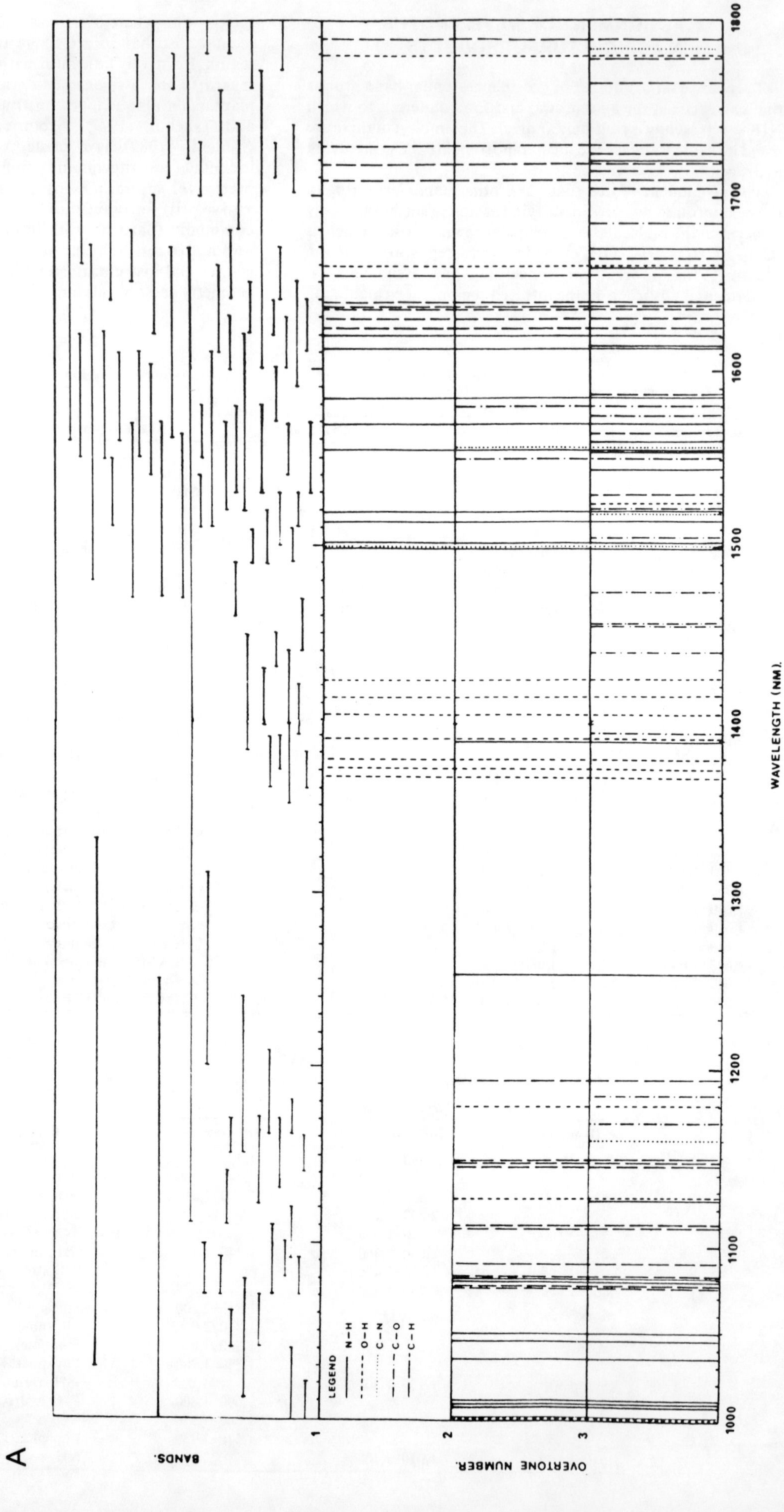
A
BANDS.
OVERTONE NUMBER.
1
2
3
LEGEND
N–H
O–H
C–N
C–O
C–H
1000
1100
1200
1300
1400
1500
1600
1700
1800
WAVELENGTH (NM).

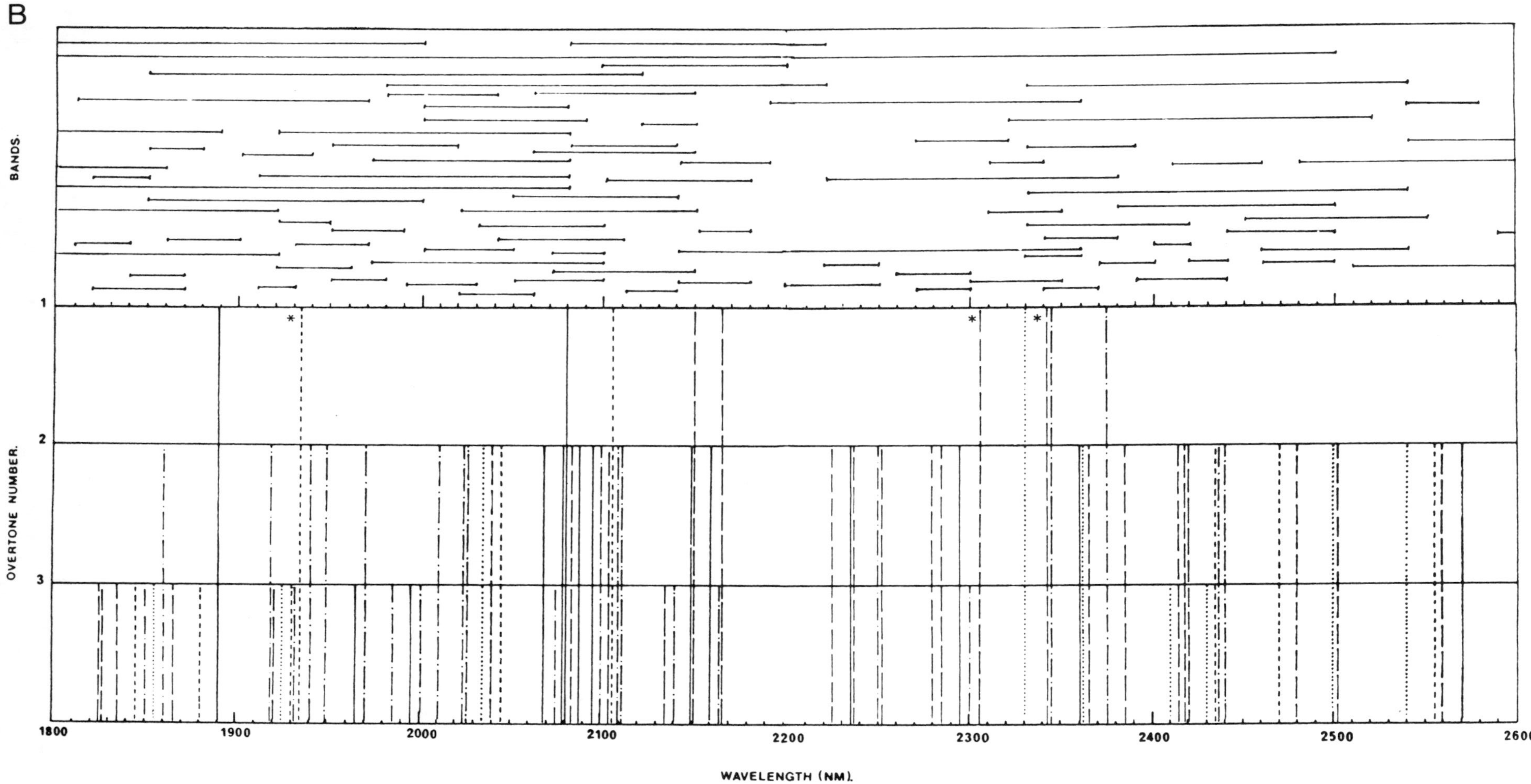

Fig. 21. Approximate positions of central wavelengths for principal absorbers in the near-infrared region. A = region from 1,000 to 1,800 nm; B = region from 1,800 to 2,600 nm. Upper section shows approximate literature values for the bands (Tables VI–X). Because each line represents a band that may extend over 100–200 nm or more, actual absorbances are practically continuous. Three combination bands, shown with asterisks, at about 1,940, 2,305, and 2,345 nm represent absorbances frequently used in determination of water (1,940 nm) and oil, respectively.

(v) temperature changes that may affect the positions of absorbers; and (vi) molecular association including the formation of dimers, hydration, crystalline state, and other forms of hydrogen bonding. Tables VI through XI give the approximate areas where overtones of some of the most common absorbers may be expected to occur. The information is based on literature values for the fundamental absorbances, and no corrections have been made for anharmonicity. The positions of combinations must be arrived at by consideration of what combinations are possible, then calculating their probable positions. Figure 21 summarizes the tables, illustrating the approximate central positions of first, second, and third overtones that occur most commonly in the near infrared. This map does not include all combination and difference bands. Bearing this in mind, together with band overlap, this figure illustrates the complexity of making assignments of absorbances in the area between 1,000 and 2,600 nm in the case of materials such as agricultural products. It also illustrates the degree to which interactions are possible, since several bond combinations may absorb in the same area. With complex molecules, such as proteins, differences in molecular structure caused by agencies such as oxidation and hydration may cause changes in the positions of the side chains (amino acid residues) relative to each other. These changes may change the overall net absorption of the molecule and consequently the optimum wavelength for making analytical measurements of absorbance.

NIR analysis should be regarded as an empirical technique in the agricultural and food industries. It may suffice to find high correlations between reflectance at some wavelength or wavelengths and the results of an established wet chemical analysis.

ACKNOWLEDGMENT

The authors gratefully acknowledge the many contributions made by Wilbur Kaye (retired, formerly of Beckman Instruments) in the assembling of this chapter.

LITERATURE CITED

ALPERT, N. L., KEISER, W. E., and SZYMANSKI, H. A. 1970. IR: Theory and Practice of Infrared Spectroscopy. Plenum Press, New York.

BAUMAN, R. P. 1962. Absorption Spectroscopy. John Wiley, New York.

COBLENTZ, W. K. 1982. A study in the near infrared reflectance spectra of protein and related compounds. M.Sc. thesis, Pennsylvania State University, University Park.

COBLENZ, W. W. 1905. Investigations of IR Spectra. Part I: Absorption Spectra. Part II: Emission Spectra. Carnegie Inst. Washington Publ. 35. Washington, D.C.

COLTHUP, N. B., DALY, L. H., and WIBERLEY, S. E. 1975. Introduction to Infrared and Roman Spectroscopy. Academic Press, New York.

KAYE, W. 1954. Near-infrared spectroscopy. I. Spectral identification and analytical applications. Spectrochim. Acta 6:257-287.

KAYE, W. 1955. Near-infrared spectroscopy. II. Instrumentation and technique. Spectrochim. Acta 7:181-204.

LAW, D. P., and TKACHUK, R. 1977. Near infrared diffuse reflectance spectra of wheat and wheat components. Cereal Chem. 54:256-265.

MURRAY, I. 1986. NIR spectra of homologous series of organic compounds. In: Proc. Int. NIR Symp. Budapest, Hungary, May.

NORRIS, K. H., and WILLIAMS, P. C. 1984. Optimization of mathematical treatments of raw near-infrared signal in the measurement of protein in hard red spring wheat. I. Influence of particle size. Cereal Chem. 61:158-165.

OSBORNE, B. G., and DOUGLAS, S. 1981. Measurement of the degree of starch damage in flour by near infrared reflectance analysis. J. Sci. Food Agric. 32:328-332.

STEWART, J. E. 1970. Infrared Spectroscopy. Marcel Dekker, New York.

WETZEL, D. L. 1983. Near-infrared reflectance analysis. Anal. Chem. 55:1165A-1176A.

CHAPTER 3

Data Analysis: Wavelength Selection Methods

WILLIAM R. HRUSCHKA
Instrumentation Research Laboratory
Agricultural Research Service
U.S. Department of Agriculture
Beltsville, Maryland, U.S.A

I. INTRODUCTION

The success and acceptance of near-infrared reflectance (NIR) analysis as a qualitative and quantitative tool in recent years are in large part the result of increases in the quality of the spectra and the power of the mathematics used in their analysis. This chapter discusses the basic factors involved in calibration and wavelength selection and describes mathematical solutions that, in commercial bench instruments, reduce to combinations of spectral measurements at a few (≤20) wavelengths. These methods account for most of those available to the users of modern commercial instrumentation. The next chapter focuses on mathematical techniques that use measurements at all of the available wavelengths. These methods are used primarily by research laboratories but are becoming available commercially as their potential strengths are verified in practical applications.

Typical near-infrared diffuse reflectance spectra are shown in Figs. 1–3. The micrometer is the preferred wavelengths unit in infrared work, although most near-infrared work has been reported in nanometers. The near-infrared region is composed of radiation with wavelengths of 700–3,000 nm, but wavelengths in the region of 1,000–2,600 nm are used in most applications. Wave numbers, also called reciprocal centimeters (cm^{-1}), are more often used in the mid and far infrared. They are units of frequency rather than wavelength. The conversion is $cm^{-1} = 10{,}000/\mu$.

The vertical axes units, log (1/reflectance)—abbreviated as log $(1/R)$, log = $\log_{10}$—are related to absorption: a higher log $(1/R)$ value means that more radiation has been absorbed (less reflected) by the sample at that wavelength. These units are used instead of percentage reflectance because there is an almost linear relation between the concentration of an absorbing component and its contribution to the log $(1/R)$ value at the wavelength absorbed. Other units for the vertical axis have been proposed but are not in general use by commercial NIR instruments. They are discussed in the chapter on Physics. The log $(1/R)$ function is used throughout this chapter with the understanding that it may be replaced by some other function of reflectance or transmittance without affecting the text, so long as attention is paid to the linearity question.

Figure 1 shows the spectra of water and ground wheat with two levels of moisture. The main features of the water spectrum are the two absorption bands that peak at 1,450 and 1,940 nm. Water is a strong absorber, and its absorption band at 1,940 nm is isolated from the bands of other absorbers, so the effect of increased moisture on the ground wheat spectrum is quite clear.

The other two major components of wheat are protein and starch (Fig. 2). These spectra have more bands and the bands are not isolated. Not only do the protein bands overlap each other, but they overlap the starch bands. Protein absorbs near-infrared radiation much more weakly than does water, and the difference between the high-protein and low-protein wheat spectra is far less noticeable than those between wheats differing in moisture content. In fact, most of the differences in Fig. 2 are traceable to small differences in the particle size distributions. The effect of large differences in particle size distributions is shown in Fig. 3, where the same sample was ground to three different mean particle sizes. A change in particle size causes a change in the amount of radiation scattered by the sample, which causes a baseline shift (additive effect). The larger particles do not change the direction of the radiation as often, so more is absorbed before leaving the sample, resulting in a higher log $(1/R)$. There is also a multiplicative effect, with the strong absorbers showing more change with particle size than the weak.

When differences in particle size are ignored, the differences caused by protein in Fig. 2 are only small changes in the shape of some of the peaks (e.g., at 2,180 nm). Although these differences are barely visible, the various mathematical methods described in this chapter enable the use of these small changes to determine protein concentration with a standard error of less than 0.2% protein.

Most research work in the near infrared has involved analysis of spectra like those in Figs. 1–3. The usual features are overlapping absorption bands with large spectral changes due to uncontrolled radiation scattering conditions masking the small spectral changes that are used to make measurements.

One set of data will be used throughout this chapter to illustrate the principles and techniques. The data are characterized in Table I and described in detail by Hruschka and Norris (1982) and Williams et al (1978, 1983). A set of 100 whole grain hard red spring wheat lots was compiled that had small within-lot variability and large between-lot variability with respect to protein and moisture content. One sample was drawn from each lot and 20 additional samples were drawn from one lot for a total of 120 samples. The 20 replicates are called the reproducibility set in Table I. The other 100 samples were divided into two sets, each of 50 samples, which are called the calibration and validation sets. This division provided roughly equal distribution of protein and moisture content in each set.

From each of the 120 samples, 20 subsamples were drawn and ground in duplicate to a mean particle size of 200 ± 30 μm. Sixteen of the subsamples were analyzed for protein by the Kjeldahl method over a 3-month period, and the average of these results is called the "definitive Kjeldahl" (KjD) for that sample. NIR spectra of the other four subsamples were obtained by an inhouse-designed, computerized spectrophotometer (Williams et al, 1983), recorded as log (1/*R*), and averaged to form a single spectrum. The spectra are 1,001 points spaced 1.6 nm apart in the region of 1,000–2,600 nm with a resolution of 7 nm. The KjDs are reported on a moisture-free basis, with moisture data being obtained from the near-infrared spectra (Norris and Hart, 1965) and then combined with the KjDs to form the "as-is" or moisture-free protein data. The 120 spectra, with the corresponding chemical data, were stored on linc-tape as 16-bit integers for computerized data analysis. Figure 4 shows four of these samples with extreme protein and moisture values.

The wide variety of methods presented here, and the use of full spectra in the examples, obscure the fact that most calibrations are developed using data from instruments that in practice have only a few preselected wavelengths available and that use linear regression on one to six spectral values. Sections II–V cover the mathematics used in at least 90% of applications so far. To help bridge the gap between this chapter and the mathematically more demanding next chapter, the conclusion contains an overview with simplified descriptions of all the mathematics used in both chapters.

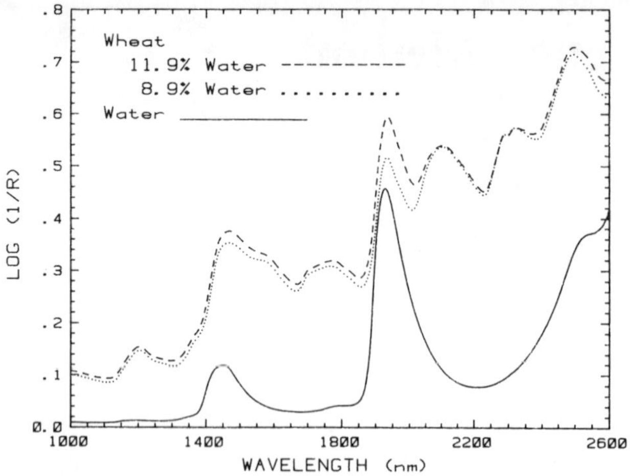

Fig. 1. Spectrum of water and spectra of wheat at two levels of moisture with constant protein content (12.4%).

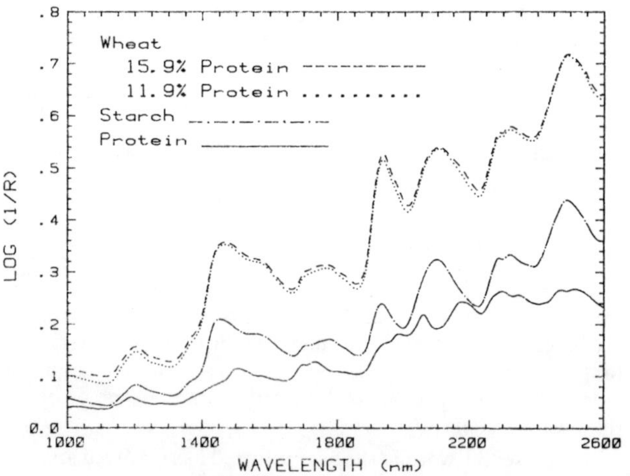

Fig. 2. Spectra of starch and protein and spectra of wheat at two levels of protein with constant moisture content (8.0%).

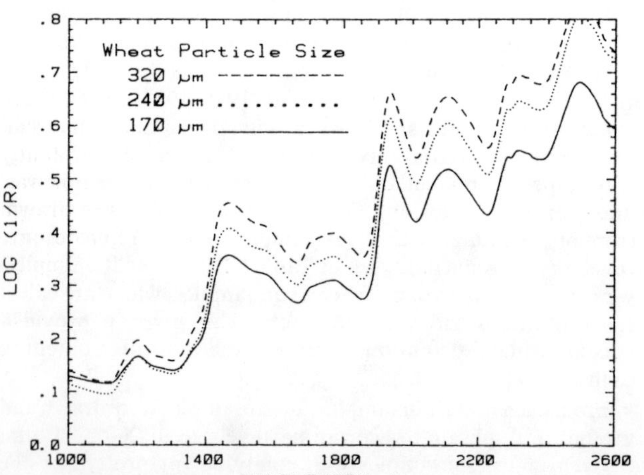

Fig. 3. Spectra of wheat (9.5% water, 11% protein) at three different particle sizes.

II. CALIBRATION, MEASUREMENT, AND VALIDATION

Measurement is a general term for quantification. The technical literature, however, uses the term only for physical quantities such as reflectance or temperature, and uses the term *determination* for chemical quantities such as percentage of protein in a wheat sample. Samples are *analyzed* for a component. When the analysis is quantitative, the component is said to be determined, and when the analysis is qualitative, the component is said to be *detected*. In the near-infrared literature,

TABLE I
Reference Data for Illustrative Hard Red Spring Wheat Samples

	Protein (%)		Moisture
Set	As Is	Dry Basis	(%)
Calibration (50 samples)			
Mean	14.15	15.98	11.47
Standard deviation	2.35	2.62	1.24
High	18.75	21.32	13.29
Low	10.13	11.35	8.97
Validation (50 samples)			
Mean	13.74	15.55	11.68
Standard deviation	2.11	2.39	1.43
High	17.71	20.05	14.32
Low	9.66	11.08	8.54
Reproducibility (20 samples)			
Mean	14.00	15.92	11.99
Standard deviation	0.061	0.065	0.273
High	14.09	16.06	12.78
Low	13.88	15.83	11.66

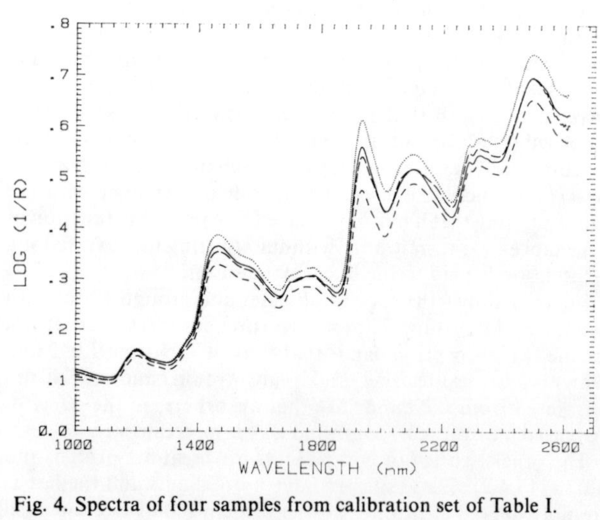

Fig. 4. Spectra of four samples from calibration set of Table I.

the term *prediction* is often used in place of measurement or determination. Although this term is routinely used by statisticians in the same way, it has connotations of looseness of method for nonstatisticians, and will probably not be an acceptable term when NIR terminology becomes standardized. In this chapter, *measurement* is used for physical quantities, *determination* for chemical quantities, and *measurement* for the general case.

A. Calibration

NIR instruments determine protein and other components by measuring log $(1/R)$ values, which must then be related to the amount of the component as determined by some other method called a *reference* or *standard* method. Establishing this relationship by using a set of samples of known composition is called *calibration* of an NIR method, whereas using the relationship to determine the amount of a component in a new sample is called an *NIR determination*. The relationship between the log $(1/R)$ values and the reference method values is expressed as an approximation, and always involves some form of regression equation. (Sections IV and V contain a review of linear regression.) The regression equation has regression constants (the Y-intercept and regression coefficients), independent variables, and one dependent variable (the reference method value). The independent variables are mathematical combinations of log $(1/R)$ values at various wavelengths. These combinations can be so complicated that they are better thought of as a series of steps, so we use the term *data treatment* to mean any mathematical process that combines log $(1/R)$ values into independent variables for use in a regression equation.

For this discussion, a distinction is made between calibration and developing a calibration model. Each data treatment has data treatment constants (such as the derivative parameters, or the amount of smoothing). Developing a calibration model involves testing different data treatments, data treatment constants, or sets of wavelengths. *Calibration* means finding the regression constants that go into the approximation once the form of the approximation, the data treatment constants, and the wavelengths have been decided upon. The following four examples illustrate the distinction.

1. One of the simplest examples of an approximation is the familiar single-term (one independent variable) regression equation:

$$Y = a + bX + e$$

or

$$\hat{Y} = a + bX$$

where Y and X are the reference and log $(1/R)$ values, respectively, a and b are regression constants, $\hat{Y}$ is an approximation to Y, and $e = Y - \hat{Y}$ is the error in the approximation for a single sample. The data treatment here is simply to take the log $(1/R)$ value as it stands, so that there are no data treatment constants. If this form of approximation were to be used, then developing the calibration model would mean finding the wavelength that provided the best approximation for all samples. If the wavelength had already been chosen, then calibration would mean finding the regression constants that gave the best approximation for all samples. An example is the use of log $(1/R)$ at 1,680 nm to measure particle size index.

2. The most common example of an approximation is a multiterm linear regression with the equation:

$$Y = a + b_1x_1 + b_2x_2 + \ldots + b_kx_k$$

where the Xs are log $(1/R)$ values at k different wavelengths. The data treatment is the same as in the first example, and there are no data treatment constants. Here, developing the calibration model involves deciding how many wavelengths to use and selecting an optimal set of wavelengths. Calibration means finding the regression constants to correspond with the set of wavelengths chosen. Numerous examples can be found using three to six wavelengths.

3. A more complicated case, but involving only single-term regression, occurs when X in the above equation is the quotient of two first derivatives (see Section VI). The data treatment constants are the derivative parameters and the degree of smoothing. If this form of approximation were to be used, developing the calibration model would involve finding the best wavelengths, derivative parameters, and degree of smoothing. Calibration would involve finding a and b in the regression. An example is in Williams et al (1983).

4. The most complex situations involve multiterm regressions where each independent variable is a combination of the log $(1/R)$ values at all the available wavelengths. These combinations may be expressed as Fourier coefficients (Giesbrecht et al, 1981), curve fit coefficients (Hruschka and Norris, 1982), factor scores (Martens and Naes, 1983), etc., and developing a calibration model involves deciding which and how many combinations to use. Calibration in this case again means finding the regression constants.

There are also mixtures of the cases. For instance, Shenk et al (1981) and Marten et al (1985) use multiterm regression, where each independent variable is a quotient of first or second derivatives. And it is possible for some independent variables in a regression to be NIR data and others to be physical or chemical data; e.g., sample temperature, grinder, variety, location, etc.

The left half of Fig. 5 outlines the calibration process, using protein as an example. Several samples (the *calibration set*) should be analyzed by both the reference method and the NIR method, although in practice the two procedures usually analyze different subsamples of each sample. The reference method values are then regressed against the data treatment values to produce calibration constants. The number of samples required and the number of near-infrared wavelengths measured depend on whether one is developing a calibration model or actually calibrating. Development of a calibration model requires more samples and more available wavelengths than calibration itself. A simple rule of thumb is to use a

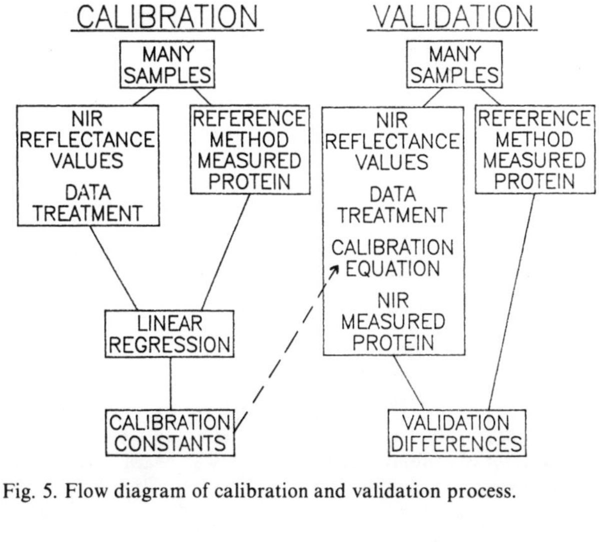

Fig. 5. Flow diagram of calibration and validation process.

minimum of 5–15 samples for each regression and data treatment constant and for any parameter of the data treatment (such as wavelength) that is allowed to vary. Examples are given in Section V.

The error of a regression approximation for a set of calibration samples can be reported as the *standard error of calibration* (SEC), the *correlation coefficient* (r), or the *coefficient of determination* (r^2). The SEC is the standard deviation (s_e) of the individual approximation errors (e) and is also called the *standard error of estimate* (SEE) for the calibration regression. The relation between r, r^2, and the SEC is illustrated in Table II, where the standard deviation of the reference data (s_Y) was assumed to be 1.0 and the SEC was calculated assuming a single-term linear regression with 50 samples. The basic relationships to notice are that the SEC decreases as r increases, r is always larger in absolute value than r^2, $0 \leq r^2 \leq 1$, and $0 \leq$ SEC. The last column of the table gives another way of understanding the correlation. In calculating its values, two things are assumed: first, that the range of the reference data is about six times its standard deviation (a reasonable ratio for sample sizes larger than 50 when the reference data are normally distributed); and second, that two samples can be correctly called different 95% of the time if their values are more than twice the SEC apart. Then the correlation permits division of the range into $6*s_Y/(2*\text{SEC})$ separate segments, and each sample is assigned to one of the segments by the calibration equation.

For example, suppose the Kjeldahl protein data for 50 samples had a standard deviation of 1.5% and a range of 9% (say, 10–19%). Then for a regression equation with a correlation of $r = 0.95$ ($r^2 = 0.90$), the corresponding entry in the fourth column of Table II (9.5) shows that the range of 10–19% could safely be divided into nine segments and one could be 95% sure of not being more than one segment off when assigning a sample to a segment. Another way of looking at sample differences is in terms of the SEC. The SEC corresponding to $r = 0.95$ (0.315 from column 3 of the table) is multiplied by the reference data standard deviation of (s_Y = 1.5%), because the table is normalized to $s_Y = 1.00$. This gives an SEC of 0.47%. Then 95% of the NIR determinations will be within twice the SEC (0.94) of the reference values, and 99% should be within 3*SEC = 1.41% of the reference values.

A second example uses the data in Table I. The as-is protein range of the calibration set is 18.75 − 10.13 = 8.62% protein, which is 3.67 times the protein s_Y of 2.35% protein, indicating that the protein data are more uniformly than normally distributed. Because each KjD value is the average of 16 individual Kjeldahl values, the sampling error is greatly reduced, permitting very high correlations with data treatments of log (1/R) values. The SEC values of 0.15% protein reported by Williams et al (1983) correspond to $r = 0.998$ and $r^2 = 0.996$, because 0.15 = 0.064 times the reference s_Y of 2.35%. Because the range is not $6*s_Y$, the fourth column entry of 47 segments must be adjusted to (3.67/6)*47 = 29 segments.

At the other extreme, correlation coefficients below $r = 0.6$ are not useful in this work because only three to four separate segments are statistically different. "Analysis" could be performed just as well by not using the regression equation at all but by declaring the determination of each unknown to be the mean value of the calibration samples. (You would still be sure of not being more than one segment off.) The situation is worse when the reference data are not normally distributed. For example, if they are uniformly distributed, then the range is approximately $3*s_Y$ instead of $6*s_Y$, and the values in column 4 of Table II must be divided by 2 to get the number of segments per range.

TABLE II
Relation of Measures of Correlation for Single-Term Linear Regression Using 50 Samples[a]

r	r^2	SEC	Segments per $6*s_Y$
0.0000	0.0000	1.010	3.0
0.2000	0.0400	0.990	3.0
0.4000	0.1600	0.926	3.2
0.6000	0.3600	0.808	3.7
0.8000	0.6400	0.606	4.9
0.9000	0.8100	0.440	6.8
0.9500	0.9025	0.315	9.5
0.9800	0.9604	0.201	14.9
0.9900	0.9801	0.143	21.0
0.9950	0.9900	0.101	29.7
0.9980	0.9960	0.064	47.0
0.9990	0.9980	0.045	66.4
0.9995	0.9990	0.032	93.9
0.9998	0.9996	0.020	148.4
0.9999	0.9998	0.014	209.9

[a] r = correlation coefficient, r^2 = coefficient of determination, SEC = standard error of calibration, and $s_Y = 1.00$ = standard deviation of reference data.

B. Measurement and Validation

Once an NIR instrument has been calibrated against a reference method, it can be used to determine the percentage of a constituent in different samples (called *unknowns*) or to measure some physical quantity of these samples. This *NIR measurement* should have a measurement error roughly equal to the SEC. However, the NIR-measurement error may, in practice, be significantly larger than the SEC. Comparison of NIR measurements and reference method measurements on a new set of samples provides a basis for calculation of the true measurement error. This comparison is called *validation* (or verification) of the calibration.

Referring again to Fig. 5 (right half), the mean ($\bar{D}$) and standard deviation (s_D) of the differences (D) between the NIR-measured and reference values for several samples (called the validation set, verification set, or prediction set) estimate the systematic and random errors, respectively, of the NIR method. The estimate $\bar{D}$ is called the BIAS and s_D is called the *standard deviation of validation differences* (SDVD = $\sqrt{[\Sigma (D - \bar{D})^2/(n - 1)]}$, where n is the number of validation samples).

The SDVD, which is a term invented for this chapter, has been called the SEP (standard error of prediction or of performance); the SEP (C) = SEPBC (bias-corrected standard error of prediction); the SDP (standard deviation of prediction); and the SED (standard error of differences). I have not found any standard term in the statistical literature, possibly because the importance of validation has been stressed only within the last 10 years. Because the terms SEP and SED have also been used for estimators different from the SDVD, readers of the literature are cautioned to look for definitions in each paper, and authors should quote definitions until this area becomes standardized.

There is some confusion about the correct way to combine the BIAS and the SDVD and how to interpret a combination. The confusion stems from two possible interpretations of the BIAS. If the BIAS is interpreted to mean a constant systematic error that is not expected to change, then a correct statement is: "If the samples used to determine the BIAS and SDVD are representative of the population to be measured, then 95% of future measurements will be within 2*SDVD of (reference value plus BIAS) and 99% will be within 3*SDVD of (reference value plus BIAS)." For example, if an instrument is calibrated with samples at one temperature and then is used to measure samples at a second temperature, the measurements may be biased. This bias is estimated by obtaining reference values for several samples, subtracting them from the corresponding NIR measurements, and finding the mean $\bar{D}$ of the differences D. Then future measurements of samples at the second temperature should be corrected for this systematic difference. If the BIAS is interpreted to mean a constant systematic error, it should not be combined with the SDVD except in statements such as the above.

On the other hand, the BIAS $\bar{D}$ may be interpreted as being caused by special characteristics of a set of samples or measurement conditions, which may change from day to day. Considered this way, $\bar{D}$ is expected to change with each new validation set and is not by itself an estimate of a constant systematic error as in the previous paragraph. Under certain assumptions, the *root mean square deviation* (RMSD = $\sqrt{[\Sigma\ D/n]}$) is then a more appropriate measure of the random error than the SDVD. To discuss this, stricter terminology than was used in the last paragraph is needed. The set of validation errors D is assumed to be representative of the "population" of all such errors. The mean (μ_D) of the error population is the true systematic error and the standard deviation (σ_D) is the true random error. The parameters μ_D and σ_D were estimated in the previous paragraph by the error sample mean $\bar{D}$ and standard deviation s_D, respectively. Some researchers assume that the measurement has no systematic error, which is the same as saying that $\mu_D = 0$. This assumption can be postulated, justified by previous experience with the measurement process or similar sample sets, or justified by showing that the sample mean is not significantly different from zero. With the assumption that $\mu_D = 0$, the RMSD is the appropriate estimator for the random error because it is an unbiased estimator for $\sqrt{[\sigma^2_D + \mu^2_D]}$.

So the decision whether to use the RMSD or the SDVD and BIAS is governed by assumptions about the systematic error of the measurement. If it is assumed to be zero, then the RMSD estimates the random error. It if is assumed to be different from zero, then the BIAS estimates the systematic error and the SDVD estimates the random error. For large n, RMSD $\cong \sqrt{[\text{SDVD}^2 + \text{BIAS}^2]}$.

RMSD is not a standardized term. It has also been used to mean $\sqrt{[\Sigma\ D^2/(n-1)]}$, which is the above RMSD with a divisor of $n - 1$ instead of n. Because this expression is not an unbiased estimator for the random error σ_D, it is not a useful estimator, except in the sense that for large n, it is very close to $\sqrt{[\Sigma\ D^2/n]}$. The RMSD is seldom found in statistics books except as a term for the ordinary sample standard deviation.

The terms *precision* and *accuracy* have been avoided in this discussion because they have not been standardized. In this monograph, precision will mean the reproducibility of a measurement under specified conditions. Accuracy will mean the closeness of the measurement to the "true" value as determined by an accepted reference method.

C. Developing a Calibration Model

Development of a calibration model is dependent on the variety and flexibility of the data treatments available and the experience of the developer. The simplest situations involve regression on log $(1/R)$ values, where the only variables are which and how many wavelengths to use from a set of six. The most complex occur in research laboratories that have most of the methods in this chapter and the next one available.

The first step is recording the spectra and reference data on sufficient numbers of calibration and validation samples. For calibration, the number of samples considered sufficient is discussed in Section V. For validation samples, no consensus exists on a minimum number and validation sets of from 6 to over 100 samples have been quoted. It is important that the validation samples cover the range of variability anticipated in future samples. This, however, is usually impossible to guarantee, and additional validations are often done routinely to detect changing situations and revise calibrations.

The second step should be the determination of the sampling and reference method errors discussed in Section III. The *reference-to-reference correlation, reproducibility*, or *laboratory error* (LE) may include sampling error and can be measured by the standard deviation of differences between reference measurements on duplicate samples. The tests should duplicate both sample preparation and analysis. An NIR method cannot correlate to a reference method better than the reference method correlates to itself, so the LE is the practical lower limit for the SEC. The LE should also be compared with the standard deviation of the reference data (s_Y). Table II can be used to convert the minimum expected SEC to the maximum expected absolute r by finding the corresponding normalized SEC = LE/s_Y in the third column. For example, if the LE is about one third of the reference standard deviation, then the fraction 0.33 gives a maximum expected absolute r of somewhat less than 0.95.

The third step is to choose data treatments that provide acceptable SECs. "Acceptable" usually means from one to two times the LE. The NIR reproducibility should also be checked for any data treatment chosen.

The fourth step is to validate the acceptable calibrations to test for *overfitting*, which happens when calibration regression approximated nonrepresentative features of the particular samples used for calibration. Several different indications of overfitting are common:

1. The SEC is much lower than the LE. This is usually the result of using too many terms in the calibration regression, or too few samples.
2. The BIAS is significant compared with the SDVD. The comparison should be the fraction $\text{BIAS}^2/\text{SDVD}^2$ because errors compare only when squared. For example, if the SDVD = 1.00, then a bias of 0.40 contributes only 16% of the variance, but a bias of 0.80 is significant. (Table III gives other examples.) If the BIAS looks significant, then further validation should be done to decide whether it is appropriate to compensate for the BIAS, or to treat it as long-term random error.
3. The BIAS is negligible, but the SDVD is more than 2*SEC. In this situation, it may be possible to get a good NIR measurement from a bad calibration by averaging the results of several NIR measurements on subsamples (essentially by averaging out sampling error).

The fifth step in developing a calibration model is to try to understand why it works: why the wavelengths are chosen and how the data treatment helps combine the log $(1/R)$ values in a meaningful way. Sometimes this may be as simple as noticing that a protein calibration uses a wavelength at which radiation is absorbed by protein. A calibration that has not been justified spectroscopically is less likely to stand up in the field even though it has been validated.

Different data treatments must be validated on separate randomly selected sample sets for the validation results to be statistically comparable, although this is rarely done in practice

TABLE III
Relation of BIAS, SDVD, and RMSD for SDVD = 1.00[a]

BIAS	RMSD	$\text{BIAS}^2/\text{SDVD}^2$
0.00	1.00	0.00
0.20	1.01	0.04
0.40	1.08	0.16
0.60	1.17	0.36
0.80	1.28	0.64
1.00	1.41	1.00
2.00	2.24	4.00
4.00	4.12	16.00
8.00	8.06	64.00

[a] BIAS = mean of validation differences, SDVD = standard deviation of validation differences, and RMDS = root mean square of validation differences.

because of the expense of reference measurements.

A table listing the individual calibration or validation errors is useful for learning about a data set, especially for detecting outliers and transcription errors. Table IV illustrates this. The number of asterisks beside each error represents the absolute size of the error in terms of the SEC. A reasonable calibration error distribution would have one third of the errors with one asterisk, 4% with two, and 1% with three. A similar distribution for validation errors would indicate no overfitting and no obvious outliers. A glance at the asterisks in Table IV suggests that the data for samples 4 and 5 have been interchanged. A reasonable error distribution results if the reference values of samples 4 and 5 are changed to 13.5 and 12.5, respectively.

III. SOURCES OF ERROR

The error in an NIR measurement is dependent on the interaction of many sources, which we will examine under three broad categories: sampling error, reference method error, and NIR method error. With the high quality of today's instruments, sampling error is usually the largest component of differences between reference and NIR measurements.

Errors are not necessarily a matter of imperfect machines or people, but they can be a result of using different measurement methods. In some ways, *difference* is a better word than *error*. For example, in the measurement of protein, the NIR method measures molecular bonding, whereas the Kjeldahl method determines total nitrogen content, which is not perfectly correlated to molecular bonding. The buyer wants to know the protein level because it relates to bread-making quality or resale price, neither of which is perfectly correlated to either protein or nitrogen level. These considerations suggest that it may be neither possible nor useful to improve the NIR-Kjeldahl correlation beyond a certain point.

TABLE IV
Portion of Table of NIR Method and Reference Method Values[a]

Sample No.	NIR Value	Reference Value	Error
1	9.6	9.5	0.1
2	10.7	10.5	0.2
3	11.4	11.5	−0.1
4	13.8	12.5	1.3******
5	12.3	13.5	−1.2******
6	14.5	14.5	0.1
7	15.1	15.5	−0.4**
8	16.8	16.5	0.3*
9	17.4	17.5	−0.1
10	18.6	18.5	0.1

[a] Asterisks show errors in terms of standard error of calibration (SEC = 0.2).

A. Sampling Error

Sampling error is caused by lack of homogeneity in the material being sampled. For example, if two 1-g samples are drawn separately from a well-mixed, 1-kg batch of ground wheat, their protein content as measured by the Kjeldahl method may vary by 0.2% protein. Most of this variation results not from lack of precision of the Kjeldahl method, but from actual protein differences between the samples. Such variation is often the effective lower limit on the SEC in an NIR calibration. Table V shows common sources of sampling error for wheat and forages.

Sampling errors occur at several places in the calibration-validation process outlined in Fig. 5:

1. The reference and NIR methods usually analyze different subsamples.
2. The methods may look at different amounts of the material. If a method used four times as much material, it will usually have about half the sampling error. In Table VI, the reproducibility for various protein determination methods are compared with the sample size used by the methods. The inverse relationship between sample size and error accounts for some of the precision differences between the methods.
3. The NIR measurement can introduce a sampling error because the radiation penetrates less than 2 mm into a sample that may be as deep as 10 mm, so only a portion of the sample is being measured. Reloading the same sample or a different subsample back into the cup several times gives a basis for calculating the loading (or reload or repack) error, which is primarily sampling error; it may, however, include such sources of error as differences in sample compression, which affects reflectance, and differences in static electricity or in particle size, which may cause certain types of materials to rise to the surface and thus introduce a bias in the measurement. Rotating the sample cup is like reloading because a different part of the sample is observed. Using diffuse transmittance instead of reflectance can reduce the errors that are caused by only part of the sample being analyzed, but other sampling errors can then occur. For example, transmittance through whole kernel grain may overemphasize the outer layers of each kernel, which may not be representative of the whole kernel; and instruments using several small, filtered sources of radiation may have different wavelengths interacting with different parts of the sample.

Although it is rarely practicable to measure accurately all the kinds of sampling error in a particular application, it is useful to know where they occur and the approximate contribution of each source to the total sampling error. This can result in

TABLE V
Sources of Sampling Error in Two Commodities

Wheat	Straw and Forage
Type of sample used	Physical composition
	Dimensions (length, etc.)
	Leaf/stem ratio
	Stage of maturity
	Moisture content
Foreign material	Subsampling for sample preparation
Blending of withdrawn sample	Blending of withdrawn sample
Sample storage	Sample identification
Sample identification	Blending of prepared (usually ground) sample
Subsampling for preparation for analysis	Subsampling for actual laboratory analysis
Blending of prepared (usually ground) sample	
Subsampling for actual laboratory analysis	

savings by indicating where to repeat the sampling. For example, suppose sampling error at the boxcar level is 0.4% protein, and reload error on subsamples, after a single sample has been ground, is 0.2%. Then the boxcar + reload error is $\sqrt{(0.4^2 + 0.2^2)} = 0.45$. Averaging the spectra of four reloads from the single ground sample halves the reload error so that the combined error from these two sources is $\sqrt{(0.4^2 + 0.1^2)} = 0.41$. But averaging the spectra of four resamplings from the boxcar would theoretically halve both errors, giving a combined error of $\sqrt{(0.2^2 + 0.1^2)} = 0.22$. In this kind of calculation, the cost of the various levels of resampling must be balanced against the benefit from the reduced error.

If the various stages of a sampling process are replicated, then the relative importance of the errors introduced at each stage can be estimated by the analysis of variance, which can also be used to measure other sources of error. I illustrate this here by using a personal communication from P. C. Williams. Some data on NIR-determined protein variability are presented in Table VII. From each of four cargoes, six lots were drawn and thoroughly mixed. From each lot, two samples were drawn and two subsamples were ground from each sample. From each grind, two repacks of the sample cup were used for duplicate NIR protein determinations. The sum of squares column shows the relative amount of variance arising from each stage of the process. The root mean square column shows the error in terms of percentage of protein. The between-cargo variability dominates the total variation, with significant variability occurring between subsamples within a cargo lot and between duplicate tests (reloads). The variance due to individual cargo lots was insignificant because sampling of cargoes is fully automatic and continuous, with low variability between lots relative to variance between cargoes. Subsampling of the cargo lot samples is manual. Variance due to grinding was also insignificant. Grinding of samples for NIR analysis at Canada's Grain Research Laboratory involves a cyclone grinder with a sample flow regulator. The ground sample is mixed very thoroughly in the sample grinder receiving jar, and also in the sample tin, to minimize the grinding error. Loading the NIR instrument cell inevitably incurs some stratification, which is reflected in the variance of duplicate testing. The relative root mean square values show that most of the variance comes from the cargoes themselves. Cargo lots showed roughly twice the variance of the subsamples, which in turn reflected about twice the variance due to grinding and reloading.

TABLE VI
Comparison of Methods for Determining Protein in Wheat

Method	Reproducibility (% Protein)	Sample Size (g)
Proton activation	0.189	1
Thermal decomposition	0.185	1
Kjeldahl nitrogen	0.154	16[a]
NIR reflectance	0.141	16[b]
Neutron activation	0.074	100

[a] Average of sixteen 1-g replicates.
[b] Average of four 4-g replicates.

B. Reference Method Error

Every reference method has a measurement error (LE or reproducibility) that is defined as the standard deviation of differences between measurements on duplicate samples. This error is also found by measuring several replicates of a sample. The standard deviation of these measurements has also been called the LE, reproducibility, or standard error of a single test (SET). The error in the first definition is $\sqrt{2}$ times the error in the second because it is the standard deviation of differences, rather than of raw measurements. The literature has not become standardized in the use of these words or definitions, sometimes even using root mean square deviations rather than standard deviations. This is no problem if the definition in use is clearly stated. In this chapter, we use laboratory error (LE) as the standard deviation of differences because it is more directly comparable to the SEC and SDVD.

For the as-is protein data in Table I, the LE using differences between duplicate samples drawn after grinding and mixing 25 g of wheat is 0.12% protein. This error contains very little sampling error because a small amount of wheat after grinding can be made very homogeneous. The KjDs, on the other hand, should have very little reference method error because the average of 16 Kjeldahl determinations should have an LE of $0.12/\sqrt{16} = 0.03\%$ protein. The standard deviation of 0.06% protein for the reproducibility set is thus a measure of the sampling error between 1,200-g portions drawn from the original lots of grain.

There are also differences between what is measured by the NIR and what is measured by the reference method. In the case of the Kjeldahl method, total nitrogen is determined and multiplied by a factor to give protein. This factor, called the protein-to-nitrogen ratio, varies between commodities and can vary within a commodity (Jones, 1931; Tkachuk, 1977; Coblentz, 1982). The NIR does not determine nitrogen, but it measures vibrations of molecular bonds that are more directly correlated to total protein. Thus, a further limit on the NIR-Kjeldahl standard error of calibration, if a constant factor is used for all samples, is the variability of the protein-to-nitrogen ratio.

In other measurements, what the reference method is measuring may not be a direct result of chemical composition. It could be a subjective value, such as a taste panel score or a physical property such as loaf volume. In cases like these, the NIR-reference method error will have a lower limit although the molecular basis of the limit may be unknown.

As discussed in Section II, NIR results can often be used to detect errors in the reference method, especially errors

TABLE VII
ANOVA of Wheat Cargo Analysis for Protein by NIR

Source	Sum of Squares	Degrees of Freedom	Mean Square	Variance Ratio	Probability	Root Mean Square
Cargoes	38.350	3	12.783	175.1	0.001	3.58
Lots	0.829	5	0.166	2.3	...	0.41
Error A	1.091	15	0.073			
Samples	0.057	1	0.057	3.0	...	0.24
Error B	0.440	23	0.019			
Grinds	0.012	1	0.012	0.5	...	0.11
Error C	1.125	47	0.024			
Duplicate NIR	0.015	1	0.015	3.9	0.5	0.12
Error D	0.359	95	0.0038			
Total	42.278	191				

involving transcription or sample mixup. This is done by treating the calibration samples as a validation set and looking for abnormally large validation errors.

C. NIR Method Error and Smoothing

NIR method errors are caused by spectral measurement errors, lack of intrinsic correlation between spectral and reference methods data, and poor choice of data treatment. A detailed discussion of the detection, measurement, and correction of errors in the measurement of spectra is given in a separate chapter, and the lack of intrinsic correlation has been discussed earlier. The relation of specific data treatments to various types of error will be mentioned in later sections.

Instruments used in NIR spectroscopy have been designed to have very low errors in both wavelength and log $(1/R)$ values because, in the applications developed for NIR instruments, the spectral variation related to variation in composition of samples is very low. For instance, Fig. 2 shows that a 4% change in protein is equivalent to a 0.01 change in log $(1/R)$ value at 2,180 nm. A 0.2% change in protein then corresponds to a 0.0005 change in log $(1/R)$. Using the slope at 2,180 nm, we can calculate roughly that a wavelength error of 0.3 nm also corresponds to a 0.0005 error in log $(1/R)$ value, which would again correspond to a protein determination error of 0.2% protein. Thus the usual specification of log $(1/R)$ and wavelength reproducibility as 0.0001 log $(1/R)$ units and 0.05 nm, respectively, ensures that the instrument-caused error is well below the usual sampling error of 0.15–0.2% protein.

Figure 6 shows a wheat spectrum with added noise of 0.0001, 0.001, and 0.01 log $(1/R)$. The 0.001 log $(1/R)$ noise is at least 10 times the maximum normal noise. Even though it is barely visible to the eye, it causes a doubling of the calibration and validation errors when added to the data sets of Table I.

Although the usual noise specifications are adequate for most NIR applications, software packages developed for research scanning instruments usually have routines for mathematically smoothing spectral data. These generally use either a running mean or a Savitzky-Golay smooth, although several other methods could be used: binomial (Marchand and Marmet, 1983), Fourier, and Gaussian (see Section VII).

A *running mean* (or boxcar smooth) simply replaces the value at each point by the mean of the values in a wavelength interval surrounding it. The interval is preferably specified in nanometers but is sometimes specified in number of points. The interval is centered at the given point, resulting in an odd number of data points per mean. In reading the literature, one should make sure of an author's definition of the running mean.

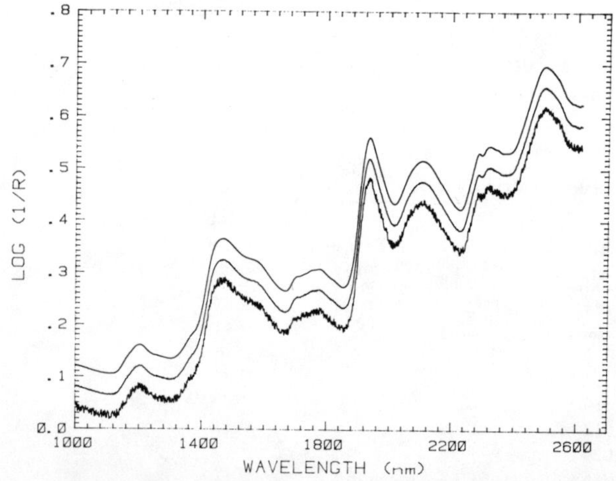

Fig. 6. Wheat spectra with added, uniformly distributed noise of 0.0001 (top), 0.001 (middle), and 0.01 (bottom) log $(1/R)$ units, respectively. Spectra are displaced for clarity.

If the number of points is $n = 2m + 1$, where m is the number of points on either side of the central point, then sometimes the "number of points" quoted is m. The advantage of this method is its ease of calculation.

The principle of the Savitzky-Golay smooth is to fit the spectrum in a wavelength interval with a polynomial by least-squares methods, and the parameters are the degree of the polynomial and the number of points to fit (Savitzky and Golay, 1964; Morrey, 1968). Then the smoothed spectrum and its derivatives are given by the polynomial and its derivatives. (See Section VI for a review of derivatives.) The method is used primarily where the bandwidth or band motion to be detected is on the order of the distance between adjacent wavelengths or smaller, where the data are very noisy, or where it is important to preserve spectral band shape.

Figure 7 illustrates the effects of the two smoothing methods on computer-generated Gaussian curves with added noise. Notice that the running mean smooth lowers the peak height of the narrow band but does not distort the wide band. The remainder of this section contains the details of the Savitzky-Golay calculation.

The Savitzky-Golay calculation involves the solution of simultaneous equations. If λ_j is the wavelength at which the smoothed or derivative value is desired, n the degree of the fitting polynomial, and $2m + 1$ the number of points to fit (giving an interval symmetrical about λ_j), then the $n + 1$ simultaneous equations are:

$$\sum_{i=-m}^{m} i^r y_i = \sum_{k=0}^{n} \left(b_{nk} \sum_{i=-m}^{m} i^{r+k} \right) ,$$

one for each r, in the $n + 1$ unknowns b_{nk}, where $r = 0, 1, \ldots, n$, and y_i is the log $(1/R)$ value at wavelength λ_{y+i}. The value b_{nk} is the coefficient of the kth term of the nth degree polynomial, b_{no} is the smoothed value at λ_j, and the value of the sth derivative at λ_j is $s!b_{ns}$.

The calculation can be thought of as a weighted average of the log $(1/R)$ values in the interval $(\lambda_{j-m}, \lambda_{j+m})$. The weights can be extracted during the calculation if a smoothed or derivative value for many points is desired. In that case, the weights for calculating the sth derivative at λ_j are $s!$ times the elements of the $(s + 1)^{\text{th}}$ row in the matrix A^{-1}B, where:

$$A_{r+1,k+1} = \sum_{i=-m}^{m} i^{r+k} \; (r, k = 0, 1, \ldots, n)$$

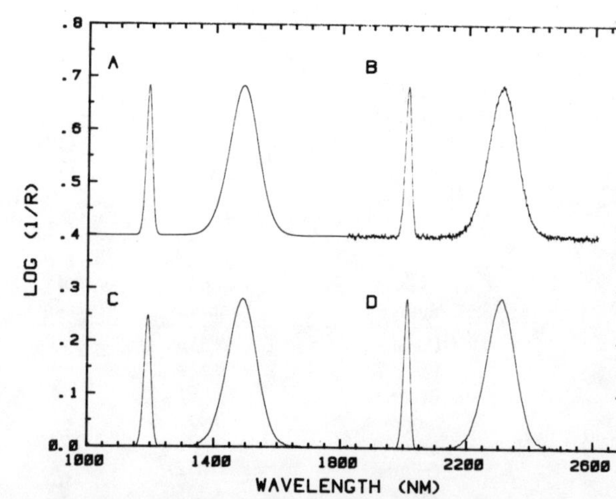

Fig. 7. Effect of smoothing: (A) Raw data, computer-generated Gaussian curves; (B) normally distributed noise (standard deviation 0.001 log $(1/R)$) added to (A); (C) running-mean smooth of (B); (D) Savitzky-Golay smooth of (B).

and

$$B_{r+1,i+m+1} = i^r \; (r = 0, 1, ..., n; i = -m, ..., 0, ..., m) .$$

The zero-order derivative is the Savitzky-Golay smooth. Savitzky and Golay (1964) give hints for simplifying the calculations, as well as tables for computing $s!b_{ns}$ for various given s and n. Users of these tables should consult the corrections by Steinier et al (1972).

IV. SINGLE-TERM LINEAR REGRESSION AND THE CORRELATION PLOT

This section gives a brief introduction to the concepts involved in single-term linear regression, the statistical procedure that answers the following question: "Given a set of data with one independent variable X and one dependent variable Y (as in Table VIII) and the corresponding scatterplot of Y against X (as in Fig. 8), what is the straight line that best fits the data?" The answer is the straight line with the equation:

$$\hat{Y} = a + bX$$

where $\hat{Y}$ is an approximation to Y, and a and b are given by the formulae in the Appendix to this chapter. Figure 9 shows the geometric meaning of Y, a, and b. For a given data point (X, Y), $\hat{Y}$ is the point on the line immediately above or below the data point. The Y-intercept (a) is the Y value at the point where the line crosses the Y axis. The slope (b) is the steepness of the line, measured by the increase in Y for each unit increase in X.

The best-fitting line is called the regression of Y on X or Y regressed against X. The regression constant a is called the constant term, and the regression constant b is called the regression coefficient. The vertical distance from a data point to this line is the residual or regression error for that point (shown in Fig. 10), and the standard deviation of all the residuals is the SEC (or SEE). The correlation coefficient (r), which is related to the SEC, ranges from -1 to 1. If $r = 1$, the data are exactly on the regression line and are perfectly correlated. If $r = 0$, then there is no correlation, and if the slope of the regression line is negative, then r is negative, giving a perfect negative correlation if $r = -1$. Figures 11–14 give examples of scatterplots of data with four levels of correlation.

In developing a calibration model using single-term linear regression, when one does not yet know the best wavelength to use, one normally finds the r value at every available wavelength. The wavelength giving the highest r value is then used for the actual calibration and subsequent validation. However, in practice, this simple approach seldom gives an adequate SEC, and a more complex calibration is usually needed. The result of successively more complex mathematical treatments in single-term regression can be illustrated with the correlation plot, which is a graph of r against wavelength.

Figure 15 shows the correlation plot for the 50 calibration samples of Table I, where the r values are calculated with as-is

TABLE VIII
Data Generated Randomly About Line $Y = 0.0 + 0.1X$[a]

Independent Variable	Dependent Variable	Residual
1.00	0.445	0.345
2.00	0.547	0.347
3.00	0.013	−0.287
4.00	0.213	−0.187
5.00	0.569	0.069
6.00	0.626	0.026
7.00	1.273	0.573
8.00	1.151	0.351
9.00	0.926	0.026
10.00	0.797	−0.203
11.00	1.312	0.212
12.00	1.258	0.058
13.00	1.723	0.423
14.00	1.200	−0.200
15.00	1.125	−0.375

[a] X = independent variable, Y = dependent variable, standard deviation of residuals = 0.289, and mean of residuals = 0.079.

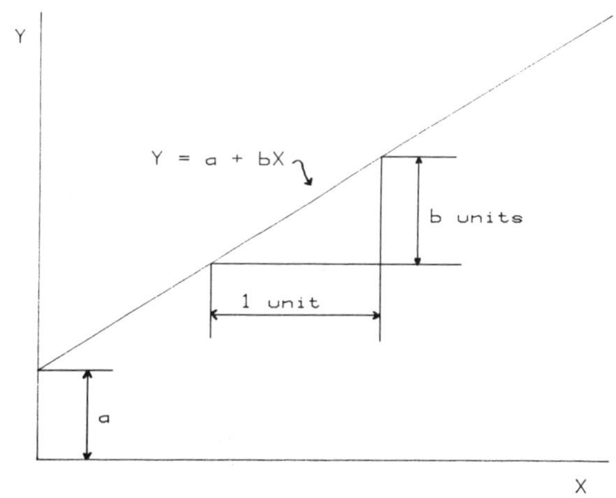

Fig. 9. Geometric meaning of constants in straight line $Y = a + bX$.

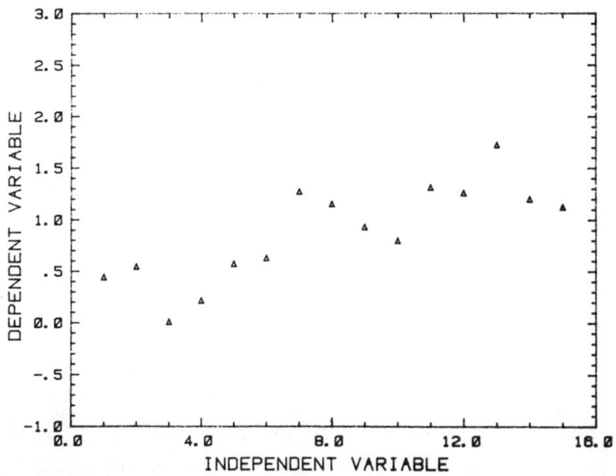

Fig. 8. Scatterplot of data in Table VIII.

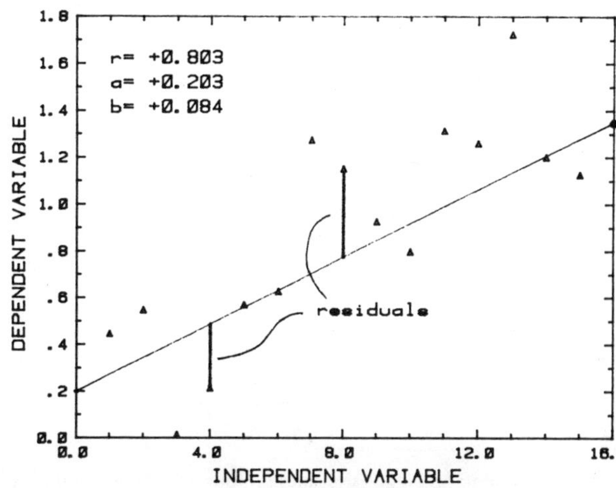

Fig. 10. Regression line through data in Fig. 8.

protein content regressed against log $(1/R)$ at each of 1,000 wavelengths. The fact that the maximum r in Fig. 15 is only 0.26 indicates that a more complicated mathematical treatment is necessary to obtain useful results.

In this example, the X in $Y = a + bX$ is log $(1/R)$ at one wavelength. One way to improve the correlation is to let X be the difference between log $(1/R)$ values at two different wavelengths. The two wavelengths can be found by an iterative process. First, the single wavelength giving the best correlation is found; then, a second wavelength is found so that the difference between log $(1/R)$ values at the first and second wavelengths gives the best correlation. Then the first wavelength is replaced with a third wavelength whose difference with the second gives the best correlation, and so on until the process converges, which means that each of two wavelengths pairs best with the other. When this technique was applied to the calibration samples of Table I, the maximum r value rose to 0.92 with the general level of the correlation plot (Fig. 16) becoming much higher. An iterative procedure does not necessarily produce the pair of wavelengths whose difference provides the highest correlation. It only provides wavelengths that would choose each other. The same process can be used with quotients instead of differences (Fig. 17), and with the quotients of differences (Fig. 18). In this last case, there are various ways of iterating the process when selecting the four wavelengths whose log $(1/R)$ values are put into the expression $(A-B)/(C-D)$. The various methods do not all yield the same choice of wavelengths.

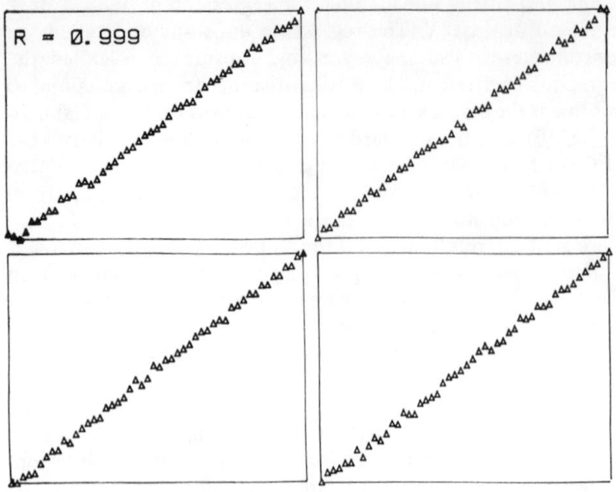

Fig. 11. Scatterplots and regression lines for four sets of data randomly generated about line $Y = 0.0 + 1.0X$ ($r = .999$).

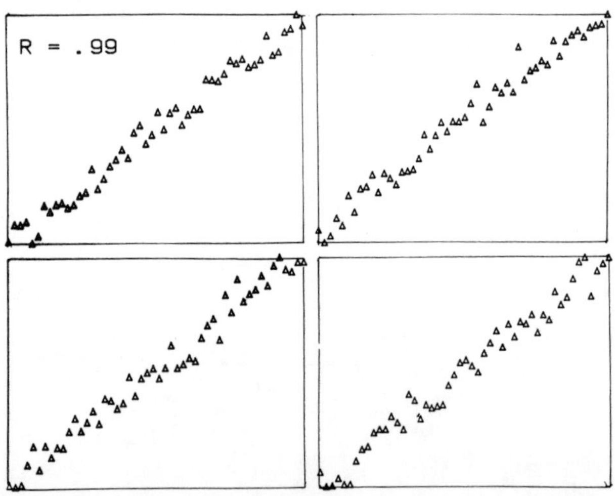

Fig. 12. Scatterplots and regression lines for four sets of data randomly generated about line $Y = 0.0 + 1.0X$ ($r = .99$).

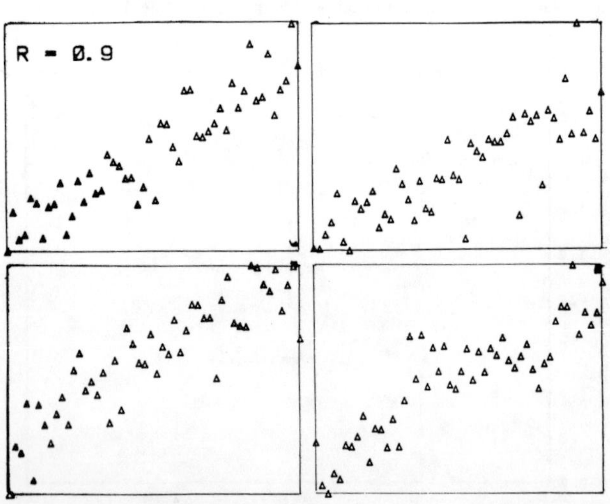

Fig. 13. Scatterplots and regression lines for four sets of data randomly generated about line $Y = 0.0 + 1.0X$ ($r = .9$).

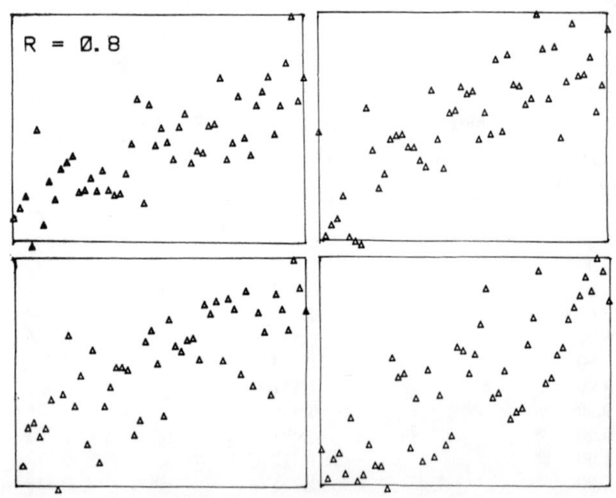

Fig. 14. Scatterplots and regression lines for four sets of data randomly generated about line $Y = 0.0 + 1.0X$ ($r = .8$).

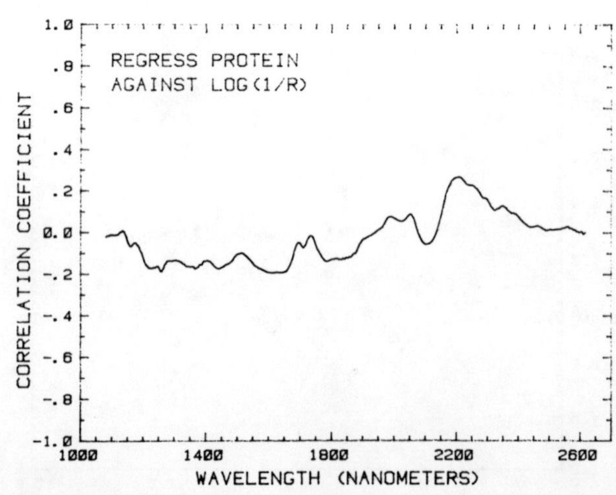

Fig. 15. Correlation to log $(1/R)$ value at each wavelength using 50 wheat samples.

These selection methods do not require a correlation plot, especially when only a few wavelengths are available; but the plot or a listing of the correlations often gives information as to why certain wavelengths are selected. Many researchers are reluctant to recommend wavelengths that are selected unless there is some spectroscopic justification. Of the various justifications offered by computer-selected wavelengths, the most common are that (i) a chosen wavelength is at the peak absorbance of the component to be determined; (ii) it is at a peak absorbance of a component whose concentration is highly correlated with that of the component to be determined; and (iii) it is part of a difference or quotient expression and serves to normalize the spectra to one level of scatter, particle size, temperature, etc. In this last case, the wavelength may be called a *reference wavelength* (see Sections V and VIII).

The correlation plot also shows which wavelengths may be unsuitable when transferring calibrations from a laboratory situation to the field. If the plot shows a sharp spike at a wavelength (as at 1,850 nm, Fig. 18), then a measurement based on that wavelength requires an instrument with very high wavelength accuracy and precision. The correlation plot can be used with multiterm regression, as discussed in Section V.

V. MULTITERM LINEAR REGRESSION

A. Basic Properties

Multiterm linear regression uses the information at a number of wavelengths to isolate the effect of a single absorber and to normalize the baseline. This technique is more difficult to describe geometrically than the single-term case, and the formulae for the regression coefficients are more complex. Given a set of data with one dependent variable Y and k independent variables $X_1, X_2, ..., X_k$, the problem is to find the set of constants $a, b_1, b_2, ..., b_k$ so that the function:

$$\hat{Y} = a + b_1X_1 + b_2X_2 + ... + b_kX_k$$

is the best fit to the data. If $k = 1$, this function is the straight line over the X_1 axis as discussed in the previous section. If $k = 2$, the function is a plane over the X_1X_2 plane. If $k = 3$, then there is no easily visualized function and no simple formulae.

Usually the X's are log $(1/R)$ values at different wavelengths. But they can be anything that will help the measurement. Examples are derivatives, quotients of derivatives, Fourier coefficients, curve fit coefficients, principal component scores, and other types of scores, all discussed in later sections and the next chapter. Auxiliary variables can also be added. For instance, the time a spectrum was measured, the instrument operator, the instrument, the wheat variety, etc., could be used in the regression. In most of this section, the term *wavelength* or *log (1/R) value* can be replaced by any other variable without changing the intended meaning.

There are various ways of choosing the wavelengths to use in a multiterm linear regression. The step-up or forward-stepwise procedure picks the wavelength giving the best single-term calibration as the first independent variable, then finds the best wavelength to add as a second variable in a two-term regression, and so on until some stopping criterion is met. The step-down procedure starts with a multiterm linear regression using all available wavelengths and eliminates variables by some criterion. The all-possible-combinations procedure tests all possible linear regressions on all subsets of the available wavelengths and reports the subsets giving the lowest SEC. This procedure is usually limited to all subsets containing only two or three wavelengths. There are also combinations of these methods. For example, the all-possible-combinations method can select two or three wavelengths, and then the step-up method can be used to add wavelengths. Alternatively, each step in the step-up method can be followed by one step of the step-down method, to check for wavelengths that can be safely eliminated when a new wavelength is added. It is important to use some criteria for selection of subsets of available variables

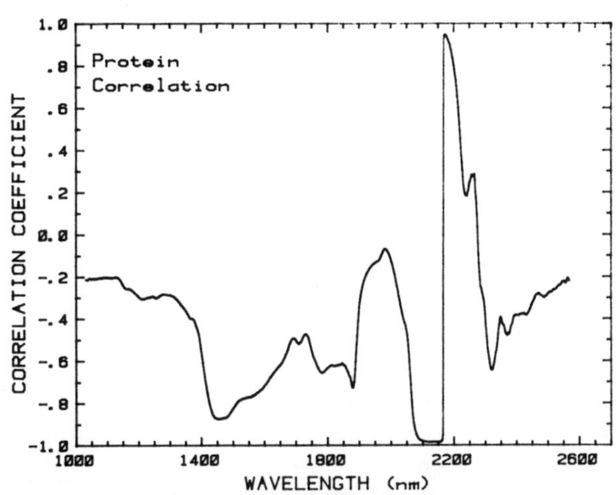

Fig. 16. Correlation to $A - B$ for 50 wheat samples. $A = \log(1/R)$ at each wavelength, $B = \log(1/R)$ at 2,166 nm.

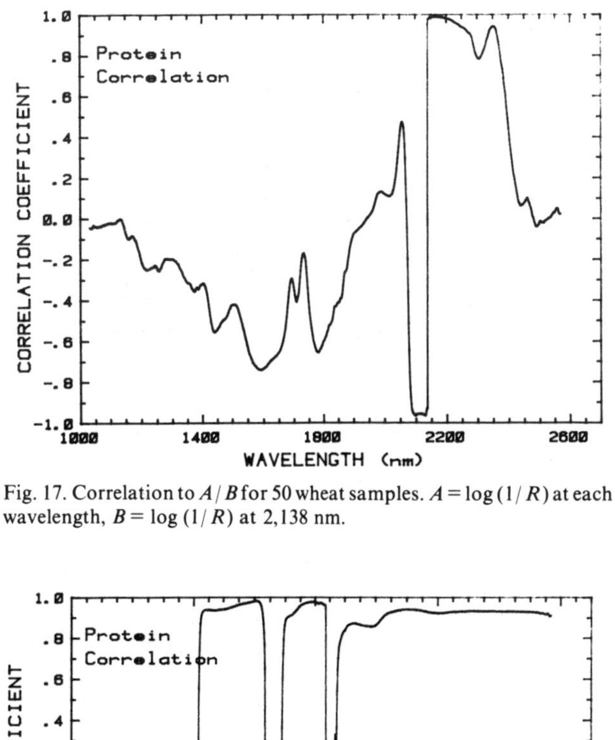

Fig. 17. Correlation to A/B for 50 wheat samples. $A = \log(1/R)$ at each wavelength, $B = \log(1/R)$ at 2,138 nm.

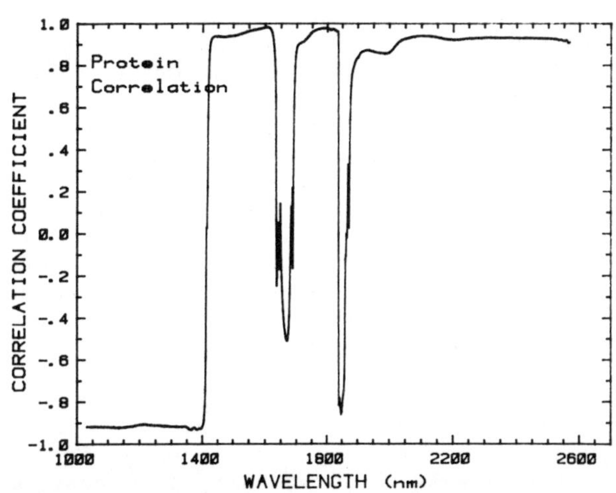

Fig. 18. Correlation to $(A - B)/(C - D)$ for 50 wheat samples. $C = \log(1/R)$ at each wavelength; A, B, and $D = \log(1/R)$ at 1,982, 1,907, and 1,838 nm, respectively.

to avoid overfitting (see Draper and Smith, 1966; Hocking, 1970, for a description of several criteria). On the other hand, the question of the validity of the various stopping techniques in the face of the multicollinearity of the log $(1/R)$ values at several wavelengths requires further research.

Nonlinear regression is not in common use by NIR workers and will not be discussed here.

A k-term regression can be reduced to a $(k-1)$-term regression by "constraining" or "forcing" the regression coefficients to sum to zero. This constraint, which is one approach to the problem of shifting baseline, is equivalent to using one of the wavelengths as a reference wavelength, as in the previous section, or to normalizing the spectra (see Section VIII), as can be shown as follows:

The constraint $\sum_{i=1}^{k} b_i = 0$ can be written $b_1 = -\sum_{i=2}^{k} b_i$. Then

$$\sum_{i=1}^{k} b_i X_i = (-\sum_{i=2}^{k} b_i)X_1 + \sum_{i=2}^{k} b_i X_i = \sum_{i=2}^{k} [b_i(X_i - X_1)]$$

and X_1 is the reference wavelength log $(1/R)$ value.

A finite-difference derivative can be considered as a multiterm linear regression reduced to a single-term regression by a constraint on the regression coefficients. For example, the finite-difference second derivative can be considered to be a three-term regression on the log $(1/R)$ values at evenly spaced wavelengths, where the regression coefficients must be in the ratio 1:−2:1. This constraint not only takes care of shifting baselines, but also separates overlapping absorption bands. Another advantage of such constraints is that they reduce the number of independent variables and thereby reduce the number of samples required for calibration.

It was mentioned above that an adequate calibration usually requires a minimum of 10 samples for each estimated regression constant and for each varied parameter (such as wavelength). The factor 10 is based on experience with validation, rather than any statistical proof. Some researchers are content with as few as 5; some demand as many as 15. Following are a few examples using the factor 10:

1. A bias adjustment on an instrument is a change in the constant term used in the regression. Only one parameter is changed, so 10 samples are enough to make the adjustment.
2. A slope adjustment of a multiterm regression consists of multiplying all regression coefficients by the same number. Since essentially only one new number is determined, 10 samples are again sufficient.
3. A three-term calibration using three preselected wavelengths would require 40 samples (including 10 for the constant term).
4. Developing a 10-term calibration requires 210 samples (1 constant term + 10 coefficients + 10 wavelength selections = 21 parameters, times 10 samples per parameter).
5. Developing a single-term calibration consisting of a quotient of second derivatives where the wavelengths and gaps for the numerator and denominator derivatives (see Section VI) are optimized would require 60 samples (10*(2 regression constants + 2 wavelengths + 2 derivative parameters)).

The SEC, r, and r^2 have the same meaning in multiterm linear regression as in the single-term case, except that r is not negative. The correlation plot should be studied after the addition of each variable. For example, Fig. 19 shows the correlation plots when the first, second, and third wavelengths are found. Notice that the maximum r in the first plot is the minimum r of the second.

B. Calculation

If there are two independent variables in a regression, the regression constants are given by:

$$D = \Sigma x_1^2 \Sigma x_2^2 - (\Sigma x_1 x_2)^2$$

$$b_1 = (\Sigma x_2^2 \Sigma x_1 y - \Sigma x_1 x_2 \Sigma x_2 y)/D$$

$$b_2 = (\Sigma x_1^2 \Sigma x_2 y - \Sigma x_1 x_2 \Sigma x_1 y)/D$$

$$a = \bar{Y} - (b_1 \bar{X}_1 + b_2 \bar{X}_2)$$

where the lowercase letters are the "centered" variables $y = Y - \bar{Y}$, $x_1 = X_1 - \bar{X}_1$, etc.

If $k \geqslant 3$, the a and b_j are best given as the solution to a set of simultaneous equations as follows: If Y_i are the Y values for n samples, and X_{ij} are the corresponding X_i values for $j = 1, 2, ..., k$, then the approximation desired is given by the least-squares solutions $a, b_1, b_2, ..., b_k$ of the set of simultaneous equations:

$$Y_1 = a + b_1 X_{11} + b_2 X_{12} + ... + b_k X_{1k}$$

$$Y_2 = a + b_1 X_{21} + b_2 X_{22} + ... + b_k X_{2k}$$

$$\cdot \quad \cdot \quad \cdot \quad \cdot \quad \cdot \quad \cdot$$

$$Y_n = a + b_1 X_{n1} + b_2 X_{n2} + ... + b_K X_{nk} .$$

A simpler set of equations results from using the centered variables $(x = X - \bar{X})$:

$$y_1 = b_1 x_{11} + b_2 x_{12} + ... + b_k x_{1k}$$

$$y_2 = b_1 x_{21} + b_2 x_{22} + ... + b_k x_{2k}$$

$$\cdot \quad \cdot \quad \cdot \quad \cdot \quad \cdot$$

$$y_n = b_1 x_{n1} + b_2 x_{n2} + ... + b_k x_{nk} .$$

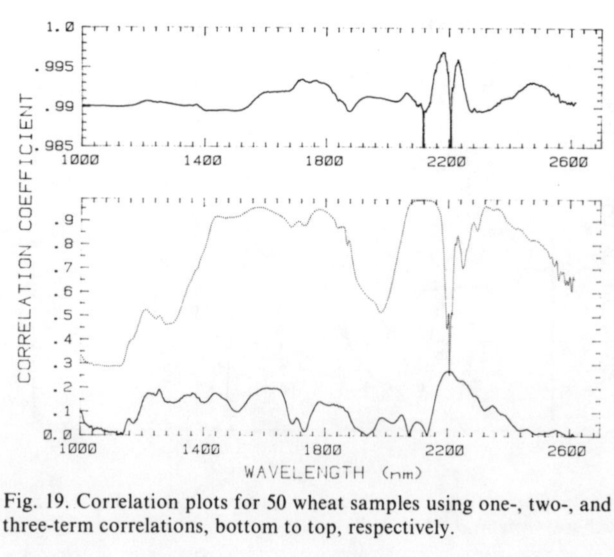

Fig. 19. Correlation plots for 50 wheat samples using one-, two-, and three-term correlations, bottom to top, respectively.

Once these are solved for the b_i, the values of a, r, r^2, and the SEC are given by the formulae at the end of this chapter.

The multiterm case is more compactly stated in matrix notation. If **y** is a column vector of n centered data points for the dependent variable Y, and **x** is the $n-$ by$-$ k matrix whose rows are the independent variable values corresponding to these data points, with each column containing the data for one independent variable; and **b** is the column vector of regression coefficients, the set of simultaneous equations with centered variable shown above can be written $\mathbf{y} = \mathbf{xb}$, and the least-squares solution is $\mathbf{b} = (\mathbf{x}^T\mathbf{x})^{-1}\mathbf{x}^T\mathbf{y}$. ($T$ = transpose, -1 = inverse.) The values of a, r^2, r, and the SEC are then found using the formulae at the end of this chapter.

VI. THE DERIVATIVE

A. Basic Properties

Derivatives are an approach to addressing two of the basic problems with near-infrared spectra: overlapping peaks and large baseline variations. The effect of derivatives is most clearly seen with the second derivative of a spectrum, which is able to separate overlapping absorption bands because it is related to curvature and has the same sign as the curvature of a spectrum. Figure 20 shows a single absorption band and its second derivative. The left and right portions of the band are curved upward, so the second derivative is positive there. The middle portion is curved downward, giving a negative second derivative with a minimum at the peak value of the spectrum where the downward curvature is greatest. The useful effect here is that the bandwidth (at half maximum) in the second-derivative spectrum is less than that of the original absorption band. The derivative will thus help resolve overlapping bands as in Fig. 21.

The second effect of the second derivative is to remove baseline shifts. Because the curvature of a straight line is zero, the derivative of a spectrum-plus-a-straight-line-baseline will be the same as the derivative of the spectrum as in Figs. 22 and 23. The quotient of two derivatives can be used to remove multiplicative as well as additive scatter effects.

The value of the second derivative in NIR work has stemmed from these two effects: resolution of overlapping peaks and removal of linear baselines. The first derivative also has these two effects, but to a lesser extent, and has a geometric interpretation as the slope of the spectrum at each wavelength (Fig. 24). Higher-order derivatives have the same two basic effects and will resolve overlapping absorption bands better

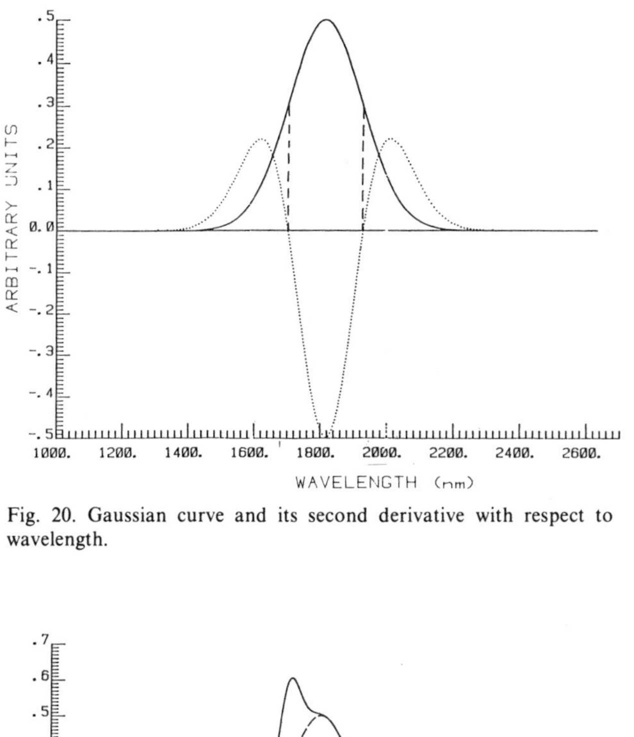

Fig. 20. Gaussian curve and its second derivative with respect to wavelength.

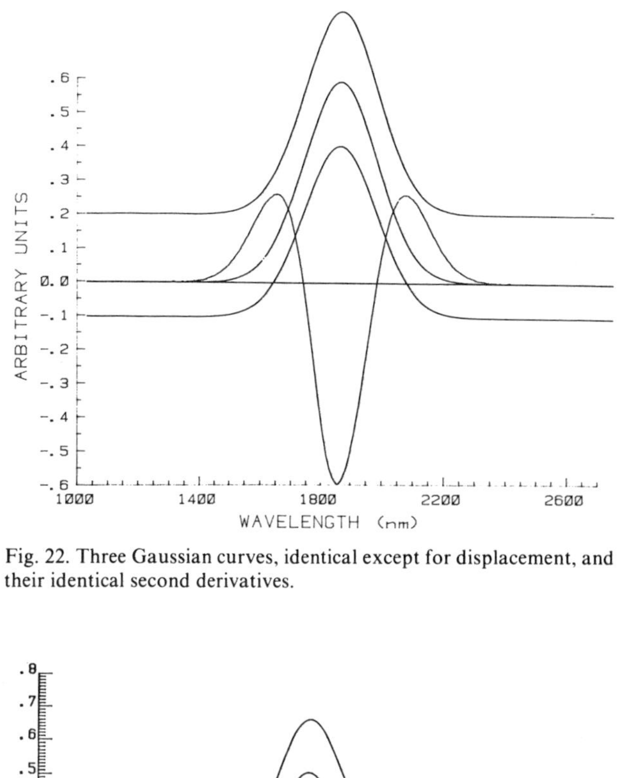

Fig. 22. Three Gaussian curves, identical except for displacement, and their identical second derivatives.

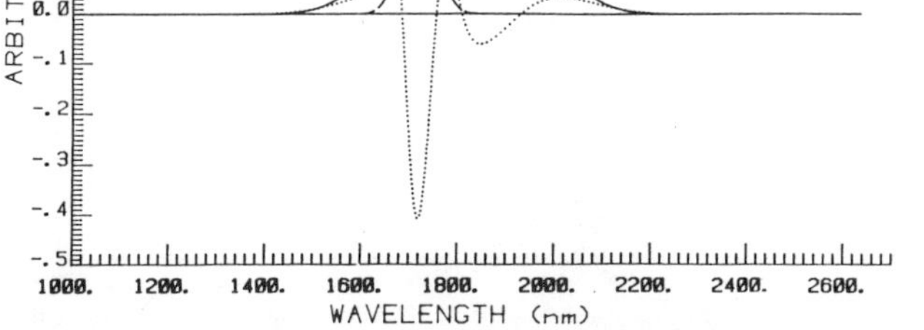

Fig. 21. Two Gaussian curves, their sum and second derivative of their sum.

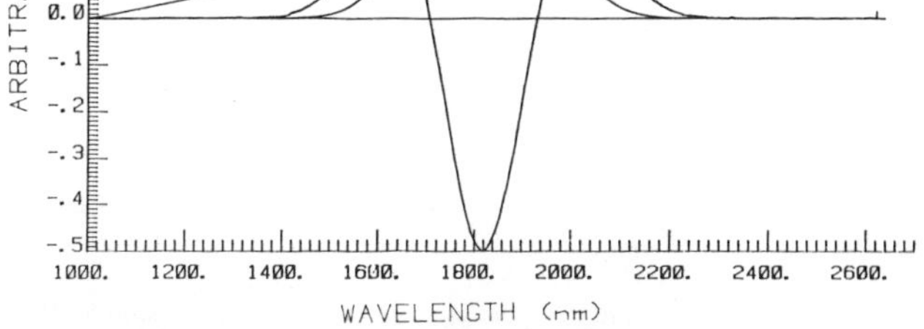

Fig. 23. Two Gaussian curves, identical except for addition of linear baseline, and their identical second derivatives.

than the lower-order derivatives, and they will remove gently curved background as well as linear background shifts. However, they are more sensitive to noise and generate more artifacts than the lower-order derivatives. They do not have any easily visualized geometric interpretation and have not been widely used in NIR work. Derivatives of orders higher than two have not been shown to have an advantage in calibration or validation.

The drawbacks of derivatives are the side lobes (see Fig. 20), which can give false information about an absorption band, and the increased sensitivity to spectral noise and wavelength instability.

In developing a calibration model using a derivative, the same search procedures are used as with linear regression methods, except that the wavelength search is done using the derivatives of the calibration spectra rather than the original spectra. Figure 25 shows the second derivatives of the spectra in Fig. 4. Figure 26 shows a correlation plot for protein regressed against the second-derivative spectra of the calibration samples of Table I together with the second derivative of a wheat protein spectrum. Notice the coincidence of correlation plot peaks with protein second-derivative peaks.

Software using more complex search techniques on multi-term quotients of derivatives where each derivative can be of a different order and have a different gap and amount of presmoothing has been developed by Shenk and Westerhaus (Marten et al, 1985).

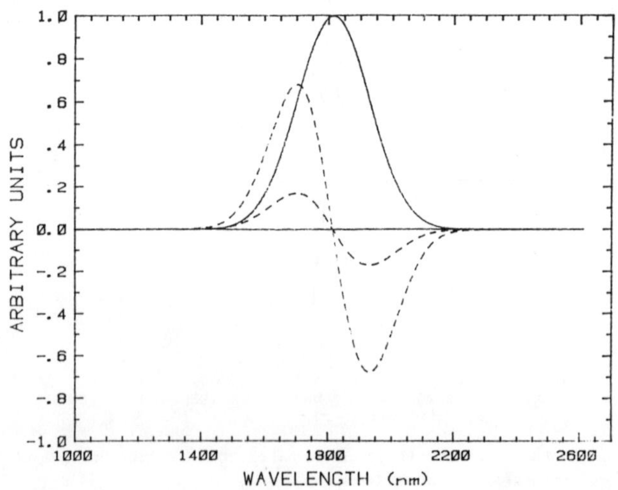

Fig. 24. Gaussian curve and its first derivative. Higher dashed line is multiple of lower.

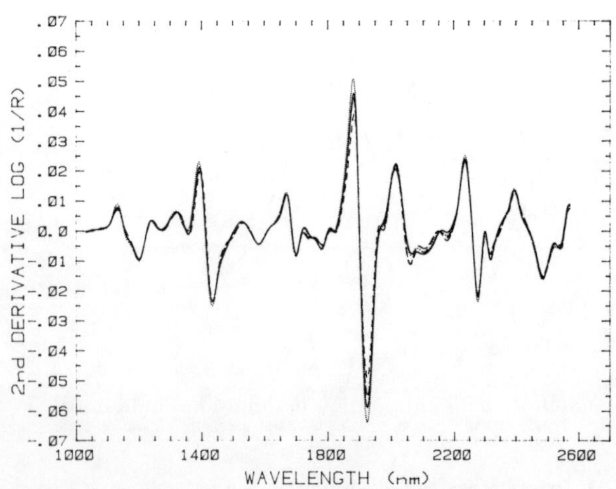

Fig. 25. Second derivatives of spectra in Fig. 4.

B. Calculation

Three common methods of calculating derivatives are the finite-difference, Fourier transform, and Savitzky-Golay methods. There is no simple relation between the derivatives as calculated by the three methods, and each has its advantages and disadvantages. The Savitzky-Golay method (see Section III) fits the spectrum with a polynomial and then takes the derivative of that polynomial. It is most useful with very sharp absorption bands, with high noise spectra, in cases where the value of the derivative at wavelengths between adjacent measured wavelengths is desired, and in detecting very small wavelength shifts. The Fourier transform method (see Section VII) is most useful when the Fourier transform of the spectrum is already available for other purposes. The finite-difference method is easiest to calculate but is more sensitive to noise than the Savitzky-Golay method because the latter includes a polynomial smooth. Both the Fourier transform and finite-difference methods are usually combined with some form of smoothing for best results. The formulae for the finite-difference calculation is as follows:

If S is a spectrum defined for evenly spaced wavelengths λ_n, $n = 0, 1, \ldots, N-1$, then the first derivatives S'_n at point n is defined by:

$$S'_n = S_{n+g} - S_{n-g}$$

where g is an integer called the gap or derivative size and S_n is the log $(1/R)$ value at point n. Similarly, the second and third derivatives (S''_n and S'''_n respectively) are defined by:

$$S''_n = S_{n+2g} - 2*S_n + S_{n-2g}$$

$$S'''_n = S_{n+3g} - 3*S_{n+g} + 3*S_{n-g} - S_{n-3g} .$$

Strictly speaking, these three equations represent first-, second-, and third-order *differences*, rather than derivatives as developed in formal numerical analysis. More important, the definitions vary from laboratory to laboratory, even between computer programs in the same laboratory. The principal variations involve whether division by some power of $(\lambda_{i+g} - \lambda_{i-g})$ is applied, and whether g is replaced by $g/2$. Transferring calibrations or replicating results between laboratories requires knowledge of the definitions in use by the systems compared. Often, the definitions are not quoted in published papers. Authors, while retaining the slightly incorrect term *derivative* for *difference*, should quote the definition being used when publishing and should quote g in nanometers rather than numbers of points. The examples in this section used the finite-

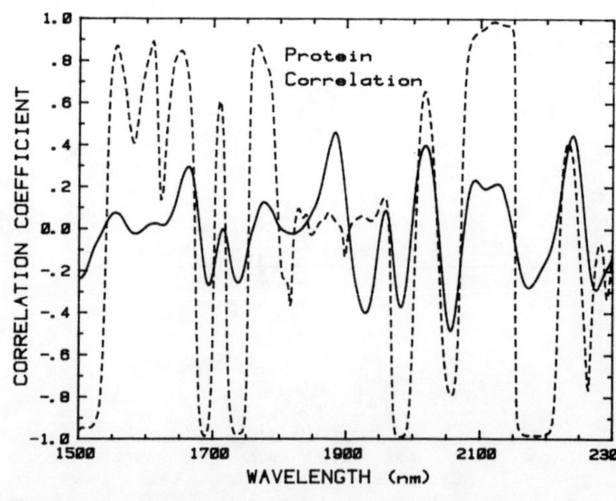

Fig. 26. Correlation of protein to second derivative spectra for 50 samples (dotted) and second derivative of protein (solid).

difference method preceded by a 5-point (8 nm) smooth and with a gap of 10 points (16 nm).

VII. THE FOURIER TRANSFORM

Although the Fourier transform requires quite complicated calculations compared with those used in the methods discussed so far, its geometric interpretation permits an appreciation of how it has been used in NIR work. Bracewell (1965) is a basic reference work on the Fourier transform; Brigham (1974) is a basic introduction to the "fast Fourier transform," which is a method of efficiently organizing the calculation of the transform; Bell (1972) and Griffiths (1975) are basic texts on Fourier transform spectrophotometers, which are used primarily in midinfrared work.

A. Basic Properties

The Fourier transform (FT) changes a spectrum into another kind of spectrum that we will call the *Fourier spectrum.* As an example, consider the two waves, a constant function, and their sum in Fig. 27. This is meant to be a very simplified example of a spectrum with two peaks and some high-frequency noise. The FT of this sum consists of the three points in Figs. 28 and 29. The point above zero in the Fourier spectrum is the FT of the constant function, and its height is proportional to the height of the constant function. The second point is the FT of the low-frequency wave, and its height is proportional to the height of that wave. The third point is the FT of the high-frequency wave, and its height is again proportional to the smaller amplitude of the high-frequency wave. Figures 28 and 29 are slightly oversimplified, but they illustrate the essential feature of the FT. The Fourier transform expresses each spectrum as a sum of waves and can be plotted as a graph whose horizontal axis is the wave frequency and whose vertical axis is the wave height.

The FT is reversible: the inverse FT of Fig. 28 is Fig. 27. Figure 30 is a flowchart showing how the FT is used. The Fourier spectrum is modified in some way and then the inverse FT is performed, giving a near-infrared spectrum that is different from the original one in some desirable way. For example, to remove the high-frequency noise from the spectrum in Fig. 27, compute the Fourier spectrum, eliminate the point corresponding to the noise (giving a modified Fourier spectrum), then compute the inverse FT of the modified Fourier spectrum. This will give the sum of the constant and the low-frequency wave without the high-frequency noise (Fig. 31).

So far the FT has been described in simplified form. The FT actually gives a pair of values for each wave frequency. The pair of values (F_k) is called a complex number; one is the real part, $\mathrm{Re}(F_k)$, and the other is the imaginary part, $\mathrm{Im}(F_k)$. The two values can be combined into the modulus $|F_k| = \sqrt{[\mathrm{Re}^2(F_k) + \mathrm{Im}^2(F_k)]}$, which is the value graphed in Fig. 28. $\mathrm{Re}(F_k)$ and $\mathrm{Im}(F_k)$ are called Fourier coefficients and the set of $|F_k|$ is called

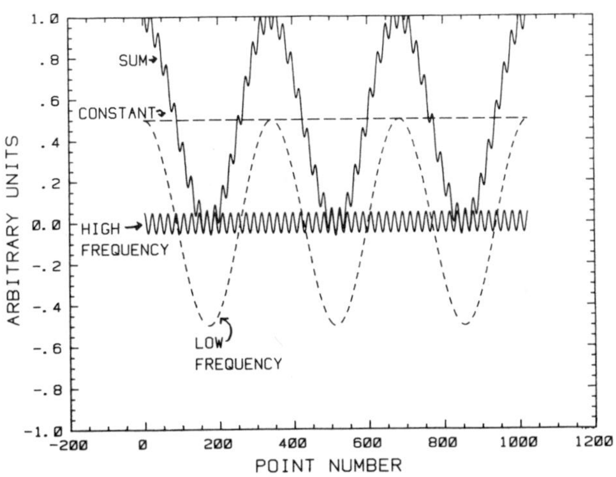

Fig. 27. Constant function, low-frequency wave, high-frequency wave, and sum of three functions.

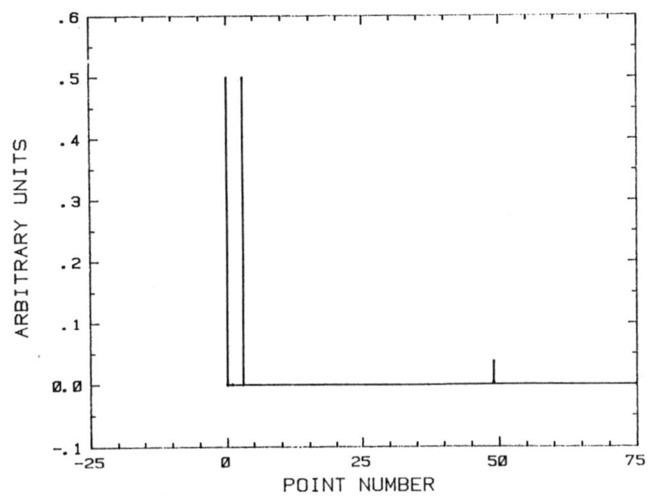

Fig. 29. Fig. 28 expanded.

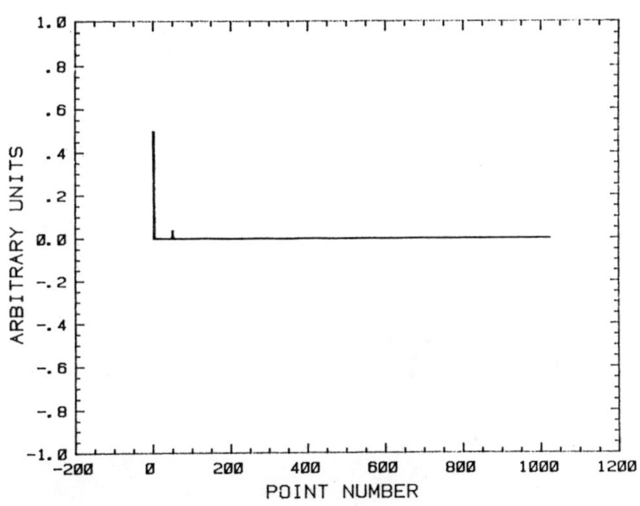

Fig. 28. Fourier transform of sum function in Fig. 27.

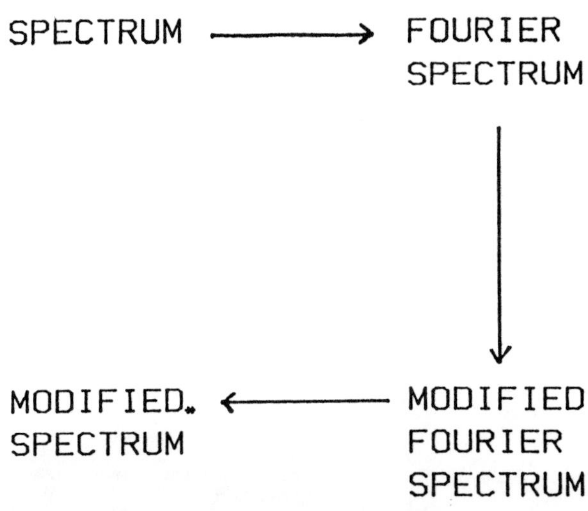

Fig. 30. Flowchart of general use of Fourier transform.

the power spectrum. The term *Fourier spectrum* is used here to mean the power spectrum except where an understanding of the application requires separation of the real and imaginary parts.

Figures 32 and 33 give a wheat spectrum of 1,024 points and the real part of its corresponding Fourier spectrum. The Fourier spectrum is symmetrical about the point 512, and point 0 is the zero-frequency point, which corresponds to the average value of the wheat spectrum. The high-frequency values are located near point 512 and the low-frequency values are located near points 0 and 1 (or near point 1,023, which is a reflection of point 1). (The symmetry occurs because the original spectrum is real, and the symmetry is actually more accurately expressed in the statement: "F_{N-k} = the complex conjugate of F_k for $k \neq 0$.")

B. Applications

Of the many applications of the FT, seven that have been used with near-infrared spectra will be described: data compression, correlation to Fourier coefficients, smoothing, normalizing, calculation of derivatives, convolution, and deconvolution. Each involves a different kind of modification of the Fourier spectrum before taking the inverse transform to get the modified spectrum, and some dispense with the inverse transform altogether.

DATA COMPRESSION

The Fourier spectrum in Fig. 32 can be modified by reducing the points near 512 to a value of zero (called "throwing out the high-frequency noise") without disturbing the essential information in the spectrum. Various measures of the lack of disturbance have been considered. One is the maximum of the difference between the spectrum and the modified spectrum. If this maximum is of the same order as the instrument noise, then probably only random instrument noise has been deleted. Another measure is the effect on calibration and validation from the resulting modified sample spectra. Zeroing all but the first 50 complex points will not disturb calibration or validation results for the samples of Table I.

Thus, 50 points of the Fourier spectra can be saved instead of 1,024 points of the original spectrum, conserving computer storage space. For example, if the original spectrum had 1,024 16-bit words (2,048 bytes) and the modified Fourier spectrum has 50 double-precision complex numbers (800 bytes), the space saving would be about 60%. If the original spectrum had 1,024 32-bit reals (4,096 bytes) and the modified Fourier spectrum has 50 single-precision complex numbers (400 bytes), the saving would be 90% (Giesbrecht et al, 1981).

CORRELATION TO FOURIER COEFFICIENTS

Sample composition can also be correlated directly to the Fourier coefficients, rather than to values in the modified wavelength spectra. If only the first 10 (low-frequency) coefficients are retained and treated as 19 separate variables in a step-up linear regression (the zero-frequency coefficient plus 9 real parts plus 9 imaginary parts), sugar and nicotine in tobacco can be determined satisfactorily (McClure et al, 1984).

SMOOTHING

In the simplified example in Section VIIA, it was demonstrated that smoothing could be accomplished with the FT. With an actual near-infrared spectrum and its Fourier transform, as discussed above, the high-frequency points were reduced to zero. Instead, the Fourier spectrum can be multiplied point by point with a weighting function of the Fourier point numbers that increases the values at the low-frequency points and diminishes those at the high-frequency points. The modified Fourier spectrum is then inversely transformed into a smoothed wavelength spectrum. The degree of smoothing and amount of band distortion are controlled by the size and form of the weighting function.

NORMALIZING

Zeroing the constant coefficient of the Fourier spectrum results in a modified wavelength spectrum whose average value is zero. This normalization can reduce the particle size effect when applied to a set of samples (see Section VIII).

DERIVATIVES

If the weighting function is designed to diminish the low frequencies and enhance the high frequencies, then the modified

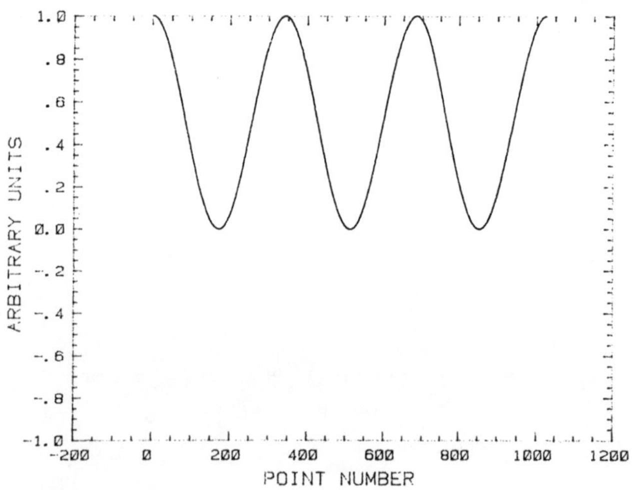

Fig. 31. Sum function of Fig. 27 with high-frequency wave removed by Fourier method.

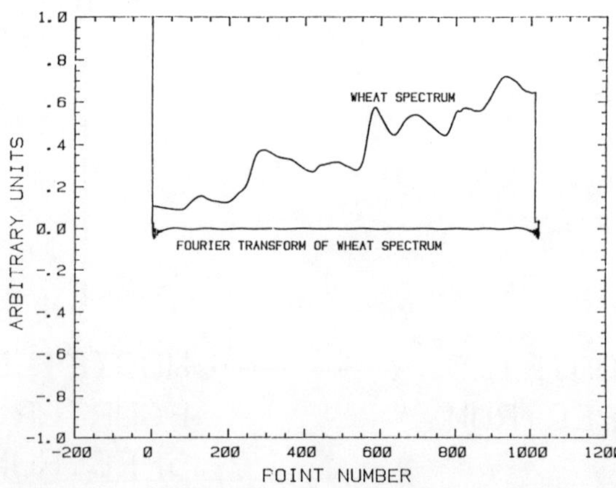

Fig. 32. Wavelength spectrum and Fourier spectrum of ground wheat.

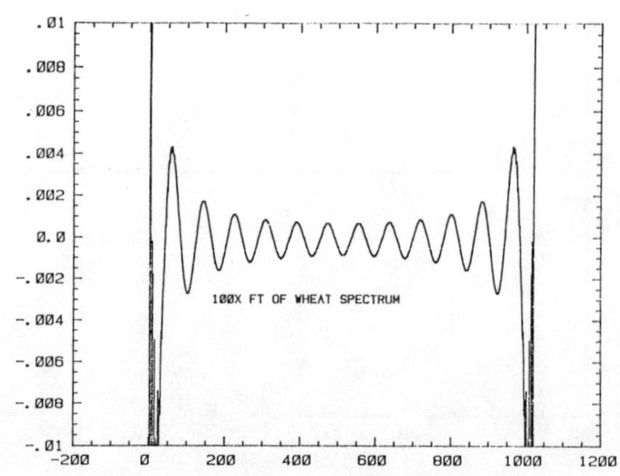

Fig. 33. Fourier spectrum of Fig. 32 expanded.

wavelength spectrum can show more detail. A specialized example is the Fourier method of calculating derivatives, which uses the relations:

$$£\,(S^{(n)}) = k^n * F_k \ (k \leqslant N/2) \text{ and}$$

$$£\,(S^{(n)}) = (N - k + 1)^n * F_k \ (k \geqslant N/2)\,,$$

where the superscript (n) denotes the nth derivative and the multiplication on the right is point by point.

CONVOLUTION

If each point in a spectrum is replaced by a weighted mean of it and its surrounding points, the spectrum is said to be convoluted with the weighting function. A simple example is the boxcar smooth (Section IIIC), where the weighting function is constant. A convolution is written:

$$S^M = S\,(*)\,W$$

where S is the original spectrum, S^M is the modified spectrum, W is the weighting function, and (*) denotes convolution. The FT can be used to accomplish convolution because of the relation:

$$£\,(S(*)W) = £\,(S) * £\,(W)$$

where the asterisk without parentheses denotes point-by-point multiplication of the FT of the spectrum and the weighting function. For example, since the FT of a Gaussian is a Gaussian, a Gaussian smooth can be accomplished by multiplying the Fourier spectrum point for point by a Gaussian, and then taking the inverse FT.

Most of the examples in this section could be discussed as some kind of convolution performed by Fourier transform methods.

DECONVOLUTION

Deconvolution is the restoration of a convoluted spectrum to its preconvoluted set of values. The convolution equation given in the preceding subsection can be reversed to read:

$$£\,(S) = £\,(S(*)W)\ /\ £\,(W)$$

where the slash denotes point-by-point division. An example of the use of convolution occurs when the slit function and the experimental spectrum are known and it is necessary to find the theoretical spectrum. Deconvolution then removes the smoothing effect of the slit function.

VIII. OTHER METHODS

A. Component Spectrum Reconstruction

Honigs et al (1984a) describe the reconstruction of the spectrum of a component of a mixture from the spectra of several mixtures with different levels of the component. Success in this would be particularly valuable in mixtures where the components are known to interact, cannot be isolated, or change their spectra when isolated. The formula for the reconstructed spectrum is:

$$S_n{}^R = (\Sigma\, xy)(1 - \bar{Y})/\Sigma\, y^2 + \bar{X}_n$$

where $S_n{}^R$ is the reconstructed spectrum at point n, $x = X_n - \bar{X}_n$ is the centered log $(1/R)$ value at point n, and $y = Y - \bar{Y}$ is the centered fractional concentration of the component whose spectrum is to be reconstructed. A closer examination of the formula shows that it essentially adds a multiple of the slope of the regression line [of log $(1/R)$ value against component level] at each wavelength to the average of all the spectra at each wavelength.

The water spectrum (Fig. 34) reconstructed from the spectra and moisture values for 50 calibration samples of Table I shows the strengths and limitations of the method. The reconstructed spectrum is remarkably like that of pure water. The exceptions are places (e.g., 2,280 nm) where water is probably correlated to starch, and some other places (e.g., 2,100 nm) where the reconstructed spectrum may be that of bound water.

B. Fast Correlation Transform

Honigs et al (1984b) describe a method sometimes called the "fast correlation transform" that can be used to select a minimal calibration set of wavelengths and/or samples from a set of data such as that described in Table I. The benefit is to reduce the number of samples needed for calculation and to select wavelengths with the most information. The method for wavelength selection is as follows:

1. Decide on the number (k) of wavelengths to select.
2. Start with the first k available wavelengths.
3. Form a matrix with $k + 1$ columns, the first k columns being the spectral data at the k wavelengths, the last column being the chemical data.
4. Row reduce the matrix on the wavelength information as described by Honigs et al (1984b), so that the largest element in each column is placed on the main diagonal before eliminating other elements of that column.
5. Using the data in the first k rows of the reduced matrix, regress the chemical data against the spectral data.
6. On repetition of steps 3–7, save these wavelengths and r values (and discard previously saved wavelengths) if r is larger than previously saved r's.
7. Discard the wavelength with the largest regression coefficient (absolute value) and replace it with the next available wavelength.
8. Repeat steps 3–7 until all available wavelengths have been tried.
9. Using the final matrix as the starting point, with all the other wavelengths being "available," repeat steps 3–8 (second pass).
10. The wavelengths saved by step 6 are then the optimal set.

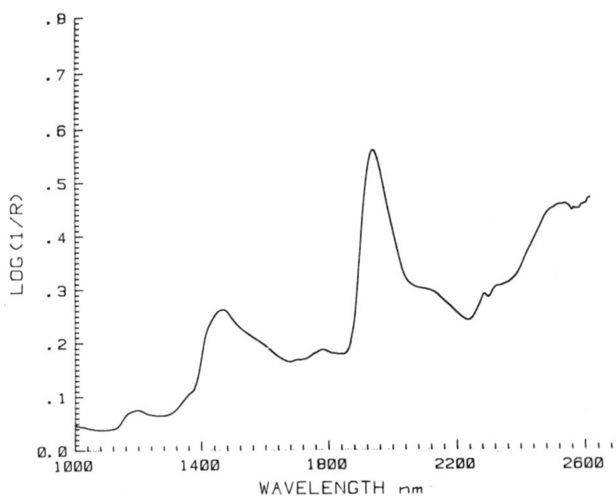

Fig. 34. Spectrum of water reconstructed from 50 wheat spectra and water data.

C. Normalizing Spectra

Normalization of spectra has been mentioned briefly in Sections V–VII. In general, it means changing a group of spectra so that the resulting ones have more features in common; or, equivalently, so that unwanted sources of variability (e.g., radiation scattering) are suppressed. This helps the visual understanding of the spectra and can reduce the complexity of a data treatment necessary to develop a calibration from the spectra. The simplest example (see Section III) was subtracting the log $(1/R)$ value at a single wavelength (called the reference wavelength) in a spectrum from all the spectral values. The result is a set of spectra with value zero at the reference wavelength. Forcing the sum of the regression coefficients to zero essentially accomplishes the same thing as was seen in Section VI. The first derivative also implies a similar normalization because it reduces a constant baseline to zero.

A more complicated normalization is accomplished by adding a different straight-line baseline to each spectrum:

$$S_n{}^M = S_n + a + b*n$$

so that the resulting spectra agree at two points instead of one (n is the point number of the wavelength). This normalization can be extended to the addition of any degree polynomial in the variable n, so that the modified spectra agree at more than two points. ("Mathematical ball-milling" [Murray et al, 1981] has been used to describe the linear case where a and b are adjusted for each spectrum so that every resultant spectrum has the same slope and Y-intercept as the average spectrum, calculated by regressing log $(1/R)$ values against wavelength.)

A similar normalization replaces S_n by some polynomial function of S_n, where the polynomial constants are the same for all log $(1/R)$ values in a single spectrum but change from spectrum to spectrum. The equation here is:

$$S_n{}^M = a + b_1*S_n + b_2*S_n{}^2 + \ldots$$

where the polynomial variable is S_n rather than n. This is more theoretically sound than the previous method, since the scatter effect on a log $(1/R)$ spectrum is dependent on absorbance rather than wavelength. Normalizing with a first-degree polynomial removes both the additive and multiplicative scatter effect. Figure 35 shows the four spectra of Fig. 4 normalized by a second-degree polynomial to agree at the indicated points. Another normalization is discussed in the subsection "Multiplicative Scatter Correction" in the next chapter.

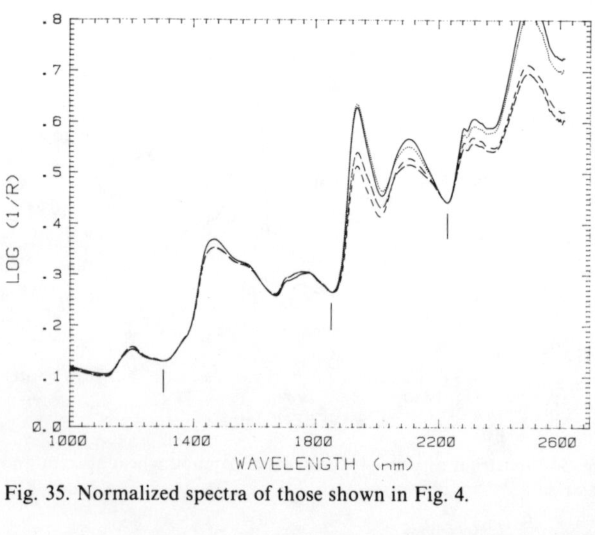

Fig. 35. Normalized spectra of those shown in Fig. 4.

D. Discriminant Analysis

The technique of discriminant analysis is like linear regression in that it uses a linear combination of variables, but the combination is used to classify a material rather than to analyze it. Mark and Tunnell (1985) give examples.

IX. CONCLUSION

The statistical processes of calibration, measurement, and validation have been described as used in near-infrared reflectance spectroscopy. Error sources have been discussed and the mathematics required for several data treatments were presented. It would seem natural at this point to compare the performances of the various data treatments in more detail. However, although researchers continue to develop and evaluate data treatments, there are too few published comparisons to permit much more than accounts of preferences and experiences. Few laboratories have the resources to devote to extensive comparisons of mathematical methods, and applications researchers have achieved success by selecting one method and using it with as much flexibility and experience as possible. Because the main sources of error when using commercially available instruments at the time of this writing are usually neither the data treatment nor the instrument error, one should raise the priority level for dealing with sampling error and reference method error and for developing a clear definition of the component or the physical property to be determined.

The next chapter is more complex mathematically than the material discussed up to now because it requires a familiarity with matrix algebra, which is reviewed in its appendixes. An overview of the mathematics discussed in both chapters is now presented, so that the reader who is unprepared for the higher level of mathematics used in the next chapter can at least get an intuitive feel for the material.

In Section II, several examples of linear regression approximations were presented. All the mathematics of both chapters can be expressed in terms of such approximations. The basic form of the approximation is:

$$\hat{Y} = a + b_1X_1 + b_2X_2 + \ldots + b_kX_k$$

where Y is the variable to be measured, such as protein concentration, loaf volume, etc.; the X_1, X_2, ..., X_k, are k independent variables, each a combination of one or more near-infrared spectral values; the b_1, b_2, ..., b_k, are k regression coefficients; and a is the regression constant term (intercept). In calibration, a set of X's and known Y's are used to estimate the b's, whereas in measurement and validation, a set of X's and the estimated b's are used to estimate an unknown Y. The main differences between the various methods discussed in this chapter are in how the near-infrared spectral values are combined into the X_1, X_2, ..., X_k. These methods are now considered in this context.

As emphasized in Section I, the most common method is for the X's to be log $(1/R)$ values at a few preselected wavelengths. Sections II–V are concerned with this method. Some other function of R values at a few preselected wavelengths may also be used, such as the Kubelka-Munk $(1 - R)^2/2R$, or the empirical scatter-correcting functions discussed in the next chapter. In what follows, the terms *spectrum* and *spectral values* refer to any of these functions of R values or to auxiliary variables. With any of the methods, spectral noise may be removed before the approximation by some form of smoothing (Section III), and X's can be indicator variables (Section V).

The next simplest method is for the X's to be derivatives, quotients, differences, or other mixtures of the spectral values at a few wavelengths (Sections IV–VI).

The following methods (data compression) are much more

TABLE IX
Data Compression Methods

Method	Latent Variables (Loadings)	Scores
Fourier transform	Sine and cosine waves	Fourier coefficients
Curve fitting	Spectra of chemical components or representative samples	Curve fit coefficients
Principal components analysis	Artificial spectra that statistically account for spectral variability	Scores
Factor analysis	Linear combinations of principal components that approximate spectra of pure components	Scores
Partial least squares	Artificial spectra that account for spectral and chemical variability	...
Extended Beer's law	Artificial spectra that account for spectral and chemical variability and approximate spectra of pure components	...

complex, but only in the sense that the X's require lengthier calculations. The calibration, measurement, and validation procedures are the same no matter how the X's are calculated. Each method begins by expressing a spectrum as the sum of elementary spectra (called loadings or latent variables), the definition of *elementary* being different for each method. Each elementary spectrum is multiplied by a "score" before being included in the sum, in order to make the sum of the elementary spectra as close as possible to the original spectrum. The scores (some or all) are then used as the X's in the approximation. What follows is summarized in Table IX.

In the Fourier transform procedure (Section VII and the next chapter), the elementary spectra are sine and cosine waves and the scores are called Fourier coefficients. In curve fitting (next chapter), the elementary spectra are actual measured spectra (of pure materials or of representative samples), and the scores are called curve fit coefficients. In principal components analysis (next chapter) the elementary spectra (called principal components) are statistically estimated to account systematically for the spectral variation. Factor analysis is similar to principal components analysis except that the elementary spectra are adjustments of the principal components to make them approximate actual spectra. The partial least-squares method (next chapter) is similar to the principal components analysis, except that the chemical data as well as the spectral data are used in the estimation of the elementary spectra. Finally, the extended Beer's law method (next chapter) is similar to the partial least-squares method in that the chemical data are used in the estimation of the elementary spectra. However, the components are designed to approximate actual spectra of pure chemical components, and the model is able to make use of assumptions about the statistical distributions of the spectral and chemical data.

Each method from either chapter can be thought of as having been developed from a regression in which the independent variables (X's) can be very simple or very complex combinations of near-infrared spectral values or auxiliary variables. In the future, computer packages will calculate these X's in a way largely transparent to the investigator, who will then be free to consider only the problems of overfitting, sampling, and the quality of spectral and chemical data.

ACKNOWLEDGMENTS

I thank Lois Harris, Dave Hopkins, Elaine Lanza, Howard Mark, Harald Martens, Dave Massie, Tom Nolan, Karl Norris, Dave Osten, John Shenk, Ed Stark, Mark Westerhaus, and Phil Williams for invaluable help in preparing this chapter.

APPENDIX

The statistical notation follows Steel and Torrie (1980), except for the use of lowercase x and y for deviations from sample means and the use of $\mathbf{M}^T$ for matrix transpose. The Fourier transform notation was adapted from several sources. Alternative notation and terminology in current use is enclosed in parentheses.

Mathematical Symbols

$\neq$	Not equal
$\leqslant$	Less than or equal
$\geqslant$	Greater than or equal
$\cong$	Approximately equal
*	Multiplication
$\sqrt{}$	Square root
$n!$	n factorial
$\lvert X \rvert$	Absolute value of X
log	Base 10 logarithm
Σ	Sum of
ΣX_i	$X_1 + X_2 + ... + X_n$

Abbreviated versions of summation are used when i and n are clear from the context:

$$\Sigma X = \Sigma X_i = \sum_i X_i = \sum_{i=1}^{n} X_i, \ \Sigma X^2 = \Sigma (X_i^2)$$

$$\Sigma XY = \Sigma (X_i Y_i), \ \Sigma (X - Y)^2 = \Sigma ((X_i - Y_i)^2) .$$

Spectroscopic Symbols

R	Reflectance
nm	Nanometer
S	Spectrum of log $(1/R)$ values
S_n	log $(1/R)$ value at spectral point number n (not wavelength n)
S'	First derivative of S with respect to wavelength
S'_n	First derivative at point number n
S''	Second derivative of S with respect to wavelength
S^M	Modified spectrum
λn	Wavelength at spectral point number n

Statistical Symbols

X	Independent variable
Y	Dependent variable
$\hat{Y}$	Estimate of dependent variable
X_i, Y_i	Individual observation
e, e_i	Individual calibration or validation error (regression error)
D, D_i	Difference between paired observations
$\bar{X}, \bar{Y}$	Sample mean (average)
μ, μ_X	Population mean
x, y	Centered variable (deviation from sample mean)
x_i, y_i	Individual centered observation (deviation)
s^2, s_X^2, s_Y^2	Sample variance

s, s_X, s_Y Sample standard deviation (SEST, standard error of a single test)

σ, σ_X Population standard deviation

n Sample size

k Number of variables in regression

a Y-intercept, constant term (b_0)

b Slope in a single-term linear regression (regression coefficient) (b_1)

b_i Regression coefficient, not including the constant term

r Correlation coefficient

r^2 Coefficient of determination

$\mathbf{M}$ Matrix

$M_{i,j}$ Matrix element in row i, column j

$\mathbf{M}^{-1}$ Inverse of $\mathbf{M}$

$\mathbf{M}^T$ Transpose of $\mathbf{M}$

$\mathbf{b}$ Column vector of regression coefficients $b_1, ..., b_k$

$\mathbf{x}$ Matrix of centered independent observations (rows = observations, columns = variables)

x_{ij} i^{th} observation of jth variable centered on mean of jth variable (element in i^{th} row, j^{th} column of $\mathbf{x}$)

$\mathbf{X}_j$ Column of data for independent variable j

$\mathbf{y}$ Column vector of centered data for dependent variable

$\bar{X}_j$ Mean of one variable

Statistical Abbreviations

BIAS Bias (mean difference)

CV Coefficient of variation or variability

RMSD Root mean square difference (RMS, root mean square)

SDVD Standard deviation of validation differences (see Section II for discussion of several other terms, including verification, prediction, measurement, analysis, SEP, SED, and SDP)

SE Standard error (used verbally to mean SEE, SEC, or SDVD when meaning is clear from context)

SEC Standard error of calibration (SEE in a calibration regression)

SEE Standard error of estimate (standard deviation of regression errors)

Statistical Formulas

$\bar{X}$ $(\Sigma\, X)/n$

x $X - \bar{X}$

s_X^2 $\Sigma\, x^2/(n-1) = (\Sigma\, X^2 - (\Sigma\, X)^2/n)/(n-1)$. The divisor $n-1$ is used when the population mean is estimated by the sample mean. The divisor is actually n minus the number of parameters estimated by sample statistics. For example, if the population mean is assumed to be zero, no parameters are estimated and the divisor is n. In a multiterm linear regression with k independent variables, $k+1$ parameters are estimated and the divisor is $n-1-k$.

CV $100{*}s_X/\bar{X}$

In any regression:

$e_i = Y_i - \hat{Y}_i$

SEE s_e. Formula depends on type of regression.

r^2 $(\Sigma\, y\hat{y})^2/\surd[\Sigma\, y^2\, \Sigma\, \hat{y}^2] = 1 - s_e^2/s_Y^2$

In a single-term linear regression calibration:

b $\Sigma\, xy/\Sigma\, x^2$

a $\bar{Y} - b\bar{X}$

r $\Sigma\,(xy)/\surd[(\Sigma\, x^2\, \Sigma\, y^2)]$

SEC $\surd[(\Sigma\, y^2 - (\Sigma\, xy)^2/\Sigma\, x^2)/(n-2)] = s_y\surd[(n-1)(1-r^2)/(n-2)] = \surd\,[\Sigma\, e^2/(n-2)]$

In a multiterm linear regression calibration:

$\mathbf{b}$ $(\mathbf{x}^T\mathbf{x})^{-1}\mathbf{x}^T\mathbf{y}$

a $\bar{Y} - \Sigma\, b_i\bar{X}_i$

r^2 $(\sum_i\ (b_i \sum_j\ x_{ji}y_j)/\Sigma\, y^2 = (\mathbf{b}^T\mathbf{x}^T\mathbf{y})/(\mathbf{y}^T\mathbf{y})$

SEC $s_Y\surd[(n-1)(1-r^2)/(n-1-k)]$
$= \surd[\Sigma\, e^2/(n-1-k)]$

In a validation, X = NIR method value, Y = reference method value (see Section IIC for alternative terminology):

D_i $X_i - Y_i$

SDVD $\surd[(\Sigma\,(X-Y)^2 - [\Sigma\,(X-Y)]^2/n)/(n-1)]$
$= s_{(X-Y)} = s_D = \surd[\Sigma\,(D-\bar{D})^2/(n-1)]$

BIAS $\bar{X} - \bar{Y} = \Sigma\, D/n = \bar{D} = \Sigma\,(X-Y)/n$

RMSD $\surd[\Sigma\,(x-Y)^2/n] = \surd[\Sigma\, D^2/n]$
$\cong \surd[\text{SDVD}^2 + \text{BIAS}^2]$

Fourier Transform Notation

j $\surd(-1)$

exp Exponential function

π Ratio of circumference to diameter of circle

N Number of points in spectrum

£ Fourier transform

$£^{-1}$ Inverse Fourier transform

S Spectrum of log $(1/R)$ values

S_n log $(1/R)$ value at point number n (not wavelength n) (points numbered $0, 1, 2, ..., N-1$)

F $£(S)$ = Fourier spectrum of S

F_k Fourier spectrum at point k (points numbered $0, 1, 2, ..., N-1$)

S $£^{-1}(F)$ = inverse Fourier transform of F

(*) Convolution

Re Real part

Im Imaginary part

$|F_k|$ Modulus of F_k

Fourier Transform Formulas

F_k $\sum_{n=0}^{N-1}\ [S_n \exp(-j2\pi nk/N)]$

S_n $[\sum_{K=0}^{N-1}\ [F_k \exp(j2\pi nk/N)]]/N$

LITERATURE CITED

In this list and in the additional reference list that follows, the statistical references all presume a "first course" in that subject, for which there are numerous equivalent texts, Neter et al (1978) being only one example.

BELL, R. J., ed. 1972. Introductory Fourier Transform Spectroscopy. Academic Press, New York. 382 pp.

BRACEWELL, R., ed.. 1965. The Fourier Transform and Its Applications. Electrical and Electronic Engineering Ser. McGraw-Hill, New York. 381 pp.

BRIGHAM, E. O., ed. 1974. The Fast Fourier Transform. Prentice-Hall, Englewood Cliffs, NJ. 251 pp.

COBLENTZ, W. K. 1982. A study in the near infrared reflectance spectra of proteins and related compounds. M.Sc. thesis, Pennsylvania State University, University Park.

DRAPER, N., and SMITH, H. 1966. Applied Regression Analysis. John Wiley, New York.

GIESBRECHT, F. G., McCLURE, W. F., and HAMID, A. 1981. The use of trigonometric polynomials to approximate visible and near infrared spectra of agricultural products. Appl. Spectrosc. 35:210-214.

GRIFFITHS, P. R. 1975. Chemical Infrared Fourier Transform Spectroscopy. John Wiley, New York. 340 pp.

HOCKING, R. R. 1970. The analysis and selection of variables in linear regression. Biometrics 32:1-49.

HONIGS, D. E., HIEFTJE, G. M., and HIRSCHFELD, T. 1984a. A new method for obtaining individual component spectra from those of complex mixtures. Appl. Spectrosc. 38:317-322.

HONIGS, D. E., HIEFTJE, G. M., and HIRSCHFELD, T. 1984b. Number of samples and wavelengths required for the training set in near-infrared reflectance spectroscopy. Appl. Spectrosc. 38:844-847.

HRUSCHKA, W. R., and NORRIS, K. H. 1982. Least-squares curve fitting of near infrared spectra predicts protein and moisture content of ground wheat. Appl. Spectrosc. 36:261-265.

JONES, D. B. 1931. Factors for converting percentages of nitrogen in foods and feeds into percentages of proteins. U.S. Dep. Agric. Circ. 183.

MARCHAND, P., and MARMET, L. 1983. Binomial smoothing filter: A way to avoid some pitfalls of least-squares polynomial smoothing. Rev. Sci. Instrum. 54:1034-1041.

MARK, H. L., and TUNNELL, D. 1985. Qualitative near-infrared reflectance analysis using Mahalanobis distances. Anal. Chem. 57:1449-1456.

MARTEN, G. C., SHENK, J. S., and BARTON, F. E., II. 1985. Near infrared reflectance spectroscopy (NIRS): Analysis of forage quality. U.S. Dep. Agric., Agric. Handb. 643. 96 pp.

MARTENS, H., and NAES, T., eds. 1983. Food Research and Data Analysis. Applied Science Publishers, London.

McCLURE, W. F., HAMID, A., GIESBRECHT, F. G., and WEEKS, W. W. 1984. Fourier analysis enhances NIR diffuse reflectance spectroscopy. Appl. Spectrosc. 38:322-329.

MORREY, J. R. 1968. On determining spectral peak position from composite spectra with a digital computer. Anal. Chem. 40:905-914.

MURRAY, I., JESSIMAN, C., and KENLEY, H. 1981. Use of a near infrared reflectance (NIR) spectrocomputer for forage evaluation. Annu. Meet Eur. Soc. Nucl. Methods Agric., 12th.

NORRIS, K. H., and HART, J. R. 1965. Direct spectrophotometric determination of moisture content of grain and seeds. Pages 19-25 in: Principles and Methods of Measuring Moisture in Liquids and Solids, vol. 4. A. Wexler, ed. Reinhold, New York.

SAVITZKY, A., and GOLAY, M. J. E. 1964. Smoothing and differentiation of data by simplified least squares procedure. Anal. Chem. 36:1627-1638.

SHENK, J. S., LANDA, I., HOOVER, M. R., and WESTERHAUS, M. O. 1981. Description and evaluation of a near infrared reflectance spectrocomputer for forage and grain analysis. Crop Sci. 21:355-358.

SHENK, J. S., WESTERHAUS, M. O., and HOOVER, M. R. 1978. Infrared reflectance analysis of forages. Pages 242-244 in: Proc. Int. Grain and Forage Harvesting Conf. ASAE, St. Joseph, MI.

STEEL, R. G. D., and TORRIE, J. H. 1980. Principles and Procedures of Statistics: A Biometrical Approach. 2nd ed. McGraw-Hill, New York.

STEINIER, J., TERMONIA, Y., and DELTOUR, J. 1972. Comments on smoothing and differentiation of data by simplified least square procedure. Anal. Chem. 44:1906-1909.

TKACHUK, R. 1977. Calculation of nitrogen to protein conversion factors for food legume breeders. Pages 78-81 in: Nutritional Standard and Methods of Evaluation. Hulse, Rachie, and Billingsley, eds. IDRC-TS7E. Int. Devel. Res. Center, Ottawa.

WILLIAMS, P. C., NORRIS, K. H., GEHRKE, C. W., and BERNSTEIN, K. 1983. Comparison of near-infrared methods for measuring protein and moisture in wheat. Cereal Foods World 28:149-152.

WILLIAMS, P. C., NORRIS, K. H., JOHNSEN, R. L., STANDING, K., FRICIONI, R., MacAFFREY, D., and MERCIER, R. 1978. Comparison of physicochemical methods for measuring total nitrogen in wheat. Cereal Foods World 23:544-547.

ADDITIONAL REFERENCES

AMERICAN SOCIETY FOR TESTING AND MATERIALS. 1982. Compilation of ASTM Standard Definitions. Philadelphia.

DAVIES, A. M. C., GEE, M. G., and FOSTER, P. W. 1984. A colour graphics display system to aid the selection of "best-pair" wavelengths for regression analysis of near infrared data. Lab. Pract. 38:78-80.

FRANK, I. E., and KOWALSKI, B. R. 1982. Chemometrics. Anal. Chem. 54:232R-243R.

JUDD, D. B., and WYZSECKI, G. 1963. Color in Business, Science, and Industry. 2nd ed. John Wiley, New York.

LANDA, I., and MOEN, R. 1979. Reduction of predictable errors in NIR measurements. (Abstr.) Cereal Foods World 24:460.

MARTENS, H. A. 1985. Multivariate calibration. Dr.techn. thesis, Technical University of Norway, Trondheim.

MOSTELLER, F., and TUKEY, J. W. 1977. Data analysis and regression. Addison-Wesley, Reading, MA.

NETER, J., WASSERMAN, W., and WHITMORE, G. A. 1978. Applied Statistics. Allyn and Bacon, Boston.

NORRIS, K. H. 1983. Extracting information from spectrophotometric curves: Predicting chemical composition from visible and near-infrared spectra. Pages 95-114 in: Food Research and Data Analysis. H. Martens and H. Russwurm, Jr., eds. Applied Science Publishers, London.

NORRIS, K. H., and WILLIAMS, P. C. 1984. Optimization of mathematical treatments of raw near-infrared signal in the measurement of protein in hard red spring wheat. I. Influence of particle size. Cereal Chem. 61:158-165.

SNEDECOR, G. W., and COCHRAN, W. G. 1967. Statistical Methods. 6th ed. Iowa State University Press, Ames.

SNEE, R. D. 1977. Validation of regression models: Methods and examples. Technometrics 19:415-428.

WENDLANDT, W. W., and HECHT, H. G. 1966. Reflectance Spectroscopy. John Wiley, New York.

WILLIAMS, P. C. 1975. Application of near infrared reflectance spectroscopy to analysis of cereal grains and oilseeds. Cereal Chem. 52:561-576.

CHAPTER 4

Multivariate Calibration by Data Compression

H. MARTENS[1]
Norwegian Computing Center
Oslo, Norway

T. NAES
Norwegian Food Research Institute
Aas-NLH, Norway

I. INTRODUCTION

A. Multivariate Calibration and Prediction

The purpose of a near-infrared instrument is to determine the concentration of chemical variables such as protein content quickly and precisely from spectrophotometric measurements. It can thereby replace slower, more expensive, or less precise methods for assessing the desired chemical constituents. To do this, the instrument must first be calibrated; i.e., a mathematical relationship for converting the near-infrared optical measurements to the desired quantity (the prediction equation) has to be established.

Diffuse reflectance spectrometry in the near-infrared wavelength region yields nonselective data because of interference from the spectra of strongly overlapping constituents and from light scatter variations. To minimize these interferences, the analytical equation of near-infrared instruments has to be multivariate; i.e. measurements on many different wavelengths must be combined. Since some of these interferences cannot be physically isolated and measured directly, the calibration of the instruments has to be done indirectly. A set of "representative" calibration samples with known chemical composition is used to program the computer of the instrument to recognize future "unknown" samples. This approach is different from that met in conventional transmission spectroscopy of clear solutions where the spectra of pure constituents can be measured and used for direct calibration.

Several indirect multivariate calibration methods are available. In the previous chapter, a review was given of the established methods for calibration and analysis presently used in commercial near-infrared instruments. This chapter summarizes some new approaches to calibration. In contrast to the established methods, they utilize as many relevant wavelengths as possible in the analysis and are therefore here called *full-spectrum calibration methods*. This chapter aims at presenting the different techniques in a unified model framework. Some of the methods will be illustrated on artificial and some on real data sets. In general, the new approaches are attempts at using the available data more efficiently by combining new statistical regression methods with an understanding of the physical phenomena involved.

B. Calibration

Calibrating a near-infrared instrument presents several problems (Table I). Figure 1 illustrates basic steps that can be used for solving them. Four different stages are required, the first two concerning the preparation of the data and the last two the statistical estimation. A calibration procedure must be reasonably simple and fast so that it can be run on small computers. Therefore it is preferable to use some kind of linear mathematical model to relate near-infrared information to chemical information. Such models are easy to compute and fast to fit to the data and can be used with a higher degree of complexity than nonlinear models. Linear statistical models are also well understood theoretically.

The recorded electrical signals from the spectrophotometer are not linearly related to the chemical composition. Over a narrow range of concentrations, most curved relationships can be approximated by a straight line, but over a wider range the error may become intolerable. To be able to use linear models over a wide variation range, some kind of initial data transformation may be required.

Experience has shown that it may be convenient to distinguish between two different data transformation (data pretreatment) aspects, with the calibration consisting of four steps:

1. The near-infrared data should first go through a response linearization step aiming to ensure that the instrument responds linearly to changes in the chemical composition of the sample.
2. The second part of the transformation, the optical correction step, eliminates such optical near-infrared interferences as light scatter and remaining specular reflectance. This step is needed to produce a model as simple as possible. Automatic warnings should be issued if abnormal levels of these optical phenomena are discovered for a sample.

Both steps may yield estimates of certain parameters. After these transformations, the near-infrared data are suitable for linear modeling. Because of the high intercorrelation between the different near-infrared wavelengths (multicollinearity), the statistical estimation is difficult. The two steps that follow, data compression and calibration regression, are useful in resolving the problem:

3. The data compression step concentrates the information from all of the available, relevant near-infrared wave-

[1] Formerly at the Norwegian Food Research Institute, Aas-NLH, Norway.

lengths in the spectrum into a small number of basic variables, which are termed *regression factors*. Data compression is important for statistical reasons (parsimony), but compressing the data also reduces computation time and computer storage requirements. Experience has shown that step 2 can reduce the number of factors in step 3. Data compression gives certain estimated parameters, *near-infrared loading spectra*, that define how the values of the regression factors are to be calculated. It gives various diagnostic checks that reveal abnormalities in the data—measurement errors and extreme samples or wavelengths—and it provides estimates of the "normal" range of variation and level of measurement noise in the near-infrared data, which helps in revealing abnormalities.

4. The final step, calibration regression, establishes linear regression relationships between the obtained regression factors and the chemical variables to be determined. It produces estimated parameters, *chemical loadings,* that define how the chemical variables are determined from the regression factors, as well as various regression diagnostics that reveal abnormalities in the chemical data in the calibration set. It may also provide estimates of the normal range of compositional variation and the normal error level when chemical data are being determined by calibrations based on the near-infrared data.

The near-infrared and chemical loadings combine mathematically to yield the calibration coefficients for analysis of unknown samples. The calibration coefficients define the weights given to the different wavelengths in a linear analytical equation, which is a simple formula that will therefore be used as a common way of expressing the analytical results from all of the different linear calibration methods treated.

Some of the methods to be presented in this chapter perform each of these four steps separately, whereas others perform two or more steps at the same time. Methods also exist that encompass all of the steps simultaneously. Transformations are considered in Section V and steps 3 and 4 in Section III.

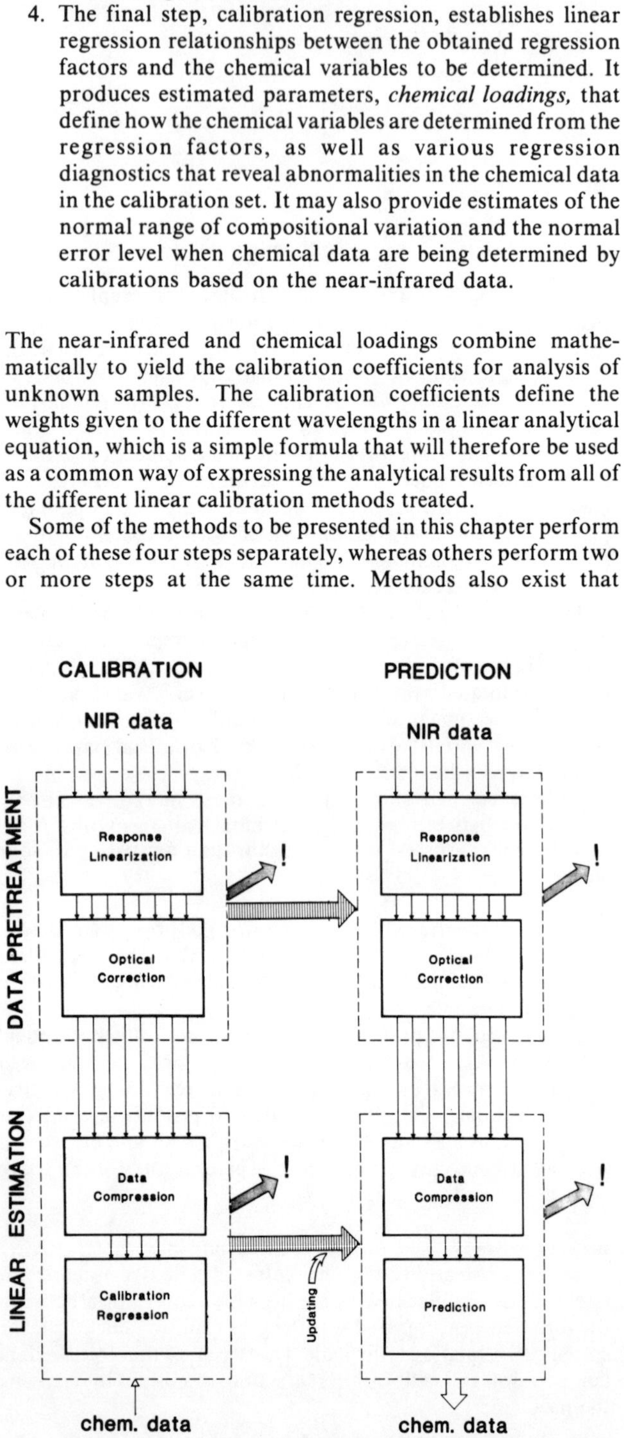

Fig. 1. Illustration of information flow in calibration and prediction.

C. Prediction

Prediction consists of predicting the chemical composition of a sample from near-infrared measurements using the calibration equation. During predicting analysis, the data must pass through the same four main steps as during calibration. The measurements from the near-infrared instrument are first submitted to response linearization. Second, the data are passed through the optical correction step, yielding quantitative information about the light scatter in each sample. Third, the values of the regression factors, which are linear combinations of the spectrum, are computed in the data compression step. This step may also give various warning diagnostics for samples that appear abnormal. Fourth, the desired chemical variables are determined linearly from the computed regression factors, using the chemical loadings found in the calibration. Diagnostics for controlling the chemical results may also be obtained.

In this chapter, standard statistical language and terminology are used with the modification that Greek letters are avoided whenever possible. Uppercase italic letters represent matrices, lowercase italics represent column vectors (or row vectors, if explicitly noted), and lowercase letters not in italics represent scalars.

The following definitions are used, given a matrix X:

X' = transpose of X
X^{-1} = inverse of X, i.e., such that $XX^{-1} = I$, the identity matrix.

The following running index system is used:

i = 1, 2, ..., I: sample number
j = 1, 2, ..., J: chemical constituent number (protein, starch, etc.)
k = 1, 2, ..., K: near-infrared wavelength channel number
a = 1, 2, ..., A: regression factor number.

If both a data set for calibration and one for prediction analysis are available, I_c may be used to denote the number of samples in the calibration set and I_p the number of samples in the prediction set.

A problem exists in describing multivariate calibration in standard statistical notation: regression is performed both over calibration samples and over wavelengths. The following chemometric convention is used in this part of the chapter:

$X = \{x_{ik}\}$: matrix of near-infrared data for sample i at wavelength k

TABLE I
Factors Involved in Calibrating Near-Infrared Instruments

Chemical and optical interferences in near-infrared data demand indirect multivariate calibration: i.e., determination of calibration coefficients for several different wavelengths.
Response of instrument to chemical variations is usually nonlinear, which can make common linear modeling difficult.
Light scatter and other optical interferences in near-infrared data are strong and nonlinear.
Measurement noise in chemical and near-infrared data creates danger of statistical overfitting.
Multicollinearity in near-infrared data creates statistical estimation problems.
High number of calibration samples is desired but expensive.
Accurate chemical or reference analysis is essential but not always available.
Comprehensive sample selection is essential, but some sample types may be difficult to obtain.

$C = \{c_{ij}\}$: matrix of chemical data for sample i and chemical constituent j.

The circumflex ("hat") above a value denotes an estimate: e.g., $\hat{C}$. Symbols and abbreviations used are summarized in Appendix A, and readers who are not familiar with matrices are referred to Appendix B.

II. LINEAR PREDICTION AND WAYS TO FIND THE CALIBRATION COEFFICIENTS

A. Linear Prediction Equation

The main emphasis here is on calibration resulting in predictors that are linear functions of the spectrum x (after a possible transformation). For constituent j in sample i, the linear prediction equation may be written:

$$\hat{c}_{ij} = \hat{b}_{0j} + \sum_{k=1}^{K} x_{ik}\hat{b}_{kj}\ , \tag{1}$$

where $\hat{b}_{0j}$ and $\hat{b}_{kj}$ are the calibration coefficients, estimated statistically by some calibration method. The coefficients $\hat{b}_{kj}$ represent proportionality constants for the chemical constituents vs. wavelength k, and $\hat{b}_{0j}$ is the intercept term.

On vector form the model is:

$$\hat{c}_{ij} = \hat{b}_{0j} + x_i\hat{b}_j\ , \tag{2}$$

where the near-infrared spectrum x_i is a K-dimensional row vector and the calibration constant vector $\hat{b}_j$ is a K-dimensional column vector as illustrated in Fig. 2.

For two or more chemical constituents, (J > 1) becomes:

$$\hat{c}_i = b_0 + x_i\hat{B}\ , \tag{3}$$

where the chemical concentration vector $\hat{c}_i$ and the intercept vector $\hat{b}_0$ are J-dimensional row vectors and the matrix $\hat{B}$ is a K × J matrix. For analyses on several samples, the equation can be written:

$$\hat{C} = 1b_0 + X\hat{B}\ , \tag{4}$$

where the matrix of predicted concentrations, $\hat{C}$, is an I × J matrix and the near-infrared matrix X is an I × K matrix. 1 is here an I-dimensional column vector with elements equal to 1.

There exist several ways to estimate the calibration coefficients in the linear prediction equation. For example, in transmission spectroscopy for clear solutions, the calibration coefficients can be determined from measurements of the absorbance spectra of the constituents present. This is called *direct calibration*. In near-infrared reflectance (NIR) and diffuse transmittance, this kind of calibration is complicated by strong interference from noise factors caused by light scatter and unknown constituents. Furthermore, some constituents do not have the same absorbance spectra in biological tissue as in their pure state. Therefore, the calibration coefficients must in this case be estimated statistically from a set of representative calibration samples with measured chemical and near-infrared data. This is called *indirect calibration*.

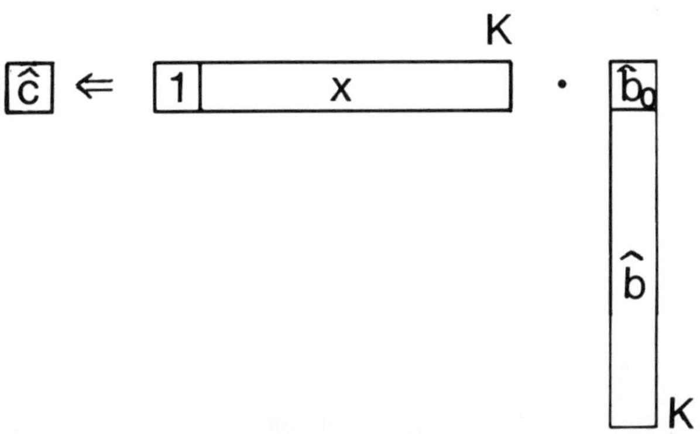

Fig. 2. Linear prediction equation, yielding concentration ĉ from measured spectrum *x* and estimated calibration coefficients $\hat{b}_0$ and $\hat{b}$ for constituent j.

B. Multiple Linear Regression as a Calibration Method

The calibration coefficients $\hat{b}_j$ for each constituent may be estimated by multiple linear regression using the following model:

$$c_j = 1b_{0j} + Xb_j + f_j\ , \tag{5}$$

where c_j and f_j are I-dimensional column vectors. Notice that whereas c_i denoted the observations of the J chemical constituents in sample i, c_j denotes the observations for constituent j in all I samples. The ordinary least-squares (LS) estimator for $(b_{0j}, b_j')'$ is:

$$[(1,X)'(1,X)]^{-1}(1,X)'c_j\ . \tag{6}$$

However, the matrix to be inverted in equation (6) is (1 + K) × (1 + K) dimensional. If K is large, inverting such large matrices requires large computer memory and long computation time. Numerical accuracy problems may also make such an operation difficult.

In this method, each calibration coefficient is treated as an *independent* statistical parameter to be estimated. Therefore, even if a solution to equation (6) could be obtained numerically for a certain calibration set, analysis of unknown samples with the resulting calibration coefficients may give inaccurate results unless the number of calibration samples is very high and all of the data very precise. The reason is sometimes called *statistical overfitting*, when too many independent parameters are estimated with too few calibration samples (see Fig. 3).

The data obtained at the different wavelengths are usually highly intercorrelated. The huge light-scatter variation introduces a strong correlation between all the wavelengths, unless some optical correction is applied on the near-infrared data. In addition, if 100 different wavelengths are measured for each sample, there are usually fewer than 100 different independent ways in which these wavelengths vary from sample

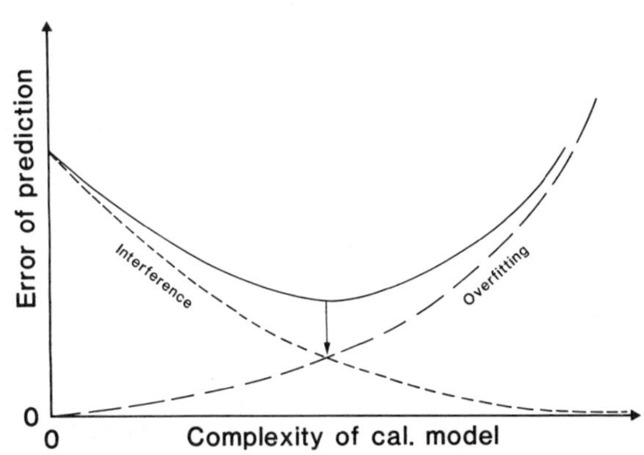

Fig. 3. Optimizing in indirect multivariate calibration. Good effect of modeling interferences and bad effect of statistical overfitting affect error of prediction in opposite directions as complexity of model increases. Complexity of model is amount of information built into calibration model relative to total amount of information available in calibration data. Amount of information built into calibration model roughly corresponds to number of parameters estimated or selected, which increases with number of regression factors, A.

to sample. This is certainly true if the number of calibration samples is less than 100! Hence, some of the wavelengths are bound to be strongly correlated to the other wavelengths.

Simultaneous intercorrelation between many different near-infrared wavelengths, which is termed *multicollinearity*, makes it difficult to obtain good calibration by the multiple linear regression method using all available wavelengths. Some type of data compression is required to correct for multicollinearity.

C. Different Classes of Calibration Methods

Several calibration methods can be used for near-infrared instruments to correct for multicollinearity. Many give similar results, but some methods behave better than others under certain circumstances. These methods can be classified in different ways.

One way of classifying calibration methods concerns whether they rely on selection or weighting of near-infrared information to "solve" the multicollinearity problem. The calibration methods described in the preceding chapter select only a small subset of the available near-infrared wavelengths, ignoring the rest. In contrast, full-spectrum methods define the regression factors as weighted linear combinations of all the relevant wavelengths and use different ways to obtain the weights that define these linear combinations of wavelengths. The spectra of certain calibration samples may be selected as wavelength weights; in other methods the weights are obtained from predefined mathematical formulae, and in still others the weights are estimated statistically. The last estimation is usually based on the near-infrared data alone, but sometimes even the chemical data are used to estimate the wavelength weights.

Another way of classifying the methods is the way they utilize information about the statistical distribution of the chemical composition in the population of samples to be analyzed. Some methods use this type of information, either implicitly or explicitly, whereas others do not (Brown, 1982).

In conventional near-infrared calibration methods (see Section IIIB), the general calibration model is:

$$\text{Chemical data} = \text{f (spectral data)} + \text{errors}, \qquad (7)$$

where f( ) means *function of*. Here the chemical data are modeled as a regression on spectral data plus errors. The conditional distribution of the chemical data given the spectral data is implicitly present in the model. Calibration methods developed in this model will also depend on this kind of information.

The direct calibration methods sometimes employed for multicomponent analysis in transmission spectroscopy in clear solutions can be extended to indirect versions suitable for near-infrared calibration. Here the model is:

$$\text{Spectral data} = \text{f (chemical data)} + \text{errors}. \qquad (8)$$

For methods developed in this model, one may choose explicitly the extent to which distribution of the chemical compositions should be incorporated in the predictor.

Another class of methods is based on regression on latent regression factors, e.g., Fourier regression, principal component regression, and partial least-squares regression; the model is:

$$\text{Spectral data} = \text{f (latent spectral regression factors)} + \text{errors}$$

and (9)

$$\text{Chemical data} = \text{g (latent spectral regression factors)} + \text{errors}.$$

In the first equation, spectral data are modeled as a function of latent factors. In the second equation, chemical data are modeled as a regression on the same latent factors. The methods developed in this model contain information about the population of samples used in the calibration and are easiest to justify if the calibration samples are representative for the distribution of future samples.

The effect of the different usage of distributional information has not yet been fully studied statistically, but theoretical considerations indicate that it is of some relevance. Sometimes a reasonably small analytical error over a maximum range of future sample qualities is required, and it is obtained by not incorporating distributional assumptions. In other cases, an absolutely minimal error over a smaller but more important range of sample variation is necessary, and it is obtained by incorporating distributional information. Naes (1985b) discusses this distinction in more detail.

A third way of classifying methods is based on the orthogonality properties of the regression factors. If the regression factors are calculated mathematically so that they are uncorrelated (orthogonal), then certain mathematical operations become simpler; i.e., no matrix inversion is needed. This method is discussed below.

III. STATISTICAL CALIBRATION METHODS FOR MULTICOLLINEAR DATA

Several indirect calibration methods are presented for multicollinear data (after a possible transformation) in the unified structure introduced in Section IB. The model applies for all the methods covered, except for the Beer's model methods to be discussed later.

For those methods where the chemical data in some way are used during the data compression stage, it is probably advantageous to calibrate for each chemical constituent separately, but possibilities of calibrating for several constituents simultaneously also exist (Sundberg, 1982). For those methods that do not use chemical data in the data compression stage, it is immaterial whether the calibration is performed for each chemical constituent separately or for several simultaneously ($J > 1$). Here main attention is given to calibration of a single constituent.

A calibration method that may be important but that is not covered herein is ridge regression (Fearn, 1983). The formula is similar to equation (6), but a diagonal matrix with the so-called ridge parameter on the diagonal is added to the matrix in brackets before inversion (Gunst and Mason, 1979). This reduces the multicollinearity problem but may distort the real information in the near-infrared data.

A. The General Model Framework

Instead of modeling only the chemical data as a function of spectral data, as is done in equations (5) and (7), i.e.:

$$c = 1\text{b}_0 + Xb + f, \qquad (10)$$

both the near-infrared data and the chemical data (one constituent) can be modeled in terms of a small set of common latent regression factors t_1, ..., t_A (I-dimensional column vectors), as illustrated by equation (9) and in Fig. 4. The model is:

$$X = 1\bar{x} + t_1p_1 + .. + t_ap_a + .. + t_Ap_A + E$$

and

$$c = 1\bar{c} + t_1\text{q}_1 + .. + t_a\text{q}_a + .. + t_A\text{q}_A + f,$$

where p_a and q_a are the near-infrared and chemical loadings of regression factor a, E and f are error terms after A factors, $\bar{c}$ is the mean of the constituent in question, and $\bar{x}$ is a K-dimensional row vector of means of X over the I calibration samples.

Using full matrix notation and subtracting the means from

both sides, this general calibration model can be written more simply:

$$(X - I\bar{x}) = TP + E \text{ (data compression)} \tag{11}$$

and

$$(c - I\bar{c}) = Tq + f \text{ (calibration regression).} \tag{12}$$

This notation (Fig. 5) will be used to illustrate the different calibration methods.

During calibration, the means $\bar{x}$ and $\bar{c}$ are calculated and subtracted from the X and c to eliminate general offset effects. Then the near-infrared loadings P, the latent regression factors T, and the chemical loadings q are estimated, together with the errors E and f. The way P and T are estimated is different for the methods treated, but in most cases q is found by least squares using $\hat{T}$ and the model summarized in equation (12). This implies that:

$$\hat{q} = (\hat{T}'\hat{T})^{-1}\hat{T}'(c - I\bar{c}) \,. \tag{13}$$

The near-infrared residuals $\hat{E}$ and chemical residuals $\hat{f}$ are found by subtracting $\hat{T}\hat{P}$ and $\hat{T}\hat{q}$ from the corresponding near-infrared and chemical observations.

The calibration model can be expressed with increasing degrees of complexity by increasing the number of regression factors, A. Testing of unknowns, cross-validation, or some other criterion with statistical significance is required to determine the optimal number of factors (Fig. 3). If too few factors are used, the model is incomplete and some interferences remain uncorrected. If too many factors are used, the estimated constants $\hat{P}$ and $\hat{q}$ (and hence coefficients $\hat{B}$, which are constructed from $\hat{P}$ and $\hat{q}$) become very noise contaminated. Both result in high analytical error.

The determination of the constituent concentrations in an unknown sample implies first a computation of the regression factors in that sample, $\hat{t}_i$ (A-dimensional row vector; t_a is used above as symbol for the I-dimensional column vector corresponding to regression factor a), from the near-infrared spectrum x_i (K-dimensional row vector), and the estimated calibration mean spectrum $\bar{x}$ and loadings $\hat{P}$, using LS and the equation:

$$x_i - \bar{x} = t_i\hat{P} + e_i \,. \tag{14}$$

The estimate of t_i, $\hat{t}_i$, is then:

$$\hat{t}_i = (x_i - \bar{x})\hat{P}'(\hat{P}\hat{P}')^{-1} \,. \tag{15}$$

(An exception is the PLS method [Section IIIF], where the $\hat{t}$'s may be found by a slightly different but more efficient procedure.)

This procedure also yields an estimate of the near-infrared residuals of the unknown sample:

$$\hat{e}_i = x_i - \hat{x} = x_i - (\bar{x} + \hat{t}_i\hat{P}) \,. \tag{16}$$

The obtained $\hat{t}_i$ and $\hat{e}_i$ may be used for detection of outliers of various types. The regression factors $\hat{t}_i$ (sometimes called *estimated scores* or merely *scores*) are used to predict the chemical variable by:

$$\hat{c}_i = \bar{c} + \hat{t}_i\hat{q} \,. \tag{17}$$

In all methods considered here, $\hat{t}_i$ is a linear combination of x_i and consequently $\hat{c}_i$ can be obtained from an ordinary linear prediction equation of the form $\hat{c}_i = \hat{b}_0 + x_i\hat{b}$ (see equation [2]). The vector $\hat{b}$ can be written as $\hat{P}'(\hat{P}\hat{P}')^{-1}\hat{q}$ and $\hat{b}_0 = \bar{c} - \bar{x}\hat{b}$.

Figure 6 illustrates the different data compression techniques, and Fig. 7 shows the subsequent use of the regression factors for the different methods to be treated.

B. Multiple Linear Regression as a Calibration Method

Figure 6 illustrates the established technique of stepwise multiple linear regression (SMLR) and the method (Norris, 1983) that may be termed *stepwise derivative linear regression* (SDLR) in the general model framework. These are not full-spectrum methods, but they may still be presented under the general framework of Section IIIA. The SDLR method is usually based on apparent absorbance, log (1/R), where R = reflectance. The log (1/R) values are used for calculating first derivatives (to eliminate spectral offset variations) and second derivatives (to eliminate linear spectral trends) with respect to wavelength. This pretreatment reduces interference from other chemical constituents. Using quotients of derivative terms serves as an optical correction.

The data compression stage for these methods consists of selecting a combination of a few wavelengths (SMLR) or

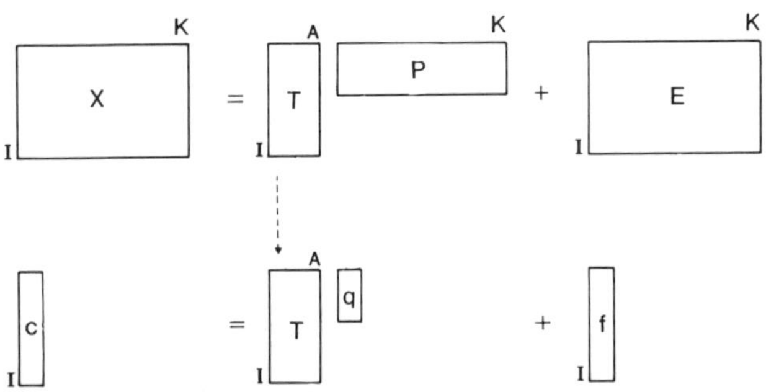

Fig. 5. General data model for calibration expressed in condensed matrix form. X and c are assumed centered.

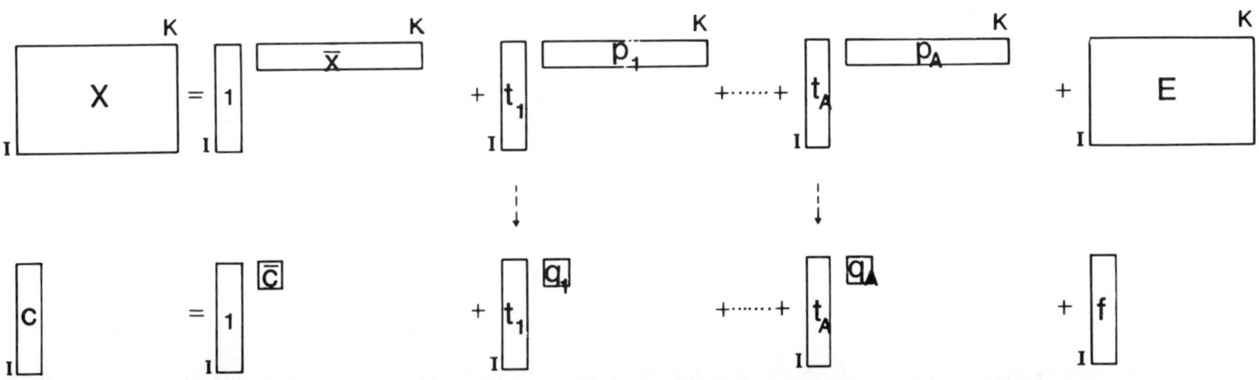

Fig. 4. General data model for calibration of one chemical variable on K near-infrared variables, using I calibration samples. Variables are concentrated to A regression factors $T = \{t_1, t_2, \ldots, t_A\}$, on which chemical variable is regressed.

quotients of derivative terms (SDLR) as regression factors. This selection can be done in a variety of ways (Draper and Smith, 1981). Many different combinations are usually tested for calibration and the "best" or most reasonable combination is adopted for the analysis.

The wavelengths or derivative terms thus selected form the regression factor matrix $\hat{T}$. This matrix is used in the estimation of the chemical loadings by ordinary least squares. The chemical residuals $\hat{f}$ are found as explained in Section IIIA. When derivatives are used, sometimes only a single regression factor is required (Norris, 1983).

For determination of the chemical constituent in an unknown sample, the same wavelength selection is used as in the calibration step to find $\hat{t}_i$. This $\hat{t}_i$ is used to estimate the chemical constituent by the equation:

$$\hat{c}_i = \bar{c} + \hat{t}_i\hat{q} \,. \tag{18}$$

For the wavelengths selected to be used as regressors in SMLR, the calibration coefficients $\hat{b}$ correspond to the chemical loadings $\hat{q}$.

An estimate of E, i.e., $\hat{E}$, does not exist in these methods. However, an extra regression step may be added to these conventional methods to provide outlier detection by regressing the unused near-infrared wavelengths in X on the selected ones (those put into $\hat{T}$). By LS solution of equation (11), the estimates of the near-infrared loadings are found to be:

$$\hat{P} = (\hat{T}'\hat{T})^{-1}\hat{T}'(X - 1\bar{x}) \,, \tag{19}$$

where X represents the unused wavelengths. The residuals $\hat{E}$ are estimated by:

$$\hat{E} = X - \hat{X} = X - (1\bar{x} + \hat{T}\hat{P}) \,. \tag{20}$$

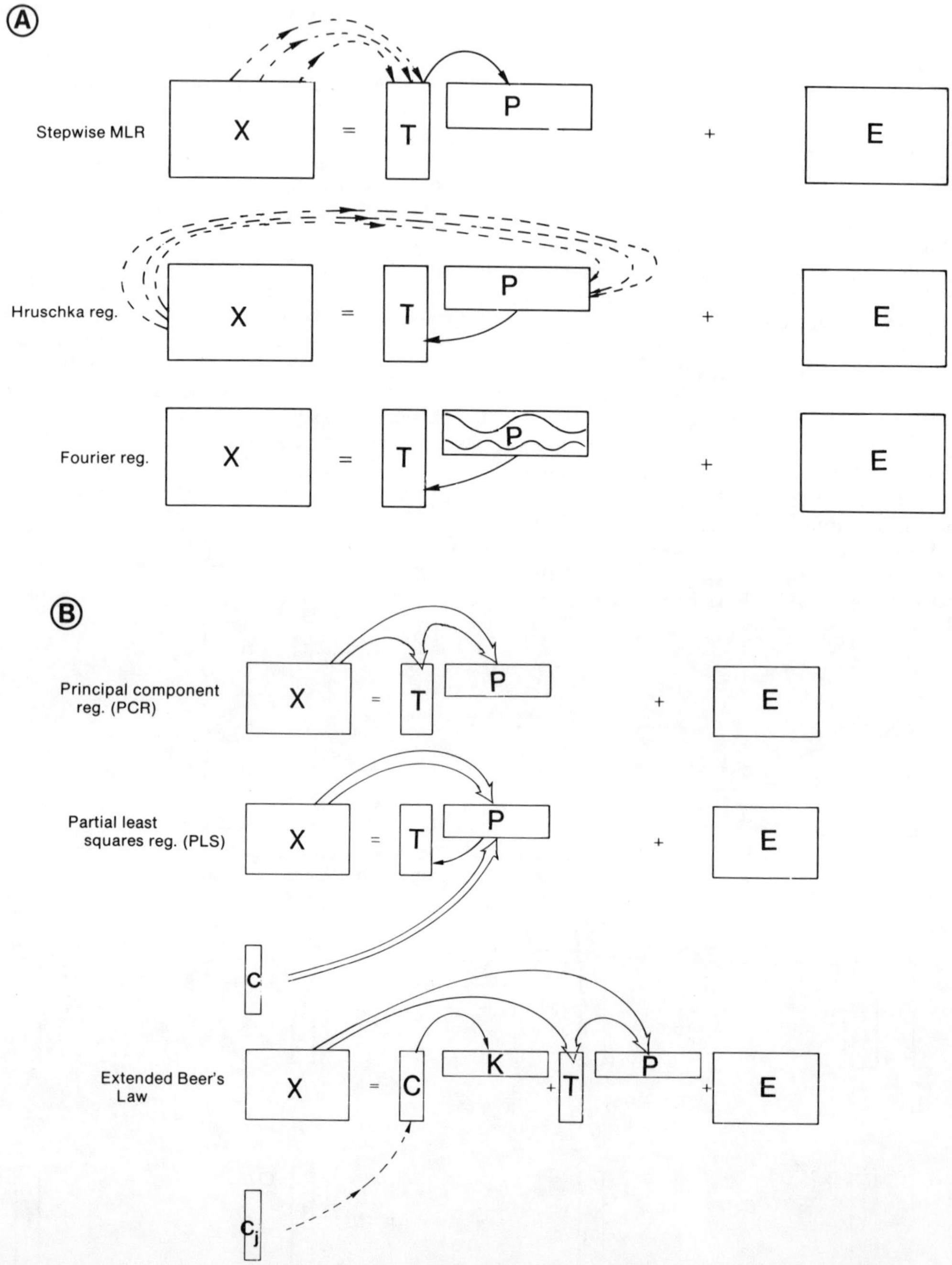

Fig. 6. Data compression stages for calibration methods. (A) Methods SMLR, Hruschka regression, and Fourier regression. Dotted arrows indicate selection and solid arrows indicate least-squares fit. X is assumed centered. (B) Methods for PCR, PLS, and extended Beer's model methods. Dotted arrows indicate selection and solid arrows indicate least-squares fit. X and c are assumed centered.

The regression factors $\hat{T}$ and the residuals $\hat{f}$ and $\hat{E}$ can be used to detect outliers in the calibration data set.

Outlier detection is also provided by the residuals of the unused near-infrared wavelengths, i.e., by:

$$\hat{e}_i = x_i - \hat{x}_i = x_i - (\bar{x} + \hat{t}_i\hat{P}) . \quad (21)$$

C. Hruschka Regression: Selecting the "Best" Calibration Samples

If there are many wavelengths K, but only a limited number, A, of relevant, independent ways in which the near-infrared spectra may vary, it is possible to use the spectra of A selected calibration samples as near-infrared loading matrix. These A samples must be selected in such a way that they include all the variation types present in the other I − A calibration samples, as well as all future prediction samples (Hruschka and Norris, 1982; H. Martens et al, 1983a).

By analogy with the previous method, where a subset of wavelengths was selected and put into the regression factor matrix $\hat{T}$, Fig. 6 shows how the spectra of a subset of samples from the calibration set are selected and put into the loading matrix $\hat{P}$. In the data compression step, the loading matrix, $\hat{P}$, is selected and used to estimate the corresponding scores T. Ordinary multiple linear regression over the K wavelengths was used by Hruschka and Norris (1982), i.e.:

$$\hat{T} = (X - 1\bar{x})\hat{P}'(\hat{P}\hat{P}')^{-1} . \quad (22)$$

Alternatively, a weighted least-squares (WLS) estimator might be used to improve the precision of the $\hat{T}$ estimates, i.e.:

$$\hat{T} = (X - 1\bar{x})V^{-1}\hat{P}'(\hat{P}V^{-1}\hat{P}')^{-1} , \quad (23)$$

where $V = \text{diag}(v_k)$ is a diagonal matrix and v_k is the expected noise level in wavelength k. In the calibration regression, $\hat{T}$ is used as regressor matrix in the ordinary linear regression equation estimating the chemical loadings, $\hat{q}$. During analysis of an unknown sample, the same near-infrared loading matrix $\hat{P}$ is used to compute $\hat{t}_i$, i.e.:

$$\hat{t}_i = (x_i - \bar{x})\hat{P}'(\hat{P}\hat{P}')^{-1} . \quad (24)$$

The residuals $\hat{e}_i$ are obtained as in equation (16) and can be used for outlier detection. The desired chemical constituents are determined as in equation (17).

Samples whose spectra are used as near-infrared factor loading matrix, $\hat{P}$, can be selected on the basis of several different criteria. If they are chosen as described by Hruschka and Norris (1982) according to their ability to reduce the calibration residuals, $\hat{E}$, this is analogous to Ottestad's (1975) component analysis. They could also be chosen according to their ability to model the chemical data from near-infrared data, i.e., by minimizing $\hat{f}$. To avoid overfitting, this could preferably be done by prediction testing or cross-validation.

In situations where it is important to limit the number of samples to be chemically analyzed, the Hruschka method for component analysis is interesting. Given cheap near-infrared data but no chemical data for a large set of samples, a small subset of samples spanning all the main near-infrared variation types can be obtained. This small sample set can then be analyzed chemically by a slower, more expensive reference method to provide a good set of data for calibration (Hruschka, pers. comm., 1983; H. Martens et al, 1983b).

The Hruschka regression method requires precise near-infrared spectra. If the measurements contain significant noise, the following methods are expected to yield somewhat improved analyses.

D. Fourier Transform Regression: Concentrating the Data to the Main Spectral Features

The Fourier transform regression method is analogous to the Hruschka one, but it uses mathematically defined rather than experimentally measured loading spectra, $\hat{P}$. It allows the concentration of the information and the isolation of certain types of noise. Each near-infrared spectrum is transformed trigonometrically from the usual wavelength "domain" to a new type of spectrum of feature "rhythms" assumed to contain all of the valuable information in the near-infrared spectra. This procedure can eliminate redundancy and erratic features that are suspected to represent random noise (the narrow peaks or high-frequency "ripple").

Figure 6 illustrates the use of Fourier transforms for multivariate calibration. During calibration, the Fourier series corresponding to 1, 2, 3, 4 ... full sine and cosine functions over the length of the relevant wavelength range can be computed according to:

$$\hat{p}_{ak} = \sqrt{\frac{2}{K}} \sin\left(\frac{2\pi ak}{K}\right),\ a = 1, 2, \ldots, \frac{A}{2};\ k = 1, 2, \ldots, K$$

$$\hat{p}_{\frac{A}{2}+a,k} = \sqrt{\frac{2}{K}} \cos\left(\frac{2\pi ak}{K}\right),\ a = 1, 2, \ldots, \frac{A}{2};\ k = 1, 2, \ldots, K . \quad (25)$$

The rows of $\hat{P}$ are then orthogonal, i.e., $\hat{P}\hat{P}' = I$.

Since factors for high a only represent noise, they are eliminated (i.e., A/2 is selected such that $A/2 \ll K/2$). The number of Fourier terms to be included, A, may be determined by, for example, significance testing, prediction testing, or cross-validation. Other types of "wavelengths" than the "harmonic" ones in the standard Fourier approach can be envisioned (Damsleth and Spjøtvoll, 1982).

Each observed near-infrared spectrum, x_i, may first be adjusted so that it meets the infinite Fourier transform requirement that the first and the last wavelength have the same value. In the data compression stage, the loading matrix $\hat{P}$ and the actual near-infrared data are used for each sample separately to estimate $\hat{T}$, according to equation (22). (For faster operation, the FFT algorithm may be used instead.) "Zero filling" can be used for higher resolution (Rabiner and Gold, 1975). In the calibration regression stage, $\hat{T}$ is used as regressor matrix in a multiple linear regression equation for estimating the chemical loadings, $\hat{q}$ (see equation [13]).

During analysis of an unknown sample, the first step slope-adjusts the data x_i (if slope-adjustment is used) and sets up the feature rhythm matrix $\hat{P}$ with the desired number of factors, A. Then the regression factors and the near-infrared residuals for outlier detection are found by using a similar formula, such as equation (16). The chemical variables c_i are determined as by

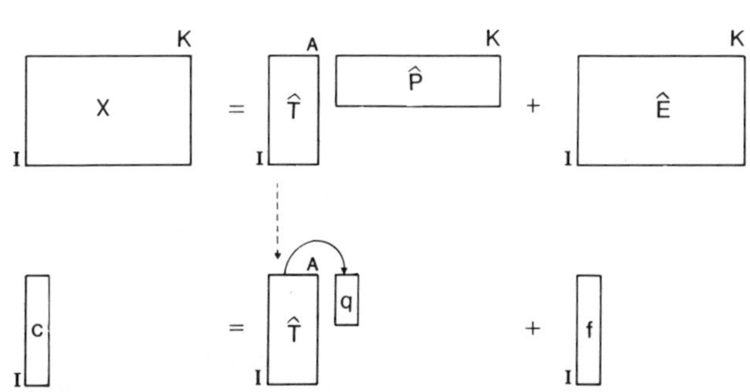

Fig. 7. Calibration regression stage in all methods covered, except those based on Beer's model. Estimated regression factors $\hat{T}$ are used for least-squares estimation of q. X and c are assumed centered.

equation (17). This corresponds to a calibration coefficient matrix:

$$\hat{b} = \hat{P}'(\hat{P}\hat{P}')^{-1}\hat{q} = \hat{P}'\hat{q}\,, \qquad (26)$$

since $\hat{P}\hat{P}' = I$.

The Fourier technique is only applicable for continuous or piecewise continuous near-infrared spectra, but for such data a great simplification can be obtained in terms of computer storage, computation time, and danger of overfitting. By so-called deconvolution prior to the back-transformation, some effects of wide-bandpass slit effects, etc., can be removed. A combination of the Hruschka or Fourier regression may prove optimal in terms of immunity to noise and computer requirements.

To summarize the previous methods, in the conventional calibration the regression factors $\hat{T}$ were selected more or less directly from the near-infrared data. In the two subsequent full-spectrum methods, the regression factors $\hat{T}$ were estimated by first selecting the near-infrared loading matrix $\hat{P}$; in Hruschka regression the spectra of some calibration samples were chosen as loadings, whereas in the Fourier transform regression the loading spectra consist of trigonometric functions. In the following three methods, the spectra P are estimated statistically.

The three methods differ in the way they use the chemical information in the data compression stage. In principal component regression (PCR), the chemical data are not used at all during the data compression; the regression factors are estimated statistically solely on the basis of the near-infrared data. This method contrasts to partial least-squares (PLS) regression, where the chemical data are used for guiding the estimation of the factors from the data in such a way that the regression factors become better regressors for the calibration regression. In the last group of methods, based on the causal Beer's model, the chemical data are used most strongly in the data compression: the chemical loading matrix is estimated so as to represent the spectra of the pure chemical constituents directly.

E. Principal Components Regression: Concentrating the Data to Their Most Dominant Dimensions

The PCR (Gunst and Mason, 1979) uses for data compression the singular value decomposition of the near-infrared data X (also called principal component analysis and eigenvector decomposition) to estimate both T and P. Given the centered data matrix $X - 1\bar{x}$, the t_1 (the first column of T, I-dimensional) and p_1 (the first row of P, K-dimensional) are estimated so that the product $\hat{t}_1\hat{p}_1$ accounts for as much as possible of the total variation in the centered X data. The vectors $\hat{p}_2$ and $\hat{t}_2$ are then chosen such that:

$$\hat{t}_2'\hat{t}_1 = 0 \qquad (27)$$

and

$$\hat{p}_2\hat{p}_1' = 0 \qquad (28)$$

and such that $\hat{t}_2\hat{p}_2$ accounts for as much as possible of the variation in the data remaining after the first factor. In this data compression stage, as much as possible of the variation in the data in the calibration samples is described by as few orthogonal factors as possible. The last factors will mainly account for nonsystematic measurement noise and redundancy in the data, and they are eliminated from the final calibration regression stage. Thereby the number of regression variables is reduced from the original K wavelengths in X to the A first PCA regression factors in $\hat{T}$.

Several different algorithms exist for computing the loadings and scores. The NIPALS or NILES method (Wold, 1966) has been found to be particularly suitable. The method is detailed in the following subsections.

CALIBRATION

The data matrix X is first centered, i.e.:

$$U = X - 1\bar{x}\,. \qquad (29)$$

For each new factor a = 1, 2, ..., A: perform steps 1, 2, and 3:

1. Estimate the near-infrared loading vector p_a (K-dimensional row vector) by LS from a preliminary estimate of the regression factor vector, $\hat{t}_a$ (I-dimensional column vector) (or from some start value, e.g., a column in U) using the model:

$$U = \hat{t}_a p_a + E\,. \qquad (30)$$

The solution is scaled to length 1 and is then:

$$\hat{p}_a = \kappa\hat{t}_a'U\,, \qquad (31)$$

where the scaling factor κ is obtained as the inverse of the length of the $\hat{p}_a$ vector obtained without scaling, i.e.:

$$\kappa = (\hat{t}_a'UU'\hat{t}_a)^{-1/2}\,. \qquad (32)$$

2. Estimate the near-infrared score vector t_a by LS from the previous estimate of $\hat{p}_a$ using the model:

$$U = t_a\hat{p}_a + E\,. \qquad (33)$$

The least-squares solution is:

$$\hat{t}_a = U\hat{p}_a'\,. \qquad (34)$$

Repeat steps 1 and 2 until convergence (typically 4 to 10 iterations). The criterion of convergence may be that the length of $\hat{t}_a$, $\hat{t}_a'\hat{t}_a$ no longer "changes."

3. Prepare new near-infrared residuals:

$$U^{(new)} = \hat{E} = U^{(old)} - \hat{t}_a\hat{p}_a \qquad (35)$$

and increase a by 1 and go to step 1.

The regression factors $\hat{t}_1, ..., \hat{t}_A$ obtained in this way are used to estimate the vector of chemical loadings, q, in the subsequent calibration regression stage. This can be done as in equation (13) or, since the scores are orthogonal, by the simple formula:

$$\hat{q}_a = \frac{t_a'(c - 1\bar{c})}{\hat{t}_a'\hat{t}_a}\,. \qquad (36)$$

The number of regression factors A to be used in the final equation can be chosen in different ways. Wold (1978) gives a cross-validation method for finding the number of significant PCA factors based on X alone, and significance tests based on distributional assumptions also exist. It is more suitable to use the ability of determination of the chemical data c as a criterion, based on true analysis or on cross-validation in the calibration regression stage. Notice that PCR as described here deletes small eigenvalues. Other versions of PCR that delete eigenvectors because of predictive relevance are discussed in Joliffe (1982).

PREDICTION

During analysis of an unknown sample, $\hat{t}_i = (x_i - \bar{x})\hat{P}'$, because $(\hat{P}\hat{P}') = I$. The near-infrared residuals of this sample are:

$$\hat{e}_i = x_i - (\bar{x} + \hat{t}_i\hat{P}) \tag{37}$$

and the concentration of the chemical constituent is determined from $\bar{c}$ and $\hat{q}$:

$$\hat{c}_i = \bar{c} + \hat{t}_i\hat{q} \; . \tag{38}$$

Alternatively, the same analysis step can be written:

$$\hat{c}_i = \hat{b}_0 + x_i\hat{b} \; , \tag{39}$$

where:

$$\hat{b}_0 = \bar{c} - \bar{x}\hat{b} \text{ and } \hat{b} = \hat{P}'\hat{q} \; . \tag{40}$$

F. Partial Least-Squares Regression: Concentrating the Data to Their Most Relevant Dimensions

The partial least-squares regression is a relatively new approach to linear regression, and the algorithm for finding the estimates is a generalization of the NIPALS algorithm for PCA. The method and its extensions are especially suitable for calibration on a small number of samples with experimental noise in both chemical and near-infrared data. It is also simple to compute, involving only a series of simple linear regressions. The method used for calibration was initially developed by Wold et al (1983) on the basis of the general PLS method (H. Wold, 1982). It has been used for near-infrared data (Martens and Jensen, 1983), fluorescence data (Jensen et al, 1982; Jensen and Martens, 1983), X-ray computer tomography data (Vangen and Martens, 1982; H. Martens et al, 1983b), and for sensory data (Hildrum et al, 1983; M. Martens et al, 1983). Computer programs for PLS analysis are available (S. Wold, 1982; Martens, 1986). Wold et al (1983a) and Lindberg et al (1983) can be referred to for applications of the method.

In both PCR and PLS, the regression factors are linear combinations of the wavelengths, estimated from the near-infrared spectra, but their estimation is different: In PCR, the data compression stage is performed independently of the calibration regression, and in such a way that the principal components account for a maximum of the total variation in the near-infrared data alone. In PLS, the two stages can be considered tied together, and the PLS factors primarily describe the near-infrared variation types that are relevant for modeling the variations in the chemical data. The resulting PLS factors describe important variations in the data themselves but at the same time are relevant for determination of the chemical data. If the different variation types in the near-infrared data have the same relative importance for modeling that data as for modeling the chemical data, then the PCR and the PLS factors are nearly identical. A comparison of PCR and PLS can be found in Naes and Martens (1985).

If the near-infrared data instead contain several major, irrelevant variation types (e.g., because of light scatter) and only minor variations relevant for modeling the chemical data, then the PCR solution may mistakenly regard the minor variations as noise and ignore them, since they come in as late factors (large a). The PLS regression has some possibility of detecting these minor, relevant variations in early factors and of clearly distinguishing them from irrelevant factors and the data noise in the later factors. This may be expected to lead to more efficient data compression and thus better calibration.

The basic PLS regression method relates one chemical variable at a time (J = 1) to the K near-infrared variables (PLS1). It can be extended in many different ways; for instance, it can be modified to calibrate for many different chemical constituents at a time (PLS2) (Martens and Jensen, 1983; Wold et al, 1983b). Since the PLS method is so new, algorithms for both PLS1 and PLS2 are described in detail. Some extensions will also be discussed.

Two different PLS1 algorithms exist, one that yields orthogonal scores and one that does not. The two algorithms yield the same final results. First, the basic version with nonorthogonal scores will be shown, to illustrate the PLS philosophy compared with PCR. The other algorithm that is less illustrative from a statistical point of view will be described afterward and used in the computations in the subsequent sections of the paper because of its computational simplicity.

Like the NIPALS algorithm, the PLS algorithms can accept missing data, but this is not shown here. In addition, the method can be used even if I < K.

BASIC PLS ALGORITHM (FIG. 8)

Calibration. The data matrix X and the vector c are first centered, i.e.:

$$U = X - 1\bar{x} \tag{41}$$

$$v = c - 1\bar{c} \; . \tag{42}$$

For each new factor a = 1, 2, ..., A, perform steps (a), (b), (c), and (d):

(a) Use the chemical residuals v and estimate the near-infrared loading vector p_a by LS from the model:

$$U = vp_a + E \; . \tag{43}$$

The solution is scaled to length 1, which gives:

$$\hat{p}_a = \kappa v'U \; , \tag{44}$$

where κ is the scaling factor that makes the length of $\hat{p}_a$ equal to 1. If desired, this estimated loading vector $\hat{p}_a$ can be modified (e.g., smoothed, as described later in this section) before the scaling to length 1.

(b) Estimate the regression factors, t_a, by LS from the estimated near-infrared loadings $\hat{p}_a$ using the model:

$$U = t_a\hat{p}_a + E \; . \tag{45}$$

The LS solution is:

$$\hat{t}_a = U\hat{p}_a' \; .$$

Steps (a) and (b) are the same as in NIPALS PCA, except that v is used as start value in (a).

(c) Estimate the chemical loadings for the previous a factors, q, by MLR from the regression factors $\hat{T} = \{\hat{t}_1, ..., \hat{t}_a\}$, using the model summarized in equation (12). Equation (36) in the PCR-NIPALS algorithm cannot be used since the scores are not orthogonal in this PLS version.

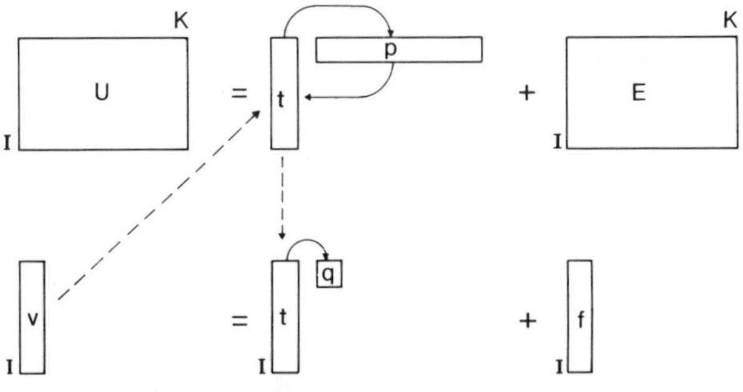

Fig. 8. Basic PLS1 algorithm. Noniterative one-regressand PLS1-regression illustrated for process of estimating parameters of PLS factor a (p_a, t_a, and q_a).

(d) Prepare new residuals, i.e.:

$$v = c - 1\bar{c} - \hat{T}\hat{q} \tag{46}$$

and

$$U = X - 1\bar{x} - \hat{T}\hat{P} \tag{47}$$

and go to step (a) if a < A.

The number of factors to be accepted in the final calibration, A, may be determined by prediction testing or cross-validation, as in PCR. Notice that a new vector $\hat{q}$ is computed by MLR for each a.

Prediction. In this basic PLS version, the analysis of an unknown sample from its near-infrared spectrum x_i proceeds just as in PCR. Equation (37) yields its near-infrared residuals $\hat{e}_i$, and (38) or (39) the concentration of the chemical constituent, $\hat{c}_i$.

As with the NIPALS algorithm, this basic PLS algorithm yields orthogonal near-infrared loading vectors $\hat{p}_1$, ..., $\hat{p}_A$, obtained by projecting the residuals U onto each loading vector $\hat{p}_a$ and subtracting this projection $\hat{t}_a\hat{p}_a$ before proceeding to the next factor. Thereby the information concerning the loading vector $\hat{p}_a$ is removed from the near-infrared data before $\hat{p}_{a+1}$ is estimated, and hence orthogonality between rows in $\hat{P}$ is ensured.

In the NIPALS algorithm a projection of U on the vector $\hat{t}_a$ was done before subtracting this effect, and hence orthogonal columns in $\hat{T}$ are also obtained. But in the basic PLS version given above, no such projection on $\hat{t}_a$ occurs, and hence the columns $\hat{t}$ are nonorthogonal.

PLS ALGORITHM WITH ORTHOGONAL REGRESSION FACTORS (FIG. 9)

A second version of the PLS regression (Wold et al, 1983b) has been designed to gain the computational advantage given by orthogonality in the regression factors. An MLR is avoided in step (c) above, and only simple regressions are necessary. The algorithm is very similar to the PLS described above with the exception of the introduction of a second type of near-infrared loadings (loading weights), w.

Calibration. The data are first centered as outlined above. For each new factor $a = 1, 2, ..., A$, perform the following steps:

(a) Estimate the near-infrared weights $w_a = (w_{a1}, ..., w_{aK})$ from the chemical residuals v using LS and the model:

$$U = vw_a + E\,. \tag{48}$$

The solution is scaled to length 1. The estimate is:

$$\hat{w}_a = \kappa v'U\,, \tag{49}$$

where κ scales $\hat{w}_a$ to length 1. If desired, this estimated loading vector $\hat{w}_a$ can be modified (e.g., smoothed, as described below) before the scaling to length 1.

(b-I) Estimate the factor score vectors t_a by LS from the estimated near-infrared loading weights $\hat{w}_a$ using the model:

$$U = t_a\hat{w}_a + E\,. \tag{50}$$

The solution is:

$$\hat{t}_a = U\hat{w}_a'\,. \tag{51}$$

(b-II) Estimate the near-infrared loadings p_a from the estimated factor score vectors $\hat{t}_a$ using LS and the model:

$$U = \hat{t}_a p_a + E\,. \tag{52}$$

The solution is:

$$\hat{p}_a = \frac{\hat{t}_a'U}{\hat{t}_a'\hat{t}_a}\,. \tag{53}$$

(c) Estimate the chemical loading element for factor a from the estimated factor scores $\hat{t}_a$ by LS using the equation:

$$v = \hat{t}_a q_a + f\,, \tag{54}$$

which gives the solution:

$$\hat{q}_a = \frac{\hat{t}_a'v}{\hat{t}_a'\hat{t}_a}\,. \tag{55}$$

(d) New near-infrared residuals are obtained by equation (35) ($\hat{p}$ is used, not $\hat{w}$!):

$$U^{(\text{new})} = U^{(\text{old})} - \hat{t}_a\hat{p}_a\,.$$

The new chemical residuals are found from:

$$v^{(\text{new})} = v^{(\text{old})} - \hat{t}_a\hat{q}_a\,. \tag{56}$$

Notice that equation (51) projects U on the vector $\hat{w}_a$ and equation (53) projects U on $\hat{t}_a$. Hence orthogonal rows in $\hat{W}$ and orthogonal columns in $\hat{T}$ are obtained! Notice also that $\hat{p}$ in the basic algorithm corresponds to $\hat{w}$ in this orthogonalized version. This may seem inconsequent, but it is designed to fit better into other notation systems (e.g., Martens and Wold, 1984).

Prediction. In this second PLS version, the analysis of an unknown sample from its near-infrared spectrum x_i proceeds as follows. The first score t_1 is found from the model:

$$x_i - \bar{x} = t_1\hat{w}_1 + e\,. \tag{57}$$

The solution is:

$$\hat{t}_1 = (x_i - \bar{x})\hat{w}_1'\,. \tag{58}$$

The next score, t_2, is found from the model:

$$[(x_i - \bar{x}) - \hat{t}_1\hat{p}_1] = t_2\hat{w}_2 + e\,. \tag{59}$$

The solution is here:

$$\hat{t}_2 = [(x_i - \bar{x}) - \hat{t}_1\hat{p}_1]\hat{w}_2'\,. \tag{60}$$

The procedure continues until the Ath factor. Notice that the procedure corresponds to the one for estimation in the calibration.

The chemical concentration is then predicted by equations

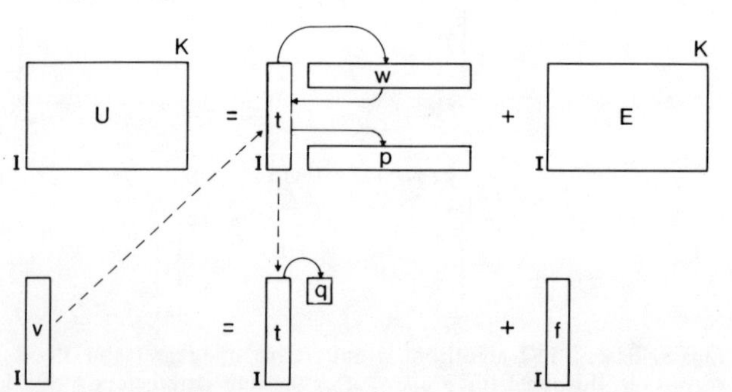

Fig. 9. Algorithm for PLS with orthogonal regression factors.

(38) or (39). In this case, vector $\hat{b}$ can be written as:

$$\hat{b} = \hat{W}'(\hat{P}\hat{W}')^{-1}\hat{q} \tag{61}$$

instead of by (40).

MODIFICATIONS OF THE PLS REGRESSION METHOD

In different calibration situations, different noise problems occur and different types of additional information can be used to counter it. The PLS regression algorithm, especially in its orthogonalized version, is well suited for incorporating different kinds of additional information.

Using Correlation Between Different Chemical Constituents. An iterative extension of the PLS algorithm given above when $J \geqslant 2$ (termed PLS2) (Wold et al, 1983b; Martens and Jensen, 1983) can in certain cases be advantageous, as when the signal-to-noise ratio in the che...ical calibration data for an important constituent is so bad that the calibration is difficult. If precise chemical data are available for other chemical constituents, and these other constituents correlate more or less to the constituent in question, then these other variables can be used to improve the calibration of the important constituent.

In conventional regression (e.g., SMLR), these other chemical variables would have to be used directly as regression factors, just like the near-infrared wavelength, and hence would have to be measured chemically in every new verification sample as well. In contrast, the PLS2 method uses the additional chemical variables for stabilizing the statistical estimation process only during calibration.

However, if there are strong nonlinearities in the data, then different linear approximations may be required in order to give optimal calibration for different chemical constituents. In such cases, the one-constituent ordinary PLS1 method is probably preferable to the PLS2 method.

The PLS2 method requires a statistical weighting of the data of the different chemical constituents in C to compensate for their differences in signal-to-noise ratio. This method also requires data from two or more chemical constituents ($J \geqslant 2$; implying the matrix notation C, F, V, q_a, and $\bar{c}$ instead of c, f, v, q_a, and $\bar{c}$). The algorithm outlined above has two extra stages between (c) and (d):

(c-II) Test whether convergence has occurred, e.g., by checking that $\hat{t}_a'\hat{t}_a$ no longer increases.

(c-III) If convergence is not reached, estimate temporary factor scores from the chemical loadings $\hat{q}_a$ by LS using the model:

$$V = t_a\hat{q}_a + F\,. \tag{62}$$

The solution is:

$$\hat{t}_a = V\hat{q}_a'(\hat{q}_a\hat{q}_a')^{-1} \tag{63}$$

Then repeat stages (a), (b-I), (b-II), (c), (c-II), and (c-III) until convergence (typically three to seven iterations).

Decreasing Sensitivity to Chemical Noise by Combined PLS/PCR. In contrast to the PCR method, the PLS regression methods use the chemical residuals v when estimating the loading spectra p_a. One would expect this to lead to more efficient data compression in PLS than in PCR, but at the expense of a slightly stronger tendency to overfitting because of errors in the chemical data. Since the PLS algorithm is a modification of the NIPALS one, it appears possible to combine the two. For a given chemical constituent ($J = 1$), one could use PLS for the first, dominating factors to ensure their chemical relevance and PCA for the last ones to eliminate the effect of chemical errors in the factor estimation. An intermediate algorithm can be used for the intermediate factors. In the PLS algorithm with orthogonal scores, a PCR corresponds to an iteration between equations (51) and (49).

Thus, PLS factors are obtained as long as steps (a), (b-I), and (b-II) are performed only once for each factor. PCR factors are instead obtained by repeating the above iteration until convergence, whereby all information about the chemical residual v is eliminated from the estimation of loading weights $\hat{w}_a$. Regression factors with properties between those in PLS and PCR are obtained by repeating the above iteration more than once but less than until convergence; each iteration will make the influence of the chemical residuals v on the estimation of loading weights w_a progressively weaker. The method has been tested successfully on near-infrared data.

Application of Extra Information about Near-Infrared Loadings. The PLS loading vector $\hat{w}_a$, obtained from equation (48) by $\hat{w}_a = v'U$, may be improved in different ways by modifying it according to additional information available. Care must be taken to retain the orthogonality of the loading vectors.

If the near-infrared data come from a continuously scanning instrument, the loading vector w_a should be a smooth function of wavelength. By applying digital filtering to $\hat{w}_a$, much of the noise in the data can be eliminated from the calibration process. A low-pass far-infrared (FIR) filter (Rabiner and Gold, 1975) is probably one of the best methods for eliminating random noise. A median filter can be used to remove spikes in the data first. This smoothing approach has given good results on a data set based on measurements from a data tomograph (H. Martens et al, 1983b). For other smoothing algorithms, see Esbensen and Wold (1983).

In conventional SMLR, only a few of the available near-infrared wavelengths are selected as regression factors. Full-spectrum methods, on the other hand, use all available, relevant wavelengths (although their relative importance can be selected by weighing them before the estimation).

A flexible intermediate method can be obtained in PLS regression (and probably for other methods described in this section) by replacing the estimated loadings $\hat{w}_{ak}$ for certain wavelengths by zero, thereby eliminating them from the estimation of that factor. If a sensible criterion is used for selecting which wavelengths to keep and which to eliminate, one would expect improved calibration results because of the reduced number of statistical parameters to be estimated and the high information content in the remaining wavelengths.

An analogue approach can be used for making the PLS regression more or less like Hruschka regression by selecting samples instead of variables.

G. Calibration Based on Beer's Model for Mixtures

Because of the multicollinearity between the different wavelengths, the MLR is dangerous to use as a calibration method. The alternative methods described above use some type of data compression to reduce the number of independent variables, followed by a calibration regression stage consisting of an LS fit of the chemical constituent to the obtained regression factors.

Since c is used as dependent variable in the calibration regression of these approaches (see equation [7]), the methods developed will depend on the distribution of the c's in the calibration sample set. In many cases it is natural to incorporate this information, but if, for example, the calibration samples are selected in a way that gives no data about the population of samples, the information should not be utilized. These two cases are sometimes called *natural* and *controlled calibration*, respectively (e.g., Brown, 1982). It can be shown that if the calibration samples are selected according to their c values in another way than that of being representative for the total population of samples, the natural methods may be difficult to justify theoretically. Generally speaking, it is safest to use samples selected to represent the variability in the population of samples. The set of calibration methods described next are based on the physical mixture model (Beer's model), and they allow an explicit choice of whether to use population information.

These calibration methods do not have a traditional calibration regression stage; therefore, the general model of Section IIIA will not be referred to in this section. Calibration of several constituents will be treated simultaneously ($J \geqslant 2$).

The Beer model for mixtures can be written as:

$$X = 1v + CK + E\,,$$

where v is an intercept (K-dimensional row vector), C corresponds to the concentration of the constituents of interest, and E consists of measurement noise with expectation zero, unknown interference effects, and other model errors (see Fig. 10 for a simplified illustration). The error terms, e, for different samples are independent. A discussion of the model can be found in Naes (1985a). Beer's model is used as an explaining concept for matrix algebra in Appendix B.

Near-infrared spectra have been shown to contain phenomena causing systematic disturbances that may be approximated by a linear factor structure. It may therefore be appropriate to assume that:

$$E = TP + E^*\,,$$

where the number of factors (rows of P) is small and the error terms in E^* for the different wavelengths are uncorrelated. The elements in E^* correspond to random measurement noise and TP corresponds to such interferences as unknown constituents and light-scatter effects. Thus the total model (extended Beer's model), including the contribution from the chemical constituents, CK, and the systematic interferences TP, may be written:

$$X = 1v + CK + TP + E^*. \qquad (64)$$

The four versions of the model that are of interest correspond to different ways of handling unknown c's and t's (Naes, 1985b). In Fig. 11, the symbol t (row vector) is the interference effect for one single sample and c (row vector) is the corresponding chemical composition of the sample.

As discussed above, we may assume that in many cases we have information in the calibration data set about the distribution of unknown c's and that in other cases no such information exists, for example because the calibration samples tell us nothing about future samples. In these two situations, it is natural to consider unknown c's as random variables and as fixed parameters, respectively. In the random case, the distribution of c corresponds to the distribution of c in the population of samples.

Concerning tP, it may sometimes be realistic to assume that both the matrix P and the distribution of future t's are well estimated by the calibration samples. In other cases, it is not so. It may for instance be realistic to assume that P, or the space spanned by the systematic interferences, is well estimated, but that the distribution of future t's is not sufficiently well described to be used in the prediction. In such cases, we may want to consider future t's as unknown fixed parameters.

When c is considered random, E(c) (the mean of c) is denoted by u and cov(c) (the dispersion matrix of c) by H. When a future t is considered random, it is assumed that $E(t) = 0$ and the rotation of P and t is chosen such that $\mathrm{cov}(t) = I$. When t is random, $\mathrm{cov}(e) = \mathrm{cov}(tP + e^*) = P'P + W$ is denoted by Σ, where e and e^* are row vectors (corresponding to one single sample) and $W = \mathrm{cov}(e^*)$.

To be useful for analysis of unknown samples, calibration methods based on this extended Beer's model require information about its model parameters. In certain applications of the Beer's model (e.g., various types of transmission spectroscopy), some model parameters (e.g., K) are known or can be measured directly, whereas others (e.g., v and P) are assumed to be zero. Such calibration (sometimes called *direct calibration*) cannot, however, be used in diffuse near-infrared spectroscopy for the reason discussed in Section I. When the model parameters are unknown, indirect versions of the

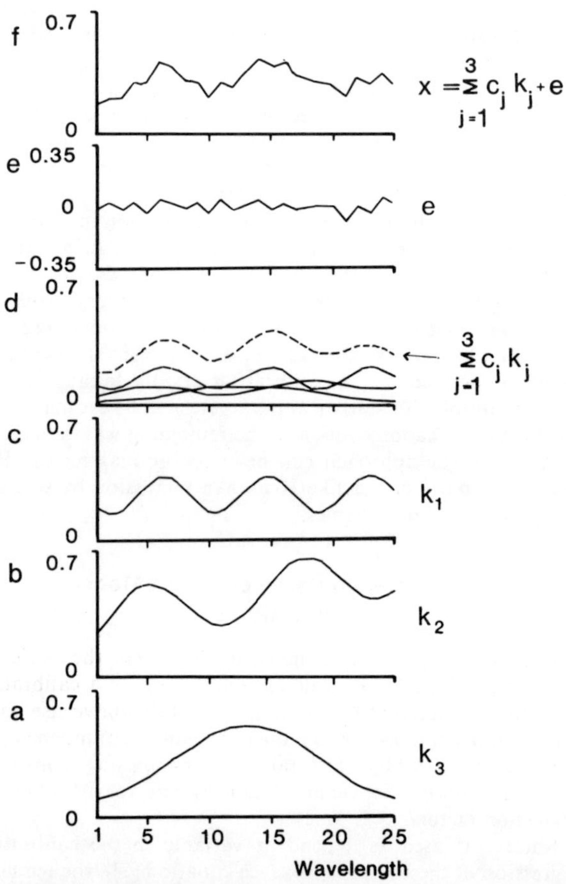

Fig. 10. Beer's model illustrated by artificial data. Three constituents have absorbance spectra k_1, k_2, and k_3 shown in a, b, and c. Mixture containing three constituents in concentrations 60%, 20%, and 20%, respectively, should yield the absorbance spectrum in d (dotted line) if no other interferences occur; this spectrum is sum of three individual contributions drawn beneath it. But if measurement process adds noise to measurement, illustrated in e, then actually measured absorbance spectrum looks like f. If this spectrum were obtained from an unknown sample, constituent concentrations and noise spectrum could be estimated by some multivariate prediction method. The k-spectra plotted are the ones used for artificial data set in Section VI.

Distributional assumption on t \ Distributional assumption on c	No	Yes
No	EGLS1	EBLUP
Yes	EGLS2	EBLP

Fig. 11. Illustration of different usage of distributional assumptions.

predictors derived from the model may be obtained by replacing the known parameters by estimates. The methods based on Beer's model differ in the way these parameters are estimated and in the way they are used in the analysis.

The model parameters in Beer's model can be estimated in different ways, dependent on robustness requirements, for example, and distributional assumptions. The maximum likelihood (ML) estimators (under normality assumption) for v, K, u, and H have been found useful (Naes, 1985a, 1986). The factor structure parameters P and W may be found by factor analysis of, e.g., the ML estimator for Σ (Naes, 1985a, 1986).

Calibration methods are given for the four different cases mentioned above and illustrated in Fig. 11. Estimates of the model parameters, which are assumed to be available, are denoted by the circumflex ("hat").

(a) No distributional assumption on c and t: EGLS1. In this case, u, H, and Σ are of no interest. It is natural to use the generalized least-squares (GLS) estimator to estimate an unknown c, treating $(\hat{K}',\hat{P}')'$ as design matrix and $\hat{W}$ as residual covariance matrix. In this case, the calibration coefficient matrix $\hat{B}$ is the first J columns of:

$$\hat{W}^{-1}(\hat{K}',\hat{P}')[(\hat{K}',\hat{P}')'\hat{W}^{-1}(\hat{K}',\hat{P}')]^{-1} \tag{65}$$

and

$$\hat{b}_0 = -\hat{v}\hat{B} .$$

The equation derived from (65) is sometimes called EGLS (estimated GLS). Here it is called EGLS1, to distinguish it from EGLS2 below.

(b) Distributional assumption on t only: EGLS2. In this case, only u and H are uninteresting. The natural predictor to use is then the GLS estimator with $\hat{K}$ as design matrix and $\hat{\Sigma}$ as covariance matrix; i.e.:

$$\hat{B} = \hat{\Sigma}^{-1}\hat{K}'(\hat{K}\hat{\Sigma}^{-1}\hat{K}')^{-1} \tag{66}$$

and

$$\hat{b}_0 = -\hat{v}\hat{B} .$$

Even this is an estimated version of GLS, but different from EGLS1. The difference is the design and covariance matrices used in the two cases.

(c) Distributional assumption on both c and t: EBLP. If both c and t are assumed to be random variables, it is natural to base the determination of c on the estimated version of the best linear predictor [BLP w.r.t. MSE $= E(\hat{c} - c)^2$] for c with $\hat{P}$ as design matrix and $\hat{\Sigma}$ as error covariance matrix. In this case:

$$\hat{B} = (\hat{K}'\hat{H}\hat{K} + \hat{\Sigma})^{-1}\hat{K}'\hat{H} \tag{67}$$

and

$$\hat{b}_0 = -(\hat{v} + \hat{u}\hat{K})\hat{B} + \hat{u} .$$

(d) Distributional assumptions on c only: EBLUP. In this case, t is an unknown fixed parameter. It is then natural to determine unknown c's by the estimated version of the best linear unbiased predictor for c (BLUP w.r.t. MSE; e.g., Henderson, 1975), treating t as a fixed effect. In this case:

$$\hat{B} = (\hat{K}'\hat{H}\hat{K} + \hat{W})^{-1}\hat{K}'\hat{H} \tag{68}$$

and

$$\hat{b}_0 = -(\hat{v} + \hat{u}\hat{K} + \hat{t}\hat{P})\hat{B} + \hat{u} ,$$

where $\hat{t}$ is the GLS estimator of t in the model treating c as random (with estimated model parameters).

Practical applications of the methods in (b) and (c) can be found in Brown (1982), Naes (1985a, 1986), and Skrede et al (1983). The methods perform well in near-infrared analysis and are multivariate generalizations of the classical and inverse methods of univariate calibration (e.g., Hoadley, 1970). Section VII presents an application of the method in (a) that is originally treated in Martens (1981). The method in (d) is not tested yet, but may be of practical interest. Although the methods treated here at first glance look fundamentally different from earlier methods, it can be shown that there is a great similarity between EBLP and, e.g., PCR. They may both be considered as estimates of the BLP for c, and the difference is how this BLP is estimated.

Other examples of calibration methods in Beer's law may be found in the works of Saxberg and Kowalski (1979) and Haaland and Easterling (1982). These methods are, however, not so well suited for near-infrared data.

The practical significance of the different distributional assumptions on c and t have not yet been fully explored (Lwin and Maritz, 1982; Naes, 1985b). It is known, however, that if the variances in $\hat{W}$ are very small, the methods will give similar results. In general, the approaches treating c as a constant unknown parameter are usually better suited than other methods for analysis outside the range of the calibration samples (extrapolation). Incorporating distributional information usually improves the precision of the analysis when the future samples are within the range of the calibration samples (interpolation). In addition, if calibration samples are representative for the population, incorporation of distributional assumptions gives the best average results (averaged over the samples analyzed).

Outlier detection is also possible in the methods presented here; Naes and Martens (1987) can be referred to for some results.

IV. PREDICTION ABILITY AND OUTLIER DETECTION

A. Evaluating Prediction Ability

The equation described in Section IIA can be applied to the samples in the calibration set as well as to new samples. By comparing the near-infrared concentrations with chemically determined values, the success of the calibration can be assessed (validation). Different types of analyses may be envisioned. Presented below are a few with special importance in practice.

Let the chemical lack-of-fit residual, $\hat{f}_{ij}$, be defined by:

$$\hat{f}_{ij} = c_{ij} - \hat{c}_{ij} \tag{69}$$

and let for a set of I samples:

$$\hat{f}_j = (c_{1j} - \hat{c}_{1j}, \ldots, c_{Ij} - \hat{c}_{Ij})' = c_j - \hat{c}_j , \tag{70}$$

where $\hat{f}_j$, $\hat{c}_j$, and c_j are I-dimensional column vectors.

The goal of a calibration is to obtain calibration coefficients that determine concentrations in future samples so that the residuals $\hat{f}_{ij}$ are as close to 0 as possible. The success can be assessed statistically by absolute values (in the same unit as the concentration itself) or by relative values (i.e., relative to the total variation in the data).

ABSOLUTE VALUES

External Validation. The most straightforward assessment of a prediction (validation) is obtained by testing on a set of new samples. The root mean square of prediction (RMSP) for each chemical constituent is defined by:

$$RMSP_j = (\frac{1}{I_p} \hat{f}_j'\hat{f}_j)^{1/2} . \quad (71)$$

This measures the total analytical error, since it includes bias as well as variation around the mean. Hruschka and Norris (1982) used the corresponding standard error of performance (SEP), where $\hat{f}_j$ is corrected for the analytical bias (see Chap. 3).

Internal Validation. When the predictor is tested on the calibration set itself, the root mean square of estimation (RMSE) is obtained in analogy to equation (71), except with I_p replaced by I_c minus the degrees of freedom used in the calibration. It may sometimes be dangerous to use this as a measure of the real analytical error because it may become much too low if the calibration model is overfitted, i.e., if measurement noise is drawn from the calibration data into the predictor. This condition may be the result of testing too many alternative wavelength combinations or of using too many regression factors in the calibration regression.

A better internal validation criterion is the root mean square of cross-validation (RMSCV) (e.g., Stone, 1974). In cross-validation, one repeats the calibration N (e.g., 4) times, each time treating one N^{th} part of the whole calibration sample set as "unknown" samples instead. In the end, all the calibration samples have been treated as unknown or prediction samples. RMSCV is then obtained in analogy to equation (71). Alternatively, RMSCV can in some cases be computed by a leverage-corrected RMSE (Cook and Weisberg, 1982), which we term RMSEL.

RELATIVE MEASURES

A measure of the success used frequently is:

$$R_j^2 = \frac{STOT_j^2 - RMS_j^2}{STOT_j^2}, \quad (72)$$

where RMS_j^2 represents the square of one of the measures RMSE, RMSP, RMSCV, or RMSEL. The term STOT signifies the total standard deviation of the constituent in the sample set, i.e.:

$$STOT_j = (\frac{1}{I}\sum_{i=1}^{I} (c_{ij} - \bar{c}_j))^{1/2} . \quad (73)$$

The quantity R_j^2 is closely related to the squared correlation between estimated and measured value (Searle, 1971). The relative ability of prediction (analysis) (RAP) (Hildrum et al, 1983) is probably a better measure of the quality of the analysis because it takes into account the level of experimental error in the chemically measured data used as "truth," represented by the chemical noise standard deviation, SCHEM. RAP is defined as:

$$RAP_j = \frac{STOT_j^2 - RMS_j^2}{STOT_j^2 - SCHEM_j^2} . \quad (74)$$

This section has discussed how the quality of analysis can be based on $\hat{f}$. Other aspects are also important when selecting a calibration method—for instance, a simple model is useful for interpretation.

B. The Importance of Outlier Detection

Measurements used for calibration and analysis usually contain random measurement noise. In general, such random noise can be compensated for by increasing the number of calibration samples or the number of wavelengths employed. However, abnormally large errors also sometimes occur that make the analytical error intolerable. Abnormalities in near-infrared data can result from a variety of factors, which are discussed in a subsequent chapter.

Near-infrared spectroscopy is often operated in routine environments. It is therefore important to have methods for automatic detection of such abnormalities or outliers, during both calibration and analysis. During calibration, outlier detection is important in order to obtain optimal calibration coefficients. When carrying out near-infrared analysis, outlier detection is primarily used as error warning.

Various types of data information are available for detecting outliers, in calibration as well as analysis:

1. Near-infrared residual, $\hat{e}_{ik}$: the lack of fit between near-infrared measurement and the calibration model. These residuals can be studied with respect to their general variation level as well as the remaining spectral features.
2. Leverage: the position relative to the rest of the calibration sample set (Cook and Weisberg, 1982).
3. Chemical residual, $\hat{f}_{ij}$: the lack of fit between chemical measurements and near-infrared values.
4. Criteria combining the variables above.

Each of these outlier information types will be discussed below, with main emphasis on their use in the general model framework of Section IIIA. Some of the statistical tests are based on rather heuristic arguments and the statistical properties are not fully explored. In addition, the tests are based on assumptions that are not always exactly fulfilled. Tests are therefore only approximate in many applications, but they give useful warnings for potential problems. The final choice of what to do when a sample or variable has been detected as outlier must rest with the user; the distinction between an erroneous sample and a correct but extreme one is a practical, not a statistical problem.

Certain types of near-infrared outliers might also be detected in the response linearization or the optical correction stages, and grave errors in chemical calibration data may be found by inspecting the input data. This aspect will not be treated here.

For a thorough presentation of outlier detection and influence in linear models based on MLR, the reader may refer to Cook and Weisberg (1982) and Beckman and Cook (1983).

C. Analysis of Near-Infrared Residuals

The multivariate near-infrared instruments usually measure more wavelengths, K, than the number of regression factors, A, required for modeling the true variability of the calibration samples. Therefore, overdetermined equations are obtained, yielding near-infrared residuals:

$$\hat{e}_{ik} = x_{ik} - \bar{x}_k - \sum_{a=1}^{A} \hat{t}_{ia}\hat{p}_{ak} . \quad (75)$$

These residuals $\hat{e}_{ik}$ are available for every calibration sample and every unknown sample, and they can be used in different ways for a variety of purposes. Both during calibration and analysis, they can be tested statistically to identify samples or wavelengths that do not fit the calibration model. The residuals of the outliers can then be studied in more detail to identify the reason for the lack of fit, to discover whether the problem stems from erratic measurement errors or from systematic errors due to some specific interference.

For scanning instruments, the presence of spectral peaks and valleys in the residuals of a sample indicates that systematic information remains in the near-infrared data. Residual features can be detected by plotting $\hat{e}_i$ vs. wavelength, but for routine operation some automatic method is preferable. One method is to study the power spectrum of $\hat{e}_i$ generated by the

Fourier transform (Rabiner and Gold, 1975). Random ("white") noise in $\hat{e}_i$ generates a flat power spectrum, whereas remaining systematic features yield abnormally high values for the "low-frequency" power terms. Fourier analysis can also be used to design optimal digital filters against erratic measurement errors in near-infrared data from scanning instruments. Cross-correlations of the spectrum of an outlier against the corresponding residuals from a data bank of near-infrared spectra may be used to identify the most probable cause of the lack of fit.

The following methods are used in detecting abnormal samples and wavelengths. In the calibration set, which contains I_c samples, the total residual standard deviation $s(\hat{e})$ after A factors is defined as:

$$s(\hat{e}) = \left(\frac{1}{df} \sum_{i=1}^{I_c} \sum_{k=1}^{K} \hat{e}_{ik}^2\right)^{1/2} , \tag{76}$$

where the number of degrees of freedom, df, depends on the calibration method and the number of factors used. For example, in PCR and PLS it may be approximated by:

$$df = I_c K - K - \max(K, I_c) A . \tag{77}$$

(This probably underestimates df somewhat [Martens and Jensen, 1983], but the problem has not been solved theoretically [Mandel, 1971].)

The near-infrared standard deviation of an individual calibration sample is defined as:

$$s(\hat{e}_i) = \left(\frac{I_c}{df} \sum_{k=1}^{K} \hat{e}_{ik}^2\right)^{1/2} , \tag{78}$$

where df is the degrees of freedom used in equation (77). Correspondingly, for an unknown sample, it is defined as:

$$s(\hat{e}_i) = \left(\frac{1}{K-A} \sum_{k=1}^{K} \hat{e}_{ik}^2\right)^{1/2} . \tag{79}$$

To detect abnormally high residuals for a sample, one may test $s(\hat{e}_i)$ against the total calibration residual, i.e., by the quotient:

$$s(\hat{e}_i)/s(\hat{e}) . \tag{80}$$

A large value of this quantity shows that the sample is abnormal in some way. Martens and Jensen (1983) used the method successfully for detecting erroneous wheat and barley samples. For calibration samples, the quantities $s(\hat{e}_i)$ and $s(\hat{e})$ are not strictly independent. Another test for a sample is obtained by replacing $s(\hat{e})$ with the total residual calculated without sample i, i.e.:

$$s(\hat{e}_{-i}) = \left(\frac{1}{(I_c - 1)} (s(\hat{e})^2 I_c - s(\hat{e}_i)^2)\right)^{1/2} . \tag{81}$$

The distribution of this criterion is unknown. However, under certain assumptions, the distribution of the square of the quotient $s(\hat{e}_i)/s(\hat{e}_{-i})$ can be approximated by a Fisher distribution with the degrees of freedom corresponding to $s(\hat{e}_i)$ and $s(\hat{e}_{-i})$. This fact can be used in practice to give a good indication of when a sample is considered to be abnormal.

In the calibration samples, the residual standard deviations for the different wavelengths are defined by:

$$s(\hat{e}_k) = \left(\frac{K}{df} \sum_{i=1}^{I_c} \hat{e}_{ik}^2\right)^{1/2} , \tag{82}$$

where df is the degrees of freedom used in equation (77). This can also be compared with $s(\hat{e})$ or $s(\hat{e}_{-k})$ (obtained in analogy to equation [81]) as above to reveal wavelengths with abnormally high noise.

Errors in individual data points, i.e., abnormal $\hat{e}_{ik}$, can be detected by testing against $s(\hat{e})$, $s(\hat{e}_{-i})$, or $s(\hat{e}_{-k})$. An F-distribution approximation can also be useful in this case.

For calibration samples, Studentized near-infrared residuals may be used instead of the obtained $\hat{e}$'s in the outlier tests. Studentized residuals are the $\hat{e}$'s divided by estimates of their standard deviations (involving leverages) to get variables with equal variance (Cook and Weisberg, 1982).

For procedures such as PCR and PLS, the data for each wavelength should generally have the same noise level to yield optimal prediction ability. Abnormally high $s(\hat{e}_k)$ indicates a need for a weighting of the near-infrared data prior to repeated calibration. If the different noise levels are unknown, the estimated $s(\hat{e}_k)$'s can be used. The weighting may be as follows:

$$x_{ik}^{(new)} = x_{ik}^{(old)} \, v_k , \tag{83}$$

where v_k is the weight corresponding to the inverse of the estimated noise level of wavelength k, e.g., $v_k = 1/s(\hat{e}_k)$. The level of prior knowledge about the information content of the different wavelengths may also be drawn into the weights v_k.

D. Leverage: Position Relative to Rest of Calibration Sample Set

Leverage of an observation is a concept developed in ordinary linear regression theory that concerns the position of the independent variables of the observation relative to the independent variables of other observations (e.g., Cook and Weisberg, 1982). Modifications of the original leverage definition are treated here as applied to the general data model of Section IIIA.

The leverage of sample i, h_i, is defined as:

$$h_i = \hat{t}_i(\hat{T}'\hat{T})^{-1}\hat{t}_i' , \tag{84}$$

where $\hat{t}_i = \{\hat{t}_{ia}, a = 1, ..., A\}$ is the row vector of regression factors for sample i, whereas $\hat{T}$ is the regression factor matrix for the whole calibration sample set (dim I × A). For calibration methods yielding orthogonal regression factors, this simplifies to:

$$h_i = \sum_{a=1}^{A} (\hat{t}_{ia}^2 / \hat{t}_a'\hat{t}_a) , \tag{85}$$

where $\hat{t}_{ia}$ is the regression factor of the sample i for factor a and $\hat{t}_a$ is the ath score vector of all the calibration samples (I-dimensional column).

The leverage of a calibration sample describes its "uniqueness" or its potential contribution to the estimated calibration solution. It is an important tool in outlier detection that complements the residuals from the near-infrared and chemical data. It reveals calibration samples that may have had very high importance for the estimation of the calibration coefficients b. It is known that $0 \leq h_i \leq 1 - 1/I$ because X is centered (Cook and Weisberg, 1982).

A leverage of zero for a calibration sample indicates that the sample has very little importance for the calibration solution. A leverage near 1 indicates a very high importance; this sample has "grabbed" a regression factor alone, which is dangerous if the sample contains much noise in its X or c data. The average value of h_i is approximated by A/I, and a warning may be given for, e.g., $h_i > 2A/I$.

The leverage can also be computed for unknown samples by using the same $\hat{T}$ in equation (84), but by using the $\hat{t}_i$'s for these samples as well. In that case, it shows how the near-infrared data of the unknown samples are positioned relative to those of the calibration samples. This h_i is useful for detecting extreme samples, those that are outside the normal concentration range. For unknown samples, a high leverage indicates a rather extreme, possibly erroneous set of concentration data.

The uniqueness of the different near-infrared wavelengths may be described by their leverages h_k defined by:

$$h_k = \hat{p}_k'(\hat{P}\hat{P}')^{-1}\hat{p}_k ,$$

where $\hat{p}_k$ is the column vector of near-infrared loadings corresponding to variable k. For methods with orthogonal P columns, h_k can be computed by:

$$h_k = \sum_{a=1}^{A} (\hat{p}_{ak}^2 / \hat{p}_a\hat{p}_a') , \tag{86}$$

where $\hat{p}_a$ is the row vector of $\hat{P}$ corresponding to regression factor a. For the PLS method with orthogonal scores, $\hat{w}$ must be used instead of $\hat{p}$.

A leverage of zero indicates that the wavelength also has had very little importance on the obtained calibration solution and may be eliminated as irrelevant. The average value of h_k is A/K; a warning for irrelevance may be given by, e.g., $h_k < 0.2A/K$; for abnormality by, e.g., $h_k > 2A/K$.

E. Analysis of the Chemical Residuals

In analogy to the near-infrared residuals, $\hat{e}_{ik}$, the chemical residuals $\hat{f}_{ij}$ reveal certain types of outliers, such as samples with errors in the chemical data c_{ij} or samples that do not fit the chemical/near-infrared relationship of the other calibration samples for other reasons, and:

$$\hat{f}_{ij} = c_{ij} - \hat{c}_{ij} = c_{ij} - \bar{c}_j - \sum_{a=1}^{A} \hat{t}_{ia}\,\hat{q}_{aj} . \tag{87}$$

A sample's $\hat{f}$-value for a certain constituent can be tested statistically to detect outliers, such as those resulting from errors in its chemical data c_{ij}. To detect abnormal samples the individual $\hat{f}_{ij}$ values can be divided by a lack-of-fit measure like RMSP, RMSE, or RMSCV and squared. Such quantities can be used as test criteria in approximate F tests (see Section IVC).

If the $\hat{f}_{ij}$ values are "leverage-corrected" (similar to Studentized; see Section IVC) before the calculation of RMSE, then this corrected RMSE can also be used. Cook and Weisberg (1982) give such a correction for residuals in conventional MLR. For methods using an explicit calibration regression stage, an analogue correction appears to be directly applicable for the chemical calibration residuals:

$$\hat{f}_{ij}^{(\text{corrected})} = \hat{f}_{ij}/(1 - h_i) . \tag{88}$$

The leverage-corrected RMSE (RMSEL) for the calibration sample set is defined by:

$$\text{RMSEL}_j^2 = (\frac{1}{I_c} \sum_{i=1}^{I_c} (\hat{f}_{ij}/(1 - h_i)))^2. \tag{89}$$

In unknown samples without chemical data c_{ij} available, the residual $\hat{f}_{ij}$ is undefined. But in certain cases when all the major chemical constituents in the sample have been calibrated for and determined, the fact that they should add up to 100% may be used for control. Let:

$$\text{DIFF}_i = 100\% - \sum_{j=1}^{J} \hat{c}_{ij} . \tag{90}$$

This can be tested against the expected mean and standard deviation of DIFF_i, estimated from the calibration sample set. The method is tested successfully in Hildrum et al (1984).

F. Combined Criteria

Expressions combining leverage and residuals may measure the influence (Cook and Weisberg, 1982) of an observation on the estimation and have the ability of detecting several different types of abnormalities at the same time. In ordinary regression, there are several such criteria, e.g., DFFITS and the Cook statistic, and analogues for the situation are immediately derived using equations (85) or (86).

A new approach for combining criteria is proposed. The criterion is designed to differentiate between influential calibration samples that are good, neutral, or bad for their analytical ability. In calibration, the leverage and the residuals may be used for detecting extreme samples. It is often difficult to determine whether an extreme sample is good or bad for the final analytical ability. A new criterion is proposed that weighs the leverage contribution from each regression factor by the positive or negative change in the analytical ability caused by that factor.

For calibration methods with orthogonal regression factors, the following criterion is proposed. For sample i, the definition is:

$$g_i = \sum_{a=1}^{A} \underbrace{\frac{\hat{t}_{ia}^2}{\hat{t}_a'\hat{t}_a}}_{\text{leverage contribution}} \cdot \underbrace{\frac{(\text{RMSP}^2(a-1) - \text{RMSP}^2(a))\cdot A}{\text{STOT}^a - \text{RMSP}^a(A)}}_{\text{weighting by relative prediction}} . \tag{91}$$

(RMSCV or RMSEL may be used instead of RMSP.) The expression in parentheses after RMSP indicates the number of regression factors incorporated. The criterion g_i is here termed *leverage relevance*, and the difference $\text{RMSP}^2(a-1) - \text{RMSP}^2(a)$ gives positive or negative weighting for the leverage contribution of each factor dependent on whether the factor has good or bad influence on the analytical ability.

$$\frac{(\text{RMSP}^2(a-1) - \text{RMSP}^2(a))\cdot A}{\text{STOT}^2 - \text{RMSP}^2(A)} \tag{92}$$

has an average value of 1. The "average" calibration sample thus has an analytical leverage g_i of A/I_c (same as h_i); "good" ones have $g_i > A/I_c$, whereas irrelevant or "bad" samples have $g_i < A/I_c$ (g_i may even be negative for exceptionally damaging samples!).

The leverage-relevance criterion has not yet been studied in detail. It has been found useful, however, in warning against possible detrimental calibration samples (see Section VI). Also, $(g_i - h_i)$ compared with the ordinary leverage has been found useful, as will be shown in Section VIF.

Further extensions of the leverage-relevance criterion for calibration samples, in analogy to DFFITS, can be envisioned, e.g., combining the g_i criterion with both chemical and near-infrared residuals in some Studentized form.

V. DATA PRETREATMENT

The transformations of the diffuse spectroscopy measurements should ideally pass through two stages, response linearization and optical correction, before being used in linear calibration modeling as treated in Section III (see IIIB). Other types of data pretreatment can be found in Young (1981).

Spectroscopic notation (unindexed capital letters) will be used in the rest of Section V.

A. Response Linearization

The near-infrared measurements from diffuse reflectance or transmission spectroscopy must pass through a linearizing stage to be suitable for the linear analytical equation:

$$\hat{c}_{ij} = \hat{b}_{0j} + x_i \hat{b}_j \tag{93}$$

over a wide enough concentration range. Different linearization functions are available. For reflectance data, R, we have, e.g.:

$$\text{Apparent absorbance: } A = \log(1/R) \tag{94}$$

and

$$\text{Kubelka-Munk transform: } K/S = \frac{(1-R)^2}{2R}. \tag{95}$$

These are discussed by Hruschka in the previous chapter. They can be extended in different ways, such as by replacing the measured reflectance R by a reflectance corrected for internal and external specular reflection, R*, according to the Saunderson theory (Saunderson, 1942):

$$R^* = \frac{R - k_1}{1 - k_1 - k_2 \cdot R}, \tag{96}$$

where k_1 and k_2 are constants defining specular reflection and internal reflection loss, respectively.

B. Multiplicative Scatter Correction

With correct response linearization, all the chemical interferences can be expected to be well modeled by the methods in Section III. Optical interferences may require different linearizations than the chemical ones. In fact, for data linearized according to the Kubelka-Munk theory for chemical interferences, the huge effect of light scatter should affect the linearized data in a strictly multiplicative way that is impossible to compensate for in linear models as treated in this chapter.

H. Martens et al (1983a) developed a multiplicative scatter correction (MSC) to eliminate this optical interference:

$$x_{ik}^{(new)} = (x_{ik}^{(old)} - \hat{u}_i)/\hat{v}_i, \tag{97}$$

where $x_{ik}^{(old)}$ is the near-infrared data after some response linearization, $\hat{u}_i$ is the estimated effect of specular reflection, and $(1/v_i)$ is the estimated scatter interference in sample i. The constants $\hat{u}_i$ and $\hat{v}_i$ were estimated from simple regressions of each spectrum x_i against the average near-infrared spectrum of the calibration sample set, $\bar{x}$ over the K wavelengths or a selected subset of them, using the model:

$$x_{ik} = u_i + v_i \bar{x}_k + e_{ik}. \tag{98}$$

This simple scatter correction was found very useful for samples with large scatter variations (wheat fractions; meat [Geladi et al, 1985]). See Section VII for applications of the method.

VI. ILLUSTRATION BY ARTIFICIAL DATA

A. Artificial Input Data

Data for 10 artificial calibration samples (1–10) and 5 test samples (a–e) were constructed to illustrate some of the phenomena encountered in near-infrared reflectance analysis of wheat flour. For simplicity, the hypothetical wheat flour was assumed to consist of protein (j = 1), starch (j = 2), and water (j = 3) so that the sum of these three constituents always is 100%. Table II shows the artificial constituent concentration data chosen, together with a comment on what each particular sample is meant to represent.

The near-infrared data were assumed to be recorded in some absorbance unit (e.g., log 1/R). To construct the spectra of the artificial samples, each sample was regarded as a linear combination of the spectra of the pure constituents (Beer's model; see Appendix B):

TABLE II
Percentage of Chemical Data in Artificial Samples[a]

Sample	Protein (j = 1)	Starch (j = 2)	Water (j = 3)	Sum	Comment
		Calibration set			
i = 1	0	80	20	100	
2	30	50	20	100	Error-free samples with constant water content
3	50	30	20	100	
4	80	0	20	100	
5	17 (−3)	50 (−10)	20	87 (−13)	Random measurement errors in chemical data
6	50 (+10)	38 (−2)	20	108 (+8)	
7	60	20	20	100	Random noise added to near-infrared data; error-free chemical data
8	50	50	0	100	
9	0	50	50	100	Error-free samples with abnormal water content
10	50	30	20	100	Abnormal light-scatter level added to near-infrared data; error-free chemical data
		Prediction set			
i = a	70	10	20	100	Error-free sample at normal water content
b	45 (+5)	44 (+4)	20	109 (+9)	Random measurement errors in chemical data
c	30	30	40	100	Error-free sample with abnormal water content
d	30	50	20	100	Random noise added to near-infrared data; error-free chemical data
e	60	20	20	100	Abnormal light-scatter level added to near-infrared data; error-free chemical data

[a] Concentrations of three chemical constituents in calibration and prediction data. Simulated chemical errors added to obtain data are given in parentheses. Simulated measurement errors added to near-infrared spectra of samples 7 and d were random, normally distributed noise with zero expectation and standard deviation of 5% of theoretical reading. Abnormal light scatter was introduced in samples 10 and e by the following, somewhat arbitrarily chosen formula:

$$x_{ik}^{(new)} = -\log(v_i^{-1} \cdot 10^{-x_{ik}}),$$

with $v_i^{-1} = 1.5$ and 0.8 for the two samples, respectively. (See Section V for further discussion on scatter.)

$$x_{ik} = \sum_{j=1}^{3} c_{ij}k_{jk} + e_{ik} \quad (99)$$

with $i = 1, 2, ..., 15$ and $k = 1, 2, ..., 25$. The symbol e_{ik} is the noise added to sample i at wavelength k. Figure 10 illustrates equation (99) for one of the samples ($i = 7$).

B. Graphical Study of the Input Data

Figure 12 shows the input near-infrared data for calibration samples 1–10 and unknown samples a–e plotted for two wavelengths expected to contain much information about protein and starch (23 = high for protein, low for starch; 17 = low for protein, high for starch). The positions of the three pure constituents are also given as the corners of the triangle inside which all error-free samples should theoretically fall. The figure shows that calibration samples 1–6, a, and b fall along a single straight line (b and 6 overlap). Water amounts to a constant 20% in these samples, so protein and starch add up to 80%.

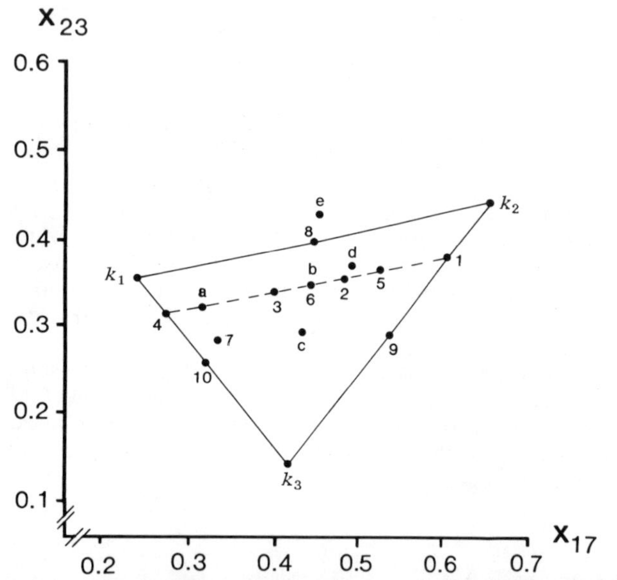

Fig. 12. Input "near-infrared" data X of artificial calibration and prediction samples plotted for wavelength 23 vs. 17. Individual samples are marked according to Table II. Triangle corners represent spectra of the three pure hypothetical constituents protein (k_1), starch (k_2), and water (k_3).

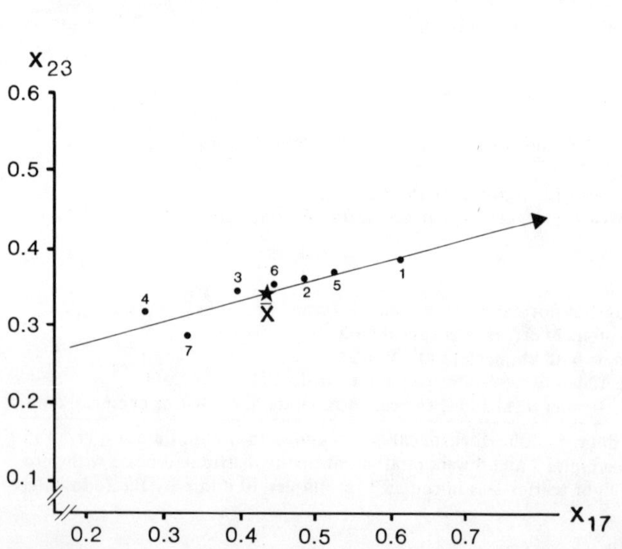

Fig. 13. Calibration samples 1–7 (illustrated as in Fig. 12) together with their mean, $\bar{x}$, and first principal component.

These samples therefore only contain one type of chemical variability (protein vs. starch) and thus form a straight line. Ideally, samples 7 and d should also fall on the line, but because of random measurement noise in the near-infrared data they lie off the line. Samples 8, 9, and c have other water contents, and samples 10 and e have abnormal light scatter.

C. Effect of Using Insufficient Range of Calibration Samples

Suppose for a moment that only the first seven samples were available for calibration; samples 8, 9, and 10 were unavailable. If nothing were known about the chemical data of the different samples, some conclusions could still be drawn from the near-infrared data. The straight line for samples 1–7 in Fig. 12 indicates that the data at the 25 wavelengths can be compressed into a few factors suitable for calibration regression. Principal component analysis (PCA) will be used for this purpose.

A PCA of samples 1–7 yielded the following results: Fig. 13 repeats the positions of the seven samples from Fig. 12. In addition, $\bar{x}$ (the average of these seven samples) is given, together with the first principal component. Because of the errors in sample 7, the estimated component line does not exactly follow the theoretically correct path passing through 1, 5, 2, 6, 3, and 4, but it is very close. Figure 14 shows the full loading spectra of the mean spectrum $\bar{x}$ and the first two eigenvectors. The mean spectrum contains peaks from both protein, starch, and water, whereas the spectrum of the first factor represents a difference spectrum between protein and starch. The second factor is strongly dominated by random noise. In this case, it is caused by noise in sample 7 (Fig. 10) since all the other six calibration samples used were made to be error free.

The five unknown samples were then submitted to analysis based on the one-factor PCR solution obtained above. The near-infrared protein values $\hat{c}$ in Fig. 15 are plotted against the "measured" protein values, c. The figure shows that sample a was analyzed accurately ($\hat{c} = 70.2\%$, i.e., $\hat{f} = 0.2\%$). This was expected, since it belongs to the same type as calibration samples 1–7 (20% water). Sample d was also accurately

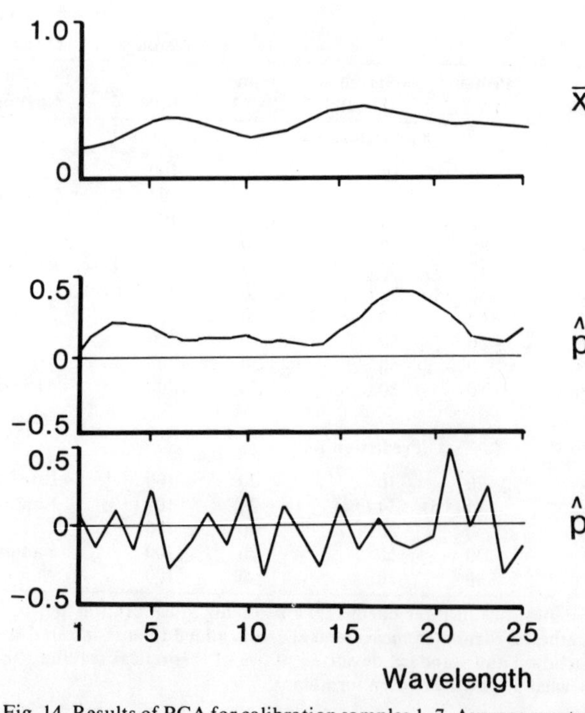

Fig. 14. Results of PCA for calibration samples 1–7. Average spectrum $\bar{x}$ and loading spectra of first two principal components are plotted against wavelength.

analyzed ($\hat{f} = 0.4\%$); the random noise in its near-infrared data did not affect the protein determination much because it was averaged over all 25 wavelengths. Sample b shows a residual of $\hat{f} = 4.5\%$ that is similar to the 5% error actually present in its chemical protein data c. This exemplifies a sample whose near-infrared protein $\hat{c}$ was much better than its chemically measured reference data c. For these samples, the analytical data were acceptable. But the samples with abnormal water content and abnormal light scatter (c and e) yielded wildly erroneous protein results ($\hat{f} = 15\%$ and 25%) because their types of variability were not included in the calibration sample set.

D. Using a Complete Calibration Data Set

Samples c and e were not accurately analyzed when using only samples 1–7 for calibration. Accuracy can be improved by including more types of samples in the calibration set. A PCR calibration using all 10 calibration samples was performed. Four components were required to give full modeling of the near-infrared calibration data. Figure 16 shows the near-infrared loading spectra for the first four principal components. Vectors 1 and 2 show smooth features and correspond to linear combinations of the protein, starch, and water. Vectors 3 and 4 are both affected by the random noise in sample 7.

The root mean square of performance (RMSP) for protein for samples a–e decreased to a minimum after four factors; that number thus appeared to be optimal for PCR with these data. Figure 17 shows the individual near-infrared results for protein with this four-factor PCR solution. Samples a and b were accurately analyzed, similar to the incomplete data set. Sample c with abnormal water content was also accurately analyzed, and so was the sample with abnormal scatter, e. Sample d, which had large random near-infrared noise, was now less accurately analyzed by the present four-factor solution than in the one-factor solution above. This was not unexpected, since there are more chances for four regression factors to find an incidental correlation to its near-infrared noise than for the one-factor solution used in Section VIC.

E. Partial Least-Squares Regression

The same 10 calibration samples were submitted to partial least-squares (PLS) calibration for protein (the PLS method with orthogonal regression factors was used). As with PCR, the PLS data compression required four regression factors to describe the near-infrared calibration data completely. The near-infrared loading weights, $\hat{w}$, are shown in Fig. 18. The first two factors are somewhat similar to those for PCA (Fig. 16), except for having opposite signs. But in contrast to PCR, the PLS displaces the random noise contributions of sample 7 to the last of these four factors, illustrating how PLS regression concentrates the near-infrared variation that has relevance for the protein modeling into the first factors. It leaves the less relevant variation behind in the last factors. PCR, on the other

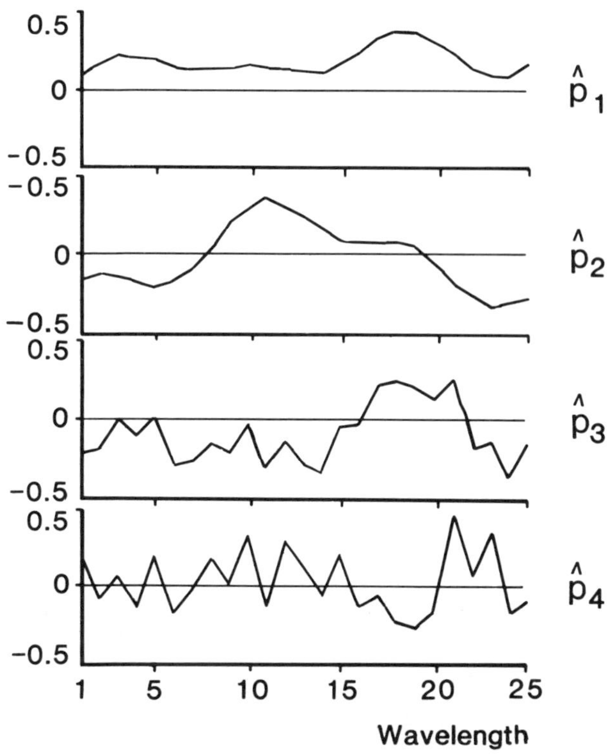

Fig. 16. Result of PCA for all 10 calibration samples. Loading spectra of first four principal components plotted against wavelength.

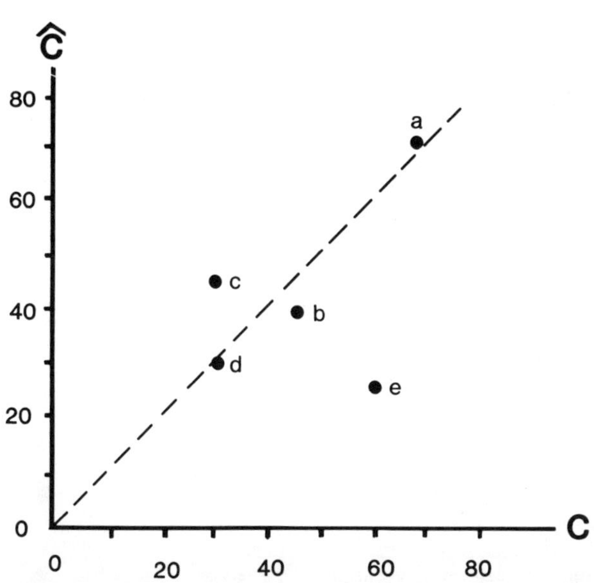

Fig. 15. Prediction of protein in samples a–e by one-factor PCR solution from calibration samples 1–7. Dotted diagonal represents ideal prediction line.

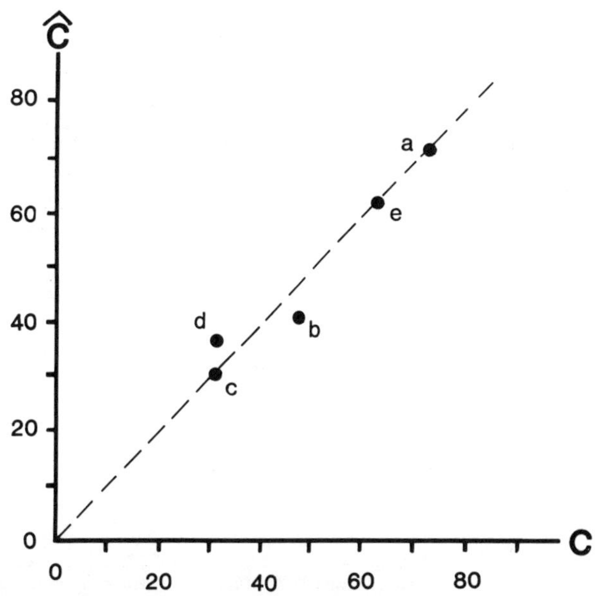

Fig. 17. Prediction of protein in samples a–e by four-factor PCR solution for calibration samples 1–10. Dotted diagonal represents ideal prediction line.

hand, takes out near-infrared factors with respect to their "near-infrared" modeling ability in the data compression stage, irrespective of their protein relevance.

The significance of this difference between PCR and PLS is shown in Fig. 19, which compares the RMSP for protein of the methods PCR and PLS on the same 10 calibration samples. The two methods show identical RMSP after zero factors (only the mean subtracted) and after four factors (full description of all near-infrared calibration data). But whereas PCR required four factors to give optimal RMSP, the PLS had optimal analytical power after only three, and this RMSP is lower than that of PCR.

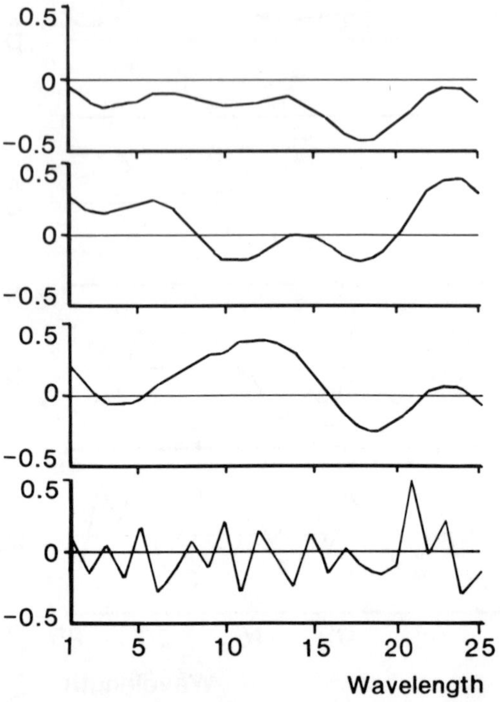

Fig. 18. Results of PLS regression (with orthogonal regression factors) for all 10 calibration samples. "Near-infrared" loading weights of first four PLS factors are plotted against wavelength.

F. Outliers

Table III shows some outlier detection results obtained by the PLS method described above, using samples 1–10 for calibration and samples a–e for determination of protein. Only the samples that yielded significant outlier detector values are given, and results for the obviously inadequate solutions (zero, one, or two factors) are ignored. For comparison, the average detector values of the calibration sample set are given.

CALIBRATION DATA SET (SAMPLES 1–10)

Calibration sample 7 (with strong random near-infrared

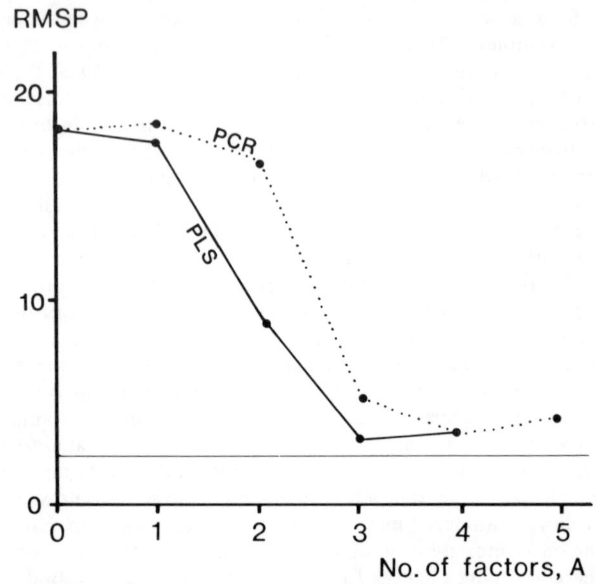

Fig. 19. Prediction error, RMSP, for protein predicted by PCR and PLS regression, plotted against complexity of calibration model (A). Both curves were calculated over prediction samples a–e, using calibration coefficients estimated using samples 1–10. Horizontal line represents theoretical lower-prediction error limit caused by protein error of 5% in sample b, resulting in a purely chemical RMSP of 2.2%. Minimal RMSP of PCR was 3.7%, whereas corresponding value for PLS regression was 3.3%. Subtracting purely chemical RMSP contribution, their "true" prediction errors came out as 1.5% and 1.1%, respectively.

TABLE III
Detection of Bad Outliers Illustrated by PLS Regression for Protein in Artificial Data Set After Three and Four Factors

	Outlier Detectors[a]							
	Three Factors				Four Factors			
	$s(\hat{e}_i)$	h_i	$g_i - h_i$	$\hat{f}_i$	$s(\hat{e}_i)$	h_i	$g_i - h_i$	$\hat{f}_i$
	Calibration set (samples 1–10)							
Average	1.1[b]	.3	0	.4[b]	0[b]	.4	0	1.1[b]
No. 5	.02	.1	−.1	−3.5	0	.1	−.1	−3.5
6	.5	.0	.0	9.6	0	.0	−.0	9.0
7	2.8	.2	−.1	−3.8	0	.9	−.8	0
	Prediction set (samples a–e)							
No. b	.4	.0		4.6	0	.0		4.0
d	2.4	.1		−5.5	2.4	.1		−6.2
e	1.5	2.2		− .4	0	2.5		−3.0

[a] "Near-infrared" residual $s(\hat{e}_i)$ defined by equation (79); leverage h_i defined by equation (84); leverage relevance g_i defined by equation (91); and protein residual $\hat{f}_i$ defined by equation (87).

[b] Root mean square of other seven calibration samples (1–4 and 8–10). The most extreme values are underlined.

noise) was detected as an outlier after three factors, both by its high near-infrared residual $s(\hat{e}_i)$ and its large protein residual $\hat{f}_i$.

After four factors (the overfitted solution), all the near-infrared residuals were approximately zero, even for sample 7, in spite of its near-infrared noise. Sample 7 also showed zero protein residual. Instead, the sample was then detected as an outlier by its high leverage h_i and as a bad outlier by its low leverage relevance g_i (shown as $g_i - h_i$). Its abnormally high leverage [$h_i = 0.900$, which is the maximum obtainable $(1 - 1/I)$] proved that it had grabbed the fourth PLS regression factor for modeling its own irrelevant noise, thereby contaminating the calibration coefficients. This overfitting effect was apparent in the erratic loading spectrum of factor 4 in Fig. 18 and in the increased RMSP at four factors in Fig. 19.

Calibration samples 5 and 6 (with errors of −3% and +10% added to their protein data) were detected as outliers with respect to $\hat{f}_i$ after both three and four factors; sample 6 was particularly conspicuous. Their estimated protein residuals $\hat{f}_i$ correspond closely to their true errors (−3.5 and +9.6% after three factors).

In addition to these apparently bad outliers, the calibration samples 8, 9, and 10 were detected as abnormal by their high leverages h_i. They were considered to be "good" outliers, however, because of their high predicting relevance g_i, which indicated that they were essential in defining regression factors important for analytical ability.

PREDICTION DATA SET (SAMPLES a–e)

Samples b, d, and e were discovered as outliers of different types. Sample b (with 5% error in its protein value) was detected after three and four factors from the large residual $\hat{f}_i$ between its near-infrared and known but erroneous protein values. Sample d (with random near-infrared noise) was detected from its high residual standard deviation $s(\hat{e}_i)$ after three and four factors. It also yielded abnormal near-infrared protein values (because of its random near-infrared noise). Sample e (with abnormal light scatter) was discovered as an outlier since its leverage h_i was exceptional. Some "near-infrared" information remained in this sample after three factors, illustrating the effect of light scatter. The scatter information in calibration sample 10 was not sufficient to model that of sample e because their light-scatter effects did not span exactly the same space. It was sufficient, however, to ensure acceptable protein prediction for e. This again illustrates that high leverage h_i in itself does not indicate an erroneous sample.

RECALIBRATION WITH OUTLIERS ELIMINATED FROM CALIBRATION DATA SET

Samples 6 and 7 were regarded as outliers and eliminated from the calibration data set before recalibration on the remaining eight samples; sample 6 was eliminated because of its extreme chemical residual $\hat{f}_i$ and 7 because of its near-infrared residuals $s(\hat{e}_i)$ and low leverage relevance, g_i.

To compare with the previous results, the detected outliers in the unknown set were not eliminated. The RMSP was now further reduced from 3.23 to 3.18% after three factors. In real situations, samples b, d, and e should have been measured again, together with 6 and 7, with respect to near-infrared and protein data. Then their errors could have been discovered and corrected.

G. Conclusions

In summary, this artificial data example illustrates how data compression can be accomplished by PCR and PLS regression, concentrating the 25 intercorrelated "near-infrared" variables to three or four regression factors without losing information. It also shows similarities and differences between the relatively well-known PCR and the new PLS method for calibration, with the latter giving the simplest solution and best analytical results. The example shows how outliers can be detected from their near-infrared residuals $\hat{e}_i$, their leverage h_i, their leverage relevance g_i, and their chemical residuals $\hat{f}_i$. Finally, it illustrates that improved analytical results can be obtained by eliminating calibration outliers.

VII. RESULTS FOR REAL DATA

A. The Real Data Sets

WHEAT SAMPLES WITH NORMAL VARIATIONS (DATA SET 1)

One hundred representative samples of Canadian hard red spring wheat were analyzed at the U.S. Department of Agriculture (Beltsville, MD) in a Cary scanning spectrophotometer with about 1,000 readings between 1,000 and 2,600 nm. The Kjeldahl-N was measured in 16 replicates for each sample. The samples were split randomly into two representative sample sets of 50 samples each (calibration set "whelgb" and prediction set "whelga"; Hruschka and Norris, 1982). A total of 155 wavelengths between 1,906 and 2,398 nm (every 3.2 nm) was chosen for a comparison of various calibration methods.

WHEAT SAMPLES WITH VERY LARGE VARIATIONS (DATA SET 2)

A single batch of Canadian hard red spring wheat was submitted to abrasive milling, resulting in 10 successive fractions ranging from almost pure hull in large flakes via aleuron-rich fractions to almost pure endosperm, finely ground. Each of these fractions was sieved into four particle-size fractions (a few fractions gave too little material to be measured). The remaining pearled endosperm and the whole-kernel sample were also included in the set after conventional milling in a Udy cyclone mill. The reflectance of the 47 samples (23 calibration, 24 prediction) was measured in a Technicon InfraAlyzer 400 with 19 fixed band-pass filters between 1,445 and 2,348 nm. Six different chemical constituents (water, ash, protein, starch, fiber, and fat) were measured on the 47 fractions in at least two replicates. Some near-infrared calibration results on these data have been reported elsewhere (Martens and Jensen, 1983; H. Martens et al, 1983a).

B. The Effect of Overfitting

A critical aspect of indirect multivariate calibration is the determination of the optimal number of regression factors, A. Using too few factors implies insufficient modeling of the chemical and optical interferences. Using too many factors implies estimating too many statistical parameters with the given calibration data. Both mistakes lead to unnecessarily large analytical errors. This is illustrated for wheat samples with normal quality range (data set 1) in log (1/R) data in the 1,900- to 2,400-nm range.

Calibrating for protein in these 50 samples by unweighted PLS1 regression, using 155 different wavelength readings in this region (every 3.2 nm), yielded the result shown in Fig. 20. The analytical error RMSP decreased sharply with the first four factors. It reached a minimum of 0.16% after seven factors, whereafter it increased again. This example illustrates the importance of selecting the optimal number of regression factors, A.

C. Comparison of Calibration Methods

Different calibration methods were applied to data set 1. Figure 21 compares the prediction error RMSP for PLS (Section IIIF) and with PCR (Section IIIE) and EGLS1 (Section IIIG) for data in log (1/R) units. To emphasize details, the RMSP is given on a logarithmic scale. The physically based

EGLS1 predictor regresses X on c with no distributional assumptions, whereas the two PCR and PLS predictors instead regress c on X and incorporate information about the calibration sample distribution. The figure shows that all three methods gave best results after seven factors for these data. PCR and PLS yielded slightly better data (RMSP = 0.16%) than did EGLS1 (RMSP = 0.19%), indicating an advantage in using population information in obtaining optimal predictors (see Sections IIC and IIIG).

PLS gave virtually the same optimal results as PCR (with minimal RMSP found after seven factors at 0.16%). The PLS analytical error approached minimum more rapidly than the PCR error did, as with the artificial data (Fig. 19).

However, Fig. 21 shows that a penalty of the more thorough use of the chemical data in PLS is a slightly increased sensitivity to overfitting when compared with PCR. The PCR-PLS intermediate method followed the PLS up to 6 factors, but its RMSP increased less than that of PLS and gave a minimum RMSP of 0.15% after 11 factors. It was thus less sensitive to overfitting than PLS, while retaining the advantage of PLS at fewer factors.

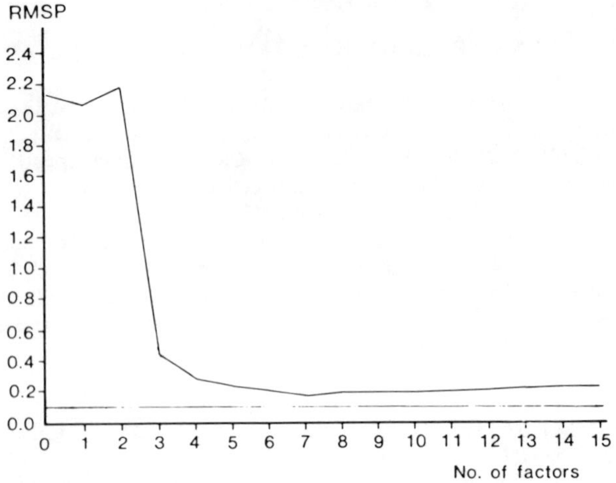

Fig. 20. Effect of overfitting. Prediction error RMSP of protein for real wheat data plotted as function of number of PLS factors, A. Near-infrared data were in log (1/R) unit and protein in percentage of sample weight. Horizontal line represents approximate level of errors in chemical data themselves.

D. Transformations of Near-Infrared Data

WHEAT DATA WITH NORMAL VARIATIONS (DATA SET 1)

Three different response linearizations and three different optical corrections were tested for protein in data set 1. The PLS regression of Section IIIF was used as a calibration method. The conventional log (1/R) linearization without optical correction yielded an optimal analytical error RMSP of 0.16% (Table IV). Paradoxically, the Kubelka-Munk linearization (K/S) gave much worse results (0.27%) in spite of being more founded on a physical modeling of the diffuse reflectance data. With the physical model further elaborated with Saunderson-corrected reflectance, using reasonably realistic parameters (specular reflectance k_1 = 0.01 and internal reflectance k_2 = 0.5), K/S gave still worse results (RMSP = 0.29%). A relatively complicated calibration solution was

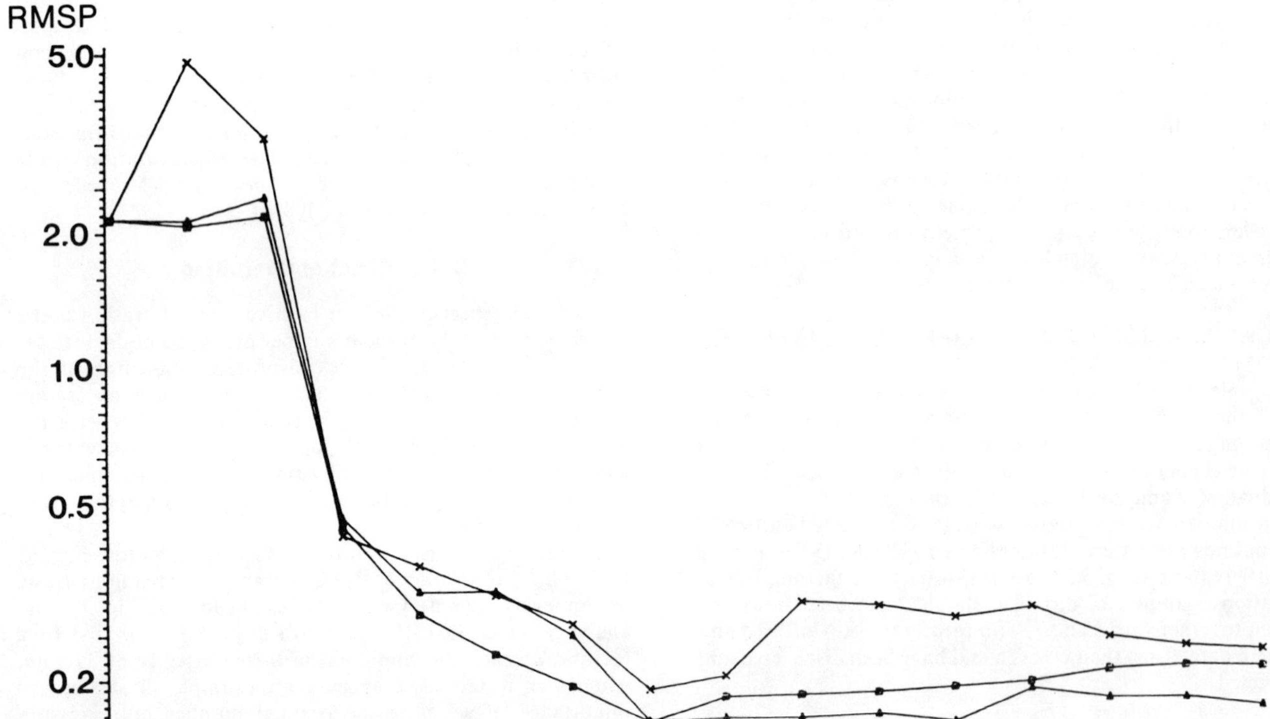

Fig. 21. Wheat protein prediction error, RMSP, plotted logarithmically against number of regression factors for three different full-spectrum calibration methods: EGLS1 (×), PCR (Δ), and PLS (□).

required in each case (five to seven PLS factors).

However, after optical correction the picture changed drastically for all three linearizations. The MSC transformation without constant term improved the optimal analysis from 0.16% to 0.15% for log (1/R) data, from 0.27% to 0.16% for Kubelka-Munk data, and from 0.29% to 0.16% for Saunderson-corrected Kubelka-Munk data. Very good results (RMSP = 0.16%) were obtained after only four factors with MSC. Reasonable data (RMSP = 0.30%) were in some cases obtained after only two factors with MSC, contrasting with RMSP = 2.2% without.

The MSC without offset is analogous to the SDLR method of dividing one derivative term by another. The MSC with constant term added may allow more general applications. For the present data, it allows a simpler modeling with fewer factors.

The strong effect of MSC on Kubelka-Munk data is not surprising, since the K/S linearization attempts to express scatter as a multiplicative effect K · (1/s).

WHEAT DATA WITH VERY LARGE VARIATIONS (DATA SET 2)

Near-infrared reflectance data measured in the fixed-filter instrument at 19 wavelengths were used to study the optimization of calibration on wheat with widely varying chemical composition and particle size. The results (Martens and Jensen, 1983) are summarized for protein in Table V, which shows that for these data (i) PLS gave better analyses than the conventional SMLR calibration method for log (1/R) data; (ii) an optical correction consisting in dividing each log (1/R) spectrum by its sum gave improved prediction; (iii) the MSC with offset term gave even better improvements; and (iv) again K/S and Saunderson-corrected K/S gave worse analytical accuracy than log (1/R) without MSC, but better than log (1/R) with MSC.

Optimizing the response linearization and the optical correction both improved the determination of protein in wheat. Together they brought the error for samples with abnormally large chemical and physical variabilities down toward that of normal wheat samples (data set 1), indicating that the importance of standardizing the sample preparation may be reduced by data pretreatment.

TABLE IV
Prediction Error RMSP for Protein in Percentage of Sample Weight in Data Set 1 Given for Different Response Linearizations, Optical Corrections, and Number of PLS Factors[a]

Response Linearization	Optical Correction	RMSP (%) a = 2	a = 4	Optimum (A)
log (1/R)	···	2.18	.28	.16 (7)
K/S	···	2.20	.33	.27 (7)
(K/S)[b]	···	2.20	.34	.29 (5)
log (1/R)	MSC[c]	.35	.16	.15 (7)
log (1/R)	MSC[d]	.30	.28	.15 (9)
K/S	MSC[c]	.35	.16	.16 (4)
K/S	MSC[d]	.43	.32	.16 (8)
(K/S)[b]	MSC[c]	.35	.17	.16 (6)
(K/S)[b]	MSC[d]	.71	.25	.16 (8)

[a] Total initial standard deviation was 2.39% and standard deviation of chemical Kjeldahl protein data was about 0.1% (Hruschka and Norris, 1982).
[b] Based on Saunderson-corrected reflectance ($k_1 = .01$, $k_2 = .5$).
[c] MSC model: $x_{ik} = u_i + v_i\bar{x}_k + e_{ik}$.
[d] MSC model: $x_{ik} = v_i\bar{x}_k + e_{ik}$.

TABLE V
Prediction Error RMSP for Protein in Percentage of Sample Weight in Wheat Data Set 2 Given for Different Response Linearizations, Optical Corrections, and Linear Calibration Methods[a]

Response Linearization	Optical Correction	RMSP (%)	Calibration Method
Log (1/R)	···	.39	(SMLR)[b]
Log (1/R)		.33	(PLS)[b]
Log (1/R)	Divided by NIR sum	.25	(PLS)[b]
Log (1/R)	MSC[d]	.24	(PLS)[c]
R	···	.24	(PLS)[c]
R	MSC[d]	.23	(PLS)[c]
K/S	···	.47	(PLS)[c]
K/S	MSC[d]	.21	(PLS)[c]
Saunderson and K/S	···	.54	(PLS)[c]
Saunderson and K/S	MSC[d]	.20	(PLS)[c]

[a] Total initial standard deviation was 2.35% and standard deviation of chemical Kjeldahl protein data was about 0.15%.
[b] Reference: Martens and Jensen (1983).
[c] Reference: Martens et al (1983a).
[d] MSC model: $x_{ik} = u_i + v_i\bar{x}_k + e_{ik}$.

INTERPRETATION OF THE KUBELKA-MUNK MSC SOLUTION

Figure 22 shows the mean spectrum $\bar{x}$ and the standard deviation spectrum of the 50 calibration samples in data set 1 for the data in Kubelka-Munk units after MSC in the range of 1.9–2.4 μm. The sharp water peak in the mean spectrum at 1,940 nm and the wide starch peak at 2,100 nm with its protein peak shoulders on both sides are, for example, clearly visible. The water peak region at 1,940 nm showed a larger variability than that of the other wavelength regions. Figure 23 shows the six first near-infrared loading spectra (PLS).

The near-infrared loading weights of factor 1 show that the scores of factor 1 correlate negatively with the water peak in the 1.94 μm region and with the starch peak near 2.10 μm and positively at higher wavelengths. Factor 2 correlates positively with wavelengths just above the water peak at 1.94 μm and negatively with starch at 2.10 μm. Factor 3 correlates negatively with wavelengths just below the water peak and with 2.25 μm, which has tentatively been ascribed to cellulose (K. Norris, pers. comm., 1981), and positively with protein peaks at 2.17 and 2.28 μm. Factor 4 correlates positively with wavelengths just below and negatively with wavelengths just above the water peak and may be a factor that compensates for wavelength shifts in the water peak caused by, e.g., temperature. Factor 5 correlates negatively with wavelengths below the water peak, coinciding with a region of strong interferences from atmospheric water vapor (K. Norris, pers. comm., 1981). The inclusion of factor 5 in the calibration solution gave decreased analytical ability, which indicates the water vapor region to have no relevance. Factor 6 was also nonsignificant, and the reason is apparent in Fig. 23: its loading weights display a very erratic wavelength dependency that may reflect random near-infrared noise and possibly some nonlinearity compensation.

The effect of these factor loading weights on the resulting calibration coefficients $\hat{b}_j$ is shown in Fig. 24. The effect of using an increasing number of factors (A = 2, 4, and 6) is illustrated in Fig. 24a. As the number of regression factors increases, the size of $\hat{b}_{kj}$ increases. Since the calibration constants amplify the random noise e_{ik} in the near-infrared measurements, it is probably preferable to have small coefficients. In addition, the detrimental effects of, e.g., the noise factor (a = 6) is apparent. The optimal calibration coefficients for protein (four factors) are given in detail in Fig. 24b. It shows positive relationship between protein and wavelengths at bands near 2.28, 2.18–2.21, 2.07, 1.98, and 1.94 μm and negative relationship at 2.25, 2.08–2.15, 2.02 μm, and at both ends of the spectral region.

E. Improvements in the Partial Least-Squares Calibration Method

WEIGHTING NEAR-INFRARED DATA

In all the calibrations reported here, the near-infrared

variables were used unweighted. Improved analysis could probably have been obtained by weighting down or eliminating the water vapor region at 1.90–1.93 μm and the apparently noisy region above 2.3 μm. An up-weighting of well-identified protein, starch, and cellulose peaks might have given further improvements.

SMOOTHING

Theoretically, one would expect the loadings of the first factors to be more precisely estimated than those of later factors because of measurement errors. Fig. 23 shows an increasing level of erratic ripple in the loading spectra of the later factors.

In an attempt at reducing this ripple in the estimates, a weighted moving average filter over ±3 neighbor variables (±9.6 nm) was applied to the loading spectra $\hat{w}_a$ inside the PLS algorithm, as described by H. Martens et al (1983b). The smoothing reduced the optimal protein RMSP from 0.164 to 0.160% after four factors.

Increasing the number of wavelength variables is another way of making the calibration more robust against random near-infrared noise. By increasing the number of wavelengths in the 1,904- to 2,400-nm range from 155 to 309, again using Kubelka-Munk linearization with MSC and smoothing over ±3 neighbor variables (±4.8 nm), the RMSP decreased further from the above-mentioned 0.160 to 0.157% after four factors. Alternatively, the individual measured spectra x_i could have been smoothed.

TRUNCATED PLS VERSION

Section IIIF discussed how each near-infrared loading vector

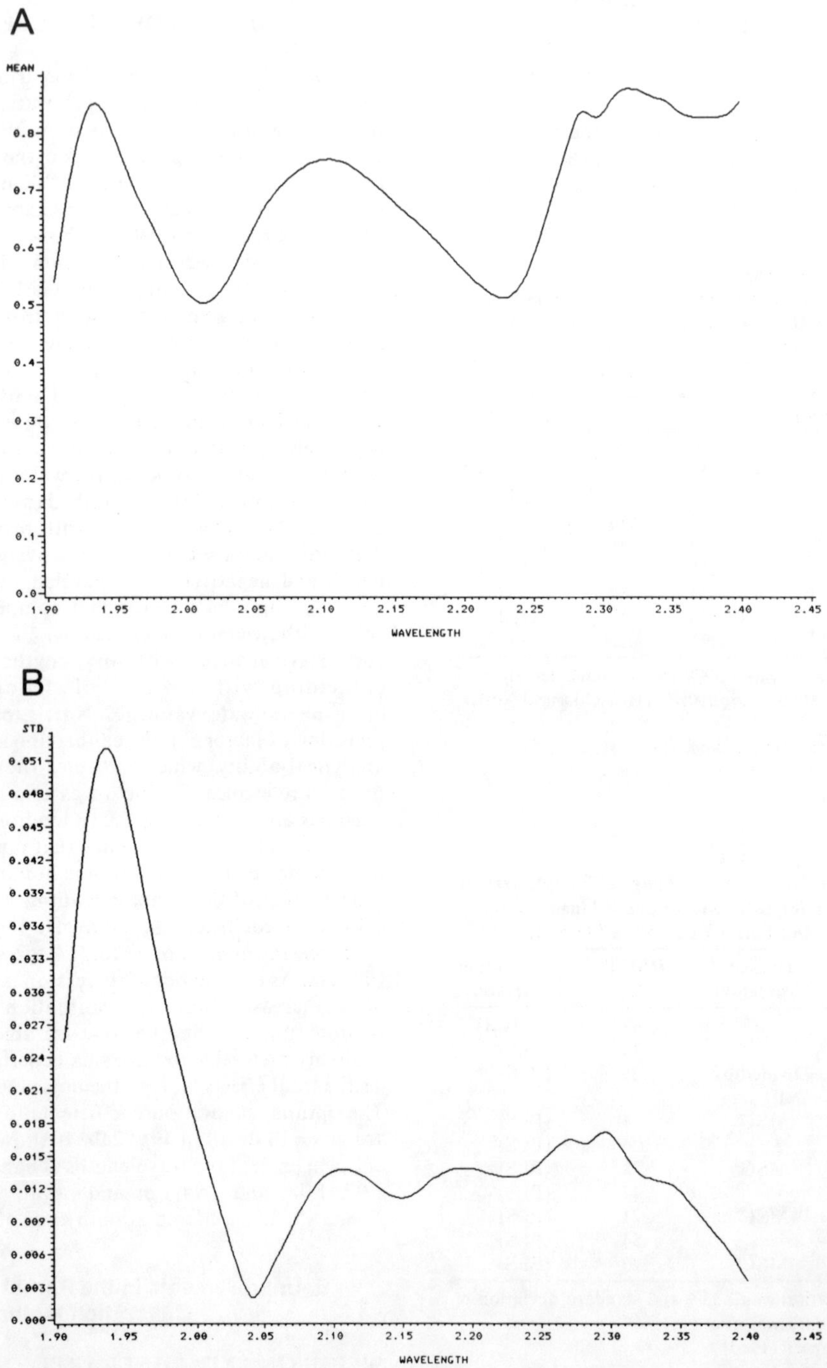

Fig. 22. Mean (A) and standard deviation (B) spectra for 50 calibration samples in data set 1 after Kubelka-Munk and MSC transformation.

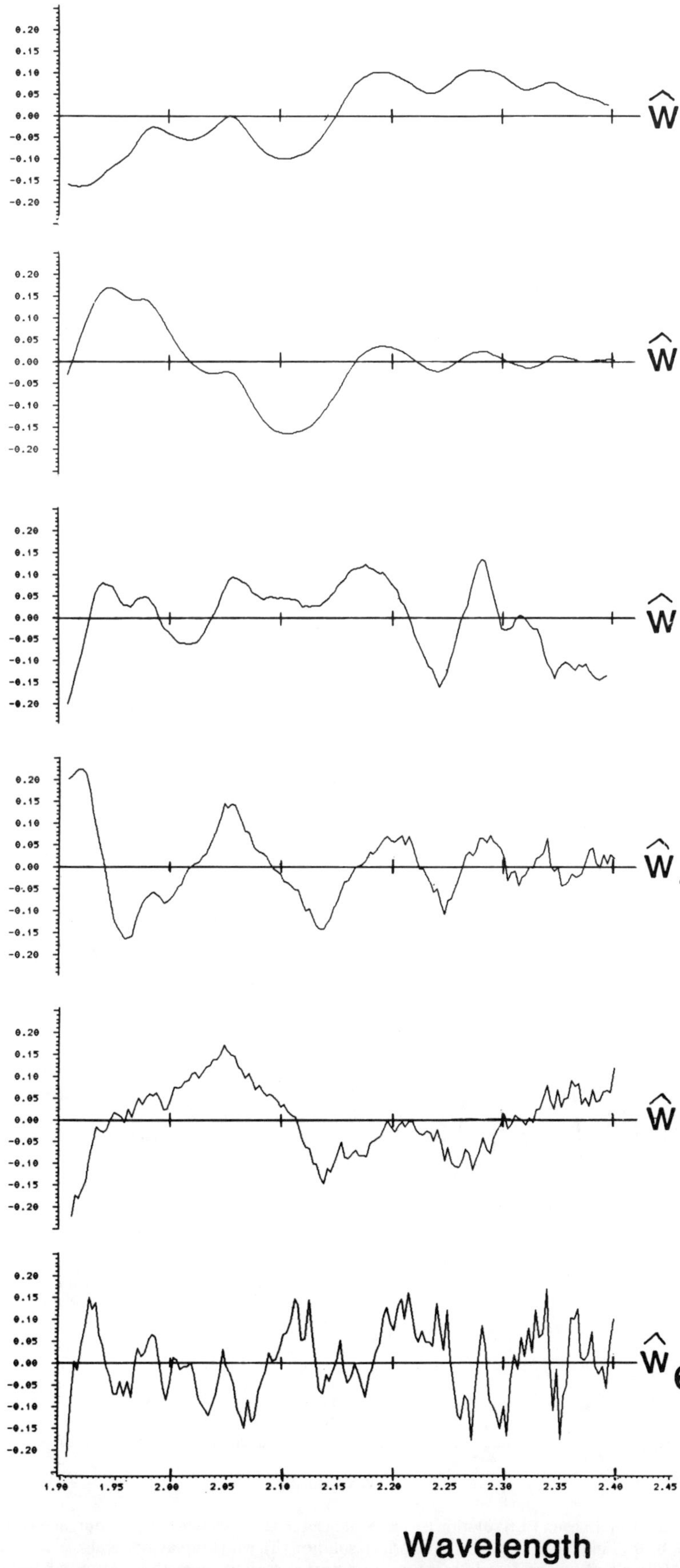

Fig. 23. Six first near-infrared loading weights for protein (PLS) for data set 1. Only four first ones improved prediction ability.

could be truncated by forcing $\hat{w}_{ak}$ of all wavelengths with small initial coefficients to zero. This was tested on the normal-range wheat data set 1 in log (1/R) unit and was found to give a small improvement compared with the full-spectrum PLS solution. The RMSP of the initial $\hat{w}_a$ vector was for simplicity used as the cutoff limit. Similarly, each regression factor vector $\hat{t}_a$ could be truncated by forcing all samples with small initial scores to zero. This was tested on the same data and likewise found to give a minor improvement in RMSP compared with the full sample set method. For truncation, the score $\hat{t}_{ia}$ of all samples was forced to zero, except the one having the largest absolute initial score, and one turn in the PLS algorithm was repeated.

VIII. DISCUSSION

A. Statistical Calibration Methods

Two of the existing calibration methods (SMLR and SDLR) use only a few wavelengths in the final calibration. Thereby the information on all the other wavelengths available or

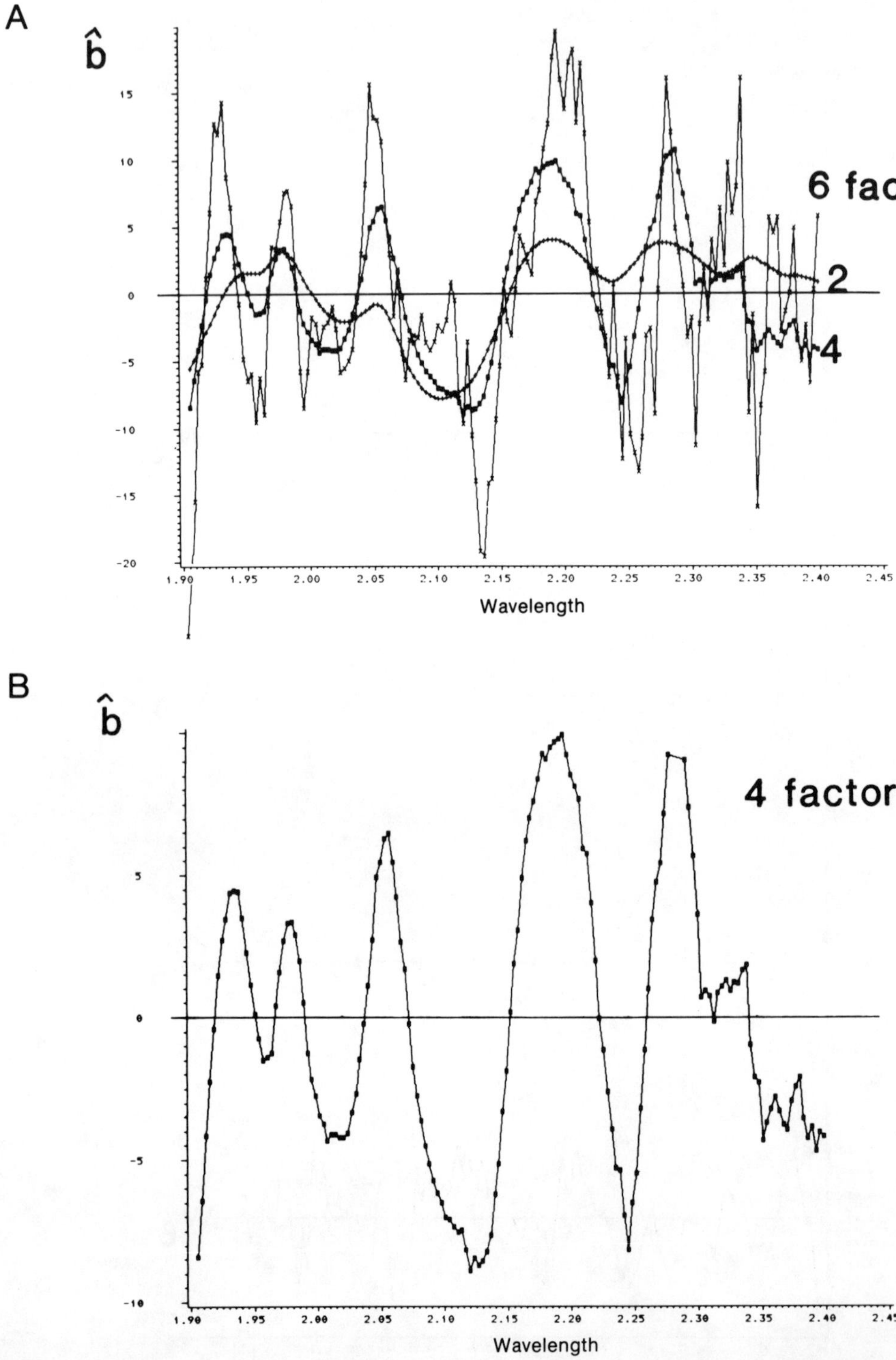

Fig. 24. Calibration coefficients $\hat{b}$ for protein from PLS solution (for K/S-data after MSC) shown for two, four, and six factors (A) corresponding to prediction errors RMSP of 0.35%, 0.16%, and 0.17%, respectively. Optimal solution (B), with four factors, is also plotted separately to show its details. Protein in sample hence is predicted by multiplying near-infrared measurement at each wavelength by corresponding $\hat{b}$ value and summing these products plus an offset; i.e., $\hat{c} = b_0 + x\hat{b}$.

potentially available is discarded in the data compression and left unused when analyzing new samples.

Several new calibration methods that use the whole spectrum x to obtain the predictor are described (full-spectrum methods). All these also rely on a data compression stage to eliminate redundancy and reduce noise in the near-infrared data. The different methods presented can be regarded as approximations to the general MLR method for estimating the calibration coefficients $\hat{b}_0$ and $\hat{b}$ in the prediction equation $\hat{c} = \hat{b}_0 + x\hat{b}$.

Figure 24 illustrates how the $\hat{b}$ spectrum may evolve as the dimensionality of the calibration model increases, from too simple (A = 2) via an optimum complexity (A = 4) to overfitting (A = 6). The different calibration methods will be summarized here, and a subjective evaluation of them given.

The conventional data compression types (SMLR, SDLR) probably require very precise near-infrared data. These methods (as well as the wavelength-truncated version of PLS regression) may be preferable if strong nonlinearities exist in the relationship between some elements of x and c. The least nonlinear wavelengths can then be chosen. By treating the unused near-infrared wavelengths as c variables, outliers can be detected even with these conventional methods.

The Hruschka regression method uses the spectra of selected calibration samples for data compression. The method may be simpler to understand than the subsequent ones and may be advantageous when the number of calibration samples is very small, since few degrees of freedom are used in the determination of the factor loadings, $\hat{P}$. But the subsequent methods that estimate the factor loadings statistically are probably better if the near-infrared data contain more noise.

The Fourier regression method has the advantage of allowing smoothing and data compression in the same step, but it is applicable only for scanning instrument data. The PCR and PLS regressions are similar to each other, but although the former is easier to understand theoretically, the latter is faster and appears more flexible and is at present the method mostly used in practice by the authors. The PLS method can easily be modified in a variety of ways to accommodate different types of restrictions or additional information, as illustrated by the multiconstituent version, the smoothed-loadings version, the truncated versions for ignoring irrelevant near-infrared information, and the PCR-PLS combination.

The extended Beer's model methods (EGLS1, EGLS2, EBLP, and EBLUP) have the advantage of being related to the physical model for the near-infrared signals. They allow the modeling of unknown interference factors as well, which the ordinary multiconstituent curve-resolution method of some UV/visual-range transmission spectrophotometers does not. Although the basic models are statistically well founded, these estimated versions are not fully understood. But they allow a choice between different usage of distributional assumptions that is theoretically appealing.

EGLS1 and EGLS2 may allow extrapolation outside the constituent concentration range of the calibration sample set. EGLS1 and EBLUP likewise seem to allow extrapolation with respect to the level of the unknown interference factors modeled during calibration. The EBLP method is expected to behave similarly to the PLS method when the error in the chemical calibration data is low.

In general, there is a great need for closer comparison of the different statistical calibration methods, both from a theoretical and a practical angle. However, for near-infrared data with very low levels of random noise, one should expect the methods to behave similarly. The apparent prediction error will often be dominated by the noise in the chemical data.

B. Factors Affecting Choice of Method

The actual choice of a statistical calibration method and its dimensionality should depend on many aspects for a particular instrument and a particular product type. Here are some important aspects:

1. Complexity of the calibration problem: number of different types of compositional variation that interfere with the near-infrared measurements in the samples (e.g., wheat protein, water, and starch constitute two interferences); degree of overlap between spectra of the different constituents and interference (protein, starch, water, etc.); degree of chemical interaction between constituents, and other phenomena modifying their spectra (e.g., water modified by protein, salt, and temperature); degree of optical scatter effect variations and other nonlinear interferences.
2. Noise level: random noise levels and probability of outliers in the chemical calibration data; random noise level and probability of outliers in the near-infrared calibration data; random noise level and probability of outliers in the near-infrared prediction data.
3. Sample distribution: extent to which the calibration sample set is representative for future samples, in terms of average quality and quality variation concerning both the constituent calibrated for and the interferences.
4. Computer storage and processing capacity.
5. Need for automatic error warnings.

C. Data Pretreatment

When light scatter varies strongly, it appears that theoretically more exact response linearizations such as the Kubelka-Munk transformation can give better linear analyses than the empirical log (1/R), but only after an optical correction. One such optical correction, the MSC, was tested and found to be beneficial, apparently yielding useful estimates for the relative scatter coefficient.

None of the response linearizations tested gives fully satisfactory linearity; new formulae should therefore be tested. The presented MSC method is likewise somewhat unsatisfactory because it ignores chemical effects on the light scatter. Ideally, the data pretreatment should be optimized iteratively together with the linear estimation.

An alternative approach is to allow nonlinear estimation. For instance, several different nonlinear transformations of the same near-infrared wavelengths can be used together as "x variables" in the data compression stage. To avoid estimating loadings for all these x variables, a wavelength selection method may then be desirable to eliminate irrelevant ones; wavelength-truncated PLS might be useful for this purpose.

Alternatively, nonlinear relationships between c and $\hat{T}$ may be used in the calibration regression stage.

D. Error Detection

Outliers of different types can be detected in the calibration sample set as well as among unknown samples, using modern regression diagnostics to assess near-infrared residuals, chemical residuals, and leverages. Samples with abnormal light-scatter levels may be detected during the optical correction stage. The outlier detection methods need more testing in practice.

In the calibration step, the elimination of outliers among the samples or among the variables should be compared with methods for down-weighting them (robust methods; e.g., Huber, 1981).

E. Updating

Finally, the problem of updating calibration results should be mentioned. Advanced instruments such as near-infrared spectrophotometers require routine checking. These control data may be used for updating the calibration constants in order to improve accuracy (reduce estimation errors from the calibration, account for instrument drift, etc.). In some methods, the updating algorithm is relatively straightforward (e.g., Cook and Weisberg, 1982, for an update of MLR). In other methods, updating algorithms are not yet available. But at least the offset and general slope (the simple corrigible errors; Cardone, 1983) can always be updated to minimize the bias. Since calibration methods may yield biased predictors, this appears to be important in practice. A weighted average of the old offset and the new bias may be used:

$$\hat{b}_o^{(new)} = w\hat{b}_o^{(old)} - (1 - w)(\bar{f}), \quad (100)$$

where $\bar{f}$ is the bias estimated in a set of new control samples and w is a weight between 0 and 1 that balances the variance of the estimated bias against the variance of the old offset. A proportional update factor, common to the coefficient of all wavelengths, $\hat{b}$, can also be made. In this way, the information in the data of analytical control samples is put to efficient use.

IX. CONCLUSIONS

This chapter has shown that indirect multivariate calibration is necessary in diffuse near-infrared spectroscopy because of extensive chemical and optical interferences at all wavelengths. It has also shown how the optimal dimensionality of such calibration methods has to be found in order to model these interferences while avoiding the noise problems from overfitting (see Fig. 3).

Several linear full-spectrum calibration methods that can be used in near-infrared spectroscopy have been described. One general group of methods relies on a data compression stage followed by a calibration regression stage, and another group relies on a joint data compression/calibration stage that allows alternative prediction methods (Beer's model methods). Methods for detection of errors in near-infrared data have been discussed, as well as errors in chemical data and abnormal sample qualities. The importance of proper linearization of the near-infrared data has been stressed, and a basic model of additive chemical interferences and multiplicative light-scatter interference has been shown to be useful.

ACKNOWLEDGMENTS

We would like to thank Ragnhild Norang and Unni Haugdahl for patient and skillful typing of the manuscript. Karl Norris, William Hruschka, Lars Munck, and Sven Åge Jensen are thanked for lending us data, and we thank them as well as Svante Wold and Kim Esbensen for valuable discussions.

APPENDIX A: SYMBOLS AND ABBREVIATIONS

Uppercase italic: matrix
Lowercase italic: vector (column vector unless otherwise stated)
Uppercase roman: upper limit for running index
Lowercase roman: scalar

^	"Hat"; statistically estimated
$1 = \begin{bmatrix} 1 \\ 1 \\ \cdot \\ \cdot \\ 1 \end{bmatrix}$	Column vector of ones used for matrix formalism
a, A	Index for regression factors, a = 1, 2, ..., A
B, *b*, b	Calibration coefficients
b_0	Calibration offset coefficient
BLP	Best linear predictor
C, *c*, c	Concentration of chemical constituents
df	Degrees of freedom
E, *e*, e	Near-infrared residuals, residuals in Beer's model
e*	Uncorrelated residuals in Beer's model
EBLP	Estimated best linear predictor
EBLUP	Estimated best linear unbiased predictor
EGLS1	Estimated generalized least-squares estimator (EGLS) with distributional assumptions on unknown interferences
EGLS2	EGLS without distributional assumptions on unknown interferences
f, *f*	Chemical residuals
g_i	Leverage relevance
GLS	Generalized least squares
h_i, h_k	Leverage
H	Covariance matrix of *c*
i, I	Index for samples, i = 1, 2, ..., I
I	Identity matrix
j, J	Index for chemical constituents, j = 1, 2, ..., J
k, K	Number of near-infrared wavelengths, k = 1, 2, ..., K
K/S	Kubelka-Munk transform
LS	Least squares
ML	Maximum likelihood
MLR	Multiple linear regression
MSC	Multiplicative scatter correction
MSE	Mean squared error
P, *p*, p	Loadings for interferences in Beer's model, near-infrared loadings
PCA	Principal component analysis
PCR	Principal component regression
PLS	Partial least squares
q, q	Chemical loadings
R	Reflectance
R*	Reflectance corrected according to Saunderson theory
RAP	Relative ability of prediction
RMSCV	Root mean square of error in cross-validation of calibration
RMSE	Root mean square of error in calibrating estimation
RMSEL	Leverage-corrected root mean square of error in calibrating estimation
RMSP	Root mean square of prediction
SDLR	Stepwise derivative linear regression
$s(\hat{e})$	Standard deviation of $\hat{e}$
SMLR	Stepwise multiple linear regression
Std	Standard deviation
STOT	Total standard deviation of c prior to modeling
T, *t*, t	Regression factor scores, interferences in Beer's model
u	Mean of c_i
U	Symbol used in PCA and PLS algorithms to replace *E*
v	Symbol used in PCA and PLS algorithm to replace *f*, intercept in Beer's model
w	Near-infrared loading weights in PLS with orthogonal scores
W	Covariance matrix of *e**
X, *x*, x	Near-infrared observations
Σ	Covariance matrix of *e* in Beer's model

APPENDIX B: MATRIX OPERATIONS ILLUSTRATED FOR MULTICOMPONENT ANALYSIS

Describing different calibration methods at sufficient precision and yet in a readable way requires matrix and vector notation. This appendix is written to help the reader who is not familiar with matrix algebra to be able to read at least the main sections of the chapter.

There are four basic matrix operations required, namely

transpose, sum, product, and inversion, corresponding roughly to turning a table 90°, summing tables, multiplying tables, and dividing tables. Beer's model will be used here to illustrate these operations.

Vectors

The near-infrared spectrum of the sample i and K different wavelengths is a row vector x_i that can be written:

$$x_i = \{x_{ik}\} = (x_{i1}, x_{i2}, \ldots, x_{ik}, \ldots, x_{iK}) . \quad \text{(B-1)}$$

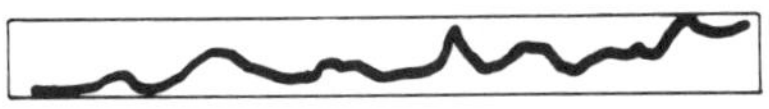

Assume that the spectrum x_i is correctly modeled by the sum of the spectral contributions from water, starch, and protein in that sample:

$$x_i = \text{water} + \text{starch} + \text{protein} . \quad \text{(B-2)}$$

The spectral contribution of constituent j (e.g., water) is defined by its near-infrared spectrum multiplied by the concentration of the constituent in that sample. Its spectrum k_j is, like the sample spectrum x_i, a row vector:

$$k_j = \{k_{jk}\} = (k_{j1}, k_{j2}, \ldots, k_{jk}, \ldots, k_{jK}) . \quad \text{(B-3)}$$

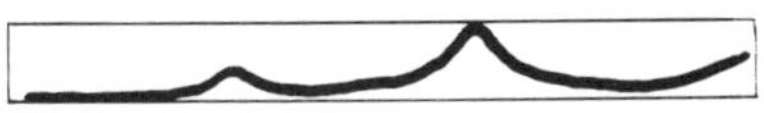

Its concentration in sample i is simply one number (a scalar) termed c_{ij}:

$$c_{ij} . \quad \text{(B-4)}$$

Hence the spectral contribution of constituent j (e.g., water) in sample i is the product of its concentration c_{ij} and spectrum k_j:

$$c_{ij} \cdot k_j = \{c_{ij}k_{jk}\} = (c_{ij}k_{j1}, c_{ij}k_{j2}, \ldots, c_{ij}k_{jk}, \ldots, c_{ij}k_{jK}) . \quad \text{(B-5)}$$

The sum of the contributions from the three constituents water, starch, and protein is then:

$$\begin{aligned} x_i &= c_{i1}k_1 \\ &+ c_{i2}k_2 \\ &+ c_{i3}k_3 . \end{aligned} \quad \text{(B-6)}$$

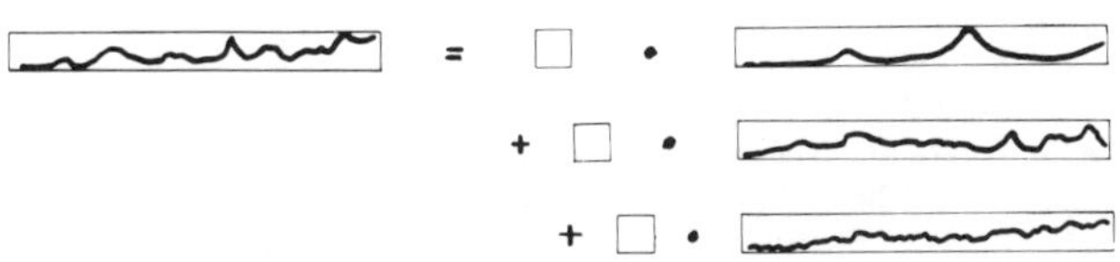

Matrices and Matrix Multiplication

In matrix algebra, equation (B-6) can be even more efficiently written by including all three concentration numbers c_{i1}, c_{i2}, and c_{i3} into the single row vector c_i:

$$c_i = \{c_{ij}\} = (c_{i1}, c_{i2}, c_{i3}) . \quad \text{(B-7)}$$

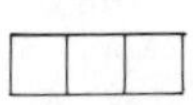

Likewise, all three constituent spectra k_1, k_2, and k_3 are included into the single table or matrix K:

$$K = \{k_{ik}\} = \quad \text{(B-8)}$$

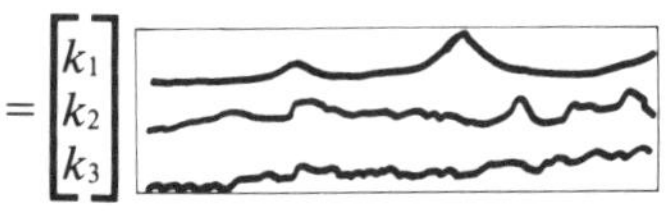

The model can then be expressed more efficiently than in equation (B-6) as follows:

$$x_i = c_iK . \quad \text{(B-9)}$$

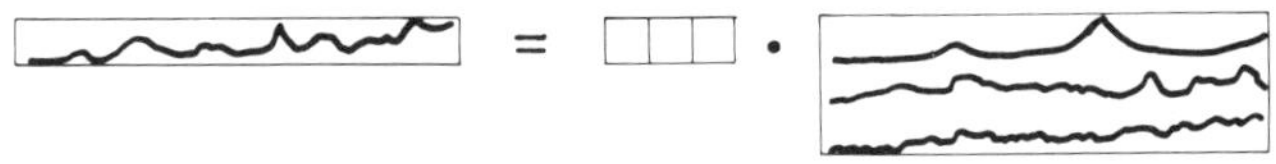

This illustrates the use of vectors (x_i and c_i) and matrices (K). Matrix multiplication is done in this way: from given c_i and K, the elements in the resulting vector x_i are obtained by:

$$x_{ik} = \sum_{j=1}^{3} c_{ij}k_{jk} . \quad \text{(B-10)}$$

If there are a number of different samples i = 1, 2, ..., I, then their spectra x_i can be collected into a single table or matrix X:

$$X = \{x_{ik}\} = \quad \text{(B-11)}$$

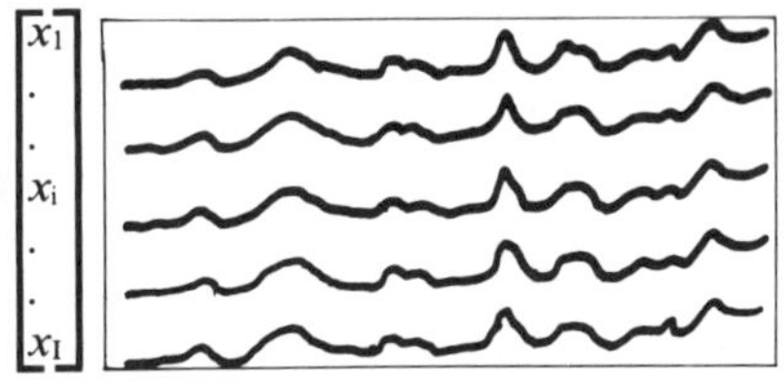

Similarly, the concentration vectors c_i are collected in the single concentration matrix C:

$$C = \{c_{ij}\} = \quad \text{(B-12)}$$

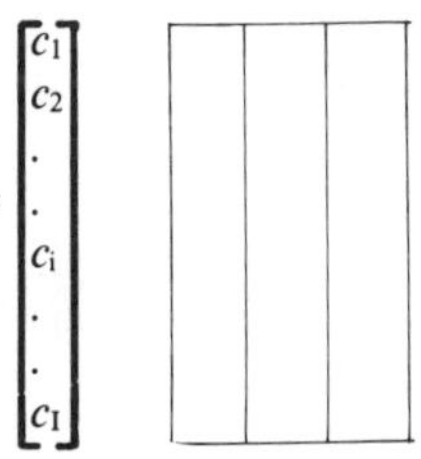

The model equation (B-10) for, e.g., I = 100 different samples can be written simply as:

$$X = CK . \quad \text{(B-13)}$$

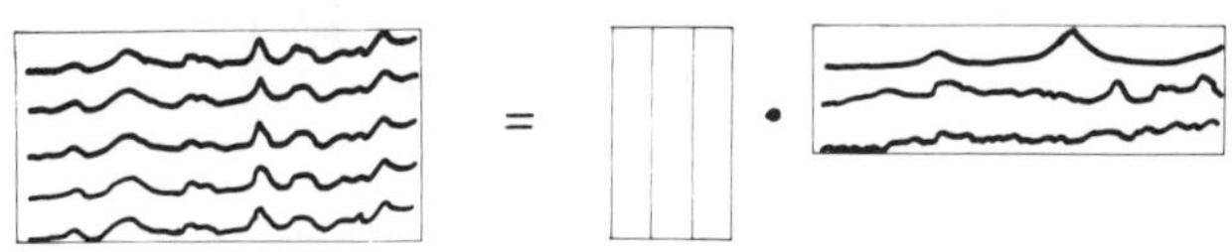

Matrix multiplication thus consists of summing up individual multiplication elements. The elements are chosen in this way:

(B-14)

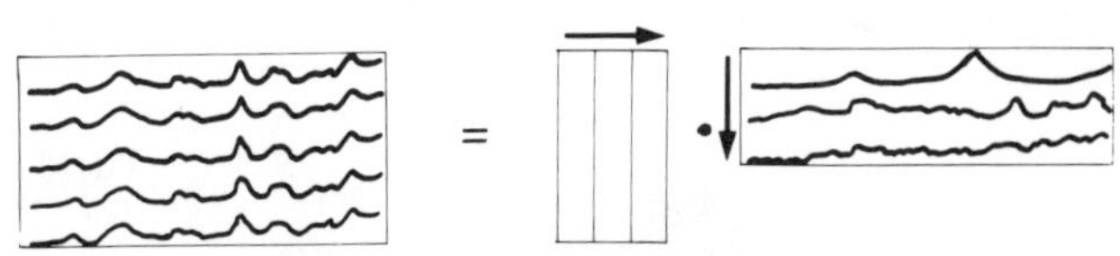

Hence, the number of columns in C must equal the number of rows in K—in this case, three (water, starch, and protein).

Multicomponent Analysis by Least-Squares Fit

Suppose now that spectra K of the pure constituents were known, that the spectra of I samples, X, had been determined. Then the concentrations, C, can be estimated. If the data behaved linearly as for transmission spectroscopy absorbance in clear solutions, the model could then be:

$$X = CK + E\,, \tag{B-15}$$

where $E = \{e_{ik}\}$ represents measurement errors. The LS fit, which averages the fitting over all the K wavelengths, would then be:

$$\hat{C} = XK'Z^{-1}\,, \tag{B-16}$$

where the matrix Z is equal to KK' and $\hat{C}$ ("C-hat") means estimated C.

Two new matrix operations are introduced in this formula: transpose $(\,)'$ and inverse $(\,)^{-1}$. These will now be treated.

Transpose of Matrix, K' (or K^T)

The transpose means that K' represents another matrix with exactly the same elements as K, but turned around so that rows become columns and columns become rows. If the matrix K is the 3 spectra with K wavelengths, i.e.:

$$K = \begin{bmatrix} k_{11} & k_{12} & k_{13} \\ k_{21} & k_{22} & k_{23} \\ k_{31} & k_{32} & k_{33} \end{bmatrix} \tag{B-17}$$

then K' is:

$$K' = \begin{bmatrix} k_{11} & k_{21} & k_{31} \\ k_{12} & k_{22} & k_{32} \\ k_{13} & k_{23} & k_{33} \end{bmatrix} \tag{B-18}$$

Hence the matrix product $Z = KK'$ implies collapsing over the wavelengths, with the result Z being the 3×3 cross-product matrix:

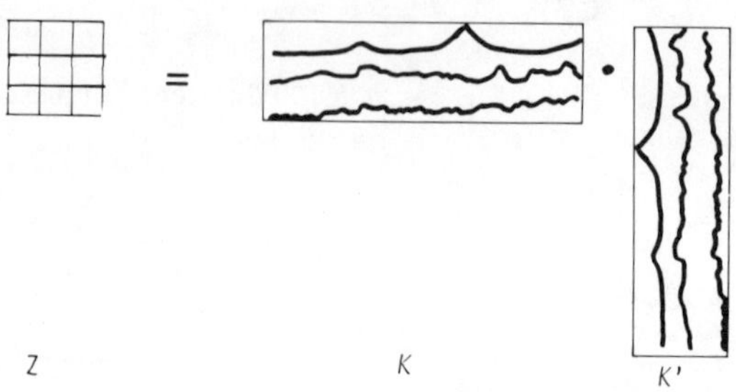

(B-19)

Inverse of Matrix, Z^{-1}

In ordinary algebra, the inverse of a single scalar number z (say z = 20) can be written either 1/z or z^{-1} (1/20 = 0.05). One definition of this scalar inverse is that a number z multiplied by its inverse z^{-1} equals 1:

$$z \cdot z^{-1} = 1$$
$$20 \cdot 20^{-1} = 1\,. \tag{B-20}$$

Matrix inversion is a straightforward extension of this; the inverse of a quadratic matrix (a matrix with the same number of rows as columns), say Z, is defined such that:

$$Z Z^{-1} = I\,, \tag{B-21}$$

where I is the "unit matrix" corresponding to the number 1 for scalars. For a 3×3 matrix, we would have:

$$I = \begin{bmatrix} 1 & 0 & 0 \\ 0 & 1 & 0 \\ 0 & 0 & 1 \end{bmatrix} \tag{B-22}$$

Given the matrix Z, a computer is required to obtain the inverse, Z^{-1}, but this is rather easily done.

However, one problem should be remembered: Just as the scalar inverse $1/z = z^{-1}$ is meaningless if z = 0, the inversion of a matrix Z can be meaningless or impossible if the matrix Z does not contain enough "information" for the computer inversion algorithm (its determinant is zero). This happens if the $J \times J$ matrix Z contains fewer than J different types of variation. In the example above, this could occur if starch and protein had the same near-infrared spectra ($k_3 = k_2$) or if the protein spectrum was equal to some combination of the water and starch spectra ($k_3 = \alpha k_1 + \beta k_2$). Then the computer cannot distinguish between protein information on one hand and water information on the other, and the inversion of $Z = KK'$ becomes impossible. It can be said that Z does not have full rank and that K does not have full row rank; the rows are so-called linearly dependent.

Provided that K has full row rank, the formula equation (B-16) for estimating the concentrations C can then be written:

$$\hat{C} = X\hat{B}\,, \tag{B-23}$$

$$\text{where } \hat{B} = K'Z^{-1} = K'(KK')^{-1}\,. \tag{B-24}$$

Equation (B-23) can be written easily without matrix algebra:

$$\hat{c}_{ij} = \sum_{k=1}^{K} x_{ik}\hat{b}_{kj},\ i = 1, 2, ..., I;\ j = 1, 2, ..., J\,. \tag{B-25}$$

How the calibration coefficients $\hat{B}$ were estimated (equation [B-24]), however, is difficult to express in traditional summation notation, whereas it is easy in matrix algebra.

LITERATURE CITED

BECKMAN, R. J., and COOK, R. D. 1983. Outliers. Technometrics 25:119-163.

BROWN, P. J. 1982. Multivariate calibration. J. R. Stat. Soc. B 44:287-308.

CARDONE, M. J. 1983. Detection and determination of error in analytical methodology. Part II. Correction for corrigible systematic error in the course of real sample analysis. J. Assoc. Off. Anal. Chem. 66:1283-1293.

COOK, R. D., and WEISBERG, S. 1982. Residuals and Influence in Regression. Chapman and Hall, London.

DAMSLETH, E., and SPJØTVOLL, E. 1982. Estimation of trigonometric components in time series. J. Am. Stat. Assoc. 77:381-387.

DRAPER, N., and SMITH, H., Jr. 1981. Applied Regression Analysis. 2nd ed. John Wiley, New York.

ESBENSEN, K., and WOLD, S. 1983. SIMCA, MACUP, SELPLS, GDAM, SPACE and UNFOLD: The way towards regionalized principal components analysis and subconstrained N-way decomposition—with geological illustrations. Pages 11-36 in: Proc. Nordic Symp. Appl. Stat. Stokkand Forlag Publ., Stavanger, Norway.

FEARN, T. 1983. Misuse of ridge regression in the calibration of a near infrared reflectance instrument. Appl. Stat. 32:73-79.

GELADI, P., McDOUGEL, D., and MARTENS, H. 1985. Linearization and scatter-correction for near-infrared reflectance spectra of meat. Appl. Spectrosc. 39:491-500.

GUNST, R. F., and MASON, R. L. 1979. Some considerations in the evaluation of alternate prediction equations. Technometrics 21:55-63.

HAALAND, D. M., and EASTERLING, R. G. 1982. Applications of new least-squares methods for the quantitative infrared analysis of multicomponent samples. Appl. Spectrosc. 36:665-673.

HENDERSON, C. R. 1975. Best linear unbiased estimation and prediction under a selection model. Biometrics 31:423-447.

HILDRUM, K. I., MARTENS, H., and LEA, P. 1984. Detection of errors in the NIR-analysis of meat components. Eur. Meet. Meat Res. Workers, 30th. Bristol, England, 9–14 September.

HILDRUM, K. I., MARTENS, M., and MARTENS, H. 1983. Research and analysis of food quality. Proc. Int. Symp. Control Food Quality Food Anal. Reading University, England, 22–24 March.

HOADLEY, B. 1970. A Bayesian look at inverse linear regression. J. Am. Stat. Assoc. 65:356-369.

HRUSCHKA, W., and NORRIS, K. 1982. Least squares curve fitting of near infrared spectra predicts protein and moisture content of ground wheat. Appl. Spectrosc. 36:261-265.

HUBER, P. J. 1981. Robust Statistics. John Wiley, New York.

JENSEN, S. Å., MUNCK, L., and MARTENS, H. 1982. The botanical constituents of wheat and wheat milling fractions. I. Quantification by autofluorescence. Cereal Chem. 59:477-484.

JENSEN, S. Å., and MARTENS, H. 1983. The botanical constituents of wheat and wheat milling fractions. II. Quantification by amino acids. Cereal Chem. 60:172-177.

JOLIFFE, I. T. 1982. A note on the use of principal components in regression. Appl. Stat. 31:300-303.

LINDBERG, W., PERSSON, J. Å., and WOLD, S. 1983. Partial least squares method for spectrofluorimetric analysis of mixtures of humic acid and ligninsulfonate. Anal. Chem. 55:643-648.

LWIN, T., and MARITZ, J. A. 1982. An analysis of the linear-calibration controversy from the perspective of compound estimation. Technometrics 24:235-242.

MANDEL, J. 1971. A new analysis of variance model for non-additive data. Technometrics 13:1-18.

MARTENS, H. 1981. Qualitative and quantitative morphometry by multivariate image analysis. Chemometric theory, illustrated by a hypothetical example: Near infrared analysis of quality in bacon. Norwegian Food Res. Inst. Rep. NR4.

MARTENS, H. 1986. "Unscrambler" program for calibration. Developed at the Norwegian Food Research Institute. Marketed by Computer Aided Modelling (CAMO), Elgsetergaten, Trondheim, Norway.

MARTENS, H., and JENSEN, S. Å. 1983. Partial least squares regression: A new two-stage NIR calibration method. Progress in cereal chemistry and technology. Developments in food science. Pages 607-647 in: Proc. World Cereal Congr., 7th. J. Holas and J. Kratochvil, eds. Elsevier, Amsterdam.

MARTENS, H., JENSEN, S. Å., and GELADI, P. 1983a. Multivariate linearity transformation for near-infrared reflectance spectrometry. Pages 205-234 in: Proc. Nordic Symp. Appl. Stat. Stokkand Forlag Publ., Stavanger, Norway.

MARTENS, H., VANGEN, O., and SANDBERG, E. 1983b. Multivariate calibration of an X-ray tomograph by smoothed PLS regression. Pages 235-268 in: Proc. Nordic Symp. Appl. Stat. Stokkand Forlag Publ., Stavanger, Norway.

MARTENS, H., and WOLD, S. 1984. Suggested recommendation for notation and terminology in chemometrics. Version 0.2. Norwegian Food Res. Inst.

MARTENS, M., LEA, P., and MARTENS, H. 1983. Predicting human response to food quality by analytical measurements. The PLS regression method. Proc. Nordic Symp. Appl. Stat. Stokkand Forlag Publ., Stavanger, Norway.

NAES, T. 1985a. Multivariate calibration when the error covariance matrix is structured. Technometrics 27:301-311.

NAES, T. 1985b. Comparison of approaches to multivariate linear calibration. Biometrical J. 27:265-275.

NAES, T. 1986. Multivariate calibration by covariance adjustment. Biometrical J. 28:99-107.

NAES, T., and MARTENS, H. 1985. Comparison of prediction methods for multicollinear data. Commun. Stat. (Sim. Comp.) 14:545-576.

NAES, T., and MARTENS, H. 1987. Testing adequacy of linear random models. Mat. Oper. Stat. (In press)

NORRIS, K. 1983. Extracting information from spectrophotometric curves: Predicting chemical composition from visible and near-infrared spectra. Pages 95-113 in: Food Research and Data Analysis. H. Martens and H. Russwurm, Jr., eds. Applied Science Publishers, New York.

OTTESTAD, P. 1975. Component analysis: An alternative system. Int. Stat. Rev. 43:83-108.

RABINER, L. R., and GOLD, B. 1975. Theory and application of digital signal processing. Prentice-Hall, Englewood Cliffs, NJ.

SAUNDERSON, J. L. 1942. Calculation of the color of pigmented plastics. J. Opt. Soc. Am. 32:727-736.

SAXBERG, B. E. H., and KOWALSKI, B. R. 1979. Generalized standard addition method. Anal. Chem. 51:1031-1038.

SEARLE, S. R. 1971. Linear models. John Wiley, New York.

SKREDE, G., NAES, T., and MARTENS, M. 1983. Visual color deterioration in blackcurrant syrup predicted by different instrumental variables. J. Food Sci. 48:1745-1749.

STONE, M. 1974. Cross-validatory choice and assessment of statistical prediction. J. R. Stat. Soc. B. 36:111-133.

SUNDBERG, R. 1982. When does it pay to omit explanatory variables in multivariate calibration? Department of Mathematics, Royal Institute of Technology, Stockholm, Sweden.

VANGEN, O., and MARTENS, H. 1982. In vivo estimation of body composition of pigs using computer tomography calibrated by PLS regression. In: Food Research and Data Analysis. H. Martens and H. Russwurm, Jr., eds. Applied Science Publishers, New York.

WOLD, H. 1966. Estimation of principal components and related models by iterative least squares. In: Multivariate Analysis. P. R. Krishnaiah, ed. Academic Press, New York.

WOLD, H. 1982. Soft modelling: The basic design and some extensions. Pages 263-271 in: Systems under Indirect Observations. Causality-Structure-Prediction. Vol. 1. Joreskog and Wold, eds. North Holland, Amsterdam.

WOLD, S. 1978. Cross validatory estimation of the number of components in factor and principal component models. Technometrics 20:397-406.

WOLD, S. 1982. SIMCA-3B manual. University of Umeå, Sweden.

WOLD, S. et al. 1983a. A multivariate calibration problem in analytical chemistry solved by partial least squares models in latent variables. Anal. Chim. Acta 150:61-70.

WOLD, S., MARTENS, H., and WOLD, H. 1983b. The multivariate calibration problem in chemistry solved by the PLS method. Pages 286-293 in: Proc. Conf. Matrix Pencils. A. Ruhe, B. Kågström, eds. Lecture Notes in Mathematics. Springer Verlag, Heidelberg.

YOUNG, F. W. 1981. Quantitative analysis of qualitative data. Psychometrika 46:357-388.

CHAPTER 5

Near-Infrared Instrumentation

W. F. McCLURE
Department of Biological and Agricultural Engineering
North Carolina State University
Raleigh, North Carolina, U.S.A.

I. INTRODUCTION

Although the theory of infrared spectroscopy has been known for more than 160 years, it was not until the 1930s that the first custom-made instruments appeared in industrial laboratories. World War II encouraged the use of infrared as an analytical tool, especially in the rubber and petroleum fields. In the 1940s the first commercial infrared spectrophotometers became available and infrared spectroscopy began significant growth. The first double-beam instrument was sold in 1947 (Archer, 1969).

Classical spectroscopists have traditionally avoided the near infrared. Working with fundamental absorption bands, many of which fall in the midinfrared between 2.5 and 15 micrometers (μm), researchers were convinced that overtones and combination absorptions that occurred in the near infrared were of little consequence.

The development of precise manufacturing techniques for fabricating photon detectors, primarily lead sulfide (PbS) detectors, revived interest in the near infrared, and the technology was attached to UV-VIS instruments. This continued the separation of near infrared from the mainstream of infrared research. However, in the 1960s, the advent of the minicomputer enabled researchers to look more closely at the visible and near-infrared spectra of various products for the purpose of sorting according to quality and composition.

In the 1950s, K. Norris of the U.S. Department of Agriculture began to investigate optical properties of dense, light-scattering materials (Norris and Butler, 1961). By 1969, his work led to the development of a computerized, near-infrared spectrophotometer that was used to obtain spectral data on meats (Ben-Gera and Norris, 1968). By 1975, McClure (McClure and Hamid, 1980) and Shenk (Shenk and Hoover, 1976) had designed and built their own computerized near-infrared systems for the analysis of tobacco and forages, respectively.

Performance of near-infrared instrumentation depends, in large part, on the components that are used in fabrication. Unfortunately, there are few, if any, components that have a constant response across that region. For example, lenses and mirrors have nonlinear characteristics. The combination of these components produces an instrument that records nonlinear spectra. Fortunately, these nonlinearities are removed in the computation of transmittance (T) or reflectance (R).

This chapter deals with the nonlinearities associated with the various components that are used in the design of near-infrared instrumentation. The discourse will lead to a discussion of design considerations used in fabricating the COMP/SPEC (McClure and Hamid, 1980). Two COMP/SPEC systems are currently in operation, one in the Bioinstrumentation Laboratory at North Carolina State University at Raleigh and the other at the U.S. Department of Agriculture Laboratory, Oxford Tobacco Research Station, Oxford, NC. Both hardware and software of the COMP/SPEC systems were designed and developed in the Bioinstrumentation Laboratory of North Carolina State University.

II. COMPONENTS OF NEAR-INFRARED SYSTEMS

A. Lenses and Mirrors: Collecting Radiation

Lenses and mirrors used in infrared instruments must often be made of special materials, since glasses normally used for the visible region are almost opaque to radiation of wavelengths longer than about 2 μm. Figure 1 shows the variation of transmittance with wavelength of various materials, along with the reflection of aluminum first-surface material.

Pyrex, although economical, has noticeable absorptions at 1.37 and 2.2 μm with a diminishing transmittance starting at 2.0 μm to almost 10% at 2.8 μm. This material is usable, but fused quartz is much superior. Aluminum first-surface mirrors are acceptable for near-infrared work, but silver and gold coatings are far superior.

There are two figures of merit that describe optical condensers used with source and dispersion systems. The principal one is the focal ratio, relative aperture, or, more commonly, the $f/$number, which is defined as:

$$f/\text{number} = \frac{f}{\phi}, \tag{1}$$

where f = focal length of the lens and ϕ = clear aperture diameter. The $f/$number is a measure of the light-gathering power (efficiency) of a lens system: the smaller the $f/$number, the greater the light-gathering power of the lens. The theoretical maximum relative aperture of an aplanatic lens is $f/0.5$, but this is practically unobtainable (Optics Guide III, 1985).

Figure 2A shows the geometric relationship between a source L and an $f/1$ condenser lens. Note here that in order to have an $f/1$ condenser:

$$F = D = 2R. \tag{2}$$

Therefore, $\gamma = \arctan 0.5 = 0.4637$ radian. Since $\alpha = 0.5$ radian, then:

$$R' = R = 0.9274R. \tag{3}$$

The area intercepted by one steradian on the surface of a sphere is:

$$A_s = (2R)^2 \,. \tag{4}$$

Hence, the area intercepted by the $f/1$ lens on the spherical surface with a diameter R' is:

$$A_f = (R')^2 \,. \tag{5}$$

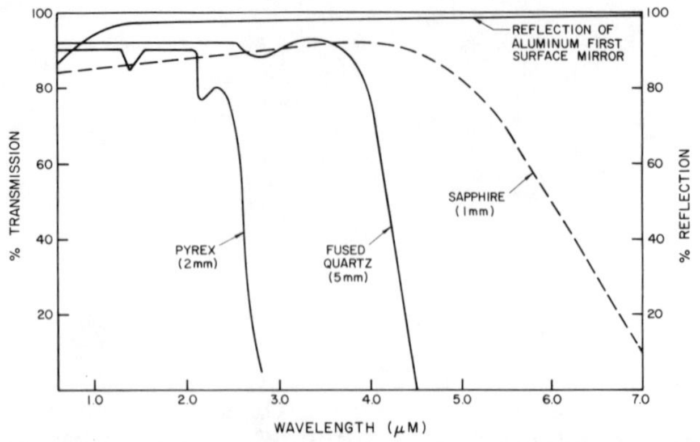

Fig. 1. Transmission/reflection of five optical materials: Pyrex, fused quartz, sapphire, and an aluminum first-surface mirror. Fused quartz and aluminum first-surface mirrors satisfy requirements of near-infrared spectroscopy.

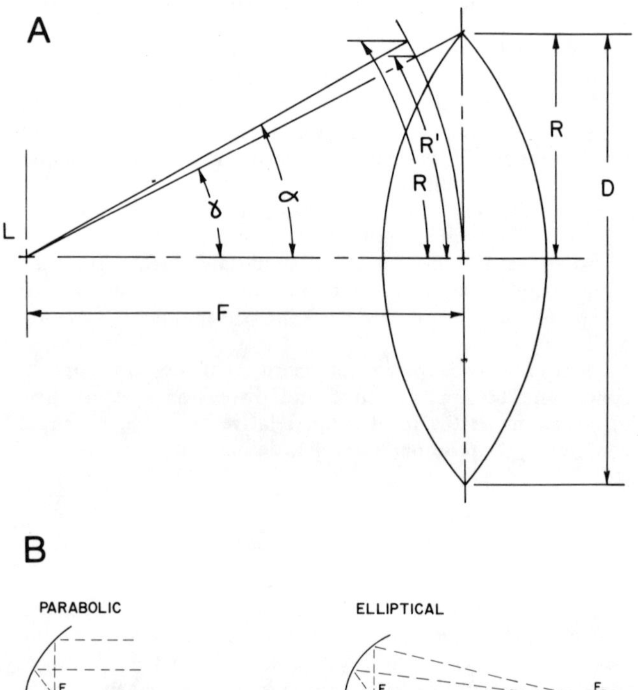

Fig. 2. Optical characteristics of $f/1$ lens (A) and first-surface mirrors (B). An $f/1$ lens collects only 5.4% of total energy emitted by light source. Elliptical and spherical mirrors are recommended when focal ratios smaller than $f/1$ are required.

Hence, the total number of steradians intercepted by an $f/1$ lens is:

$$\Omega = \frac{(0.9274R)^2}{4R^2} \tag{6}$$

$$\Omega = 0.6755 \,. \tag{7}$$

The equivalent solid angle is an important aspect of the relative aperture of a condenser lens because lamp irradiance is usually given in watts per steradian. Here it is seen that an $f/1$ lens would collect only a small portion of the total energy emitted from a point source, or:

$$\%\ \text{collected} = \frac{1}{12.57\ \text{Ster/sphere}} \times 0.6755 \times 100 \tag{8}$$

$$\%\ \text{collected} = 5.4\% \,. \tag{9}$$

This illustrates one means of increasing the illumination available to a monochromator from a given source. However, to achieve f numbers below 1.0, the designer should resort to parabolic or ellipsoidal mirrors (see Fig. 2B) (Birth, 1969).

Degree of collimation of a lens system is defined (Birth, 1969) by the size of the source and the focal length of the condenser (Fig. 3A), or:

$$D = \frac{\bar{d}}{2f} \,, \tag{10}$$

where $\bar{d}$ = source size and f = lens focal length. Note that as D becomes smaller, divergence of the beam is smaller (i.e., the collimation is better). Since the filament of sources used in the monochromator is rectangular in shape (to match the shape of the entrance slit of the monochromator), the degree of collimation in the horizontal plane differs from the degree of collimation in the vertical plane.

To increase the power of the source, the dimensions of the source must also increase. Thus, to maintain a high degree of

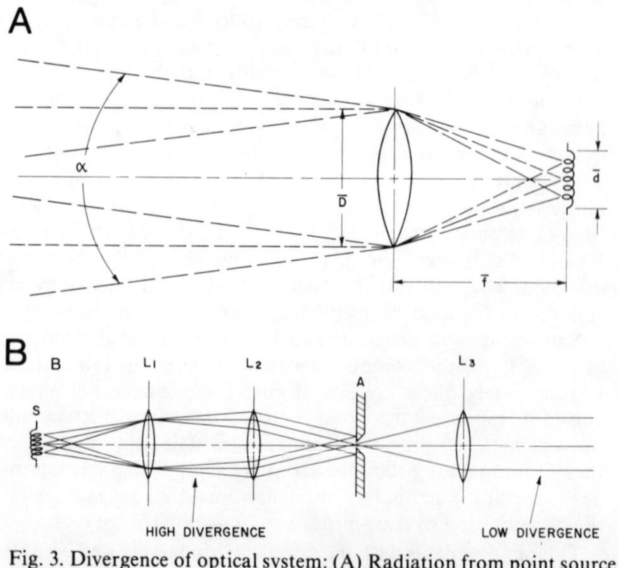

Fig. 3. Divergence of optical system: (A) Radiation from point source located at focal point of condenser yields parallel beam of radiation $\bar{D}$. Since real sources are not point sources but have finite dimensions, resultant beam has finite dimensions. (B) In instruments where low divergence is required, an aperture A must be used. In dispersion systems, A takes form of slit that also determines, at exit, bandpass of spectrometer.

collimation, a long-focal-length condenser is necessary (see Fig. 3B). Large-diameter optics are required to secure a low *f*/number. Therefore, users of near-infrared systems cannot increase the energy output of a monochromator simply by using a higher-powered source. Extra energy comes only by widening the entrance and exit slits or by replacing the current input optics with optics having a lower *f*/number.

B. Radiation Sources

A good radiation source for near-infrared spectroscopy should have the following characteristics:

A

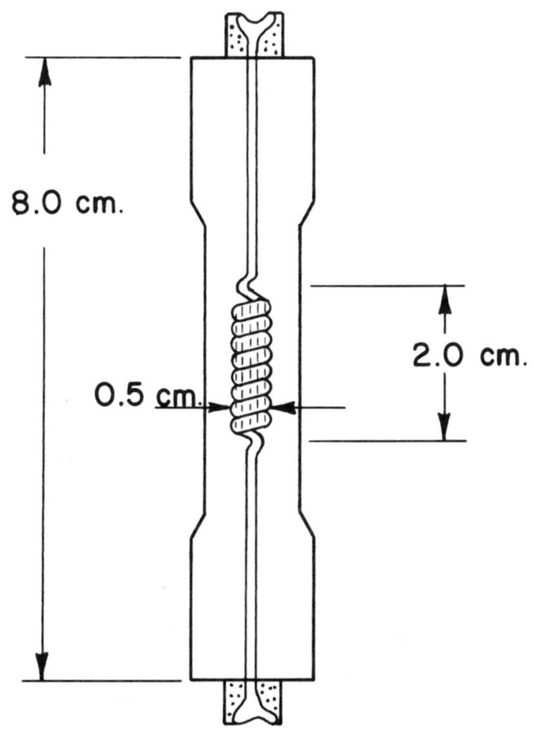

B

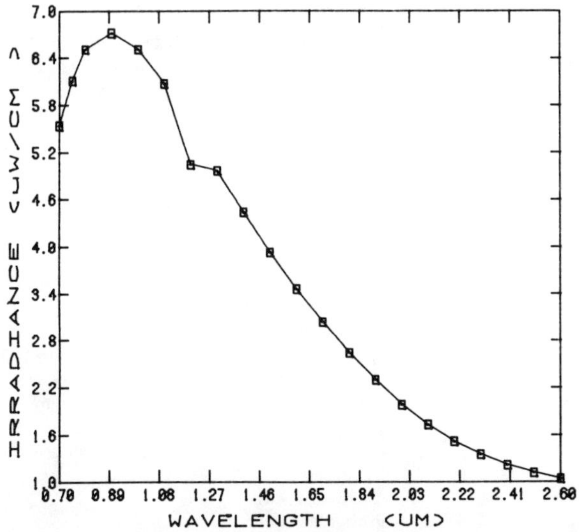

Fig. 4. Physical size (A) and spectral irradiance (B) of model DXM tungsten halogen lamp.

1. The beam should emit a continuum of radiation over the range from 0.9 to 2.6 μm.
2. The intensity of the source should be high enough so that extensive signal conditioning (amplification, cross-correlation circuitry, etc.) is not needed.
3. The overall size of the lamp should be small to take advantage of low *f*-number optics.
4. The filament of the lamp should be compact, with a projected area equivalent to the slit of the spectrometer.
5. The envelope of the lamp should contain iodine vapor, which reduces envelope darkening and extends the useful life.
6. The source should not fluctuate over long periods of time or flicker over short periods of time.

Tungsten-halogen lamps, such as those coded DXM and FAD by the American Standards Association (see Fig. 4A), meet the above criteria. Any tungsten lamp with a quartz bulb is satisfactory, but tungsten-halogen provides a longer life and more stability because of the cleaning action of the halogen. Stabilized DC power supplies, rather than AC power supplies, are used extensively to provide radiation that varies little with time. In single-beam systems like the COMP/SPEC, it is extremely important that this power supply be of the highest quality possible, having good regulation and low ripple. Operating the lamps from a 60-Hz source is not recommended unless the data-recording circuitry is synchronized with the AC source. The irradiance (μW/cm^2 per μm at a distance of 42 cm) of a typical DXM lamp is shown in Fig. 4B. The lamp output peaks at 0.9 μm (6.72 μW/cm^2) and tapers off to 1.04 μW/cm^2

TABLE I
Spectral Irradiance of Three Quartz-Iodine Tungsten-Filament Lamps[a]

Wavelength (nm)	Lamp QL-2	Lamp QL-5	Lamp QL-10
250	0.0051	0.0052	0.0051
260	0.0093	0.0093	0.0090
270	0.0158	0.0159	0.0155
280	0.0253	0.0252	0.0244
290	0.0380	0.0380	0.0369
300	0.0545	0.0548	0.0532
320	0.104	0.105	0.102
350	0.237	0.242	0.234
370	0.366	0.374	0.363
400	0.643	0.647	0.630
450	1.26	1.26	1.23
500	2.04	2.04	2.02
550	2.93	2.96	2.91
600	3.88	3.94	3.88
650	4.79	4.91	4.80
700	5.54	5.72	5.58
750	6.11	6.32	6.14
800	6.51	6.69	6.49
900	6.72	6.94	6.71
1,000	6.51	6.73	6.53
1,100	6.07	6.25	6.11
1,200	5.53	5.67	5.55
1,300	4.97	5.09	4.98
1,400	4.44	4.52	4.44
1,500	3.93	4.00	3.93
1,600	3.46	3.51	3.45
1,700	3.03	3.06	3.01
1,800	2.63	2.65	2.61
1,900	2.29	2.28	2.26
2,000	1.98	1.97	1.95
2,100	1.73	1.71	1.70
2,200	1.52	1.51	1.50
2,300	1.36	1.34	1.33
2,400	1.22	1.21	1.21
2,500	1.12	1.10	1.11
2,600	1.04	1.03	1.04

[a]Source: Stair et al (1963).

at 2.6 μm. The lamps also provide sufficient radiation for operation over the visible spectrum down to 0.4 μm without extensive signal conditioning of the detector signal. Emission (Table I) varies little from lamp to lamp. These lamps are unsurpassed to date as economical sources of radiation for near-infrared spectroscopy (Birth, 1969).

Not all the energy of a lamp can be made available at the entrance slit of the monochromator. Assuming $f/1$ entrance optics, the energy available from a 200-W tungsten-halogen lamp is 5.5 $\mu W/cm^2$ nm. The area of one steradian at 43-cm radius is $(43\ cm)^2$. Using a nominal bandpass of 10 nm and recognizing that the solid angle of an $f/1$ lens is 0.676, we see that the energy available at the monochromator becomes (Birth, 1971):

$$E_m = (5.5\ \mu W/cm^2\ nm)(43\ cm)^2\ (10\ nm)(0.676) \quad (11)$$

$$= 68.74\ mW/10\text{-nm bandpass}\ . \quad (12)$$

Hence, we can see that only a small fraction of the energy ever reaches the entrance to the monochromator—in this case, only 0.034%.

C. Monochromators

The function of a monochromator is to disperse, or spread out, the radiation according to wavelength. Prisms and gratings are the most popular dispersion elements used in monochromators. Neither type is perfect, and the user must take trade-offs in order to achieve an optimum optical system.

There are two types of gratings: the classical ruled-plane grating and the concave holographic grating. The ruled-plane grating has a series of triangular, parallel grooves cut into some solid material (such as glass) with a diamond-shaped tool mounted on a ruling engine. One of the major problems in making a ruled-plane grating is the inaccuracies, or errors, made with the mechanical engine, which results in "ghosts" and "grass" (stray light).

The holographic grating was developed to overcome many of the imperfections in ruled-plane gratings. The term *holographic* refers to the method used for producing this type of grating. Two intersecting laser beams are used to produce interference fringes in a photosensitive material deposited on optically flat glass (Fig. 5A). The interference fringes recorded on the photosensitive material are then developed, thus creating triangular-shaped grooves. Holographic technology has improved to the point where precision grooves can be made on concave plates, making it possible to produce "flat-field" monochromators without the use of auxiliary optics (Fig. 5B). The simplicity of the optical and mechanical parts of such systems has made possible the manufacturing of simple, high-throughput, narrow bandpass, low-stray-light spectrophotometers at costs commensurate with filter-based systems. Table II compares the ruled and holographic gratings.

A problem with gratings that does not exist with prisms is that light with several wavelengths leaves the grating at the same angle of dispersion (Fig. 6A). This phenomenon is called *overlapping orders*, and (as noted in Fig. 6B) the orders vary in their efficiencies. Therefore, when using any kind of grating system, filters or prism monochromators must be employed to eliminate the overlapping orders. For example, if a grating is used to scan the near-infrared region from 1.0 to 2.5 μm (Fig. 7), the spectrum beginning at 1.0 μm would appear at 2.0 μm as the second-order spectrum. Thus, a high-pass cutoff filter inserted at 1.8 μm would eliminate the second-order spectrum. The third-order spectrum starting at 3.0 μm is of no consequence since scanning stops at 2.5 μm because of lack of sensitivity of the PbS detectors.

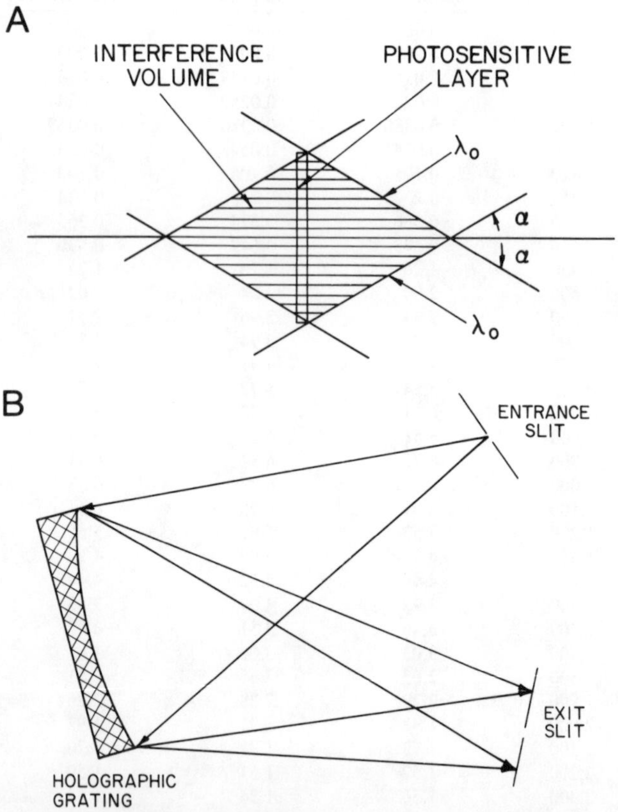

Fig. 5. Typical recording (or grooving) of plane, holographic grating using laser/etching techniques (A). Simplified monochromator using single concave holographic grating. In this case, grating, the only inside optical component, provides very linear dispersion at the exit slit (B).

TABLE II
Comparison of Ruled and Holographic Gratings[a]

Property	Classically Ruled Gratings	Holographic Gratings
Efficiency	At blaze 60–99%	At maximum 35–99%. Efficiency curve can be made flatter than ruled gratings to cover wider spectral domain. In blazed holographic gratings, efficiency curve is identical to ruled gratings
Ghosts	Approximately 10^{-5} of line	No ghosts
Scattered light	At best 10^{-5} to 10^{-6} at 5Å of laser line in the visible	10^{-6} to 10^{-8} at 5Å of laser line in the visible
Size	Generally limited to 8 in. × 8 in. for groove spacings of more than 600 g/mm	Up to 17-in. diameter in all groove spacings. Much larger size, up to 1 m, is possible
Groove density	Maximum of 3,600 g/mm; in general, stray light increases drastically with groove density	Up to 6,000 g/mm. Stray light does not increase with groove density
Aberration correction	Not possible	Possible

[a]Source: Hayat (1978).

Prism monochromators may be used in conjunction with gratings as order sorters. The system is then called a *double-monochromator* system. Figure 8 is an optical diagram of the Cary 17 monochromator. McClure and Hamid (1980) built the COMP/SPEC around a Cary 17 monochromator. Note from Fig. 8 that the radiation passes through the prism on its way to the grating. Hence, only a very narrow band of wavelengths reaches the grating (i.e., the orders are sorted out by the prism). The stray-light specification of the Cary 17 is as good as or better than any available (less than 0.1% at 2.65 μm). Unfortunately, Varian Instruments no longer manufactures this monochromator. However, there are a number of the Cary 14 and 17 instruments still in use, and when these are retired from service, the monochromators would provide the basis for a COMP/SPEC design.

The characteristics of monochromators vary from one design to another. If monochromator characteristics were constant, and a single-beam recording of relative lamp energy vs. wavelength were made, then the resultant curve would be a function only of the lamp output and detector-response curves. Experimentally, this is not found to be the case, as monochromator characteristics contribute significant spectral features to the output.

The reciprocal dispersion (nm/mm) of any monochromator is not constant but varies from one end of the range to the other. For instance, Fig. 9 is a plot of the reciprocal dispersion for the Cary 17. Over the near-infrared range from 0.9 to 2.6 μm, the reciprocal dispersion varies from 3.8 down to 2.8 nm per millimeter of slit width. Therefore, if the monochromator is operated over this range with a fixed 2-mm slit width, the half-intensity bandwidth of the radiation exiting the monochromator will vary from 7.6 nm at 0.9 μm to 5.6 nm at 2.6 μm. The bandpass can be held constant by programming the slit function into the controlling computer.

Observed responses of monochromators are influenced by Wood's anomalies, absorption characteristics of other optical elements (lenses, mirrors, etc.) both inside and outside the monochromator, and water vapor in the radiation path. Figure 10 shows an experimental response curve obtained for a Cary 17 in the single-beam mode. The grating for this system was blazed at 1.6 μm. *Blazing* a grating increases the energy efficiency of the grating over a wide range of angles on both sides of the angle at which the diffracted energy is maximized, which is called the *blaze angle*. The blaze angle C is the angle between the normal to the broad face of the groove and the grating normal (Fig. 11). A grating can be blazed at an angle specified by the purchaser to maximize efficiency in any part of the spectrum. For example, if the operator intends to work mainly in the region near 2,200 nm, the grating can be blazed at a wavelength in this area. A blazed grating will break up or diffract up to about 85% of the incident energy at angles close to the blaze angle and will maintain high (up to 50%) efficiency at angles of 15°–20° beyond the blaze angle in the first-order spectrum.

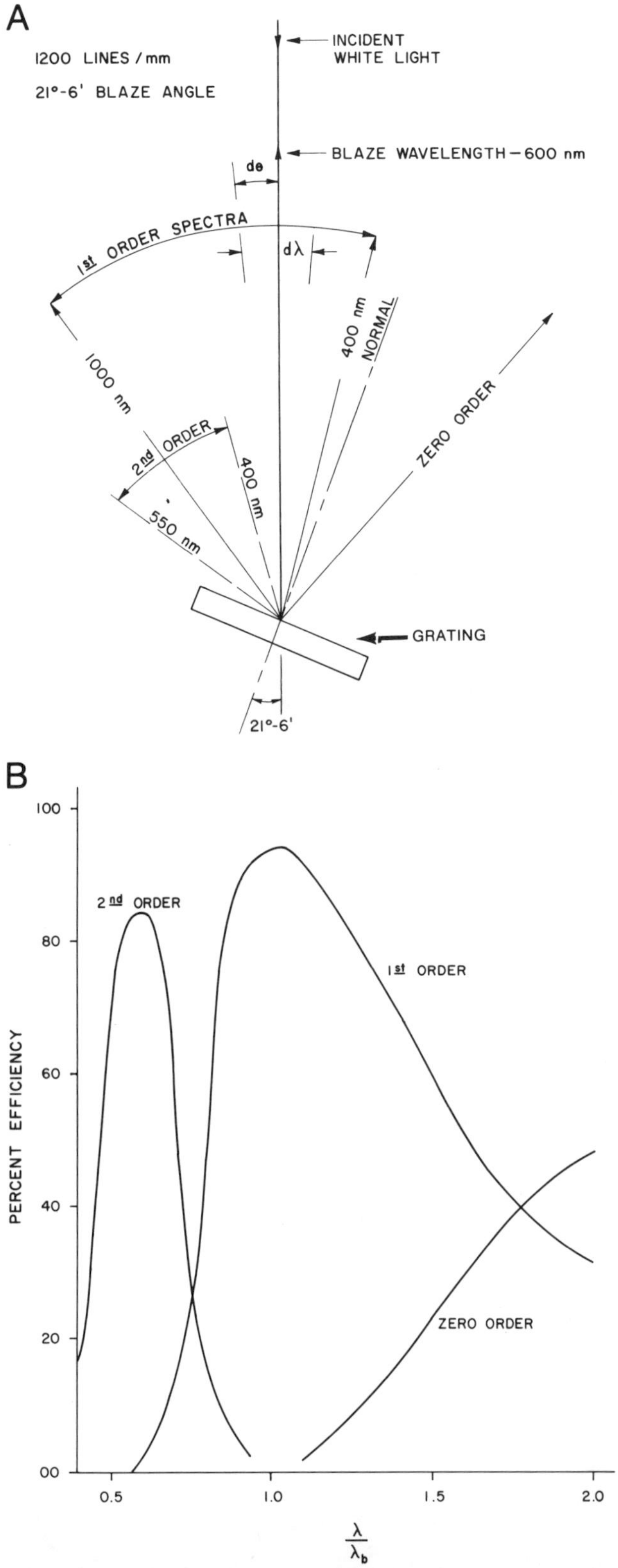

Fig. 6. Grating characteristics: spectral distribution (A) and efficiency (B). First-order spectrum is one usually used because of its higher-efficiency broad wavelength coverage.

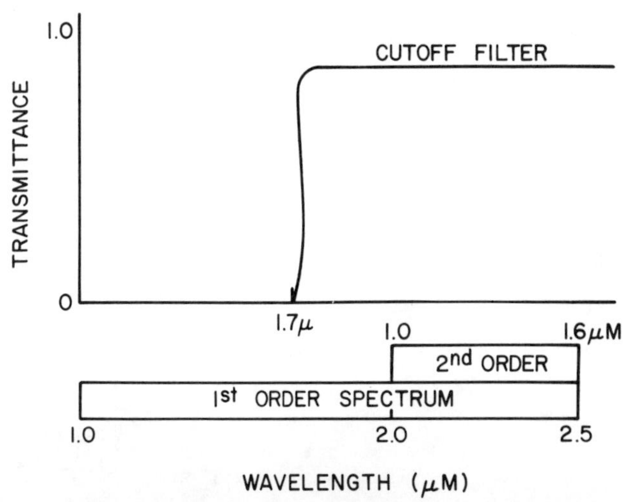

Fig. 7. Overlapping of orders of near-infrared grating. Cutoff filter can be used to block second-order spectrum.

There are only two Wood's anomalies in the near-infrared, at 0.959 μm and 1.259 μm. *Wood's anomalies* are spurious absorbances, usually fairly sharp spikes, that are associated with any grating monochromator operating in the single-beam mode and covering a wide wavelength range. They are caused by redistribution of the energy of the primary order among the other orders, including the one leaving the exit slit, and depend partly on the material from which the grating, and particularly its surface, is made. Most gratings are made from glass, with an aluminum or silver coating layer, into which the grooves are cut. More recently, the coating is a photographic film. Silver has a strong absorption band at about 300 nm and is not recommended for gratings intended for measurements in the UV area.

There is to date no way of avoiding Wood's anomalies completely, although the grating can be blazed so that they occur in an area that is of lower importance to the operator. This may cause some loss in efficiency of the grating, but it is important to effect a compromise to the extent that the Wood's anomalies do not interfere with the measurements to be made. For example, measurements made in the area of 1,259 nm may be affected. The sharp dips at 1.38, 2.2, and 2.5 μm are the result of OH absorption by traces of water vapor in the air path and water in the silica optics. The Cary 14 with Infrasil optics or the Cary 17 with Superasil W optics minimizes the effects of this water absorption. All of these anomalies occur only in the single-beam system response curves, and they are corrected by the computer in the computation of reflectance or log $(1/R)$. If the above anomalies do appear in the reflectance or log $(1/R)$ curves, the system is not performing properly. Gratings, mirrors, detector (PbS), and lamp (tungsten-halogen) all contribute to the nonlinearity of a near-infrared system (see Fig. 10).

D. Filters

The first instrument designed to measure reflected light used filters to define the wavelength range (Taylor, 1919). As the demand for narrow-band interference filters grew, so did the technology of filter manufacturing, so that today narrow-band filters for almost any region of the optical spectrum can be obtained.

FIXED FILTERS

The quality of narrow-band interference filters is determined by the percentage of transmittance at the peak wavelength, the

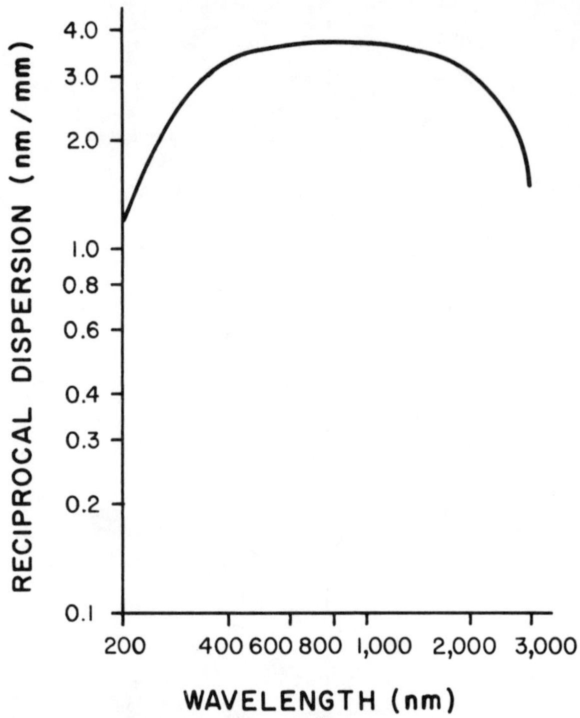

Fig. 9. Reciprocal dispersion of Cary 17 monochromator. Note that dispersion is not constant over wavelength but varies from 1.1 nm/mm to 3.7 nm/mm.

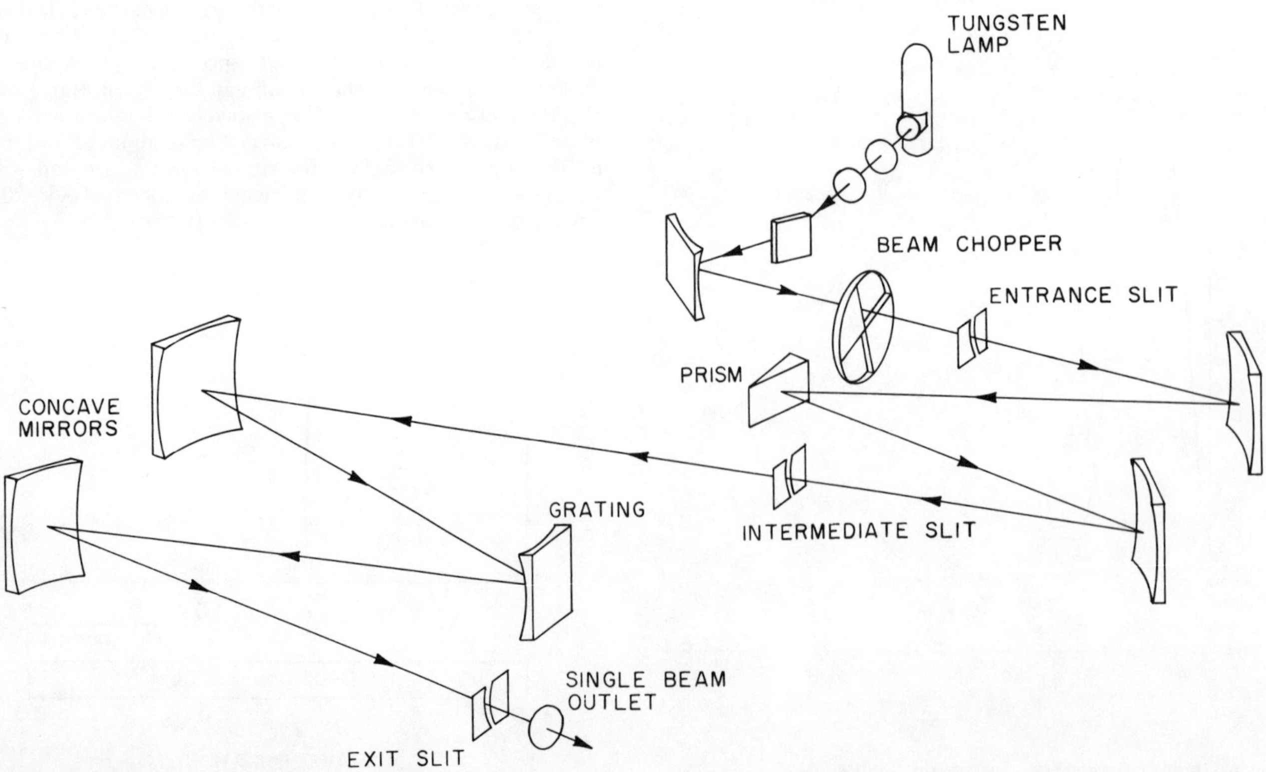

Fig. 8. Diagram of Cary 17 monochromator, a double monochromator with light passing first through prism monochromator and then reflected from grating. Prism thus removes higher-order spectra.

half-intensity bandpass, and the upper (UCO) and lower cutoffs (LCO) (Fig. 12). Filters suitable for near-infrared instrumentation should have a half-intensity bandwidth (HIBW) of 10–20 nm and a peak transmittance of at least 30%.

Narrow-band interference filters that do not have LCO and UCO pass more than one wave band. This phenomenon is similar to grating monochromators. Maxima of the same height at λ_0 occur also at $\lambda_0/2$, $\lambda_0/3$, etc. For instance, if the longest wavelength at which maximum intensity occurs (first-order band λ_0) is found at 2.0 μm, maxima will also occur at 1.0 μm (third-order band), etc. The other orders have the same peak transmittance but a narrower and steeper bandpass.

In some cases it may be desirable to make use of these narrower and steeper orders. For example, if the second order fell at 1.5 μm, the first order would be at 3.0 μm and the third at 1.0 μm. The influence of the first-order bandpass could be ignored since the PbS does not respond in this region, but the third-order band would have to be masked in order to sense only the second-order wavelengths. If the third order is used to isolate the 1.5 μm band, the second order appears at 2.75 μm and the fourth order at 1.125 μm. This increasing proximity of the side bands acts as a limitation regarding the order of interference that can be made use of for practical purposes.

The literature records many applications of the planar-fixed filter configuration. Birth and Norris (1965) used two filters mounted in a rotating wheel to study the quality of horticultural products; different pigment systems could be studied by changing the filters. McClure (1968, 1969b) designed a two-filter moisture meter for determining the water content of tobacco leaves in the transmittance mode; Fig. 13 is a simplified diagram of this instrument. Energy from a tungsten lamp passed alternately through two filters (1.82 and 1.93 μm) and through a tobacco leaf. The energy reaching the detector was decoded with photodiodes and the difference in the logarithm of the two detector signals was computed for scaling and display on the panel meter.

An obvious disadvantage of the fixed-filter arrangement is that the number of filters that can be mounted on a given filter wheel is limited. However, there is evidence that indicates that the precise positioning of the bandpass of the filters is not necessary to get good estimates of sample chemistry (McClure, 1969a).

VARIABLE FILTERS

Variable interference filters (sometimes called wedge interference filters) can be made by pressing the interference material into the shape of a wedge between two pieces of glass. Norris (1964) constructed a spectroradiometer using a linear wedge-interference filter for the range 0.4–1.2 μm. This instrument was later used (Downs et al, 1964) for measuring irradiance in plant growth chambers.

Circular variable interference filters are available for the near-infrared range (Birth, 1971; Optical Components, 1983). These are made in circular segments, each covering only a portion of the near-infrared region. However, with the advent of the holographic grating, both linear and circular variable filters became too expensive for most applications. Much better resolution with less stray light is obtainable with the gratings and at lower costs.

TILTING FILTERS

Filter parameters are generally expressed with the incident beam collimated and normal to the surface of the filter. When the filter is tilted from the perpendicular, or the beam is other than normal to the surface of the filter, the passband broadens, its center wavelength shifts toward shorter wavelengths, and its peak transmittance is reduced (Fig. 14). The extent of these

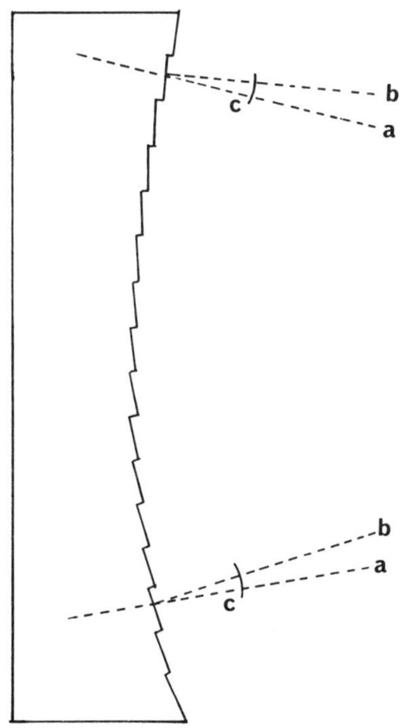

Fig. 11. Illustration of blaze angle of holographic grating: (a) = grating normal, (b) = blaze normal, and (c) = blaze angle.

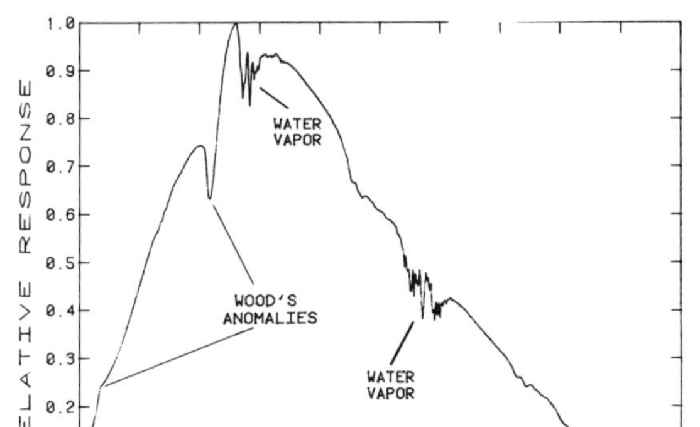

Fig. 10. System response of COMP/SPEC. This response curve includes nonlinearities of source, detector, and monochromator optics. Wood's anomalies and water vapor absorption bands are also present.

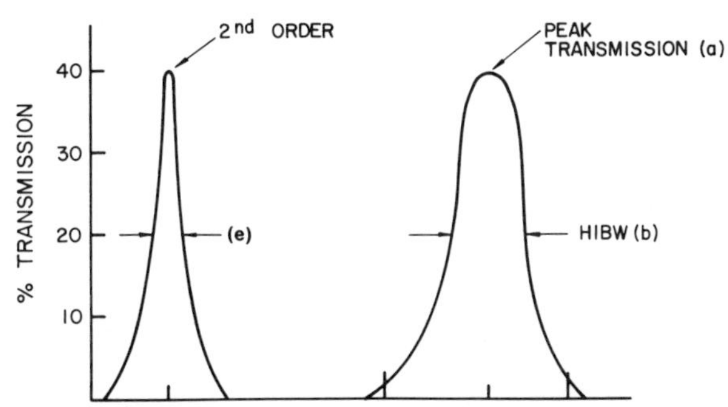

Fig. 12. Characteristics of narrow-band interference filters: (a) peak transmission, (b) half-intensity bandwidth (HIBW), (c) lower cutoff (LCO), (d) upper cutoff (UCO), and (e) second-order bandpass.

changes is ultimately determined by the number of degrees of tilt from the normal angle of incidence and/or the cone angle of the beam. The formula for determining the shift of the center wavelength with the angle ($\leqslant 30°$) of incidence is:

$$\lambda_\phi = \lambda_0 \sqrt{\frac{n^2 - \sin^2\theta}{n}}, \tag{13}$$

in which λ_ϕ = center wavelength at angle, λ_0 = center wavelength at normal incidence, θ = incident angle in degrees, and n = effective index of refraction of the total filter.

Of course, the narrower the bandwidth of the filter, the more noticeable are the effects of the angle of tilt. Although the wavelength shift and bandwidth change can generally be theoretically predicted by use of the formula above, exact wavelength and bandwidth changes for a given angle of incidence and/or cone angle are best determined empirically for each individual filter as in Fig. 14.

In electro-optical systems, the most common off-normal situation encountered is one where the radiation source is incident to the filter surface in either a convergent or divergent configuration. The effect encountered in such an arrangement is substantially less than that produced by collimated off-normal rays of equivalent angles.

E. Detectors

Infrared detectors may be characterized by three basic parameters: their spectral response, speed of response, and the minimum amount of radiant power they can detect. Other parameters, such as temperature of operation, magnitude and type of signal produced, and recommended circuit that best couples the detector to a readout device, are of less importance.

Infrared detectors may be classified into two broad categories according to principle of operation: thermal detectors and photon detectors. Thermal detectors have been available since the early 1800s. Their principle of operation involves the absorption of infrared energy by a temperature-sensitive surface. Generally, the response time of thermal detectors is too long (~10 ms) to be considered for near-infrared applications. One unique advantage of thermal detectors is their uniform sensitivity at all wavelengths.

Although thermal detectors have been around since the early 1800s, photon detectors are relatively new, having been developed during the last 35 years. The operation of photon detectors requires that the incident radiation be strong enough to liberate charge carriers from either the crystal lattice (intrinsic detectors) or from impurities intentionally added to the host crystal (extrinsic detectors).

The minimum detectable power, called *noise equivalent power*, is usually measured by placing a black body of known temperature (usually 500° K) at a convenient distance from the detector. Its radiation is interrupted periodically (1,000 Hz) by means of a rotating sector disk, such that a detector signal of constant frequency is obtained. The detector noise is measured by calculating the radiant power received at the detector, usually with a narrow-band amplifier, when the detector is shielded from the radiation source. The noise equivalent power (NEP) is then calculated from the relation:

$$\text{NEP} = \frac{P}{(S/N)}, \tag{14}$$

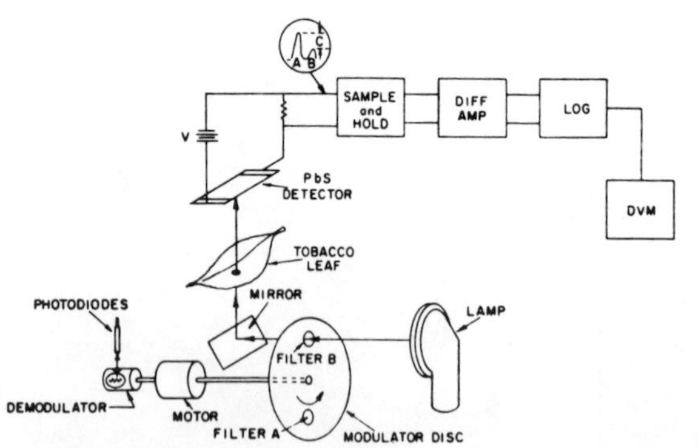

Fig. 13. Schematic/block diagram of near-infrared moisture meter for tobacco, a dual interference-filter design.

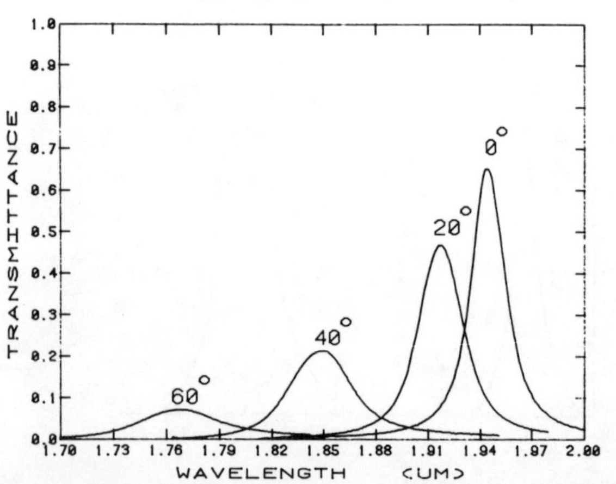

Fig. 14. Effect of tilting narrow-band interference filter from normal to light beam. Note that peak wavelength shifts toward shorter wavelengths (regardless of directions of tilt from normal) and that half-intensity bandwidth broadens.

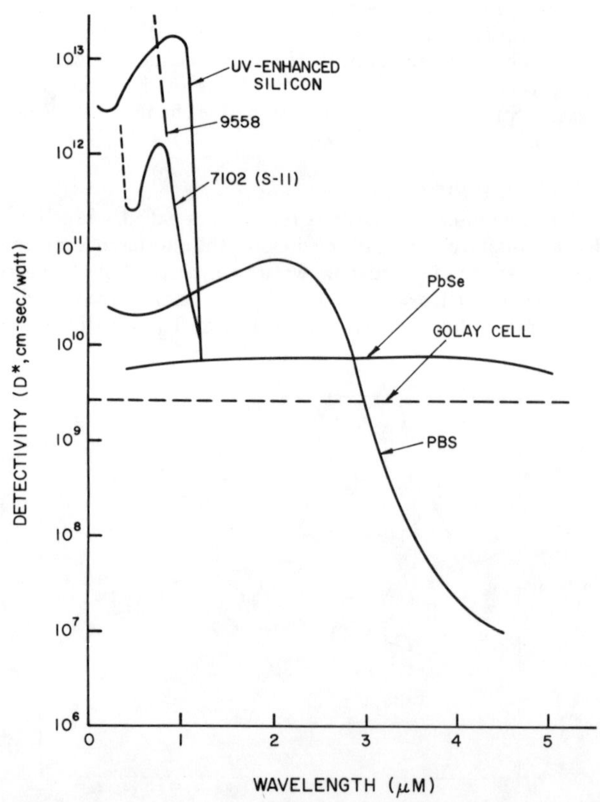

Fig. 15. Detectivity (D^*) of more popular near-infrared detectors at room temperature. These curves were obtained at chopping frequency of 90 Hz with amplifier bandwidth of 1 Hz. The PbS detector has adequate detectivity for near-infrared work.

where P = power received by the detector, S = signal voltage, and N = noise voltage.

The more commonly used term for characterizing the sensitivity of infrared detectors is detectivity, D, where:

$$D = \frac{S/N}{P} \tag{15}$$

or, in comparison of detectors, a normalized quantity, D^*, which is the detectivity of a detector of a 1-cm^2 area whose noise is reduced to that obtained with an amplifier of 1-Hz bandwidth, or:

$$D^* = \frac{S/N}{P_D}\left(\frac{f}{A}\right)^{(1/2)}, \tag{16}$$

where f = amplifier bandwidth, A = detector area, and P_D = the power density at the detector. The spectral curves for the more popular near-infrared detectors are shown in Fig. 15.

Both the 7102 and the 9558 PMTs (photomultiplier tubes) have higher D^* than lead sulfide (PbS), but both have limited operating ranges. In the COMP/SPEC, for example, the operating range of the 9558 is from 0.4 to 0.85 μm. UV enhancement of silicon (Longerich and Ramaley, 1974) also improves its performance in the near-infrared, but its range of operation in the COMP/SPEC is limited to 0.55–1.15 μm. PbS can be operated over the entire region from 0.5 to 4.0 μm with a D^* that is only slightly larger than lead selenide (PbSe), but the characteristics of other elements in the system (i.e., lamps and monochromator) limit its usefulness to the range 0.9–2.6 μm.

Silicon, unlike PbS, is not especially sensitive to temperature change and need not be thermoelectrically cooled, especially when operated in an air-conditioned laboratory. From 0.4 to 0.95 μm, the deviation of absolute spectral responsivity with temperature is less than 0.5% per degree centigrade.

Cooling can be accomplished using thermoelectric elements. The PbS detector arrangement, described later, is cooled with four Borg-Warner 930-17 thermoelectric cooling elements. The circuitry for controlling temperature is given in Fig. 16. This circuit controls the detector temperature to within ±0.01° C for a laboratory temperature variation of ±5° C.

The thermistor is mounted in the cooled block of aluminum (the circular piece in Fig. 17A) within 1 mm of a detector. The hot block (the square piece of aluminum in Fig. 17B) is mounted against the surface of the sample compartment to provide additional heat conduction to the outside.

III. COMPUTERIZED SPECTROPHOTOMETRY: THE COMP/SPEC

An instrument that reaches the user today without incorporating some type of digital computer technology is probably obsolete before it is pressed into service. Near-infrared spectrophotometry, with the exception of Fourier transform infrared spectroscopy (FTIR), is probably more dependent on computers than any other phase of spectrophotometry.

The literature reveals several instruments that were interfaced to a computer to enhance their function and to improve ease of operation. Longerich and Ramaley (1974) described the circuitry needed to connect a Cary 14 spectrophotometer to a 2114B Hewlett-Packard computer. An optical tachometer was attached to the wavelength drive mechanism of the spectrophotometer for wavelength encoding.

Mattson and Smith (1977) interfaced a Nova 1220 to a Perkin-Elmer 180 dual-beam spectrophotometer. The abscissa (wave number) carousel was stepper-motor driven; an electromechanical encoder sent signals to the wave-number readout as the carousel rotated. Ordinate values were converted to ratiometer counts between 1 and 12,000. Both abscissa and ordinate values were transmitted in serial ASCII format from the spectrophotometer to the computer. A 10-bit digital-to-analog converter was used to drive an X-Y recorder for displaying the spectra.

Norris and Barnes (1976) briefly described a computerized spectrophotometer incorporating a Cary 14 monochromator and a Nova 2/10 computer. Shenk and Hoover (1976) discussed a computer-controlled spectrophotometric system that utilized a Jobin-Yvon HRS-2 monochromator. Technicon Industrial Systems and Neotec Corporation currently are building computer-controlled instruments, but detailed designs are proprietary. McClure and Hamid (1980) described in detail the design of the COMP/SPEC. Its design will serve as a means of drawing together the components discussed above into one system.

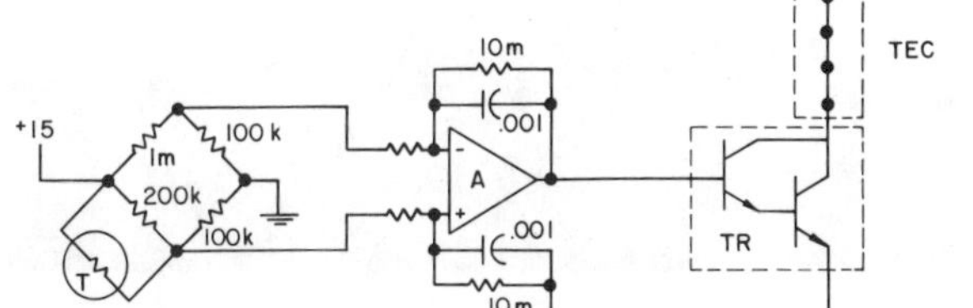

Fig. 16. Thermoelectric cooling control circuit for PbS detectors in COMP/SPEC (T = Fenwal GA51P32, 100K at 25° C; A = Signetics 5556; TR = Motorola HEP-S9102; TEC = four Borg-Warner 930-17 thermoelectric elements.

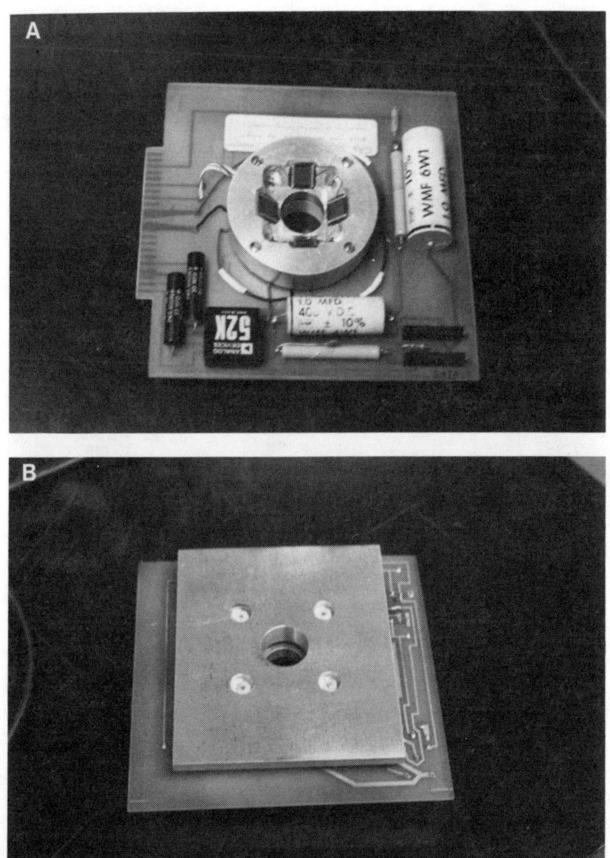

Fig. 17. Mounting arrangement for PbS reflectance detectors in COMP/SPEC: (A) circular cold block to which PbS cells are bonded with thermally conductive glue and (B) square heat sink that conducts heat to instrument cabinet.

A. General Design

A block diagram of the computerized spectrophotometer is shown in Fig. 18 and a pictorial view in Fig. 19. The system is constructed around a Cary 17 monochromator that is controlled by a Nova 2/10 minicomputer with 64K bytes of core memory. Spectra are displayed on a Tektronix 4010 storage display console; hard-copy spectra are recorded on a Tektronix 4662 X-Y plotter. Computer peripherals consist of (i) a Data General 6027 tape drive; (ii) two disk drives consisting of a Data General 6050 10-megabyte disk and a 6030 floppy disk; and (iii) an OKIDATA 2350 line printer. The computerized spectrophotometric system is mounted on rollers so that it can be conveniently positioned within the laboratory.

B. Optomechanical

The Cary 17 is a double monochromator consisting of a prism and a grating monochromator in series, permitting measurements in excess of six optical density units (OD) before stray light becomes a problem. High energy output in the near-infrared region of the spectrum was achieved by specifying the blaze wavelength of the grating to be 1.6 μm. Even though the scanning range of the Cary 17 is from 0.18 to 3.2 μm, the use of lead sulfide detectors for near-infrared measurements restricts the useful scanning range from 0.9 to 2.6 μm with a nominal bandpass of 7 nm.

The Cary 17, unlike the Cary 14, is driven by a stepping motor with a stepping resolution of 200 steps per revolution. The gear train coupling the motor to the cam screw of the monochromator provides further reduction, giving a wavelength resolution of 0.01 nm/step. The advantages of the stepping-motor drive over the conventional synchronous motor include its simplicity, resettability, reversibility, and superior flexibility for operating under computer software control. Both the scanning rate and data interval can be controlled by software. Furthermore, wavelengths can be encoded by a software counter without additional hardware. The work of the author thus far has been at a scanning rate of 10.8 nm/s with a return speed of 18 nm/s. Controls on the front of the monochromator allow operator intervention for setting the starting wavelength before going to computer-controlled scanning.

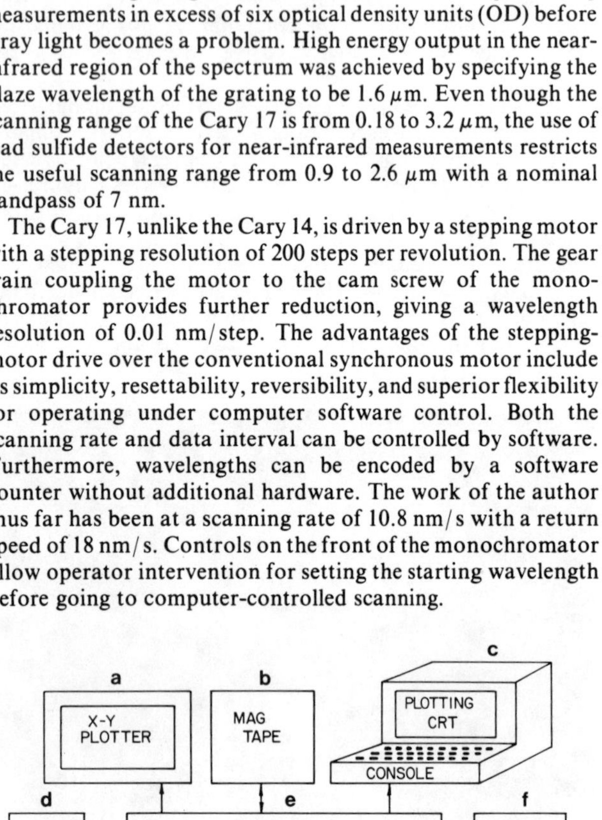

Fig. 18. Block diagram of COMP/SPEC, custom-built computerized spectrophotometer: (a) Tektronix 4662 plotter, (b) Data General (DG) 6027 magnetic tape drive, (c) Tektronix 4010 plotting CRT, (d) OKIDATA 2350 printer, (e) DG Nova 2/10 computer, (f) two disk drives: DG 6030 and DG 6045, (g) DXN tungsten-halogen lamp, (h) Cary 17 monochromator, and (i) sample compartment with PbS detector (M = mirror, D = detector, S = sample).

The light source is a Sylvania DXM tungsten-halogen lamp rated at 250 W at 30-V input. The light is chopped mechanically at 60 Hz before entering the monochromator. Slits of the monochromator are manually adjustable to give a bandpass of up to 11.0 nm for a maximum width of 3 mm.

Light from the monochromator vertically enters the sample compartment (32 × 23 × 23 cm) by being reflected at a right angle by an ellipsoidal first-surface mirror. The mirror has a short radius of 14.98 cm and a long radius of 24.38 cm. When the mirror is positioned with the short axis in the horizontal plane, the slit of the monochromator is focused near the top of the sample compartment for reflectance measurements. Turning the mirror 90° focuses the beam near the bottom of the sample compartment for transmittance measurements. A permanently mounted card holder at the top and a removable card holder at the bottom of the sample compartment make it easy to convert from transmittance to reflectance measurements, respectively, and vice versa.

The system currently operates in the single-beam mode. As compared with the double-beam mode, the single-beam mode has the advantage of high energy throughput, which is important for optically dense light-scattering materials. Furthermore, signal-to-noise ratio is higher in this mode because of the absence of cross-talk noise encountered in double-beam systems. Cross-talk noise, a characteristic of double-beam instruments, is produced when the light source is divided to illuminate the sample and the internal standard. A signal from one beam can reach the other and interfere with the measurement, producing the cross-talk noise.

C. Optoelectronic

Separate detectors are required for the visible and near-infrared regions of the spectra. An EMI 9558 photomultiplier tube covers the region from 0.3 to 0.90 μm for high OD applications in the transmittance mode. Silicon solar cells work well over the region from 0.4 to 1.2 μm in both the reflectance and transmittance modes, and PbS photoresistive cells cover the region from 0.9 to 3.0 μm. Because of the dynamic range of the energy throughput of the monochromator and the response of the detectors, the practical scanning range of the system is limited in the single-beam mode to 0.4–0.9 μm for the visible and 0.9–2.6 μm for the near infrared (using PbS detectors). Four detector circuits are shown in Fig. 20. With the exception of circuit (B), each is constructed on an 18- × 18-cm, custom-etched printed circuit card that plugs directly into prewired 44-pin pc-edge connectors mounted in the sample compartment. Circuit (B) was mounted directly on the bottom of the 9558

Fig. 19. Author prepares sample for scanning on COMP/SPEC I located in Bioinstrumentation Laboratory at North Carolina State University. Another COMP/SPEC (COMP/SPEC II) is owned and operated by U.S. Department of Agriculture scientists at Oxford Tobacco Research Station, Oxford, NC.

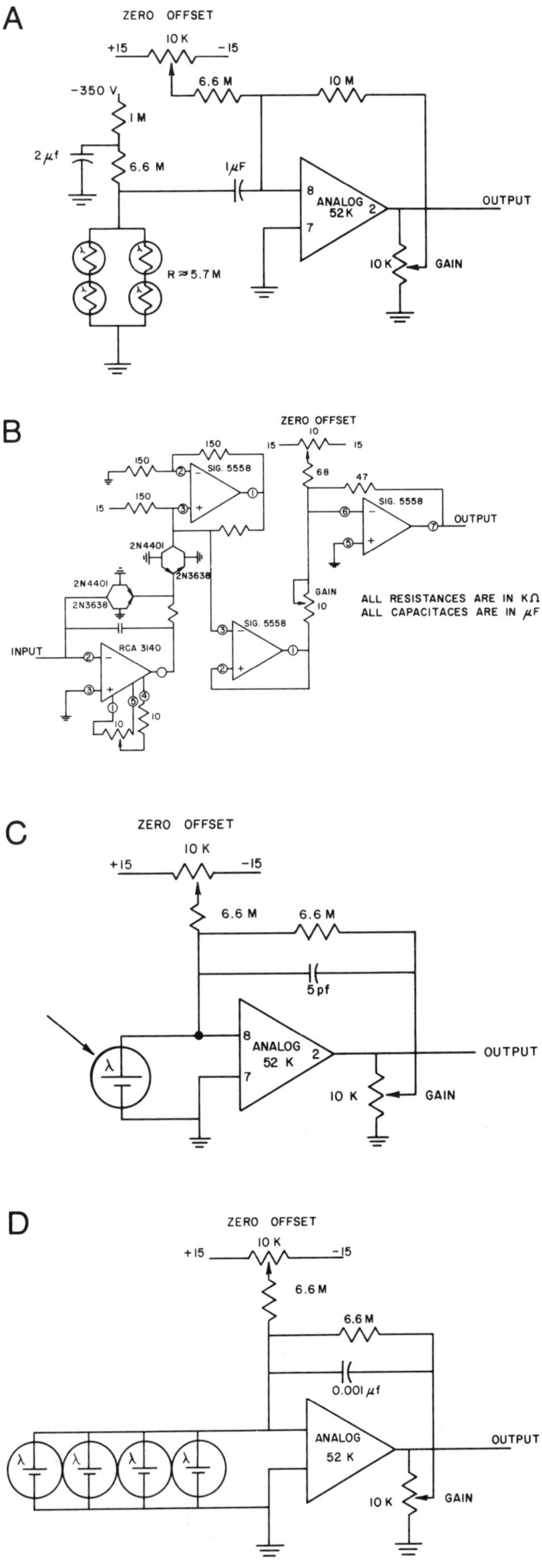

Fig. 20. Schematics of four detector circuits designed for COMP/SPEC: (A) PbS reflectance, (B) photomultiplier-tube log transmittance, (C) silicon transmittance, and (D) silicon reflectance.

PMT housing. An AC synchronous 1,800-rpm motor drives a disk with two open sectors for modulating the light beam entering the monochromator. As shown in Fig. 21, the detector preamplifier output is a square waveform, the peak-to-peak magnitude of which is proportional to full light (A) and no light (B) and is in phase with the AC line voltage.

Demodulation of this signal is accomplished by the circuit in Fig. 22. The circuit consists of two sample-and-hold modules (Analog Devices SHA-5) and a difference amplifier (Analog Devices 52 K). The inputs of both SHA-5s are connected to the output of the detector-preamp. If the sample-and-hold pulses given to the SHA-5s are properly timed, the output of the difference amplifier will be the voltage difference between points A and B of the detector output signal.

The sample-and-hold pulses, TIM-1 and TIM-2, are generated by the circuit in Fig. 23. It is essentially a phase-synchronized rectifier that gives two 30-Hz, 20-s pulse strings at 180° apart that are used to trigger the SHA-5s. TIM-1 drives SH-1 and TIM-2 drives SH-2 (Fig. 22). The output of the difference amplifiers is, therefore, A-B, the signal that is proportional to the intensity of the light reaching the detector.

D. Digital Interface

The digital interface was designed using the Data General Corporation general-purpose interface board with 4040 basic interface and 4041 input/output data registers (Archer, 1969). The 4040 basic interface contains the prewired circuitry that connects the interface to the data and control lines on the bus, networks for passing the interrupt and data channel priority signals along the bus, and a device selection network that allows

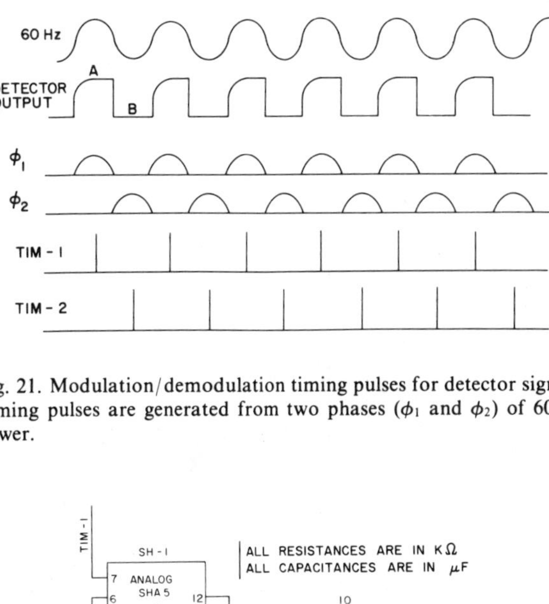

Fig. 21. Modulation/demodulation timing pulses for detector signals. Timing pulses are generated from two phases (ϕ_1 and ϕ_2) of 60-Hz power.

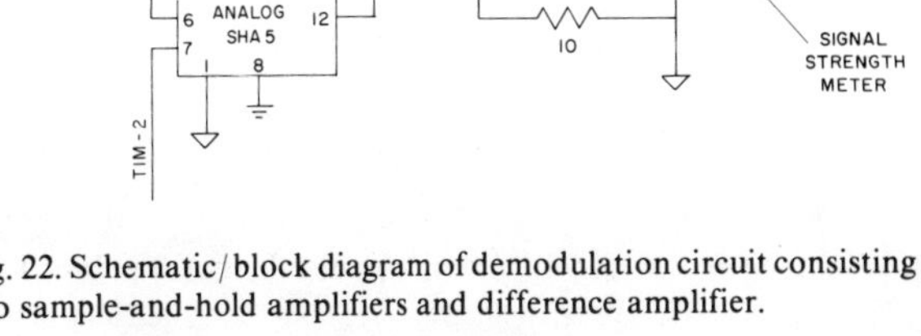

Fig. 22. Schematic/block diagram of demodulation circuit consisting of two sample-and-hold amplifiers and difference amplifier.

a choice of any device code by putting in appropriate jumpers. The 4041 data registers act as buffer to control the data flow between custom logic and the internal data bus. All the above circuitry is mounted on a standard 38- × 38-cm board that is divided in the middle by two rows of approximately 200 wire-wrap pins, of which 48 are available for custom interfacing. Half the board is reserved for customer logic and is configured for mounting 65 14- or 16-pin ICs.

Commercially available components were mounted on the general-purpose board for control of the monochromator and the analog-to-digital converter. The device selection network was jumpered for a device code of 21 for both the analog-to-digital converter and the monochromator control. Scanning direction of the monochromator was provided via the output buffer according to the status of DATA 14 and DATA 15 (Fig. 24).

A mechanical chopper synchronously rotating at 1,800 rpm in conjunction with the H-100 interrupter V gives a forward scanning pulse rate of 1,080 pulses per second. The output of the interrupter is buffered with a transistor (W) and shaped by a Schmidt trigger (X) before going to the synchronizer (Y). The computer initiates forward scanning by raising DATA 15 to logic 1 and lowering DATA 14 to logic 0. This releases the clear of the flip-flop J so that it can be set by the next TIM-2 pulse. Since the flip-flop J is negative edge triggered, the presence of TIM-2 when DATA 15 goes high will not set the flip-flop. The synchronizer (Y) shapes the output of the interrupter, which is timed with the one shot (Z). The output of the one shot is passed through gates (Q) and (R) to the driving transistor (S) for the translator module. Pulses at the input of (S) move the stepping motor in the forward (UP/SCAN) direction.

The output of (Z) also triggers one shot (E′), which triggers the one shot (F′), which furnishes the convert command to the A/D converters. The OR-gate (D′) allows additional START pulses from the computer to generate additional conversions as required by the data-taking scheme.

The end of convert signal from the A/D converter is double-buffered through gates (I′) and (H′) and taken inside the computer. The output of the one shot (G′) transfers the binary conversion of the A/D converter to the output of the data input buffer and at the same time sets the DONE flip-flop and raises the interrupt flag. At this time the data are ready to be brought into the computer with the DIA instruction.

Reverse scanning of the monochromator is initiated by setting DATA 14 to 1 and DATA 15 to 0. There is no requirement for synchronization in the reverse mode because no data are taken. Pulses generated by IOPLS from the computer can be varied as the system requires. For example, reverse scanning begins at a rate of 1,000 pulse per second for 1,000 steps and then changes to 1,800 steps per second until the scan reaches within 1,000 steps of the starting wavelength, whereupon the speed is reduced to 1,000 steps per second for the last 1,000 steps.

The stepping motor of the monochromator is bidirectionally controlled by a SLO-SYN STM 1800 DV translator module with logic inputs that can be easily interfaced to a computer. A single 10- to 15-V negative exchange of voltage with a minimum pulse width of 50 μm (fall and rise time not to exceed 2 μs) at pins (R) or (S) of the translator module will cause the motor to step down and up the wavelength scale, respectively. Two transistors pulled up to +15 V are adequate to interface the translator device to the TTL logic of the computer. The module is capable of driving the stepping motor in a four- or eight-step switching sequence. The eight-step sequence was employed in the COMP/SPEC, giving a scanning resolution of 0.01 nm.

The manual control circuitry is shown in detail in Fig. 24. In

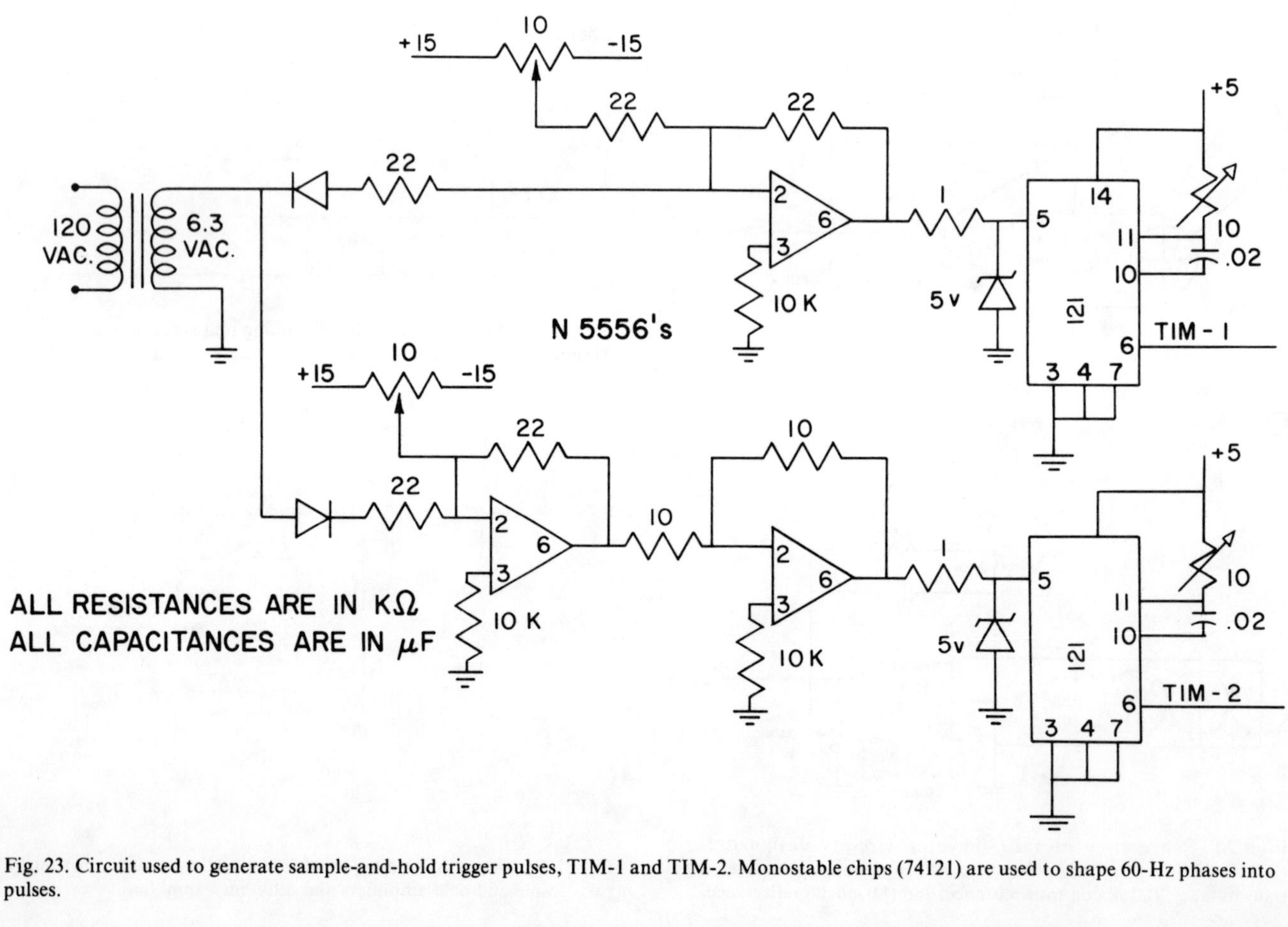

Fig. 23. Circuit used to generate sample-and-hold trigger pulses, TIM-1 and TIM-2. Monostable chips (74121) are used to shape 60-Hz phases into pulses.

the manual mode, DATA 14 and DATA 15 are a logic 0. The output of NAND gates (C) and (Q) is high, allowing outputs of gates (F) and (P) to activate the driving transistors (H) and (S), respectively. The oscillator OSC is on the translator pc-board and is activated by switch S_3. Switch S_2 changes the operation from computer to manual, and S_1 determines the direction of manual scanning. Manual scanning is asynchronous because no data are taken.

IV. PERFORMANCE OF THE COMP/SPEC

A summary of the specifications and certain performance indices for the COMP/SPEC is given in Table III. Data for computing performance indices were taken at a scanning rate of 10.8 nm/s and a return rate of 18 nm/s. Thus, the forward-scanning (data-taking) time was 2.62 min and the return (reset) time was 1.57 min, for a total time per spectra of 4.19 min. Each spectrum scanned contained 1,700 data points, with each point being the average of 1,126 analog/digital conversions.

A. Photometric Noise

Photometric noise in a computerized spectrophotometer is a measure of total system noise inherent in the detectors, amplifiers, and A/D converter. Photometric noise for the COMP/SPEC was computed from two reflectance spectra, from 900 to 2,600 nm, of a Halon reflectance standard run as a sample with a 20% attenuation (0.7 log [1/*R*] units). A log (1/*R*) spectrum was computed for each reflectance spectrum, and the two log (1/*R*) spectra were subtracted so that the noise data are clustered about zero. This noise spectrum S_n was used to compute the photometric noise. No smoothing was used in any computation.

Peak-to-peak noise was computed by subtracting the minimum and maximum data points in S_n. Bias *b* was obtained by averaging all the points in S_n, or:

$$b = \frac{\sum_{n=1}^{N} d}{n}, \tag{17}$$

TABLE III
Specifications and Performance Indices of COMP/SPEC

General

Light source: Sylvania DXM (250 W at 30 VDC)
Slits: adjustable 0–3 mm
Bandpass: 0–11 nm for range 900–2,600 nm
Wavelength drive: Slo-Syn TYPE MO91-FD06
Scan speed: 10.8 nm/s
Return speed: 18 nm/s
Optical path: single beam
Range: 0.4–2.6 μm
A/D conversion time: 10 μs max
A/D word size: 15 bits plus sign bit

Performance

Absorbance limits: 7 OD with 9558 photomultiplier tube
Silicon: 5 OD
PbS: 4 OD
Stray light: 0.0001% for range .24–.5 μm
0.001% for range .21–1.8 μm
0.1% at 2.65 μm
Wavelength precision: 0.005 nm at 1682.4 nm
Wavelength resolution: 0.01 nm
Photometric noise: 162 μOD

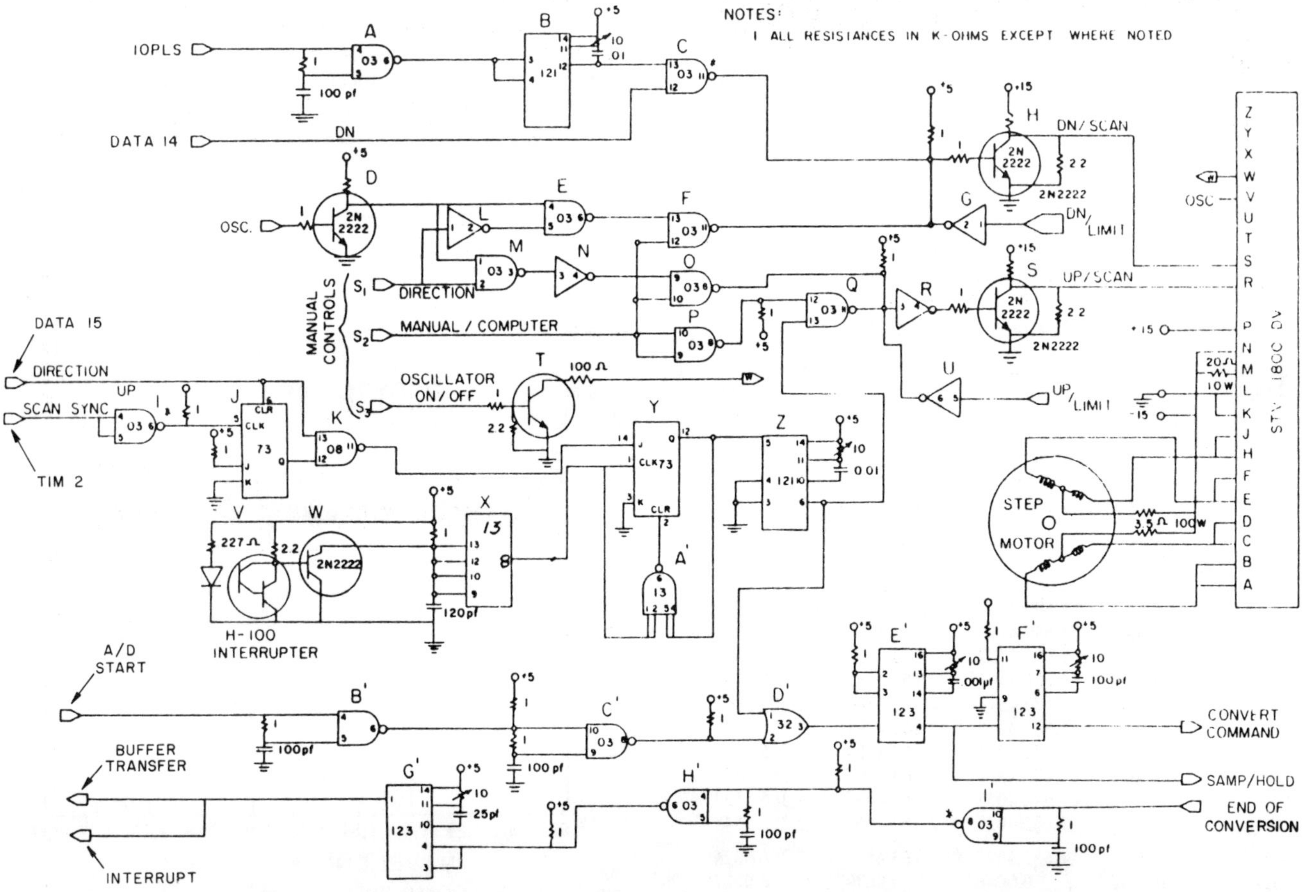

Fig. 24. Circuit diagram of scan-control interface and A/D converter control circuit. Data taking was timed to occur between stepping-motor pulses.

where b = bias, d = value of the nth data point, and N = number of data points in S_n.

The RMS noise was the computed square root of the average of the square of the data points in S_n, or:

$$S_{\mathrm{RMS}} = \sqrt{\frac{(d)^2}{n}}, \tag{18}$$

where S_{RMS} = RMS noise.

Plotting S_n as a function of wavelength (Fig. 25) shows the RMS noise to be 162 μOD with a peak-to-peak noise of +1,150 and −1,112 μOD at 2.536 μm and 2.576 μm, respectively. The average offset was 44 μOD. The noise was higher on each end of the spectrum where the signal-to-noise ratio is lower.

B. Wavelength Precision

Wavelength precision is extremely important in near-infrared spectroscopy, much more important than accuracy.

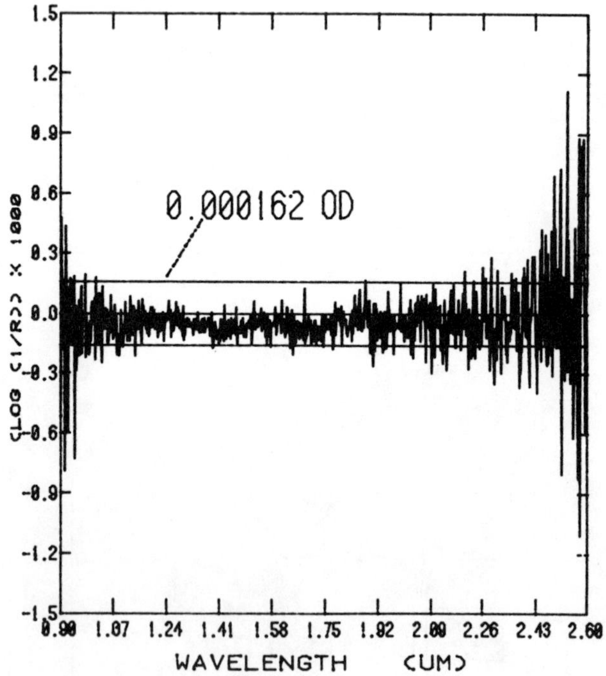

Fig. 25. Noise spectrum of COMP/SPEC I. The RMS noise is 162 μOD.

TABLE IV
Wavelength Precision of COMP/SPEC

Reading Number	Wavelength Number 1	2	3	4
1	1,143.9643	1,682.4187	2,163.4509	2,306.5693
2	1,143.9640	1,682.4129	2,163.4722	2,306.5598
3	1,143.9634	1,682.4129	2,163.4722	2,306.5600
4	1,143.9779	1,682.4103	2,163.4635	2,306.5657
5	1,144.1007	1,682.4078	2,163.4540	2,306.5807
6	1,144.0965	1,682.4026	2,163.4593	2,306.5578
7	1,143.9675	1,682.4074	2,163.4599	2,306.5594
8	1,143.9642	1,682.4061	2,163.4507	2,306.5661
9	1,143.9671	1,682.4057	2,163.4495	2,306.5527
10	1,143.9643	1,682.4035	2,163.4566	2,306.5548
AVG	1,143.9929	1,682.4087	2,163.4559	2,306.5626
SD	0.0548	0.00462	0.00797	0.00775
CV	4.8×10^{-3}	2.7×10^{-4}	3.7×10^{-4}	3.4×10^{-4}

Correlograms, showing the relationship between wavelength information (log [1/R], etc.) and chemistry of all samples in a calibration set, often exhibit sharp peaks. A shift in the wavelength mechanism from one spectrum to another would cause excessive errors for both calibration and prediction. Wavelength accuracy, on the other hand, can be off by several nanometers without causing any problems, since the calibration and prediction sets are usually run on the same machine and wavelength inaccuracies are compensated for in the collection of the data. Moving of calibration equations from one instrument ("master") to another ("slave") requires that the slave be allowed to deviate from prescribed wavelengths, picking its own optimum wavelengths. The latter procedure is necessary in recognition that the wavelength mechanism may not register exactly with that of the master.

Wavelength precision (Table IV) for the COMP/SPEC varied slightly with wavelength. The standard deviation was determined to be 0.055, 0.005, 0.008, and 0.008 for the polystyrene wavelengths of 1,143.9, 1,682.4, 2,163.4, and 2,306.5 nm, respectively. These values were determined by calculating (by interpolation) the zero crossing of a quadratic polynomial first derivative (through 9 points) for each of the unsmoothed log (1/R) 1,700-point spectra.

This procedure is extremely sensitive to treatment of the data. Smoothing the data before computation of the derivatives will give different zero crossings. However, the precision figures should vary very little from one treatment to another.

C. Fourier Analysis of Instrument Performance

Fourier analysis of near-infrared spectra obtained with dispersion (monochromator) systems is a quick, reliable way to monitor instrument anomalies. Generation of smoothed spectra (for determining noise), derivative spectra (for determining wavelength precision), and power spectra (for monitoring scanning anomalies) can be accomplished from the Fourier domain with much less distortion than that encountered in wavelength space.

Figure 26 is a noise spectrum generated with the aid of Fourier analysis. It was obtained by transforming the log (1/R) spectrum, $S1$, to Fourier space retaining only the first 200

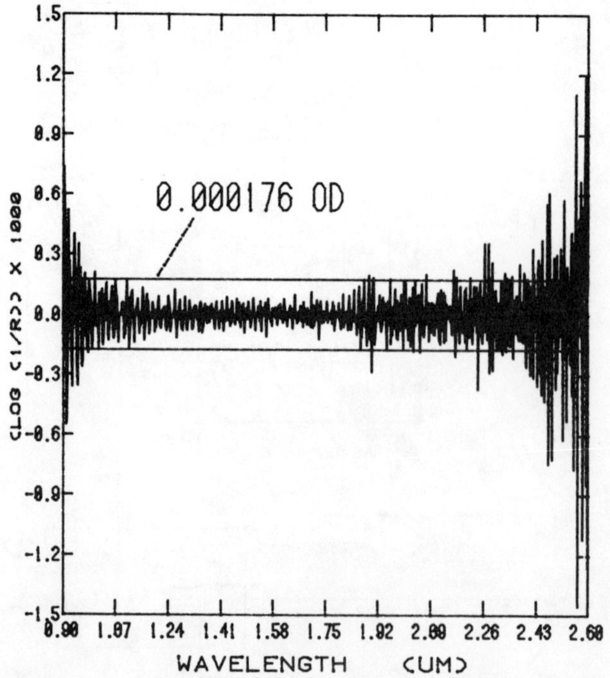

Fig. 26. Noise of COMP/SPEC I determined by Fourier analysis is 176 μOD.

Fourier coefficient pairs. Another log $(1/R)$ spectrum, $S2$, was computed from these coefficients. The difference spectrum ($S1$-$S2$) is the noise data in Fig. 26. It is obvious that generation of $S2$ was a smoothing operation invoked by dropping Fourier pairs 201 and higher. The RMS noise in Fig. 25 is 176 μOD with peak-to-peak noise of 1,190 μOD and −1,450 μOD at 2.597 and 2.566 μm, respectively. Offset for the noise spectrum in Fig. 25 was less than 1.0 μOD.

Figure 27 shows the power spectrum for two different COMP/SPEC systems. Both power spectra were computed from a single tobacco spectrum. Note that in Fig. 27A there is a very powerful spike at index 757, one that never appeared in the power spectrum of COMP/SPEC II (Fig. 27B). This spike appeared in all spectra recorded on COMP/SPEC I regardless of the material being scanned. The index represents a frequency of 4.8 Hz and is thought to be associated with a bad gear in the drive train of the monochromator.

V. SOFTWARE FOR COMP/SPEC

Results obtained with any computerized spectrophotometric system depend on the software packages developed for its support. COMP/SPEC software was developed in-house independently of all other packages that may be available. A brief discussion of COMP/SPEC software packages is included to give the reader an appreciation of the type of programs needed to take advantage of the near-infrared spectrophotometric hardware.

COMP/SPEC software comes in three packages: Scanning/Analysis (SCANL), Analytical Software Programs (ASP), and Computerized Spectrophotometric Analytical System (CSAS). Before discussing these packages, it is helpful to take a look at the organization of the spectra data files for COMP/SPEC.

A. COMP/SPEC File Structure

All spectra files are organized into three parts: data directory (DD), spectra, and chemistry (sometimes called Y-data). The data directory and space for spectra are established on a diskette before scanning of a set of samples begins. Spectra files are organized into blocks of 64 16-bit words each. The first block in the file is devoted to the DD; only 40 locations are currently used. The first two words of each spectrum array denote the sample number. Position of the chemistry arrays is computed when a spectra file is established.

A typical DD is given in Table V. Much of the information in DD is self-explanatory. The DD is updated when samples are

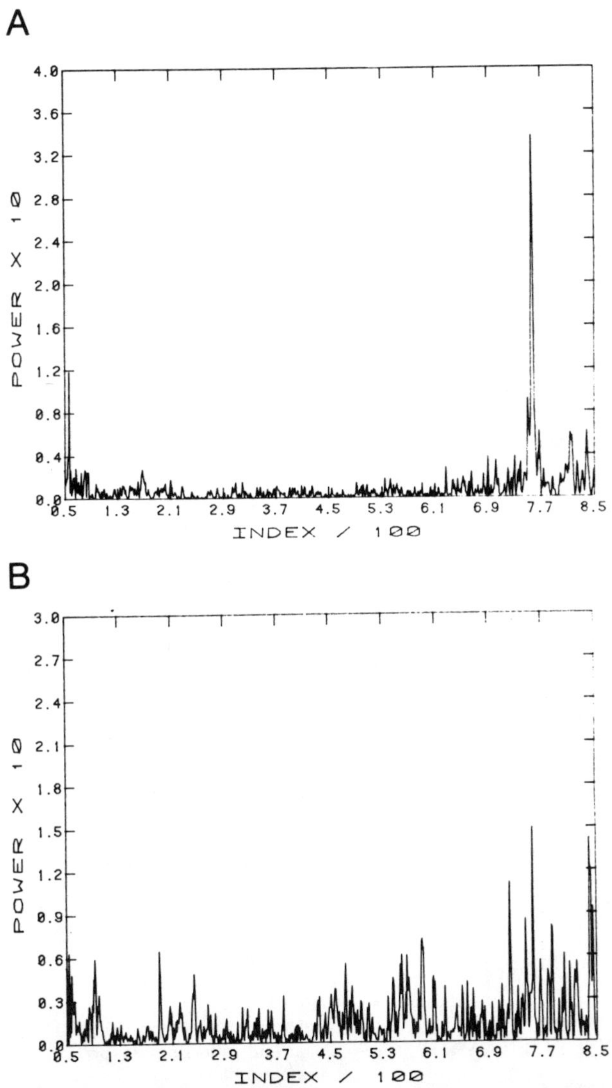

Fig. 27. Power spectrum from COMP/SPEC I (A) and COMP/SPEC II (B) providing information relative to instrument anomaly in COMP/SPEC I, a 4.8-Hz spike at point 759. This peak never appears in power spectrum of COMP/SPEC II.

TABLE V
Typical Data Directory for Spectral Files Recorded with COMP/SPEC

DIRECTORY INFORMATION
June 11, 1984
File: LR1

Displacement/Definition	Value
1 DISC NUMBER	1
2 SECTION	0
3 FORMAT	4
4 INITIAL WAVELENGTH	910
5 WAVELENGTH INTERVAL	200
6 POINTS IN EACH SPECTRUM	840
7 NUMBER OF SPECTRA	60
8 INITIAL SAMPLE NUMBER	1
9 SAMPLE NUMBER INCREMENT	1
10 DATA TYPE	1
11 SPECTRUM SIZE IN RECORDS	27
12 WAVELENGTH OFFSET	10
13 Y-SIZE IN RECORDS	3
14 LOCATION OF FIRST Y	1621
15 NUMBER OF Y DATA	3
16 TYPE OF SMOOTH	1
17 SMOOTH INTERVAL	21
18 TYPE OF DERIVATIVE	0
19 DERIVATIVE INTERVAL	0
20 DIVIDE BY R?	0
21 DELTA OD?	0
22 DELTA FOR OD	0
23 ANTILOG?	0
24 LOG?	1
25 SHRINK FACTOR	2
26 SCALE	0
27 SHIFT	0
28 PSHIFT	0
29 WINC (WAVELENGTH INCR.)	2
30 PLOTTER SPEED	10
31 CURSOR SPEED	10
32 OD OFFSET	0
33 NUMBER OF SCANS	0
34 LAST SAMPLE # ON FILE	60
35 PULSES PER POINT	100
36 RECORD NUMBER	2188
37 SPECTRUM NUMBER	60
38 APNP1	1700
39 FOURIER TRANSFORMATION	0
40 NUMBER OF FOURIER PAIRS	0

added to the file (D7), when Y-data (chemistry arrays) are added to the file (D13–D15), and when transformations (smoothing, derivative, Fourier) are performed (D16–D25 and D39–D40). SCAN also interfaces to an analog X-Y plotter that uses special DD information D27–D32; this feature is used in displaying spectra on the Tektronix 611 storage display. D33 controls the number of scans to be averaged for one sample.

B. Scanning/Analysis

The SCAN software package contains 15 programs that are accessible from the console. The programs (Table VI) permit scanning or collection of spectral data, making changes to the DD, making data transformations (smoothing, derivatives, differences, etc.), printing of spectral data, and display of spectra on the storage scope. The programs are intended for qualitative spectral analyses during the scanning process; hence, it is not necessary to exit the scanning process to determine trends in the data.

C. Analytical Software Program

The ASP for COMP/SPEC is a set of programs, written in Fortran, that provide the user with more powerful tools for making detailed spectral analyses. Table VII lists the major programs in this package. A detailed discussion is beyond the scope of this chapter on hardware, but a brief description of each program is given in Table VII.

D. Computerized Spectrophotometric Analytical System

Software for COMP/SPEC has evolved over the last 14 years. During the course of software development it became apparent that many spectra "housekeeping" functions should be brought under the roof of one program, and CSAS was developed to meet this need. It is a program written in 132 overlays so that it will run on a 64-kilobyte computer. Table VIII is a list of the functions that can be performed with CSAS. Only the major functions will be discussed.

Using CSAS, not only can chemical data arrays be edited and attached to spectral data, but also external X-Y files can be established. Spectra and external X-Y files may be plotted. Statistical data with respect to spectra and chemistry can be obtained. Pooled variance of chemistry data (multiple analyses) is easily obtained. Correlations between single wavelength and wavelength ratios vs. chemistry can be generated. Finally, the two commands NOTE and PRINTF are useful for labeling graphs and making slides.

TABLE VI
Command Options in SCAN Software Package for COMP/SPEC

Command Displacement	Program	Description
0	DATAQ	Data acquisition (scanning)
1	SANLS	Spectrum analyzer
2	PNCH	Print spectral data
3	UCALB	Write data directory (DD) to diskette
4	RDWTD	Read/write spectrum to/from diskette
5	UDDLR	Update DD
6	DDLR	Display DD
7	PCAL	Calibrate analog plotter
8	DSPLY	Display spectra on scope
9	SMDR	Smooth spectrum/display
10	CDERV	Compute derivative spectrum/display
11	DFSP	Compute difference of two spectra
12	CALB	Set up DD
13	TRSF	Transform entire spectra file
14	MNUL	Releases for manual control
15	EODS	Return to CLI

TABLE VII
Programs in Analytical Software Package for COMP/SPEC

Program Type/Name	Description
1. Transformations	
A. FTRSF	Conventional transformations (smoothing, derivative, shrinking, etc.)
B. FOTRSF	Fourier transformation
C. FOPWR	Power spectra transformation
2. Spectrum handling	
A. FDFSD	Difference spectroscopy
B. SHUFFLE	Shuffle any spectrum of spectra from one file to another
C. SPLIT	Split file into two or more parts
D. SUBADD	Correct set of spectra from one standard (say ceramic) to halon reference standard
E. STAT	Print spectrum statistics (max/min, std dev, etc.)
F. AVER	Compute average and std dev spectra of any number of spectra
G. CWAVE	Change wavelength range of spectrum
H. SHIFT	Shift spectrum along wavelength axis for plotting, difference spectroscopy, etc.
3. Calibration and prediction	
A. SMLR	Stepwise multiple linear regression program
B. LRRD	For any chemical constituent attached to spectra file, compute gap −2nd derivatives and determine best ratio of derivatives for range of wavelengths specified
C. PRED	Take calibration equations from equation files established by SMLR and predict unknowns or knowns. If known, pred statistics are provided
D. MCANS	Multiple constituent analyses. Take any number of established equations (set up by program EQUATION) and predict constituent from set of spectra
E. EQUATION	Set up established equations for product; MCANS uses these equations for predicting unknowns

TABLE VIII
Computerized Spectrophotometric Analytical System Command Summary

Command Code	Function
Handling Spectrum and Chemistry Arrays	
RS #3	Read spectrum numbered 3
RY #3	Read Y-data set number 3
TY #-# ...	Type all (TY) or range of Y-data values
TW #-# ...	Type range of spectrum wavelength values
TI #-# ...	Type range of spectrum index values
TXDS [/X/Y] "FILENAME"	Type contents of external data set
LY #-# ...	List (type) range of Y-data set names
LYP	Print all Y-data set names
PY #-# ...	Print all (PY) or range of Y-data values
PW #-# ...	Print range of spectrum wavelength values
PI #-# ...	Print range of spectrum index values
EDITS [I#/W#]	Edit current spectrum (at I-index W-WL)
EDITY [#]	Edit current Y-data set (at index #)
DIF WL1 WL2 ...	Difference of two wavelengths (WL1-WL2)
RAT WL1 WL2 ...	Ratio of two wavelengths (WL1/WL2)
PV[P] #-# ...	Peaks and valleys report (P-PRINT)
PVX[/P] "FILENAME"	Peaks and valleys report for external file
YST [P/L]	Y-data statistics (P-PRINT L-LIST ALL)
VARY [/P] #-# ...	Pooled variance of Y-data sets

(*continued on next page*)

TABLE VIII (continued)

Command Code	Function
CORW [P] WL WL ...	Correlation statistics between WL pairs
CORWC [P] WL/C#	Wavelength/constituent correlation
CORWRC [/P] WL/C#	Wave ratio/constituent correlation
POSS [P]	List/print all positive spectra in file
NEGS [P]	List/print all negative spectra in file
MM #-# ALL	List max and min values from spectra
SCALE?	List current spectrum scale factor
SET SCALE #	Set scale factor to number #
SCALE [/M/D/A/S] #-#	Scale specified spectra
FRESET	Reset current file (remove scale)
CS	List current spectrum number
CY	List current Y-data set name
DIR [P/M]	Directory list (print/maintenance)
Plotting Spectra and External (X-Y) Files	
PLOT [#-#,#]	Plot current spec or range from in file
PLOTF "FILENAME"	Plot plot file named "FILENAME"
PLOTY X# Y# ...	Plot two Y-data sets as X and Y coordinates
REGLF "FILENAME"	Draw regression/error/std deviation line
SYMF "FILENAME"	Draw symbols at plot file points
CHART	Create grid pattern on plotter
BOX	Create grid pattern without tick marks
LABEL	Label plotting grid
TIC# WL ...	Tic mark with rotation # at wave-length WL
LINE X Y X Y ...	Draw line (XY to XY etc.) on plotter
HL # ...	Draw horizontal line on plotting grid
VL # ...	Draw vertical line on plotting card
NOTE# "MESSAGE"	Print "MESSAGE" with rotation on plotter
PRINTF# "FILE"	Print contents of ASCII file on plotter
PDIR [M]	List/maintenance plotting parameters
File Commands	
GET FILE XX	Get input file named: XX
FILE	List current input file name
INITIALIZE[/F] "DIR" ...	Initialize directory (/F - FULL)
RELEASE "DIR" ...	Release directory
FLOAT "FILE" INTO "FILE"	Convert file to floating point
COPY "FILE" INTO "FILE"	Copy file into another file
CXYPF X# Y# ...	Create external plot file from Y data
YINPUT "FILENAME"	Attach Y-data set to current file
NEWY "Y-LABEL"	Attach dummy Y-data set to current file
? [/P/S]	List (P-PRINT S-SORT) available commands
STOP	Exist CSAS. Return to CLI
	End of summary

ACKNOWLEDGMENTS

I would like to thank these people for their contributions: Brenda Mason for typing the manuscript, M. L. McLester for his technical assistance and proofreading, and my many other colleagues for their patient listening as I talked about ideas. Especially, I thank my wife, Judy, for her understanding during all those Saturdays when I was away from home.

LITERATURE CITED

ARCHER, E. D. 1969. Infrared analysis—I. Lubrication 55:13-24.

BEN-GERA, I., and NORRIS, K. H. 1968. Direct spectrophotometric determination of fat and moisture in meat products. J. Food Sci. 33:64-67.

BIRTH, G. S. 1969. Development of a high intensity optical system for non-destructively evaluating the interior quality of agricultural products: Engineering analyses of a high energy spectrophotometer. Dep. of Agric. Eng., Purdue University, Lafayette, IN.

BIRTH, G. S. 1971. Spectrophotometry of biological materials. Ph.D. thesis, Purdue University, Lafayette, IN.

BIRTH, G. S., and NORRIS, K. H. 1965. The difference meter. U.S. Dep. Agric. Agric. Res. Serv. Tech. Bull. 1341. Washington, DC.

DOWNS, R. J., NORRIS, K. H., BAILEY, W. A., and KLUETER, H. H. 1964. Measurement of irradiance for plant growth and development. Am. Soc. Hortic. Sci. 85:663-671.

HAYAT, G. S. 1978. Holographic grating update. Electro-optical Sys. Des. April, pp. 50-54.

LONGERICH, H., and RAMALEY, L. 1974. Digital interface for a Cary 14 spectrophotometer. Anal. Chem. 46:2067-2071.

MATTSON, J. S., and SMITH, C. A. 1977. An on-line minicomputer system for infrared spectrophotometry. Chap. 2 in: Computers in Chemistry and Instrumentation. Vol. 6. J. S. Mattson, H. B. Mark, Jr., and H. C. McDonald, Jr., eds. Marcel Dekker, New York.

McCLURE, W. F. 1968. A rapid method for measuring moisture content of cured tobacco: Research report. Dept. of Biol. and Agric. Eng., North Carolina State University, Raleigh.

McCLURE, W. F. 1969a. Affect of shrinking spectral data on estimating chemical composition. Dept. of Biol. and Agric. Eng., North Carolina State University, Raleigh.

McCLURE, W. F. 1969b. Design of a near infrared moisture meter for tobacco. Int. Res. Rep. North Carolina State University, Raleigh.

McCLURE, W. F., and HAMID, A. 1980. Rapid NIR measurement of the chemical composition of foods and food products. Part I: Hardware. Am. Lab. 12:57-69.

NORRIS, K. H. 1964. Simple spectroradiometer for the 0.4 to 1.2 micron range. Trans. ASAE 7:240-242.

NORRIS, K. H., and BARNES, R. F. 1976. Infrared reflectance spectrocomputer design and application. Proc. Technicon Int. NIRA Conf., 7th. December. Tarrytown, New York.

NORRIS, K. H., and BUTLER, W. L. 1961. Techniques for obtaining absorption spectra on intact biological samples. IRE Trans. Bio-Med. Electron. 8:153-157.

OPTICAL COMPONENTS. 1983. Corion Corp., Holliston, MA.

OPTICS GUIDE III. 1985. Melles Griot, Irvine, CA.

SHENK, J. S., and HOOVER, M. R. 1976. Infrared reflectance spectrocomputer design and application. Pages 122-125 in: Advances in Automated Analysis. Technicon, Tarrytown, NY.

STAIR, R., SCHNEIDER, W. E., and JACKSON, J. K. 1963. A new standard of spectral irradiance. Appl. Opt. 2:1151-1154.

TAYLOR, A. H. 1919. Measurement of reflectance using filters. J. Opt. Soc. Am. 4:9-10.

CHAPTER 6

Commercial Near-Infrared Reflectance Analyzers

P. C. WILLIAMS
Grain Research Laboratory
Canadian Grain Commission
Winnipeg, Manitoba, Canada

I. INTRODUCTION

The term *commercial near-infrared analyzers* refers to the small bench-type instruments used by grain elevator companies, flour and feed mills, government laboratories, food processing companies, and other organizations for the routine analysis of a wide range of commodities for protein, moisture, oil, fiber, and other constituents. For the purpose of this chapter, a *constituent* is defined as a discrete factor such as oil or lysine. A *parameter* is a factor associated with end-product utilization, such as wheat hardness or forage digestibility. The commercial instrument is the result of the new near-infrared technology that was launched in 1971 when the DICKEY-john Grain Analysis Computer (GAC) and the Neotec Grain Quality Analyzer (GQA) were introduced at a press conference at Springfield, IL, as instruments for the estimation of moisture, protein, and oil in soybeans. Shortly afterward, the Neotec GQA was demonstrated at the 1971 annual meeting of the Kansas Wheat-Growers' Association at Hutchinson, KS (Rosenthal, 1971). Since that time, significant modifications made to the data-reduction systems and calibration practice have resulted in an array of bench-type near-infrared analyzers marketed by at least seven companies. Those listed in Table I include past and present models.

At the time of their introduction, the most attractive features of near-infrared analyzers were the speed of analysis, very low labor costs, simple sample preparation, nondestructive analysis of the sample, and omission of chemicals (pollution-free analysis). In recent years digital microprocessors, improved optical systems, and calibration practices have added the dimensions of high accuracy and precision. Modern near-infrared analyzers, fitted with filters capable of using the wavelengths optimum for individual commodities and constituents, provide routine analysis of an accuracy and precision equal to wet chemical methods (Williams et al, 1983). Near-infrared technology is a method for quantitative analysis, and it therefore comes under the category of applied analytical chemistry.

A near-infrared instrument detects electronic signals diffusely reflected from a sample surface that is irradiated by the instrument with light of many wavelengths. The signals at wavelengths specific to the constituents to be determined are amplified and translated by the instrument into composition data such as protein or moisture. The instrument compares all signals received from the sample with an internal reference standard (usually ceramic or sandblasted gold) that enables it to distinguish among samples of different composition. The instrument has to be programmed to recognize samples of different composition by calibration against a standard reference method of chemical or physicochemical analysis.

As explained in Chap. 2, when molecular vibrations coincide with specific wavelengths of the irradiating light, some of the molecules absorb energy at those wavelengths and are excited to a higher energy level. The absorbances of energy are called fundamental absorbances, and excitations to higher levels are referred to as first- or second- or higher-order overtones. The degree or intensity of absorbance can be measured by a spectrophotometer that compares the intensity of energy transmitted or reflected with the incident energy at all wavelengths throughout the range over which the instrument is set up to operate. In its simplest form, the Beer/Lambert law states that the more molecules of a certain type present in the sample, the more energy will be absorbed at wavelengths specific to those molecules, or to groups within the molecule; as a result, the degree of absorbance can be used to determine the concentration of those molecules present in the sample. The successful application of near-infrared spectroscopy depends on measuring the absorptions at the optimum wavelength for a given constituent and relating the measurements statistically to the concentration of the constituents or to functional properties of the commodity on which the measurements are made.

II. TECHNICAL PRINCIPLES AS THEY APPLY TO INSTRUMENTATION

A. Reflectance: Specular, Fresnel, and Diffuse

Near-infrared analyzers represent a marriage of convenience between spectroscopy in the near-infrared area of the spectrum and applied statistics. The near-infrared area in relation to the complete electromagnetic spectrum is illustrated in Chap. 2. The area between 1,000 and 1,900 nanometers (nm) is sometimes referred to as the *overtone area* and the area between 1,900 and 2,500 nm as the *combination area* because many of the absorption bands observed in this area arise as combinations of the first overtones of at least two fundamental vibrations. When a surface is irradiated by a source of energy such as light energy, some of that energy is reflected directly from the "outside" of the surface. This is called *specular* reflectance (Fig. 1) or sometimes *Fresnel* reflectance (the subject is treated in more detail in Chap. 1). The main difference between specular and Fresnel reflectance is that the former refers to reflectance directly from an uneven surface, such as that of powdered grain, whereas the latter is generally associated with planar surfaces. This type of reflected light has not acquired information from the sample and possesses essentially the same energy as it originally carried. Probably a better term than either "specular" or "Fresnel" is *regular*

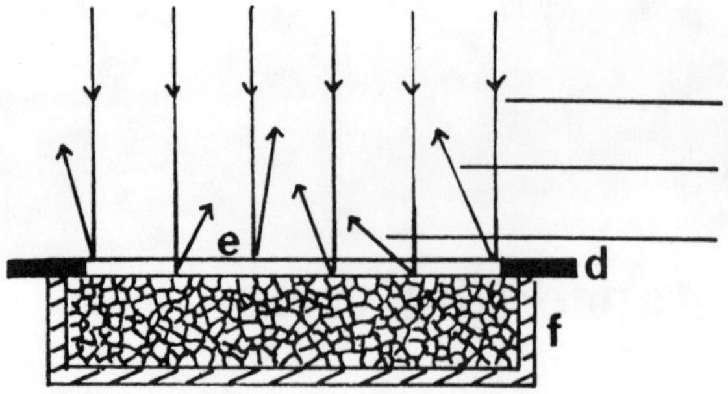

Fig. 1. Regular reflectance: a = incident light, b = reflected light from surface of cell (no absorbance/no information), c = reflected light from sample surface (no absorbance/no information), d = plastic cell cap, e = cell window, and f = sample cell.

reflectance, which embraces both. Regular (specular or Fresnel) reflectance is really a type of noise that interferes with the detection of the diffuse near-infrared signal from the sample. The irradiation strikes the surface and is reflected back at random because of irregularities in the surface. The reflectance has a normal or Gaussian distribution with regard to direction, so most of it is oriented toward the energy source. Other light energy penetrates the surface and is reflected from within the sample. When this light emerges from the surface, where it has been in contact with millions of molecules, it carries less energy than it originally did. This type of reflectance is *diffuse reflectance*.

Diffuse reflectance describes reflected energy that escapes from the surface of a sample after having penetrated the sample (Fig. 2). The diffusely reflected light also emerges at random angles through 180°. Some of the energy that has entered

TABLE I
Commercial Near-Infrared Instruments

Company	Instrument	Model	System of Wavelength Selection[a]	Wavelengths Available[b]	Maximum No. of Terms in Calibration[c]	Type of Computer	Status[d]
Anacon	NIR Analyzer	...	DF	2	2	Analog	NM
DICKEY-john	Grain Analysis Computer	GAC I	DF	6	6	Analog	S
		GAC II	DF	6	6	Analog	S
	Instalab	GAC III 800	DF	10	10	Digital	A
		GAC III 880	DF	8	8	Digital	A
	Grain Analysis Computer	GAC III 660	DF	6	6	Digital	A
		GAC III 650	DF	5	5	Digital	A
		GAC III 640	DF	4	4	Digital	A
Labormim	Infrapid	31	TF	300 (6)	3	Analog	A
		61	M	...	...	Micro	A
LT Industries	IR Analyzer	7000	M	1,200	10[e]	Mini	A
		Quantum 1200	M	1,200	10[e]	Mini	A
Neotec	Grain Quality Analyzer	GQA I	TF	300 (6)	3	Analog	S
		GQA II	TF	300 (6)	3	Analog	S
		GQA 21	TF	300 (6)	3	Analog	S
		GQA 31	TF	300 (6)	3	Analog	S
		GQA 31EL	TF	300 (6)	3 (1)	Analog	S
		Ultraspeed	TF	300 (6)	3 (1)	Analog	S
		GQA 41	TF	300 (8)	4	Micro	S
	Feed Quality Analyzer	FQA 51	TF	768 (12)	6	Micro	A
Pacific Scientific	NIR Analyzer	101	TF	300 (3)	3 (1)	Micro	A
		102	TF	300 (3)	3 (1)	Micro	A
		4250	TF	300 (3)	20+	Mini	A
		3000	TF	300 (3)	3 (1)	Mini	A
	Feed Quality Analyzer	FQA 51A	TF	768 (12)	6	Micro	S/A
		7250	TF	700 (7)	9	Mini	A
	Research Composition Analyzer	6100	M	950	6	Mini	A
		6350	M	950	6	Mini	S
		6250	M	950	6	Mini	A
	Dynascan	7000	TF	700 (7)	9	Mini	A
Percon Prufgërate	Inframatic	8100	DF	6	6	Micro	A
		8120	DF	20	20	Micro	A
Technicon/DICKEY-john	InfraAlyzer	2.5A	DF	6	6	Analog	S
Technicon	InfraAlyzer	200	DF	6	6	Micro	A
		300	DF	6	6	Micro	A
		400	DF	19	19	Micro	A
		250	DF	6	6	Micro	A
		500	M	700	9	Mini	A
Trebor	NIR Analyzer	GT 70	M	1,900	14[f]	Micro	A
		GT 7700	M	1,900	14	Mini	A
		GT 80	IRLED	12	14	Micro	A
		GT 90	IRLED	12	14	Micro	A
		GT99	IRLED	12	14	Micro	A

[a] DF = discrete filters, TF = tilting filters, M = monochromator, and IRLED = infrared light-emitting diodes.
[b] Figures in parentheses refer to maximum number of wavelengths used in normal calibration.
[c] Figures in parentheses refer to number of wavelength terms used in quotient mathematics mode.
[d] NM = not marketed, S = superseded, A = available, and S/A = superseded but available on special request.
[e] Programming can be modified to accept more variables on request.
[f] Includes two terms for temperature compensation.

becomes absorbed by chemical constituents present in the sample. Individual constituents absorb energy at specific wavelengths, and it is the light diffusely reflected from the sample that carries information as to the energy absorbed at these wavelengths. The energy entering the sample from the irradiation is compared by near-infrared instrumentation to that escaping by diffuse reflectance. The difference between energy entering and leaving the sample, relative to that of a constant background with essentially zero absorbance (such as a ceramic tile or the internal surface of an integrating sphere) is computed into the concentration of specific constituents by the instrument, provided it has been calibrated to recognize what these differences mean in terms of concentration. The amount of energy entering and leaving the sample is affected by the particle size and shape and hence the bulk density and packing performance of a powdered material, all of which affect the effective path length. Pressure on the sample caused by placement of a spring-loaded cell cover also affects path length and energy interchange. Changes in the numbers of particles and their proximity to each other cause radical changes in the number of faces upon which specular and Fresnel reflectance can occur, which in turn affect the dissipation of irradiating energy from the sample. These changes also increase the pattern of diffuse reflectance. This subject is discussed more fully in Chap. 1.

B. Band Overlap

When a substance contains two or more constituents, such as protein, starch, and moisture, it is possible for absorbance bands caused by vibrations of different atomic combinations (e.g., C-H and N-H stretching) to occur so closely together that the bands overlap each other. Although a constituent may display several prominent bands, all of which can theoretically be employed in analysis, in practice many of these bands may be masked by the presence of stronger bands from other constituents. As a result, the wavelengths actually selected as optimum by computer for the analysis of the original constituent may not coincide with any of the more prominent bands displayed by the pure constituent. Other constituents (e.g., moisture and starch) and absorbing vibrations such as the peptide-link grouping may direct the computer to select wavelengths for actual determination that coincide with areas of least interference from other absorbers.

Moisture is usually present at levels of about 4–90% in food materials, depending upon whether the materials are fresh, as in meat, or mainly dehydrated, as in grain products. It has very strong absorbers in the areas of 1,470 and 1,940 nm. The water content of fresh products has profound effects on the state of other constituents and, as a result, on the wavelengths selected for analysis of these constituents. It also affects the method of sample presentation. Cereal products contain 50–70% starch, which has strong absorbers in the areas of 2,100 and 2,275 nm. Oil and cellulose have powerful absorbers in the area from 2,310 to 2,350 nm. Water, starch, cellulose, and oil can all interfere in the determination of constituents present in lesser concentrations. For example, in the determination of protein in cereal products, the most widely used bands are at 2,150–2,180 nm. These are relatively low-intensity protein bands, but they coincide with an area of lowest interference from starch and moisture (Williams and Krischenko, 1986).

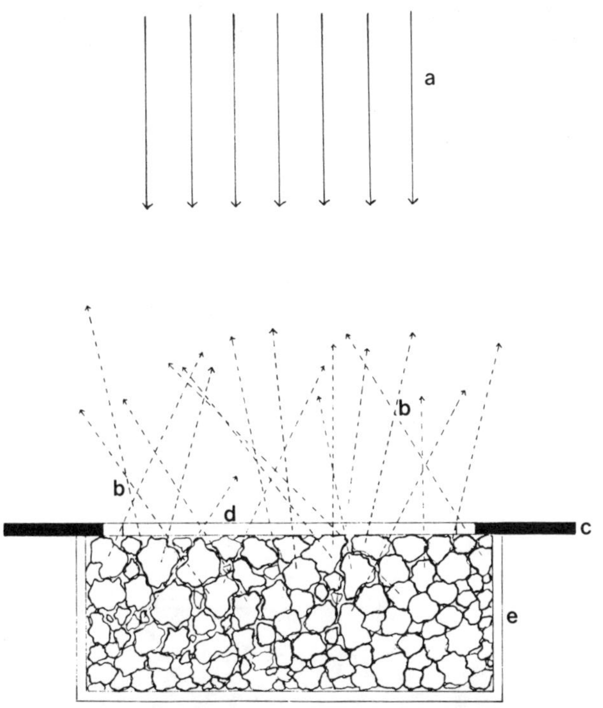

Fig. 2. Diffuse reflectance: a = incident light, b = diffusely reflected light from within sample body (some of light is absorbed and diffusely reflected light carries this information), c = plastic cell cap, d = cell window, and e = sample cell. Light is diffused between and through sample particles.

C. Transmittance and Absorbance

Commercial near-infrared instruments measure signals in the region between 1,100 and 2,500 nm in reflectance mode. Reflectance, as it is usually referred to in connection with commercial instruments, is analogous to absorbance. Band intensities in spectroscopy are referred to in terms of transmittance or absorbance. *Transmittance* is the ratio of the radiant energy transmitted through a sample to the radiant energy impinging upon the sample. *Absorbance* is the logarithm of the reciprocal of the transmittance, or $A = \log_{10} (1/T)$. Reflectance in terms of the radiant energy reflected from the surface of the sample is analogous to transmittance through the sample, in that it represents the ratio of the radiant energy diffusely reflected from a surface to the radiant energy impinging on the sample; i.e., reflectance $(R') = \log_{10} (1/R)$, where R is the amount of light energy emerging from the sample.

The Trebor GT-90 whole-grain near-infrared transmittance instrument probably receives its signal partly as truly transmitted energy and partly by a type of multiple reflectance, where the energy passes through the sample and consists of energy reflected from the outside layers of many individual grains rather than truly transmitted energy that has passed through the grains themselves. Pacific Scientific models 6250 and 7000 are capable of true transmittance measurements in the region from 600 to 1,100 nm. Using a cell several centimeters thick, this technique is particularly suitable for analysis of slurries and liquids. The new LT 7000 and Trebor 70 series are also capable of true transmittance measurements.

When the frequency of a vibration coincides with the frequency of the irradiating energy, absorption of the energy occurs at that frequency. If energy of only a certain frequency is allowed to irradiate the material containing molecules that vibrate at that frequency (for example, by use of interference filters), the amount of energy absorbed is a measure of the number of molecules of that type that are present in the sample. The log $(1/R)$ signal conforms closely but not strictly to the Beer/Lambert law. The absorbance/concentration relationship is not completely linear throughout the near-infrared range, and the nonlinearity is compensated for in computation by the instrument. This is the basis of the accuracy of quantitative near-infrared analysis.

D. Optical Density

Transmitted or reflected light energy is sometimes recorded in terms of *optical density* (OD) values. Optical density, which is the log of the reciprocal of the transmitted energy [log (1/*T*)], relates to transmittance rather than absorbance. For near-infrared reflectance spectroscopy, the energy reflected from the surface is recorded as the log (1/*R*), which is analogous to log (1/*T*), and the OD has come to be adopted for the expression of reflected energy (reflectance). An OD value of 1 means that 10% of the energy used to irradiate the sample is transmitted through the sample or, in terms of log (1/*R*), reflected from the sample. Table II illustrates other OD values.

The log (1/*R*) is not strictly equivalent to log (1/*T*). When a transparent sample is irradiated, the energy passes through the sample or is absorbed by it. If the sample is not transparent, the signal will be diffusely reflected, although part of it may pass through the sample. In the reflectance mode the energy reflected back from the sample is reflected by regular and diffuse reflectance, and not all of the energy of the signal can be accounted for in terms of absorbance or "pure" reflectance. Most of the measurements involved in quantitative near-infrared analysis are made in the area of about 0.5–0.8 OD, so that a fairly high percentage of the incident energy is utilized. Measurements in the OD range of 4 or 5 or higher are based on relatively small fractions of the incident energy and are proportionately more susceptible to the influences of the chief noise generators, temperature fluctuation and vibrations. Some instruments use light sources of lower energy levels, which tend to be more sensitive to internal fluctuations in temperature and to fluctuations in the temperature of the sample. In near-infrared reflectance and near-infrared transmittance instruments operating in this high OD range, extra precautions have to be taken to compensate for the significantly lower signal-to-noise ratio (see Section IIE).

E. Instrument Geometry

Instrument geometry means the geometry of the lines followed by the irradiating energy in its path to the sample and reference standard and from the sample or standard to the detector. The spatial and geometric relationships of the sample, energy source, and detector are fundamentally important to the efficiency of a near-infrared instrument. There are four main types of geometry in simple bench-type commercial NIR instruments: overhead illumination with detection by a single central detector, overhead illumination with detection by a single offset detector, overhead illumination with multiple offset detectors, and the integrating sphere. An example of the first type of geometry is the original DICKEY-john GAC, Model I (Fig. 3). The optical system is illustrated in Fig. 4.

Energy from the source was focused onto the sample through a circular aperture. The single lead-sulfide detector was centrally placed between the light source and the sample surface so that the actual area of sample illuminated was an annular ring. Energy reached the detector from the surface of the sample by diffuse reflectance. This area was much larger than the detector, and the system was designed to minimize regular reflectance by placing the detector very close to the surface. The later versions of the GAC (e.g., the GAC III) have a single detector offset to one side, which greatly reduces interference caused by the specular component. To provide even detection of the reflected signal, the sample cell is rotated during reading. Irradiation is from overhead as with the earlier model (Fig. 4).

All of the Neotec/Pacific Scientific bench instruments have used overhead irradiation, with four offset detectors, set at equal distances around the irradiated area and at an angle of 45° to the horizontal. This technique minimizes interferences caused by regular reflectance and maximizes the reception of reflected energy. For samples with wide ranges in particle size and shape, the cell is rotated to improve precision.

TABLE II
Optical Densities (OD)

OD	Transmittance Reflectance (%)
0	100
0.1	79.4
0.2	63.1
0.3	50.1
0.4	39.8
0.5	31.6
0.6	25.7
0.7	20.0
0.8	15.9
0.9	12.6
1.0	10.0
2.0	1.0
3.0	0.1
4.0	0.01
5.0	0.001
6.0	0.0001

Fig. 3. DICKEY-john Grain Analysis Computer Model I. (Courtesy DICKEY-john Corp., Auburn, IL))

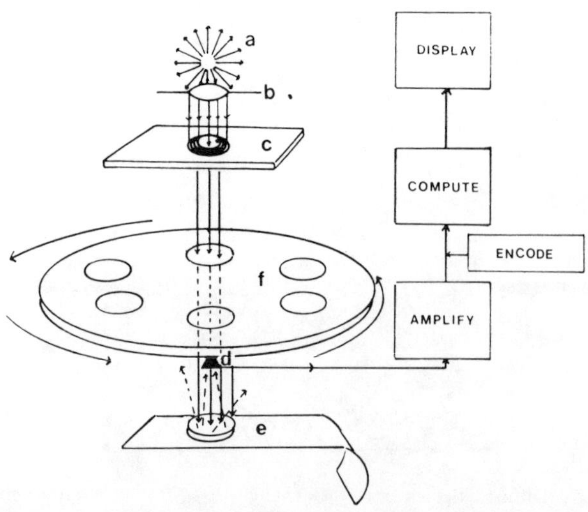

Fig. 4. Optical system of DICKEY-john Grain Analysis Computer Model I: a = light source, b = collimating lens, c = slit system, d = detector, e = sample cell, and f = horizontally rotating filter wheel.

Irradiation from source to sample may be either of the single-beam or double-beam type. Single-beam irradiation allows more energy to be applied to the sample; as a result, the intensity of reflectances and absorbances is enhanced and, theoretically, the degree of resolution can be improved. To ensure this, and the commensurate improvement in signal-to-noise ratio, all of the components must be of high stability. (The signal-to-noise ratio is that of the signal reaching the detectors from the sample at specific wavelengths to that of all unrelated background signals reaching the detectors.) Highly stable components and design are essential because single-beam illumination involves separate irradiation of sample and standard. Even the minute amount of drift that might occur in the 20 sec during which the sample or standard is being irradiated could cause significant errors, since the differences in the actual energy reflected from the surface of samples differing in composition are also minute. Instrument drift must be absolutely minimal to enable precise (and therefore accurate) analysis using single-beam irradiation. Modern electronics allows this. Before the last two decades, electronics had not reached the present state of excellence in manufacture, and the double-beam mode, which is more tolerant of minute deficiencies in components and design, was preferred. The disadvantage of the double-beam mode is that the illumination/irradiation is split and only half of the energy emitted by the source actually reaches the sample. When an instrument operates in double-beam mode, the geometry becomes more critical and the accuracy of measurement is more sensitive to minute differences in the distances between energy source, sample, and detector. Double-beam mode irradiation is facilitated by the use of an integrating sphere.

The Technicon InfraAlyzer models 300 and 400 operate in the double-beam mode and use an integrating sphere. Light energy enters the sphere via an oscillating mirror at the top that causes the irradiation to strike the sample or deflects it to impinge on the walls of the sphere. Energy returned from the sample by diffuse reflectance rebounds from the internal reflecting walls of the sphere and reaches one of two detectors. The internal surface of the integrating sphere is coated with a film of gold that has been sandblasted to provide a diffusely reflecting surface. This serves as a standard, and the irradiation that enters the sphere without irradiating the sample creates the reference signal. The signal from the sample is continuously compared with the standard, and as a result the sample can be inserted at any time. A ceramic standard is also located at the rear of the sample drawer, mainly for reference purposes, and is not used in routine analysis. The arrangement is illustrated in Fig. 5.

A variation on the InfraAlyzer geometry is the "transflectance" concept, which is commercially exploited by Technicon (Fig. 6). This method analyzes liquid samples by irradiating through the sample onto a ceramic or other diffusely reflecting backing plate. The energy diffusely reflected from the backing plate passes through the liquid sample and accumulates information en route to the integrating sphere. Transflectance is most useful in the analysis of very thin layers of liquids—0.1 to 1.0 mm in thickness in the 1–2.5 μm region. In the case of nontransparent liquids, as the thickness or degree of opacity of the fluid layer increases, the proportion of light energy reflected through the fluid via the reflectance standard decreases relative to that diffusely reflected from particles within the fluid, until true diffuse reflectance is taking place. An effective method of analyzing this type of material is to use a lower wavelength region, from 600 to 1,100 nm in transmittance mode, in conjunction with a silicon detector, thus permitting higher-energy irradiation and use of cell thicknesses of several centimeters. Both methods are very useful for the analysis of milk, other opaque liquids, and clear solutions, and the use of larger cells enables the analysis of thicker slurries, since flow properties and cleanup are facilitated. Pacific Scientific models 6250 and 7000, the recently introduced Trebor models 70 (7700) and 99, and the LT 7000 use this approach.

F. Reflectance Standards

Near-infrared analysis is based on the concept of comparing the information emanating from the sample by diffuse

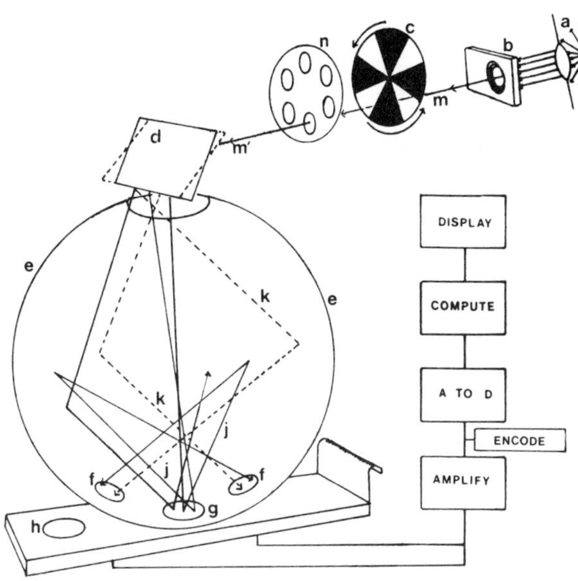

Fig. 5. Technicon InfraAlyzer system: a = light source/collimating lens, b = slit system, c = chopper (light/dark) wheel, d = oscillating mirror, e = integrating sphere, f = detector, g = sample, h = reflectance standard (not involved in analysis), j = diffusely reflected light from sample to detector, k = diffusely reflected light from standard (interior surface of sphere) to detector, m—m′ = light path, and n = filter wheel (up to 20 filters).

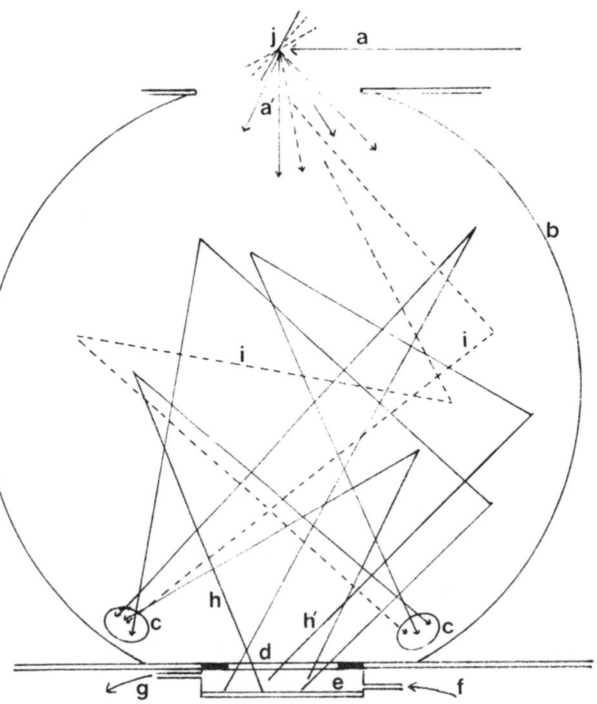

Fig. 6. Transflectance principle. a = incident light, b = integrating sphere, c = PbS detectors, d = flow-through cell, e = diffusely reflecting backing plate (e.g., ceramic), f = liquid inlet port, g = outlet port, h, h′ = "transflected" lgiht carrying information from liquid cell contents, i = light diffusely reflected from internal surface of integrating sphere, and j = oscillating mirror.

reflectance with energy reflected from a suitable standard. The absorbance from the standard is subtracted from that of the sample, at given wavelengths, and the difference in absorbance is computed into constituent composition. An ideal standard should be optically "flat." It should also be readily available, durable, easy to keep clean, possess high reflectance properties, and be capable of a high degree of uniformity between instruments. Several standards have been proposed and used in instruments. Practically any substance could serve as a standard, and substances like Teflon that have absorption bands in the near-infrared region can be used as standards because the instrument subtracts the signal of the standard from that of the sample and corrects for any absorbance from the standard in the areas used for analysis. An experimental calibration was once set up at the Grain Research Laboratory, Winnipeg, using ground wheat as a standard in a Neotec GQA Model 31 fitted with a special sample drawer that held two sample cells. The "standard" was a sealed "permacell" filled with ground wheat. Forty samples were used in the calibration, and the results differed from a normal calibration in that the observed optical data were greatly reduced. The calibration constants could be employed to compute protein using optical data from "unknown" samples in the usual way, provided the same ground wheat was retained as the standard. Difference curves generated by subtracting a sample low in concentration in a constituent from a sample high in that constituent can be useful in identifying potential wavelengths for analysis, since the "spectrum" created represents areas of maximum difference in absorbances.

The earliest generation of DICKEY-john instruments used no standard at all. Pacific Scientific/Neotec, and subsequently DICKEY-john, used Teflon standards in their earlier models. Although the instruments gave reasonable results, Teflon was easily scratched during cleaning and had a high propensity to accumulate static electricity. Dust was thus attracted to the Teflon plate, which needed cleaning at least daily and more frequently in very dusty locations. In addition, Teflon was found to be very temperature sensitive, and standards differed between instruments. Compressed resublimed sulfur was proposed by Kronstein and co-workers (1963) and by Tkachuk and Law (1978). This standard has excellent reflectance properties but is not durable and causes difficulty with cleaning off dust. Ceramic was first used as a diffuse-reflectance reference standard by Technicon/DICKEY-john in the InfraAlyzer 2.5A, which was introduced during 1975. Shortly afterward, Neotec replaced its Teflon standard with ceramic in the GQA and FQA series. Ceramic has proved to be a most practicable standard. It provides a serviceable diffuse-reflectance signal and has practically no absorption bands in the near-infrared region. It is readily available in the form of inexpensive and easily handled disks. Its introduction improved analytical precision, and therefore accuracy, and by mid-1976 all commercial near-infrared instruments used ceramic as a standard. The Technicon InfraAlyzer models 250, 300, 400, and 500 use gold as a reference standard, which is an excellent reflective surface if it is sandblasted to provide diffusion. It is very durable, and because of its location as the inner surface of the integrating sphere it is essentially immune from a serious problem with all standards—namely, dust accumulation on the standard. The instrument geometry has to be very exact with double-beam/integrating sphere optics, and the irradiation/detection system tends to be more sensitive to sample dust on the surface of the sphere cover glass than with systems having external standards.

G. Wavelength Selection

Commercial near-infrared analytical equations require a relatively small number of wavelengths and can accommodate the measurement of constituents such as protein, starch, moisture, and oil (fat) with this set of wavelengths. These four constituents, and in particular protein and moisture, were the most important ones measured in the earlier generations of near-infrared instruments. During the past three to five years, the use of near-infrared technology has been expanded to the measurement of many factors including amino acids, fiber components, and associated parameters such as grain hardness, particle size, and digestibility in a wide range of foods, commodities, and raw materials. Accurate measurement of these more diverse parameters calls for a wider selection of wavelengths and mathematical treatments.

The most critical part of the system is the selection of the correct wavelength at which to make the measurements. Studies at the U.S. Department of Agriculture Instrumentation Research Laboratory, Beltsville, MD, (Williams and Norris, 1983) showed that use of a wavelength for analysis of protein only 3 nm from optimum resulted in a significant decline in accuracy. Wavelengths are selected by comprehensive spectrophotometers that usually scan between 1,000 and 2,600 nm. The spectrophotometer or monochromator is interfaced to a minicomputer that uses multiple regression analysis to select the optimum wavelengths for the measurement of specific constituents. There are four principal systems for wavelength generation. These are (i) discrete filters with a half-height bandpass of about 10 nm; (ii) tilting filters, each of which "scans" a region of about 120–130 nm and works in conjunction with an encoder system that selects wavelengths by microprocessor; (iii) the monochromator, which can scan a wavelength range of about 2,000 nm in the near-infrared and also employs a microcomputer to select the actual wavelengths at which measurements are made for analysis: and (iv) the use of near-infrared light-emitting diodes at specific wavelengths.

Three main types of wavelengths are selected. The first is wavelengths that correspond to known absorbances of the specific constituents to be determined. A good example is the 2,308- to 2,310-nm wavelength for oil determination, which coincides with a major band displayed by oil and has been assigned to a C-H stretching combination of a -CH_2- grouping (Holman and Edmondson, 1956; Law and Tkachuk, 1977). The second type is a wavelength corresponding to an absorber on a constituent that is inversely related in concentration to the constituent to be determined. For example, a computer may select a wavelength near 2,100 nm for determination of protein in a set of flour samples. This is really a starch wavelength, but there is a high negative correlation between the starch and protein contents of flour. The third type of wavelength occurs when the optimum wavelength for measurement of a constituent corresponds to an area of least interference from other constituents, rather than to a band caused by a specific absorber. This occurs particularly in the case of constituents present in minor concentrations as compared with other constituents. An example of the "least-interference" concept is the 2,180-nm wavelength for determination of protein in wheat. It is not the most prominent protein band, but it occurs in an area of lowest interference from starch, cellulose, and moisture, all of which interact to mask the more prominent protein bands at 1,980 and 2,056 nm.

H. Mathematical Treatment

Commercial near-infrared instruments record optical signals in the form of the log $(1/R)$, or apparent reflectance. The Technicon InfraAlyzer, DICKEY-john GAC, and Percon Inframatic series use the log $(1/R)$ signal directly to compute constituent concentration. The advantage of the log $(1/R)$ format lies in its simplicity. Provided the instrument has access to even a single wavelength point suitable for the measurement of a constituent, satisfactory analysis is possible. A wide range of wavelengths, or even the influence of particle size itself, can serve as a reference for the prediction wavelength. The original Pacific Scientific/Neotec GQA series used the log (1/*R*) signals in the form of a delta log (1/*R*), which is attained by subtraction of the signal at wavelength point *B* from that at the primary wavelength point, *A*. As the delta value approaches zero, it becomes analogous to a first derivative of the log (1/*R*) signal at the primary wavelength point *A*. Later models of the GQA employed the first derivative of the log (1/*R*) signal. Derivatives require instruments that scan a fairly large region of wavelength to implement the derivatization, which requires making measurements at two wavelengths (in the case of the first derivative). The first derivative algorithm normally used is:

$$D_1 \log (1/R) = \log (1/R)(A) - \log (1/R)(B) \qquad (1)$$

for measurements taken at wavelength points *A* and *B*. The *wavelength point* or *primary* wavelength is the *A* wavelength, and the signal at wavelength *B* is subtracted from the *A* signal. For example, if a first-derivative wavelength point is selected at 2,156 nm, this is the *A* wavelength, and the *B* wavelength occurs at a wavelength shorter than the *A* wavelength by a distance in nanometers specified by the derivative or "gap" size. The algorithm can be interpreted as a "binocular" observation, taken from two wavelength points rather than at a single one.

Another form of mathematical treatment is the extraction of the second derivative of the log (1/*R*). The formula most commonly used for the computation of the second derivative is:

$$D_2 \log (1/R)_A = \log (1/R)(A) - 2 \times \log (1/R)(B) + \log (1/R)(C) \qquad (2)$$

so that three wavelength points are used for every term. This can be likened to a "stereoscopic" or three-dimensional measurement taken at the central wavelength, *B*. Figure 7 illustrates the format of the log (1/*R*) and first and second derivatives of the log (1/*R*). The subject is treated in more detail by Morrey (1968). The primary reason for using a derivative is to reduce the baseline shifts caused by particle-size changes between samples, sample positioning, and other factors concerning the sample and its presentation. Derivatization resolves anomalies in the spectra, or areas of the spectra, that are caused by variations in particle size, shape, bulk, density, and other factors, which means that differences between spectra are limited to those caused by chemical composition. Theoretically, this procedure simplifies the identification of wavelengths specific to chemical composition. Other mathematical treatments, including third and fourth derivatives of log (1/*R*), the Kubelka/Munk treatment, curve-fitting, and the partial least-squares methods, have all been investigated (W. Hruschka and K. Norris, pers. comm., 1983). They appear to offer no significant advantages over the simple log (1/*R*) and first and second derivatives thereof, other than the fact that quotient treatment of first- and second-derivative data generally improves the stability of calibrations. This feature has been discussed in Chap. 3.

No single mathematical treatment is ideal for all applications. It is advisable to optimize mathematical treatments of raw log (1/*R*) for the measurement of individual constituents in a given commodity at the same time as wavelength selection, since certain constituents are measured better with, for example, the first derivative of the log (1/*R*) than with the log (1/*R*) itself. *Sophistication* in computerized spectrophotometers usually means more comprehensive computerization. Computerized spectrophotometers range from the highly refined equipment installed at the U.S. Department of Agriculture, Agricultural Research Center, Beltsville, MD, and at North Carolina State University, Raleigh, to the more compact, commercially available research composition analyzer models 6250 and 6350 offered by Pacific Scientific, Gardner/Neotec Division, and the InfraAlyzer 500 offered by Technicon Industrial Systems, both of which can operate as computerized spectrophotometers and as analytical instruments. These instruments are optically equal in performance to the one-of-a-kind instruments and are supported with excellent software packages. LT Industries and Trebor Instruments have recently introduced their own versions of the computerized spectrophotometer.

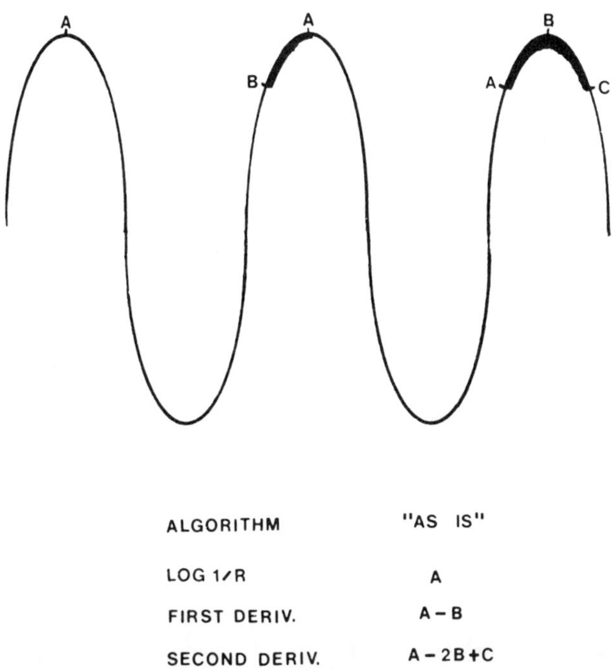

Fig. 7. Format of log (1/*R*) and first and second derivatives of log (1/*R*) algorithms.

I. Optical Systems

The two main methods by which most commercial near-infrared instruments generate wavelengths are the discrete filter and the tilting filter principles. Both systems give excellent results for a wide range of analyses. The filters used in near-infrared instruments are called narrow band-pass interference filters. The bandwidth at half height of these interference filters allows the passage of light over a narrow wavelength region (usually about ±5 nm) only. The efficiency of transmittance abruptly increases to and recedes from a maximum (Fig. 8).

The wavelength at which peak efficiency in transmission is observed occurs when the filter is perpendicular to the source of irradiation, which is sometimes referred to as the *central* or *starting* wavelength point of the filter. Instruments fitted with discrete filters can make effective measurements only at the central wavelength points of the filters.

When an interference filter is tilted at an angle to the primary energy source, the central wavelength of the filter shifts toward the shorter wavelength. This tilting filter phenomenon was introduced to near-infrared technology by Neotec Corporation after preliminary research had been conducted by K. H. Norris and co-workers at the U.S. Department of Agriculture at Beltsville. The principle is illustrated in Fig. 9. The degree to which the wavelength shifts is given approximately by the formula:

$$\lambda_2 = \lambda_1 \frac{(n^2 - \sin^2\theta)^{1/2}}{n} \qquad (3)$$

where θ = angle of filter relative to "normal" to light source, λ_1 = central wavelength, λ_2 = wavelength at angle θ, and n = refractive index of the filter.

Example 1: A narrow band-pass filter has a central wavelength of 2,192 nm. What is the effective wavelength at an angle of 27°31′ to the vertical?

The refractive index of filters with peak efficiency at wavelengths of up to 2,220 nm is 1.45 and 1.95 for filters of longer wavelengths (2,220–2,500 nm).

$$\lambda_{27°31'} = 2{,}192\,\frac{(1.45^2 - \sin^2 27°31')^{1/2}}{1.45}$$

$$= \frac{2{,}192\,(1.45^2 - 0.21345)^{1/2}}{1.45}$$

$$= 2{,}078 \text{ nm.}$$

During the passage of the filter through the area of tilt there is some loss of efficiency of transmittance toward the extreme angle of tilt. The filter is usable for analysis over a range of about 9% of the wavelength of the starting filter, and thereafter the degree of transmittance is not sufficiently reliable for use in quantitative analysis. For example, a filter with starting wavelength of 2,190 nm effectively scans an area of about 175 nm down to 2,015 nm. A tilting filter system with three filters allows an effective scanning area of about 520 nm (separated into three regions of the near-infrared spectrum), and a system with six or seven tilting filters provides a potential scanning area of over 1,000 nm. In this type of system, the available wavelength per filter is usually quantized into intervals (pulse points) corresponding to about 1.7 nm. The specific wavelengths at which the instrument takes measurements are selected by its computing device via a coding system.

J. The Basic Near-Infrared Analytical System

The near-infrared analytical system consists of a source of irradiation, a focusing system to direct the energy onto the surface of the sample or reference, a nonabsorbing diffuse reflectance reference standard, a sample cell with a quartz or glass window through which the sample is irradiated, a filtering process to ensure that only energy of specific wavelengths reaches the sample, and a detector to measure the energy reflected from the surface. Most instruments incorporate some form of encoder to ensure that only signals emitted from the sample at specific wavelengths are processed into analytical data. The signal received by the detector is amplified and relayed, usually via a storage system, to a data-reduction device, and a meter displays the analytical result. These components are illustrated in Fig. 10. The sequence of irradiation, wavelength selection or filtering, detection, amplification, computation, and display is used in all types of near-infrared reflectance analyzers.

III. COMMERCIAL BENCH-TYPE INSTRUMENTS: FIRST GENERATION

A. DICKEY-john Grain Analysis Computer Model I

The Grain Analysis Computer (GAC I) was the first near-infrared reflectance instrument offered by the DICKEY-john Corporation. It was commissioned by the Illinois Department of Agriculture under a sharing program with the U.S. Department of Agriculture for the estimation of protein, oil, and moisture in soybeans. The instrument used the basic system described in Fig. 10 except that it used no internal reference standard. It employed six discrete narrow band-pass interference filters with central wavelengths at 1,680, 1,940, 2,100, 2,180, 2,230, and 2,308 nm, which were used to measure all constituents. The filters were mounted in a horizontally revolving wheel. Computation was by means of analog computer circuitry. The wavelengths were chosen on the basis of spectral studies of the principal constituents isolated from

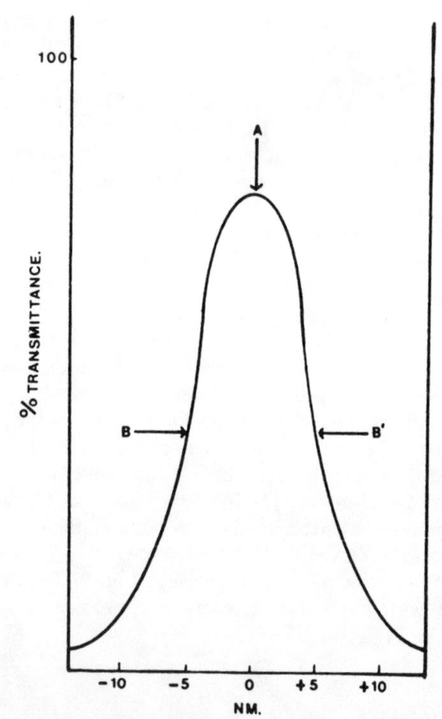

Fig. 8. Efficiency of transmittance of narrow band-pass filter: A = peak absorbance; B, B′ = half-height band width.

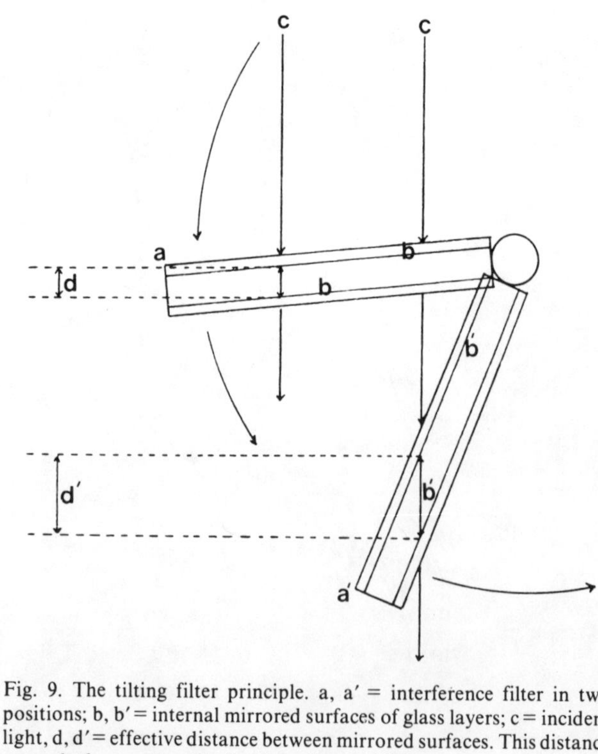

Fig. 9. The tilting filter principle. a, a′ = interference filter in two positions; b, b′ = internal mirrored surfaces of glass layers; c = incident light, d, d′ = effective distance between mirrored surfaces. This distance controls the wavelength of an interference filter.

grains, such as wheat and soybean protein, starch, oil, and water. Figure 11 gives the log (1/*R*) spectrum and the second derivative of the log (1/*R*) spectrum for these four principal plant constituents. The second-derivative trace highlights areas of maximum and minimum absorbance, with the "valleys" being areas of maximum absorbance. An area that is relatively insensitive to absorptions from constituents occurs at about 1,680 nm, and fluctuations about this area are caused mainly by variance in particle size. As a result, 1,680 nm is a valuable reference wavelength point for measurements of several constituents.

The GAC I carried one product board, which could be calibrated for three constituents. The instrument could carry calibrations for the determination of up to three constituents, such as oil, protein, and moisture, in one commodity. At the time of its introduction, the GAC I was used only to test soybeans. The original product boards were hard-wired, and for calibration the operator was provided with tables to enable selection of the correct resistors to solder into the board. The sample cup was open, with no cover. The cup was loaded with ground sample and leveled to provide a smooth surface. The operator calibrated the instrument by reading 30–50 samples into it and recording the log (1/*R*) data. These were regressed against the chemical data for protein, moisture, oil, or other constituents, using stepwise multiple linear regression. After calculating the regression coefficients, the operator used the tables to select resistors to wire the boards. The regression coefficients were also used to calculate a log reference value for each product/constituent, which was wired into the board by the operator using diodes. The GAC I used only one lead-sulfide detector, which was centrally located above the sample cup and protected from dust by an optical glass window. The sample cup entered the instrument via a drawerlike access system.

The GAC Model I was soon followed by the Model II, which used a Teflon internal reference standard at the back of the sample drawer. When the drawer was pulled out, the reference standard was positioned beneath the irradiation detection system. Pushing in the drawer replaced the standard with the sample. The GAC II carried three product boards, each of which could carry calibrations for three constituents. The instrument was used in its original form at the Grain Research Laboratory, Winnipeg, to test wheat, barley, soybeans, and rapeseed. Typical results are summarized in Table III. The results were reasonably comparable with those from more modern near-infrared instruments, and the Model II can be said to have been the forerunner of such instruments as the DICKEY-john GAC III, 400, 600, and 800 series and the Technicon InfraAlyzer series.

B. Neotec Grain Quality Analyzer

The Neotec Grain Quality Analyzer (GQA) Model I (Fig. 12) was also originally designed under contract for the state of Illinois to analyze soybeans. This instrument used the tilting-filter principle with three interference filters mounted in a vertical, spinning paddle-wheel arrangement. The actual wavelengths used were factory-wired into the optical system as "pulse points" corresponding to the angular position of the filter with respect to the light beam, which in turn produced the desired wavelength. (A pulse point is a specific wavelength.) The region scanned by the filters used in each individual instrument assembled in the factory was compared spectrally with a set of reference filters. The scanning region of each filter was divided into 128 pulse points, starting at the central wavelength, and the exact wavelengths (pulse points) at which the instrument actually took readings were stipulated by means of a computer-controlled encoder.

The other innovation of the Neotec GQA was the use of a form of derivatized treatment of the log (1/*R*) signal. The GQA used the delta log (1/*R*) algorithm, which involved two wavelengths (pulse points) per wavelength term. Three wavelength terms (six pulse points) were wired into the instrument and used for the estimation of oil, protein, and moisture. The GQA was also equipped with three analog computer boards into which constants could conveniently be set via screw potentiometers, and the instrument was capable of analyzing three commodities for three constituents. This instrument was the first to employ screw potentiometers for calibration. In calibration mode, the instrument recorded the diffuse reflectance signals in the form of the log (1/*R*) from the sample surface. The GQA used four lead-sulfide detectors, which were offset at an angle of 45° to the sample surface to maximize the reception of diffuse, relative to specular, reflectance and surrounded the sampling area of the instrument at intervals of 90°. The system layout of the GQA I is diagrammatically represented in Fig. 13.

The Neotec GQA I used Teflon as a reference standard against which the signal from the sample was compared. The sample cell of the Neotec GQA was a round plastic cup, and the sample surface was protected by a removable optical glass window. The cup held about 6–8 g of ground grain, which was compressed against the window by means of a spring-loaded disk and a threaded cap. The relatively uniform surface presented to the optical system improved the precision of measurement. The cell entered the instrument via a drawer, which at its inner extremity carried the Teflon standard. When the drawer was pulled out, the standard was positioned under the optical system/detectors. Pushing in the drawer replaced the standard with the sample. The log (1/*R*) signal was computed into the delta log (1/*R*) immediately. The three filters of the instrument were compared spectrally to standard reference filters during factory assembly, and a factor to standardize all filters was hard-wired into each instrument. The

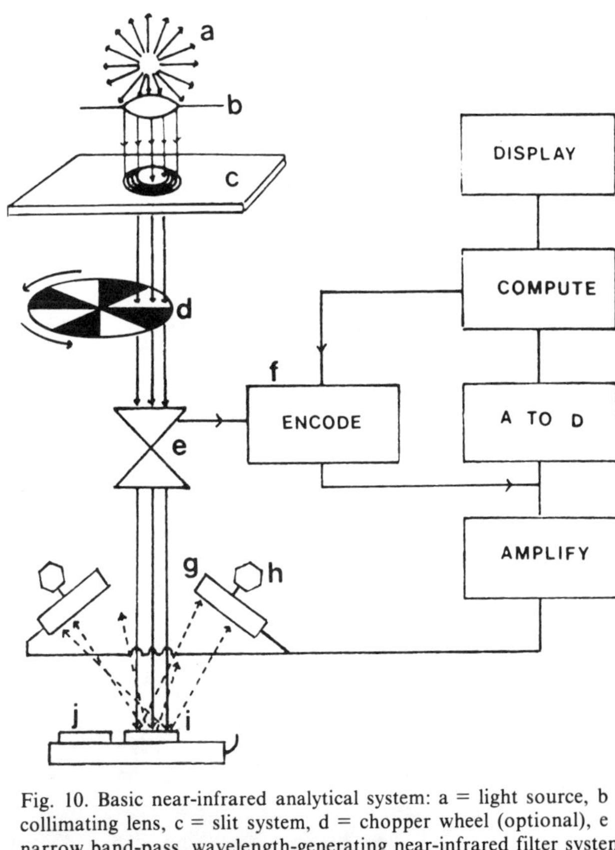

Fig. 10. Basic near-infrared analytical system: a = light source, b = collimating lens, c = slit system, d = chopper wheel (optional), e = narrow band-pass, wavelength-generating near-infrared filter system, f = wavelength-generating encoder, g = detector, h = detector cooling system (optional), i = sample, and j = diffuse reflectance standard.

delta log (1/*R*) was multiplied by these factors to give three optical values, called *C* values, for each sample. Calibration constants (*K* values) were developed by reading 40–50 samples in the instrument and recording the three *C* values for each. The *C* values were then regressed against standard chemical or physicochemical information. The four *K* values (three regression coefficients and an intercept) were then set into the instrument analog computer circuitry via three series of screw potentiometers on each of the three product boards. The instrument consisted of two boxes, the larger of which contained the optics, detectors, amplification, and data storage systems, and the smaller, upper box housed the analog computer and display. Results have been reported for analysis of several commodities by the GQA (Williams, 1975).

C. Anacon NIR Analyzer

The Anacon NIR Analyzer was developed during 1973–74. It possessed two rotating interference filters, and the sample was presented to the optics in an open cell. A high and a low calibration standard were provided for use by the operator in setting up the instrument. Calibration was extremely simple. A set of readings was obtained for a series of samples, and these readings were regressed against chemical data, such as protein, and a chart prepared using simple regression. The meter could be calibrated to read protein (for example) directly by preparing an overlay for the meter.

The results from the original instrument were not satisfactory because precision was very poor. The sample presentation involved the operator holding the sample in the open cup underneath the optics and detection system. The result was affected by the position of the sample relative to the detector, and errors of over 2% in protein could occur. The sample presentation system was redesigned, with the sample cell positioned in a recessed depression at the bottom of the sample presentation block. This fixed the position of the sample relative to the detectors, and the improvement in the results was dramatic. Some results for the analysis of hard red spring wheat, barley, and oats are presented in Table IV, together with the precision of check sample testing. Data for wheat were encouraging. The instrument was very fast in operation (practically instantaneous), but the very simple optical system coupled with other electronic factors resulted in significant daily baseline shifts. The respective coefficients of variation for the readings of low and high calibration standards were 6.2 and 4.6. The instrument was not marketed commercially, although it had the potential of being the first low-cost, simple near-infrared instrument applicable to grain analysis.

D. Pacific Scientific/Neotec Automated Digital Analyzer

The early-generation near-infrared instruments (1970–73) were beset with numerous operational factors that led to frequent breakdowns, poor precision and accuracy, and general unreliability. The design and introduction in 1974 of the Automated Digital Analyzer (ADA; Fig. 14) by the original

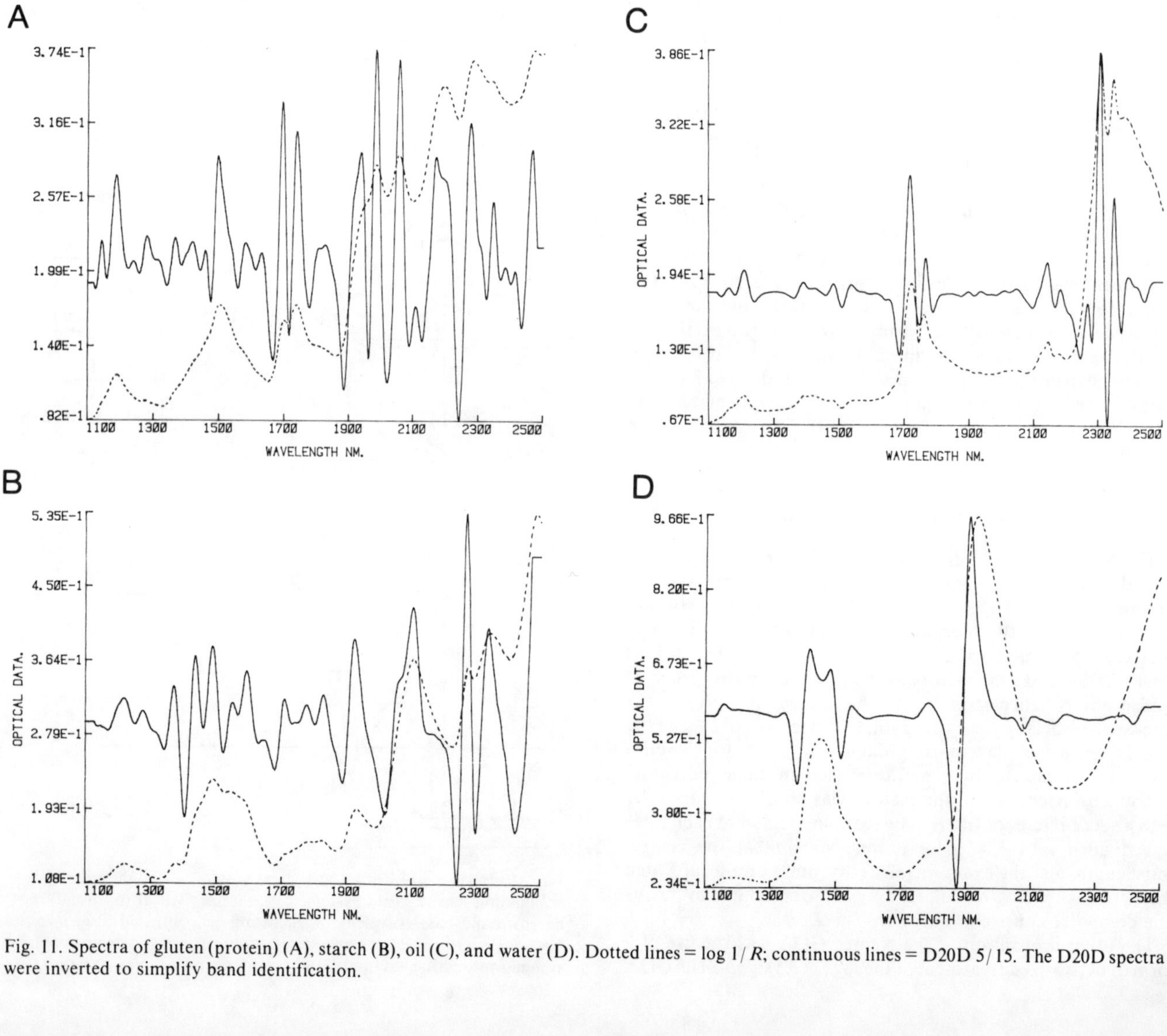

Fig. 11. Spectra of gluten (protein) (A), starch (B), oil (C), and water (D). Dotted lines = log 1/*R*; continuous lines = D20D 5/15. The D20D spectra were inverted to simplify band identification.

Neotec Corporation became a significant factor in the development of modern instrumentation (Williams et al, 1978). The instrument possessed four unique features: its automated sample presentation; the inverted optical system, which irradiated the sample from underneath; its pioneer use of a digital computer to transpose the optical signals into protein and moisture; and the first operator-controlled wavelength selection system. Analysis of a set of 10 samples could be completed in about 70 sec. The instrument performed a simultaneous analysis for protein and moisture and was the first near-infrared instrument to compute the protein results to a constant moisture basis (13.5%).

The ADA was also the forerunner of the modern spectrocomputer in that its software enabled the instrument to select the optimum wavelengths at which to make its measurements from six limited, but operator-designated, ranges of wavelengths, two in each of the three tilting filters. The algorithm used was $\Delta R/R$. This was analogous to the delta log $(1/R)$, but the log $(1/R)$ signal was converted to the antilog before further processing. The delta was set up by the selection of the second wavelength in each filter and divided by the value of R at the primary wavelength. Up to six wavelength points (pulse points) could be selected for each calibration. Computation of constants was carried out by a Data General NOVA 1220 digital minicomputer using forward stepwise multiple linear regression.

A wavelength "term" is the complete form in which an instrument uses wavelength. A term may involve any number of wavelengths. For exmple, a single log $1/R$ term involves only one wavelength (e.g., A in Fig. 7). A single second derivative term involves three wavelengths (e.g., A, B, and C in Fig. 7), whereas a quotient second derivative term involves six wavelengths, (e.g., A, B, C, A′, B′, and C′; see the discussion of quotient mathematics in section IV-C). A wavelength "point" differs from a "term" in that the point usually denotes the starting wavelength of a term, e.g., A in the first derivative or B in the second derivative (Fig. 7). A wavelength point may or may not be smoothed but usually is. In the ADA, the wavelength terms were displayed on a teletype, and the operator had the option of limiting the number of terms used by eliminating those that the regression coefficients indicated were not contributing to the accuracy of prediction. Usually four to five terms were adequate for the measurement of protein and three for the measurement of moisture.

TABLE III
Accuracy of Analysis by DICKEY-john GAC Model II

	Hard Red Spring Wheat		Barley	Soybean		Rapeseed	
Parameter	**Protein**	**Moisture**	**Protein**	**Oil**	**Protein**	**Oil**	**Protein**
Mean Std[a]	13.63	11.46	11.10	19.5	40.48	44.2	20.55
Mean NIR	13.71	11.52	11.05	19.7	40.55	44.3	20.32
SD Std	1.04	0.70	0.98	0.92	1.33	2.27	0.96
SD NIR	0.97	0.63	0.92	0.53	1.33	2.86	1.29
SEP[b]	0.27	0.32	0.46	0.50	1.09	1.99	0.80
r	0.93	0.89	0.88	0.90	0.90	0.80	0.80

[a]Standard methods of analysis were protein by Kjeldahl (AACC, 1983, Method 46-12), moisture by air oven (Method 44-15A), and oil by Goldfish extraction using anhydrous ether (Method 30-20).
[b]Standard error of performance.

Fig. 12. Pacific Scientific/Neotec Grain Quality Analyzer Model I.

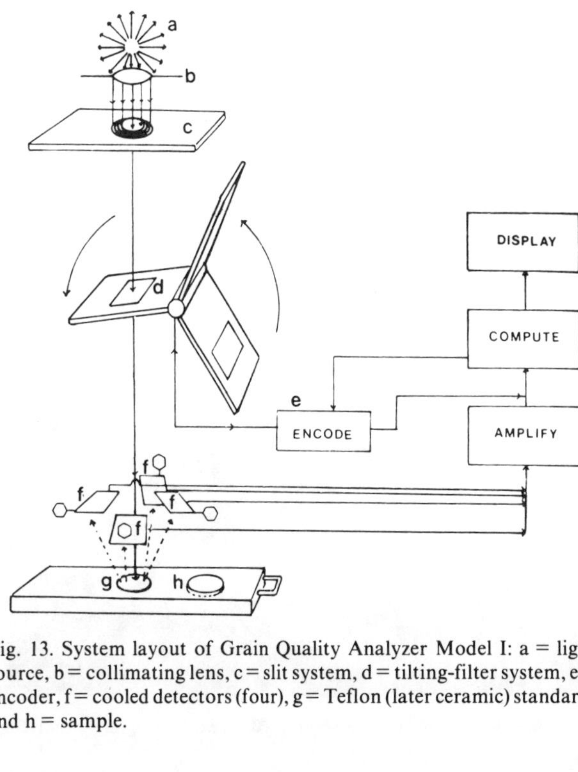

Fig. 13. System layout of Grain Quality Analyzer Model I: a = light source, b = collimating lens, c = slit system, d = tilting-filter system, e = encoder, f = cooled detectors (four), g = Teflon (later ceramic) standard, and h = sample.

The ADA is still in daily use at the Canadian Grain Commission, and it has now performed more than five million tests for protein and moisture. An optical CRT display has been added and the software updated. The algorithm in use now is the first derivative of the log $(1/R)$. The punched paper-tape method of calibration has been updated and the original Teflon standard replaced with ceramic, but the optics and sample presentation systems have not been changed. Sample cells are mounted in trays of 10. The bottoms of the cells are of optical glass, cemented into plastic cell frames with threaded lids. Sample trays are fed into the instrument automatically, and each individual cell is stopped over the sensing area by a photoswitch. After all 10 samples have been read, the instrument reads the ceramic standard and computes the results. The original spring-loaded plastic compression plate was never used with the sample compartments of the ADA. The plastic disk is cemented to a circle of foam rubber 8 mm thick and 35 mm in diameter, which is in turn cemented to the threaded cell cover. Cell closure is thus a single operation. Cells are overfilled and struck level to provide a uniform amount of sample in each cell. The cells are capped and residual meal vacuumed away. The first cell in each tray is offset 2 mm from the end, the last offset 1 mm. This difference is sufficient to trigger a fail-safe microswitch that prevents trays from being placed in the instrument backwards and helps in preserving sample identity. The sample presentation, analysis, and ejection systems are all controlled by microswitches (Fig. 15).

The ADA threw new light on the design and functionality of the early-generation near-infrared instrumentation manufactured by the Neotec Corporation. Its trouble-free operation during early months indicated that the optical and detection systems of the original GQA were not responsible for malfunctions, which were most likely attributable to the analog computer. The ADA was originally designed specifically for work with hard red spring wheat, but for the last five years it has been used to test hard red spring and winter wheats, soft white spring and winter wheats, durum wheat, and barley. Typical results for these surveys are given in Table V.

E. Technicon/DICKEY-john InfraAlyzer 2.5

The DICKEY-john GAC Model II was followed by the Technicon/DICKEY-john InfraAlyzer Model 2.5. This instrument incorporated a newly designed, closed sample cup with a quartz window instead of glass and analog computer "grain boards" that could be programmed by means of screw potentiometers from the outside of the instrument. The diode values were still manually entered into the grain boards using thumb switches, but the potentiometers greatly improved the ease of operation of the InfraAlyzer 2.5. "Diode" values were really log reference values or starting points from which the computation of constituent concentration derived. Both the original GAC I and the InfraAlyzer 2.5 used a light source that was underpowered electronically, increasing the life of the lamp

TABLE IV
Accuracy of Analysis for Protein by Anacon NIR Analyzer

Parameter	Hard Red Spring Wheat	Barley	Oats
Mean Std[a]	13.69	11.69	10.28
Mean NIR	13.90	11.62	10.48
SD Std	2.48	0.51	1.23
SD NIR	2.64	0.81	1.24
SEP[b]	0.38	0.55	0.85
r	0.99	0.75	0.76
N	12	12	12
	Precision		
Mean check[c]	13.88	12.05	12.12
CV	1.6	1.5	4.2
N	12	12	12

[a]Standard analysis for protein by Kjeldahl (AACC, 1983, Method 46-12).
[b]Standard error of performance.
[c]Mean and coefficient of variability (CV) of respective check samples.

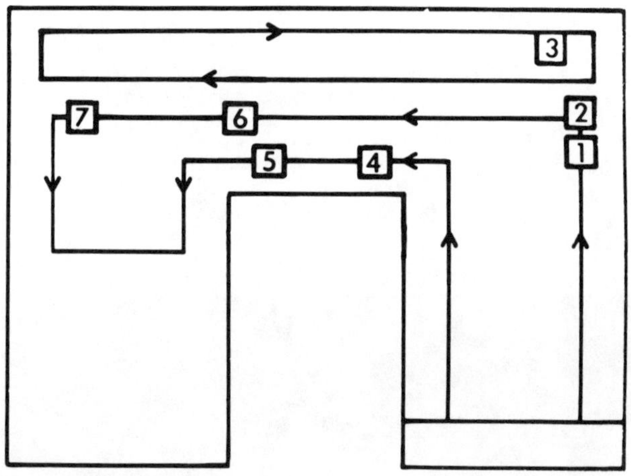

Fig. 15. Microswitching controls of Automated Digital Analyzer: 1 = fail/safe switch, 2 = stops inward and initiates lateral movement of sample cell trays, 4 = initiates reading first sample, 5 = alerts instrument to position of samples, 6 = counts sample cells (light-emitting diode), 7 = activates ejector motor, 3 = arrests lateral movement and is triggered by driving lug on chain drive.

Fig. 14. Pacific Scientific/Neotec Automated Digital Analyzer.

TABLE V
Accuracy of Analysis of New Crop Samples of Different Cereals by ADA

Parameter	Hard Red Spring Wheat	Durum Wheat	Soft White Spring Wheat	Barley
Mean Std[a]	14.27	14.48	10.91	12.32
Mean NIR	14.28	14.42	10.91	12.30
SD Std	1.76	1.87	1.08	1.58
SD NIR	1.83	1.82	1.06	1.59
SEP[b]	0.28	0.24	0.19	0.34
r	0.99	0.99	0.98	0.98
N	883	149	452	172

[a]Standard analysis by Kjeldahl (AACC, 1983, Method 46-12).
[b]Standard error of performance.

to the extent that in three years of continuous operation of 11 InfraAlyzers at the Canadian Grain Commission there were no instances of lamp failure.

IV. COMMERCIAL BENCH-TYPE INSTRUMENTS: SECOND GENERATION

The Neotec GQA I (including a version called the GQA II that had a motor-driven, drawer-type sample cell that exposed a very large area of sample to the detecting system), the DICKEY-john GAC I, and the InfraAlyzer 2.5 were the first generation of near-infrared instruments, together with the Anacon NIR Analyzer and another instrument designed by E. J. Brach at the Government of Canada Engineering and Statistical Support Division, Central Experiment Farm, Ottawa (E. J. Brach, pers. comm., 1974). The Ottawa instrument incorporated an integrating sphere and also promised to be a low-cost instrument, but development was halted because of the higher priorities of remote-sensing equipment design at the Ottawa institution. Feedback from the users, together with the availability of high-quality components and digital microprocessors at low cost, stimulated the development of a second generation of near-infrared instruments.

A. Other Neotec GQA Models

The second generation of instruments included the Neotec models GQA 21, 31, and 41. The GQA 21 was an updated version of the original Model I. Model 31 originated during early 1974, with a Teflon standard and improved analog computing. It incorporated thermoelectrically chilled, lead sulfide detectors, which stabilized the signal and output. The 1976 version employed a ceramic standard. All of these instruments, as well as those produced by other companies, incorporated a fan to introduce airflow to and from the instrument and took other precautions to control the temperature. Model 41 (Fig. 16) was introduced during 1974 as the first bench-type instrument to incorporate digital computing and the first to allow the operator to change wavelengths without changing filters. The instrument also featured the first completely keyboarded operation, whereby the operator was able to enter constants, change intercepts, and change wavelengths by pressing keys rather than by using a screwdriver. The Model 41 carried three tilting filters, but the operator could select up to four delta R/R terms per constituent, two of which were in one filter. The two terms were selected for the filter covering the area that had most significance in the measurement of specific constituents; for example, two terms (pulse points) would be selected in filter number two (the "protein" area) for the measurement of protein. Wavelengths were again entered as pulse points, via a drawer-type keyboard that was also used to enter calibration constants. The GQA 41 could be interfaced to a teletype or CRT terminal for displaying $\Delta R/R$ values. It was capable of simultaneous analysis of up to 16 parameters (four components in four commodities). The results of analysis of the common constituents were superior to those of contemporary instruments, and standard deviations of difference of 0.18–0.22 were normally obtained for determination of protein in wheat. The instrument was also useful for the analysis of mixed feeds and their constituents (Williams and Starkey, 1980). A combination of higher cost per unit, together with advances in technology, rendered the GQA 41 obsolete by about 1977 when the third generation of instruments made its appearance.

During 1977 two events of significance took place that affected future near-infrared bench instrumentation. The first was the availability at reasonable prices of digital microprocessors, which stimulated a large-scale change to the use of these highly efficient components for data reduction in bench-type instruments. The second was the development of division or quotient mathematical treatments of the log $(1/R)$ signal at the Instrumentation Research Laboratory of the U.S. Department of Agriculture, Beltsville (Norris and Williams, 1984).

B. Neotec Feed Quality Analyzer

In 1976, Neotec introduced its Feed Quality Analyzer Model 51 (FQA 51; Fig. 17) at the 61st National Convention of the American Association of Cereal Chemists in New Orleans. This instrument was an extension of the GQA Model 41 and was the first near-infrared instrument to have automatic self-calibration capability. It was fitted with a vertically rotating filter wheel carrying six narrow band-pass filters (Fig. 18). By use of the tilting-filter principle, each filter was subdivided into

Fig. 16. Pacific Scientific/Neotec Grain Quality Analyzer Model 41. (Courtesy Pacific Scientific Co., Silver Spring, MD)

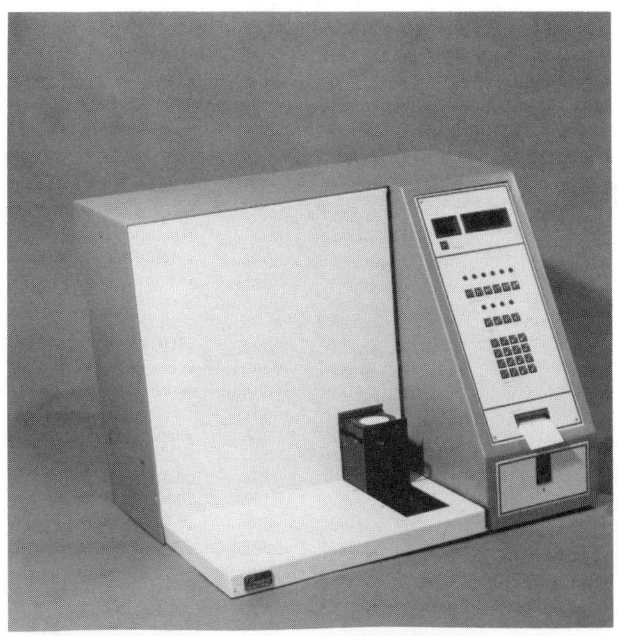

Fig. 17. Pacific Scientific/Neotec Feed Quality Analyzer Model 51. (Courtesy Pacific Scientific Co., Silver Spring, MD)

128 wavelength points, giving a possible total of 768. Table VI gives an abbreviated listing of wavelengths corresponding to the pulse points of all six filters used in the FQA 51A. At the starting or central wavelength of a filter—when the filter is perpendicular to the light source—the filter is at its longest effective wavelength, and as the filter is tilted in either direction, the effective wavelength becomes shorter. The FQA 51A also gave the operator the option of seven mathematical treatments of the log $(1/R)$ signal, including the log $(1/R)$ itself and the first and second derivatives thereof. Quotient treatment could be applied to the first- and second-derivative algorithms. The software enabled the operator to select pulse points (wavelengths) and mathematical treatments. Having specified these, the operator read the optical values of up to 50 calibration samples. The optical data were automatically stored by the FQA. In its "autocal" mode, the instrument then computed its own calibration constants using a forward stepwise multiple linear regression. Although pulse points could be inserted for up to six wavelength terms, the system usually selected only two to five of these for analysis. The operator could view the standard error of estimate progressively as each sample was added to the calibration and could delete samples that apparently had gross errors in their chemical analysis. The FQA 51A could also be used up to a point as a scanning instrument. Wavelengths could be changed by increments of up to five pulse points, and by observing the changes in the standard error of calibration the operator could optimize the wavelengths. The FQA 51A could accommodate up to 96 calibrations.

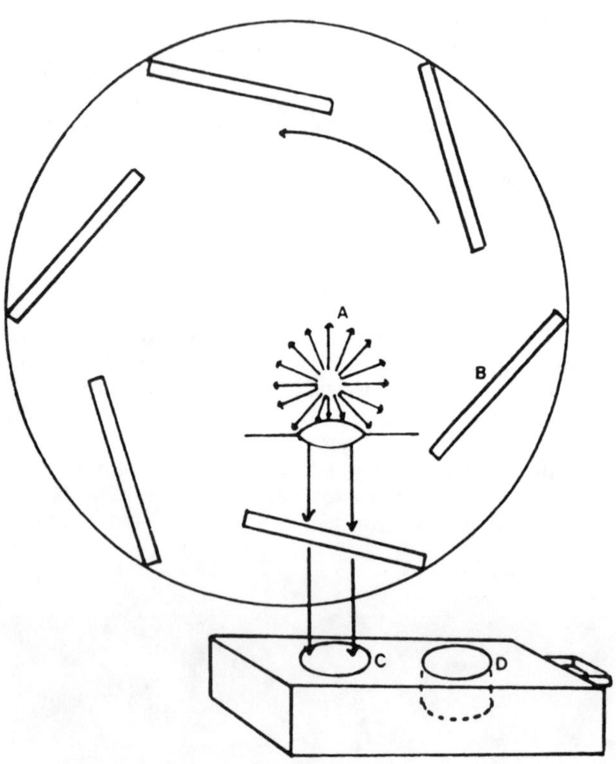

Fig. 18. Vertical filter-wheel system carrying six tilting filters: A = light source, B = narrow band-pass tilting filter, C = ceramic standard, and D = rotating sample cell.

C. DICKEY-john GAC III Series

The DICKEY-john Corporation introduced its new GAC III series (Fig. 19) at the fall meeting of the American Association of Cereal Chemists in San Francisco in 1977. The instrument featured digital data reduction and a rotating sample cell. The optical system consisted of up to 10 discrete filters and used the log $(1/R)$ signal directly with no further processing. Four extra filters enabled the use of a quotient delta log $(1/R)$ algorithm, but for most applications the raw log $(1/R)$ algorithm using the original six filters (the same wavelengths as the original GAC I) was preferred. The filters were carried in a filter wheel that moved by a stepping process—each filter was positioned over the sample for a set period and then moved to a new position. The sample was constantly illuminated although the irradiation was intermittently interrupted by a chopper wheel, creating alternate light/dark periods. The original GAC Model I horizontal, rotating filter wheel rotated at 450 rpm. By leaving the filter wheel in one position during many rotations of the chopper wheel, the instrument was enabled to make multiple readings either of the sample or the internal standard. The chopper wheel whirled at 1,800 rpm, which increased the

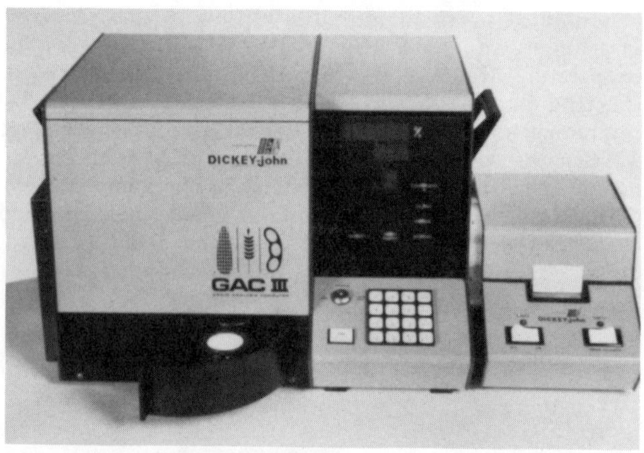

Fig. 19. DICKEY-john Grain Analysis Computer Model III. (Courtesy DICKEY-john Corp., Auburn, IL)

TABLE VI
Wavelengths/Pulse Points (PP) for FQA 51A

Starting Wavelength											
1,680 nm		1,760 nm		2,100 nm		2,220 nm		2,310 nm		2,360 nm	
PP	λ	PP	λ	PP	λ	PP	λ	PP	λ	PP	λ
10	1,608	140	1,687	265	2,009	395	2,126	520	2,254	650	2,305
25	1,622	155	1,702	280	2,027	410	2,145	535	2,265	665	2,316
40	1,635	170	1,715	295	2,043	425	2,162	550	2,275	680	2,326
55	1,647	185	1,727	310	2,057	440	2,177	565	2,284	695	2,334
70	1,656	200	1,737	325	2,070	455	2,190	580	2,291	710	2,342
85	1,665	215	1,745	340	2,080	470	2,200	595	2,298	725	2,348
100	1,671	230	1,751	355	2,088	485	2,209	610	2,303	740	2,353
115	1,676	245	1,756	370	2,094	500	2,215	625	2,306	755	2,357
122[a]	1,677	249	1,757	376	2,096	504	2,216	632	2,308	760	2,358

[a] Last figure in each column represents last usable pulse point on filter for derivative math.

number of readings taken in unit time and hence the precision of the analysis. The GAC III was the first instrument to use a stepping mechanism for the filter wheel. The single lead-sulfide detector was offset to minimize specular reflectance and cooled for high sensitivity. Use of the high-speed chopper gave better precision and a better signal-to-noise ratio. The rotating sample cup further improved analytical precision by minimizing error caused by loading of the sample cell. The optical system is illustrated in Fig. 20. The energy source was a tungsten/halogen lamp, with the light energy emanating from the lamp and passing through a collimating lens that focused the light into parallel rays. The light beam was interrupted periodically by means of the rotating chopper disk. The chopper enabled measurements to be taken with the light on and off, generating an A/C signal from the detector that reduced noise by compensating for minute changes in the detector. The light passed through narrow band-pass near-infrared interference filters so that only light of certain wavelengths reached the sample surface and thence the detector. The signal from the detector was amplified and converted to digital form for computation.

The precision of analysis by all near-infrared bench-type instruments is enhanced by the taking of multiple readings of both sample and standard. This was achieved by illumination of the sample or the standard for a set period of time, during which, as a result of the speed of rotation of the disk carrying the interference filters, the filter-wheel assembly, or a stepping filter wheel with a chopper mechanism, a set number of readings was taken and averaged. The DICKEY-john instruments made 120 readings of the sample before averaging and displaying the result, and the Neotec GQA series took 84 readings. The actual precision of reading a sample can be verified by making a number of separate determinations of both the optical values and the concentration of, for example, protein, using a sealed cell filled with a commodity such as ground wheat (Williams, 1975). The cell is placed in the sample drawer and simply read a number of times—12–20 readings are adequate. The precision determined in this way depends solely on the capacity of the instrument to make readings of a constant sample surface (repeatability). The results of such a study for a DICKEY-john

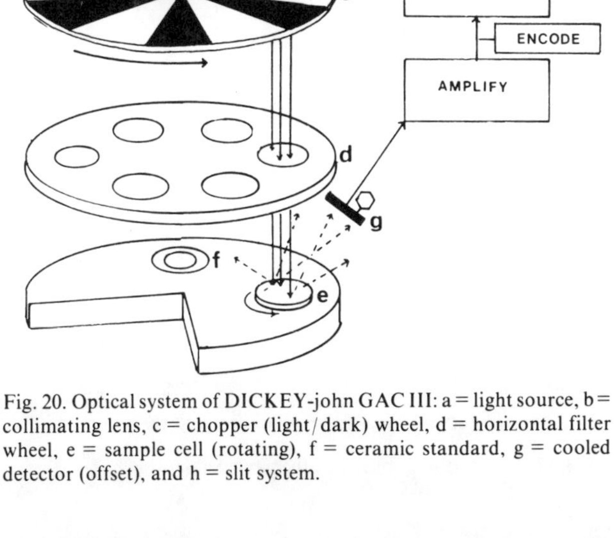

Fig. 20. Optical system of DICKEY-john GAC III: a = light source, b = collimating lens, c = chopper (light/dark) wheel, d = horizontal filter wheel, e = sample cell (rotating), f = ceramic standard, g = cooled detector (offset), and h = slit system.

GAC III and a Neotec GQA Model 31 EL are presented in Table VII. For this study, the drive belt of the GAC III sample cell was removed so that the sample cell could not revolve. Both instruments were clearly capable of a very high degree of precision. The principle of using a sealed sample cup to assess instrument precision was adopted by a number of users, and for a time special threaded cells called permacells were marketed by the Neotec Corporation for use with their instruments. There was less demand for the permacells by users of the Technicon/DICKEY-john instruments, since daily readings of the signal from the ceramic standard were adequate for the establishment of instrument stability.

Another feature of the GAC III series was the convenient push-button operation. The front panel carried a clearly illuminated display activated by a keyboard by which all constants were entered and changed. Calibration still involved multiple linear regression of optical against chemical data. This was achieved by a separate computer, or by direct interfacing of the GAC with an auxiliary desktop computer such as a Hewlett-Packard Model 9815. The GAC III was also the first portable bench instrument to convert analytical data to constant moisture basis. The basis for reporting was programmable, but in North America most results are reported on a 13.5% or 14% moisture (or moisture-free) basis.

The introduction of near-infrared reflectance analysis focused renewed attention on the accuracy and precision of routine laboratory analysis (Miller et al, 1978). Discrepancies between the pioneer instruments and wet chemical analysis had been attributed solely to failure of the instruments. Most of the earlier efforts to adopt near-infrared instrumentation were directed to the determination of protein in cereals, particularly in wheat. Variations in moisture content of wheat often resulted in significant discrepancies in the apparent protein content (Williams and Sigurdson, 1978), and laboratories gradually came to realize the importance of reporting the results of testing for protein and other constituents on a constant moisture basis. The extent of errors in wet chemical analysis also became apparent, and the establishment of the technology was accompanied, not without some soul-searching, by a progressive improvement in the analytical precision and accuracy of routine laboratory analysis by wet chemistry methods.

The introduction of quotient mathematical treatments in 1977 provided the basis for a further improvement in the precision and accuracy of near-infrared testing for protein, particularly in the presence of a high degree of variance in particle size or moisture content, which also affects the mean particle size and particle-size distribution of ground grains and seeds. Quotient mathematics were developed at the Instrumentation Research Laboratory at the Agricultural Research Center of the U.S. Department of Agriculture at Beltsville. The principle (Fig. 21) involves division of a primary wavelength term by a second, or reference, wavelength term. Based on use of the raw log $(1/R)$ signal, each wavelength point represents a single central wavelength that has been subjected to the specific smoothing technique preferred by the manufacturer. If the first or second derivative of the log $(1/R)$ is used, each wavelength term represents respectively two or three individual wavelengths. The primary, or "starting", wavelength of each term is selected by multiple regression after reading samples. The reference wavelength is selected by matrix inversion using

TABLE VII
Precision of Analysis of Wheat by Near-Infrared Instruments

	DICKEY-john GAC III		Neotec GQA 31EL	
Parameter	Protein	Moisture	Protein	Moisture
SD	0.05	0.06	0.06	0.06
CV	0.36	0.52	0.43	0.52

the same samples. To improve the integrity of the primary wavelength, the reference wavelength may be reentered as a primary wavelength to reselect a reference wavelength. This procedure is repeated until neither wavelength changes further. By using very large numbers of samples to maximize the variance in both particle size and moisture level and by testing the quotient (division) procedure with different mathematical treatments, it was concluded (Williams and Norris, 1983; Norris and Williams, 1984) that the first derivative of the log $(1/R)$ provided the most reliable form of mathematical treatment for both particle size and moisture variance. Raw log $(1/R)$, Kubelka/Munk, and first- and second-derivative treatments were all evaluated, and it was apparent that the division step effectively improved analysis by any of these algorithms. Late in 1977, the Neotec Corporation introduced its GQA 31EL, modified to accommodate quotient mathematics. The instrument possessed two circuit boards for the ordinary delta log $(1/R)$ algorithm and one for the quotient version of the first derivative of the log $(1/R)$. Access to all boards was by screw potentiometers. During the early part of 1978, the DICKEY-john GAC III was also made available with optional quotient mathematics.

D. Neotec GQA 101 Series

The first instruments specifically designed for the use of quotient mathematics were the Neotec GQA 101 series. These instruments were low cost and simple, with two product circuits capable of carrying calibrations for the determination of protein and moisture in two commodities, such as wheat and barley. The instruments incorporated a push-button control panel. The drawer-type sample presentation was replaced with a cell mounted in a rigid plastic plate. An optical glass window was cemented into the bottom of the cell. The cover of the cell was fitted with a plastic plate backed with foam rubber to provide compression for the sample. The sample was loaded into the cell and capped with the special compression cap, and the entire cell assembly was inserted into the instrument in the vertical position. The ceramic standard was placed behind the sample cell compartment so that when the sample cell assembly was removed for reloading, the standard was irradiated. The 101 series also incorporated the tilting-filter principle but with the paddle-wheel, three-filter assembly rotating in the horizontal plane rather than the vertical. The operator was able to change wavelengths (pulse points) by means of a push-button coding system. Calibration was a very simple procedure because the GQA 101 used only a single delta log $(1/R)$ quotient term for protein and a single nonquotient term for moisture. As a result, simple linear regression, rather than multiple regression, could be used to compute calibration constants.

The instrument could be calibrated for protein, or any other constituent for which quotient mathematics was used, by a two-point system that did not require a computer. Calibration with two samples calls for the preparation of two very carefully blended samples that should represent the upper and lower limits of concentration anticipated. The samples are ground and read in the instrument. Optical values are recorded for both samples, and the calibration constants are calculated according to the formulae:

$$K_1 = \frac{(\text{high composition \%} - \text{low composition \%}) \times 100}{(\text{high optical signal} - \text{low optical signal})} \quad (4)$$

$$K_0 = \text{low composition \%} - \frac{(K_1 \times \text{low optical signal})}{100} . \quad (5)$$

Example 2: Protein of high sample is 16.6, low sample is 10.6. Optical signal (C value) of high sample is −33.3 and of low sample is −45.1:

$$K_1 = \frac{(16.6 - 10.6)}{[-33.3 - (-45.1)]} \times 100 = \frac{(6.0)}{(11.8)} \times 100 = 50.8$$

$$K_0 = 10.6 - \frac{(50.8 \times -45.1)}{100} = 33.5 .$$

The constants K_1 (50.8) and K_0 (33.5) are set into the instrument by push buttons.

Based on a quotient algorithm, calibrations for constituents other than protein involved only a simple linear regression rather than the customary multiple linear regression of the earlier instruments. The GQA 101 was also the first instrument to use a wavelength in the neighborhood of 2,020 nm for determination of moisture. The original version of the instrument used the more familiar 1,940-nm point, but research with the Pacific Scientific (then Neotec) Research Composition Analyzer Model 6350 interfaced directly to a GQA 101 revealed that a wavelength near 2,020 nm gave more accurate analysis for moisture (R. C. Moen and P. C. Williams, unpubl. data, 1978).

E. Neotec GQA Model 102

The Model 101 was soon joined by the 102 (Fig. 22), which was equipped with more memory and was capable of carrying calibrations for four constituents and four commodities. It also had the option of using the first or second derivative of the log $(1/R)$ as a three-term algorithm or as single-term-quotient, first- or second-derivative mathematics.

The Model 102 used the standard Neotec sample cell, which fitted into a vertical drawer. Both the Neotec 101 and 102 were compact instruments. They possessed three narrow band-pass filters, and by using the tilting-filter principle the wavelength area scanned by each filter was divided into 100 wavelength points, or pulse points, so that 300 wavelength points were available to the operator. This allowed for changing wavelengths to accommodate a range of commodities and constituents. The wavelength range was from about 1,900 to 2,340 nm. Calibration was easier than with any of the preceding

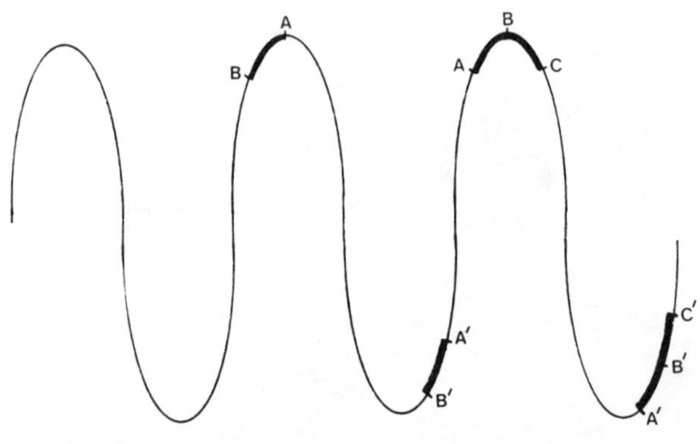

ALGORITHM	AS IS	QUOTIENT
FIRST DERIV.	A−B	(A−B)/(A′−B′)
SECOND DERIV.	A−2B+C	(A−2B+C)/(A′−2B′+C′)

Fig. 21. Principle of quotient mathematical algorithms.

instruments because the single wavelength term meant that regression (calibration) constants could be computed using the simple $y = a + bx$ equation, where x was the optical datum point for each sample. Since most of the variance in particle size and moisture content was compensated for by the quotient algorithm, the calibration samples needed only to represent the range in composition, and as a result fewer samples were necessary. About 20 were recommended, in contrast to the 50 samples recommended for calibration of other bench instruments. Recently, the company introduced its Model 3000, which is a version of the Model 102 with an auxiliary IBM PC computer and a full range of sample presentation systems.

F. Labormim Infrapid 31

Simultaneously with the introduction of the Neotec GQA 31EL, the Hungarian government's Labormim Corporation, Budapest, commenced the manufacture of the Infrapid, an instrument that was essentially the European counterpart of the GQA 31EL, built under license to Neotec by the Budapest-based corporation. The instrument was identical with the GQA 31EL in essentially every aspect, and it is still available.

G. Neotec GQA Ultraspeed

An innovative instrument designed and introduced by Neotec at the same time as the Model 101 was the Ultraspeed, which was a GQA 31 EL fitted with a built-in, Falling Number KT-30 burr-type grinder. The theory was that the use of quotient mathematics in conjunction with the grinder would allow the instantaneous determination of protein in wheat and barley by grinding the whole grains and scanning the ground whole meal directly in a gravity-fed, vertically oriented cell. The sample of wheat or barley (20–25 g) was poured into a small hopper at the top of the instrument. When the grinder was activated by a switch the sample was ground, passing directly into the detecting area. A small impeller kept the ground sample passing through the system without clogging, and the result was displayed when all of the sample had passed through the detection area. Only a single-quotient, first-derivative term was used for protein measurement, and calibration was again achieved using a simple $y = a + bx$ linear regression. The instrument was not very successful, primarily because of poor analytical precision, and it is no longer marketed. The algorithm could compensate for wide variations in particle size, but the ground whole meal from the grinder was not blended before entering the detection area. Differences in sample constitution caused by texture were magnified by stratification and static electricity, creating significant discrepancies. These were accentuated when the grain was high in moisture, since moisture content influences the manner in which grains are reduced in all grinders, particularly the burr-mill type.

H. Technicon InfraAlyzer Model 400

The Technicon Corporation introduced its first wholly Technicon-built instruments during 1978. The InfraAlyzer 400 was the first of these, and the most comprehensive (Fig. 23). It used the discrete filter approach, with the log $(1/R)$ signal used directly. The filter assembly could accommodate up to 19 filters and offered the option of filters at additional wavelengths. The standard instrument was equipped with a filter wheel carrying six filters with wavelengths at 1,680, 1,940, 2,100, 2,180, 2,230, and 2,308 nm, which were the wavelengths originally selected at the U.S. Department of Agriculture Instrumentation Research Laboratory for use with the untreated log $(1/R)$ algorithm. Other filters were presented to the light source and sensing system by a stepping mechanism, rather than the familiar continuously rotating filter wheel. The action of the stepping motor and chopper wheel has been described earlier (Section IVC). Precision of the reading of the optical system was further improved by the incorporation of an integrating sphere. The beam of light alternately irradiated the inside of the sphere, which was used as an internal reference standard, and the sample. This was facilitated by an oscillating mirror system. A ceramic reference standard was provided to enable the operator to check instrument stability. Twin lead-sulfide detectors were also located in a position to receive the signal for both sample and reference. The optical system is represented in Fig. 5. Wavelengths (filters) available in the standard 19-filter InfraAlyzer 400 corresponded to 1,445, 1,680, 1,722, 1,734, 1,759, 1,778, 1,818, 1,940, 1,982, 2,100, 2,139, 2,180, 2,190, 2,208, 2,230, 2,270, 2,310, 2,336, and 2,348 nm. Other filters are available and in use in the Model 400R (Research) instruments. The signal from the detector was amplified in the usual manner before being computed into protein, moisture, etc., by the microprocessor.

The InfraAlyzer 400 was joined by the 300 and 200. These models were lower cost, simplified versions of the Model 400. The chief difference was that these instruments were only

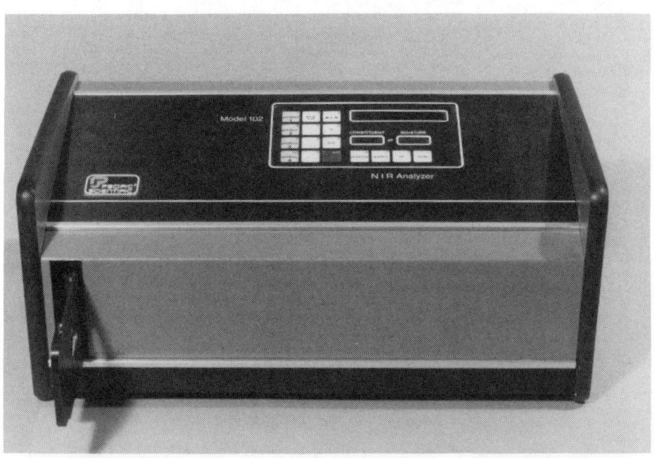

Fig. 22. Pacific Scientific/Neotec Grain Quality Analyzer Model 102. (Courtesy Pacific Scientific Co., Silver Spring, MD)

Fig. 23. Technicon InfraAlyzer Model 400. (Courtesy Technicon Corp., Tarrytown, NY)

available with six filters. Other than this, the Model 300 was essentially the same as the Model 400, with the integrating-sphere optics. The Model 200 carried the same optics but was programmed by means of a removable cartridge into which the constants were manually inserted. It was marketed by the International Stanley Corporation as the International Stanley Grain Analyzer. Models 400 and 300 were programmable with a push-button operation, with the push-button keyboard being incorporated into the actual face of the instruments. Calibration was facilitated in all near-infrared instruments offered by Technicon, DICKEY-john, and Neotec by an interfaced Hewlett/Packard Model 9815 or similar desktop computer, which was an optional accessory. The Model 9815 calculator was itself programmable by means of cassette tapes, the repertoire of which included multiple linear regression. Technicon engineers developed a novel, easy-to-load sample cell with quartz window, which was used with models 400, 300, and 200. The InfraAlyzer is now available in several versions, some of which are custom produced for specific applications. The parent company researches the wavelengths for individual commodities and constituents. Several versions of the InfraAlyzer 400 are marketed for use in the dairy, textile, and pharmaceutical industries, in addition to the grain and food industries where bench-type near-infrared instruments first saw service as quantitative analytical instruments. Special cells are available for the analysis of liquids.

A recent introduction to the Technicon series is the InfraAlyzer Model 250, which is a redesigned version of the 300. The instrument offers precalibrations and transferability of calibrations to other instruments of the same type. Large-scale monitoring of instruments by telephone (modem) hookup is also featured. The optics, detection system, and internal software are essentially the same as in the well-established models 300 and 400.

I. Pacific Scientific Feed Quality Analyzer Model 51A

The Model FQA 51A, which became available in 1978, was an updated version of the FQA 51 that featured a rotating sample cell, updated software that added the option of quotient first- and second-derivative treatments of $\log(1/R)$, and ability to be interfaced with an auxiliary computer that allowed the operator to select pulse points for new constituents and commodities and to optimize mathematical treatments. The computer came equipped with twin flexible disks for storage and processing of several hundreds of samples and of many calibrations. The use of flexible disks, each of which could store spectral data for over 400 samples, enabled comprehensive research into wavelength selection and optimization of mathematical treatment for the analysis of new commodities and constituents.

J. Percon Inframatic 8100

The most recent introduction to the European near-infrared field is the Inframatic 8100 (Fig. 24), a well-engineered, compact instrument offered by Percon Prufgërate GmbH of Hamburg, West Germany. The 8100 uses the $\log(1/R)$ signal directly, with six narrow band-pass interference filters. One of the most innovative features is that it has no sample cell. The ground sample is packed into a built-in area that is confined by a hinged, doorlike attachment. The packed sample is irradiated through a quartz window, which must be brushed clean between samples. The Inframatic is calibrated with factory-developed constants and is marketed ready for use in the measurement of protein, moisture, and kernel hardness in wheat and of protein and moisture in barley. It can be calibrated for other commodities and constituents in the same way as the DICKEY-john GAC or Technicon InfraAlyzer series. It is the first instrument marketed with a factory calibration for kernel hardness, which is becoming a very important parameter in the wheat industry. European grain differs from North American, Australian, South American, and South African grains in that it usually is of much higher moisture content. Moisture exerts a strong influence on the manner in which the grain is reduced in any grinder, and the calibrations established for the Inframatic 8100 were developed using European commodities. Use in North America or any of the other areas referred to would require new calibration constants based on regionally produced grain.

K. Pacific Scientific Model 7000

The FQA 7000 series is a tilting-filter instrument available with seven filters, each with an effective scanning area divided into 100 pulse points. The format of the filter wheel is essentially the same as that of the FQA51 (Fig. 18). The instrument, which is used mainly for the analysis of more complex commodities and constituents, is interfaced to an auxiliary computer that gathers and stores data, selects pulse points, and optimizes mathematical treatments. The operator may work in both reflectance and transmittance modes and has available a range of sample presentation cells that enable the analysis of whole grains, powdered (ground) commodities, liquids, and slurries. The instrument moves the sample while the measurement is being made and allows the scanning of from 1 to 100 cm^2 in area for reflectance and up to 200 cm^3 for transmittance. Another version of the FQA 7000 is the Dynascan, which is designed specifically for on-line analysis and is equipped with the same range of software and sample presentation facilities.

V. COMMERCIAL BENCH-TYPE INSTRUMENTS: THIRD GENERATION

The third generation of near-infrared instruments offered by all seven of the instrument manufacturers is programmed to provide the operator with basic information as to internal instrument troubleshooting. The operator can verify whether the instrument is functioning normally by reading the optical signals from the reference standard or a check sample. The instruments themselves detect their own electronic failings and most obvious errors and indicate the inconsistencies with documented error messages. If the instruments should fail to provide the accepted levels of accuracy and precision, time can be saved in servicing if the operator is able to inform the parent company of the symptoms displayed by the instrument. Troubleshooting software provides the operator with the necessary

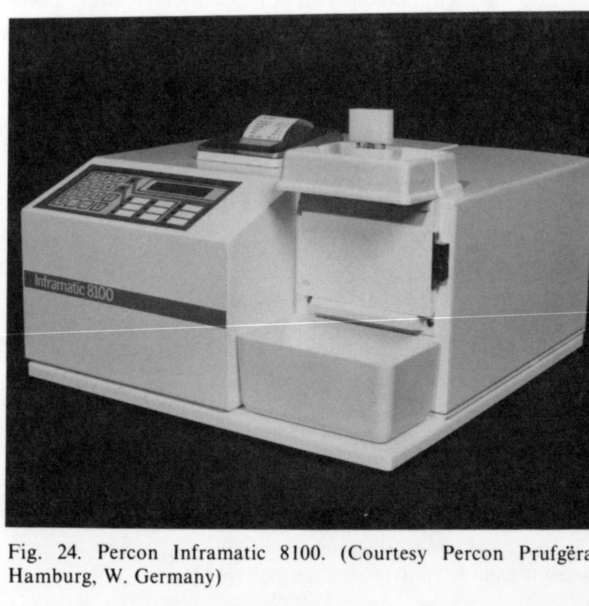

Fig. 24. Percon Inframatic 8100. (Courtesy Percon Prufgërate, Hamburg, W. Germany)

information to assist the factory or service center to diagnose the faults from long distances, and thereby saves time in ensuring that the correct parts are available. In many instances the service center can instruct the operator as to how the faulty parts may be replaced, saving considerable time and expense in travel and labor.

During the past seven to eight years, most near-infrared bench-type instruments have had the capacity for interfacing with auxiliary terminals or computers, thus allowing automatic computation and storage of large numbers of calibration constants. In the case of tilting-filter instruments, this capacity also enables the operator to optimize wavelengths (pulse points). The chief drawback to an instrument wherein the wavelengths can be changed is that the operator needs to know which wavelengths to use. On the other hand, the chief advantage of the tilting-filter approach is that the operator has access to a large wavelength range. All wavelengths used in an near-infrared instrument must be optimized relative to each other, and the manual changing of even three wavelength terms to optimize them all is a formidable and time-consuming process. An auxiliary computer can select wavelengths from any series of samples, and modern instruments can be interfaced to auxiliary computers to optimize wavelengths directly as pulse points for the Pacific Scientific instruments. Pulse points or wavelengths selected on any near-infrared instrument can be transferred to other instruments of the same model, which can then be calibrated by manual methods or by automated software. Scanning of samples by computerized spectrophotometers working in the log (1/*R*) mode enables the identification of the optimum discrete filters to use with instruments having this algorithm.

A. Percon Inframatic 8120

During 1984, Percon Prufgërate expanded the capability of its six-filter Inframatic 8100 to include 20 filters. The new Percon 8120 has the same compact size as the original model, and it also uses a stepping mechanism/chopper wheel to achieve wavelength selection and precision. The 8120 can be interfaced to an auxiliary computer and is provided with software that enables calibration, calibration storage, and statistical analysis of analytical data in a manner similar to the Pacific Scientific, Technicon, DICKEY-john, and Trebor instruments. The Inframatic is also available with special sample-presentation cells for analysis of liquids and other products, such as butter and cheese. Percon offers calibrations for more than 60 commodities and 20 constituents or parameters.

B. Trebor GT-90 Whole Grain Analyzer

During 1980, the introduction of the Trebor GT-90 (Fig. 25) marked a new phase of near-infrared technology in that it was the first instrument available to the industry that was capable of testing wheat for protein and moisture without grinding the sample (Williams et al, 1985b). Grinding the sample had always been regarded as one of the most critical sources of error in near-infrared testing of grains and seeds. Earlier work at the U.S. Department of Agriculture Instrumentation Research Laboratory had demonstrated the feasibility of testing whole wheat for protein and moisture by transmittance spectroscopy in the near-infrared region (K. H. Norris and P. C. Williams, unpubl. data, 1979), and Tkachuk (1981) reported on the determination of protein in whole wheat by near-infrared reflectance. The GT-90 uses near-infrared light-emitting diodes (NIRLEDs) at discrete wavelengths as the source of irradiation. Interference filters are incorporated with the NIRLEDs to complement the optical system. Energy passes through the sample-sensing area and is detected by a silicon detector (Fig. 26). The optical system is unique in the use of NIRLEDs and also in that they are illuminated one at a time. The flow of energy is turned on and off between readings. Total time for a complete pass of all 12 NIRLEDs is 0.25 sec, and the instrument takes 100 readings of each increment of the sample. Whole grain passes through the sensing area in increments of 3–3.5 g, and the sample is stationary when the transmitted signal is being sensed. Amplification and computation of the signal into constituent composition are achieved in a manner analogous to near-infrared reflectance instruments. Energy signals are sensed at 12 wavelengths, and all of the signals are used to compute 12 calibration constants. Multiple linear regression analysis is used to generate the constants (regression coefficients and an intercept). Calibration is facilitated by interfacing with a minicomputer. Two additional constants are included (13 and 14) to compensate respectively for temperature of the instrument and the sample. The mathematical algorithm used is the log (1/*T*), but the instrument can also be adapted to use the

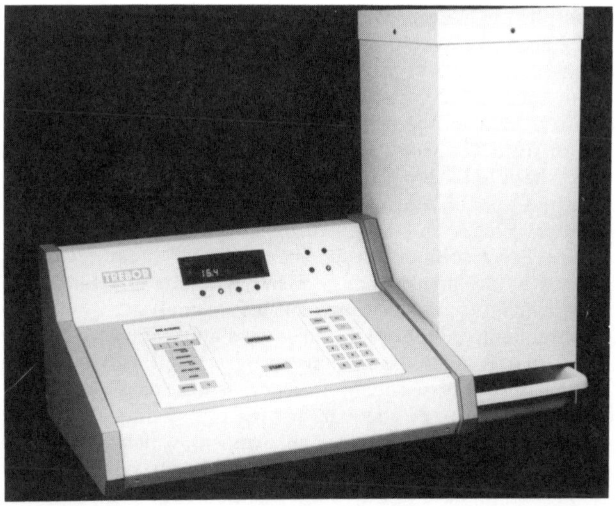

Fig. 25. Trebor GT-90 whole-grain NIR analyzer. (Courtesy Trebor Instruments, Gaithersburg, MD)

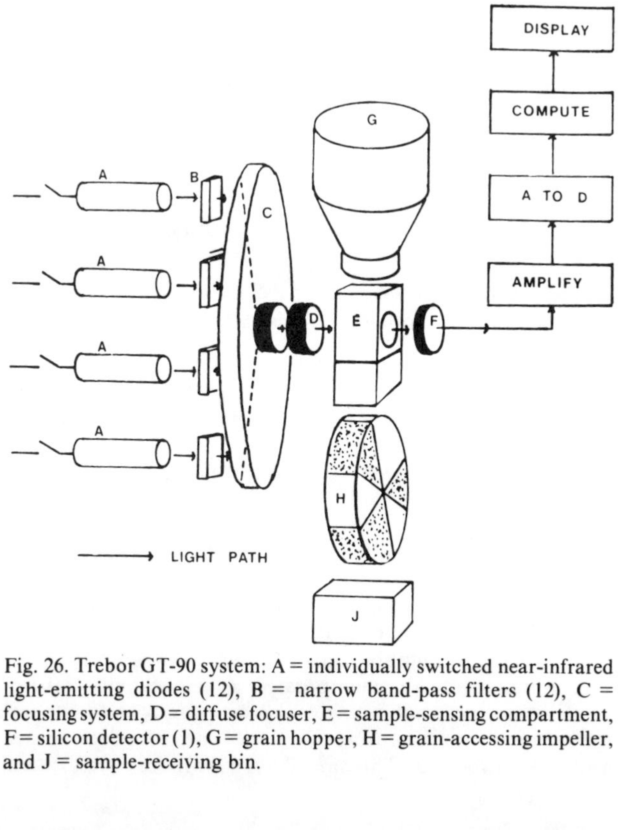

Fig. 26. Trebor GT-90 system: A = individually switched near-infrared light-emitting diodes (12), B = narrow band-pass filters (12), C = focusing system, D = diffuse focuser, E = sample-sensing compartment, F = silicon detector (1), G = grain hopper, H = grain-accessing impeller, and J = sample-receiving bin.

second derivative of log (1/*T*) alone or in the quotient version. The GT-90 uses all 14 terms in its operating algorithm.

The forte of the GT-90 is its capacity to give an accurate moisture test on whole grains, as well as protein. Moisture testing on ground samples requires correction for the moisture loss incurred during grinding (Williams and Sigurdson, 1978). Moisture results are reliable after only 1 hr of tempering to preselected moisture levels, which makes the instrument particularly suitable for work in flour mills and other areas where tempering wheat is necessary. The software includes the facility for changing the slope of a calibration. Using the Teach mode, the accuracy of a calibration can be checked by testing a series of samples of known protein and moisture content in the usual way. The instrument calculates the slope of the regression line of GT-90 against standard results, and the slope can be automatically adjusted by the operator. Activation of the Teach slope correction enables the GT-90 to multiply all calibration constants by the slope. Reanalysis of the samples verifies that the results are in accordance with the standard data.

The accuracy of testing is affected by grain size and shape, and for a given commodity separate calibrations are recommended for varieties with a size or shape that differs significantly from the more "normal" sizes and shapes. This is true of durum wheats, where grains can range from spring wheat shape to the long, angular shape typical of the old variety Senator Cappelli, and to barley, where the North American varieties, Klages and Harrington, have distinctive sizes and shapes. The results of testing wheat and barley are seriously affected by the presence of excessive amounts of small weed seeds, which should be cleaned out before analysis.

Subsequently, Trebor has introduced a modification to the GT-90 that enables the analysis of very small samples. The GT-80 can produce protein results on as little as 5 g of whole wheat, with accuracy adequate for use in breeding and other screening programs. A further improvement, which is available on the GT-99, an updated version of the GT-90, is the use of a moving cell in place of the rotating sample access mechanism. The modification was incorporated to enable analysis of liquid milk, meat, and other commodities. The sample size is large (over 100 g) and the cell is fashioned from polycarbonate, a durable, clear plastic. The sample cell stops an operator-directed number of times during a single pass of the cell. At each position the sample is stationary, thus improving the precision of analysis. The replacement of the wheel-type sample access mechanism also provides the potential of near-infrared transmittance analysis of powdered (ground) samples, including flour, using a thinner cell.

VI. SIMPLE COMPUTERIZED SPECTROPHOTOMETERS

A. Pacific Scientific/Neotec Research Composition Analyzer Model 6350

An earlier chapter described the design and operation of the sophisticated computerized spectrophotometers used in such centers as the Instrumentation Research Laboratory of the U.S. Department of Agriculture at Beltsville, MD. During the past five years, computerized spectrophotometers have become available at prices that make them realistic and even essential for large-scale operations requiring continuous monitoring of a wide range of commodities and parameters and the applied research that is often necessary to obtain the most precise results. The first of these was the Pacific Scientific/Neotec Research Composition Analyzer (RCA) Model 6350. The system, which could be operated in either reflectance or transmittance modes, incorporated a holographic grating/single-beam spectrophotometer controlled by a Data General Nova 3 minicomputer, using twin flexible disks for data storage and manipulation. The cam-driven holographic grating enabled very rapid scanning (about 2.5 complete scans per second) from 1,100 to 2,500 nm in the near-infrared or from 600 to 1,100 in the visible/very-near-infrared region. More recently, the RCA 6350 has been updated to incorporate a Northstar minicomputer in place of the Nova 3, more sophisticated software, a Winchester 5-megabyte hard disk, an auxiliary flexible disk, and a printer. The newer version is marketed as the RCA 6250 (Fig. 27). The RCA 6350/6250 enables the high-speed optimization of wavelengths, mathematical treatments of the log (1/*R*) signal, and the smoothing and derivatization necessary to obtain the most accurate and precise analyses possible with commercially available near-infrared instrumentation.

The software, which includes a very comprehensive selection of data processors, carries out statistical evaluation of analytical results automatically. It can also be used directly as an analytical instrument, which allows the operator to verify the accuracy and efficiency of wavelengths and calibrations realized. Additional peripherals allow the production of hard copies of data and plots of spectra and multiple correlation traces. The RCA 6350 and 6250 have both silicon and lead sulfide detectors. The silicon detector is used in the visible/very-near-infrared region, from 600 to 1,100 nm, for transmittance measurements. Lead sulfide detectors are used from 1,100 to 2,500 nm for both transmittance and reflectance. The optical data are recorded at 2-nm intervals, so that data for 700 wavelength points are accumulated. The spectral data accumulation program allows the operator to add up to eight sets of chemical data to the spectral data accumulated for individual samples. Spectral and chemical data can be listed to verify the accuracy of chemical data, and the spectra can be displayed and recorded. The wavelength range displayed can be expanded for close inspection of localized areas of the spectrum and for recording of the log (1/*R*) signal at any wavelength point. Samples can be deleted from or restored to data files. Data files can be created from existing files, and files can be merged to provide more comprehensive sample variance for calibrations. Additional chemical constituents can be added, and chemical data can be replaced with different data.

Wavelengths are selected by multiple linear regression, which selects up to nine individual wavelength terms by any of nine mathematical treatments. After the selection of the wavelength terms, the correlations and multiple correlations can be displayed across the scanned spectrum. Correlations generated during selection of the first term are the most significant. Both the magnitude and shape of the correlation "bands" are valuable to the operator. This principle is not applicable to the log (1/*R*) algorithm, since at least two wavelengths are essential to analysis using that algorithm. Ideally, the areas of highest

Fig. 27. Pacific Scientific/Neotec Research Composition Analyzer Model 6250. A = monochromator, B = computer/terminal, C = printer.

correlation should appear as a broad band with a relatively wide area of wavelength. This means that a calibration based on such a wavelength will tolerate some fluctuation in optimum wavelength outside of the selected point.

A typical correlation plot is illustrated in Fig. 28. The point of highest correlation occurred at 2,132 nm, but the correlation was high from 2,120 to 2,160 nm. Earlier work (Williams and Norris, 1983) had shown that variation of as little as 3 nm in the wavelength selected for prediction could result in significant changes in the accuracy of analysis. Differences in factors such as the moisture content of the sample could force the computer to select an optimum wavelength in a completely different area of the spectrum. The broader the wavelength range of the highest correlation band, the more tolerant the calibration based on such a wavelength would be of fluctuations in the optimum wavelength for measurement of an individual sample and of factors that can affect the optimum wavelength, such as moisture, particle size, and bulk density. The high correlation at 1,942 nm (Fig. 28) is a far sharper band and less reliable for use as a primary wavelength in a calibration. A small shift in optimum wavelength at such a point would cause a large change in the multiple correlation between optical signal and constituent composition. Table VIII illustrates the changes in the coefficient of correlation at wavelength points surrounding the 2,132- and 1,942-nm peaks, together with the efficiency of single-point calibrations based on these wavelength points in determination of protein in wheat. The wavelength points at which the next highest correlations occur should be recorded after the first (primary) wavelength has been selected. Often two or more of these will be selected by multiple linear regression as secondary, tertiary, etc., wavelength points to accompany the primary point in a calibration set. If the only wavelength points at which high correlations occur have sharp correlation peaks, this type of calibration requires careful evaluation. Notice that the correlation plot for a single quotient-mathematics wavelength point (Fig. 28B) has many more broad bands, which indicates the increased stability of calibrations based on quotient mathematics.

To verify the efficiency of a calibration, the 6350 and 6250 (and other computerized spectrophotometer systems) incorporate programming to enable the operator to use the instrument as an analyzer. The calibrations generated can be stored and used to analyze samples that have not previously been used in the calibration. Samples used for analysis or verification of calibrations must be of known chemical composition. The mean difference between standard and 6350 or 6250 results (bias) is calculated, as well as the standard deviation of the difference (standard error of performance). An editing program is available for changing intercepts or slopes, and many other options.

The 6350 and 6250 computers can be directly interfaced to Pacific Scientific filter instruments such as models 101, 102, 51A, and 7000. The optics of the respective instruments can be used to record the signals from calibration samples directly into the computers of the 6350 or 6250 instruments and then transposed into different mathematical forms, such as first or second derivative or log $(1/R)$, and used to select wavelengths, which are displayed as pulse points. The pulse points can then be set into the instruments by the operator and the same samples used to calibrate the instruments. The stand-alone

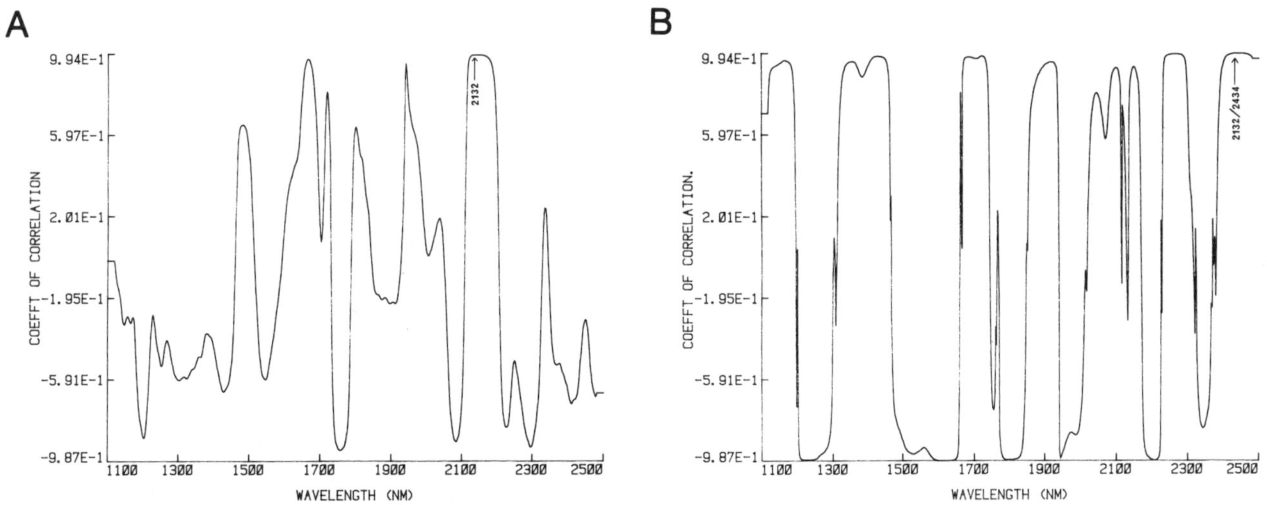

Fig. 28. Wheat protein correlation plots for 1,100–2,500 nm. A, first derivative of log $1/R$; segment = 5 wavelength points, derivative (gap) = 25 points. B, same, but for quotient algorithm.

TABLE VIII
Changes in Coefficient of Correlation Close to Peak Correlation for Broad and Sharp Correlation Peaks

	Wavelength (nm)							
Parameter	**2,122**	**2,132**	**2,142**	**1,932**	**1,942**	**1,952**	**2,132 + 2,434[a]**	**1,942 + 2,136**
Coefficient of correlation[b]	0.988	0.990	0.988	0.158	0.975	0.953	···	···
SEP[c]	0.199	0.182	0.257	2.25	0.423	1.70	0.199	0.181
$\bar{d}$	−0.302	−0.12	0.01	0.39	−0.72	−0.32	−0.15	−0.21
r^d	0.998	0.997	0.995	0.563	0.994	0.731	0.998	0.998

[a]2,434 and 2,136 are respectively the second wavelength points selected to give a two-wavelength point calibration using 2,132 or 1,942 as primary points.
[b]Coefficient of correlation for calibration.
[c]Standard error of performance of calibrations based on the respective single wavelength terms.
[d]Coefficient of correlation for analysis, corresponding to SEP.

Model 101 is limited to a first-derivative algorithm, but models 102, 51A, and 7000 can use several different mathematical treatments. If the instruments are interfaced with an auxiliary computer, all models can use any of several mathematical treatments featured in the software, since the instrument optics are simply being used to record the log (1/*R*) data from the samples. This is theoretically applicable to any near-infrared instrument, including stand-alone instruments that operate in the log (1/*R*) mode, providing the wavelength range is available. The accuracy of the calibrations can be verified by analysis of further samples of known chemical constitution. The forte of the 6350 or 6250 computerized spectrophotometers lies in the selection of pulse points (or wavelengths) and the optimization of mathematical treatments of optical data for the analysis of all types of commodities and constituents. Wavelengths selected for analysis of common constituents, such as protein, moisture, and oil, frequently fall into similar patterns, such as the region at 2,140–2,160 nm for protein by first-derivative mathematics, or 2,170–2,185 nm for protein by second derivative. The presence of high concentrations of such constituents as fiber, oil, or moisture may change the optimum wavelength for determination of such constituents as protein or oil to different areas of the spectrum, typically in the region of 1,600–1,800 nm. These constituents can be revealed by displaying the spectrum of one or two samples. The most recent version of the 6250, 7000, and 3250 series enables interfacing to IBM PC-AT or PC-XT minicomputers, which use the Pacific Scientific software. Among other improvements, one of the new features is the incorporation as optional peripherals of hard disks with far larger capacity (up to 20 megabytes) and increased computing speed.

B. Labormim Infrapid 61

The Labormim organization in Hungary introduced its simplified computerized spectrophotometer during 1982. The Infrapid 61 is a lower cost, less sophisticated version of the spectrocomputer, and it does not conform to spectral and electronic specifications of the Pacific Scientific or Technicon InfraAlyzer 500 systems. The optical system of the Infrapid 61 depends on a latticed monochromator that resolves the near-infrared region into specific wavelength areas over the range of from 1,300 to 2,400 nm with a reproducibility of ±1 nm. The light source is a 12-V tungsten/halogen lamp, and four lead-sulfide detectors are used. Mathematical transformations are limited to log (1/*R*) and first and second derivatives thereof as simple or quotient wavelength points. The optical and software systems enable optimization of wavelength points, and up to 20 calibrations can be accommodated simultaneously. The instrument is compact, and by the use of its monochromator provides simple, uncomplicated access to a comprehensive array of wavelengths. Wavelength selection can be sped up by interfacing the Infrapid 61 to a minicomputer. The wavelength range, selection capability, and derivatization capability allow the Inframatic 61 to be used for analysis of a wide range of commodities and constituents.

C. Technicon InfraAlyzer Model 500

Technicon Industrial Systems announced the development of the Model 500 computerized spectrophotometer (Fig. 29) at the 1979 Pittsburgh Conference. The instrument incorporates a stand-alone, scanning, holographic grating monochromator operating in the double-beam mode and a Hewlett-Packard Model 1000 (HP-1000) minicomputer with associated software. The monochromator can also be interfaced with other external computers. Optics include the monochromator supported by a light/dark, sample/reference chopper and an integrating sphere. Sample holders are available to accommodate liquid and powdered samples. The analysis of liquids is enabled by the transflectance concept, by which a thin layer of liquid is irradiated. The irradiation passes through the liquid and is diffusely reflected back from a reference standard to impinge upon the detectors via the internal wall of the integrating sphere. This system, including the liquid sample compartment, can also be used with models 400 and 300. Transflectance is essentially nonlinear and does not conform to the Beer/Lambert law in the near-infrared region. This shortcoming is overcome by computerization, but the technique tends to be limited by the viscosity of the liquid and thickness of the liquid layer. A built-in microprocessor carries out analytical computations after regression coefficients have been derived. The HP-1000 is used for the more comprehensive computations involved with wavelength selection. The wavelength range of 1,100–2,498 nm is scanned by the holographic grating monochromator at a rate of 8 nm/sec. The 700 wavelength points can be scanned in about 90 sec. The detection system utilizes an integrating sphere and twin lead-sulfide detectors. This system is identical to the InfraAlyzer 400 and 300 models, and consequently the calibrations are readily transferable.

The chopper wheel creates three states—sample, reference, and dark. The cycle from sample and reference to light is 40 Hz, and the cycle from light to dark is 80 Hz. The objective of this difference in illumination is to eliminate the need for cooling the detectors. Each sample and reference signal is flanked by two dark signals. A large number of both sample and reference signals are generated during each analysis to provide analytical precision. The instrument possesses 10 kilobytes of memory for constituent storage, and up to 98 products can be accommodated, depending upon the number of regression constants (*F* values) assigned. This depends, in turn, on the number of wavelengths used for each product/constituent combination. Wavelength selection in commercially available instruments, using 700 data points, involves some form of stepwise multiple linear regression. An HP-1000 minicomputer complements the optics of the monochromator unit. Two disks, a micro floppy and a hard disk, are used in conjunction with the HP-1000. The hard disk provides at least 15 megabytes of storage, and each micro diskette can store data from 700 wavelength points for 75 samples. The computer has a resident memory of up to 512 kilobytes. The interfaced computer, and a full array of user-orientated software, complete the system. Experienced users have the additional facility of being able to program the HP-1000 in BASIC, FORTRAN, or Pascal. The manufacturer supplies software capable of performing all of the more familiar functions of data acquisition, access of chemical information, mathematical transform of optical signals, visual display, data reduction, and data analysis. Up to 20 wavelength terms can be selected, although in practice it is unusual for more than four terms to be required.

D. LT Analyzer Model 7000

The LT Analyzer 7000 is another instrument that combines the ability to research and optimize wavelengths and mathematical treatments. The spectral response is from 1,200 to 2,400 nm, with possible expansion down to 400 nm. It is a fast-scanning spectrophotometer, capable of five scans per second over the whole range. The LT-7000 is a product of LT Industries of Rockville, MD, the newest company to offer a relatively low-cost, comprehensive near-infrared analyzer. The instrument is primarily intended as an on-line analyzer, and it can make measurements at distances up to 3 m (10 ft) from the sample or sample stream using fiber optics extension. Its design incorporates flexibility in that it can operate as an ordinary general-purpose analytical spectrophotometer in the ultraviolet, visible, and near-infrared spectral ranges, in both reflectance and transmittance modes.

E. Trebor Models GT-70 and 7700

The Trebor GT-70 is a departure from the NIRLED concept used in Trebor's earlier instruments in that it incorporates a 90-mm, holographic grating monochromator and can operate in the very-near-infrared from 600 nm to 1,100 nm, as well as up to 2,500 nm. It is a high-energy, single-beam spectrophotometer that uses a silicon detector and four lead-sulfide detectors set at 45° to the reflecting surface. It carries out a complete scan in 20 sec, taking 256 readings at 1-nm intervals across the selected wavelength. Its sample presentation system allows measurements to be made in reflectance or transmittance modes, and the sample can be scanned in up to 18 operator-controlled positions. This feature is included in lieu of rotating the sample cell.

The Trebor 70, which is operated by an internal microprocessor, includes twin, 3-in. floppy disks, each with 1 megabyte of storage. These enable the operator to process data on a remote minicomputer while reading more samples and recording the optical data on the floppy disks for future processing. The instrument can be converted to the Model 7700 (Fig. 30) by interfacing to an auxiliary computer, such as the IBM PC-AT or PC-XT, or a KAYPRO. Of these, the PC-AT offers the most comprehensive package, with a 20-megabyte hard disk in combination with a 1.2-megabyte floppy disk system. The PC-AT has a high-resolution video terminal with a color graphics adapter. In its 7700 format, this Trebor model can be used as a computerized spectrophotometer to research the optimum wavelengths, mathematical option, and mode (transmittance or reflectance) in which to make the analysis. When this has been carried out, the calibration can be stored and used in the operational mode for analysis. In common with all other computerized spectrophotometers, the Trebor 7700 is fortified with a full range of software options allowing data and sample manipulation, regression, editing, and the facility for the operators to assemble their own programs.

F. LT Quantum 1200

The most recent addition to the selection of near-infrared instruments available is the LT Quantum 1200 (Fig. 31), offered by LT Industries. This updated version of the LT-7000 incorporates full-range wavelength selection between 1,200 and 2,400 nm and the facility for analysis of powders, pastes, and liquids in transmittance and reflectance modes. The Quantum 1200 can be used for laboratory as well as on-line operation, and calibrations are not subject to the normal difficulties associated with on-line/laboratory systems. It incorporates all of the

Fig. 29. Technicon InfraAlyzer Model 500. (Courtesy Technical Corp., Tarrytown, NY)

features of the computerized spectrophotometer and near-infrared analyzer together with on-line and off-line applicability (using fiber-optics remote measurement) and compatibility with PC computers, which gives great flexibility in software. The specifications are identical to those of the LT-7000, but the instrument is more compact and both its design and software have been updated. The Quantum 1200 can also be used for slurries or solid samples, and improved fiber optics have extended the range of its off-line application beyond the original 3 m. In common with all modern near-infrared instruments, the Quantum 1200 also carries comprehensive diagnostics.

G. Applications of Computerized Spectrophotometers

Several types of applications are available to the operator of the computerized spectrophotometer. These include normal wavelength and mathematical treatment optimization, which is the accepted, traditional forte of the computerized spectrophotometer. The instrument can also be used as an analytical tool in quantitative analysis. This feature is particularly attractive to operations that are obliged to analyze a wide diversity of commodities for several constituents or parameters. Such constituents include amylose, fiber components, lignin, β-glucan, amino acids, and simple sugars. Parameters include digestibility, wheat hardness, and water absorption. Wavelengths for those constituents can be searched and the instrument can then be used to analyze samples. Recent work (Norris et al, 1977; Rubenthaler and Bruinsma, 1978; Williams et al, 1984; Williams et al, 1985a) has verified the possibility of estimating amino acids by means of near-infrared spectroscopy. It was necessary to use up to four wavelength terms for each individual amino acid; on the basis of second-derivative mathematics, this involved 12 individual wavelengths per amino acid. Although two wavelength terms were common to most of the amino acids, four amino acids called for about 30 individual wavelengths in a single commodity, which is beyond the scope of any of the simple commercial near-infrared instruments available. A computerized spectrophotometer can utilize 700 wavelength points or more and is the type of equipment that makes the rapid, simultaneous analysis of several amino acids in different commodities feasible. Alternatively (Williams et al, 1985a), an instrument such as the Pacific Scientific FQA 51A, in which the tilting filters enable the simultaneous use of many wavelengths, can be employed. In practice, it is seldom necessary to determine more than five (the so-called limiting) amino acids—lysine, threonine, methionine, tryptophan, and isoleucine.

The above applications are all quantitative, but there are several useful qualitative applications. These are discussed in Chap. 15.

Fig. 30. Trebor Model GT 7700. (Courtesy Trebor Instruments, Gaithersburg, MD)

Fig. 31. LT Industries Quantum 1200. (Courtesy LT Industries, Inc., Rockville, MD)

VII. OPERATION OF COMMERCIAL BENCH-TYPE INSTRUMENTS

The principles of successful operation are applicable to all instruments, and they can be resolved into three parts: calibration, operation, and monitoring.

A. Calibration

Calibration of near-infrared instruments has historically been the most formidable obstacle to their overall acceptability in industry. It is necessary to calibrate any instrument separately for each individual constituent, and for most commodities. The most important aspects of a calibration are the accuracy and precision of analysis of which it is capable and its stability, or capacity to provide good analytical results for samples with a wide range of variance and over a long period.

There are several prerequisites to successful calibration (Table IX). Identification of the sources of variance and the assembly of samples representing this variance are at the same time the most onerous and the most important tasks. All samples must be freed of foreign material and ground or otherwise suitably prepared for analysis. Some preliminary treatment may be necessary. For instance, cottonseed should be delinted before grinding. The other steps in Table IX are self-explanatory and are included in the operator's manual supplied

TABLE IX
Prerequisites for Calibration

1. Identify all sources of variance, including composition, growing location, etc.
2. Assemble samples with range of composition within each variance bracket.
3. Prepare sample—cleaning, documentation, grinding, etc.
4. Carry out standard analyses.
5. Record optical data.
6. Compute calibration constants.[a]
7. Set constants into instrument.[a]
8. Verify validity of calibration by analysis of unused samples.
9. Carry out slope/bias corrections where necessary.
10. Reverify calibration by reanalysis of unused sample set.
11. If exercise is to be carried out with computerized spectrophotometer, repeat steps 5–10, changing mathematical treatments, derivative size, etc., to optimize wavelength and mathematics.

[a] Some instruments compute and store their own calibration constants.

with new instruments. All modern instruments have push-button methods for entering constants and changing from calibration to operational modes.

Calibrations may be classified as *closed* (enumerative) or *open* (analytical) (Deming, 1975). Either type of calibration can be applied to simple instruments or computerized spectrophotometers. With a simple instrument fitted with six discrete filters and operating with the log $1/R$ algorithm, such as the InfraAnalyzer 300 or DICKEY-john GAC III 660 series, an auxiliary computer is used to compute constants for whichever of the six wavelengths represent the optimum set for the population of samples used in calibration. This principle applies equally to the InfraAlyzer 400 with 19 filters, the Inframatic 8120 with 20 filters, or the more comprehensive instruments with up to 1,200 available wavelengths, and built-in microcomputers, when wavelength selection becomes more complex. The more wavelengths used in a calibration, the more complicated wavelength selection becomes, and the more critical is sample selection.

A closed calibration represents an ideal situation for near-infrared application. The operator has a fixed or closed population of perhaps 2,000–4,000 samples to analyze. A typical closed calibration is a breeding program that has several thousand samples to analyze rapidly in one season. The characteristics of the crop may change the following season, which will then represent a new closed population. The samples represent a large range of variance, but since no more samples will arise, it is only necessary to analyze that population. The samples are all ground and read on the instrument and the optical data stored. Next, the optical data are all derivatized. Derivatization reduces error components caused by particle size and shape and bulk density. The spectra can then be studied to identify samples that differ in absorbance in the area where absorbance is anticipated for the constituent to be measured—for example, in the 2,180-nm region for protein. One method for this type of investigation is the determination of Mahalanobis distances at several wavelength points (Mark and Tunnell, 1985). The Mahalanobis distance (Mahalanobis, 1936) for a spectrum is the distance in absorbance units by which the individual spectrum differs from the mean for all spectra in the population at a given wavelength point (or points). A set of 30–40 samples is then selected with universal distribution of Mahalanobis distances, or simply differences in log $(1/R)$ or derivatized log $(1/R)$ signal. The samples are carefully analyzed chemically and used to generate the wavelengths and calibration constants. Selection of samples using Mahalanobis distances or any other method on underivatized spectra can be misleading since significant differences in spectra can be induced simply by sample texture and preparation, which combine to cause differences in particle size and shape. The derivatization step minimizes these sources of error. The calibration is verified with 20–30 further samples of known composition and used to analyze the remaining large population of samples. The chief advantage of this procedure is to minimize the amount of chemical analysis, but if more than one constituent is to be determined, the process of sample selection must be repeated to include samples representative of the range of each constituent.

The identification of samples on the basis of spectral characteristics such as Mahalanobis distances becomes increasingly difficult for analyses 1) of components that are present in relatively low concentration, such as amino acids, 2) where the presence of one or more constituents interferes with the constituent to be measured, or 3) when the optimum wavelength for the analysis of a particular constituent or parameter is due to an absorber of relatively low intensity compared to other absorbers. Under these circumstances, it is necessary to perform a wavelength search to identify the optimum wavelengths for calibration before the calibration step is possible. This requires preliminary chemical analysis. A guideline to selection of samples for chemical analysis for constituents other than protein is to select on the basis of those that have already been tested for protein. Growing conditions and genetic factors that have resulted in significant differences in protein usually lead to significant differences in other constituents, even though they may not be correlated with protein content.

Open or analytical calibrations are used for the analysis of indefinite populations, where samples with different variance components continue to arrive for analysis ad infinitum. In situations of this type, the most successful approach is to accumulate large numbers of samples with chemical data. An analysis of variance can then be carried out in which all of the sources of variance likely to be encountered in future samples are identified and their relative importance categorized. Groups of samples are then accumulated corresponding to these sources of variance, each group carrying a range of composition with respect to the constituents to be analyzed. The samples are then assembled into calibration and verification files and the calibration set used to identify wavelengths and optimize mathematical treatment for the analysis. The efficiency of the calibration is then tested, and the most successful calibration used in future analysis.

The most likely sources of variance are the range in composition, texture, growing conditions (location and season), moisture content, and primary processing procedure (grinding, cooking, drying, etc.). Chapter 7 describes the sources of variance in detail. Samples should be accumulated that represent a uniform distribution of composition across the anticipated range. The reason for this uniform distribution of samples, as distinct from samples with a normal or a Gaussian distribution, is to increase accuracy at extremes of composition. More variance is included in the samples and the calibration constants are more stable over a long period and give better accuracy at extremes of composition. The chief difficulty is in assembling samples with a range in composition of more than one constituent when analysis for several constituents is necessary. Intercorrelation between constituents then becomes more important. If a calibration is based on samples with Gaussian distribution with respect to composition, when the calibration is used to analyze unknown samples the results of analysis will be regressed toward the mean (Fig. 32). As a result of this phenomenon, which is discussed in more detail by Dunne and Anderson (1976), accuracy will be maximized close to the mean. However, the results of samples at high extremes appear to be lower than the chemical analysis would indicate, whereas the results of samples at low extremes appear to be higher than they should be. This type of error can be serious in situations

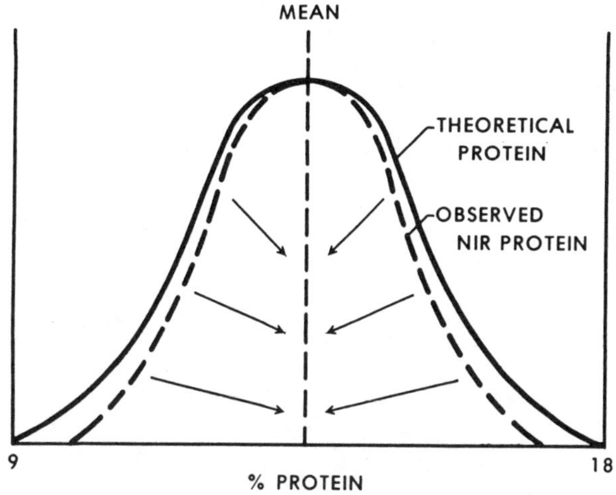

Fig. 32. "Dunne effect": regression of NIR protein results toward the mean.

where accuracy is critical at extremes, for example at grain silos where prices are higher for high-protein wheat and low-protein barley and where payment of protein premiums depends on a verification analysis. The "Dunne effect" can result in discrepancies that may prove costly to the producer of the wheat or barley. The slope of a simple regression is given by:

$$b_{xy} = r \times \frac{SD_y}{SD_x} \quad (6)$$

where SDy = standard deviation of the dependent variable (standard results such as Kjeldahl protein). In the ideal situation, the slope is equal to 1 or adjusted to 1 (according to the method described below in the section Fine-Tuning of Calibrations), so that:

$$SDx = r \times SDy \,, \quad (7)$$

which means that the standard deviation of the independent variable (near infrared) is equal (if r is very close to 1, e.g., 0.998) or less than SDy and is reduced by the factor r. This is the mathematical basis for regression toward the mean:

$$\text{Since } SDx^2 = \frac{\Sigma\,(x - \bar{x})^2}{N - 1} \text{ and } SDy^2 = \frac{\Sigma\,(y - \bar{y})^2}{N - 1}\,, \quad (8)$$

$$\text{then } \Sigma\,(x - \bar{x})^2 = r^2 \cdot \Sigma\,(y - \bar{y})^2 \,. \quad (9)$$

For a single sample, this reduces to:

$$(x - \bar{x}) = r \cdot (y - \bar{y}) \,.$$

Example 3: Using values of 13.7 and 0.91 for $\bar{x}$ and r, and a Kjeldahl value of 15.4:

$$\begin{aligned} x - \bar{x} &= 0.91 \cdot (15.4 - 13.7) = 0.91 \times 1.7 \\ &= 1.55 \\ x &= 1.55 + \bar{x} \\ &= 15.25 \,. \end{aligned}$$

The magnitude of the effect is related to the accuracy of sampling, sampling preparation, and both near-infrared and standard chemical analysis. An estimate of the effect can be arrived at by calculating the coefficient of correlation between near-infrared and standard chemical analysis and the mean of the chemical analyses. The extent to which the chemical analysis of an individual sample differs from the population mean is multiplied by the coefficient of correlation. This value is added to the population mean to give a result that approximates the result obtained by near-infrared (NIR) analysis on that calibration. The formula for this is:

$$\text{NIR result} = \text{population mean} + (\text{deviation from mean} \times r) \,, \quad (10)$$

where r is the coefficient of correlation between near-infrared and chemical analysis of past samples using that calibration. An example is given:

Example 4: Mean = 12.2, r = 0.87, chemical result of sample (barley protein) B is 9.8.

$$\begin{aligned} \text{NIR}(B) &= 12.2 + [(9.8 - 12.2) \times 0.87] \\ &= 12.2 + (-2.4 \times 0.87) \\ &= 12.2 - 2.09 = 10.11 \,. \end{aligned}$$

The lower the coefficient of correlation, the more serious will be the discrepancies. This type of error is minimized by calibration with samples having a uniform distribution of the concentration of constituents.

There are four types of calibration used in near-infrared instruments, and they are summarized in Table X.

UNIVERSAL CONSTANTS

Universal constants are those that combine the optical and chemical data from very large sample populations into one calibration using one set of constants. The samples represent all known and anticipated sources of variance and are ground on several grinders, and the optical data are accumulated from several instruments of the same type. The concept of universal constants was developed at the Grain Research Laboratory, Winnipeg, as a result of the observation of similarities in the regression coefficients from over 200 calibrations prepared for Neotec and Technicon/DICKEY-john instruments of the same type. Originally the constants from 12 calibrations (about 500 samples) were combined into one comprehensive calibration and used to determine protein in hard red spring wheat. Standard errors of performance (SEP) of 0.37% protein and 0.29% moisture were obtained using a Neotec Model 31. Subsequently, universal constants were computed directly from optical data recorded from over 1,000 samples. The SEPs were 0.29 for protein and 0.27 for moisture.

The chief difficulty with universal constants lies in the accumulation of a sufficient number of samples representing all anticipated sources of variance. A useful guideline is to include three to nine samples for each source of variance. The sample subsets should represent the range in composition, with the minimum number (three) representing high, low, and intermediate composition. For example, if a calibration is to be set up to determine protein in wheat from 100 locations, this will require about 300 samples, which will then represent the high, low, and intermediate protein contents from all locations. If a variance in moisture is also anticipated, the number of samples will increase in practice to about 600–900 to accommodate the range in both protein and moisture. The calibration process should be repeated each season, with data for the new season being added to that for the previous seasons until no significant changes in efficiency of the analysis occur, indicating that all of the variance likely to affect the results has been incorporated into the calibration. This open type of calibration is applicable to very large-scale operations that analyze large numbers of samples on an ongoing basis using a number of instruments of the same type. It is also applicable to instrument manufacturing companies who do relatively little analysis themselves, yet are responsible to many customers using their instruments for the analysis of similar commodities for the same constituents. Operations that do not have access to enough samples may find one of the other types of calibrations more suitable to their

TABLE X
Types and Applications of Near-Infrared Calibrations

Type	No. of Samples[a]	Application
Universal constants	200–2,000 (cumulative)	Analysis of any commodity for parameters for which large numbers of samples with standard analyses are to be analyzed over period of time, using several instruments of same type.
Simple	30–60	Analysis for single parameter in single commodity with minimum sources of variance.
Quotient mathematics	12–20	Analysis for single parameter in commodity with wide variance in particle characteristics.
Two-point	2	Analysis of commodity with wide variance from several sources.

[a] Number of samples used in calibration.

operation. Instrument companies can accumulate enough samples with enough variance in characteristics and composition by acquiring relatively small numbers of samples (up to 50) from individual customers, recording the optical and chemical data, and storing it. (Hard disks can store over 6,000 spectra and floppy disks can accommodate over 100. Data storage is no longer a limiting factor, and the capacity of modern minicomputers fitted with hard and floppy disks is far in excess of the ability of most organizations to accumulate samples with associated standard analytical data.) All of the data can then be combined to produce comprehensive calibrations.

Questionable chemical data should be verified by analysis at a reliable laboratory. If the standard data indicate that the sample is truly different from others, it should be retained in the calibration set. Outliers of this type represent new sources of variance and improve the integrity of the calibration (Williams and Antoniszyn, 1986). If a true outlier is found, it will bias the calibration toward itself, but it is likely that others are lurking in the vicinity. Incorporation of true outliers adds to the variance incorporated into the calibration and improves its long-term efficiency. By storing data from samples from different growing locations and seasons and recombining the data into a single calibration, the Canadian Grain Commission presently operates near-infrared instruments with universal constants compiled from over 4,000 samples. The SEPs for protein and moisture are respectively 0.22 and 0.20. Operations such as flour mills and food manufacturing plants can also accumulate samples from mill streams, patent flours, etc., or food and feed types and ingredients with chemical data over a period of several months, to develop and expand a universal calibration using samples processed at different times. The advantages of such constants are stability in application to any new material and the capability of transfer to other instruments. Slope/bias corrections may be necessary. Theoretically, a slope correction means that a calibration is not strictly transferable. A slope correction by application of a simple regression equation (see section below on Fine-Tuning of Calibrations) is much faster and simpler than developing a new calibration.

The number of instances where calibration constants are truly transferable over a wide range of concentration is relatively small. Constants may appear to be transferable if they are found to operate successfully over the working range of concentration required by the user. At extremes of concentration, such calibrations are usually not so reliable, with discrepancies becoming progressively bigger at greater extremes. The distribution of the errors reflects the distribution of concentration in samples used for calibration, so that for Gaussian assembly of calibration samples, the distribution of errors will be Gaussian and large discrepancies relatively rare over the working range. This is an argument in favor of this method for the assembly of calibration samples.

SIMPLE CALIBRATIONS

Simple calibrations, the most frequently used system for preparing a near-infrared instrument for analysis, is the type described in operator's manuals prepared by the instrument manufacturers. Simple calibrations are satisfactory provided the observed and anticipated variance is not excessive. First, it is essential to define the sources of variance in the material to be analyzed. Next, 50–60 samples should be accumulated with a uniform distribution of composition. The actual steps for calibration differ for individual instruments, and they are described in operator's manuals prepared by the respective companies.

CALIBRATIONS USING QUOTIENT MATHEMATICS

It is practically impossible to include examples bearing all identifiable sources of variance simultaneously with a full range of composition in 40 or 50 samples. Unless all of the variance to be encountered in future analysis is included in the calibration samples, significant discrepancies will inevitably occur in subsequent analysis, particularly when changes in season or growing location or processing procedure occur. Consequently, simple calibrations are sample sensitive, and periodic recalibration is necessary to ensure continued accurate analysis. The development of *quotient mathematics* resulted in a simpler and more efficient system for calibration. Quotient treatment of a derivatized log $(1/R)$ signal means taking the reading at primary wavelength point A and dividing it by the derivatized signal at a second wavelength point B. This treatment was found to minimize errors caused by variations in particle size and moisture content, two of the most significant sources of variance. Analysis using a single-quotient wavelength term gave better accuracy and precision than up to six simple log $(1/R)$ wavelength terms. With quotient mathematics, calibration constants can be calculated using a simple linear regression ($y = a + bx$), since only two variables (quotient optical signal and standard data) are involved. The calibration constants are generally much smaller than those of other types of calibration, which is usually an indication of stability. About 12–20 samples are sufficient for the calibration provided they represent a uniform distribution across the expected range of composition. The samples are ground, mixed, and read in the instrument in the usual way to record the optical signals, which are then regressed against the chemical data and the simple regression coefficient b (K_1) and intercept a (K_0) are computed. The K_1 and K_0 are set into the instrument and accuracy is verified using a different set of samples of known composition.

TWO-POINT CALIBRATION

Since the chief factors affecting accuracy by near-infrared analysis can be controlled by quotient treatment, only two samples are theoretically necessary to establish the slope and intercept for calibration. These two samples must represent the range of concentration anticipated—for example, 10–17% protein in wheat. The optical signals are read for the two samples and are used to calculate the K values (calibration constants). The formulae for K_1 and K_0 are given above in equations (4) and (5). The instrument is ready for analysis after the calibration constants are entered, and the accuracy of the instrument can be verified by the analysis of new samples of known composition, as for all other types of calibration. This type of calibration has been used in a Neotec GQA 31EL for three consecutive seasons without change in the intercept (Williams and Cordeiro, 1985). For routine analytical work the two-point calibration is likely to give the most consistent results when variation is anticipated from growing location and season, processing procedure, or other, similar sources that affect the particle characteristics of the finished commodity. For best results it is advisable to prepare the two high- and low-composition samples by blending samples from several sources. Several kilograms should be prepared, since the two samples can be used over a long period.

Methods of sample selection for calibration have been the subject of much study in recent years. These can be summarized into two approaches: selection on the basis of spectral characteristics and on the basis of chemical or physicochemical characteristics. The objective underlying the first method, which is typified by the work of Honigs and co-workers (1985), is to provide a sample set with a comprehensive variance in spectral characteristics with the minimum of chemical analysis. The sample set is selected on the basis of Gaussian or uniform distribution of optical (spectral) data taken at specific wavelengths across the spectrum scanned by the instrument or by a system representative of all of the wavelengths. If the operator wishes to maximize accuracy close to (within one standard deviation of) the mean, selection on the basis of Gaussian distribution is preferred. If maximum possible accuracy is needed at extremes of composition as well as near the mean, universal distribution of samples will give better data.

A variation on this theme is the selection of samples by computation of Mahalanobis distances (Mark and Tunnell, 1985), although both of these approaches lead to essentially similar sample sets.

The Gordian knot in this approach is that differences in underivatized spectral characteristics are not necessarily associated with differences in chemical composition. Samples that differ widely in a constituent such as protein may only differ minutely in the optical signal at the optimum wavelength point for making the measurement, whereas, conversely, samples that are of similar chemical composition may differ widely in optical signal (Williams and Krischenko, 1986). Variation in moisture content can introduce very significant differences in the spectra. Physical hardness, sample preparation technique, and fiber, oil, and moisture contents are other factors that may bear no close relation to the constituents to be measured but can cause variance in spectra. Chemical constituents such as protein, fiber, and oil may vary considerably independently of one another. If samples are selected on the basis of spectral differences in, for instance, the oil band areas, they will not necessarily be suitable for the analysis of other constituents, and the selection step will have to be repeated. Sample sets selected solely on the basis of spectral characteristics may also result in samples with an excellent distribution of spectra but a range in chemical constitution that either does not cover the full range in composition of future samples to be analyzed or has an uneven distribution of composition. To a certain extent, derivatization (first or second order) tends to improve this type of sample selection. It minimizes differences in spectra caused by particle size and other characteristics so that differences observed in the spectra are more likely to be the result of differences in composition. This approach is affected mainly by constituent interaction, since a positive difference between two spectra caused by variation in one constituent may be counterbalanced by a negative or positive difference in another constituent, whereas differences in other constituents may not be apparent in the spectra, particularly if the range is small. A further criticism is that large numbers of samples may need to be scanned to select a set with a suitably wide range of spectral characteristics. Since the accumulation step is often the most difficult, the actual saving in time and convenience may be minimal. On the positive side, selection of samples on the basis of spectral characteristics could be useful in many industrial applications where interactions between constituents and major fluctuations in particle size and packing characteristics are less likely.

Sample selection on the basis of chemical or physicochemical analysis is the approach most likely to provide stable calibrations, taking all factors into consideration, including types of material and parameters to be determined. It has the advantage of computing constants on the basis of regression of optical data directly against the parameters to be determined. The ideal method is the combination of selecting samples on the basis of both spectral and composition differences, which is possible with a computerized spectrophotometer, particularly where the spectral data are derivatized. Access to sufficient numbers of samples representing all anticipated sources of variance remains the most important prerequisite. The variance necessary to the development of stable calibrations can only come from the samples.

A novel type of two-point approach that has recently been found to give encouraging results using a computerized spectrophotometer is to develop a difference curve between two samples differing widely in the constituents to be measured (Williams, unpubl. data, 1986). Described as *differential calibration*, this is achieved by using the "low" sample as a reference standard, then reading the "high" sample; the technique is thus also referred to as "high/low" calibration. The instrument subtracts the low from the high sample and the peak absorbances should represent the wavelengths most likely to be successful in analysis. It is advisable to develop a difference curve between two samples with a small range in concentration as well, to ensure that the difference "bands" represent areas where spectral differences are truly associated with differences in composition. The difference curve should be derivatized to minimize variance caused by particle characteristics, using the derivative parameters intended for use in future analysis. The strongest two or three difference bands are then stipulated in a calibration equation. Some typical results are given in Table XI.

In all near-infrared applications, one of the most important factors is intercorrelation between constituents. The growing conditions that result in differences in protein content in seeds, forages, and other commodities are likely to cause variability in other constituents. When this happens, correlations may occur between the constituents to the extent that the operator may believe that a calibration is successful in analysis of a particular constituent, whereas in fact the determination may be based on changes in concentration of another that is highly correlated with the one being determined. A useful indication that this is not happening is the wavelength set selected, which should be different for different constituents and should conform to absorption bands known to be associated with the constituent to be determined, although this will not always be the case. All types of calibrations are vulnerable to this source of error. Table XII gives an indication, by no means exhaustive, of some constituents likely to be intercorrelated.

FINE-TUNING OF CALIBRATIONS

Two other principles can be employed to improve the efficiency and flexibility of near-infrared instruments. These are instrument normalization and the *t*- or *F*-value wavelength exclusion system.

Normalization of Instruments. Near-infrared instruments of the same model can be aligned with each other by regressing the signals from individual channels from one instrument against those of other instruments. This process, which was introduced by Neotec/Pacific Scientific in 1978 (R. C. Moen, pers. comm., 1978), is known as *normalization* of instruments. About 20 samples should be read and the optical signals recorded from one instrument, which is thereby established as the "master" instrument. The optical signals from all others (called "slave" instruments) are regressed against those of the master instrument for each wavelength point. A single sample is then read on each slave instrument, and the optical signal is adjusted for each wavelength point, using the respective regression coefficients. This process is facilitated in Pacific Scientific models by a calibration transfer program. The master instrument is then calibrated by standard calibration or by means of universal constants. By adjusting the basic optical signals, using the regression coefficients obtained from the optical data, the slopes of all slave instruments are in accordance with that of the master instrument, so that one set of calibration constants operates in all similar instruments after the adjustment of the intercepts. The accuracy of each slave instrument is verified by analysis of 20–30 samples, and slope/bias adjustments are made as necessary. Table XIII illustrates the principle of slope/bias adjustment. At least 12 samples should be reanalyzed after slope/bias adjustment to ensure that the calibration is accurate.

The t- *or* F-*Value Exclusion Method.* Most multiple regression programs compute *t* values for each wavelength point (calibration constant) used in the calibration. These values may be used to fine-tune a calibration. For example, in the original measurement of ash by the Technicon InfraAlyzer, seven wavelengths were found to be necessary (Williams et al, 1981). These included the six wavelengths (filters) of the standard InfraAlyzer together with a 2,345-nm filter. Constants were developed for the original 19 wavelengths of an InfraAlyzer 400 using forward stepwise multiple linear regression. The *t* values were used to exclude wavelengths.

TABLE XI
Accuracy of Protein Determination in Wheat Using High/Low Calibration

	D10D[a]				D20D[b]			
	Selected		High/Low		Selected		High/Low	
	HRS[c]	SWS[d]	HRS	SWS	HRS	SWS	HRS	SWS
SEP	0.211	0.233	0.159	0.196	0.221	0.180	0.184	0.158
r	0.996	0.940	0.998	0.955	0.949	0.962	0.997	0.972
$\bar{d}$	0.07	−3.42	−0.08	−0.21	−0.21	3.41	−0.17	0.76

[a] Five wavelength points smoothing, 15 points derivative (gap) size.
[b] Five wavelength points smoothing, 10 points derivative (gap) size.
[c] Hard red spring wheat.
[d] Soft white spring wheat, analyzed using calibration derived from HRS samples.

TABLE XII
Possible Intercorrelations Between Constituents

Commodity	Constituents	Coefficient of Correlation[a]
Cereal grains	Protein/starch	−0.7 to −0.9
	Protein/moisture	−0.3 to −0.5
Barley	Protein/fiber	+0.6 to +0.8
Oilseeds	Protein/oil	−0.2 to −0.9
	Oil/fiber	+0.4 to +0.7
	Protein/fiber	+0.5 to +0.8
Soybeans	Oil/protein	−0.6 to −0.9
	Oil/fiber	+0.5 to +0.8
	Protein/fiber	−0.4 to −0.8

[a] Approximate values from which individual comparisons may vary widely.

TABLE XIII
Example of Slope/Bias Adjustment

Protein Kjeldahl	Protein NIR	Protein NIR (after adjusting *K* values)
17.1	18.0	17.1
16.6	17.3	16.5
15.9	16.7	15.6
15.0	15.7	15.1
14.1	14.7	14.1
13.0	13.5	13.2
12.3	12.6	12.6
11.6	12.0	11.5
10.9	11.2	10.8
10.3	10.5	10.0
9.6	9.7	9.7
8.8	8.7	8.8

Statistical Analysis[a]

r	0.999	...
b	0.9046	...
a	0.83	...
k_0	0.1174[b]	0.1062[c]
k_2	0.9716	0.8789
k_3	−1.1043	−0.9989
k_5	−0.3872	−0.3503
A[d]	12.16	11.83
SEP	0.29	0.18
$\bar{d}$	0.54	−0.01

[a] r = coefficient of correlation, b = coefficient of regression (slope), a = intercept, k_0–k_3 = calibration constants, A = calibration equation intercept, *SEP* = standard error of performance, and $\bar{d}$ = mean difference between NIR and Kjeldahl values.
[b] Original regression coefficients (calibration constants).
[c] Adjusted constants. Note that original intercept A has been multiplied by regression coefficient (0.9046) before addition of intercept, 0.83.
[d] Calibration intercept, set into instrument (GAC III).

Usually, wavelengths responsible for constants with the lowest t or F values of the wavelength/regression coefficient series can be dispensed with. This is not necessarily true, since some such wavelengths/constants can still contribute toward the final calibration, so that the effect of eliminating a wavelength must be tested at each stage of the elimination process. The multiple regression is recalculated and the calibration used to analyze verification samples. If the SEP improves, more wavelengths can be excluded until there is no further improvement in the apparent efficiency of the calibration. Another indication of efficiency is the corrected multiple correlation coefficient, which starts to decrease when wavelengths are not contributing. These wavelength elimination steps must all be verified by analysis of test samples and derivation of the SEP and bias. In this way, the forward stepwise regression process can be employed as a backward stepwise regression, with the operator, instead of the computer, controlling the elimination of wavelengths. The 2,345-nm filter for flour "ash" determination was researched by this method. For calibrations involving derivatized mathematics, the F values are preferable to t values for this purpose and, by using forward stepwise regression, can be employed in the same way. Both t- and F-value exclusion methods can be featured in computerized calibration programs. As a further refinement, wavelengths that have been eliminated in an early step can be reinserted and retested at a later stage, when they may become useful in analysis.

The final stage in a calibration is verification of its effectiveness in analysis. Several methods have been suggested for evaluation of the efficiency of the calibration/verification exercise, but methods that relate the SEP to the standard deviation of the initial data are the most practicable. The ratio of the SEP to the standard deviation of the original data (i.e., SD_x/SEP_{xy}) is called the RPD and should be as high as possible, preferably above 10. This topic is discussed in more detail in the next chapter.

The regression coefficient of slope (b) and the intercept (a) should be calculated between NIR and standard values. If the slope exceeds 1.05 or is less than 0.95, a slope/bias correction should be carried out. First the bias is removed by adjusting the original intercept (K_0, F_0, or A). Then the calibration constants (including the intercept) are all multiplied by the regression coefficient (b) and the new intercept (a) is added to the original intercept (see Table XIII), thereby adjusting the slope and improving the accuracy of the calibration over the range. The samples must be reanalyzed to verify accuracy. The instrument can then be used for analysis. When the regression coefficient (b) is calculated, the coefficient of correlation (r) is also generated. This should be above 0.95. If r is less than 0.90, the accuracy of future analysis using the calibration will require careful monitoring, since slope/bias correction will not improve the overall relationship of NIR to standard values but will ensure that the deviations will be more evenly distributed across the range of concentration.

B. Operation

The operation of individual instruments in analysis is fully described in the operation manuals issued by their manufacturers and will not be further covered in this monograph.

C. Monitoring

PRECISION

Precision is defined as reproducibility of the analytical result. The daily performance of a near-infrared instrument should be monitored by the regular analysis of a reference sample. Results of these analyses will indicate the precision of testing. Ideally, three samples should be used, with one of them tested every 20th sample. The samples should represent the normally anticipated range in composition. (Every 10th sample is recommended for new operations.) If sums of X, X^2, and N are recorded, the precision can be updated by day, week, month, and year. Precision of a single sample, including sampling and sample preparation error, should be no more than 1.0–1.5% of the mean result. This corresponds to ±0.16–0.19% for protein in wheat. Repeat analysis of the same ground sample should be less than 1% of the mean, or about 0.08–0.13% for wheat protein. Precision of reading an already loaded cell (repeatability) should be less than 0.7% of the mean for both protein and moisture provided the cell is located in the instrument in exactly the same orientation each time.

ACCURACY

Accuracy is defined here as the degree of closeness of the near-infrared result to the true chemical composition as determined by the standard (reference) method. Since many near-infrared analyses deal with constituents such as protein and fiber, which have not been well defined in terms of actual chemical composition, the accuracy of the analysis is referenced to the most reliable chemical or physicochemical analysis available. Verification of accuracy of near-infrared analysis can only be assessed by analysis of blind duplicates, or check samples of known constitution that are interpolated (and are unidentifiable by the laboratory), for retesting at random. The time-tested method of spiking samples by adding additional constituents to the sample cannot be applied in near-infrared analysis of grains, forages, or related commodities. The configuration of "free" constituents will not necessarily be the same as the constituents when they are bound together in a sample, and the wavelengths suitable for measurement of a particular constituent in the sample may not be suitable for measuring the free constituent. This is illustrated in spectra 94 and 241 in the section on spectra following Chap. 15. Spectrum 94 is that of a wheat gluten preparation containing over 90% protein ($N \times 5.7$) on a dry matter basis, and spectrum 241 is that of a sample of wheat with 13% protein. Many of the bands apparent in the gluten are not prominent in the wheat spectrum, and the wavelengths selected for protein measurement are not the strongest "protein" bands in the wheat spectrum. Prominent gluten bands that are not noticeable in the wheat spectrum include the strong gluten absorbances at 1,185, 1,495, 1690, 1,905, and 2,170 nm, which appear only as minor absorbances, shoulders, or not at all in the wheat spectrum. This indicates that the association of the gluten with starch, oil, fiber, and moisture in the wheat has caused significant modifications to the absorbances in these areas. Several other strong gluten bands do appear in the wheat spectrum, but they have apparently shifted by as much as 15 nm.

The results of near-infrared analysis of samples of known composition provide a practicable method of verification of accuracy, which can be expressed in terms of the SEP or the RPD. A second verification of the accuracy of analysis lies in the operation for which the analyses are being used. For example, if the instrument is being used for the buying and selling of wheat or barley on the basis of protein content at a grain elevator, the ongoing financial operation of the elevator depends on the accuracy of the analysis, which is verified by standard analysis of parcels of grain shipped from the elevator by the buyers. The elevator receives information on protein content back from the buying agency. This serves as an alternative and very effective monitoring and verification system. Standard deviations of differences (SEP) for an industrial operation such as a grain elevator or a flour mill may be as high as 0.4. Provided there is no bias, a high SEP in such an operation is not serious, since the two sets of sampling, sample preparation, and analytical error involved in analysis of both producers' samples and shipments are likely to result in errors of this magnitude. Coefficients of correlation should be 0.9 or higher. Accuracy evaluation is discussed further in the next chapter.

VIII. OPERATION OF COMPUTERIZED SPECTROPHOTOMETERS

It is convenient to divide the operation of a computerized spectrophotometer into preparatory, exploratory, and application phases.

A. Preparatory Steps

Six operations should be carried out before commencing wavelength or other searches. These are summarized in Table XIV and discussed in the following subsections.

VERIFY ACCURACY AND PRECISION OF WAVELENGTHS OF ABSORPTION PEAKS

Should the instrument have received rough treatment during shipping, installation, by accident, or by wear and tear, the optical system may become slightly out of alignment. The wavelengths of peak absorptions recorded by the instrument may not then correspond to the true wavelengths at which measurements should be made. This can be misleading to the operator, particularly when the information is to be used to set wavelengths into a smaller commercial instrument or to purchase filters for use in a discrete filter instrument. The accuracy of wavelengths can best be verified by scanning a pure substance, for which the absorption maxima of the wavelengths have been comprehensively established elsewhere. Absorption maxima of the substance as scanned by the instrument should conform to the known maxima. A convenient standard

TABLE XIV
Summary of Preparatory Operations

Step[a]	Operation	Comments
(i)	Verify accuracy and precision of wavelengths of absorption peaks using standard substances	Polystyrene is convenient test substance.
(ii)	Select samples	Preferably 100–200, at least 50.
(iii)	Prepare samples	Clean out foreign material and grind.
(iv)	Perform standard tests	Duplicate.
(v)	Read optical data on computerized spectrophotometer	Optimize instrument parameters such as derivative size. Make at least two files: one for calibration and one for verification, covering range of variance.
(vi)	Add standard data and check	Verify identification of samples and chemical data.

[a]Some instrument companies recommend use of sample selection program to decide which samples from population are best to use for calibration. This step would be inserted between steps (iii) and (iv).

substance is polystyrene, which can be purchased in sheets or prepared by cutting a section from a clean polystyrene petri dish or disposable polystyrene cup, or by using polystyrene powder obtained by grinding a piece of the material in an impeller mill and sieving the powder through a 75-mesh sieve. Polystyrene characteristically displays eight well-established absorption maxima or peaks at 1,144, 1,206, 1,680, 1,761, 2,145, 2,167, 2,307, and 2,354 nm. Four of these, together with data from a typical monochromator scan of polystyrene, are given in Table XV. The figures are the means of six runs representing eight 50-scan tests on a commercially available computerized spectrophotometer. The precision of the wavelength peak was better at the higher wavelengths. Precision of wavelength measurement by a spectrophotometer is critical. A difference of 1 nm in reading can be equivalent to 0.3% protein in wheat (W. R. Hruschka, pers. comm., 1985). Wavelength accuracy is not so important in future analyses, provided the instrument always takes its readings at the same wavelength.

SELECT SAMPLES

Sometimes a period of several weeks may elapse before enough samples can be accumulated to complete calibration/verification sets. During this period, the samples acquired early should be carefully stored and preferably refrigerated. The age of the samples may become important. Aging of some commodities is often accompanied by changes in certain constituents as a result of moisture changes, oxidation, agglomeration of particles, particularly in oilseeds (J. A. Panford, pers. comm., 1984), or deterioration. These changes may produce bands that are not present in fresh material, and, conversely, bands present in fresh material may disappear or diminish with age. Such physical properties as bulk density may also change with age.

PREPARE SAMPLES

Preparation of samples includes removal of foreign material, documentation, grinding, mixing, and storage. These steps are discussed in the next chapter.

PERFORM STANDARD TESTS

Standard (reference) tests must be carried out by reliable and accessible procedures. A useful procedure is to pass the samples through the test laboratory, then to repass them through the laboratory under reassigned sample numbers on a different day. Samples whose results differ by more than 1.5× the standard error of the test should be repeated in duplicate or discarded. Accurate and precise chemical or physicochemical analysis is vitally important to successful application of near-infrared spectroscopy.

READ OPTICAL DATA ON SPECTROPHOTOMETER

At least two sets (files) of samples should be recorded. One should contain about 15–25 samples for use as test samples (sometimes called unknown or unused samples) to evaluate the accuracy of the final calibration. They should be assembled completely independently of the samples used in wavelength selection and calibration—for example, on different days. The other file or files are used in wavelength selection, mathematical optimization, and calibration. Samples may be recorded as separate files of, say, 30–50 samples, corresponding to series with different types of variance. Files can be combined later on a separate disk to give a single large file. File editing enables the removal of samples from a file and the readdition of these samples to the same file or any existing file, and most computerized spectrophotometers enable new files to be created by transferring samples from an existing file or files. Calibrations, correlation plots, and files of mathematically treated data can be rapidly regenerated from raw data by computerized spectrophotometers so that only raw data need be stored. Raw data used in calibrations should never be discarded. Storage of other data uses up memory and restricts the efficiency of the system. It is good practice to copy raw optical data to a second diskette to protect against damage or loss. Data recorded as raw log $(1/R)$ data may be retained indefinitely as such. The raw data may be manipulated as described above, but it cannot be changed in terms of the number of scans taken, the area scanned, or the precision (single or double), which are all fixed at the time of recording.

ADD STANDARD DATA

Several chemical constituents or physicochemical parameters can be accommodated per sample. They are usually added as an option of the data recording program.

B. Exploratory Steps

The five exploratory steps are listed in Table XVI.

DISPLAY SPECTRA

The spectra of all samples should be displayed to ensure that they are smooth throughout the wavelength range and that no noise peaks are present. A noise spike indicates an electronic malfunction in the instrument or in the power supply to the instrument, and the source must be traced and corrected before any further data are recorded or processed. This step will also illustrate the variability in scatter caused by particle

TABLE XV
Precision of Reading Polystyrene by RCA Model 6350

Standard Value		1,144 nm	1,682 nm	2,165 nm	2,305 nm
Run 1	λ	1,144.16	1,682.58	2,165.55	2,305.14
	SD	0.030	0.014	0.024	0.012
2	λ	1,143.87	1,682.32	2,165.49	2,304.98
	SD	0.024	0.021	0.015	0.009
3	λ	1,143.94	1,682.40	2,165.46	2,305.01
	SD	0.015	0.012	0.015	0.009
4	λ	1,143.78	1,682.19	2,165.50	2,305.23
	SD	0.024	0.025	0.011	0.016
5	λ	1,143.55	1,681.98	2,165.44	2,305.10
	SD	0.067	0.045	0.032	0.024
6	λ	1,143.46	1,681.96	2,165.31	2,304.67
	SD	0.016	0.022	0.008	0.016
Mean λ		1,143.79	1,682.23	2,165.46	2,305.02
Range in λ		0.70	0.62	0.24	0.56
SD of λ		0.258	0.243	0.082	0.194
Mean SD		0.029	0.023	0.020	0.016
SD of SD		0.019	0.012	0.010	0.005

TABLE XVI
Summary of Exploratory Operations

Step	Operation	Comments
(i)	Display	Examine system response (ceramic or other standard signal), raw log $(1/R)$ spectra, and derivatized traces to detect irregularities such as noise spikes or incorrect mathematical treatment.
(ii)	List	List standard data of calibration and verification samples to detect errors.
(iii)	Perform regression	Select number of wavelength points for optimum performance. Optimize smoothing and derivatization.
(iv)	Display correlation plot	Document areas of high correlation for future reference.
(v)	Verify	Analyze test samples.

characteristics and will show the type of material being tested and its rough composition by the presence of oil bands, protein bands, and bands for other constituents.

LIST DATA

Following the completion of the operation for recording, the chemical or other standard data should be checked to ensure that the correct standard data have been entered. A single grossly erroneous chemical result can increase standard errors of calibration and cause a significant wavelength shift, particularly for the second, third, and subsequent wavelength points. This will increase sample sensitivity of the calibration.

PERFORM REGRESSION

Regression is the method used to select wavelengths and to optimize mathematical algorithms, including derivative ("gap") size. All calibrations should be examined by analysis of files of unknown samples. The lowest SEP and highest coefficient of correlation will be the optimum combination of wavelength and derivatization. This may occur with even a single wavelength term with derivatized signals. A single derivatized term involves at least two wavelengths, producing a type of stereo effect when used in a calibration, and often gives satisfactory and stable results with no further wavelength points. This outcome is particularly likely if sample selection has been done carefully to maximize representation of variance in composition and physical sample characteristics. It is good practice to display the correlation plot after selection of the first wavelength by derivatized or quotient mathematics, the operation that is listed as step (iv) in Table XVI. Both positive and negative correlations will be displayed, enabling the operator to observe and document the wavelengths where broader areas of correlation occur. These wavelengths can subsequently be specified in a calibration if they are not selected by the computer. Regression programs do not always select the best wavelengths for analysis—they simply select those with highest correlations (see Section VI-A). The first log $(1/R)$ wavelength selected by multiple regression always gives a poor correlation and high standard error.

In stepwise regression, the first wavelength chosen has the highest individual correlation against the dependent variable. This wavelength may not appear in the final calibration. The log $(1/R)$ algorithm requires at least two wavelengths, one of which acts as a reference to the other. The regression program selects wavelengths in sequence, then rechecks to verify whether the earlier wavelengths are those most suitable for the subsequent ones. In doing so, alternative wavelengths may be selected and the earliest wavelengths eliminated in favor of those that are more suitable. As more wavelengths are added, the multiple correlation increases and the standard error of calibration (SEC) improves (becomes smaller). This process does not necessarily mean that the SEP (accuracy) also improves, and the addition of too many wavelengths may actually reduce the accuracy of analysis. The corrected multiple correlation may become smaller and the SEP may increase. This is referred to as overfitting of data (see Chaps. 3 and 4). The t or F values indicate the usefulness of a term, but analysis of unknown samples is the best test of the efficiency of a calibration. As a general rule, the fewer the number of terms in a calibration, the better. For example, although six terms are possible in each of the Technicon InfraAlyzer models 250 and 300, the DICKEY-john GAC and Instalab series, and the Percon Inframatic 8100, all three companies recommend only three terms for protein and two for moisture, for testing grains. Calibrations with small numbers of terms are generally regarded as being more stable. In some cases, addtion of extra terms may be necessary to accommodate an extra variable.

The form of multiple regression to use has been the subject of some controversy. For all practical purposes, both forward and backward stepwise multiple regression can be employed. There are pitfalls in both methods. Forward stepwise multiple linear regression (FSR) can be misleading, in that mathematically as more terms are added, the coefficient of multiple correlation increases and the standard error of estimate or calibration (SEC) decreases. This does not necessarily mean that the SEP will improve.

The backward stepwise regression (BSR) method works by calculating the total multiple regression for a stipulated number of wavelength points and then eliminating points on the basis of the t-test. The chief argument against the BSR system is that once a wavelength has been eliminated it will not be reintroduced, allowing a wavelength to be eliminated early in the computation that would have been more useful in analysis than one of the wavelengths actually used in the final equation. As mentioned above, wavelengths that have been eliminated in earlier stages can be reintroduced by the operator and tested at a later stage, at which time they may become useful in analysis.

Automatic exclusion or inclusion of wavelengths on the basis of F or t values is largely responsible for the shortcomings of both FSR and BSR, since no actual analysis is performed. The shortcomings of both types of regression can be overcome by operator analysis of a set of verification samples, whether the wavelengths are included on the basis of the F-test (for FSR) or the t-test (BSR). The evaluation of a wavelength/mathematical treatment combination for a calibration should be based on the SEP, bias, and coefficient of correlation of observed near-infrared vs. standard values. Used in this way, both FSR and BSR are satisfactory methods of regression for the optimization of wavelengths and mathematical treatments for calibration. If log $(1/R)$ values are highly correlated to each other, regression constants tend to increase and become more variable (sample sensitive). Ideally, log $(1/R)$ values should be orthogonal (uncorrelated). The addition of an extra term then has no effect on the other constants. This is seldom encountered in practice. The use of single-term (quotient) mathematics, or the two-point high/low system, avoids this source of error, since with only a single set of optical data no intercorrelation is possible.

The best policy is to keep the number of wavelength terms as low as practicable, preferably one or two (with derivatized mathematics), which can best be achieved by testing unknowns at each step. The size of the regression coefficients is also a useful indication of the potential stability of a calibration. "Size" in regression coefficients must be assessed with regard to the algorithm employed or the format in which the log $(1/R)$ data are recorded. Regression coefficients for derivatized log $(1/R)$ values are always higher than those for the untreated log $(1/R)$ data at specific wavelengths. Similarly, Technicon InfraAlyzer coefficients are usually higher than DICKEY-john GAC coefficients, since the InfraAlyzer works with the raw log $(1/R)$ signal, whereas the log $(1/R)$ signals are multiplied internally by a factor in the GAC. Bearing this in mind, there is still very large variation in the size of regression coefficients for a given algorithm; as a general rule, the lower the coefficients are, the better. Very high regression coefficients usually indicate that the optical signals are very small and highly correlated to each other. They are usually associated with high sample sensitivity and generally unstable calibration. Very large regression coefficients still work satisfactorily in many instances, but the analytical results require careful monitoring. Regression coefficients for quotient mathematics or high/low calibrations are usually the lowest (Table XVII). The constants were used in the computation of the data presented in Table XI.

A curious feature of the log $(1/R)$ algorithm is that, with certain commodities and constituents, a single log $(1/R)$ wavelength may apparently provide good analytical data with no reference wavelength. Under these circumstances, the particle-size characteristics of the sample are probably acting as a reference for the single wavelength. Such systems are also likely to be highly sample sensitive. This can be tested by repeating the wavelength selection using another sample set (such as the analytical, rather than the calibration, set). Sample sensitivity will immediately be revealed on reciprocal analysis. Normally, log (1/*R*) algorithms require at least two wavelengths. The only parameter that can truly be measured using a single log (1/*R*) wavelength is particle size itself. Intervariability between grinders and instruments prevents transfer of particle-size data between instruments unless calibration is carried out. This involves more than one wavelength (Williams and Sobering, 1986).

Another method of wavelength selection is the multiple-range selection system. The wavelength range is divided into 10–15 subranges, for example from 1,100 to 1,200, 1,200 to 1,300, ..., 2,400 to 2,500 nm. Single wavelength points are selected from each of these subranges, and the most effective points on the basis of correlation to the constituent and SEC are combined into a single calibration by stipulating these wavelengths. The wavelengths used may be intercorrelated but are all correlated to the protein or other constituent. Typical results of this multiple-range selection approach are illustrated in Table XVIII. With use of the multiple-range selection calibration, the SEP and bias were both lower than when direct-selection calibrations were used, despite a higher SEC.

A novel adaptation of near infrared is that of fiberglass-disk near-infrared reflectance (FDNIR). By this technique, substances in solution or true suspension are spotted onto fiberglass disks. Pure solvent is spotted onto another fiberglass disk, which is used as a standard in place of the ceramic standard. The disks are air-dried or dried with warm air (e.g., from a hair dryer) and read in the near-infrared instrument, using the solvent-spotted disk as the standard. The instrument subtracts the standard from the sample and corrects for bands in the sample caused by the solvent. Calibration is achieved by spotting regular increments of the substance to be determined onto disks. About 12–20 increments are recommended. Wavelengths can be searched and mathematical treatments generated in the usual way. The technique is very fast: a single sample takes about 5 min per test, but assembly-line systems provide samples that can be read as fast as the instrument can read powdered dry samples. It allows the storage of very large numbers of samples in the form of fiberglass disks compressed in tins, and it is also applicable to discriminant analysis. Figure 33 shows spectra of bovine serum albumin developed by FDNIR spotted from aqueous solution; the smallest concentration shown in the figure represented 2 mg of albumin. Figure 34 shows two FDNIR spectra of maltose, with one spectrum developed from a 15-min incubation of a flour high in starch damage and the other spectrum developed from a flour low in starch damage. The technique has been applied to the analysis of aqueous solutions by Meurens (1984).

If the computerized spectrophotometer is to be used to optimize wavelengths for a commercial instrument, the mathematical treatment must be identical with that used by the instrument from the aspects of algorithm, smoothing, and derivatization, where applicable. The selection of filters for a discrete filter instrument requires the log (1/*R*) treatment. In the selection of pulse points for the Pacific Scientific series of instruments, in addition to the log (1/*R*), the first- or second-derivative treatments are used in simple and normalized form, and the company provides software that enables direct identification of wavelengths as pulse points by using the mathematical treatment specific to the instrument. The operator needs information from the instrument manufacturers with regard to the number of points summed (smoothing) and derivative or gap size. The number of readings taken in unit time is used to reduce system noise in commercial instruments. This can be simulated in some computerized spectrophotometers by specifying the number of scans per reading. If a bench instrument is to be supported by an auxiliary computer, the optics and detection systems of the instrument enable it to work with any mathematical treatment that can be accommodated by the computer. For example, if a Percon Inframatic 8120, 20-filter instrument were interfaced with a computer, it could employ log (1/*R*) or first or second derivatives instead of the fixed first-derivative algorithm with which it is furnished on delivery, provided that the derivative employed wavelengths corresponding to the filters installed.

Instruments are commercially available that employ at least five different mathematical treatments of data. Mathematical treatments, including least squares and principal component regression, curve-fitting, derivative and quotient mathematics, log 1/*R* (untreated), two-point calibration, and others, all represent options available to the operator for the optimum use of the log 1/*R* signal for the analysis of individual constituents in individual commodities. No single mathematical treatment is superior to all others for all applications, and careful optimization is advisable for all applications. It is good practice when using quotient mathematics to stipulate the selected denominator wavelength as a primary wavelength (numerator), then to use it as a reference to select a new denominator. The wavelength should be close to the original primary wavelength, which is then reused as numerator to reselect the denominator. This procedure ensures the best wavelength set to act as numerator and denominator for the samples used in the search. Another feature of quotient mathematics is that the second or third wavelength selected in a stepwise wavelength search is often the same as or close to the wavelength selected as denominator. "Best-pair" programs select the two wavelengths by a different process. The computer compares all wavelengths two at a time and selects the best pair simultaneously, comparing each pair of wavelengths from the whole array. This system is different from the normal computer program, which selects the best single wavelength and then selects the second wavelength by comparing all other wavelengths to find the best one to accompany the first. Similarly, "best-triple" programs search through all available wavelengths three at a time. Best-pair and best-triple selections are lengthy procedures. The best-pair program often selects two wavelengths that are very close together. This is analogous to development of a first derivative, although the recorded log (1/*R*) data will be of larger

TABLE XVII
Calibration Constants for Selected and Specified (High/Low) Calibrations

D10D[a]				D20D[b]			
Selected		**Specified (High/Low)**		**Selected**		**Specified (High/Low)**	
λ	κ	λ	κ	λ	κ	λ	κ
2,152	841.336	2,150	921.331	1,984	862.393	1,984	1,581.181
2,442	374.335	1,960	−120.692	2,166	1,700.327	2,182	547.479
1,774	−438.436	1,484	166.382	1,144	782.810	2,054	65.919
Intercept	6.69		24.25		19.56		17.93

[a]Five wavelength points smoothing, 15 points derivative (gap) size.
[b]Five wavelength points smoothing, 10 points derivative (gap) size.

magnitude than the first-derivative data.

A note on objective analysis by near infrared. Certain parameters can be determined directly by the near infrared because of the relationship of the parameter to other chemical constituents in the material to be analyzed. These include flour "ash," which depends on a near-infrared determination of absorbance at 2,345 nm, or an oil band (Williams et al, 1981), and digestibility in forages, which is determined by near-infrared measurements of absorbance at cellulose bands (Norris et al, 1976). *Objective analysis* means defining the objective of the analysis, then determining the end result directly. *Quality* in terms of acceptability by the consumer or primary processor is usually a combination of several features, including texture, color, and other factors. The quality of a fruit, for example, can be evaluated by its color, texture, taste, and keeping quality (among other factors). Chemical assessments usually involve determinations of sugars, starch, pH, moisture, pectins, and other constituents, followed by computer-assisted assembly of the data into an overall quality evaluation. An equally objective assessment of "quality" by near-infrared analysis in, for example, fruit can be carried out by assembling a large number—200 or more individual fruits—and then ranking them in order by each factor involved in the overall definition of quality. The individual rankings are then totaled to give an overall ranking based on all quality factors. The near-infrared instrument is then calibrated directly to the quality ranking and used to assess quality in future samples. The principles of this form of discriminant analysis are discussed fully in a paper by Molnár (1982). The readout gives a total quality rank from 1 to 100. An example of such a ranking is given in Table XIX.

TABLE XVIII
Comparison of Calibrations Based on Forward Stepwise Regression (FSR) and Multiple-Range Selection (MRS) for Determination of Protein in Wheat

FSR			MRS		
Wavelength	SEP[a]	$\bar{d}$	Wavelength	SEP	$\bar{d}$
		D10D[b]			
2,152	0.244	0.03	2,152	0.244	0.03
2,447	0.148	0.07	1,666	0.244	0.03
1,774	0.211	0.07	1,752	0.184	−0.05
		D20D[c]			
1,984	0.150	−0.17	1,984	0.150	−0.17
2,166	0.207	−0.25	1,736	0.171	0.09
1,144	0.221	−0.21	2,164	0.212	−0.22

[a]Standard error of performance.
[b]Five wavelength points smoothing, 15 points derivative (gap) size.
[c]Five wavelength points smoothing, 10 points derivative (gap) size.

VERIFICATION

Standard errors of performance should be less than 0.15–0.20 times the standard deviation of the concentrations of the constituent under investigation in the verification samples. If the coefficient of correlation is low (0.7–0.8), the near-infrared analysis has not been effective and correction of the slope by application of slope/bias correction (see Section VII-A) will not improve accuracy. The reasons for the inaccuracies should be

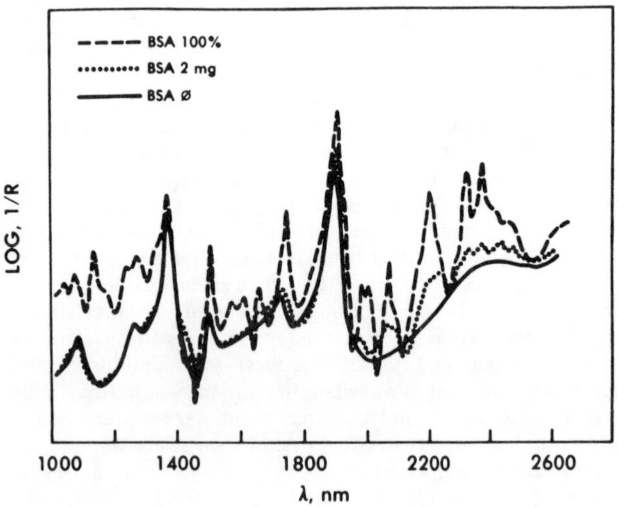

Fig. 33. Spectra of bovine serum albumin by fiberglass-disk near-infrared analysis.

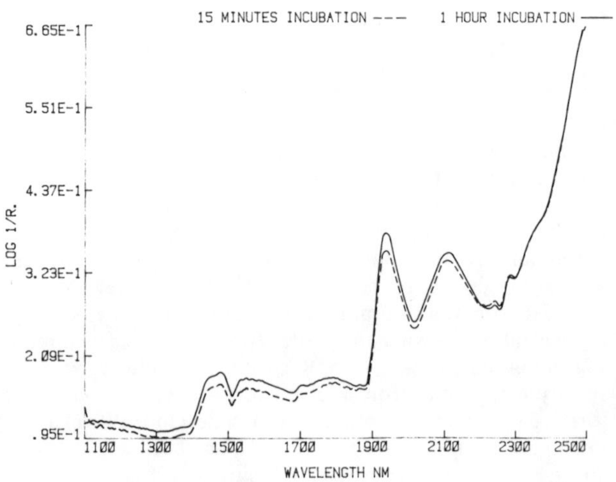

Fig. 34. Fiberglass-disk near-infrared analysis spectra of autolysis of hard red spring flour with high starch damage.

TABLE XIX
Specimen Ranking of Fruit for Objective Near-Infrared Calibration for Overall Quality

Fruit	Color	Texture	Taste	Storage Texture	Storage Taste	Overall	Rank[a]
A	4	1	5	3	7	20	4
B	7	11	6	8	6	38	8
C	1	6	4	2	3	16	2
D	3	8	7	11	8	37	7
E	12	9	9	12	10	52	12
F	9	12	8	6	11	46	10
G	10	5	3	4	1	23	5
H	2	4	2	1	2	11	1
I	11	7	11	9	12	50	11
J	6	3	1	5	4	19	3
K	5	2	10	7	9	33	6
L	8	10	12	10	5	45	9

[a]Near-infrared instrument is calibrated directly against "rank" values.

located and corrected for, usually by improving the integrity of the calibration samples or optimizing the mathematical treatment. It is important to remember that NIR is not universally applicable, and certain constituents or parameters may be determined more accurately by other analytical methods.

C. Application Steps

ALL NEAR-INFRARED INSTRUMENTS

1. Calibrate with samples used in exploration.
2. Analyze unknowns.
3. Ideally, repeat steps (1) and (2) using different set of samples.
4. Repeat exploratory steps using different mathematical treatments, etc., to optimize treatment.

PACIFIC SCIENTIFIC INSTRUMENTS ONLY

5. Convert wavelength to pulse points. The Pacific Scientific IMATH program automatically carries out this step for instruments in the GQA, FQA, and 7000 series. Remember that the first derivative of the log $(1/R)$ algorithm uses two pulse points per wavelength term. Similarly, a second-derivative term uses three wavelengths, and the A and C wavelengths are equidistant above and below the selected primary B wavelength. The FQA 51A and 7000 series operate in log $(1/R)$, first-, and second-derivative modes. The segment and derivative (gap) sizes for the FQA 51A are hard-wired and the insertion of the B wavelength is sufficient. Both A and B pulse points must be inserted for first-derivative algorithms. The gap size in the 7000 series is programmable.
6. Check on possible filter crossing. Derivatized mathematics may not allow the use of the full filter because of the size of the derivative. If the selected pulse point is too close to the end of the range covered by the filter, display the individual coefficients of correlation and select the pulse point with the highest correlation concomitant with the minimum distance from the end of the filter that will allow the full derivative size. This usually gives better analytical accuracy.
7. Enter pulse points to instrument. Neotec models 101, 102, 51, and 7000 series can be pulse-pointed by keyboard, whereas Model 31 EL requires a simple soldering process.

Steps (1) through (4) are described in the operator's manuals provided by all instrument companies, and steps (5) through (7) are described in the manuals for the respective Pacific Scientific models.

This monograph was prepared mainly to illustrate the basic principles of near-infrared technology and its application to raw materials and finished products of the agricultural and food industries. However, there is a very large scope for applications in other industries, such as plastics, pharmaceuticals, petrochemicals, and textiles. The application of near-infrared analysis to these industries often involves unique sampling and sample preparation steps, sometimes with very specific conditions of temperature and humidity. These types of applications may be simpler than agricultural and food applications in that many of the constituents to be analyzed are pure chemical compounds, often allowing calibrations to be carried out simply by preparing samples with varying concentrations of the substance. Selection of samples for calibration on the basis of spectral characteristics may be very useful in these situations. Also, accuracy can be evaluated by spiking with known amounts of the sought-after constituent, which makes life a little easier, whereas applications to agricultural, food, and biological materials in general are fraught with the complexity of their individual constituents and their mutual interactions.

ACKNOWLEDGMENTS

I wish to acknowledge the helpful contributions of John Judge of Technicon Industrial Systems, Dave Funk of DICKEY-john Corporation, Don Webster of Pacific Scientific, Peter Perten of Percon Prufgërate, Bob Rosenthal of Trebor Industries, Isaac Landa of LT Industries, Karl Norris of the U.S. Department of Agriculture, and Ed Stark of KEC Analysis.

LITERATURE CITED

AACC. 1983. Approved Methods of the AACC. Method 44-15A, approved October 1975; revised October 1981. Method 46-12, approved October 1976; reviewed October 1982; revised November 1983.

DEMING, W. E. 1975. On probability as a basis for action. Am. Stat. 29:146-152.

DUNNE, W., and ANDERSON, J. A. 1976. A system for segregating Canadian wheat into subgrades of guaranteed protein content. Can. J. Plant Sci. 56:433-450.

HOLMAN, R. T., and EDMONDSON, P. R. 1956. Near-infrared spectra of fatty acids and related substances. J. Anal. Chem. 28:1533-1538.

HONIGS, D. E., HIEFTJE, G. M., MARK, H. L., and HIRSCHFELD, T. B. 1985. Unique sample selection via near-infrared spectral subtraction. Anal. Chem. 57:2299-2303.

KRONSTEIN, M., KRAUSHAAR, R. J., and DEACLE, R. E. 1963. Sulfur as a standard of reflectance in the near-infrared. J. Opt. Soc. Am. 53:458-465.

LAW, D. P., and TKACHUK, R. 1977. Near infrared diffuse reflectance spectra of wheat and wheat components. Cereal Chem. 54:256-265.

MAHALANOBIS, P. C. 1936. On the generalized distance in statistics. Proc. Natl. Inst. Sci. India 2:49-55.

MARK, H. L., and TUNNELL, D. 1985. Quantitative near-infrared reflectance analysis using Mahalanobis distances. Anal. Chem. 57:1449-1456.

MEURENS, M. 1984. Analysis of aqueous solutions by NIR reflectance on glass fiber. Proc. Annu. Users Conf., 3rd. Pacific Scientific, Rockville, MD, February.

MILLER, B. S., POMERANZ, Y., THOMPSON, W. O., NOLAN, T. W., HUGHES, J. W., DAVIS, G., JACKSON, N. G., and FULK, D. W. 1978. Interlaboratory and intralaboratory reproducibility of protein determination in hard red winter wheat by Kjeldahl and near infrared procedures. Cereal Foods World 23:198-201.

MOLNÁR, P. 1982. The design and practical use of an overall quality index for food products. Pages 115-147 in: Food Research and Data Analysis. H. Martens and H. Russwurm, Jr., eds. Applied Science Publishers, London.

MORREY, J. R. 1968. On determining spectral peak positions from composite spectra using a digital computer. Anal. Chem. 40:905-914.

NORRIS, K. H., BARNES, R. F., MOORE, J. E., and SHENK, J. S. 1976. Predicting forage quality by infrared reflectance spectroscopy. J. Anim. Sci. 43:889-897.

NORRIS, K. H., and WILLIAMS, P. C. 1984. Optimization of mathematical treatments of raw near-infrared signal in the measurement of protein in hard red spring wheat. I. Influence of particle size. Cereal Chem. 61:158-165.

NORRIS, K. H., WILLIAMS, P. C., and MATTERN, P. 1977. Rapid estimation of amino acids in wheat by near infrared reflectance spectroscopy. (Abstr.) Cereal Foods World 22:461.

ROSENTHAL, R. D. 1971. Introducing the Grain Quality Analyzer. Presentation at Annu. Meet., Kans. Wheat-growers' Assoc., Hutchinson.

RUBENTHALER, G. L., and BRUINSMA, B. L. 1978. Lysine estimation in cereals by near-infrared reflectance. Crop Sci. 18:1039-1042.

TKACHUK, R. 1981. Protein analysis of whole wheat kernels by near infrared reflectance. Cereal Foods World 26:584-587.

TKACHUK, R., and LAW, D. P. 1978. Near infrared diffuse reflectance standards. Cereal Chem. 55:981-995.

WILLIAMS, P. C. 1975. Application of near infrared reflectance spectroscopy to analysis of cereal grains and oilseeds. Cereal Chem. 52:561-576.

WILLIAMS, P. C., and ANTONISZYN, J. 1986. The significance of outliers. NIR/NIT 86. Proc. Hung. Sci. Soc. Food Ind., Budapest.

WILLIAMS, P. C., and CORDEIRO, H. M. 1985. Effects of calibration practice on correction of errors induced in near-infrared protein testing of hard red spring wheat by growing location and season. J. Agric. Sci. Camb. 104:113-123.

WILLIAMS, P. C., and KRISCHENKO, V. P. 1986. A comparison of two near-infrared computerized spectrophotometers for the determination of protein in wheat. Appl. Spectrosc. 40:785-790.

WILLIAMS, P. C., MacKENZIE, S. L., and STARKEY, P. M. 1985a. Determination of methionine in peas by near-infrared reflectance spectroscopy (NIRS). J. Agric. Food Chem. 33:811-815.

WILLIAMS, P. C., and NORRIS, K. H. 1983. Effect of mutual interactions on the estimation of protein and moisture in wheat. Cereal Chem. 60:202-207.

WILLIAMS, P. C., NORRIS, K. H., GEHRKE, C. W., and BERNSTEIN, K. 1983. Comparison of near-infrared methods for measuring protein and moisture in wheat. Cereal Foods World 28:149-152.

WILLIAMS, P. C., NORRIS, K. H., and SOBERING, D. C. 1985b. Determination of protein and moisture in wheat and barley by near-infrared transmission. J. Agric. Food Chem. 33:239-244.

WILLIAMS, P. C., PRESTON, K. R., NORRIS, K. H., and STARKEY, P. M. 1984. Determination of amino acids in wheat and barley by near-infrared reflectance spectroscopy. J. Food Sci. 49:17-20.

WILLIAMS, P. C., and SIGURDSON, J. T. 1978. Implications of moisture loss in grains incurred during sample preparation. Cereal Chem. 55:214-229.

WILLIAMS, P. C., and SOBERING, D. C. 1986. Attempts at standardization of hardness testing of wheat. II. The near-infrared reflectance method. Cereal Foods World 31:417-420.

WILLIAMS, P. C., and STARKEY, P. M. 1980. Influence of feed ingredients upon the prediction of protein in animal feed-mixes by near-infrared reflectance spectroscopy. J. Sci. Food Agric. 31:1201-1213.

WILLIAMS, P. C., STEVENSON, S. G., and IRVINE, G. N. 1978. Testing wheat for protein and moisture with the Automated Digital Analyzer. Cereal Chem. 55:263-279.

WILLIAMS, P. C., THOMPSON, B. N., WETZEL, D., McLAY, G. W., and LOEWEN, D. 1981. Near-infrared instruments in flour mill quality control. Cereal Foods World 26:234-237.

CHAPTER 7

Variables Affecting Near-Infrared Reflectance Spectroscopic Analysis

P. C. WILLIAMS
Grain Research Laboratory
Canadian Grain Commission
Winnipeg, Manitoba, Canada

I. THE PHILOSOPHY OF ERROR

A. What Is Error?

Errors are defined as differences between computed or measured values and the true or correct values. Before discussing the sources of error in near-infrared reflectance (NIR) testing, it is necessary to define error and the philosophy of error. There are three concepts concerning errors: unless errors are detected, no one knows they are there (this also applies to so-called "outliers"); the significance of the error is directly proportional to the true result; and the significance of the error is also directly proportional to the significance or impact of the true result. The most usual method of detection of errors is by some kind of system for monitoring accuracy of analysis—by using check samples, "blind" duplicates (duplicate tests carried out on different batches of analysis unknown to the laboratory), or, in the case of NIR analysis, by checking the results against standard laboratory data. The other way of finding out that errors exist is the "hard" way, where errors lead to missed specifications, improperly formulated feeds, or some other industry-related outcome. The significance of the error is proportional to the true result; for example, an error of 0.4% in protein testing is considered high for wheat analysis but not for soybean analysis, since the respective means are about 12% and 40%. Similarly, an error of 0.4% in protein testing is important when the accuracy demanded is ±0.2%. By the same reasoning, an NIR error of 0.3% is acceptable where the accuracy of standard analysis by a reference method (see Chap. 3) is no better than 0.3%, since NIR testing is cheaper and faster. Errors of ±3% in digestibility of a forage crop would be more than adequate for screening genotypes in a breeding program. Notwithstanding this type of practical philosophy, it is always preferable to achieve the highest accuracy possible in any test procedure.

Errors can only be detected by comparison with reliable known values. There are two types of errors: random, or unbiased errors, which usually reflect precision as well as accuracy; and biased errors, which are all higher or lower than the true values. These reflect a persistent change in the system and may not reflect changes in precision.

B. Definitions

For the purpose of this chapter, *accuracy* is defined as closeness to the true or correct result and *precision* as day-to-day or batch-to-batch reproducibility. Precision, which is often confused with accuracy, is determined by computing the standard deviation of the results of check sample analysis. It involves all steps of the test process, including sampling and sample preparation, and should not be confused with *repeatability*, which only involves the actual test, with none of the preliminary steps. For example, repeatability of an NIR test is taken to mean the agreement between successive readings of a sample by the instrument after loading the cell. The accepted method of expressing accuracy in classical chemical analysis is in terms of comparison with analysis of known calculated chemical constitution and by "spiking" samples with known amounts of pure substances and determining the recovery (usually theoretically 100%). Expression of accuracy in this manner is difficult in NIR analysis. Near-infrared reflectance spectroscopy is applied to testing for constituents such as moisture, protein, oil, starch, fiber components, and other organic substances of simple and complex nature. Recent applications, such as the estimation of amino acids, which themselves are relatively simple compounds, were complicated by the fact that neither the amino acids themselves nor the proteins of which they form a part exist in simple forms in the foods, feeds, and commodities under investigation. The most common method of expressing accuracy is in terms of the standard deviation of differences (SDD) between NIR and standard analyses of "true-life" or actual "unknown" samples and by the use of check samples, which have been analyzed by standard chemical methods to a degree that establishes confidence in their composition.

During the earlier period of quantitative NIR testing, the SDD came to be referred to as the *standard error of prediction*, but the term preferred nowadays is the *standard error of performance* (SEP). *Prediction* implies guesswork, and we like to think that NIR technology has progressed beyond that point, particularly with regard to the determination of constituents such as oil, protein, moisture, starch, fiber components, and even amino acids. Tests for parameters such as hardness or in vitro digestibility are prediction tests, whether by NIR or wet chemistry, and in the case of such parameters the SEP may be regarded as the standard error of prediction. The SEP statistic, which is illustrated in Tables I and II, will be referred to later.

II. STATISTICAL TERMS NECESSARY TO THE EVALUATION OF ACCURACY

Several terms are necessary and essential to the interpretation of statistical analysis of the results of NIR testing. Unless all of these aspects are properly appraised, the analyst can draw conclusions that are incorrect and frustrating, and often costly discrepancies can result. A word of explanation about dependent and independent variables. The x-axis data are usually referred to as the independent and the y-axis data as the dependent variables, since all values of y depend on the sampling and analytical efficiency of x. In computing a set of calibration constants, or in evaluating accuracy by computing the SEP, the standard values (Kjeldahl protein, oven moisture-extracted oil, etc.) are the independent variables (x_i) and the optical data (log [1/R], derivatized log [1/R], or NIR analytical data) the dependent variables (y_i).

The statistics that must be determined and interpreted (Table II) include (i) the standard error of performance, SEP; (ii) the mean difference or bias, $\bar{d}$; (iii) the mean result of testing the samples by the standard method (chemical or physical) of analysis, $\bar{x}$ (the mean of the NIR figures is $\bar{y}$); (iv) the standard deviation of both standard and NIR series of samples used in checking, SD_x and SD_y; (v) the coefficient of variability between SEP and $\bar{x}$, CV; (vi) the coefficient of correlation between standard and NIR results, r; (vii) the regression coefficient, b, and intercept, a for the same figures; and (viii) the root mean square deviation, RMSD. Formulae for calculating all of these parameters are given in Table III. The distribution of deviations from standard data of predicted NIR data also form an important part of the interpretation of the analytical efficiency of an NIR calibration.

The examples given in Table II were calculated from Example 1 in Table I, and the third section of Table II illustrates the application of the respective regression equations to predict the values of x (standard data) from the observed values of y.

III. PRECISION

Precision, which is part of accuracy, means reproducibility or consistency in the analytical data. It should be determined before accuracy is evaluated, since long-term accuracy is impossible to achieve if a procedure is not capable of giving good precision.

Precision is evaluated by repeated analysis of samples (preferably at least three, with a range in composition), using all

TABLE I
Example 1: Wheat Protein

Standard Protein %	NIR Protein % (A)	Differences (A)	NIR Protein % (B)	Differences (B)
x = 13.1	y_A = 13.2	$y_A - x$ = +0.1	y_B = 13.4	$y_B - x$ = +0.3
12.6	12.6	±0	12.8	+0.2
10.7	11.0	+0.3	10.8	+0.1
15.9	15.6	−0.3	16.3	+0.4
9.4	9.9	+0.5	9.3	−0.1
12.1	12.2	+0.1	12.2	+0.1
14.3	14.1	−0.2	14.6	+0.3
17.8	17.2	−0.6	18.4	+0.6
16.5	15.8	−0.7	17.0	+0.5
10.2	10.6	+0.4	10.2	±0
14.7	14.3	−0.4	15.0	+0.3
11.4	11.7	+0.3	11.5	+0.1

TABLE II
Statistical Treatment of Data from Example 1

Statistics of Raw Data

Statistic	Abbreviation	Standard	NIR (A)	NIR (B)
Mean result	$\bar{x}$ or $\bar{y}$	13.23% ($\bar{x}$)	13.18% ($\bar{y}_A$)	13.46% ($\bar{y}_B$)
Standard deviation	SD (σ)	2.65	2.27	2.85
Number of observations	N	12	12	12
Sum of x^2 (y^2)	Σx^2 (Σy^2)	2,178.11	2,142.24	2,263.07
Sum of x (y)	Σx (Σy)	158.7	158.2	161.5

Statistics of Comparisons

Statistic	Abbreviation	A (x:y_A)	B (x:y_B)
Mean difference between x and y values (bias)	$\bar{d}$	−0.042	+0.233
Standard deviation of differences	SEP	0.396	0.206
Sum of differences squared	Σd^2, $\Sigma(x-y)^2$	1.75	1.12
Sum of differences	Σd, $\Sigma(x-y)$	−0.50	2.80
Coefficient of variability	CV	2.99	1.56
Coefficient of correlation	r	0.972	0.999
Regression coefficient	b	1.1357	0.929
Intercept	a	−1.739	+0.725
RPD[a]		6.7	12.9

Application of Regression Statistics, $x = a + by$

Observed y	Predicted x (NIR.A.)	Predicted x (NIR.B.)
9	8.48	9.09
11	10.75	10.94
13	13.03	12.80
15	15.30	14.66
17	17.57	16.51

[a] Ratio of (standard error of) performance to (standard) deviation.

of the steps necessary to the final analytical procedure, including sample preparation. It is expressed as the standard error of a single test (SET). The establishment of precision involves identification of factors influencing the procedure and development of methods for avoiding or correcting for them.

The SEP and the RMSD are amalgamations of the combined errors of both standard and NIR tests and the sampling, sample preparation, and variability factors inherent in testing an array of randomly distributed, unknown samples. The SEP and RMSD differ in that the former is bias corrected (see the discussion in Chap. 3). The separate SEP and bias figures are preferred for reporting accuracy since the RMSD by itself includes the bias but does not indicate its magnitude or sign. This is immaterial if both mean x and mean y are quoted. For example, in the evaluation of the accuracy of testing a relatively simple commodity such as ground wheat for protein, the SEP is a combination of the errors of both standard (Kjeldahl) and NIR tests, sampling error, sample preparation error, the physical constitution and chemical composition of the test material, and residual error. All aspects of each test procedure (standard and NIR) must be included in the evaluation of precision. Over 30 sources of error have been identified in the Kjeldahl test (Williams, 1974), and later in this chapter an even larger number of factors will be described that affect the NIR procedure. The NIR and Kjeldahl test procedures both include sampling and preparation of the sample, but the Kjeldahl test for total nitrogen or protein also involves reagent preparation, weighing, digestion, and distillation. The extent to which precision is itself affected by sampling, sample preparation, and test procedure can be identified by testing under different conditions. The precision of the NIR test has been determined for wheat that was ground separately for each test, for the same ground wheat but separate sample cell loading, and by using a sealed sample cell ("permacell") that was placed in the sample cell holder in the same position for every reading. This series was used to determine the precision of both NIR protein and moisture testing (Table IV).

The permacell values represent the actual or basic error (repeatability) of the NIR instrument. The aspects of error contributed by all loading and sample preparation can be extracted from the total error by squaring the individual standard error (SEP) statistics and subtracting the squared value for sample loading from the figure for separate grinds to determine the error resulting from grinding, and the squared value for permacell or instrument error from the value for separate loading to determine the proportion of total error resulting from sample loading. The square root is then extracted from the residual figure. In this way, the proportions of error for the different operations can be identified. The influence of sampling on the error of Kjeldahl testing, which can also be isolated from the total error, has been reported to be as high as 34% of the total error or precision of testing wheat (Hildebrand and Koehn, 1944). This procedure is illustrated in Table V using the wheat data of Table IV.

These figures suggest that the proportion of total test (or precision) error from different causes varies between constituents. For example, cell loading is a bigger source of error than grinding in the case of protein, but grinding is by far the biggest factor in NIR testing for moisture. This is logical, since the grinding action affects both the moisture content and the moisture log $(1/R)$ signal. The effect of grinding increases directly as the original moisture content of the sample increases. As the grinder is used for more samples, it heats up and causes more moisture loss. The moisture log $(1/R)$ value in the near-infrared region is stronger than that of protein over the same concentration range, so the precision of the moisture signal is likely to vary more from sample to sample than that of protein because of differences in the diffuse reflectance from the surface.

TABLE IV
Precision (SET) of Near-Infrared Reflectance Testing of Protein and Moisture in Wheat ($N = 12$)

Procedure	% Protein	CV[a]	% Moisture	CV[a]
Separate grinds	0.168	1.18	0.371	3.28
Separate sample loading	0.135	0.95	0.117	1.04
Permacell	0.065	0.46	0.051	0.45

[a]CV = coefficient of variability.

TABLE III
Formulae for Calculating Statistical Parameters

Parameter	Abbreviations	Formula
Standard error of performance	SEP	$\left(\dfrac{\Sigma(x-y)^2 - ((\Sigma(x-y))^2 \div N)}{N-1}\right)^{1/2}$
Mean difference $(x - y)$	$\bar{d}$ (bias)	$\Sigma(x-y) \div N$
Mean x	$\bar{x}$	$\Sigma x \div N$
Mean y	$\bar{y}$	$\Sigma y \div N$
Standard deviation of x	SDx	$\left(\dfrac{\Sigma x^2 - ((\Sigma x)^2 \div N)}{N-1}\right)^{1/2}$
Standard deviation of y	SDy	$\left(\dfrac{\Sigma y^2 - ((\Sigma y)^2 \div N)}{N-1}\right)^{1/2}$
Coefficient of variability	CV	$(\text{SEP} \times 100) \div \bar{x}$
Coefficient of correlation	r	$\dfrac{(\Sigma x^* y) - ((\Sigma x^* \Sigma y) \div N)}{[(\Sigma x^2 - ((\Sigma x)^2 \div N)) \times (\Sigma y^2 - ((\Sigma y)^2 \div N))]^{1/2}}$
Coefficient of regression (slope)	b_{xy}	$\dfrac{(\Sigma x^* y) - ((\Sigma x^* \Sigma y) \div N)}{(\Sigma y^2 - ((\Sigma y)^2 \div N))} = r * \dfrac{\text{SD}x}{\text{SD}y}$
Intercept of regression	a_{xy}	$\bar{y} - (b^* \bar{x})$
Ratio of SEP to SDx	RPD	SD$x \div$ SEP

In a similar way, the absolute precision of the standard tests for protein, moisture, and oil has been determined on wheat (protein and moisture) and rapeseed (oil). The standard tests used were the Kjeldahl procedure (Method 46-12), single-stage air oven (Method 44-15A), and Goldfisch extraction with anhydrous diethyl ether (Method 30-20) (AACC, 1983). The results and the proportion of error are summarized in Table VI. All tests were carried out by experienced technicians. The testing error in standard analysis includes mixing and weighing of the ground sample. The sample size for standard testing for moisture and oil is larger than that for protein, and the contributions of larger sampling error and more sources of error in the actual testing (mixing and weighing the ground sample and digestion conditions are the most important sources of error in Kjeldahl testing) result in a proportionally higher basic testing error for protein than for moisture and oil. The best way of improving overall analytical precision is to improve sampling, subsampling, and grinding techniques and to develop a consistent weighing technique in the case of standard analysis and cell-loading procedure in the case of NIR.

For interpretation of the SEP in terms of the efficiency of the NIR instrument and its calibration, the precision of both standard and NIR procedures should be extracted from the SEP. This is illustrated in Table VII, which uses data from a comparative study of NIR instruments (Williams et al, 1983). The true test error (TTE) was derived by squaring the standard error of a single test (SET) or precision of both the Kjeldahl and the respective NIR instruments, subtracting both from the square of the SEP to obtain the variance of the test error, and extracting the square root, i.e.:

$$\text{TTE} = \sqrt{(\text{SEP})^2 - (\text{SET}_{\text{Std}})^2 - (\text{SET}_{\text{NIR}})^2}\,. \qquad (1)$$

The very low TTE of the Beltsville universal computerized spectrophotometer is an indication of the potential of the NIR technique using a very-high-resolution monochromator and a powerful computer with comprehensive interfacing and software. Often a high SEP is associated with a high SET, and the TTE may be much lower, indicating to the operator that there is scope for improvement in sampling, sample preparation, and cell loading before attempting on-line analysis of important material. It is theoretically impossible for the SEP to be lower than the combined SET for the standard and NIR techniques used. Accordingly, the minimum SEP likely for NIR determination of protein in wheat is about 0.175 by a commercial bench-type instrument.

TABLE V
Distribution of Error for Near-Infrared Reflectance Testing of Protein and Moisture in Wheat

Error Source	Proportion of Error (%) Protein	Moisture
Grinding and sampling	35.4	90.1
Cell loading	50.4	8.0
Residual (instrument test error)	15.0	1.9

IV. INTERPRETATION OF STATISTICAL EVALUATION OF NIR ANALYSIS

Table II illustrated two different types of data. The comparison between NIR (A) and standard data showed that although the overall mean NIR data agreed closely with that of the standard data, the SEP of nearly 0.4 indicated that the NIR determinations were rather variable. The CV was 3.0, and should ideally be between 1% and 1.5% of the mean value for the standard data. The SEP of the NIR (B) series was about one half that of the (A) series, and the CV of 1.6 was acceptable. The earlier statement that the magnitude of the error is proportional to the true result means that the SEP should be compared with the mean standard result to evaluate the significance of the figures. This is indicated by the CV. In these two examples, the SEP of 0.396 was high, since the mean standard result was 13.23%, giving a CV of 3.0. If the test material had been ground undefatted soybeans with a protein content of 41%, the CV would reduce to 0.97% of the mean, and the results would be regarded as excellent. The mean and CV are adjuncts to the SEP and essential to its interpretation.

The mean deviation or bias is also very important, since it indicates the magnitude and direction in which the NIR results deviate from standard results. This information enables the operator to adjust the intercept of the instrument to maintain the accuracy of analysis. There are four prerequisites before taking the step of adjusting the intercept. The first is to examine the SET, since if the precision of the NIR instrument is poor, adjusting the intercept will only improve the mean result but not the efficiency of analysis. The second is evaluation of the variability of standard data, or standard deviation (SD*x* of Table III). The variability of the NIR determinations is indicated by the magnitude of deviations from standard analyses (SEP). If the SEP value is similar to the SD*x*, the instrument is not predicting the standard values any better than the standard deviation of the original data. The SEP should be much lower than the SD*x*, and ideally the ratio of the SD*x* to

TABLE VI
Distribution of Error in Standard Testing for Protein, Moisture, and Oil ($N = 24$)

Sample Source	Standard Error Protein	Moisture	Oil	Proportion of Error (%) Protein	Moisture	Oil	Error Source
Separate grind	−0.178	0.256	0.870	70.0	92.8	92.4	Sampling and grinding
Single grind	0.098	0.069	0.240	30.0	7.2	7.6	Testing and instrumental

TABLE VII
True Test Error (TTE) of Some Near-Infrared Instruments for Testing Wheat for Protein

Parameter	Kjeldahl	Infra-Alyzer 400 F4[a]	Infra-Alyzer 300 F4	GAC III F7	GQA 31EL	GQA 101	BUCS[b] D20D/D20D[c]
SEP	0.171	0.186	0.275	0.218	0.244	0.241	0.145
SET[d]	0.089	0.123	0.138	0.116	0.118	0.105	0.110
TTE	0.146	0.107	0.221	0.162	0.194	0.198	0.032

[a] F4 = intercept + 3 constants; F7 = intercept + 6 constants.
[b] Beltsville universal computerized spectrophotometer.
[c] Second derivative of log $(1/R)$ ÷ second derivative of log $(1/R)$ at reference wavelength.
[d] Standard error of duplicate testing.

SEP should be 10 or higher. This ratio can be expressed as the RPD (ratio of [standard error of] performance to [standard] deviation). It is calculated by dividing the SDx by the SEP. In the case of NIR (A), the RPD = 2.65 ÷ 0.396 = 6.7, and for series (B) the RPD is 12.9. For a series of fairly uniform samples where the SDx is not high, the RPD cannot be as high as 10, but it should indicate that the SEP is appreciably lower than the SDx. For example, RPD values of 3 or 4 will verify accurate analysis if the SDx is only 0.4–0.5.

Assuming the slope of the relationship between x and y is 1.0, or has been corrected to 1.0, the RPD is mathematically related to the coefficient of correlation through the formulae:

$$\text{RPD} = \text{SD}x \div \text{SEP} \tag{2}$$

and

$$\text{SEP} = (N - 1)/(N - 2) \times (1 - r^2)\text{SD}y\,. \tag{3}$$

A corollary of this is that for a given value of r, the RPD will be constant regardless of the SDx. This is illustrated in Table VIII for a limited number of values. A second corollary of this relationship is that SEP values are limited by the degree of correlation between standard and NIR testing. Very low values reported for the SEP must be suspect unless they are accompanied by the r and SDx data.

Another method of standardizing the SEP is to relate it to the range of the standard data, as proposed by Starr et al (1981). This value, called the RER, represents the range divided by the SEP. It should also be as high as possible, but it can be inflated by a single sample at an extreme of concentration, whereas the standard deviation is not so markedly affected by a single sample. This is illustrated by the example given in Table IX. The RPD of 3.8 is not good. The RER of 14.0 looks satisfactory, but this appearance is misleading because the deviations indicate a low-high/high-low slope change.

The third prerequisite is the pattern of distribution of the deviations between standard and NIR data. The series (B) SEP data showed that the results at the lower protein level agreed closely with the standard data and the bias was +0.2. If the detailed results are examined, it is apparent that the NIR results became progressively higher relative to the standards at higher protein levels. The eight patterns that usually characterize this type of relationship (Table X) indicate that the slope of the regression line of NIR on standard data is greater or less than 1.0. Ideally, the slope (b) should be 1.0 and the intercept (a) should be 0. A slope change may be accompanied by a positive or negative overall bias, in which case a slope/bias correction is necessary. Biases can also occur with no slope difference.

Types 7 and 8 can be simply corrected with an intercept adjustment. All of the other types of bias require a slope/bias adjustment if the errors at either extreme are greater than 1.5% of the mean (e.g., 0.21 at 14% protein or 0.6% at 40% oil, etc.). This can be achieved by multiplying all of the calibration constants (K or F values), including the intercept, by the regression coefficient b_{xy}. The intercept a is then added to (or subtracted from) the original calibration intercept. An example is given in Table VIII of Chap. 6.

The fourth, and one of the most important prerequisites, is the coefficient of correlation, r, which is a measure of the closeness of fit between NIR and standard data. The coefficient should be as close to 1.0 as possible. The importance of this statistic has been discussed earlier, and in the example given in Tables I and II. NIR series (A) had a coefficient of correlation of +0.972. This means that $(0.972)^2 \times 100$ or 94.5% of the variance in the y values (NIR) can be attributed to variation in the x or standard values. The parameter r^2 is referred to as the coefficient of determination. In the NIR (B) series, $r = 0.999$, or 99.8% of the variance in y can be accounted for by variance in x. If the coefficient of correlation is low (for example, 0.65), slope/bias correction will not effect an improvement in analytical accuracy and the reasons for the poor agreement between standard and NIR results must be sought. In many cases, this can be traced to poor sample selection for calibration

TABLE IX
Illustration of RPD and RER[a]

	Kjeldahl Protein		NIR Protein	*d*
	10.0		10.6	0.6
	10.4		11.6	0.6
	11.3		11.7	0.4
	11.9		12.2	0.3
	12.4		12.2	−0.2
	12.9		12.9	0
	13.4		13.2	−0.2
	13.6		13.6	0
	13.9		13.2	−0.7
	14.3		13.6	−0.7
	18.0		16.9	−1.1
Mean	13.07		12.93	−0.14
SD	2.16		1.64	0.56 (SEP)
r		0.992		
RPD (SDx ÷ SEP)		3.9		
RER (Range ÷ SEP)		14.3		

[a]RPD = ratio of SEP to SDx; RER = ratio of SEP to range in x (independent variable).

TABLE X
Different Types of Bias and Slope Differences in Near-Infrared Reflectance Analysis

No.	Low Standard Result	High Standard Result
1	Accurate	High
2	Accurate	Low
3	High	Low
4	Low	High
5	High	Accurate
6	Low	Accurate
7	High	High
8	Low	Low

TABLE VIII
Relationship of Standard Error of Performance to Coefficient of Correlation and SDx[a]

r	SDx	2.5	2.0	1.5	1.0	0.5	50	40	30	20	10	5
0.998	SEP	0.158	0.126	0.095	0.063	0.032	3.16	2.529	1.897	1.265	0.632	0.316
0.995		0.250	0.200	0.150	0.100	0.050	5.00	4.000	3.000	2.000	1.000	0.500
0.99		0.353	0.282	0.212	0.141	0.071	7.06	5.648	4.24	2.82	1.41	0.706
0.98		0.498	0.398	0.299	0.199	0.100	10.00	7.97	5.98	3.98	1.99	1.00
0.95		0.781	0.625	0.469	0.312	0.156	15.60	12.50	9.37	6.25	3.12	1.56
0.90		1.090	0.872	0.654	0.436	0.218	21.80	17.44	13.08	8.72	4.36	2.18
0.85		1.317	1.054	0.791	0.527	0.264	26.40	21.07	15.00	10.54	5.27	2.64
0.80		1.500	1.200	0.900	0.600	0.300	30.00	24.00	18.00	12.00	6.00	3.00
0.75		1.654	1.323	0.992	0.662	0.331	33.10	26.46	19.85	13.23	6.62	3.31
0.70		1.786	1.429	1.072	0.714	0.358	35.72	28.58	21.43	14.29	7.14	3.58

[a]E.g., for $r = 0.98$ (RPD = SDx ÷ SEP) = (2.5 ÷ 0.498) = 5.02, (40 ÷ 7.97) = 5.02, etc.

(poor range in composition and/or excessive variability in physical constitution), poor sample preparation and blending, or inaccuracies in standard data.

To summarize the interpretation of these simple statistics, the SEP indicates the variability in deviations of y from x, and the bias indicates the average amount by which the results differ. The SEP together with the bias pinpoint the accuracy of the test procedure. Ideally, the SEP and bias should both be equal to zero. The coefficient of correlation indicates the closeness of fit between the NIR and standard data over the range of composition. A high correlation coefficient with a low SEP and bias means that the NIR results are accurate over the anticipated range, and likely to remain so, provided these statistics were based on a sufficient number of observations. A high correlation coefficient with a bias at one or both extremes of composition means that these discrepancies are also practically certain to persist, as illustrated by NIR series (B); regression statistics b and a can be used in the equation $y = a + bx$ to correct the slope and bias and to effect improvement in the SEP. A low coefficient of correlation between NIR and standard data means that the NIR analysis has not been successful, and it is possible that NIR spectroscopy may not be applicable to that particular analysis. The coefficient of correlation (r) and the coefficient of determination (r^2) are the most meaningful statistics for "instant" appraisal of analytical efficiency. Provided the coefficients of correlation/determination are high (above 0.9), the accuracy can always be fine-tuned by a slope/bias adjustment, despite an apparently high initial SEP. If the coefficient of correlation is low (0.8 or less), the SEP may be capable of improvement, but it is not possible to obtain really high accuracy unless the coefficient of correlation is high (at least 0.90).

All of these statistics must be interpreted in the light of the regression coefficients, which should be as low as possible for a given instrument or mathematical treatment, and the standard deviation of the original population with regard to the parameter being measured when tested by the standard method—i.e., SDx and the associated CV. As the SDx increases, the coefficient of correlation can also be expected to increase, provided the precision of testing by both standard and NIR methods is satisfactory. Accurate testing becomes more critical to an operation as the variance (SDx) in the population increases. After all, if the standard deviation of the anticipated samples is approaching zero, there will be no need to test more than one sample.

V. SOURCES OF ERROR IN NIR TESTING

At first glance, the NIR technique appears to be relatively free from error sources. It is a simple process, involving grinding and mixing a sample, loading a cell, and reading the result. No weighing is necessary, and there are no reagents to prepare. In practice, however, nearly 40 sources of error have been identified. All of them can affect the results, but the most important thing to remember is that excellent results, with coefficients of variability of from 1 to 1.5, are consistently achieved, and the precision of even the most simple of NIR instruments is comparable to that of the reference method with which it is usually compared. Furthermore, NIR has been successfully applied to the determination of relatively minor constituents, such as methionine, which is present only up to 0.3% in the commodity (field peas) for which its application has been reported (Williams et al, 1985). Error sources can be divided into factors associated with the instruments, those associated with the samples, and those associated with the operator. The most troublesome sources of error are likely to be sample selection for calibration, sampling and sample preparation, wavelength selection, and standard laboratory analysis. The sources of error are summarized in Table XI. Factors marked with an asterisk also affect precision.

This list is not exhaustive, and individual operators may identify other sources of error. The factors are not listed in order of importance, since all of them contribute toward the total error. Individual error sources will be described briefly in the remainder of the chapter. More information on individual sources can be gleaned from the references cited.

A. Factors Associated with the Instrument

INSTRUMENT NOISE

Noise in connection with any type of spectroscopy is defined as any disturbance, especially a randomly distributed (higher or lower than normal) and persistent disturbance that influences the quality and clarity of the electronic signal. The efficiency of an NIR spectrophotometer is markedly affected by the ratio of the electronic signal from the sample to the background signal or noise. Referred to as the signal-to-noise ratio (SNR), this should be as high as possible. There are two main types of

TABLE XI
Error Sources in Near-Infrared Reflectance Testing[a]

Instrument Factors
*Instrument noise
Stray light
Nonlinearity of signal
*Wavelength selection
*Mathematical treatment of raw log (1/R) data
Segment size
*Derivative size
*Smoothing of raw optical signal
*Static electricity
*Instrument temperature control
*Power supply
*Instrument geometry
Cell window
Aging of components
Sample Factors
Chemical composition, including absorbing groups present
Interactions between chemical constituents
Influence of chemical constituents on physical condition
Moisture status of sample before and after grinding
*Bulk density of ground sample
*Physical texture of sample
External factors affecting sample physical nature, such as weather damage (frost, bleach, disease, etc.)
*Sample temperature
*Ambient temperature
Sample type, including variety
Conversion factors
Operational Factors
Calibration practice
Number of samples used in wavelength selection or calibration
*Sample selection with particular reference to distribution of samples with respect to composition
Accuracy of laboratory analysis
*Wavelength selection
*Mathematical treatment
Sample preparation
*Sampling and subsampling procedure
*Grinder type
*Condition of grinder
*Mean particle size
*Particle-size distribution
*Blending after grinding
Chemical or physicochemical analysis (standard analysis)
Sample storage
Before use
After grinding
Sample cell loading
*Packing
*Cleanup
*General carelessness

[a] Asterisks indicate factors that also affect precision.

instrument noise: noise that affects the instrument output on a standard background, such as an optically "flat" standard, and noise caused by interaction between instrument and sample. A certain amount of noise is inevitable, even in the best of smooth-running instruments.

Pure instrumental noise can be caused by several agencies. One of the most important is instrument temperature. Temperature affects wavelength stability, and changes in the temperature of the instrument can cause the optimum wavelength at which the instrument actually takes measurements to differ slightly from the wavelength at which the instrument has been set up to take the measurements. For example, an instrument that has been designed or programmed to take readings at certain optimum wavelengths may, because of changes in ambient or instrument temperature, take its readings 1 or 2 nm above or below the optima. Since the magnitude of this type of change is not constant throughout the NIR area, the wavelengths at which the instrument is taking its measurement may deviate to a greater or lesser extent from the preselected wavelengths. Because the calibration constants are already set into the instrument, this source of error causes a change in the slope of instrument response to the sample. Correction of the slope by the recognized statistical procedure effectively restores the accuracy of the instrument until another temperature change occurs, which may cause the results to revert to their original pattern. This typical occurrence can become very frustrating to the operator who seeks efficiency in accuracy and precision.

Small fluctuations in temperature also affect the output of the lead-sulfide detectors. The lead sulfide (PbS) is deposited as a film, which acts as a resistor. When photons (light energy beams) impinge on the detector, its resistance is changed. The changes in resistance are minute and are amplified before storage and computation into concentration. Factors other than light photons can cause changes in the resistance. These include temperature changes that themselves can cause changes in the resistance of PbS detectors. Any changes in resistance generated by external sources are amplified and recorded as normal changes in the output of the detector and can therefore give rise to spurious signals. In modern NIR bench instruments, the lead sulfide detectors can be fitted with thermoelectric elements to control the temperature as closely as possible by their cooling effect.

Factors that cause the room or benching, or moving parts of the instrument itself, to vibrate can cause or increase instrument noise. These sources are referred to collectively as mechanical noise. Wear of moving parts of the instrument can cause increases in vibration and slight shifts of components relative to each other. Another source of mechanical noise may be, for example, a minute shift in the position of the shaft of a motor. The instrument stabilizes itself to the new circumstances, causing changes both in optimum wavelength and in detector output, with the net effect that all results change. Changes in the slit aperture can also occur as a result of wear and aging, changing the total amount of energy impinging upon the sample and again causing differences in the output.

Changes in temperature and vibrations are intermittent and can affect precision of instrument output both within and between days of operation. Variations resulting from wear and aging of mechanical and electronic components usually effect a gradual change in the signal. This process applies to moving parts such as bearings and to the optical, detection, amplification, and computing components. It is detectable only by the daily analysis or reading of a stable check sample, the instrument ceramic standard, or a reflectance standard such as polystyrene, which has characteristic bands. If both the magnitude of the signal and (in the case of comperterized spectrophotometers) the position of the bands are recorded, changes in wavelength and output can easily be detected. Operators of bench instruments such as the DICKEY-john GAC III, Technicon InfraAlyzer, and Percon Inframatic 8100 can detect the changes by reading the ceramic standard every day. The readings will be taken at the same wavelengths, and small fluctuations in the signal (log values) will indicate the type of influences referred to. Table XII illustrates typical data for a ceramic standard taken over a three-month period for an NIR bench instrument. Throughout the time, the instrument gave consistently good results for protein and moisture in wheat, barley, and durum wheat.

Precautions that can be taken to minimize instrument noise include stabilization of the mounting of the sensitive detectors and amplification components so as to minimize vibration, cooling of the detectors, and use of more than one detector, which tends to average out the influences of noise-inducing agencies.

This type of problem is more common with computerized spectrophotometers. The smaller bench instruments take from 80 to 120 readings from the sample and the standard at only a few selected wavelength points. Spikes of abnormal power tend to be averaged out without causing major discrepancies in the actual electro-optical data used to compute the signal. Computerized spectrophotometers scan 1,000–1,500 nm of wavelength, usually at 2-nm intervals, so that 700 or more wavelength points are represented in each spectrum. They take several seconds to do this, and as a result are rather more vulnerable to spikes or other sources of noise. Most computerized spectrophotometers take a complete scan in about 0.2 sec, and are usually operator-directed to take at least 50 scans per sample. This procedure also tends to average out minor power fluctuations. Figures 1A and 1B illustrate a noisy spectrum of ground wheat caused by a faulty detector and the corresponding spectrum after the fault had been rectified. All of these agencies cause an overall interaction between the components of the instrument, the power supply, and the external and internal factors to give a more or less stable background noise level for the instrument, which is likely to vary slightly from instrument to instrument of the same type. It is the efficiency with which the instrument can differentiate between its own background noise and the significant changes in the signal to the detector induced by the insertion of a sample (the SNR) that is the most important feature of the instrument to the operator.

STRAY LIGHT

Another term familiar to those involved with NIR instruments is *stray light*, which can be a serious source of error. It is defined as light energy that reaches the detector from wavelengths other than those at which the instrument is designated to take readings, or from sources other than the sample. There are three main types of stray light. The first is simply extraneous room light that reaches the detection system of the instrument. It can be caused by incomplete closing of the drawer or a fault in the instrument design that does not exclude room light. Most bench-type NIR instruments enclose the optical system in a "black" chamber to protect the system as far as possible. The second type of stray light is associated mainly with the light source. Ideally, the light or energy source should be a lamp having the energy emissions of its radiation occur mainly in the

TABLE XII
Log Values from Ceramic Standard from DICKEY-john Instrument over Three-Month Period

Date	Log 0	Log 1	Log 2	Log 3	Log 4	Log 5
5 September	150.71	165.09	133.59	95.501	60.791	140.46
17 September	149.97	164.45	132.89	94.781	60.210	140.05
25 September	150.03	164.52	133.00	94.770	60.618	140.18
14 October	146.60	161.02	129.39	90.861	55.490	135.00
18 November	147.09	161.57	130.02	91.609	56.912	135.51
10 December	151.72	166.29	134.77	96.468	61.445	142.23

NIR region. The tungsten filament lamp used in most NIR instruments peaks in energy emission at about 900 nm. It would be better if the peak emission occurred at about 2,000 nm. Other energy sources (lamps) exist that do peak closer to the optimum area, but the energy of their emissions is much lower than the tungsten-halogen lamp. The third type of stray light is associated with grating monochromators. This phenomenon of *overlapping orders* is discussed in Chap. 5.

The monochromatic light that is sought for making a quantitative measurement (i.e., light of a single wavelength) is seldom pure, and usually energy (light) of other wavelengths will intrude. Gratings or interference filters allow the passage of radiations of more than one wavelength. Radiations reaching the detector at wavelengths other than the one at which the measurement is to be made are technically stray light. Sometimes this extraneous energy is well outside the bandpass of the filter or wavelength point at which the measurement is to be made. For example, if the measurement is being made at 2,100 nm, the light source may also be emitting energy at other areas, such as in the 1,200-nm area. The radiation from this latter area that reaches the detector is stray light. It is normally compensated for in instrument design by the use of secondary filters and exclusion of other types of radiation via a prism or other device. Another method of restricting energy emission to certain wavelengths is the use of light-emitting diodes—in this case, near-infrared light-emitting diodes (NIRLEDs). They exist in a wide range of wavelengths, but the emissions that are most readily available are in the range of 800–1,100 nm. At this

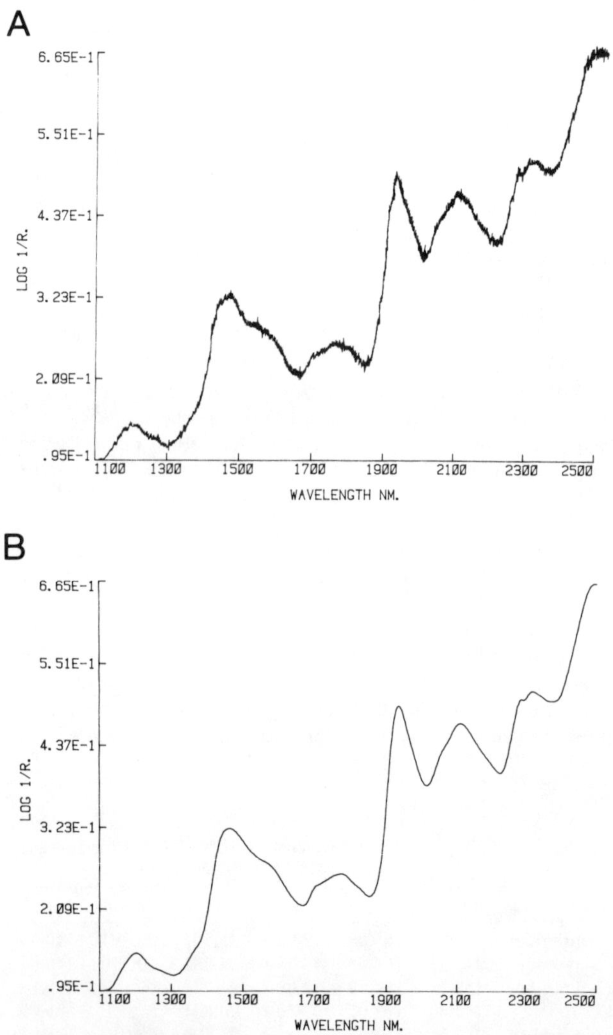

Fig. 1. Influence of detector noise on spectrum of ground wheat before (A) and after (B) correction of fault.

time, NIRLEDs emitting in the range of 1,700–2,300 are not economically feasible for use in a commercial instrument.

NONLINEARITY

The log $(1/R)$ signal does not vary strictly in a linear manner with concentration. In this aspect, NIR differs from classical transmission spectroscopy, where transmittance essentially follows the Beer/Lambert law and the absorbance of energy by a transparent sample is proportional to the concentration of absorbing molecules. The slope of the curve of reflected energy in the NIR may not increase as fast as it should and may tend to level off at higher concentrations of composition. This factor becomes more important with constituents having stronger absorbances, such as moisture and oil, and is less important with constituents such as protein. For example, a protein difference of from 8% to 16% may only result in a change in reflected energy of about 0.002 optical density (OD). On the other hand, a similar change in moisture may result in changes of over 1 OD, depending upon the commodity.

The nonlinearity factor has three aspects. Two are the aspects of basic intensity of the absorbing constituent (for example, the effect is stronger with moisture than with protein) and of the wavelength where the measurement is to be made. At either end of the spectrum, the signal tends to be more nonlinear. The third aspect is the relationship between the detector and the light source. Lead-sulfide detectors peak in efficiency at about 2,500 nm. The tungsten-halogen lamp peaks in efficiency at about 900 nm. These two factors tend to become balanced out, and the system incorporating both peaks in efficiency at about 1,700 nm; theoretically, the best SNR should occur in the 1,700-nm area. In practice, this is not necessarily true, since some designs of computerized spectrophotometers choose to optimize their instruments at other areas of the spectrum (usually toward the higher wavelength) by changing the blaze angle of the grating (see Chap. 5). Interferences caused by moisture and nonlinearity often result in changes in the optimum area for taking measurements, as is apparent to many users of computerized spectrophotometers who find that their instruments select wavelengths for measuring constituents such as oil or protein in the 1,650- to 1,750-nm region, rather than the "traditional" areas of 2,308 or 2,180 nm, respectively.

There are two main groups of nonlinearity sources, one associated with the instrument and the other with the sample. Lead-sulfide detectors and amplification devices are not strictly linear over the range of energy/wavelength experienced in monochromator instruments. This nonlinearity is not a significant source of error, but it does exist. Stray light can cause nonlinearity; for example, if a reading is being taken at a wavelength of 1,940 nm and the instrument is "leaking" light at another wavelength, the stray light will influence the signal reaching the detector. This light will cause a nonlinear response to the variance at 1,940 nm and can constitute a serious source of error. The logarithmic amplifiers used on older instruments were nonlinear, but they are no longer used in modern instruments and therefore are not involved in nonlinearity.

The gain setting on monochromators and other instruments can be responsible for nonlinearity. In terms of an amplifier, *gain* is the ratio of the output to the input. In NIR instruments, the input signal is typically in the form of microvolts and the output in the form of volts. The maximum signal level that an amplifier in present-day NIR instruments can tolerate is about 10 V. Ideally, the instrument gain should be set to provide a signal level of as close to 10 V as possible to maximize efficiency. But the reflectance characteristics of the sample influence the signal reaching the amplifier/detector system. An instrument can be set up to analyze material such as ground wheat with high efficiency. If now the instrument is used for flour analysis, nonlinearity can occur. The signal reaching the detector/amplifier depends on the reflectance of the sample. White flour has much higher reflectance than ground wheat and may saturate the capacity of the detector/amplifier, resulting in serious nonlinearity. Organizations intending to purchase an NIR instrument should inform the instrument manufacturer as to the chief type of material for which they will be using it. For example, if the purchaser wishes to analyze mainly sunflower seeds, the instrument company can maximize the gain to accommodate the dark-colored sunflowers. If a company wishes to analyze mainly ground cereals, the gain can be adjusted accordingly. If an instrument is purchased to analyze one or two similar commodities and then is used to analyze a different type of material, it may be necessary to get a technician from the instrument company to adjust the gain. Saturation of the detector/amplifier signal can occur at any part of the spectrum where the signal from the sample is strong—for example, in the 1,400- to 1,800-nm region.

Interaction between the instrument and the sample can cause nonlinearity. If the sample is completely light absorbing so that theoretically no signal comes off, there will still be a small amount of reflectance from the surface. This phenomenon can cause nonlinearity when absorbance is very high, as with material that varies widely in moisture content. At 1,940 nm, the log (1/*R*) or OD may be as high as 1.2, so that only about 6% of the irradiated energy is reflected. Low-moisture samples may give an OD signal as low as 0.4, with about 40% of the energy being reflected. The "dark-surface" reflectance is usually about 0.5%, but the 0.5% error in 6% reflectance (from the high-moisture sample) has a much more marked influence than the 0.5% error in the 40% reflectance from the low-moisture sample and thus causes nonlinearity, with the effect being most marked at high moisture levels. Differences in particle size can also influence the OD signal very significantly. Generally, the coarser the particles, the higher the OD, so dark-surface reflectance has a more significant effect on nonlinearity than with finer-ground samples. Dark-surface reflectance can be measured by taking readings from substances such as carbon black, and a correction can be applied via the software to eliminate that reflectance as a source of nonlinearity.

Nonlinearity can also be caused by the samples. The amount of change in the reflectance signal (OD) reaching the detector/amplification system is not necessarily related to the differences in concentration of specific constituents. For example, a much greater change in OD occurs from a change of 5–10% in moisture content than of 30–35%, and the OD changes even less for a fluctuation of 75–80% moisture, which is a typical moisture level for meat products. The chief reason is the changes in scatter properties of the sample that are induced by changes in the water content. The refractive index (RI) of water is about 1.3 and that of a commodity like ground wheat is about 1.45. Changes in moisture content change the RI so that at high moisture content it becomes closer to that of water, whereas at low moisture content it is closer to that of wheat. These changes in scatter are caused by both refractive index and particle-size changes. At higher moisture levels, the instrument may be forced to use a completely different wavelength area, such as the 1,450- or 1,150-nm area instead of the more familiar 1,940-nm band, for the determination of moisture. The higher moisture level may cause different bands to be selected for the determination of other constituents, such as oil, protein, or fiber. To a certain extent, software can compensate for many cases of minor deviations from linearity, but some nonlinearity will remain and cause errors in testing, particularly at extremes of concentration of certain constituents.

WAVELENGTH SELECTION

Wavelength selection is one of the most critical aspects of NIR analysis. For the purpose of this monograph, the term *measurement* is used to denote the assessment of the limits and magnitude of a physical factor, such as hardness, or, in the case of NIR spectroscopy, the intensity of energy emitted at definite wavelengths. *Determination* is used to denote the assessment of chemical composition. NIR bench instruments take measurements at previously determined wavelengths and use their interpretation of the intensity of the signal in the determination of composition. This is achieved by multiplying the energy signal by appropriate calibration constants and adding an intercept. Ideally, the wavelengths at which measurements are to be made are optimum for the constituent and the commodity, and the constants, which are regression coefficients, are also optimum. In practice, wavelengths are seldom exactly optimum for all samples to be analyzed, and it is logical that the more wavelengths that are used, the less likely that all wavelengths will be optimum; i.e., the fewer the number of wavelengths used in a calibration, the better.

As a wavelength becomes farther away from the optimum point, the associated regression coefficient computed for it changes and tends to compensate for the wavelength difference. Above a certain deviation from optimum, the regression coefficient no longer compensates for the deviation and the accuracy of the measurement and determination suffer. The regression coefficients may become very big, and since they are used to multiply the reflected signal, a small deviation between the optimum sample/wavelength relationship results in a correspondingly large deviation from the true result. In general, the smaller the regression coefficients the better, and large coefficients tend to be very sample sensitive. Symptoms include an increase in the frequency of large errors, which may be biased in direction, since regression coefficients may be positive or negative depending on the wavelength.

In NIR analysis of several series of samples of the same commodity (e.g., wheat), a computerized spectrophotometer usually selects primary wavelengths that agree closely. Using log (1/*R*) analysis with a single wavelength is not common. The log (1/*R*) algorithm normally requires at least two wavelengths. Instances do occur where a calibration with a single log (1/*R*) wavelength gives satisfactory accuracy with certain sample sets, but such calibrations are usually very sample sensitive and unstable. A single wavelength term often gives satisfactory results with first- or second-derivative mathematics, especially if the samples do not represent a wide range of variances other than in the constituent to be determined. The first wavelength and its associated regression coefficient embrace most of the variance in sample sets of the same commodity, and stable calibrations may be obtained. If only a single wavelength point is used when large variance in extraneous factors is present, the error of testing may be fairly high, since such other factors as variations in growing location, season, and in moisture, particle-size distribution, and other constituents will induce interferences. These can be compensated for by using more than a single wavelength. Wavelength terms subsequent to the first can be regarded as "supporting" terms. The selection of the second, third, and subsequent terms by the same computerized spectrophotometer usually reveals that these vary to a greater extent than the primary wavelength; i.e., they are more sample sensitive, as is reflected in the variability of the regression coefficients. The regression coefficients associated with the most important wavelengths show less variability between sample sets than those of the other wavelengths. Table XIII gives variation in regression coefficients for 20 sample sets of hard red spring wheat at different wavelengths.

The regression coefficients associated with 2,180 nm and its reference wavelength at 2,100 nm showed the least variability. Column 6 illustrates the contribution made at each wavelength to the overall computation of protein. The contribution, which is a combination of the intensity of the signal and the magnitude of the regression coefficient, is calculated from the fraction contributed by each wavelength point of the product of the regression coefficient and the log (1/*R*) values at that wavelength. On the basis of the contribution of each wavelength point to the computation of protein, it is clear that the wavelengths at 2,100 and 2,180 nm were the most important, followed by those at 2,230 and 1,680 nm.

Column 7 of Table XIII, which relates the coefficient of variability to the percentage contribution of each wavelength point, is a combination of high percentage contribution and low variability in the regression coefficient. The data were taken from 20 randomly selected calibrations, each using 40 samples of hard red spring wheat, for protein content. A low value in column 7 indicates a low ratio of variability to percentage contribution (i.e., low variability in regression coefficient between different calibrations, together with a consistently high contribution to the total protein result). Although the wavelengths at 2,100 and 2,180 nm are essentially equivalent and highly significant in the computation of protein, the third most important wavelength is at 1,680 nm. The 1,940- and 2,230-nm wavelengths are less stable, and the "oil" filter at 2,308 nm contributes little to the equation. A similar analysis can be applied to the moisture constants, where the 1,940-nm wavelength is the most significant, followed by those at 2,308 and 2,230 nm. The wavelengths at 2,100, 2,180, and 1,680 nm are of much less significance in the computation of moisture in fixed-filter instruments. When the least significant wavelengths are eliminated from the calibration system, the contribution of the most significant wavelength increases at a greater rate than the others. This type of study is possible only when a large number of individual calibrations with widely different sample sets of the same commodity have been completed to enable the estimation of the variability of each individual regression coefficient.

A similar analysis can be applied to data from a computerized spectrophotometer. The increased computing capability allows the operator to identify the most important wavelengths, then to create calibrations using these to analyze files of additional "unknown" samples. Extensive studies of wavelengths can be completed in a very short time and stable calibrations developed. The precision of reading the same samples is excellent. Table XIV gives the precision of reading 12 samples of hard red spring wheat drawn from two totally different areas in duplicate on a computerized spectrophotometer at six different wavelength points.

Column 5, which relates the mean difference to the mean log (1/*R*), shows that the mean differences between duplicate cell loadings of the same 12 samples of wheat are less than 1% of the mean log (1/*R*) at the six wavelength points studied and that they are randomly distributed with respect to wavelength area.

The software of the computerized spectrophotometer includes the capability of displaying the correlation coefficients for all wavelengths referred to. The array of coefficients across the scanned area for the selection of the first wavelength provides the most useful information because these are simple coefficients of correlation and may be positive or negative. It gives the point of highest correlation, but also all other points of high and low correlation. It also shows the shape of all correlation peaks. Those with broad bands are most likely to provide stable calibration constants with low coefficients of variability (although there is no certainty in this assumption). The second wavelength selected to accompany the first, whether as a second wavelength or as denominator, often coincides with other bands of high correlation. This procedure should not be confused with the "best-pair" program used by some to select the best two wavelengths to accompany each other; the wavelengths are selected two at a time and not singly. A "best-triple" program is also available.

The multiple correlation bands displayed after the selection of the second and subsequent wavelengths become progressively sharper, illustrating why tertiary and other wavelengths are often more sample sensitive than the primary wavelength. There are three ways of improving the stability of calibrations. The first is by using very large numbers of samples incorporating all sources of variance in the calibration (the concept of universal constants), and the second is by using *division* or *quotient* mathematics. Correlation plots for single numerator/denominator (quotient) wavelength points usually show several broad areas of high correlation, most of which are capable of acceptable stable calibrations. The concept of quotient mathematics was developed by Norris and Williams (1984) and has been discussed in earlier sections of this monograph. The third method is that of basing calibrations on wavelengths identified by difference spectra. It is discussed in more detail in Chap. 6.

The chief difficulties in the development of universal constants are the accumulation of sufficient samples with the necessary variance in physical characteristics and chemical composition and the chemical analysis that is necessary. Protein, moisture, and hardness are relatively rapid tests to perform, and approved standard methods are available. When the field is extended to oil and particularly fiber components or amino acids, difficulties occur not only with the expense and time involved with the analysis but also with deciding which chemical or physicochemical method to use. Table XV gives recommended standard procedures for the most commonly needed parameters.

The cost of chemical testing may act as a deterrent to the development of universal constants, but it is a relatively low price to pay for stability in ongoing analysis. Several attempts have been made to circumvent the chemical analysis by selecting calibration samples on the basis of spectral

TABLE XIII
Variability of Regression Coefficients for Protein at Different Wavelengths (n = 20)[a]

Wavelength (nm)	Regression Coefficient	CV[b]	% Contribution to Protein Computation	Ratio CV to % Contribution
2,308	0.0142	2,204	3.2	689
2,230	−0.1480	391	8.1	48
2,180[c]	1.3304	35	43.8	0.8
2,100[c]	−1.0860	25	35.9	0.7
1,940	0.0308	105	1.8	58
1,680	−0.2072	127	4.6	28

[a]Data taken from DICKEY-john GAC III.
[b]Coefficient of variability.
[c]Most important wavelengths.

TABLE XIV
Precision of Duplicate Log (1/*R*) Readings at Different Wavelengths

Wavelength (nm)	Mean Log (1/*R*) of Duplicates[a]	Mean Difference Between Duplicates	CV[b]	(Mean Difference × 100) / Mean Log (1/*R*)
2,316	0.5104	0.0043	1.14	0.84
2,228	0.3984	0.0032	1.31	0.80
2,180	0.4332	0.0032	1.31	0.74
2,102	0.4843	0.0031	1.30	0.64
1,934	0.4695	0.0023	1.47	0.49
1,675	0.2351	0.0020	1.50	0.85

[a]Duplicate cell loadings.
[b]CV = (standard deviation of difference between duplicates × 100) ÷ mean log (1/*R*).

characteristics only, the theory being that sample sets can be selected with sufficient variance to provide a stable calibration. Chemical analysis would then be applied to the relatively small number of samples (20–30). There are three main criticisms of this approach. First, variance in the optical signal is not necessarily associated with variation in chemical composition, and it is just as likely to be caused by differences in particle size and shape. For example, a large difference of protein content in wheat may be associated with quite variable differences in optical signal. Conversely, large differences in spectral data are not necessarily associated with differences in chemical composition. This observation is illustrated by the data of Tables XVI and XVII. Table XVI summarizes data for three pairs of samples used on two different computerized spectrophotometers, one at the Grain Research Laboratory, Winnipeg, the other at the All Union Institute for Agrochemical Services at Moscow, USSR (Williams and Krischenko, 1986). Two of the pairs differed in protein by over 5%. The differences in log $(1/R)$ at 2,180 nm, a widely used wavelength for protein determination, were not highly correlated to protein on either instrument.

The instrument/sample interaction was clearly apparent, in that the difference in log $(1/R)$ for the second pair of samples (B) was over 10 times that of the first pair (A) despite an 18% reduction in the difference in protein. Also, the difference in log $(1/R)$ for the third pair (C) differed widely between instruments, unlike the first two pairs.

Table XVII summarizes the relationships between log $(1/R)$ values and protein content at 12 wavelengths for 12 samples of hard red spring wheat. The average coefficient of correlation between protein content and log $(1/R)$ for the 12 wavelength points was only 0.35, for both the Winnipeg and Moscow instruments. At the widely used log $(1/R)$ point of 2,180 nm, the coefficients of correlation were respectively 0.52 and 0.53, which were the highest for the wavelength points studied.

The second criticism of sample selection on the basis of spectral characteristics is that in order to select a set of 20 or 30 samples with maximum variance in optical signal and chemical composition, it may be necessary to assemble and scan a large number of samples so that the actual saving in time and trouble may not be appreciable. A third factor is that differences between samples in composition of constituents, such as fiber components, oil, protein, and most amino acids, or simple sugars, are not necessarily associated with the same differences in optical characteristics.

MATHEMATICAL TREATMENT OF THE log $(1/R)$ SIGNAL

Several NIR instruments have been used successfully for the analysis of different commodities for protein, moisture, and other constituents using the log $(1/R)$ signal directly after amplification but without further mathematical treatment. Mathematical treatment of optical signals was first suggested by Morrey (1968) as a method for correction of band overlap before analysis. Treatments were used in NIR instruments to resolve band overlap and baseline drift. The most successful mathematical treatments used to date are the smoothed log $(1/R)$ itself; the first or second derivative of the log $(1/R)$; the delta log $(1/R)$, as used in the early Neotec/Pacific Scientific instruments; the quotient version of the first or second derivative (Norris and Williams, 1984); and the Kubelka/Munk equation. All of these have been discussed in earlier chapters. Other treatments, such as curve-fitting (Hruschka and Norris, 1982) and the full-spectrum procedures, also described earlier, have been used in computerized spectrophotometers with some success; in general, however, the raw log $(1/R)$ and the first or second derivative of it in standard or quotient form have proved to be most practicable to use in bench-type instruments and, as well, the most consistent in performance. Bands that appear as peaks or shoulders in the log $(1/R)$ spectrum occur as crossovers from positive to negative values (zero points) in a first-derivative spectrum and as negative ("valleys") but much sharper points in a second-derivative spectrum (Fig. 2).

SEGMENT SIZE

The term *segment size* was introduced by Pacific Scientific to describe the degree of smoothing for use with the computerized spectrophotometer. Segment size refers to n (see section below on Smoothing of Raw Optical Signal), the number of data points involved in the smoothing. Recent research (Williams et al, unpubl. data, 1986) has verified that there was no

TABLE XV
Recommended Procedures for Standard Analysis

Parameter	Method	Reference
Protein	Kjeldahl	AACC Method 46-12
	Kjelfoss	Williams and MacEachern, 1981
	KjelTec	Mindel, 1986
Moisture	Single-stage air oven	AACC Method 44-15A
Oil	Goldfisch extraction with anhydrous ether	AACC Method 30-20
Crude fiber	Extraction	AACC Method 32-10
Acid detergent fiber	Extraction	Van Soest, 1963
Neutral detergent fiber	Extraction	AACC Method 32-20
Lignin	Extraction	Van Soest and Wine, 1968
In vitro digestibility	Enzymic (pepsin/cellulase/amyloglucosidase)	Jones and Hayward, 1975
Lysine	Ion-exchange chromatography	Spackman et al, 1958
Methionine	Gas-liquid chromatography	Mackenzie, 1977

TABLE XVI
Log $(1/R)$ (2,180 nm) Values for Paired Wheat Samples Differing Widely in Protein

	(A)			(B)			(C)		
	High	Low	Difference	High	Low	Difference	High	Low	Difference
Protein	17.2%	10.4%	−6.8%	16.3%	10.7%	−5.6%	15.5%	10.6%	−4.9%
Winnipeg 6350	0.41844	0.41538	−0.00306	0.42577	0.38542	−0.04035	0.39509	0.38876	−0.00633
Moscow 6350	0.45500	0.45116	−0.00384	0.44601	0.40466	−0.04135	0.43031	0.31825	−0.01206

significant improvement in analytical efficiency beyond a value of 6 for *n* for log (1/*R*), or the first or second derivative thereof.

DERIVATIVE SIZE

In derivative mathematics, the size of the derivative or gap is an important factor. There is also an optimum size for a derivative (Williams et al, unpubl. data, 1986). If the derivative is too small, the algorithm may not minimize system noise efficiently, since the derivatized signal may be sensitive to interactions. If it is too large, the sensitivity will be "damped out" and the calibration will be lower in analytical efficiency because of its lower sensitivity to variations in composition. The size of the derivative is also related to the degree of resolution of the instrument and the smoothing procedure. For instruments with very high and efficient resolution, smaller derivatives are more effective because they are more sensitive to changes in composition. In working with computerized spectrophotometers, both the derivative size (gap) and the segment of wavelength that comprises each wavelength point need to be optimized. The accuracy of analysis is affected less by the segment size than by the derivative size (Williams et al, unpubl. data, 1987). For instruments with lower resolution power, larger derivatives are generally more effective. The influence of derivative size on the form of the correlation plot is illustrated in Fig. 3.

Table XVIII illustrates the effects of changing segment and derivative size with a computerized spectrophotometer. The test substance was hard red spring wheat. Forty-eight samples were used for calibration and 20 for analysis. The accuracy of analysis tended to improve with increased smoothing (segment size) for both first and second derivatives, although there was no significant difference between segment sizes of 4 and 10 points. The optimum derivative or gap sizes were respectively 15 and 10 points (about 30 and 20 nm).

The Pacific Scientific Research Composition Analyzer 6250 records optical data at intervals of approximately 2 nm across a spectrum of 1,100–2,500 nm, and the spectrum is divided into about 700 points. If a segment for a derivative is composed of 5 points, this means that each wavelength point used in the algorithm consists of about 10 nm starting at the wavelength point selected. If the algorithm is log (1/*R*) and the wavelength point selected is 2,180 nm, the actual signal regressed against protein for the chemical data is a combination of the signals recorded at 2,180–2,188 nm. Similarly, if the first derivative is used, with segments of 5 points and a derivative size or gap of 20 points, for a selected wavelength of 2,150 nm the two wavelength points use the signals recorded from 2,150 to 2,158 and from 2,190 to 2,198 nm. This factor is important when a tilting filter is used because, depending on the derivative size, a given starting wavelength point may result in the end of the derivative being past the end of the filter. If this situation develops, which it may do in connection (for example) with the Pacific Scientific Feed Quality Analyzer 51A, good analytical results can still be obtained by studying the multiple correlations and selecting wavelengths (pulse points) corresponding to the highest correlation that will avoid crossing from one filter to the next. The exercise is a simple one if the instrument is interfaced to a terminal.

TABLE XVII
Coefficients of Correlation Between Log (1/*R*) and Protein Content at 12 Wavelength Points for 12 Samples of Hard Red Spring Wheat

Wavelength (nm)	Winnipeg 6350 *r*	Moscow 6350 *r*
1,162	0.40	0.30
1,240	0.29	0.24
1,392	0.20	0.20
1,496	0.24	0.26
1,680	0.31	0.25
1,940	0.23	0.31
2,100	0.29	0.32
2,180	0.52	0.53
2,230	0.52	0.52
2,308	0.43	0.45
2,346	0.43	0.45
2,450	0.33	0.31
Mean	0.35	0.35

SMOOTHING OF RAW OPTICAL SIGNAL

Smoothing is a means of reducing instrument/sample noise. The simplest method, and one that is widely practiced by

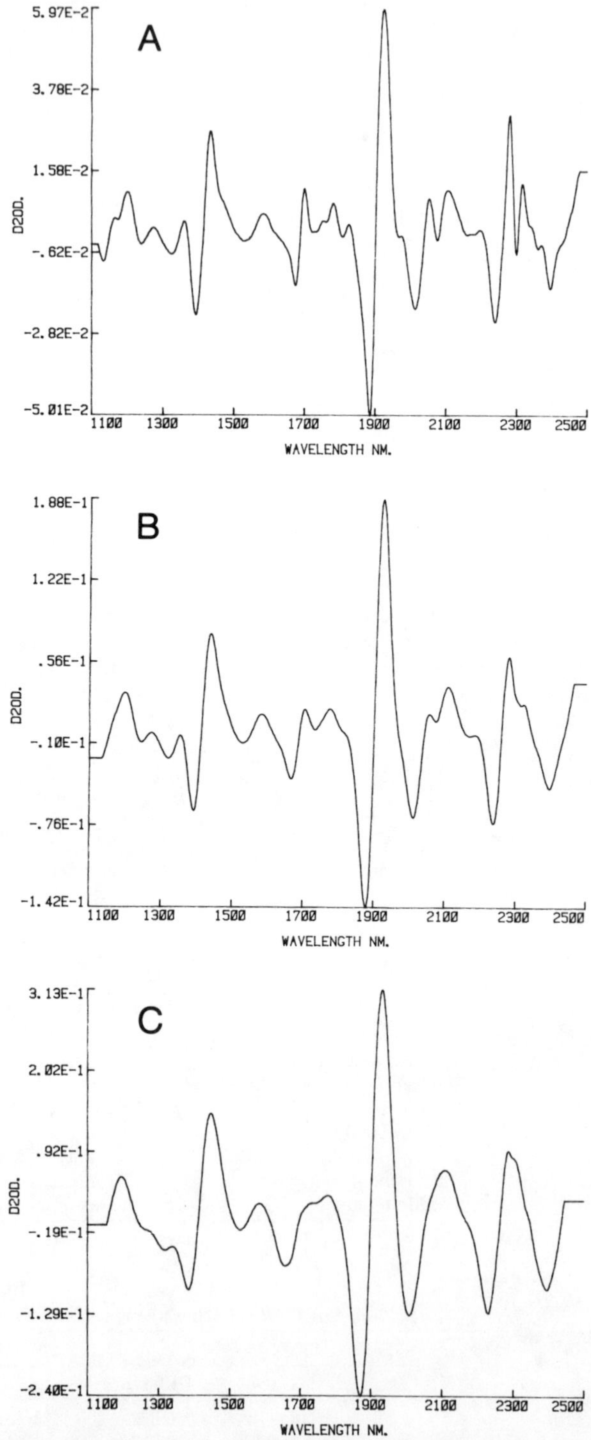

Fig. 2. Influence of derivatization (A, 20 nm; B, 50 nm; C, 90 nm) on band appearance in near-infrared region.

computerized spectrophotometer users, is the "boxcar" method, which involves summing data from a certain number of wavelength points (n) and then dividing by the number of wavelength points. The optical data for the first wavelength point are then replaced with the mean of the optical data of the number of points used in the smoothing. The second point is treated in the same way, until the signals recorded at every wavelength point in the spectrum have been replaced with the means of the set number of points. The last few wavelength points remain unsmoothed. They are at both ends of the spectrum (around 1,100 and 2,500 nm), a region where wavelengths are hardly ever used in NIR analysis. The other common methods of smoothing include the Fourier transform and Savitzky and Golay (1964) methods, which have been described in Chap. 3. Commercial bench-type instruments use another method of smoothing. They achieve essentially the same effect by taking multiple readings of both sample and standard, then dividing the total summed signals by N, the number of readings. The divided or mean readings for standard and sample are used in computation of composition. About 40 readings are enough to give a stable result. The Pacific Scientific instruments usually take 84 readings per sample, and the DICKEY-john, Technicon, and Percon instruments take 120. This method increases the time per test slightly because of the longer reading time.

STATIC ELECTRICITY

Static electricity is an accumulation of electrical charge. It is particularly apparent in areas where the relative humidity is low, such as western regions of the United States and Canada, Australia, the Middle East, the Indian subcontinent, Africa, and parts of South America—i.e., in most of the areas where cereals other than rice are grown. The buildup of charge is affected by several agencies, including relative humidity of the air and fibrous materials such as nylon and wool that are capable of conducting charges; these materials may be present in the clothing of the operators, in carpeting, and so forth. The most noticeable influence of static electricity in NIR analysis is its effect on the orientation of particles of ground material in sample cells. Cells with quartz rather than glass covers are more susceptible to static electricity. Particles of dust or sample are attracted and become closely associated with the surface, making it difficult to keep clean and allowing sample-to-sample contamination. Also, the orientation of the sample at the surface during sample loading contributes to sample/instrument noise. Static electricity also affects buildup of dust particles on the quartz windows located on the inside of NIR instruments to protect the detectors from the environment. It is essential to keep these clean, since dust accumulation affects the accuracy of analysis. Ideally, the windows should be cleaned by gentle dusting with optical tissues, since brushing may impart enough charge to increase the static effect on dust buildup. The shape and composition of particles affect the accumulation of static electricity. Long fibrous particles tend to become charged more readily than rounder or cubic particles. Particles of substances such as powdered straw may become strongly orientated at cell surfaces. Orientation of particles at the sample/instrument interface affects the intensity and pattern of the diffuse-reflectance signal and, as a result, the precision and accuracy of NIR analysis.

INSTRUMENT TEMPERATURE CONTROL

Control of instrument temperature is essential because fluctuations affect the performance of detectors and other components. Faulty temperature control results in gradual increases in positive or negative drift as the temperature of the instrument fluctuates. Control is achieved by factory installation of thermoelectric elements during instrument assembly. These stabilize the temperature of the instrument over a small range. NIR instruments usually operate efficiently at an ambient temperature of about 25° C (77° F), with a maximum of about 40° C (104° F), a temperature that is rarely attained in buildings where testing is carried out. Cooling in most NIR instruments is achieved by internal fans. If these stop working for any reason, results can become erratic as a result of temperature fluctuation and overheating of some components. This problem can occur over fairly short periods of time and will be revealed if regular precision tests are carried out. On the other hand, results may also be affected if ambient temperature

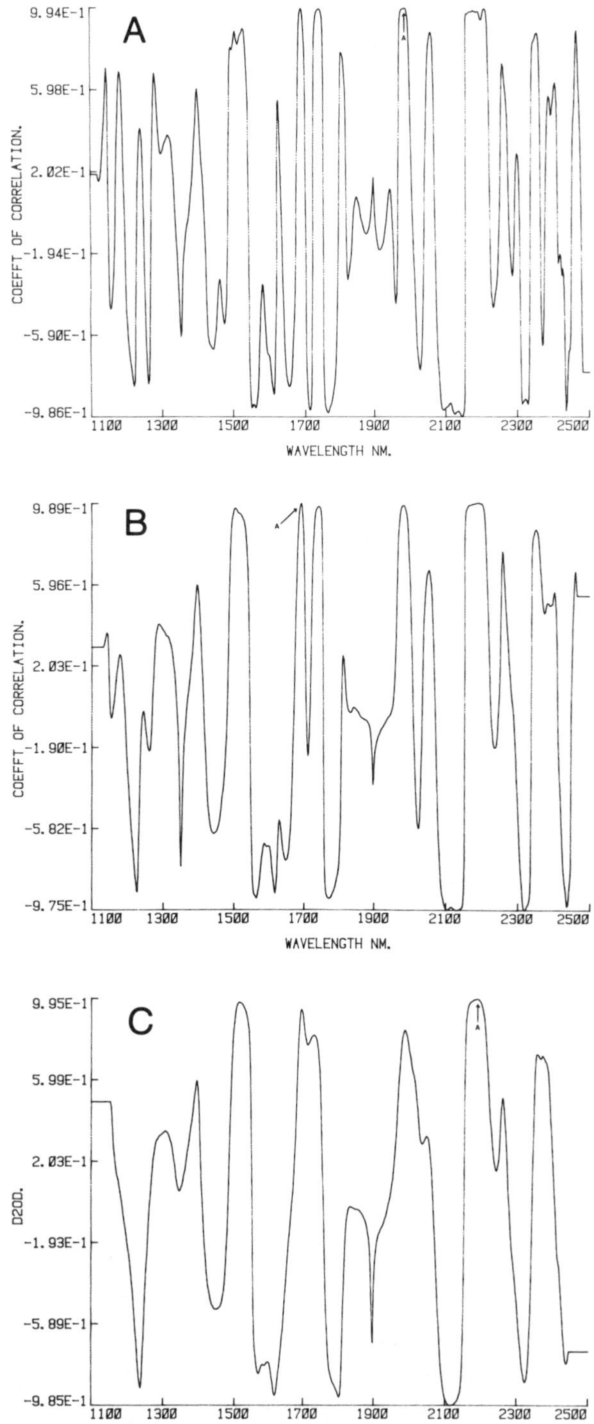

Fig. 3. Influence of derivative size on correlation plot for wheat protein determination (A = primary wavelength selected; all D2OD): (A) second derivative, segment 10 nm, derivative (gap) 20 nm; (B) second derivative, segment 10 nm, derivative (gap) 50 nm; (C) second derivative, segment 10 nm, derivative (gap) 90 nm.

falls too low. Generally speaking, if the atmosphere is comfortable for the operator, it is also suitable for the instrument.

POWER SUPPLY

Electricity is supplied to the instrument in the form of pulses (alternate current) that are usually delivered at 60 cycles per second for 110- to 115-V supplies and 50 cycles per second for the 220- to 240-V service typically supplied in Europe, Africa, Asia, and Australia. Voltage is electrical pressure, and fluctuations in voltage mean that the instrument still receives its power as 50 or 60 cycles per second but that the actual amount of power, or amperage, varies. Voltage fluctuations of ±5% are common and may vary during the time of day or, on a more local basis, when the power supply to a building is in high demand for short periods. This occurrence is common at large grain terminals when conveyers and automatic dumpers are in intermittent use. In some rural areas and in developing countries, "brownouts" may occur, periods of low and insufficient power caused by heavy overload and partial failure in supply. They can cause significant malfunction of NIR instruments.

The occurrence of spikes of high or low electrical intensity constitutes a potent source of noise. They are distinct from periodic variations in line voltage, which are an everyday occurrence in many areas. Spikes can have many causes, including faulty relays on refrigerators and other appliances, vacuum cleaners, grinders, car induction coils, and other equipment. Laboratory gas chromatographs have ovens that are activated by relays. Faulty brushes on electrical motors can cause arcing, another potential source of spikes. Radio frequency noise can also affect amplifiers and detectors. This type of noise is sometimes referred to as inductive noise. Inefficient grounding can cause ground loops, wherein stray current seeking a ground cannot find adequate egress and disperses in different directions, causing sudden changes in the supply. If any such sources of spikes or interruptions in the normal power supply occur adjacent to an NIR instrument, or even in the same building when there is a common source of power, results can be spurious. The interruption in the power supply may last a fraction of a second or for several seconds. In that time, an instrument may take several readings from a sample and the final result can be biased by 10% or more. A line filter can assist in providing a constant source of power to the instrument.

Modern NIR instruments are designed to tolerate voltage fluctuations of up to ±10% of line supply. Building power supplies are generally stable to about ±5%. Power fluctuations can cause differences in the speed of rotation of motors and associated components, such as filter wheels. Filter wheels commonly rotate at about 6 revolutions per second or higher, and power fluctuations can change the number of readings taken in unit time, in turn affecting the precision of measurements. Commercial bench-type NIR instruments normally take 80–120 readings per sample, and a small reduction in the number of readings does not have a marked influence on the results. Total power failure may cause an instrument to lose its calibration constants and can also affect performance of instruments that require fairly extensive warm-up time. Most NIR instruments require about 1–2 hr of initial warm-up time after a deenergized period of 2 hr or longer.

INSTRUMENT GEOMETRY

The precision engineering of NIR instruments results in very accurate and finite distances between the energy source (lamp) and the sample surface. The computation of composition from the optical signal assumes that this distance is constant, and minute changes in the distance can cause differences in the magnitude of the signal reaching the detectors from the sample relative to that from the standard. Mirrors are often interposed between energy source and sample. These all form part of the geometry, and the more complicated the pathway of light from source to sample, the more critical do small differences in the distances become. Minute shifts in the position of a mirror may cause partial deflection of the signal. Changes in path length, the distance from sample to detector, can be caused by small amounts of sample preventing the cell from resting completely flat on the sample cell carrier. This source of error, which can affect accuracy and precision, can be avoided by keeping the cell carrier clean. Such changes can also be caused by minute alterations in the positions or angles of mirrors, collimating lenses, and other ancillary devices that affect the light path. This problem is particularly important with instruments that operate in the double-beam mode.

CELL COVERS (WINDOWS)

The sample cells of most NIR instruments interpose an optical glass or quartz window between the sample and the detector to ensure uniformity of the surface. The thickness and refractive index of the window can affect the accuracy of testing because they are rarely completely planar or of uniform thickness. Often a change of window, caused by breakage, can cause a linear offset in analytical results. It is particularly important to determine consistency of results between sample cells when more than one cell is to be used in high-throughput operations. It is also important to maintain clean quartz or glass cell windows, especially when analyzing oilseeds or other commodities having more than 8% oil. A case could be made for

TABLE XVIII
Influence of Segment and Derivative (Gap) Size on Accuracy of Determination of Protein in Wheat Using Pacific Scientific Research Composition Analyzer 6250[a]

	Segment									
	2[b]		4		6		8		10	
	SEP	$\bar{d}$	SEP	$\bar{d}$	SEP	$\bar{d}$	SEP	$\bar{d}$	SEP	$\bar{d}$
D10D	0.20	0.11	0.16	0.07	0.16	0.08	0.14	0.07	0.14	0.07
D20D	0.29	0.46	0.22	0.13	0.21	0.13	0.20	0.14	0.20	0.16

	Derivative (Gap)													
	2[b]		5		10		15		25		35		45	
	SEP	$\bar{d}$	SEP	$\bar{d}$	SEP	$\bar{d}$	SEP	$\bar{d}$	SEP	$\bar{d}$	SEP	$\bar{d}$	SEP	$\bar{d}$
D10D	0.20	0.04	0.19	0.04	0.16	0.05	0.14	0.04	0.15	0.08	0.16	0.24	0.16	0.19
D20D	0.29	0.13	0.21	0.12	0.19	0.22	0.27	0.11	0.24	0.15	0.21	0.56	0.21	0.25

[a] Figures represent mean results, e.g., for D10D segment size, SEP of 0.16 and $\bar{d}$ of 0.08 are mean values for SEP with segment size of 6 points summed and derivative sizes of 2–45 points summed. For D20D derivative size, SEP of 0.21 and $\bar{d}$ of 0.12 are mean values of SEP and $\bar{d}$ for derivative size of 5 points and segments of 2–10 points, etc.

[b] Number of wavelength points.

an open cell with no windows, in that such a sample-presentation system would theoretically be less susceptible to regular reflectance. The changes in the microorientation and degree of hydration of particles at the immediate surface caused by evaporation are likely to incur errors that would tend to counterbalance the benefits of an open reflectance surface.

B. Factors Associated with the Sample

Most sources of error associated with the sample concern particle size and shape and bulk density, all of which are a function of the relationship between the texture of the commodity and the grinding procedure. The texture of plants and of commodities derived from them is affected by over 20 factors, as summarized in Table XIX.

CHEMICAL COMPOSITION

The chemical composition includes the constituents of the commodity and the absorbing groups present in them, any of which may interact with each other. The constituents in agricultural and food commodities that are most likely to cause interactions are moisture, protein, oil, and fibrous carbohydrates. Although a constituent may be present in small concentrations, it is important to remember that the irradiating light energy penetrates the surface of a ground sample to a depth of about 1 mm, and in doing so it passes through particles, cells, and molecules. To the irradiating light photons, the sample consists not of particles but mainly of space. In that area, of 1 mm by about 2 cm, there is approximately 0.5–2 g of sample, depending on the bulk density. By Avogadro's Law, there are 6.025×10^{23} molecules in one gram molecule, so in 1 g of ground wheat with 12% moisture there are about 4×10^{21} molecules of water and (assuming a very arbitrary value of 1,000,000 for the molecular weight of wheat protein) about 7.8×10^{16} molecules of protein—plenty of molecules to cause absorbances, but proportionately more of those of smaller molecular weight. Since most of the molecules in plant- and animal-derived materials possess dipoles of varying strength, there are many sources of interference in NIR measurement.

The degree of interaction depends on imbalances of composition and the intensity of absorbances. In such commodities as feathermeal, fishmeal, meatmeal, and blood meal, protein is the major constituent; in most other commodities, however, starch, fibrous carbohydrates, oil, and moisture are the most important causes of interactions. In fresh foods such as meat, fish, vegetables, and fruit, water is the most important factor. There are three main types of interaction. The first is the influence of one or more constituents on changing the optimum wavelength for determination of another. The second is an effect of the chemical constituents on the physical nature of the commodity and on the pattern of reduction during grinding. Moisture, oil, fiber, and protein can all affect comminution. Under some circumstances, mineral content can also cause significant interactions. For example, plant species such as *Atriplex* (saltbush) may accumulate over 20% mineral constituents. This level affects comminution and may also cause interferences. The third type of interaction is the effect of moisture alone on the hydration status of all other constituents.

Interactions Between Chemical Constituents. The most common interactions occurring in food, feed, and forage commodities include those between oil and protein, starch and protein, fiber and protein, oil and fiber, and starch and fiber. Very few commodities have both oil and starch present in high concentration. The most likely influence of an interaction is that the presence of the constituent that has either the highest concentration, the strongest absorber, or the combination of both of these tends to mask the absorbers of the constituents present in lower concentrations or with weaker absorbers in the area where its own strongest absorbers occur. As a result, the instrument becomes less sensitive to variations in concentration of the "lesser" constituent. In instruments with discrete filters, the accuracy of analysis may be reduced because the wavelengths are not optimum. In a computerized spectrophotometer this often results in the selection of wavelength points in alternative areas of the NIR for determination of the lesser constituent.

Computerized spectrophotometers may select wavelengths that correspond to higher overtones of the original wavelength. For example, an absorber caused by a second overtone of a $-CH_2-$ stretching frequency at 2,275 nm, which may be optimum for a fiber determination in a commodity low in oil, may be ignored by a computerized spectrophotometer in favor of a higher overtone of the same absorber in the 1,700- to 1,720-nm area if the search for "fiber" wavelengths is conducted in a commodity high in oil. This type of behavior by a computerized spectrophotometer serves as a pointer to the sample/instrument relationship that exists between the samples, the constituent, and the instrument and can be useful in explaining random or biased errors. When a constituent such as protein is determined at a wavelength that does not correspond to a major absorber (e.g., the 2,150- to 2,160-nm area for determination of protein using first-derivative mathematics), the wavelength may be associated with an area of minimum interference by other constituents, rather than with the absorber of the constituent being determined. The determination of oil, protein, and moisture in many commodities has been well researched and the optimum wavelengths and mathematical treatments established. Table XX illustrates wavelengths that have been used successfully for the analysis of some common constituents. For expansion of the technique to new commodities and constituents, flexibility in wavelength and mathematical treatments is essential.

Influence of Chemical Constituents on Physical Condition. Oil, moisture, fiber, protein, and mineral matter may all affect the physical state of a commodity. Most commodities must be reduced to a smaller and more uniform particle size before analysis by NIR. In general, the easiest commodities to reduce to a powder are starchy, low to medium (8–12%) in protein, and low in moisture and oil. This observation is apparent in the ease with which wheat is reduced to fine particles, particularly wheat of low (8–10%) moisture content. High-protein wheat reduces in particle size in a different manner than lower-protein wheat because of a type of binding effect exerted by the protein.

TABLE XIX
Factors Affecting Physical Texture of Agricultural and Food Commodities

Chemical composition, especially fiber and oil content
Moisture content
Plant species, particularly those making up forage
Plant variety
Stage of maturity, particularly of forages, grasses, and straws
Lignification
Growing environment, particularly climate during growing season
Soil fertility
Weather damage during harvest period
Frost during maturation period, especially of cereal grains
Storage conditions
Length of time
Temperature
Atmospheric conditions
Type of container
Moisture content of material at beginning of storage
Fungal disease
Insect damage
Processing practice, e.g., heat treatment or freezing
Drying conditions
Temperature of sample that is affected by:
Ambient temperature
Temperature of grinder, which is affected by number of samples ground in sequence, time between samples, and construction of grinder

Wheats of a certain inherent texture, those that are medium hard to hard, may become softer and easier to grind at higher protein levels. Softer wheats, durum wheats, and common wheats of very hard texture tend to become even more resistant to grinding at higher protein levels.

Oil acts as a pasting agent, and when a commodity high in oil is pulverized it tends to form a paste rather than a powder. Oil contents up to 5% have relatively little influence, but when the level reaches 7–8% the effect of the oil becomes noticeable and the resultant ground-up grains and seeds are no longer free flowing. The degree of agglomeration increases directly as the oil content increases until the oil content reaches 20–30%, when few discrete particles exist. At oil contents of 45% and higher, the comminuted seeds may take on the texture of a paste rather than a powder, depending on the fiber content. The manner of comminution of the commodity affects the accuracy and precision of analysis, since the packing of sample cells is affected. The physical nature of the surface, and hence the diffuse reflectance from it, are strongly influenced by oil content. Accuracy of analysis of commodities high in oil is also affected by the extrusion of oil during grinding and at the quartz or glass windows of sample cells. These windows must be carefully cleaned with alcohol and then with water and thoroughly dried between samples. When ground-up oilseed samples are stored, even under refrigeration, they tend to agglomerate and the particle characteristics change (J. A. Panford, unpubl. data, 1985). Reanalysis may result in large discrepancies. It is recommended that oilseeds, or any type of seed with oil content above 10%, be analyzed within a day or two of grinding for the most precise results. Whole-seed residues may be stored and reground if further NIR analysis is necessary.

Fibrous constituents may cause significant changes in the pattern of grinding and bulk density of a commodity. The most fibrous substances are mature straws, particularly of crops such as faba beans, cotton, flax, and even wheat, particularly the part of the straw nearest the ground. These materials are relatively easy to reduce to finer particles, but the particles tend to be long and thin and quite different from those of ground grains and seeds. Fibrous commodities have low bulk densities and generally form open surfaces with many deep cavities. Consequently, the amount and pattern of diffuse reflectance from such surfaces are quite different from those from the surfaces of ground grains. The orientation of light fibrous particles is affected to a greater degree by static electricity at the sample surface than is that of the denser, more uniformly shaped particles of ground grains. Seeds that are relatively high in fiber, such as oats and barley, are more difficult to analyze with precision than those lower in fiber, and the SEPs for the analysis of fibrous seeds are higher than those for wheat, for example. With a seed such as oats, the presence of high fiber, starch, and up to 5–6% oil causes particular difficulties in grinding and cell packing. Food legumes such as the lupin, which may approach 10% oil, may also cause difficulties in grinding, mixing, and cell loading. Wheat flour, particularly soft wheat flour, and pulses low in oil content such as peas, faba bean, and lentil have high bulk densities and very fine particle sizes and pack densely in the sample cell.

In NIR instruments, the round area viewed by the detectors is usually about 2–2.5 cm in diameter. Consequently, the actual surface area of ground material contributing signal toward the analysis may be quite large. One gram of wheat flour is believed to have a surface area of between 1.5–4 m^2. Because of the differences in physical nature of ground commodities caused by chemical constitution, very wide differences in effective surface area exist between commodities, affecting the path length and the pattern of diffuse reflectance from the sample and thus both the accuracy and precision of analysis. Very fibrous samples, particularly samples with a high degree of agglomeration caused by high oil content, have the smallest effective surface area. Ground seeds low in oil, such as wheat and peas, have the highest surface area and generally give good results by NIR spectroscopy.

Interactions between constituents that are highly correlated to each other can result in measurements of a constituent being highly dependent on variations in the composition of another. This dependence can occur to the extent that the wavelengths selected for analysis of the first constituent may have been selected on the basis of the second constituent. The operator may believe that constituent *A* is being determined, whereas it is actually constituent *B* that is responsible for the measurement. This problem is not critical unless the concentration of both needs to be determined. It becomes more important in the case of constituents such as amino acids, when determinations may be reflecting variations in protein rather than the actual amino acids. Wavelengths selected will assist the operator in determining whether the wavelengths conform to the required constituent or to an associated constituent, such as protein. Other examples are the NIR determination of damaged starch in flour and hot-water extract by direct determination on ground barley. Both of these parameters are likely to be highly influenced by variations in grain hardness and consequently particle size.

Moisture Status of the Sample. The moisture status of the sample affects the accuracy of analysis before and after grinding. Samples such as meat or fresh fruit, which are very high in moisture, are impossible to grind and are best presented to the instrument in the original form, as a minced sample or a slurry. Samples that are merely fairly high in moisture (14–20%) are difficult to grind. The term *uniform particle size and shape*

TABLE XX
Wavelengths That Have Been Used Successfully for Analysis of Some Common Constituents

Constituent	Commodity	Wavelength[a] (nm)	Algorithm
Protein	Wheat	2,180	log (1/*R*)
	Wheat, others	1,680	log (1/*R*)
		2,310	log (1/*R*)
	Barley	2,180	log (1/*R*)
	Wheat	2,150	D1OD
	Forages	2,164	D2OD
	Forages	2,100	log (1/*R*)
	Faba beans	2,310[a]	log (1/*R*)
	Rapeseed	2,156	D1OD
	Soybean	2,060	log (1/*R*)
	Pork	2,226/2,470	D1OD
	Fish	1,684	log (1/*R*)
Moisture	Wheat	1,940	log (1/*R*)
	Wheat, forages	2,020–2,026	D1OD
	Wheat, others	2,230	log (1/*R*)
	Rapeseed	1,390	D2OD
	Soybeans	2,010	D1OD
	Pork	1,752/1,468	D1OD
Oil	Rapeseed	2,306	D1OD
	Fish	2,310	D1OD
	Soybeans	2,302	D1OD
		2,334	D2OD
	Sunflower	2,393	D2OD
Fat	Pork	1,750	D1OD
NDF[b]	Forage	2,294	D2OD
		1,674	D2OD
ADF[c]	Forage	1,666	D2OD
		1,555	D2OD
IVDMD[d]	Forage	1,555	D2OD
		1,674	D2OD
Ash (oil)	Wheat flour	2,345	log (1/*R*)

[a] Some of these wavelengths are not the primary wavelengths but have been found to be significant in determination of protein in some commodities.
[b] NDF = neutral detergent fiber.
[c] ADF = acid detergent fiber.
[d] IVDMD = in vitro dry matter digestibility.

when applied to most commodities is mythical. Particles even of wheat, ground in the best of grinders, contain particles of an infinite variety of shapes and range in size from 1 μm or less to particles with a diameter of over 500 μm. Uniform particle size is even more difficult to achieve with highly fibrous commodities, and the best that can be hoped for in grinding is to achieve, consistently, a powder with the finest particle size commensurate with fastest grinding time and lowest heat buildup, in which the particles (albeit of widely different size) are randomly and evenly distributed. Variability in moisture between samples of the same commodity causes significant differences in mean particle size, particle shape, and particle-size distribution. Table XXI summarizes the influence of moisture on the mean particle size and particle-size distribution of three types of wheat when tempered to different moisture contents and ground on a cyclone grinder fitted with a 1.0-mm (0.04-in.) screen.

These data represent the means of several samples each of hard red spring, soft white spring, and durum wheats. The second section of Table XXI illustrates the more extreme differences in particle-size characteristics that occur. There is a strong sample/moisture interaction with respect to the influence of moisture on the comminution of wheat and other commodities, and changes in moisture status affect the generation of particles of different sizes in different ways. The effects were more noticeable with the harder wheats, the most significant being a considerable increase in the percentage of coarse particles. The moisture of the "low" soft white spring line was not so low as either the durum or hard red spring lines. Also, soft white spring wheat is not so soft as soft white or soft red winter wheats, and it is likely that these softer types would exhibit even greater variance at greater extremes in moisture level. High moisture content before grinding has the effect of increasing particle size. On the other hand, low moisture content facilitates grinding, and wheat and many other commodities pulverize more readily at low moisture. Very low moisture levels (6–8%) have the effect of causing cereals (except soft wheats or rye) and pulses to pulverize even more readily than at levels of 9–10%. Soft wheats and other soft cereals are slightly more resistant to grinding at 6–7% than at 9–10% moisture. High moisture (above 18%) makes it next to impossible to reduce fibrous material to a reproducible particulate material.

The moisture content of a sample also affects powdered material in that the hydration status of all components is directly affected by the amount of moisture present, as well as by the relative affinity of each constituent for water. The degree of hydration may influence the optimum area of the spectrum where the absorbers of specific constituents occur. This is of particular importance in the case of fresh products, such as meat or fruit. Some absorbers of substances such as starch and cellulose become intensified at very low moisture levels. Since components of many agricultural and food commodities contain constituents that differ widely in affinity for water (for example, cellulose has a much higher affinity for water than starch), it is likely that when two constituents occur at low moisture levels, the constituent with highest affinity for water reduces the hydration status of other constituents. This may cause changes in the spatial orientation of the molecules, with the net effect of different absorbers becoming optimum for the moisture-deprived constituents. This area has not been widely researched, but it has been observed that testing of commodities such as wheat and barley, which are very low in moisture (7–8%), is usually fraught with errors much larger than can be attributed simply to the mathematical difference in moisture content. The statement made in the chapter on chemistry that molecules of the same substance all have identical vibrations may not be true in the case of large molecules of plant origin. For example, molecules of the same protein in different samples with different moisture content may become hydrated to different degrees. When this differential hydration occurs, the molecular vibrations may be changed to the extent that wavelengths optimum for determination of protein in some samples may not be optimum for others. Proteins, starch, pentosans, and some fiber components are very complicated molecules, and this area of their chemistry is not fully understood.

Because moisture content also affects the relative proportions of all other constituents present, it is essential to determine the moisture contents of sample sets used for precision or accuracy determinations. For example, a difference in moisture content of 2% between two soybean samples may mean an apparent difference of nearly 1% protein.

BULK DENSITY

Bulk density of ground or otherwise powdered materials influences the amount and orientation of material that can be packed into a sample cell. Since the tightness of the packing is affected, the surface from which reflectance occurs is also affected. Bulk density is affected mainly by chemical composition, and in general the higher a commodity is in constituents with low density (such as cellulose and pentosans) the lower will be the bulk density, and the more difficult it will be to attain an even and reproducible surface. Table XXII gives the bulk densities of several commodities, as determined in triplicate by a crude but reproducible method involving the filling by gravity of a 25-cc measuring cylinder with a known constant weight (10 g) of the cyclone-ground materials.

The wide differences in bulk densities between commodities result in commensurate differences in the surface characteristics

TABLE XXI
Influence of Moisture on Mean Particle Size and Particle-Size Distribution in Wheat

Wheat Type	Moisture	500–1,000 μm	500 μm	354 μm	210 μm	149 μm	74 μm	H_2O%	Mean Particle Size (μm)
				Mean Results (n = 12)					
Durum	Normal	1.6	4.4	15.9	15.1	28.0	34.9	12.6	193
	High	1.4	6.4	15.1	16.9	23.5	36.7	16.2	197
HRS	Normal	2.0	4.6	11.6	11.6	22.0	48.8	12.2	178
	High	2.2	6.3	11.6	13.6	21.0	45.2	17.0	185
SWS	Normal	3.4	5.7	8.7	8.5	15.8	57.4	12.7	176
	High	4.8	5.0	9.4	8.4	15.9	56.4	15.4	189
				Extreme Results					
Durum	Low	1.6	5.0	17.4	19.6	29.4	27.0	7.2	202
	High	5.1	11.7	22.6	18.0	20.6	22.0	15.9	267
HRS	Low	2.2	6.6	15.2	16.0	21.4	38.7	8.7	203
	High	4.4	11.5	18.7	14.9	21.3	29.2	15.4	253
SWS	Low	3.8	5.8	8.3	8.5	15.1	58.4	10.0	181
	High	5.2	6.4	9.4	9.4	15.3	54.2	15.7	200

and the optical signal from the surface. Errors in NIR testing of commodities such as straws, dried grasses and other forages, wheat bran, soybean hulls, and similar substances can be improved by two-stage grinding. The dried material should be reduced to a fairly coarse particle size by grinding with a hammer mill or a mill with cutting action (such as the Wiley No. 4 mill) using a coarse screen. The material should then be reground through a cyclone or similar grinder, using a 1.0-mm screen. The two-stage grinding procedure is more time-consuming, but the effort is more than justified by the improvement in accuracy of NIR analysis.

PHYSICAL TEXTURE OF THE SAMPLE

Factors that affect physical texture also affect grinding; particle size, shape, and distribution; and, therefore, diffuse reflectance (see Table XIX). Physical texture means the composition or structure of the commodity. The texture of a substance influences the way it can be reduced to a fine particle size. It was pointed out above that the faba bean has the highest bulk density of all agricultural commodities tested for that parameter. Another attribute of the faba bean is that it is one of the hardest of all seeds physically, and it is very difficult to reduce to a fine particle size. The large seed size in terms of volume means that the seeds of many faba bean genotypes will not freely enter the grinding chambers of grinders such as the cyclone sample mill. To obtain a uniform sample, it is necessary to grind the beans with a two-stage reduction process. The best type of grinder for the first stage is the Christy-Norris 8-in. hammer mill, using a special screen with 8- to 10-mm holes. This grinding produces a meal with coarse consistency, which can easily be reduced to a fine powder by passage through a cyclone, Retsch, Falling Number, or similar grinder in the same way as the fibrous material referred to above. Seeds such as palm kernels are similar in their need for this type of two-stage grinding. Durum wheat is equal in hardness, but the smaller grain size allows it to be reduced by a single passage through a cyclone grinder.

In general, the harder the material, the more uniform the particle size of the final ground material (assuming the use of a cyclone or similar grinder). Materials such as soft wheat and forages may be more readily reduced, but these and other commodities (such as dried grass or straws), although they are not physically hard, do not reduce to give the same degree of uniformity in mean particle size and particle-size distribution.

FACTORS AFFECTING PHYSICAL TEXTURE

The species of the plant affects the texture of a powdered forage sample by changing the leaf-to-stem ratio and the overall degree of lignification. Stems are more highly lignified than leaves, and a higher proportion of leafier species means that a dried forage will reduce to a fine powder more readily. Stage of maturity also affects texture, since lignification usually increases during the maturation stage. The chief reason is the effort of the plant to stiffen its stems to enable them to hold up the heads that are becoming heavier with the developing seed. As a result, plants with larger seeds tend to have stiffer, more highly lignified stems. Powdered mature plants thus contain a preponderance of larger, irregular particles and have a lower bulk density than powdered, dried immature forages.

Lignin reinforces the strength of structural elements and the protective coats of plants and seeds. Field maturation and certain storage conditions are usually associated with lignification. Plant species and even varieties differ widely in their degree of lignification, which again has the net result of causing resistance to comminution and causes changes in particle characteristics and packing pattern.

Plant variety has an important effect on the texture of grains and seeds. Both wheat and barley varieties differ widely in physical hardness and give powdered meals that differ in particle characteristics sufficiently for the optical signals in the NIR to be used to determine their texture with a high degree of precision. Growing environment affects the texture of grains and forage plants, as does soil fertility by affecting the chemical composition and, in the case of forages, the ratio of leaf to stem. Generally there is a higher ratio of leaf to stem and less lignification under conditions of cool to medium (18–25°C mean ambient) temperatures and adequate soil moisture than under hot, dry conditions (30°C and upwards).

Wet weather during the harvest period can affect texture by causing bleaching and possibly sprouting, both of which tend to soften grains. Wet weather can affect texture of forages by inducing the development of molds and other fungi. Frost during the maturation of grains causes them to become significantly harder. Poor storage can also affect texture by encouraging the development of molds or of bacterial colonies with associated temperature buildup. Storage effects are related to temperature, length of time, and atmosphere. Ideal storage conditions combine low temperature with good aeration, conditions that have no adverse effects on grains, forages, or any other type of commodity. When conditions are other than ideal, the shorter the storage period the better. Samples before or after preparation should be reduced to 12% moisture or lower (8% for high-oil-containing seeds) and stored in airtight plastic or metal containers. Glass containers are also suitable but are expensive, breakable, and not as conveniently stacked or stored. The recent introduction in Australia of bulk storage under high carbon-dioxide concentration is an effective method of protection against infestation with insects, but it is not so effective against anaerobic fungal and bacterial development unless the moisture content is low. Fungi and insects can exert significant harmful effects on the composition and texture of any commodity. Both types of lesions can result in changes in composition, usually at the expense of the more soluble carbohydrate and protein fractions, resulting in changes in the texture of the material. Similarly, preprocessing practices such as drying, sterilization, or freezing can affect the cellular nature of materials, which in turn affects their comminution.

TABLE XXII
Bulk Densities of Some Common Powdered Commodities Ground in Cyclone Grinder (1.0-mm screen)

Substance	Bulk Density (g/cm^3)
Faba bean	0.639
Lentil	0.542
Wheat flour (HRS)	0.514
Corn (maize)	0.506
Wheat (HRS)	0.495
Soybean hulls	0.489
Barley	0.390
Wheat bran	0.349
Oats	0.345
Defatted rapeseed	0.338
Dried alfalfa	0.240
Barley straw (mature)	0.142

SAMPLE TEMPERATURE

Higher temperatures during grinding have a small but persistent effect in reducing resistance to comminution, so that commodities such as grains reduce more readily. Grinders usually heat up to their maximum operating temperature after grinding five to eight samples in sequence. To avoid possible differences in particle characteristics, the grinder should be allowed to run for 1–2 min before use, then three or four samples of dispensable wheat or other material should be ground before the authentic samples are processed.

Samples of animal origin present a further variety of textures and packing factors, many of which are affected by temperature (e.g., butter). These samples include dairy products (butter, cheese), liquid and dried milk, ice cream mix, cream, meat and derived products, blood, bone, fish and meat meal, and many others. Many of the products need no sample preparation beyond careful blending, but they do require specialized systems for presenting the sample to the instrument. This aspect has been well researched by instrument manufacturing companies, all of which have devised ingenious and practical cells for the analysis of special products including powders, slurries, liquids, stiff pastes, and such materials as butter and margarine. For example, of the more than 60 commodities for which Percon has marketed the Inframatic 8100 or 8120, more than half have required a sample-presentation system other than the regular ground-grain system.

The influence of sample temperature on the accuracy of NIR analysis of wheat has been documented (Williams et al, 1982). The chief reason for the effect observed on wheat was attributed to sensitivity to temperature of wavelengths used in measurement of protein. The main effect on wheat protein determination was a nearly linear decrease in observed protein content as sample temperature rose from −30 to 45°C. The effect of temperature on wavelength is related to the intensity of the signal and the data treatment applied to the log $(1/R)$ signal. At certain wavelengths, derivatized log $(1/R)$ data are affected more by temperature than the underivatized data. The optimum wavelength for protein determination tends to be more sensitive to external factors than the wavelengths used for determination of some other constituents. One of the most important temperature effects is on the principal moisture band at 1,925–1,945 nm. Temperature affects the degree of hydrogen bonding and the hydration status associated with all constituents, and these changes influence the wavelengths at which overtones or combinations actually occur. When instruments are set up to use specific wavelengths by discrete filters or by encoder, temperature fluctuations can cause differences in the phenomenon referred to above as instrument/sample noise. The temperature effects can largely be offset by careful selection of wavelengths during the design of instruments, which has been achieved in the development of the more modern NIR instruments. Interactions between mathematical treatment, wavelength, and temperature can largely be offset by optimization using a computerized spectrophotometer. Other useful precautions are to use more than one grinder and interchange grinders to minimize temperature increase; to allow the samples to cool before analysis, when practicable; and to calibrate by using samples of different temperatures representing the anticipated temperature range.

Cooking methods represent yet another variable. Cooking temperatures vary from the 40°C traditionally used in drying pasta products, through the 80–100°C used in modern pasta drying, through boiling and steam cooking to high-temperature baking. There is a strong interaction between cooking temperature and texture of the commodity, and individual operators should carefully research the effects of processing temperature on the texture and sample-preparation technique best suited to their products.

AMBIENT TEMPERATURE

The ambient temperature affects the temperature of the instrument, commodity, subsample, and rate of temperature increase during grinding. Grain samples that have been stored in bulk during a North American or northern European winter usually reach a low temperature and do not warm up quite so quickly during grinding. The reverse is true in areas such as Australia, the Middle East, the Indian subcontinent, or even North America during the peak harvest season of July–September when temperatures reach 30–40°C. The effects of temperature on texture, which have been referred to above, differ from the temperature effect described in the preceding subsection. Those observations were made on the same ground samples, the only difference being the temperature of the actual sample at the time of presentation to the instrument, and the results were attributed to an effect on wavelength rather than texture. The ambient temperature can also affect instrument performance, since detectors and other components are sensitive to temperature. This applies particularly to computerized spectrophotometers, which may not be equipped with the same degree of temperature control as commercial bench-type instruments.

SAMPLE TYPE

Because of their physical and chemical constitution, some commodities are more subject to error caused by instrument/sample noise than others. The materials that are most difficult to analyze with accuracy are those with an imbalance of one or two constituents, including raw oilseeds, some of which are rich in both oil and fiber; forages, particularly mature plants; and materials such as dried wheat gluten, fishmeal, or sugar-beet pulp. It is essential that calibrations for the analysis of such materials be established with a very comprehensive array of samples, but it is often difficult to accumulate samples with sufficient variance. This is discussed in section C of this chapter.

CONVERSION FACTORS

Conversion factors are used to convert the constituent that is actually measured into the constituent sought. The best known of these is the nitrogen-to-protein factor arbitrarily established at 5.7 for wheat and flour, 6.25 for other grains, pulses, seeds, and forages, and 6.38 for dairy products. Other factors exist or are an integral part of the test, such as the crude fiber test, which does not give a true indication of the true indigestible or digestible fibrous components present in a fibrous food or feed. Such factors were established by our predecessors, chemists who, had they had the facilities of modern laboratories to work with, would doubtless have been as skeptical of the accuracy of these factors as we are. The factors are a source of error in NIR testing, since an NIR instrument is really a "supermicroscope" that "sees" beyond the sample surface, theoretically detecting every molecule present.

The nitrogen-to-protein factor is probably the most important of these conversion factors, since protein is the constituent most frequently determined by NIR. The reference method is the Kjeldahl test in some form, which determines total nitrogen with accuracy. It is unlikely that the amount of nitrogen in the protein fraction of every sample of every commodity is constant. A variation of as little as ±0.2% nitrogen in the protein would cause the nitrogen-to-protein conversion factor to change to the extent of introducing an error of about ±0.16% protein in wheat at 13.5% protein or about ±0.5% protein in soybeans at 40% protein. Tkachuk (1977) noted changes of up to 0.13 in the nitrogen-to-protein factor for wheat and 0.09 for chickpea. These changes would be sufficient to cause errors of over 0.2% protein in wheat and 0.3% in chickpea at respectively 13% and 20% protein. Conversion factors are hidden errors. From a philosophical viewpoint, the excellent accuracy that it is possible to achieve with NIR may imply that the nitrogen-to-protein factor is practically constant. In practice this is not necessarily so, since it is likely that the NIR and Kjeldahl techniques measure all of the nitrogen, with the result that both techniques are equally affected by changes in the nitrogen-to-protein factor and neither is truly measuring "protein."

A similar argument can be presented in the case of the nonprotein nitrogen, most of which exists in the form of simple peptides and amino acids, and the main absorbers responsible for NIR measurements are likely to be the same as those in protein. Accordingly, there is no reason why the NIR technique would not encompass this type of "nonprotein" nitrogen in the same way as the Kjeldahl test.

C. Operational Factors

CALIBRATION PRACTICE

Number of Samples Used. Traditionally, NIR calibrations have been established using anywhere between 20 and 50 samples. During the early years of NIR technology, instrument manufacturers were reluctant to advocate the use of more than 24 samples for calibration, mainly because of an apprehension on the part of the purchaser of embarking on a calibration procedure that involved the assembly and chemical analysis of a larger number of samples. Early research verified that calibrations carried out with too few samples were sample sensitive and did not give reliable results over a long period. Subsequently, it was shown that larger numbers of samples possessing a wider range of variance in the parameters likely to be encountered in routine analysis gave more reliable results, and calibrations of this type came to be referred to as *robust* or *stable*. The minimum number of samples necessary to establish a stable calibration of the "simple" type (see Chap. 6) is 35–40; preferably, however, more samples should be used, depending on the nature of the material to be analyzed, the constituents sought, the sources of variance likely to affect the analysis, and the instrument to be used.

Sample Selection. The two principles followed in sample selection are identification of the likely sources of variance in the commodity to be tested and assembly of a set of samples representative of all of the variance, including chemical and physical composition and spectral identity. The actual number of samples used in a calibration depends upon the anticipated variance. If the variance in the available samples is limited, the efficiency of the calibration in terms of accuracy and precision will not be improved by simply increasing the number of samples. For example, if an NIR instrument is to be used to monitor the output of a flour mill, where particle size is relatively uniform, and the range of protein is unlikely to exceed 0.5%, a calibration based on 100 samples is not likely to be any more effective than one based on 30.

For a simple commodity such as soft white spring wheat, the variables most likely to affect the wheat are growing season, growing location, planting time, irrigation and fertilizer practice, wheat variety, fungal and insect disease, and weather effects during maturation and harvest periods. For a more complex commodity such as a forage, the variables of plant species and stage of maturity at harvest may be added. For an even more complicated commodity such as a feed mix, not only must the individual components of the mix be considered, but also their proportions in the mix and the interaction effects of all of the above on the individual components such as grains and dried grasses or forage legumes that are often used in the mix (Williams and Starkey, 1980).

For commodities of animal origin, such as powdered milk, cheese, butter, meat, and wool, the chief factors affecting NIR analysis include sample composition, temperature, and processing procedure. Fat, protein, and moisture content are the most important constituent variables, and temperature affects the texture and general physical state of most of these commodities. For example, presentation of butter to an NIR instrument using a sample cell that depends on a cut, vertical planar surface would be difficult if the ambient temperature were high enough to melt the butter even slightly, since the surface would not remain flat. Species and stage of maturity (age) affect commodities such as fish, fishmeal, meat, and meatmeal. Storage is also a factor, since materials of animal origin tend to deteriorate much more rapidly than those of plant origin. Sample preparation is more complicated with fish and meat products because of their peculiar texture and high moisture content.

All variables should be included at least three times in any calibration, and accumulation of enough samples for a stable calibration is often the most critical aspect. A practical consideration in sample accumulation for calibration is that identification of a variable makes it easy to accumulate samples with a range of that variable. If an unexpected, "new" element appears, such as the effects of a new season or an epidemic of a different fungal disease, samples bearing that variable to different degrees can be accumulated and used to update the calibration at the time of its appearance. New elements include a change in processing procedure in a factory or the proposed use of different grain types as the basis for feed mixes, such as a switch from barley to feed-grade wheat because of price differentials. The operator should be aware of the likelihood of such differences in commodity. A uniform distribution in composition across the anticipated range minimizes errors caused by regression toward the mean from extremes of concentration. This phenomenon is particularly apparent when the calibration samples are selected with Gaussian distribution of composition, as has been discussed in detail in the previous chapter. It is good practice to invest time in accumulating as comprehensive a set of calibration samples as possible before recording any optical data. Samples collected in the early days of the assembly period should be protected from change by careful storage until use.

Accuracy of Standard Laboratory Analysis. Accuracy in standard (reference) analysis is essential to setting up efficient NIR calibrations and for assembling samples to be used in monitoring an operation. The early days of NIR analysis saw many cases of NIR instruments blamed for poor accuracy, when in most of these instances it was the standard laboratory analyses used in monitoring that were responsible for high standard errors. The 14 years of NIR application in North America have witnessed significant improvements in the attention paid to standard laboratory analyses, and this is reflected in improved NIR statistics. These are partly the result of improvements in instrumentation and partly improvements in standard wet-chemical analyses. Standard laboratory accuracy is a critically important factor in the efficiency of NIR analysis, since the SET of standard analysis is a component of the accuracy of the overall test.

SAMPLE PREPARATION

Sampling and Subsampling Procedure. Sampling is the most important single source of error in any chemical or physicochemical analysis of agricultural commodities and of most food products and ingredients. The sample must be truly representative of the total population in every sense, including chemical composition, physical constitution, and the presence of foreign material. Table XXIII illustrates the number of kernels of wheat in different weights, assuming an average weight per kernel of 35 mg. Similarly, a 70-lb bale of hay may represent 2,000–10,000 plants, depending on species and stage of maturity of the plants at harvest.

The larger the population, the more comprehensive the sampling method needs to be. For example, to gain a representative sample of the progeny of a single F_2 wheat plant, planted as a single row, all that is necessary is to divide the entire yield in half manually or by using a sample splitter, then one half can be used for analysis and the other half replanted as necessary. At the grain elevator or farm, the practice of diverting by hand portions of the stream of grain entering or leaving a truck into a pail gives a satisfactory sample, provided the entire truckload is sampled in this way every 15–20 sec throughout the loading or

TABLE XXIII
Approximate Number of Kernels of Wheat in Different Weights

Weight	No. of Kernels
Kilogram	29,000
Tonne (1,000 kg)	29,000,000
Carlot (50 t)	1,428,570,000 (1.43×10^9)
Vessel (10,000 t)	285,714,230,000 (2.85×10^{11})

unloading. To sample a vessel, the stream of wheat passing into it should be automatically sampled continuously, using a Woodside or diverter-type sampler (Bauwin and Ryan, 1974). The sample stream should be divided and a small percentage diverted into a container that will eventually contain the sample that represents the entire shipment. The Kostur sampling system used by the Canadian Grain Commission at terminal elevators is typical of this method (Bryan and Elvidge, 1977). The sample has to be thoroughly blended before analysis and subsampled by being passed through a sample divider such as the Boerner divider repeatedly until the final half-samples are small enough for analysis. Similarly, a single row of grass from an experiment may all be dried, ground, and analyzed, whereas bulk quantities of hay and forages may be sampled during the loading of a truck or building of a stack by withdrawing a handful from every 10th or 20th bale (or simply every 5 min) and placing these subsamples into a large plastic bag. The final bagful of forage or hay is well mixed by hand and two or three large handfuls placed in a separate plastic bag and labeled. This is the sample for analysis.

Grinder Type. There are two main types of grinding, namely "normal" grinding for sample analysis and two-stage grinding, which involves prereduction of such materials as forages, feed pellets, or large seeds that cannot enter the grinding chamber, followed by further reduction on a cyclone or similar grinder. Grinding and grinders have formed the subject of a recent publication (Williams, 1984). For normal grinding, be it for standard or NIR analysis, it is important to use the grinder that is best for that commodity. Grinders suitable for different commodities are summarized in Table XXIV. The Udy cyclone, Cyclotec, Falling Number KT3100, and Retsch (Brinkman) centrifugal grinders, all fitted with 1.0-mm screens, are most suitable for seeds low in oil, and the cyclone grinder with forage head is suitable for grinding up to 20-g samples of forages, or even bigger samples if the receiving jar is replaced with a modified sample receptacle.

For oilseeds, an impeller-type mill is preferred, such as the Krups Model 75, Braun, Moulinex, or similar ones. For forages and feeds and for very hard seeds such as faba beans, the Christy/Norris 8-in. hammer mill and the Falling Number KT3100 hammer mill are most suitable. For prereduction, the Christy/Norris 8-in. mill fitted with a special screen with round holes 8 mm in diameter, or the Wiley Model 4 grinder with a 1.0-mm screen, are both useful. The coarse material should then be further reduced in a cyclone or similar grinder with a 1.0-mm screen. Oilseeds can conveniently be ground in a high-speed impeller mill (Krups, Braun, or Moulinex). The seeds should be ground in short 15-sec bursts, with the sample being blended manually between bursts until a uniform meal is obtained. About six to eight bursts are sufficient. Specialized grinding, as in the use of 50% diatomaceous earth or snap freezing in liquid nitrogen, has been suggested for oilseeds. These methods are more time-consuming and are impracticable for large-scale applications, but they are useful for low-throughput operations. Soybeans can be ground successfully using a cyclone grinder with no screen. The particle size is more variable, but cleanup is satisfactory. Other high-oil commodities, such as maize, oats, and lupin, can also be ground in this way. The precision of analysis is not as good as when the seeds are ground in an impeller grinder, and oil extraction may be reduced (P. C. Williams and D. C. Sobering, unpubl. data, 1986).

Grinder Condition. The mean particle size, particle-size distribution, and therefore the diffuse-reflectance signal can be markedly affected by the condition of the grinder used in sample preparation. Cyclone sample mills need periodic replacement of the Carborundum ring around the grinding chamber and of the 1.0-mm screen. The surface of the Carborundum becomes progressively polished by the passage of grain samples. At the same time, the holes of the 1.0-mm screen tend to increase in size and the net result is an increase in mean particle size. This process can be monitored and the grinder maintained by periodically replacing the Carborundum disk or screen. It is important to ensure that screens are replaced accurately: sometimes a small piece of grain lodges underneath the screen and prevents it from seating properly, which can result in differences in particle characteristics and dramatic errors in testing. The burrs of burr mills such as the KT3303 become worn with time and need to be replaced. Under normal use, the Carborundum disk and screen of a cyclone grinder

TABLE XXIV
Suggested Grinders for Specific Laboratory Testing

Workload	Commodity	Analysis[a]	Method[b]	Grinder
High	Grains, cereals, pulses	Proximate	Standard[c]	KT 3303, cyclone,[d] Cyclotec[d]
High	Grains, cereals, pulses	Proximate, hardness	NIR	Cyclone, Cyclotec
Medium	Grains, cereals, pulses	Proximate	Standard, NIR	KT 3303, KT 3100, cyclone, Cyclotec, Retsch,[c] Christy/Norris 8-in.[d]
Medium	Grains, cereals, pulses	Moisture	Standard	Wiley intermediate
Medium	Oilseeds	Oil, protein, moisture, fiber	Standard, NIR	Krups, Culatti, Moulinex, Braun
Medium	Cereal grains	Physical hardness	Standard (particle-size index)	KT 3303, Cemotec[e]
Medium	Forages	Proximate	Standard, NIR	Christy/Norris,[d] cyclone (forage head), Wiley No. 4[d]
Medium, Prereduction	Forages, hard seeds	Proximate	Standard	Christy/Norris,[f] Wiley No. 4[f]

[a] Proximate analysis includes protein, moisture, oil, fiber, minerals, carbohydrate (by difference).
[b] Standard analysis means wet chemistry methods (Table XV).
[c] Retsch centrifugal grinder is marketed under other names, e.g., Brinkman centrifugal grinder.
[d] Fitted with 1.0-mm screen.
[e] Cemotec grinder is available from Tecator/Fisher Scientific.
[f] Fitted with 8- to 10-mm screen.

should be replaced about every 50,000 samples and a set of burrs about every 100,000 samples. Passage of stones or pieces of steel through grinders causes serious damage to screens and burrs, and, in the case of the cyclone grinder, the sparks generated are potentially hazardous. Hammer mills are generally the most rugged of grinders; barring accidents, they need very little attention besides lubrication. The screen should be replaced about every 100,000–250,000 samples, depending on the material ground. Small impeller mills need fairly frequent replacement of the impellers, especially if they are used to grind soybeans or similar fairly hard commodities.

Mean Particle Size and Particle-Size Distribution. The two parameters of particle size and particle-size distribution are closely associated, and both are affected by factors that include grinder type and revolutions per minute, screen size, texture (i.e., physical hardness) of the commodity, chemical composition, and moisture content, all of which have been discussed above. Particle size, shape, and distribution affect the surface presented to the instrument. These parameters are among the most important causes of instrument-sample interaction or system noise. Figure 4 illustrates spectra of wheat samples of different hardness/particle size that have been ground on an Udy cyclone grinder using a 1.0-mm screen. It is possible to obtain successful NIR results with a wide range of sample sizes, provided the instrument is calibrated accordingly, but in general the finer and more uniform the particles, the more effective the analysis. Coarse and uneven particles can change the path length to varying degrees, which affects precision rather than accuracy, and two-stage grinding is recommended when particles are large and irregular in shape, particularly with forages, straws, and very hard seeds.

Blending After Grinding. The step of blending after grinding is a simple but essential one in improving precision. The ground sample stratifies in the grinder receptacle and must be thoroughly blended before testing. The sample jar should be capped and shaken and the contents stirred 15–20 times with a spatula before loading the sample cell (NIR) or weighing (reference method).

SAMPLE STORAGE

Storage Before Use. The most important factors to consider when samples need to be stored for a significant period (more than two weeks) before analysis are changes in chemical composition and damage caused by insects, molds and other fungi, and bacteria. Storage of samples in kraft paper envelopes can change moisture levels significantly. In the northern states of North America and in Canada, as well as in dry regions such as Australia, the Indian subcontinent, the Middle East, northern and southern Africa, and parts of Latin America, samples stored in this way lose moisture very quickly and reach an equilibrium moisture level at about 8% in three or four weeks. In other areas, such as northern Europe, parts of the eastern United States and Canada, and the northwest coast of North America, samples stored in envelopes may gain moisture. Fungal and bacterial infestations are more likely to occur in areas where the relative humidity is high, whereas insect infestations are prevalent in hot areas. Storage in sealed, thick, 4-mil (0.1-mm) plastic bags will protect material from moisture loss but not from prolonged exposure to insects or from mold or fungal infestation if samples are high in moisture. Storage in metal, glass, or rigid plastic containers effectively protects against any deterioration if temperature and humidity are low. If temperature and humidity are high, infestation from fungi and bacteria may occur, and if insect eggs are present, infestations will tend to develop. Generally, forages are less likely to deteriorate in storage because their chemical composition is not so conducive to insect or fungal infestation.

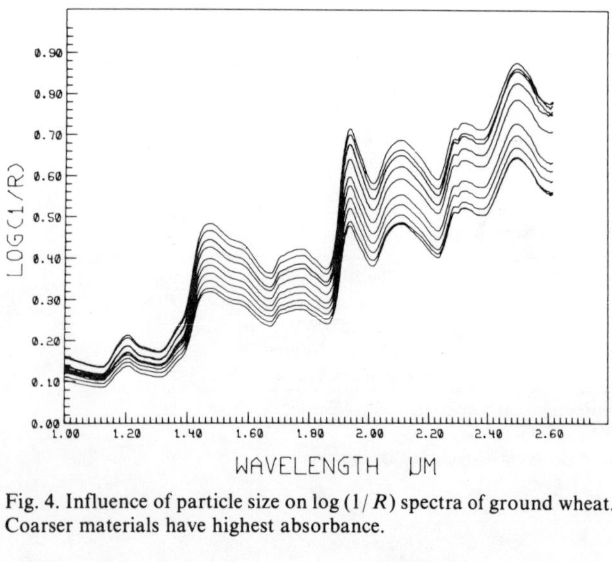

Fig. 4. Influence of particle size on log $(1/R)$ spectra of ground wheat. Coarser materials have highest absorbance.

Aeration is also an important aspect of storage, and well-aerated storage can protect many types of material for long periods. Ideally, samples should be stored refrigerated or frozen, but numbers and bulk usually make this impracticable. If samples are tempered to different moisture levels before grinding, they should be allowed to rest in the tempered state for at least two days and should be stored in metal or plastic containers, taped to prevent moisture loss (or gain), or in thick plastic bags. If the standing period is likely to be longer than two days, the tempered samples should be refrigerated to prevent bacterial or mold development or insect infestation. Samples stored in refrigerated conditions for long periods may increase significantly in moisture content, to the extent that they may require air-drying before use as NIR, calibration, or monitoring samples.

Sample Preservation After Grinding. The same principles for sample storage before grinding are applicable to preservation of the ground, well-mixed sample after grinding. The main differences are that the samples are usually smaller and that the storage period after grinding is usually shorter since samples are frequently ground, analyzed, and thrown away, and a 24-hr storage period is usually sufficient. Under conditions where longer storage is necessary, precautions must be taken to preserve the samples—for example, over weekends. The storage method depends on the size and chemical composition (particularly moisture content) of the sample, length of storage, normal atmospheric conditions, and space and finances available for storage. Large samples are more difficult and expensive to store for long periods than small ones; other than for samples to be used as check samples, or those that may be needed for verification of earlier test results, such as cargo samples, the long-term storage of large ground samples should not be encouraged.

Where storage is essential, samples should be kept in metal containers with tightly fitting lids and the containers themselves should be refrigerated to inhibit the development of insect infestation and molds. The lids may also be taped to prevent moisture interchange with the atmosphere. Samples of intermediate size (100–1,000 g) are also best stored in metal containers, but since there are usually more of them (for example, flour samples awaiting physical dough-testing and baking tests), round, plastic pails are acceptable. Round (cylindrical) plastic containers are preferred to square ones because they are easier to clean. Square containers have corners and also stimulate stacking in storage. Stacking of plastic containers is not advisable because of the possibility of cracking. The ideal storage area is a refrigerated room with air circulation and fitted with movable ("rollaway") shelf units. In this type of storage room, several tons of samples can be stored with easy access. Other types of areas include air-conditioned rooms, preferably with no windows, or simply ordinary rooms with air-circulating fans. Aeration, or frequent air change, is very important in sample storage, particularly in the absence of refrigeration.

For samples of 15–25 g, the size normally ground for use in analysis, small, round, flat metal containers of about 20 mm in depth and 60 mm in diameter (2-oz ointment tins) are the most suitable for short-term sample preservation. They are easy to clean and to label and stack in trays, and they can be taped to protect against moisture interchange. They are also convenient for mixing the sample in before weighing. The best alternatives to these containers are plastic vials with tight-fitting caps or plastic bags. Small paper envelopes are useful for conveying ground samples to the laboratory, but they can only be used once and are completely porous to water movement inward or outward. Plastic vials can be stored in boxes but are not so durable as metal tins, are subject to buildup of static electricity, and are more difficult to clean. For large-scale analytical operations, turnaround time of sample containers is a significant factor that occupies many staff-hours. Storage of analytical samples in envelopes or plastic bags makes it difficult to mix the sample before weighing or loading sample cells, and the static electricity accumulation by some plastics may cause further difficulties in transferring samples from the bags.

Atmospheric conditions are mainly ambient temperature and relative humidity. In dry climates, samples lose moisture very quickly and the equilibrium moisture level for grain with the air may be as low as 6–7%, so that samples stored in paper envelopes or 2-mil plastic bags in the open air lose moisture down to that level. This loss can cause significant inaccuracy in reporting the results. Even ground samples stored in the small metal containers referred to above can lose moisture more or less rapidly, depending on the initial moisture level of the sample after grinding. Taping the lid to the can with masking tape protects against moisture loss for short periods, but at high levels of moisture and room temperature, molds develop rapidly. Table XXV illustrates the rate of moisture loss from taped and untaped tins of cyclone-ground wheat. The study was carried out in Vancouver, British Columbia, where at the time of testing the relative humidity was very high. It is likely that moisture losses could be even greater in drier areas.

SAMPLE CELL LOADING

Packing. Cell-loading error is influenced mainly by mixing of the ground sample before analysis, by the bulk density of the sample, by particle size and shape, and by the amount of sample loaded into the cell. The mixing of the ground sample before loading the cell has been examined (Williams, 1975). No improvement in precision was obtained after 15 mixes of the ground sample in the sample container of an Udy grinder. A spatula was used with a rotating movement. Cell-loading error has been determined for several commodities including red spring, soft, and durum wheats, barley, oats, and rye. The error resulting from grinding increased directly as the amount of fiber in the sample, being highest in the case of oats. The sample-loading error was significantly greater for the more fibrous materials. In some cases, the sample cell-loading error was higher than the grinding error. There was also a sample/cell-loading interaction. The instrumental error was consistently low for all commodities.

The amount of material packed into the cell can affect the results by influencing the compression of the sample at the surface, which will affect the diffuse reflectance. This variable is easily controlled by overfilling the cell and striking the surface level before putting on the cap for all samples. In this way, the average weight of material actually used varies from about 2 g for straws and other fibrous materials to nearly 6 g for maize (corn) and pulses, but it is consistent among samples of a single type. Table XXVI illustrates the variance in log $(1/R)$ signal for several commodities as a result of underfilling and overfilling the sample cell of a DICKEY-john GAC III. The log $(1/R)$ signals increased for cereal grains as a result of overfilling but decreased for the straw sample, which had a much lower packing density. The marked increase in log $(1/R)$ signal in sunflower was probably caused in part by extrusion of oil at the surface. This can be avoided by use of an open sample cup with no window. Pacific Scientific instruments give similar results, and their sample cell uses about twice as much sample. The cell-loading error of the Technicon InfraAlyzer series is also comparable, and the sample cell uses about the same amount of material as the DICKEY-john GAC III. The Percon Inframatic 8000 series presents a different type of loading since it has no sample cell as such. The sample is packed by tamping into a cell formed by a door-type attachment to the front of the instrument. The tamping must be uniform to ensure precision. Fairly large errors (up to or over 1% in wheat protein) can occur if the sample is not sufficiently well packed. Cleanup between samples should include the magnetic catch of the sample chamber door. If it is dust coated, the door may fall open during tamping. It should not be necessary to hold the door during tamping.

Cleanup. Sample-to-sample contamination can occur if cells are not efficiently cleaned between samples. For materials low in oil, such as cereals, pulses, and forages, brushing away fine material after emptying the cell is usually sufficient. This can become more laborious when the sample cell has a quartz window instead of optical glass. Quartz windows are more subject to buildup of static electricity, and fine particles adhere more closely to quartz than to glass. It is necessary to wipe the window with lens tissues periodically. For oilseeds and some mixed feeds (e.g., poultry feeds that have been sprayed with tallow), the cell window must be lightly brushed free of particles of meal, wiped free of oil using a tissue moistened with alcohol, wiped free of alcohol using a tissue soaked in water, and finally dried with another tissue. This seemingly tedious routine, however, only occupies about 30 sec and is essential to prevent contamination.

Cleanup is not limited to the sample cell. The cell carrier and

TABLE XXV
Moisture Loss from Ground Wheat Stored in Metal Cans With and Without Taping

Days from Initial Moisture Test	Initial Moisture Level[a] 14.3% Moisture Loss (%)		12.6% Moisture Loss (%)		10.4% Moisture Loss (%)	
	Taped	Untaped	Taped	Untaped	Taped	Untaped
1	0	0.1	0	0.1	0	0
2	0	0.9	0	0.4	0	0.1
4	0	3.3	0	1.4	0	0.3
6	0	4.0	0	2.1	0	0.3
21	0.4	···	0.2	···	0.1	···
42	0.6	···	0.5	···	0	···

[a]Single-stage air oven (130°C ±1 for 1 hr).

TABLE XXVI
Influence of Packing Cell on DICKEY-john Log (1/*R*) Values

Commodity	Log 0	1	2	3	4	5	Protein	Moisture	Hardness
Barley									
Under	132.39	99.73	107.55	118.42	114.92	52.38	10.7	9.5	...
Normal	134.14	101.60	109.50	119.92	116.11	54.11	11.4	9.6	...
Over	139.62	107.48	115.07	125.22	120.87	59.75	11.3	10.1	...
HRS Wheat									
Under	180.92	139.65	150.73	162.51	155.18	73.34	12.1	9.1	46.7
Normal	183.67	142.04	153.17	164.98	157.40	76.90	12.1	9.1	45.2
Over	184.92	143.63	154.49	165.98	158.35	76.54	12.1	9.2	41.5
Straw									
Under	132.78	104.10	106.19	113.84	100.94	54.97	...	...	...
Normal	128.78	100.88	102.72	110.14	97.53	53.19	...	...	...
Over	127.91	100.27	101.99	109.29	96.45	52.75	...	...	...
Sunflower									
Under	369.05	238.90	238.46	226.90	226.18	148.70	...	...	...
Normal	370.38	240.78	240.17	228.76	227.74	150.23	...	...	...
Over	434.28	291.36	290.18	276.92	274.32	183.34	...	...	...
Normal/ Over[a]	404.10	268.94	268.02	255.68	252.49	170.12	...	...	...

[a] Because of extremely high change in sunflower readings, overfilling was repeated with lower degree (about one half) of "overfilling" amount.

the glass or quartz window that protects the detectors from the atmosphere also accumulate dust (see the section above on Static Electricity) and should be cleaned daily or after every 50 samples. Buildup of dust on the interior window can cause progressive development of bias, and an abrupt change in results can occur following cleanup after a lengthy accumulation of dust.

CARELESSNESS

Carelessness, an annoying, ever-present, and significant source of trouble, includes mistakes in numbering samples and generally incorrect identification, reading samples out of sequence, recording the wrong data, misplacing decimal points, generally poor-quality bench-work, and other lapses. It is usually associated with lack of concentration on the part of the operator. Routine analysis is a boring operation, and it is advisable to make the task as interesting as possible by keeping the operators *au fait* with the reasons for and implications of the analysis. Good supervision helps by detecting anomalous results, maintaining good morale, and generally running an efficient operation.

LITERATURE CITED

AACC. 1983. Approved Methods of the AACC. Method 30-20, approved April 1961; revised October 1975; reviewed October 1982. Am. Assoc. Cereal Chem., St. Paul, MN.

AACC. 1983. Approved Methods of the AACC. Method 32-10, approved April 1961; revised October 1975 and October 1981. Am. Assoc. Cereal Chem., St. Paul, MN.

AACC. 1983. Approved Methods of the AACC. Method 32-20, first approval October 1977, final approval October 1982. Am. Assoc. Cereal Chem., St. Paul, MN.

AACC. 1983. Approved Methods of the AACC. Method 44-15A, approved October 1975; revised October 1981. Am. Assoc. Cereal Chem., St. Paul, MN.

AACC. 1983. Approved Methods of the AACC. Method 46-12, approved October 1976; reviewed October 1982; revised November 1983. Am. Assoc. Cereal Chem., St. Paul, MN.

BAUWIN, G. R., and RYAN, H. L. 1974. Sampling, inspection, and grading of grain. Pages 115-134 in: Storage of Cereal Grains and Their Products. C. M. Christensen, ed. Am. Assoc. Cereal Chem., St. Paul, MN. 549 pp.

BRYAN, J. M., and ELVIDGE, J. 1977. Mortality of adult grain beetles in sample delivery systems used in terminal grain elevators. Can. Entomol. 109:209-213.

HILDEBRAND, F. C., and KOEHN, R. C. 1944. Sources of error in the determination of the protein content of bulk wheat. Cereal Chem. 21:370-374.

HRUSCHKA, W., and NORRIS, K. H. 1982. Least squares curve-fitting of near-infrared spectra predicts protein and moisture of ground wheat. Appl. Spectrosc. 36:261-265.

JONES, D. I. H., and HAYWARD, M. V. 1975. The effect of pepsin pretreatment of herbage on the prediction of dry matter digestibility from solubility in fungal cellulose solutions. J. Sci. Food Agric. 26:711-718.

MacKENZIE, S. L. 1977. Analysis of methionine in peas. Improved gas chromatographic practice. J. Chromatogr. 130:399-402.

MINDEL, B. 1986. Kjeltec Auto Plus: An intelligent Kjeldahl system. Tecatur. In Focus 9:12.

MORREY, J. R. 1968. On determining spectral peak positions from composite spectra with a digital computer. Anal. Chem. 40:905-914.

NORRIS, K. H., and WILLIAMS, P. C. 1984. Optimization of mathematical treatments of raw near-infrared signal in the measurement of protein in hard red spring wheat. I. Influence of particle size. Cereal Chem. 61:158-165.

SAVITZKY, A., and GOLAY, M. J. E. 1964. Smoothing and differentiation of data by simplified least squares procedures. Anal. Chem. 36:1627-1638.

SPACKMAN, D. H., STEIN, W. H., and MOORE, S. 1958. Automatic recording apparatus for use in chromatography of amino acids. Anal. Chem. 31:1190-1205.

STARR, C., MORGAN, A. G., and SMITH, D. B. 1981. An evaluation of near-infrared reflectance analysis in some plant breeding programmes. J. Agric. Sci. 97:107-118.

TKACHUK, R. 1977. Calculation of the nitrogen to protein conversion factor. Pages 78-82 in: Nutritional Standards and Methods of Evaluation for Food Legume Breeders. J. H. Hulse, K. O. Rachie, and L. W. Billingsley, eds. IDRC-Ts7e. International Development Research Centre, Ottawa.

VAN SOEST, P. J. 1963. Use of detergents in the analysis of fibrous foods. I. Preparation of fiber residues of low nitrogen content. J. Assoc. Off. Anal. Chem. 46:825-829.

VAN SOEST, P. J., and WINE, R. H. 1968. Determination of lignin and cellulose in acid detergent fiber with permanganate. J. Assoc. Off. Anal. Chem. 50:50-55.

WILLIAMS, P. C. 1974. Errors in protein-testing and their consequences. Cereal Sci. Today 19:281-286.

WILLIAMS, P. C. 1975. Application of near infrared reflectance spectroscopy to analysis of cereal grains and oilseeds. Cereal Chem. 52:561-576.

WILLIAMS, P. C. 1984. A study of grinders used for sample preparation in laboratory analysis of grains. Cereal Foods World 29:770-775.

WILLIAMS, P. C., and KRISCHENKO, V. P. 1986. A comparative study of two computerized spectrophotometers for the near-infrared analysis of wheat for protein. Appl. Spectr. 40:785-790.

WILLIAMS, P. C., and MacEACHERN, L. W. 1981. A note on the use of titanium oxide catalyst in an automated Kjeldahl-type instrument. Lab. Pract. 30:125.

WILLIAMS, P. C., MacKENZIE, S. L., and STARKEY, P. M. 1985. Determination of methionine in peas by near-infrared reflectance spectroscopy. J. Agric. Food Chem., 33:811-185.

WILLIAMS, P. C., NORRIS, K. H., GEHRKE, C. W., and BERNSTEIN, K. 1983. Comparison of near-infrared methods for measuring protein and moisture in wheat. Cereal Foods World 28:149-152.

WILLIAMS, P. C., NORRIS, K. H., and ZAROWSKI, W. S. 1982. Influence of temperature on estimation of protein and moisture in wheat by near-infrared reflectance. Cereal Chem. 59:473-477.

WILLIAMS, P. C., and STARKEY, P. M. 1980. Influence of feed ingredients upon the prediction of protein in animal feed mixes by near-infrared reflectance spectroscopy. J. Sci. Food Agric. 31:1201-1213.

CHAPTER 8

Analytical Application to Fibrous Foods and Commodities

F. E. BARTON II
Russell Research Center
U.S. Department of Agriculture
Athens, Georgia, U.S.A.

I. INTRODUCTION

The analysis of highly fibrous feeds by near-infrared reflectance (NIR) analysis is different from that of grains in several respects. The components of the plant matrix are more complex and involve numerous discrete interactions. Recent work by Hruschka and Norris (1982) showed that for ground wheat, the summation of spectra of the chemical components (protein, starch, cellulose, moisture, and simple sugars) did not adequately reflect the total spectral composition when curve-fitting techniques were applied to the spectra. When the complexities of a forage sample are considered, it becomes obvious that the interaction of protein with lignin and carbohydrate along with minor constituents would make analyses by "pure components" impossible. Therefore, the analyses must be made on the basis of the spectra correlated to empirical results. Again, the amount of fiber is much greater—i.e., the fiber is the matrix as opposed to being a component, as in most foods and feeds.

There has been very little work completed on fiber content and fiber structure in foods and grains. Lund and Smoot (1982) studied the dietary fiber content of tropical fruits and vegetables. They found that not only was the amount of fiber small and variable, but the fibrous fraction differed between species and from that of forage plants. Baker (1977, 1978) and Baker and Holden (1981) examined cereals and grains for fiber content and evaluated several methods of fiber analysis to see which one would be most suitable for cereals. Baker (1977) found that a buffered acid detergent fiber determination apparently improved recovery of cellulose. The general viewpoint of these authors was that no one method seemed to be best as none gave a number that corresponded to an identical fraction in all cereals and grains. The new enzymatic procedures of Asp et al (1983) may provide both an assay and a means of characterizing fiber for monogastrics and ruminants. This chapter will be devoted to the following subjects: the structure and composition of highly fibrous materials, with forage grasses as the examples; methods of analyses; and NIR analysis specifically.

II. STRUCTURE AND COMPOSITION OF FORAGES

The microanatomical differences between warm- and cool-season (C-3 vs. C-4) grasses were examined by Akin and Burdick (1973). One of the major differences noted was that warm-season grasses had a more highly developed parenchyma bundle sheath than the temperate grasses. In this study, the sections were stained for lignin with acid phloroglucinol to identify the sites of lignification. In a subsequent study (Akin and Burdick, 1975), the percentage of each type of tissue was reported for temperate and tropical grass species. The warm-season (tropical) grasses contained far more of the less easily degradable tissues than the cool-season (temperate) grasses. Figure 1 shows the cross section of a typical warm-season grass, coastal bermudagrass (*Cynodon dactylon* (L.) Pers.) (CBG). Note the large thick-walled parenchyma bundle sheath on the outside of the large vascular bundles of CBG. This tissue is degraded slowly by rumen microorganisms in CBG and, in some cases, can show a positive reaction for lignin with histochemical reagents, such as chlorine sulfite. The mesophyll of the warm-season grass is composed of small, densely packed cells, whereas a typical cool-season grass, Kentucky-31 tall fescue (*Festuca arundinacea* Schreb.) (KY-31) has more loosely arranged cells. As shown by Akin and Burdick (1975), the cross-sectional area of fescue that comprises the mesophyll is 60% whereas that of bermudagrass is only 27%.

Warm-season grasses are recognized to have higher fiber contents than cool-season grasses. The values in Table I taken from Barton et al (1976) reflect average increases of 10 percentage units in neutral detergent fiber (NDF) and 4 percentage units in acid detergent fiber (ADF) for the warm-season grasses. These differences persist when only the four-week warm-season grasses are compared (NDF average 65.0, ADF average 34.6). The lignin data are of particular interest, in that the higher-fiber-content, less-digestible warm-season grasses also have a higher lignin content (4.9 vs. 4.3%). If one considers only the four-week regrowth samples, the higher-fiber, less-digestible warm-season grasses are identical in lignin (4.3%) and virtually identical in digestibility (61.6 to 61.5%) to the cool-season grasses, while maintaining an average of nine percentage units NDF and four percentage units ADF higher fiber content. Clearly, compositional differences do not answer the question of quality and animal performance differences for warm-season vs. cool-season forages.

One way to try to resolve some of the obvious differences between digestibility data and compositional analysis is to study both by the same method. Akin et al (1973) showed the pattern of forage digestion by rumen microorganisms (Fig. 2). This and subsequent studies (Akin and Burdick, 1975, 1981; Akin et al, 1975, 1977, 1983a, 1983b, 1984a, 1984b; Akin and Barton, 1983) have shown that the indigestible residue is made up of lignified vascular bundles, sclerenchyma, and cuticle for both warm- and cool-season species and that the parenchyma bundle sheath of the warm-season grass is very slowly degraded and not fully digested with a 48-hr in vitro incubation. Microscopic (light and electron) examination of the digested leaf sections allows direct observation of the plant tissues that were attacked and digested by rumen microorganisms. If samples are viewed after specific periods of incubation, the relative rate and mode of microbial attack can be ascertained. This qualitatively describes the fiber digestion of a given species grown under the specific environmental and management conditions for a given sample, but is not necessarily related to composition. To relate microscopic evaluation to a particular fiber analysis, the material represented by the residue after digestion by rumen microorganisms must be determined.

III. THE ANALYSIS OF FORAGES

Fibrous materials traditionally have been analyzed by the Weend Proximate Analysis Procedures as a means of estimating total digestible nutrients. In the proximate analysis procedures, percentage dry matter is determined by oven drying, percentage crude protein is expressed as 6.25 × percentage nitrogen from the Kjeldahl analysis, percentage fat by ether extraction, percentage crude fiber by alternate base and acid treatments, and percentage ash by incineration. These procedures continue to be the standard methods in use by many state testing laboratories. The analyses are empirical, based on the assumption that the reagents or experimental conditions affect each sample in an identical manner. They are all gravimetric procedures, and the calculated results are relative percentages. Moore and Mott (1973) and Martens and Russwurm (1983) have published excellent reviews that detail the status of gravimetric forage analyses. Since the molecular weight of a forage sample or any constituent therein cannot be determined, these percentages are the only way to express compositional properties quantitatively. The analyses are very dependent on sampling techniques, technician experience, the environment in which the sample is analyzed, and all of the procedures are destructive.

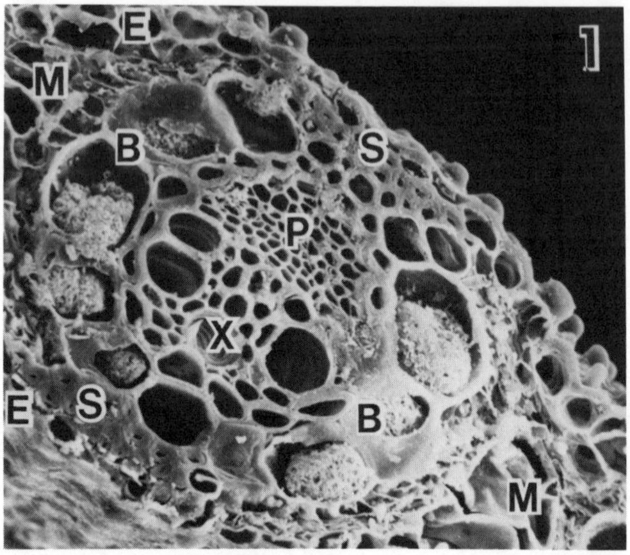

Fig. 1. Cross section of coastal bermudagrass showing various tissues within plant: sclerenchyma (S), epidermis (E), parenchyma bundle sheath (B), mesophyll (M), xylem (X), and phloem (P) (×500).

Peter J. Van Soest published procedures on detergent fiber analyses (Van Soest, 1963a, 1963b; Van Soest and Wine, 1967, 1968) and a means of classifying forage fractions (Van Soest, 1967, 1973). The detergent analysis system defines the sample by separating it into two fibrous fractions, an NDF fraction obtained by extraction in a boiling, buffered 2% sodium lauryl sulfate solution and an ADF fraction obtained by extraction with a boiling 2% solution of hexadecyltrimethyl ammonium bromide in 1.0*N* sulfuric acid. The NDF test (Van Soest and Wine, 1967) is considered to yield data on total fiber or cell walls, and the ADF test (Van Soest, 1963b) gives the lignified portion of the plant cell walls. The ADF procedure is primarily used as a preparatory step to lignin determinations, which are carried out either with 72% sulfuric acid (Van Soest, 1963b) or permanganate (Van Soest and Wine, 1968). The detergent analysis procedures, ADF and NDF in particular, have been used with varying degrees of success to predict respectively the extent of digestion and relative intake (Rohweder et al, 1978).

TABLE I
Percentage Compositional Analysis of Grasses[a]

Grass	IVDMD[b]	Protein	Ash	NDF[c]	ADF[d]	Hemi-cellulose	Holo-cellulose	PML[e]
Coastal: 4 week	66.1	19.2	7.8	61.0	29.1	31.8	61.2	4.1
Coastal: 8 week	50.4	11.0	5.4	71.2	40.0	31.2	66.8	6.0
Coastcross-1: 4 week	66.1	18.7	7.2	60.0	31.9	28.1	53.6	3.5
Coastcross-1: 8 week	54.9	13.9	7.2	62.9	39.0	23.9	55.6	5.5
Bahia: 4 week	59.6	15.7	6.2	71.0	35.7	35.3	60.8	3.4
Bahia: 8 week	53.2	9.2	5.9	67.5	35.0	32.5	76.0	5.3
Pangola: 4 week	54.5	7.0	4.7	69.4	41.7	22.7	57.3	6.3
Pangola: 8 week	48.6	5.9	5.1	67.0	29.5	37.6	47.4	4.8
Average tropical	57.8	12.6	6.2	66.2	35.2	30.4	59.8	4.9
Kenhy: 4 week	65.6	13.2	8.3	58.2	33.6	24.7	41.6	3.2
Ken-Blue: 4 week	58.1	15.5	7.1	54.0	30.6	23.5	43.5	4.3
Brome: 4 week	64.2	14.3	8.8	56.2	34.3	21.9	51.1	4.8
Orchard: 4 week	62.8	14.8	8.2	57.9	33.3	24.6	43.9	4.1
Kentucky-31: 4 week	62.7	14.2	8.8	58.4	31.4	25.5	45.2	3.4
Timothy: 4 week	66.8	13.4	8.8	55.6	34.7	20.9	42.0	4.1
Kentucky-31: 4 week (fall)	55.0	12.6	8.4	59.8	31.6	28.2	46.0	5.6
Kenhy: 4 week (fall)	60.1	12.6	8.0	57.3	30.7	26.6	43.8	4.2
Orchard: 4 week (fall)	58.8	17.8	9.7	54.0	29.4	24.6	45.5	4.8
Ken-Blue: 4 week (fall)	61.0	17.3	6.3	57.6	27.5	30.1	40.3	4.1
Average temperate	61.5	14.6	8.2	56.9	31.7	25.1	44.3	4.3
Average standard deviation	2.62	0.13	0.13	1.08	1.00	1.04	1.00	0.30

[a]Source: Barton et al (1976); used by permission.
[b]In vitro dry matter digestibility.
[c]Neutral detergent fiber.
[d]Acid detergent fiber.
[e]Permanganate lignin.

In recent years, these procedures have begun to replace the crude fiber procedure as the preferred methods of analyses for fiber in forages. The ADF procedure is now a method approved by the Association of Official Analytical Chemists (AOAC). The literature values for many forages in the above references are comparable to those in Table I that were obtained by the detergent analyses procedures. However, the assumption is made that the reagents are effecting the same treatment on all samples regardless of species, environment of growth, and agricultural management practices.

One way to examine this question is with microscopy, i.e., evaluation of leaf sections before and after microbial digestion. Akin et al (1975) examined by scanning electron microscopy the tissues that comprise the residues of NDF and ADF of leaf sections from CBG, a warm-season grass, and KY-31, a cool-season grass. In these experiments, 5-mm sections of the leaf blades were treated with the boiling reagents, prepared for microscopy, and viewed with a scanning electron microscope (Fig. 3). The mild treatment with NDF left the cell walls virtually intact in CBG and slightly distorted the mesophyll in some KY-31 samples. The amount of tissue removed from the leaf section was determined gravimetrically (Table II). It was found that much less tissue was removed from the sections than from Wiley-milled, ground leaf blades. The NDF conditions were such that the fragile cell wall membranes were not ruptured and cell contents not removed unless the cell was opened by the knife when the sections were cut (Table II).

Treatment with ADF reagents revealed differences both between species and for all species when compared with digestion. For the warm-season CBG, the residue contained portions of the parenchyma bundle sheath. This tissue (Fig. 3b), which resisted the acidic treatment, is slowly degraded by rumen microorganisms. The opposite is true for KY-31: the only tissues remaining after a 1-hr treatment were cuticle, sclerenchyma patches, and pieces of vascular tissue, far exceeding the digestion of KY-31 by rumen microorganisms. Thus, as a measure of extent of digestion (Rohweder et al, 1978), ADF would overestimate the digestion of KY-31 and underestimate the digestion or nutritive value of CBG. Direct comparisons of quality estimated from ADF values between temperate and tropical (cool- and warm-season) grasses must be made with caution. The differential response of the plant cell wall to these analytical reagents reflects differences in their availability to rumen microorganisms, and a linear response suitable for all species should not be expected.

Barton et al (1976) determined correlations between rumen bacterial digestibility and chemical data of warm- and cool-season grasses (Barton et al, 1976). It was found that protein and lignin content were the most significant factors. There was no correlation between digestibility and NDF if both the temperate and tropical grasses were considered as one set. Clearly, the empirical analyses do not truly measure quality when different and diverse species are being compared.

The digestibility of ADF by rumen microorganisms was examined by electron microscopy (Barton et al, 1981). The patterns of degradation or removal of tissue by ADF reagents and of digestion by rumen microorganisms paralleled those observed by Akin et al (1975). In these experiments, the isolated acid detergent treatment, followed by bacterial digestion, removed more fiber than either alone (Fig. 4a). The reverse dual treatment (Fig. 4b), i.e., digestion followed by acid detergent treatment, consistently removed even more tissue. The gravimetric results given in Table III showed about a 5–12 percentage unit difference for the two treatments when the order was reversed. The qualitative data from the micrographs greatly enhanced the interpretation of the numerical gravimetric results. Moreover, it showed where overinterpretation of apparent fiber digestion values can be made if no consideration is given to the effect of the dual treatment on the plant cell wall.

Traditionally, the major inhibitors to forage quality and therefore animal performance are considered to be the amount of cell wall fiber and the extent of lignification (Moore and Mott, 1973). Barton and Akin (1977) examined the effect of lignin removal from the cell wall on the remaining tissues and the digestion of those tissues by rumen microorganisms. Potassium permanganate, used to oxidize lignin from ADF, was used to delignify the cell walls of temperate and tropical grasses (Barton and Akin, 1977). The results of those experiments showed that the lignified tissues were disturbed (Fig. 5). The parenchyma bundle sheath was separated from the highly lignified inner bundle sheath and the sclerenchyma was, in some cases, separated into individual cells. In the temperate grasses, some of the mesophyll was removed. When the permanganate treatment was applied to NDF residue of ground forage, about 17% of the material was removed. Only about a fourth of this amount was lignin, and the rest of the material removed was carbohydrate. The rate and extent of digestion increased when the delignified leaf sections were incubated with rumen microorganisms. Tissues that were usually degraded between 24 and 48 hr were extensively degraded after only 1 hr (Fig. 6). Some tissues (sclerenchyma) that usually are not degraded also

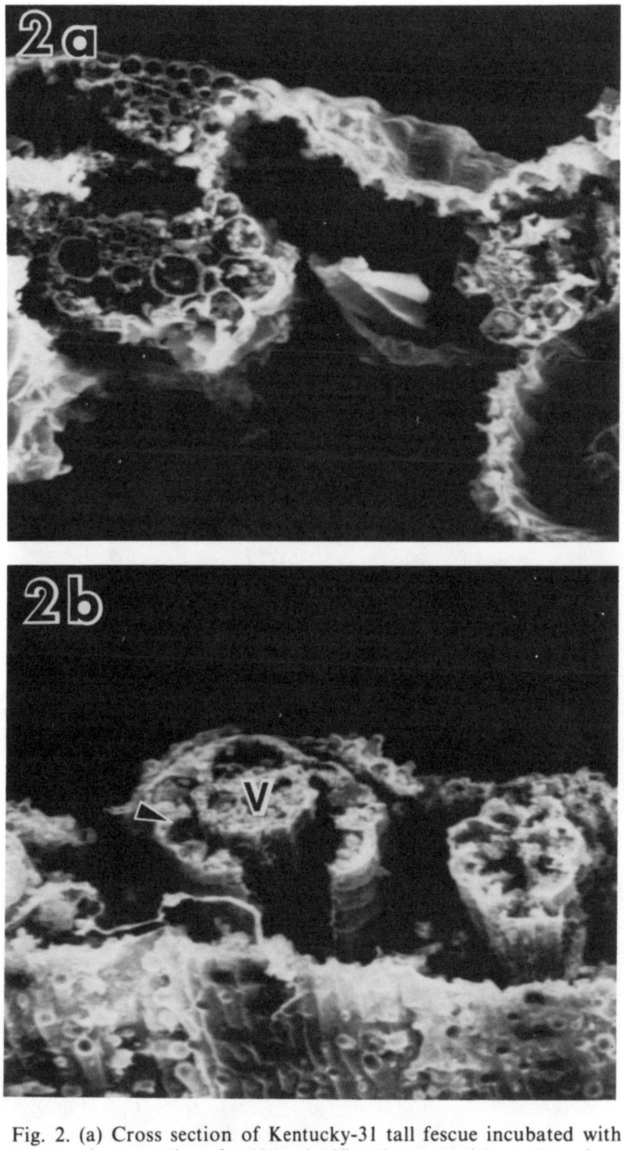

Fig. 2. (a) Cross section of Kentucky-31 tall fescue incubated with rumen microorganisms for 48 hr; lignified tissues, cuticle, and portions of bundle sheaths remain (×150). (b) Cross section of coastal bermudagrass incubated with rumen microorganisms for 48 hr; lignified vascular tissues (V) are partially degraded (×150).

were digested. Thus, it is not just the amount of lignin but the extent to which it is tied to the plant cell wall that determines the rate and extent of digestion and forage quality. These subjective criteria are not obvious to the analyst in the usual empirical analyses. It is quite possible to overinterpret the results of those analyses. Other methods give analytical results that are more definitive, but the cost in time and resources is high and the number of analyses that could be done is quite limited.

Two techniques that give more definitive results are high-pressure liquid chromatography (HPLC) and carbon-13 nuclear magnetic resonance spectroscopy (^{13}C-NMR). These techniques have been used to characterize hemicellulose, lignin, and lignin-carbohydrate complexes. Each has some distinct advantages over the other, but they also complement each other.

Albersheim et al (1967) published the alditol acetate procedure for the analysis of plant cell wall hydrolysates. The standard gas-liquid chromatograph (GLC) method had one disadvantage in that sample preparation was laborious and time-consuming. The HPLC method of Barton et al (1982) and the modifications by Windham et al (1983) have decreased the number of steps and time required as well as providing an improved recovery and good materials balance.

The procedure is essentially a 2.0*N* trifluoroacetic acid hydrolysis of NDF followed by HPLC analysis of the hydrolysate. The method is precise—i.e., a 1–2% error in each component sugar (Table IV)—and small differences in component sugars are readily apparent ($P < 0.05$) (Table V). Perhaps the most significant factor of these analyses is the materials balance. When simple sugars or commercial polysaccharides were hydrolyzed and analyzed, the recovery through HPLC was approximately 96.7–98.5%. For a sample of NDF, about 50% of the fiber is hydrolyzed and present in the hydrolysate and 50% is present in the fibrous residue (Windham et al, 1983). Of the fraction hydrolyzed, only 40–70% (20–35% of the NDF) was recovered. That fraction not recovered as sugar either precipitated in the round bottom flask during washing or was retained in the Whatman C-18 Sep-Pak that

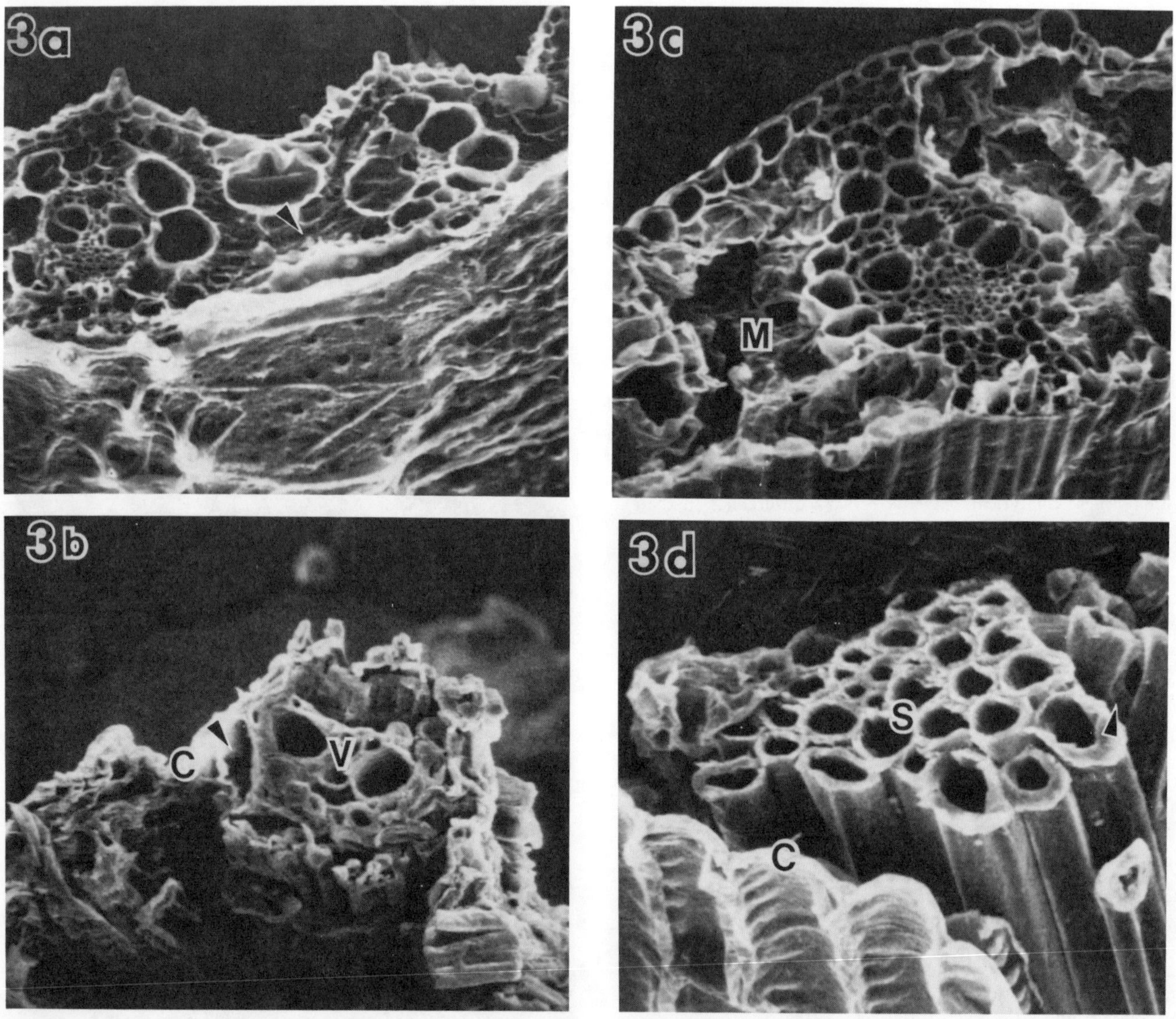

Fig. 3. (a) Cross section of coastal bermudagrass leaf extracted for 60 min with neutral detergent reagent; tissues, including mesophyll (arrow), are intact as in control sections (×240). (b) Cross section of coastal bermudagrass extracted for 60 min with acid detergent reagent; lignified vascular tissue (V) and cutinized epidermis (C) remain, and portions of outer bundle sheath (arrow) are infrequently observed to remain (×256). (c) Cross section of Kentucky-31 tall fescue leaf extracted for 60 min with neutral detergent reagent; tissues are similar to those in control samples, and all are intact except for mesophyll (M) (×224). (d) Portions of tissues remaining in Kentucky-31 tall fescue after 60-min extraction with acid detergent reagent; cuticular layers of epidermis (C) and sclerenchyma (S) separated into individual cells (arrow) are seen (×650).

was used as a final cleanup step. The acetone soluble material, which was yellowish brown, resembled a lignin carbohydrate complex (LCC). The yellowish-brown material has been analyzed by ^{13}C-NMR. The work of Barton et al (1981) on ADF digestion and Akin et al (1975) on the NDF and ADF of forage grasses and earlier work on the grasses themselves (Akin et al, 1973) suggests that the readily hydrolyzable fraction could be the slowly degraded portion of the plant cell wall, and the extent to which it is tied to lignin or plant phenolics could determine its rate of digestion. The highly lignified portion of the plant cell wall that is essentially indigestible establishes by difference the maximum energy a ruminant can derive from a given forage sample. The largest compositional difference between tropical and temperate grasses is the amount of "hemicellulose." Tropical grasses contained 30–35% more hemicellulose, as estimated by the percentage NDF minus the percentage ADF of the respective forage (Goering and Van Soest, 1970; Barton et al, 1976). Van Soest (1967) also stated that hemicellulose appears to have a special relation to lignin. The above results would support that statement.

For 40 years, NMR spectroscopy has been one of the most definitive structural elucidation tools and has provided chemists with a wealth of structural and analytical information. During the last 15 years, the utility of this spectroscopic technique has been extended to more complicated and biological samples by ^{13}C-NMR. The technique was used by Barton et al (1982) and Windham et al (1983) to assist in the analysis of the hydrolyzable fraction of plant cell walls previously discussed. In this case, the hemicellulose of forages was obtained by alkali extraction (Routley and Sullivan, 1958). The NMR spectra (Fig. 7) show structures that have been interpreted as glucoarabinoxylans. The spectrum contained no resonance (signal) for the C-6 carbon of glucose, the only carbon different from the first five carbons of the sugar xylose. It is known from HPLC analysis that the hemicellulose fraction contains 5–10% glucose. One may reasonably assume that the C-6 of glucose is bound and shifted under another resonance. This assumption was verified by adding the totals for xylose and glucose obtained from the NMR (Table VI) and comparing the

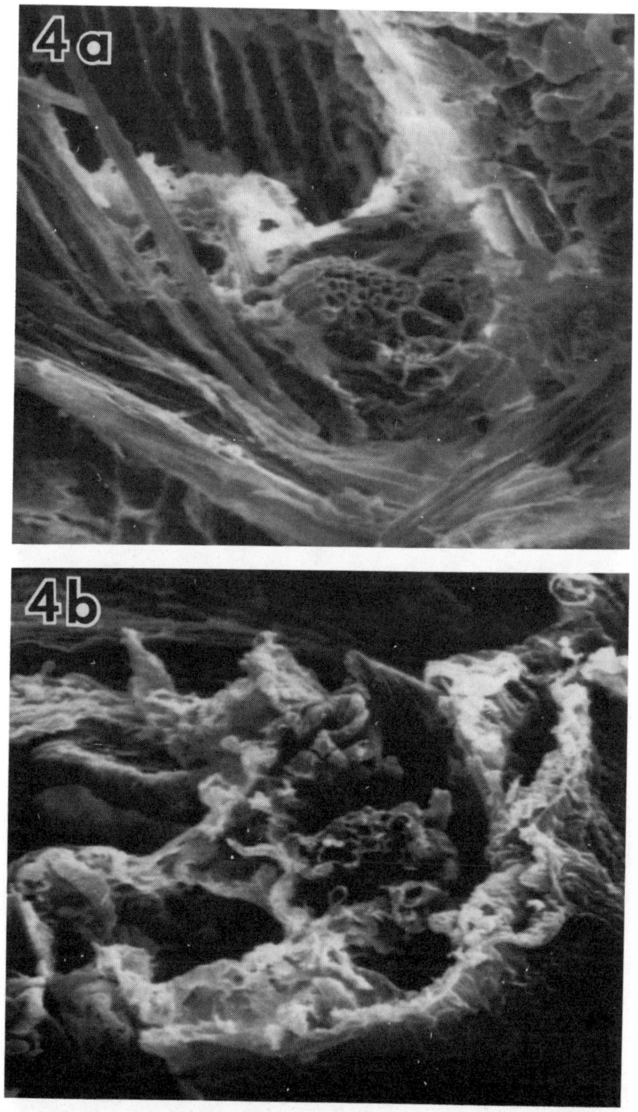

Fig. 4. (a) Cross section of Kentucky-31 tall fescue extracted for 60 min with acid detergent reagent and then incubated with rumen microorganisms for 48 hr; only cuticle, portions of vascular bundles, and sclerenchyma remain (compare with Figs. 2a and 3d). (b) Cross section of Kentucky-31 tall fescue incubated with rumen microorganisms and then extracted with acid detergent reagent for 60 min; small pieces of vascular bundles, cuticle, and some sclerenchyma remain (compare with Figs. 3a and 4a).

TABLE II
Percentage Residue of Neutral Detergent Fiber and Acid Detergent Fiber from Whole, Wiley-Milled Forage and Intact Leaf Samples of Coastal Bermudagrass and Kentucky-31 Tall Fescue[a]

	Neutral Detergent Fiber		Acid Detergent Fiber	
Grass	**Whole**[b]	**Leaf**[c]	**Whole**	**Leaf**
Coastal bermudagrass	59.4 ± 0.3	78.3 ± 1.6	29.1 ± 0.8	25.3 ± 0.7
Kentucky-31 tall fescue	50.7 ± 0.6	79.1 ± 2.3	28.6 ± 0.2	27.8 ± 1.3

[a] Source: Akin et al (1975); used by permission.
[b] Average of 12 determinations plus standard deviation for whole, ground samples.
[c] Average of 3 determinations plus standard deviation for leaf samples.

TABLE III
Percentage Digestibility of Five Grasses, Acid Detergent Fibers, and Combined Dry-Matter Disappearance (CDMD)[a]

Grass	% IVDMD	% IADFD	% ADER	CDMD-1[b]	CDMD-2
Coastal bermudagrass	57.41	5.66	46.56	77.24	67.08
Coastcross-1 bermudagrass	61.17	5.91	41.17	77.16	68.53
Kentucky-31 tall fescue	64.54	5.76	44.73	80.40	76.93
Kenhy tall fescue	66.87	7.18	47.15	82.49	75.50
Orchardgrass	64.53	6.98	53.86	83.63	71.35

[a] Source: Barton et al (1981); used by permission. Isolated acid detergent fiber (ADF) disappearance (IADFD) is dry matter removed from isolated ADF by rumen microorganisms; acid detergent extracted residue (ADER) is percentage dry matter removed by extraction with acid detergent reagent of in vitro residue.
[b] CDDF-1 = (NDF × ADER/100) + IVDMD; CDMD-2 = (% ADF × IADFD/100) + (100 − ADF), where NDF is neutral detergent fiber.

compositional values to the HPLC results (Table V). The agreement was within ±5% when each was compared as percentage of whole grass. From these data, a structure for grass hemicellulose that has an arabinoxylan attached to a 1:4 glucan (Fig. 8) was suggested. The work of Albersheim et al (1967) describes the hemicellulose of sycamore as a xyloglucan and an arabinoxylan. The evidence described above from the ^{13}C-NMR indicates a single structure.

The NMR method has been used to elucidate the structure of forage lignin (Himmelsbach and Barton, 1980). Lignin, which is a polymer of phenylpropanoid (C-9) units, is quite complex. It differs only in the 3 and 5 position substitution of the phenyl ring by methoxyl groups, i.e., *p*-coumaryl (none), coniferyl (monomethoxyl), and syringyl (dimethoxyl). The work of Ludemann and Nimz (1973, 1974a, 1974b), Nimz (1974), Nimz and Ludemann (1974, 1976), and Nimz et al (1974a, 1974b, 1975) has characterized the lignins of hardwoods and softwoods. Acetal lignins extracted and described by Bolker and Teraschima (1966) contained some carbohydrate, as did the lignin extracted from bermudagrass by the milled wood lignin procedures (Fig. 9). Grass lignins cannot be extracted free of carbohydrates. There were differences noted by Himmelsbach and Barton (1980) in the spectra of the lignins from different species of grass. The cool-season (temperate) fescue contained about equal amounts of the three lignin moieties, whereas the warm-season (tropical) bermudagrass contained predominantly the unsubstituted *p*-coumaryl moiety. This unsubstituted lignin would be the one considered most likely to be attached to carbohydrates (Sarkanen and Hergert, 1971) and would have the greatest chance to tie up more carbohydrates. This tie-up of more of the warm-season grass carbohydrates through the *p*-coumaryl unit would render them less available for digestion by rumen microorganisms than the cool-season grass carbohydrates. The nature of this binding of carbohydrate and lignin had not been discretely determined. Harkin (1973) has suggested a covalent linkage that would be consistent with the work of Himmelsbach and Barton (1980). Evidence can be found for the existence of an ether or a carbonyl ester resonance in the lignin spectra and in those of the LCC shown in Himmelsbach and Barton (1980). The large resonances at 162 and 172 ppm (Fig. 1 in Himmelsbach and Barton, 1980) indicated the presence of carbonyls in esters.

Examination of the spectra of the LCC isolated from the HPLC analysis of the NDF showed an arabinoxylan associated with an unsubstituted lignin moiety and the presence of an ester linkage (Fig. 10). Subsequent HPLC analysis of the carbo-

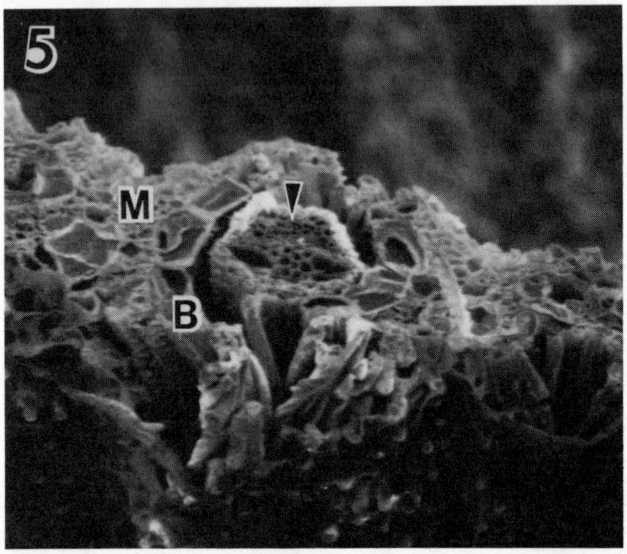

Fig. 5. Cross section of CBG-NDF treated for 30 min with buffered $KMnO_4$ reagent and demineralized for 10 min. Mesophyll (M) and phloem (arrow) are intact but parenchyma bundle sheath (B) is almost completely separated from rest of vascular bundle. Sclerenchyma has been divided into individual cells and separated from vascular bundle (×384). (Reprinted, with permission, from Barton and Akin, 1977)

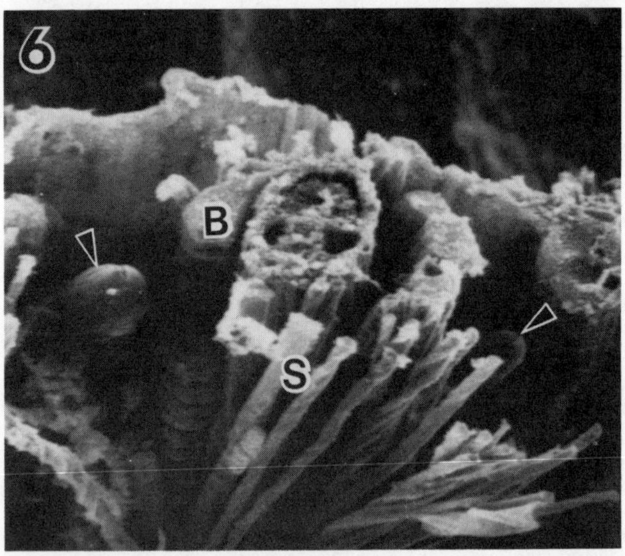

Fig. 6. Delignified CBG-NDF digested. Mesophyll is completely removed and phloem and parenchyma bundle sheath (B) are beginning to be removed. Sclerenchyma (S) is broken apart into fiber cells, and rumen protozoa (arrows) are present between vascular bundles (×384). (Reprinted, with permission, from Barton and Akin, 1977)

TABLE IV
Precision of Method[a]

	Sugar Components (mg) After Repeated Injections[b]			
Injection No.	**Xylose**	**Arabinose**	**Glucose**	**Galactose**
1	37.13	11.57	13.88	TR[c]
2	37.28	11.74	13.73	TR
3	37.82	11.53	14.45	TR
4	38.14	11.76	14.16	TR
5	37.98	11.46	14.09	TR
6	37.19	11.08	13.91	TR
7	37.35	11.76	13.75	TR
8	37.89	11.55	14.47	TR
9	37.13	11.32	13.51	TR
10	37.64	11.17	13.43	TR
Mean	37.55	11.49	13.94	TR
SD	0.38	0.24	0.35	
CV	1.02	2.09	2.55	

[a]Source: Barton et al (1982); used by permission.
[b]Milligrams recovered per 100 mg of coastal bermudagrass neutral detergent fiber hydrolyzed.
[c]Trace; amount present was inadequate for quantitative purposes.

TABLE V
Precision of Sample Preparation for Coastal Bermudagrass (CBG) Neutral Detergent Fiber[a]

Sample	Xylose	Arabinose	Glucose	Galactose
CBGa	36.67 ± 0.44[b]	11.58 ± 0.37	14.06 ± 0.26	TR[c]
CBGb	37.47 ± 0.36	11.46 ± 0.34	14.04 ± 0.36	TR
CBGc	37.35 ± 0.26	11.64 ± 0.39	13.92 ± 0.34	TR

[a]Source: Barton et al (1982); used by permission. Data presented as milligrams recovered per 100 mg of fiber hydrolyzed.
[b]Each value is mean of five observations (i.e., five injections of three sample preparations).
[c]Trace; amount present was inadequate for quantitative purposes.

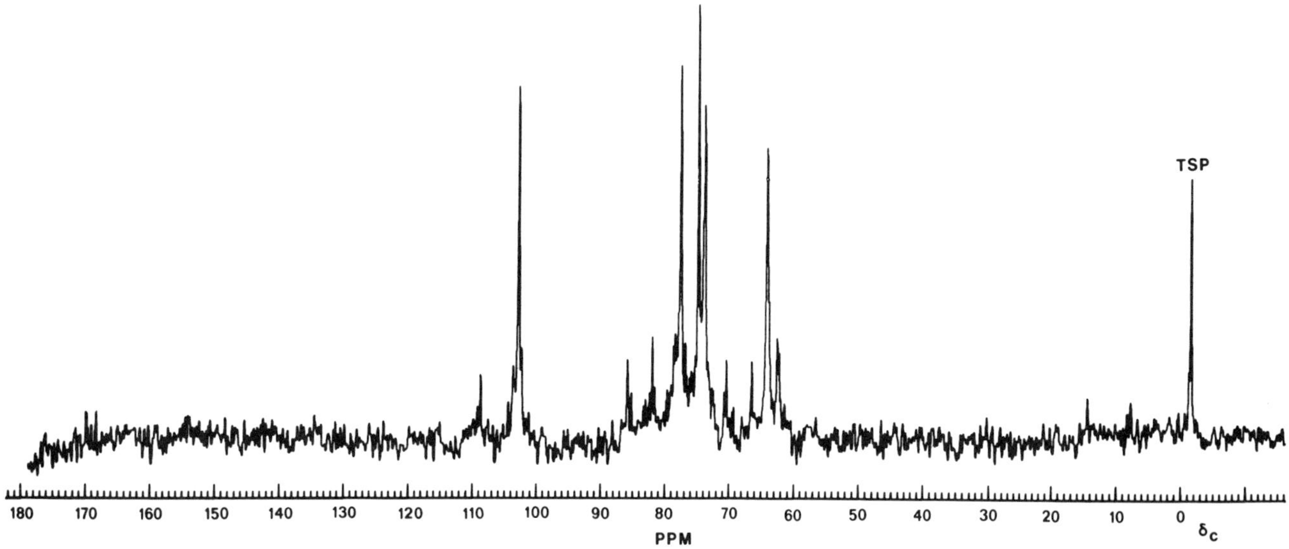

Fig. 7. Carbon-13 nuclear magnetic resonance spectrum of coastal bermudagrass hemicellulose (in D_2O, 70° C) at 25.2 MHz.

TABLE VI
Relative Percentage Composition of Coastal Bermudagrass (CBG) and Orchardgrass (OG) Forage Hemicelluloses by C-13 Nuclear Magnetic Resonance

Carbohydrates	CBG	OG
Xylose + glucose	84.1	77.5
Arabinose	15.9	22.5

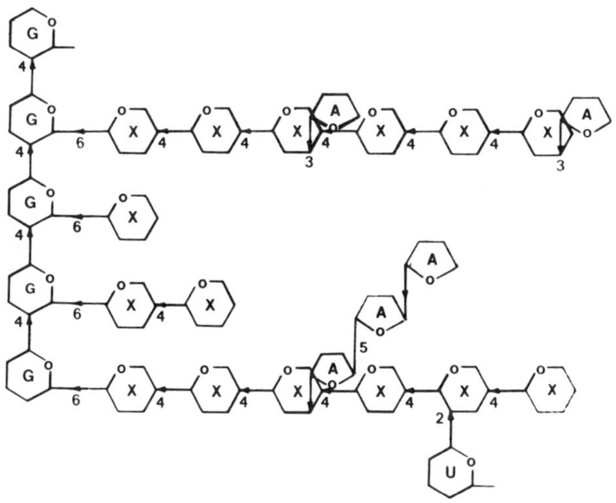

Fig. 8. Possible structure for polysaccharide portion of grass hemicellulose.

hydrate portion gave a xylose-to-arabinose ratio of 1:1 (F. E. Barton II, W. R. Windham, and D. S. Himmelsbach, unpubl. data, 1983). This suggested that the lignin was preferentially associated with the arabinose moiety because a normal xylose-to-arabinose ratio is in the range of 3–4:1.

The use of ^{13}C-NMR and the interpretation of the spectral data along with HPLC data gave a fairly clear picture of the structure of the molecular entities that comprise the more fibrous and lignified portions of the plant. It can be said with some certainty what comprises the plant cell wall and how some of the pieces fit together, but the methods lack several important criteria for good routine analytical methods.

By definition, a good analytical method is one that is accurate, precise, rapid, low cost, easily interpretable, and simple in sample preparation and that examines the structure in situ with no chance for sample preparation artifacts to alter the results. Certainly the NMR method is accurate, precise, and interpretable, but it is neither rapid nor low cost. The HPLC method is an improvement in speed over the gas chromatography alditol acetate procedures and reasonable in cost, but still not fast enough to handle large numbers of samples. The standard gravimetric methods are inexpensive, and large sample numbers are accommodated by tailoring laboratories. Many state testing laboratories run 2,000–3,000 samples a year. What is needed is a faster, cheaper method. Three recent texts have given updated and expanded treatments of the standard methodology (Ulbricht and Theander, 1980; Wheeler and Mochrie, 1981; Martens and Russwurm, 1983). A recent article by Asp et al (1983) described an enzymatic method of fiber analysis and a method of describing soluble and insoluble dietary fiber. These procedures may be refined and automated, but they still retain the high labor costs involved with gravimetric analyses.

IV. NIR AS AN ANALYTICAL METHOD

Near-infrared reflectance (NIR) analysis is a rapid, inexpensive method to analyze forages for compositional, bioassay, or quality estimates. The method was developed and used for the rapid determination of oil, protein, and moisture in grains and oilseeds (Norris and Hart, 1965; Ben Gera and Norris, 1968; Hymowitz et al, 1974; Rinne et al, 1975). It has been used to predict the nutritive value of feedstuffs (Norris and Barnes, 1976; Shenk et al, 1977). More recently, the composition of forages has been investigated (Norris et al, 1976; Shenk and Hoover, 1976; Shenk et al, 1977, 1978; Barton and Burdick, 1979, 1980, 1983; Barton and Coleman, 1981; Burdick et al, 1981).

The NIR method requires a near-infrared spectrometer equipped for diffuse reflectance, a minicomputer for data handling and calculation of calibration equations, and suitable samples to be used for calibration. A set of samples (n) is selected for calibration. There are three basic ways to select samples for the calibration: (i) Y% or some number of samples from the set taken randomly; (ii) every i^{th} sample; or (iii) a subset that covers the analytical range and is qualitatively similar to the entire sample population to be analyzed. In the case of the particular system that is part of the National Near-Infrared Research Project (U.S. Department of Agriculture), spectra are taken and 700 data points for each sample are stored in a computer file. The data for several compositional factors (X) are added to the spectral information. The file then contains n samples of 700 data points and X constituents. Multiple linear regression techniques are used to generate an analytical equation as shown in equation (1):

$$\% X = C_0 + C_1(I_1) \pm C_2(I_2) \pm C_j(I_j)\,, \qquad (1)$$

where C_0 is a constant representing the intercept and C_j are the constants multiplied by the intensity function at j. The regression techniques can be stepwise or multiple stepwise (Shenk et al, 1977); they can involve derivatives of the logarithm of reciprocal reflectance (Norris et al, 1976; Shenk et al, 1978) and combinations of techniques such as ratios of derivatives (Norris, 1983a; 1983b). A recent paper describes the use of Gauss-Jordan linear algebra as a means of obtaining calibration equations (Honigs et al, 1983). An equation can be derived for each compositional factor. The spectra of unknown samples are then taken and the analytical result calculated. It is possible to obtain up to 10 analyses per sample in a few minutes. Although the instrument is expensive, the per-sample cost is quite low. Mobile systems are available to take the analysis to the samples (J. S. Shenk, pers. comm., 1982). Once properly calibrated, the instruments can be used to analyze several hundred samples a day.

Calibration of NIR instruments has been, and will continue to be, the source of much research (Barton and Burdick, 1979; Barton and Coleman, 1981; Hruschka and Norris, 1982; Honigs et al, 1983). Calibration depends on three factors: the appropriateness of the calibration set, the quality of the calibration data, and the mathematical treatment of the spectral data. The appropriateness of the samples chosen for calibration has been the subject of considerable statistical debate. As mentioned above, some researchers believe that the selection should be random or at least arbitrary from the total population

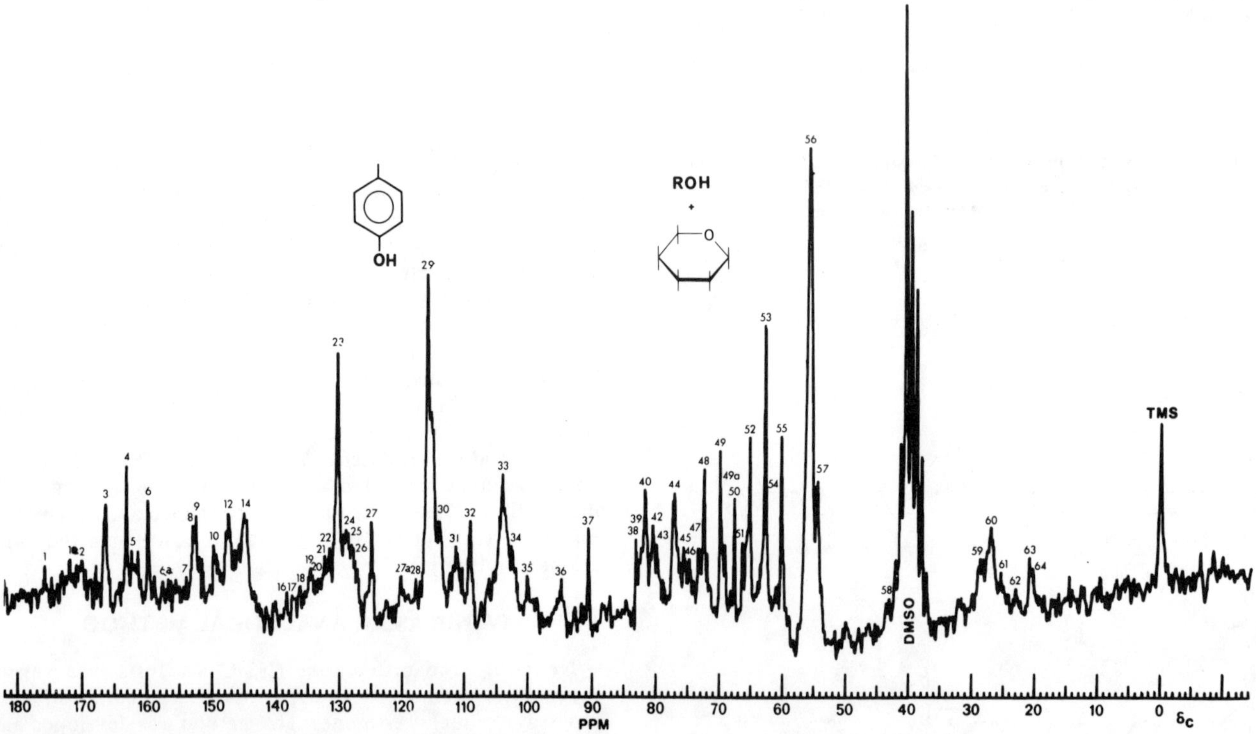

Fig. 9. Carbon-13 nuclear magnetic resonance spectrum of coastal bermudagrass lignin at 25.2 MHz. Unsubstituted aromatics predominate (signals 23 and 29). Presence of carbohydrate is shown in 60–90 ppm region.

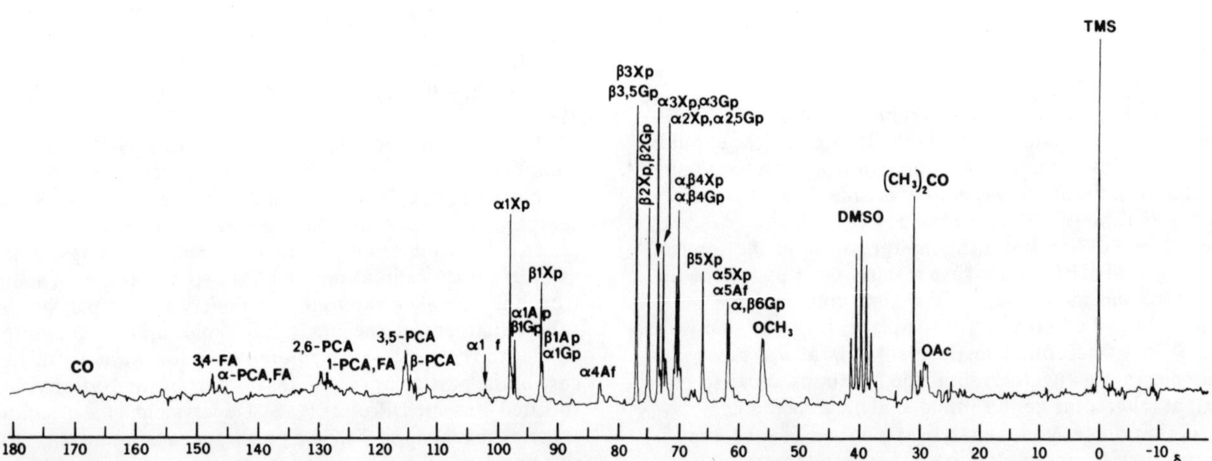

Fig. 10. Carbon-13 nuclear magnetic resonance spectrum of nonpolysaccharide residue hydrolyzed from coastal bermudagrass neutral detergent fiber. Appears to be lignin-carbohydrate complex.

(Shenk and Hoover, 1976); others would choose the samples to reflect a wide range of compositional data (Barton and Coleman, 1981). However a set is chosen, once its spectra are recorded, the choice of the mathematical data treatments must be considered next.

Norris et al (1976) reported that the logarithm of reciprocal reflectance and its second derivative were the two most promising mathematical treatments for forages. The logarithm of reciprocal reflectance gives a spectral result analogous to absorbance. The second derivative (D2OD) can be obtained by successive derivatization over typically 20–30 nm with some smoothing, or by using three equal spectra lengths (A, B, and C) of 20–30 nm and the equation:

$$\text{D2OD} = \log(1/R)\,A - 2\log(1/R)\,B + \log(1/R)\,C\,. \quad (2)$$

Maddams (1980) reviewed curve-fitting and derivative techniques to show both their utility and their limitations. Most of the forage work has been done using second derivatives. This mathematical treatment generally gives better-resolved spectra with minima occurring for each maximum that was resolved. The data treatment can also remove differences resulting from a nonuniform particle size from the spectra. If the appropriate derivative segment is selected, a multicomponent absorption signal (Fig. 11) can be resolved into its individual absorption with a minimum at the wavelength where the maximum in log ($1/R$) (absorbance) spectra would have been had the band resolved.

The spectra in the near infrared are broad, featureless, often overlapping absorbances. The signals arise from carbon-hydrogen stretching vibrations, their harmonics, overtone, and combination bands. Early work by Kaye (1954, 1955) characterized the spectra for simple molecules and described basic instrumentation necessary for acquiring spectra.

The signals in this region, 0.8–2.5 μm, are weak (100 to 1,000 times less intense than the midinfrared), but the radiation sources are more intense and the detectors more efficient. Since the amount of absorbed light is small, the distortion is also small. The NIR is an optically thin region—i.e., dispersion is weak and refractive indices are relatively constant. The weak signals, which would normally be considered a disadvantage, are in reality more of an advantage, since there is less variation in signal intensity compared with either the midinfrared or ultraviolet.

The third calibration consideration is the quality of the analytical data of the samples themselves. The data in Table VII were obtained from 100 Old World Bluestem grass samples that were used to calibrate an NIR instrument (Coleman and Barton, 1982). These data were obtained from three separate calibrations. First, the laboratory gravimetric values from the samples that were used in the calibration were recorded two to three years before the spectra were taken. The results were not good, as indicated by low R^2 values and high standard errors of calibration (SEC) (Table VII, rows corresponding to run I). The R^2 and SEC improved when all the samples were reanalyzed and the regression made to recalibrate the instrument (Table VII, rows corresponding to run R). There were still some samples that were identified as having suspect data—i.e., they failed the t test, or had spectra that differed from those of other samples with similar data. These samples were reanalyzed and their new data, if different, substituted in the calibration file to give the third set of R^2 and SEC values (Table VII, rows corresponding to run C), which were more acceptable. Good standard analytical data are absolutely essential to calibration. The subsequent NIR analyses can only be as good as the total of all laboratory errors.

The calibration and analyses of forages have been accomplished in our laboratory on three instrument systems. The first was a tilting-filter reflectance spectrometer; the second was an early commercial scanning monochromator; the third was the spectrophotometer common to the U.S. Department of Agriculture Near-Infrared Network (Barton and Burdick, 1979; Barton and Coleman, 1981; Burdick et al, 1981; Barton and Burdick, 1983). Although these instruments differ in design and levels of sophistication of the spectrometer, they all yield spectral data with 300–2,000 data points per spectrum. The tilting-filter and the network spectrometer use 700 data points (the network spectrometer produces 1,873, but only 700 data points are centered and averaged data points are saved). The older monochromator initially took only 300 data points, but currently takes 2,000. This capability is important when multiple analyses and mathematical transformations (derivatives, ratios, etc.) of the spectral data are required.

Fixed-filter instruments give one data point per filter. As such, the scanning systems are capable of producing two orders of magnitude more spectral data. With this increase in available data comes the requirement for computational hardware and software to utilize it. The inherent computational capabilities of the tilting-filter instruments are quite limited. Barton and Burdick (1979) and Burdick et al (1981) showed that although one "product" (file) would hold six equations based on six wavelengths with a possibility of using one or more of five mathematical treatments, it was necessary to use a different set of wavelengths for each constituent to be analyzed (Table III in Burdick et al, 1981). To change derivative segment width, a new EPROM (Erasable Programable Read Only Memory) chip had to be installed. This constituted a major revision (Table V in Barton and Burdick, 1979). The current network instrumenta-

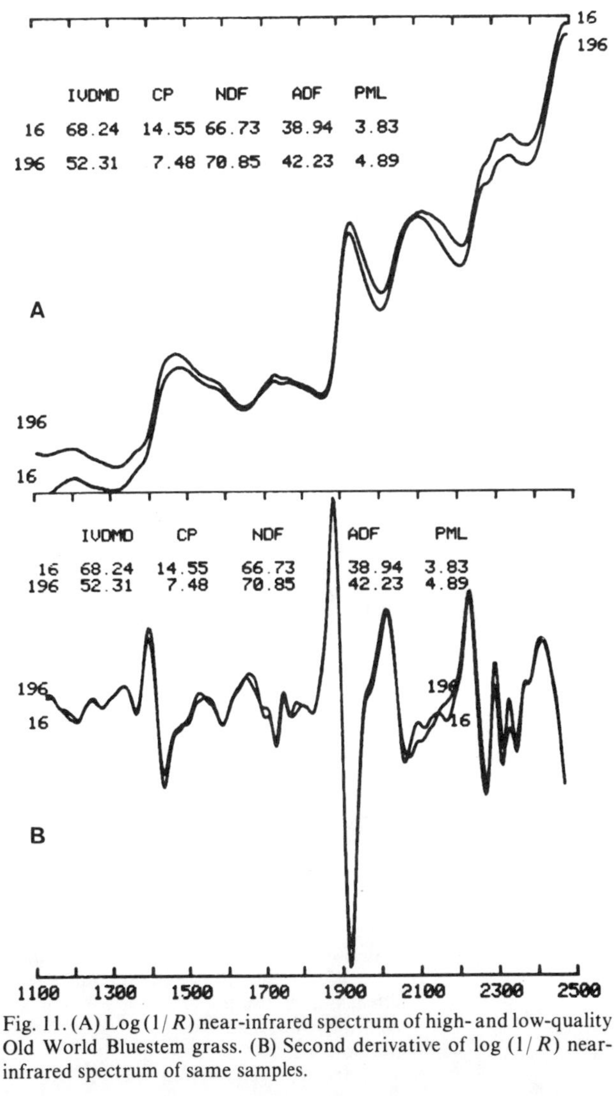

Fig. 11. (A) Log ($1/R$) near-infrared spectrum of high- and low-quality Old World Bluestem grass. (B) Second derivative of log ($1/R$) near-infrared spectrum of same samples.

tion, which incorporates a holographically ruled grating monochromator, is interfaced to, in our case, a Digital Equipment Corporation PDP 11/34A minicomputer. This computer and its attendant software (which was written at Pennsylvania State University) run either the monochromator or the tilting-filter spectrometer. The computer stores all the spectral data points and performs smoothing and mathematical transformation. In addition, all of the laboratory data are stored so that the spectral transformation parameters can be changed and the regressions rerun without having to reenter data. The data files can also be edited, which has led to improvement in SEC and SEP (standard error of performance) as shown in Barton and Coleman (1981, Table IX). Norris (1983a) has prepared an excellent review of the process of extracting compositional data from spectral data.

Two areas of the NIR method need further study. First, the precision, and with it the accuracy of the fiber and other laboratory measurements, must be improved. Second, the physical significance of the analysis must be determined. It is essential that methods be developed that measure components in the plant that influence its digestion and subsequent utilization by ruminant animals.

The Tecator Fibertec System is a semiautomated extraction apparatus that enables improvements in accuracy and precision of gravimetric fiber determinations. The word *extraction* is used here in the context of solubilization of materials from the sample whether they are hydrolyzed, digested, solubilized by reflux, or complexed. The accurately weighed sample is introduced into a sintered glass crucible and placed into the Fibertec hot-extraction unit (up to six crucibles at a time). The reagents are pumped into the crucibles (preheated) and boiled in the crucible extraction column. After the allotted time, the samples are filtered and washed with no sample transfer. The Fibertec has an excellent system for washing the sample free of all detergent, which eliminates one large source of error. The vacuum filtration and washing are very uniform for all samples, further helping to reduce errors. The precision of analysis is improved overall, as shown in Table VIII. The standard deviation for triplicate determinations is roughly three-fourths that of the conventional boil-and-stir procedure.

Improvement of calibration data can be demonstrated by using the data to calibrate an NIR spectrophotometer. Table IX shows the results of three different wet-chemical calibration data sets on the same sample. The first (A) is the average of triplicate analyses by four different research laboratories (12 replicates); the second (B) is a single-laboratory triplicate determination from the author's laboratory. The third is triplicate determination using the Fibertec in our laboratory. The precision (as determined by lower SD of triplicate) of the Fibertec (F) results is generally better than the average of the four laboratories and always better than the single-laboratory determination. This is reflected in the calibration of an NIR instrument by the R^2 and SEC values. Note the large between-laboratory errors as indicated in the standard deviation column for the average of the combined results from four laboratories. The Fibertec results are obviously more precise, but since the results appear as good or better than the averages for the data for calibration from the four laboratories, the results may also be more accurate and yield a better calibration. Data in Table IX provide an internal check for within-laboratory consistency. Since crude protein was not run on the Fibertec, those values labeled "R" are a repeat of the single-laboratory analysis done at a different time and used for an internal check.

The second major limiting factor to the NIR method is the relationship of the compositional data to the spectra. Recent work by Himmelsbach et al (1983) has shown that solid-state ^{13}C-NMR can be used to compare the relative ratio of plant constituents. This method may be able to provide accurate data on constituents as they are in situ. Historically, the one great limitation to the use of ^{13}C-NMR for structural and quantitative results is that the sample must be dissolved in a

TABLE VIII
Variation Resulting from Technique

Constituent	Conventional		Fibertec	
	Mean	SE[a]	Mean	SE
Neutral detergent fiber	68.5	0.30	64.9	0.24
Acid detergent fiber	39.1	0.39	36.4	0.34
PML	6.54	0.45	5.43	0.74

[a] Between replicate mean squares.

TABLE VII
Effect of Laboratory Data on Calibration[a]

Analysis	Run[b]	Mean + SD	λ's[c]	R^2[d]	SEC[e]	Repeat[f]
Dry matter	I	93.3 ± 1.82	3	0.63	1.11	0.03
	R	95.3 ± 1.16		0.73	0.61	0.06
	C	94.7 ± 1.06		0.84	0.43	0.02
Protein	I	12.2 ± 1.96	3	0.84	0.80	0.03
	R	12.4 ± 2.00		0.87	0.71	0.04
	C	12.3 ± 2.01		0.94	0.49	0.05
Neutral detergent fiber	I	67.7 ± 3.12	5	0.65	1.86	0.16
	R	68.2 ± 2.91		0.73	1.52	0.08
	C	67.6 ± 2.71		0.82	1.15	0.27
Acid detergent fiber	I	38.8 ± 2.55	3	0.63	1.55	0.10
	R	38.6 ± 2.95		0.81	1.27	0.25
	C	38.8 ± 2.90		0.87	1.04	0.09
Permanganate lignin	I	4.9 ± 1.14	3	0.38	0.90	0.05
	R	3.7 ± 1.23		0.66	0.71	0.01
	C	3.7 ± 0.87		0.61	0.54	0.01
IVDMD[g]	I	60.4 ± 4.10	3	0.68	2.33	0.21
	R	60.2 ± 5.78		0.65	3.40	0.33
	C	62.5 ± 3.34		0.83	1.36	0.11

[a] Source: Coleman and Barton (1982); used by permission.
[b] I = file data were several years old; R = samples were all reanalyzed routinely; C = samples from R that were statistical outliers were reanalyzed and new data incorporated into file.
[c] Number of wavelengths used in equation.
[d] Coefficient of determination.
[e] Standard error of calibration.
[f] Repeatability error.
[g] In vitro dry matter digestibility.

suitable solvent before the spectrum can be taken. This limitation means that solid samples that are insoluble (i.e., whole grass) are not compatible with the NMR technique. Again, the extraction of hemicellulose, lignin, or LCC modifies the in situ structure. Lignin once isolated is very susceptible to condensation reactions upon exposure to moisture and air. Interpretations must be made with the knowledge that artifacts of isolation are likely to exist. The new NMR technique of cross-polarization/magic-angle spinning (CP/MAS) (Schaefer and Stejskal, 1979) has made it possible to record the spectra of rigid materials, including fibrous residues and even whole-grass samples.

Basically, there are three important constituents of plant fiber: carbohydrate, lignin, and protein. Solid-state ^{13}C-NMR spectra of these constituents have been obtained, separately or in combination, by several groups of researchers in the past three years (Atalla et al, 1980; Earl and VanderHart, 1980, 1981; Rutar et al, 1980; Dixon et al, 1981; O'Donnel et al, 1981; Schaefer et al, 1981) with the CP/MAS technique (Schaefer and Stejskal, 1979). Several researchers have also reported that the ^{13}C-NMR signals caused by protein carbonyls were proportional to values obtained for protein by standard chemical methods.

Using this technique, Himmelsbach et al (1983) reported the application of CP/MAS ^{13}C-NMR for the determination of the ratio of carbohydrate, lignin, and protein between grass species from single spectra of whole-plant material. Figure 12 gives the spectra of two common forage grasses, K-31 and CBG. Both spectra contain relatively narrow lines that correspond to the carbohydrate, lignin, and protein in the plant. The spectra in Fig. 13 show how the individual constituents in the whole grass can be divided into their individual components for quantitative analysis.

Since no molecular weights can be determined for the grasses, and the packing densities and a number of physical parameters cannot be easily obtained or held uniform, direct measurement of a constituent is not possible by this technique. Ratios were determined to compare the quantitative results for these two grasses. The data in Table X are results obtained from several instruments and at two field strengths. The 50-MHz carbon

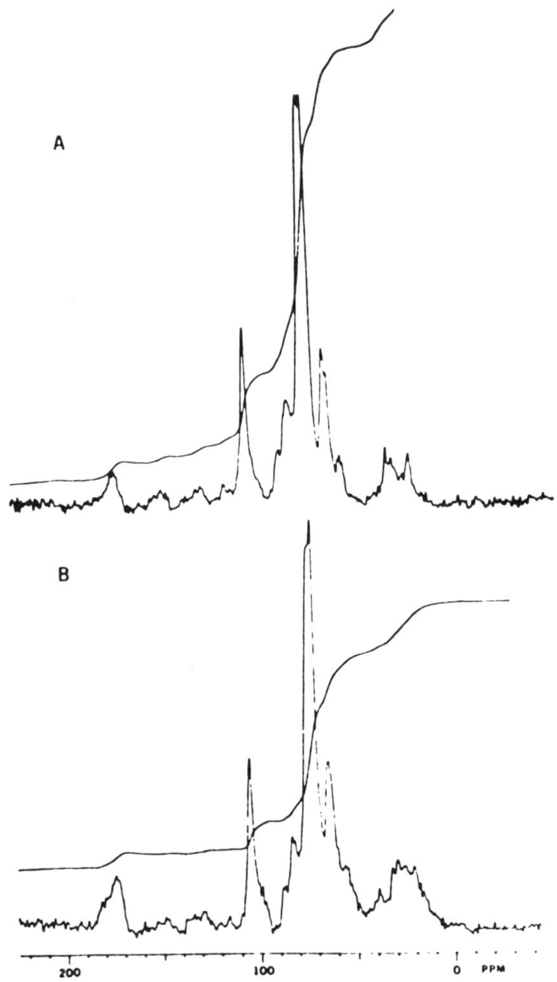

Fig. 12. Solid-state carbon-13 nuclear magnetic resonance spectrum of coastal bermudagrass (A) and Kentucky-31 tall fescue (B) at 50 MHz.

TABLE IX
Effect of Laboratory Data on Calibration

Analysis	Type[a]	Range	$\overline{SD}$	λ's[b]	R^2[c]	SEC[d]	Repeat[e]
Protein	A	7–21	0.62	1	0.94	1.07	0.1676
	B	8–23	0.18	1[f]	0.95	1.05	0.1434
	R	7–22	0.15	1	0.94	1.10	0.1789
Neutral detergent fiber	A	43–76	1.99	3	0.97	1.45	0.2750
	B	43–75	0.45	3	0.96	1.76	0.2693
	F	44–75	0.42	3	0.98	1.24	0.2458
Acid detergent fiber	A	30–44	2.03	3	0.93	1.18	0.2627
	B	29–45	0.25[g]	3	0.85	1.84	0.2994
	F	28–43	0.43	3	0.87	1.49	0.2684
Permanganate lignin	A	3–12	0.51	3	0.92	0.71	0.1662
	B	2–12	0.21	2	0.71	1.50	0.1029
	F	...	...	...	...	...	...
Acid detergent lignin	A	3–10	0.56	3	0.87	0.70	0.1552
	B	3–10	0.11	3	0.90	0.67	0.1201
	F	1.5–10	0.30	3	0.94	0.57	0.1386
In vitro dry matter digestibility	A	42–76	2.57	4	0.95	2.00	0.6426
	B	47–69	1.04	4	0.95	1.38	0.5251
	F	...	...	...	...	...	...

[a]A = average of four best labs (conventional method); B = conventional method data from Athens laboratory; R = rerun percentage protein in Athens laboratory; F = Fibertec analysis data.
[b]Number of wavelengths used in equation.
[c]Coefficient of determination.
[d]Standard error of calibration.
[e]Repeatability error.
[f]Using three degrees, $R^2 = 0.98$, SEC = 0.64.
[g]Six replications per sample.

data appeared to be the best and agreed closely to NIR and quite well with the wet chemical analyses data (CA). A probable explanation is that both spectral methods, the NIR and NMR, are measuring responses that result in signals from the constituents in the entire plant material, whereas the wet chemical methods measured entities that may not have been representative of the entire plant material. In the chemical methods, as in all empirical methods, it has been assumed that all the plant materials respond to the chemical reagents identically, which is unlikely to be the case. It is possible that more low-molecular-weight carbohydrates may be washed away as cell solubles from a cool-season grass species such as KY-31 than from a warm-season grass such as CBG. In the case of NDF, the result would be a higher value for the CBG/KY-31 carbohydrate ratio. It could be caused by structural differences between the plants and not true differences in carbohydrate contents of the intact plants, or differences in response of soluble carbohydrate to the solid-state technique.

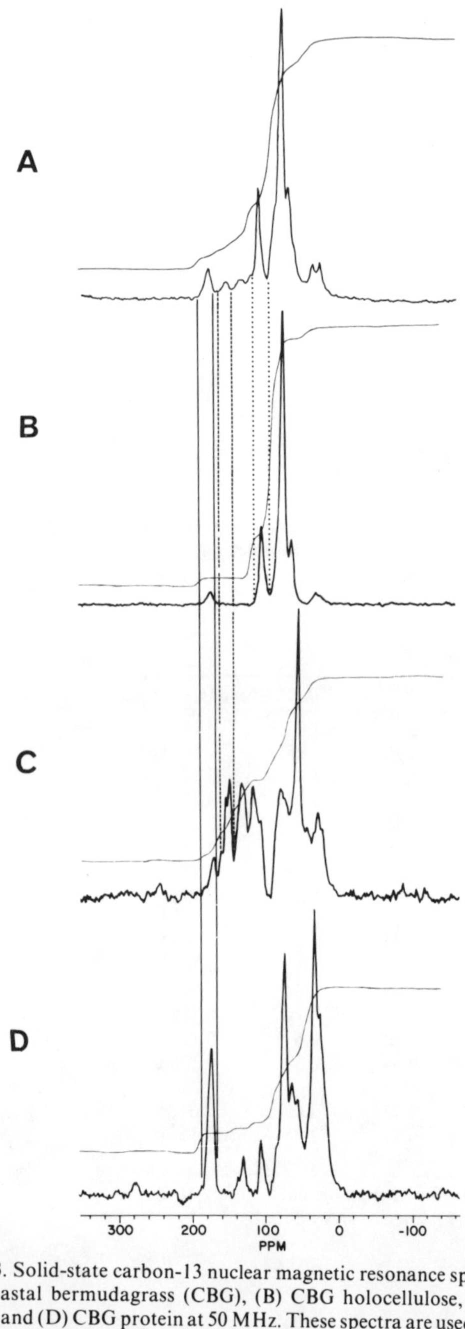

Fig. 13. Solid-state carbon-13 nuclear magnetic resonance spectrum of (A) coastal bermudagrass (CBG), (B) CBG holocellulose, (C) CBG lignin, and (D) CBG protein at 50 MHz. These spectra are used to define regions used for quantitative determinations of carbohydrate, lignin, and protein.

Structural differences in lignin could also cause "lignin" in one species to be more susceptible to permanganate treatment. This susceptibility would be especially true for the type of "lignin" in a warm-season plant like CBG where *p*-coumaryl units have been suggested to occur as side chains on polymeric lignin, or even on polysaccharides as α-, β-unsaturated esters (Himmelsbach and Barton, 1980). These C-9 phenolic units would be more susceptible to hydrolysis by the acid detergent reagent and would be removed to a greater extent with the hydrolyzable carbohydrate than would the true polymeric lignin. As a result, there would be a slightly lower value for the CBG/KY-31 ratio for "lignin" by the chemical method than was actually present in the plant.

For protein, the instrumental techniques might produce a different ratio of protein value because the standard Kjeldahl chemical method actually determines total reduced nitrogen. The Kjeldahl method could be influenced by nonprotein nitrogen in the plant, such as reduced nitrogen in free amino acids or as ammonium ion in ammonium nitrate that had been absorbed into the plant from fertilization. The NIR and CP/MAS ^{13}C-NMR methods would tend to discriminate against free amino acids vs. protein. NIR discriminates by picking different wavelength sets for protein than for amino acids (Williams et al, 1984, 1985). CP/MAS ^{13}C-NMR can be made to discriminate against smaller and generally more mobile species (Stejskal and Schaefer, 1975) and would not respond to residual ammonium nitrate.

The acquisition of solid-state ^{13}C-NMR spectra is faster than solution phase for these high-molecular-weight entities (3–4 vs. 48 hr per sample). However, the cost of these high-quality research instruments is high ($175,000 to more than $300,000). It is currently the only structurally definitive (easily interpretable) method of obtaining quantitative data on plant cell walls.

V. ADVANTAGES OF THE CHEMOMETRIC METHOD

Chemometrics as a discipline within chemistry can be defined as the development and application of mathematical and statistical methods to extract useful chemical information from chemical measurements (Kowalski, 1977). The extraction of compositional information from spectral curves can be considered as a chemometric method (Norris, 1983a, 1983b). The basic impetus for the development of *predictive analyses* is the increasing cost of performing laboratory analyses and the time required to obtain the results. Chemometric methods have been used for decades. Whenever a standard curve is constructed from a series of standard solutions assuming linearity from Beer's Law and used to read the concentration of unknowns directly from the scale knowing only its absorbance,

TABLE X
Ratio of Fiber Constituents in Coastal Bermudagrass vs. Kentucky-31 As Determined by CP/MAS ^{13}C-NMR,[a] Near-Infrared Reflectance, and Chemical Analysis[b]

Technique	Constituent		
	Carbohydrate	Lignin	Protein
Nuclear magnetic resonance			
37.7 MHz	1.27	1.57	0.68
50.0 MHz	1.27	1.66	0.59
Near-infrared reflectance[c]	1.27	1.70	0.58
Chemical analysis[c]	1.33	1.60	0.65

[a]Cross-polarization/magic-angle spinning carbon-13 nuclear magnetic resonance.
[b]Source: Himmelsbach et al (1983); used by permission.
[c]Each ratio is result of at least triplicate replications with standard deviations of ±1 in percentage composition.

a chemometric method has been used. Standard curves used in analyses accomplished with ion-selective electrodes are another example. The great decrease in the cost of computers has made possible the NIR technology to which this text is directed. Because of the statistical routines, calibration equations can be developed and used to provide an analytical result without having performed the analysis in question, simply from taking the spectrum. NIR is essentially a chemometric method.

Currently, commercial laboratories charge from $15 to $30 for a protein and fiber analysis. With current NIR technology, the completion of 1 or 10 analyses on a sample entails the same analytical input, which realizes far more information at the same cost. The cost of NIR equipment ranges from under $10,000 to over $100,000; the cost of the associated computer would range from less than $5,000 to systems over $50,000. Computer capacity and software flexibility are inherently important. The computer must be capable of storing all spectral data, the calibration (chemical) data, and regression equations. The software must be capable of performing derivative calculations and varying the derivative size and smoothing of log $(1/R)$ data. The system requires a sophisticated multiple regression analyses program to ensure that the optimum wavelengths are selected for the calibration equations. It should also contain internal statistical tests to examine the spectra and data of the samples for identification of erroneous data, or of samples that may be considered outliers. As such, the systems can also be used to manage laboratory data bases and provide internal as well as external checks on accuracy. With the savings in labor resulting from the ability to run several hundred samples in one day, NIR becomes very cost-effective even for the smaller laboratories (Williams, 1977).

The advantages of the NIR method for managing laboratory data can best be illustrated by examples. Table XI gives the results of NIR analysis of a series of samples, five of which were blind duplicates. The column labeled SDP (standard deviation between like samples by NIR) shows that the NIR instrument recognized the identical samples more closely than did the laboratory analyses, as shown by the column labeled SDA (standard deviation of like samples by laboratory methods) for everything but crude protein. The numerical difference between SDP and SDA for protein is probably very close to the instrumental error for the method. Overall, it is apparent that precision as well as speed can be improved. These analyses required five weeks in the laboratory by gravimetric methods, but only one hour by NIR.

The data in Table XII result from in vitro digestibility determination on a set of Old World Bluestem grasses. The samples in the middle with the large negative biases (i.e., predicted values are larger than measured values) were run as triplicates, with acceptable laboratory standard deviations. When these samples were rerun, the new values agreed with the NIR predicted values. This error could have been caused by a restriction in the automatic pipette or in the straining of the rumen inoculum, which lowered the number of rumen microorganisms, or by lowering of the activity of the inoculum because of residual detergent in the three racks of tubes. It is difficult to check standard laboratory results for an indeterminant error such as this. The results would normally have been accepted and the error of 10–30% would not have been detected. This example illustrates that accuracy can be improved through the use of a chemometric procedure. Table XIII contains data obtained from the determination of dry matter in which both positive and negative biases were found. In the laboratory, the tray used to carry "dry" crucibles from the oven to the desiccator to the balance held 20 crucibles. The data reflect differential moisture absorption among trays of crucibles in a desiccator over time. The chemometric procedure is very sensitive to laboratory practices that involve consistent errors and makes these practices quite easy to check, whereas before the errors went undetected.

If one is going to implement a chemometric procedure in the laboratory, several factors need to be considered. Equations to produce the analytical result can be generated that give excellent calibration statistics but do not provide acceptable analysis. It is possible to overfit the data, and one must be aware of the sources of laboratory error and the "true errors" associated with the procedure that produced the data to be used for calibration. This is discussed in detail in Chapters 3 and 4. Since the spectra from solid-state ^{13}C-NMR can be readily interpreted and quantitative data are present in the spectra, it seems logical to use the spectroscopic data as input for the chemometric measurement. NIR instrumentation gives analyses in about 2 min, whereas the solid-state ^{13}C-NMR

TABLE XII
Actual vs. Predicted IVDMD[a]

Sample No.	Average Residual Size	Bias
1–26	2.5	Positive and negative
27–36	10.0	All negative
37–40	4.4	All negative
41–60	1.7	Positive and negative

[a] In vitro dry matter digestibility.

TABLE XIII
Actual vs. Predicted Dry Matter

Sample No.	Average Residual Size	Bias
1–20	1.74	All negative
21–40	0.42	0.0
41–60	1.78	All positive

TABLE XI
Calibration of Near-Infrared Reflectance Instrument for Broad Spectrum of Samples[a]

Analysis	N[b]	SEC[c]	R^2[d]	Repeatability	SDP[e]	SDA[f]
In vitro dry matter digestibility	3	1.74	0.92	0.16	0.37	1.72
Crude protein	2	0.72	0.98	0.02	0.22	0.09
Acid detergent fiber	3	1.6	0.88	0.08	0.24	0.28
Acid detergent lignin	3	0.78	0.87	0.06	0.12	0.29

[a] Source: Barton and Coleman (1981); used by permission. Set of 30 samples included 13 bermudagrass, 5 alfalfa, 4 orchardgrass, and 1 timothy grass. Twenty were hays and 10 were fresh-frozen and freeze-dried.

[b] Number of wavelengths.

[c] Standard error of calibration.

[d] Coefficient of determination.

[e] Average standard deviation of predicted values of El Reno samples that were duplicated; i.e., five different samples.

[f] Average standard deviation of actual laboratory values for 10 El Reno samples that were duplicated.

requires 2–4 hr as the principal analytical method, which is much too slow. It is also possible to use NMR data to calibrate the NIR instrument. The same "chemical" information that is obtained from NMR is present in NIR spectra and would provide the accuracy and precision needed to measure quality in forages. Chemometrics can be used for basic research into forage, plant, and fiber structure as it relates to microbial degradation. Although NIR instrumentation is currently the front end of the method, other portions of the electromagnetic spectrum and other techniques may be coupled to the computer for chemometric analyses in the future.

LITERATURE CITED

AKIN, D. E., AMOS, H. E., BARTON, F. E., II, and BURDICK, D. 1973. Microanatomical differences of warm-season grasses revealed by light and electron microscopy. Agron. J. 65:825-828.

AKIN, D. E., and BARTON, F. E., II. 1983. Rumen microbial attachment and degradation of plant cell walls. Fed. Proc. 42:114-121.

AKIN, D. E., BARTON, F. E., II, and BURDICK, D. 1975. Scanning electron microscopy of coastal bermuda and Kentucky-31 tall fescue extracted with neutral and acid detergents. J. Agric. Food Chem. 23:924-927.

AKIN, D. E., BARTON, F. E., II, and COLEMAN, S. W. 1983a. Structural factors affecting leaf degradation of Old World Bluestem and Weeping Love Grass. J. Anim. Sci. 56:1434-1446.

AKIN, D. E., BROWN, R. H., and RIGSBY, L. L. 1984a. Digestion of stem tissues in *Panicum* species. Crop Sci. 24:769-773.

AKIN, D. E., and BURDICK, D. 1973. Microanatomical differences of warm-season grasses revealed by light and electron microscopy. Agron. J. 65:533-537.

AKIN, D. E., and BURDICK, D. 1975. Percentage of tissue types in tropical and temperate grass leaf blades and degradation of tissues by rumen microorganisms. Crop Sci. 15:661-668.

AKIN, D. E., and BURDICK, D. 1981. Relationships of different histochemical types of lignified cell walls to forage digestibility. Crop Sci. 21:577-581.

AKIN, D. E., RIGSBY, L. L., and BROWN, R. H. 1984b. Ultrastructure of cell wall degradation in *Panicum* species differing in digestibility. Crop Sci. 24:156-163.

AKIN, D. E., ROBINSON, E. L., BARTON, F. E., II, and HIMMELSBACH, D. 1977. Changes with maturity in anatomy, histochemistry, chemistry, and tissue digestibility of bermudagrass plant parts. J. Agric. Food Chem. 25:179-186.

AKIN, D. E., WILSON, J. R., and WINDHAM, W. R. 1983b. Site and rate of tissue digestion in leaves of C_3, C_4, and C_3/C_4 intermediate *Panicum* species. Crop Sci. 23:147-155.

ALBERSHEIM, P., NEVINS, D. J., ENGLISH, P. D., and KARR, A. 1967. A method for the analysis of sugars in plant cell-wall polysaccharides by gas-liquid chromatography. Carbohydr. Res. 5:340-345.

ASP, N. G., JOHANSEN, C. G., HALLMER, H., and SILJESTEM, M. 1983. Rapid enzymatic assay of insoluble and soluble dietary fiber. J. Agric. Food Chem. 31:476-482.

ATALLA, R. H., GAST, J. C., SINDORF, D. W., BARTUSKA, V. J., and MACIEL, G. E. 1980. Carbon-13 NMR spectra of cellulose polymorphs. J. Am. Chem. Soc. 102:3249-3251.

BAKER, D. 1977. Determining fiber in cereals. Cereal Chem. 54:360-365.

BAKER, D. 1978. Fiber in wheat foods. Cereal Foods World 23:557-558.

BAKER, D., and HOLDEN, J. M. 1981. Fiber in breakfast cereals. J. Food Sci. 46:396-398.

BARTON, F. E., II, and AKIN, D. E. 1977. Digestibility of delignified forage cell walls. J. Agric. Food Chem. 25:1299-1303.

BARTON, F. E., II, AKIN, D. E., and WINDHAM, W. R. 1981. Scanning electron microscopy of acid detergent fiber digestion by rumen microorganisms. J. Agric. Food Chem. 29:899-903.

BARTON, F. E., II, AMOS, H. E., BURDICK, D., and WILSON, R. L. 1976. Relation of chemical analysis to in vitro digestibility for selected tropical and temperate grasses. J. Anim. Sci. 43:504-512.

BARTON, F. E., II, and BURDICK, D. 1979. Preliminary study on the analysis of forages with a filter-type near infrared reflectance spectrometer. J. Agric. Food Chem. 27:1248-1252.

BARTON, F. E., II, and BURDICK, D. 1980. Prediction of crude protein in dehydrated coastal bermudagrass by NIR reflectance. Pages 103-106 in: Proc. Res. Ind. Conf. Coastal Bermudagrass Processors Assoc., 10th, Athens, GA.

BARTON, F. E., II, and BURDICK, D. 1983. Prediction of forage quality with NIR reflectance spectroscopy. Pages 532-534 in: Proc. Int. Grassl. Congr., 14th, Lexington, KY. Westview Press, Boulder, CO.

BARTON, F. E., II, and COLEMAN, S. W. 1981. Potential of near infrared reflectance spectroscopy for measuring forage quality. Okla. Agric. Exp. Stn. Misc. Publ. 108.

BARTON, F. E., II, WINDHAM, W. R., and HIMMELSBACH, D. S. 1982. Analysis of neutral sugar hydrolysates of forage cell walls by high-pressure liquid chromatography. J. Agric. Food Chem. 30:1119-1123.

BEN-GERA, I., and NORRIS, K. H. 1968. Determination of moisture content in soybeans by direct spectrophotometry. Isr. J. Agric. Res. 18:125-132.

BOLKER, H. I., and TERASCHIMA, N. 1966. Infrared spectroscopy of lignins. Pages 110-124 in: Lignin Structure and Reactions. J. Marton, ed. Am. Chem. Soc., Washington, DC.

BURDICK, D., BARTON, F. E., II, and NELSON, B. D. 1981. Prediction of bermudagrass composition and digestibility with a near infrared multiple filter spectrophotometer. Agron. J. 73:399-403.

COLEMAN, S. W., and BARTON, F. E., II. 1982. Calibration of a near infrared reflectance spectrometer for prediction of forage quality. Okla. Agric. Exp. Stn. Misc. Publ. 109.

DIXON, W. T., SCHAEFER, J., SEFCIK, M. D., STEJSKAL, E. O., and McKAY, R. A. 1981. Quantitative chemical composition of materials such as humic soils, lignins, and coals by high resolution carbon-13 NMR. J. Magn. Reson. 45:173-176.

EARL, W. L., and VANDERHART, D. L. 1980. High resolution magic angle sample spinning carbon-13 NMR of solid cellulose. J. Am. Chem. Soc. 102:3251-3252.

EARL, W. L., and VANDERHART, D. L. 1981. Observation by high-resolution carbon-13 nuclear magnetic resonance of cellulose. 1. Related to morphology and crystal structure. Macromolecules 14:570-574.

GOERING, H. K., and VAN SOEST, P. J. 1970. Forage fiber analysis (apparatus, reagents, procedures, and some applications). U.S. Dep. Agric. Agric. Handb. 379.

HARKIN, J. M. 1973. Lignin. Pages 323-373 in: Chemistry and Biochemistry of Herbage. G. W. Butler and R. W. Bailey, eds., Vol. 1. Academic Press, New York.

HIMMELSBACH, D. S., and BARTON, F. E., II. 1980. ^{13}C nuclear magnetic resonance of grass lignins. J. Agric. Food Chem. 28:1203-1208.

HIMMELSBACH, D. S., BARTON, F. E., II, and WINDHAM, W. R. 1983. Comparisons of carbohydrate, lignin, and protein ratios between grass species by cross polarization-magic angle spinning carbon-13 nuclear magnetic resonance. J. Agric. Food Chem. 31:401-404.

HONIGS, D. E., FREELIN, J. M., HIEFTJE, G. M., and HIRSCHFELD, T. B. 1983. Near-infrared reflectance analysis by Gauss-Jordan linear algebra. Appl. Spectrosc. 37:491-497.

HRUSCHKA, W. R., and NORRIS, K. H. 1982. Least-squares curve-fitting of near-infrared spectra predicts protein and moisture content of ground wheat. Appl. Spectrosc. 36:261-265.

HYMOWITZ, T., DUDLEY, J. W., COLLINS, F. E., and BROWN, C. M. 1974. Estimations of protein and oil concentration in corn, soybean, and oat seed by near-infrared light reflectance. Crop Sci. 14:713-715.

KAYE, W. 1954. Near-infrared spectroscopy. I. Spectral identification and analytical applications. Spectrochim. Acta 6:257-287.

KAYE, W. 1955. Near-infrared spectroscopy. II. Instrumentation and technique. Spectrochim. Acta 7:181-204.

KOWALSKI, B. R. 1977. Chemometrics: Theory and Applications. Am. Chem. Soc. Symp. Ser. 52. Washington, DC.

LUDEMANN, H. D., and NIMZ, H. 1973. Carbon-13 nuclear magnetic resonance spectra of lignins. Biochem. Biphys. Res. Commun. 52:1162-1169.

LUDEMANN, H. D., and NIMZ, H. 1974a. ^{13}C-Kernresonanzspektren von Ligninen, 1: Chemishe Verschiebungen bei monomeren und dimeren Modellsubstanzen. Makromol. Chem. 175:2393-2407.

LUDEMANN, H. D., and NIMZ, H. 1974b. ^{13}C-Kernresonanzspektren von Ligninen, 2: Buchen-und Fichten-Bjorkman-Lignin. Makromol. Chem. 175:2409-2422.

LUND, E. D., and SMOOT, J. M. 1982. Dietary fiber content of some tropical fruits and vegetables. J. Agric. Food Chem. 30:1123-1127.

MADDAMS, W. F. 1980. The scope and limitations of curve fitting (to vibrational band systems). Appl. Spectrosc. 34:245-267.

MARTENS, H., and RUSSWURM, H., Jr., eds. 1983. Food Research and Data Analysis. Applied Science Publishers, New York.

MOORE, J. E., and MOTT, G. O. 1973. Structural inhibitors of quality in tropical grasses in anti-quality components of forages. In: Anti-Quality Components of Forage Crops. A. G. Matches, ed. Crop Sci. Soc. Am., Madison, WI.

NIMZ, H. 1974. Beech lignin—proposal of a constitutional scheme. Angew Chem., Int. Ed. Engl. 13:313-321.

NIMZ, H., EBEL, J., and GRIESBACH, H. Z. 1975. Structure of lignin from soybean cell suspension cultures. Z. Naturforsch. Teil C 30:442-444.

NIMZ, H., and LUDEMANN, H. D. 1974. ^{13}C-Kernresonanzspektren von Ligninen, 5: Oligomere Lignin modellsubstanzen. Makromol. Chem. 175:2577-2583.

NIMZ, H., and LUDEMANN, H. D. 1976. Kohlenstoff-13-NMR-Spektren von Ligninen, 6: Lignin-:und DHP acetate. Holzforschung 30:33-40.

NIMZ, H., LUDEMANN, H. D., and BECKER, H. Z. 1974. Kohlenstoff-13-Kernresonanzspektren von Ligninen, 4: Die Lignine der europaischen mistel (*Viscum album* L.). Pflanzenphysiol. 73:226-233.

NIMZ, H., MORGHARAB, I., and LUDEMANN, H. D. 1974. ^{13}C-Kernresonanzspektren von Ligninen, 3: Vergleich von Fichten lignin mit kunstlichem lignin nach Freudenberg. Makromol. Chem. 175:2563-2575.

NORRIS, K. H. 1983a. Extracting information from spectrophotometric curves. Predicting chemical composition from visible and near-infrared spectra. Pages 95-114 in: Food Research and Data Analysis. H. Martens and H. Russwurm, Jr., eds. Applied Science Publishers, New York.

NORRIS, K. H. 1983b. Multivariate analysis of raw materials. Pages 527-535 in: Chemistry and World Food Supplies: The New Frontiers. CHEMRAWN II (Chemical Research Applied to World Needs). L. W. Shemilt, ed. Pergamon Press, New York.

NORRIS, K. H., and BARNES, R. F. 1976. Infrared reflectance analysis of nutritive value of feedstuffs. Pages 237-241 in: Proc. Int. Symp. Feed Composition, Anim. Nutr. Requir. Comput. Diets, 1st. P. V. Fonnesbeck, L. E. Harris, and L. C. Kearl, eds. Utah Agric. Exp. Stn., Logan.

NORRIS, K. H., BARNES, R. F., MOORE, J. E., and SHENK, J. S. 1976. Predicting forage quality by infrared reflectance spectroscopy. J. Anim. Sci. 43:889-897.

NORRIS, K. H., and HART, J. R. 1965. Direct spectrophotometric determination of moisture content of grain and seeds. Pages 19-25 in: Principles and Methods of Measuring Moisture in Liquids and Solids. A. Wexler, ed. Vol. 4. Reinhold, New York.

O'DONNEL, D. J., ACKERMAN, J. J. H., and MACIEL, G. E. 1981. Comparative study of whole seed protein and starch content via cross polarization-magic angle spinning carbon-13 nuclear magnetic resonance spectroscopy. J. Agric. Food Chem. 29:514-518.

RINNE, R. W., GIBSON, S., BRADLEY, J., SEIF, R., and BRIM, C. A. 1975. Soybean protein and oil percentages determined by infrared analysis. U.S. Dep. Agric. Agric. Res. Publ. ARS-NC-26.

ROHWEDER, D. A., BARNES, R. F., and JORGENSEN, N. 1978. Proposed hay grading standards based on laboratory analyses for evaluating quality. J. Anim. Sci. 47:747-759.

ROUTLEY, D. G., and SULLIVAN, J. T. 1958. The isolation and analysis of hemicellulosis of bromegrass. J. Agric. Food Chem. 6:687-692.

RUTAR, V., BLINC, R., and EHRENBERG, L. 1980. Protein content determination in solid organic materials by proton-enhanced magic angle sample spinning carbon-13 NMR. J. Magn. Reson. 40:225-227.

SARKANEN, K. V., and HERGERT, H. L. 1971. Classification and distribution. Pages 874-876 in: Lignins, Occurrence, Formation, Structure, and Reactions. K. V. Sarkanen and C. H. Ludwig, eds. Wiley-Interscience, New York.

SCHAEFER, J., SEFCIK, M. D., STEJSKAL, E. O., McKAY, R. A., and HALL, P. L. 1981. Characterization of the catabolic transformation of lignin in culture using magic-angle carbon-13 nuclear magnetic resonance. Macromolecules 14:557-559.

SCHAEFER, J., and STEJSKAL, E. O. 1979. High resolution ^{13}C-NMR of solid polymers. Pages 282-324 in: Topics in Carbon-13 NMR Spectroscopy. G. C. Levy, ed. Vol. 3. Wiley-Interscience, New York.

SHENK, J. S., and HOOVER, M. R. 1976. Infrared reflectance spectrocomputer design and application. Pages 122-125 in: Advances in Automated Analysis. Technicon International Congress (1976), Vol. 2. Industrial Symposia, Mediad.

SHENK, J. S., NORRIS, K. H., BARNES, R. F., and FISSEL, G. W. 1977. Forage and feedstuff analysis with infrared reflectance spectro/computer system. Pages 454-463 in: Proc. Int. Grassl. Congr., 13th, Leipzig, German Democratic Republic.

SHENK, J. S., WESTERHAUS, M. O., and HOOVER, M. R. 1978. Infrared reflectance analysis of forages. Pages 242-244 in: Proc. Int. Grain Forage Harvesting Congr. ASAE, St. Joseph, MI.

STEJSKAL, E. O., and SCHAEFER, J. 1975. Removal of artifacts from cross-polarization NMR experiments. J. Magn. Reson. 18:560-563.

ULBRICHT, T. L. V., and THEANDER, O. 1980. OECD Workshop at Uppsala, Sweden, 17–19 Sept. 1980. Agric. Environ. 6:87-348.

VAN SOEST, P. J. 1963a. Use of detergents in the analysis of fibrous feeds. I. Preparation of fiber residues of low nitrogen content. J. Assoc. Off. Agric. Chem. 46:825.

VAN SOEST, P. J. 1963b. Use of detergents in the analysis of fibrous feeds. II. A rapid method for the determination of fiber and lignin. J. Assoc. Off. Agric. Chem. 46:829-835.

VAN SOEST, P. J. 1967. Development of a comprehensive system of feed analyses and its application to forages. J. Anim. Sci. 26:119.

VAN SOEST, P. J. 1973. The uniformity and nutritive availability of cellulose. Fed. Proc. Fed. Am. Soc. Exp. Biol. 32:1804-1808.

VAN SOEST, P. J., and WINE, R. H. 1967. Use of detergents in the analysis of fibrous feed. IV. Determination of plant cell-wall constituents. J. Assoc. Off. Anal. Chem. 50:50-55.

VAN SOEST, P. J., and WINE, R. H. 1968. Determination of lignin and cellulose in acid detergent fiber with permanganate. J. Assoc. Off. Anal. Chem. 51:780-785.

WHEELER, J. L., and MOCHRIE, R. D., eds. 1981. Forage Evaluation: Concepts and Techniques. Am. Forage Grassl. Council, Lexington, KY, USA, and CSIRA, East Melborne, Victoria, Australia.

WILLIAMS, P. C. 1977. The economics of near infrared reflectance spectroscopy in testing grain for protein and moisture. Pages 131-136 in: Advances in Automated Analysis. Technicon International Congress (1976), Vol. 2. Industrial Symposia, Mediad.

WILLIAMS, P. C., MacKENZIE, S. L., and STARKEY, P. M. 1985. Determination of methionine in peas by near-infrared reflectance spectroscopy (NIRS). J. Agric. Food Chem. 33:811-815.

WILLIAMS, P. C., PRESTON, K. R., NORRIS, K. H., and STARKEY, P. M. 1984. Determination of amino acids in wheat and barley by near-infrared reflectance spectroscopy. J. Food Sci. 49:17-20.

WINDHAM, W. R., BARTON, F. E., II, and HIMMELSBACH, D. S. 1983. High-pressure liquid chromatographic analysis of component sugars in neutral detergent fiber for representative warm and cool-season grasses. J. Agric. Food Chem. 31:471-475.

CHAPTER 9

Near-Infrared Research in Europe

BRIAN G. OSBORNE
TOM FEARN
Flour Milling and Baking Research Association
Chorleywood, Hertfordshire, U.K.

SUSAN G. STEVENSON
Grain Research Laboratory
Canadian Grain Commission
Winnipeg, Manitoba, Canada

I. INTRODUCTION

A. History and Extent of NIR Usage in Europe

The foundations for near-infrared reflectance (NIR) analysis of materials such as cereal grains were laid by Norris and Butler (1961), Hoffmann (1963), Norris (1964), Massie and Norris (1965), and Norris and Hart (1965), but it was not until about 1971 that the first commercial instruments for NIR analysis became available. The two instruments available at that time (Neotec Grain Quality Analyzer and DICKEY-john InfraAlyzer) were both of U.S. manufacture and were not marketed in Europe until 1973-74. 1973 saw the appearance of the first NIR instrument in Europe, when the prototype Neotec GQA Model I was marketed for the first time. In 1975 the first InfraAlyzer in Europe was purchased by the Flour Milling and Baking Research Association (FMBRA) in the United Kingdom and, following the successful demonstration of the instrument's ability to measure protein and moisture in wheat, several flour-milling firms also invested in NIR. Interest in the technique gradually grew, with instruments becoming used for the analysis of wheat and flour in Finland, France, Hungary, Norway, Spain, Sweden, Switzerland and West Germany. The early 1980s saw the introduction of the Inframatic 8100, a low-cost, six-filter NIR instrument with the capacity for 18 calibrations. The 8100 is manufactured in Hamburg, West Germany, by Percon Prufgerate GmbH. This instrument was designed to determine protein, moisture, and hardness in grain products using samples ground on a Falling Number mill (KT 120 or 3100) and was found to perform as well for wheat as any other NIR instrument available in Europe at that time.

By 1979, grain trading to a protein specification using NIR analysis was accepted practice in the United States and Canada, and research workers in those countries had begun to discover many new applications for the technique. This research was facilitated by the appearance of a low-cost computerized spectrophotometer (Neotec 6350 Research Composition Analyzer) in 1978 (Landa, 1979). This instrument enabled spectra to be studied, wavelengths selected, and mathematical treatments optimized so that novel applications could be made that widened the scope of NIR research beyond the few laboratories that possessed one-of-a-kind computerized spectrophotometers. In 1979 the first Neotec 6350 in Europe was installed at FMBRA and was quickly followed by several others in the U.K. By 1981, NIR had become established in Eire, Greece, Italy, and the Netherlands and a study group of the International Association for Cereal Science and Technology (ICC) had been inaugurated. In 1984, more than 60 U.K. flour mills were using a total of approximately 100 NIR units. About 500 units were being used in the whole of the British food and agricultural sectors (Anonymous, 1984).

The present position concerning the use of NIR in the milling and baking industries in Europe is that the greatest use has been made in those countries where wheat is traded with a protein content specification. For example, in 1982 the entire milling wheat crop in Eire and about 50% of the milling wheat crop in the U.K. were traded using NIR measurement of protein. There was a smaller but significant utilization of NIR in France, West Germany (Beuch, 1983; Höhne, 1983; Schorch, 1983; Zwingelberg, 1983), and Sweden (Allvin, 1980), whereas applications in the other countries in Europe were mainly confined to the use of NIR in plant breeding and milling research establishments. The use of NIR in the baking industries in Europe has not yet become well established.

The current major use of NIR in Europe in terms of number of instruments is still therefore the measurement of protein content for the trading of wheat and flour. A standard method for carrying out such analysis has now been recommended by FMBRA and adopted by the ICC.

B. Some Early Problems and Their Solutions

INFLUENCE OF GRAIN TEXTURE ON NIR ANALYSIS

Pioneering work by Williams (1975) established the principal factors that influence the performance of NIR instrumentation. He recognized that the most important factor was the particle-size variation in individual ground samples as presented to the instrument. This in turn is influenced by the method of grinding. Williams also noted that different types of wheat required different calibrations, again because of differences in particle size when ground. By the time NIR instruments were introduced into Europe, it was already known that both precision and accuracy of analysis were affected by the method of sample preparation and by the procedure of loading the sample into the cell. It was also expected that the instruments would require calibrations specific to European wheat varieties. The first attempt at such calibrations in the U.K. was made with an InfraAlyzer 2.5A using samples from the 1974 wheat harvest, and it was discovered that standard errors of performance (SEP) of about 0.2% could be obtained for both protein and moisture using a closed-cup sample presentation technique and separate calibrations for hard and soft varieties. It was observed that, using the InfraAlyzer 2.5A, the log reflectance values for hard wheats were higher than those for soft wheat (Fig. 1); by assigning an arbitrary value of 10 to hard wheats and 0 to soft (classified by variety), a calibration for the differentiation between hard and soft wheats was obtained. When the hardness value was 5.0 or less, the soft calibration was used, and when 5.1 or more, the hard calibration was switched into operation.

Other researchers in Europe approached the problem of the influence of grain texture in different ways. Bolling and Zwingelberg (1979) also noted the influence of hardness on log reflectance values but preferred to grind samples such that a meal of consistent particle size was obtained, which enabled a single calibration to be used for both hard and soft wheats. Later work by Zwingelberg (1983) reported the use of NIR to determine the hardness of wheat and classification of wheats of various quality grades into hard or soft categories. Saurer (1978) found that separate calibrations for hard and soft wheats were unnecessary when using a Grain Quality Analyzer but that separate calibrations for spring and winter wheat were desirable. Starr et al (1981) reached a similar conclusion using an InfraAlyzer 400. Osborne et al (1982d) showed that although the improvement in accuracy in separate calibrations for hard and soft wheats was statistically significant at the 5% level using 220 samples, the practical difference was very small. The SEP for protein was 0.20% for hard wheats, 0.23% for soft wheats, and 0.25% for hard and soft combined. Although it could be shown statistically that separate calibrations tended to be slightly more accurate, it is arguable whether the extra effort involved in obtaining them could be justified.

EFFECT OF GRINDING METHOD

The initial work in the U.K. was carried out using the KT 120 grinder as used for the Hagberg Falling Number test, but the accuracy of the NIR determination of protein and moisture contents of soft wheats when ground with the KT 120 was poor. It was found that a finer grinding using the Udy cyclone mill gave satisfactory results for soft wheats. The Udy cyclone mill had been recommended by Williams (1981) and Williams and Thompson (1978), but they and others have pointed out that samples are subject to overheating and moisture loss (Maddix, 1976; Fearn, 1981; Bolling and Zwingelberg, 1982). In addition, accumulation of high-protein fines within the grinder can cause contamination of subsequent samples unless the grinder is correctly used. For these reasons it was considered likely that a hammer mill such as the KT 120 should be a more suitable grinder for wheat, and therefore the use of this and the KT 3100 was reexamined. It was found that, using an InfraAlyzer 300B, calibrations for protein and moisture in wheat ground using a KT 3100 mill with a slightly finer (0.8-mm) screen were as good as those using a Udy cyclone mill. The other problem with either the KT or Udy mills is the loss of moisture on grinding. When protein and moisture are read simultaneously by NIR on the same ground sample, protein content may be reported on the basis of any given moisture content or on a dry-weight basis. When it is necessary to have a measure of the actual moisture content of the sample, this can be achieved either by calibrating for moisture loss (Bolling and Zwingelberg, 1979) or by calibrating for direct analysis of the whole grain (Osborne et al, 1982c).

INFLUENCE OF TEMPERATURE ON NIR ANALYSIS

The action of grinding a wheat sample prior to NIR analysis is accompanied by a rise in sample temperature (Williams et al, 1982). Likewise, commercial roller milling of wheat into flour may result in temperatures of up to about 30° C or more in the freshly milled flour. In addition to these variations in sample temperature, grain intake points and flour bins are often located in positions subject to ambient temperature change. An experiment was carried out to investigate whether a fluctuation in NIR protein readings of nine different U.K. wheats occurred with time after grinding sequentially on a Udy cyclone mill. The instrument used was an InfraAlyzer 300B, and the results are displayed in Fig. 2. There was no evidence of bias or trend with time, except that there was more variability in the first half of each series of readings than in the second. In this experiment, the samples were returned to their bulk sample after each reading and the whole stirred again before resampling for the next reading, so this effect could be attributed to a reduction in

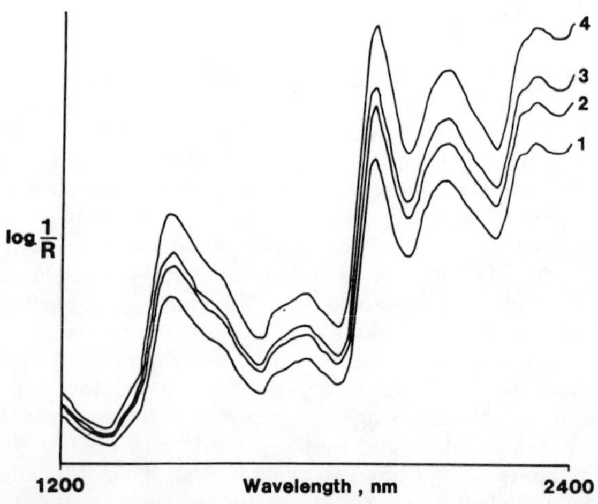

Fig. 1. Near-infrared reflectance spectra of four U.K. homegrown wheat samples ranging in kernel hardness from very soft (1) to very hard (4).

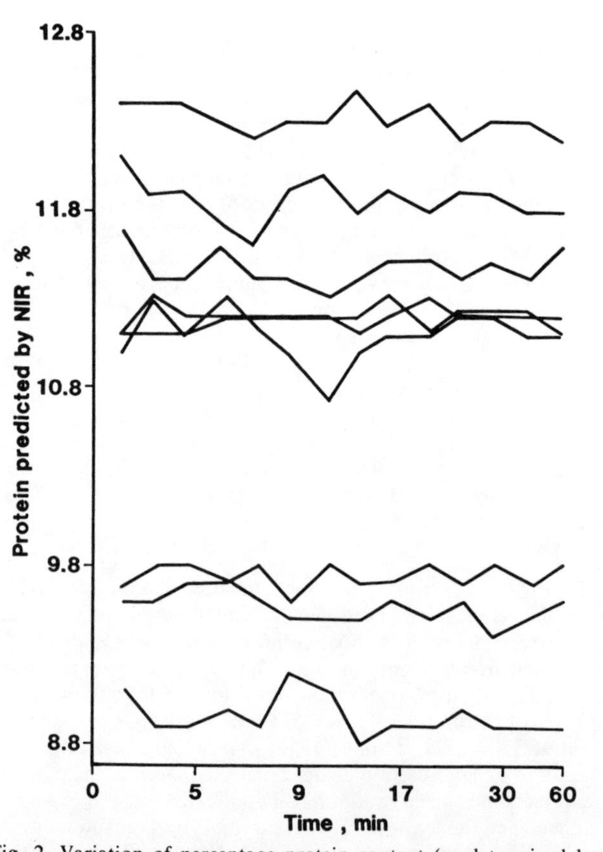

Fig. 2. Variation of percentage protein content (as determined by near-infrared reflectance) of nine different wheat samples with time after grinding (Udy cyclone mill).

sampling variance with extra stirring (Williams, 1975). With the instrument and grinder used, the NIR reading taken immediately after grinding was not significantly different from that taken after 60 min, so no waiting period between grinding and analysis was required.

To test for the effect of flour sample temperature and ambient temperature on protein readings made using an InfraAlyzer 300B, 20 samples of flour were adjusted to 5, 15, 20, and 40° C and read with the InfraAlyzer in a constant temperature room set first at 15° C and then 20° C. At both ambient temperatures, a bias in the protein readings of 0.024% protein per ° C change in sample temperature was observed (B. G. Osborne, unpubl. data). Thirty-nine flour samples were read first in an ambient temperature of 15° C, then 20° C on consecutive days, and the experiment was repeated with another 46 samples. For this experiment, the samples were at the same temperature as the instrument in each case. The pooled results for 85 samples showed a bias of 0.02% protein per ° C change in ambient temperature. These experiments verified that temperature both of the sample and the instrument (ambient) exerted a small but possibly significant influence on the NIR results.

FREQUENCY OF CALIBRATION

Initially, NIR was mainly used in Europe for harvest survey work and as an aid to selection of cultivars in plant-breeding studies. Since harvest surveys are carried out because the composition of wheat samples, climatic conditions, and the varieties grown vary from year to year to a considerable extent in Europe, it was possible that NIR instruments would require a new calibration for each harvest. During the earlier years of NIR application, it was necessary to recalibrate each year until sufficient data were accumulated to enable a more general calibration to be achieved. At that time, it was found possible to obtain single calibrations for protein and moisture in wheat for four U.K. harvest years, but annual bias adjustment factors were necessary (Osborne et al, 1982d). These bias factors were caused by instrumental drift associated with the older type InfraAlyzer 2.5A. An InfraAlyzer 300B was calibrated for protein in KT-120-milled wheat in May 1981 using wheats from the 1980 U.K. harvest. This calibration was used to carry out protein determinations in wheats from the 1981 and 1982 harvests, and bias checks against Kjeldahl were carried out at approximately six monthly intervals using 20–40 samples each time. Up to March 1983, there had been no significant bias changes. Individual bias checks had revealed mean differences of NIR-Kjeldahl, none of which was significant at the 5% level, and the standard deviation of differences between NIR and Kjeldahl data (SEP) on a total of 132 samples was 0.28% (Osborne, 1983b).

If calibration of an InfraAlyzer-type system is carried out by means of the traditional method of multiple regression of all six log reflectance values for the calibration samples against reference data, then the resulting seven-term calibration equation is likely to be unstable over long periods of time. This is because terms that have no significant contribution to the analysis but that may on the other hand be sensitive to other variables in the samples or the instrument are included in the equation. Modern NIR instruments use equations with two or three log reflectance terms. Calibrations of this type are stable from year to year, and this, together with improved instrumental design (Day and Fearn, 1982a, 1982b), has removed the necessity for annual recalibration.

DEVELOPMENT OF CALIBRATION METHODS

NIR is an empirical technique in that no rigorous quantitative theory has yet been developed. For this reason, it is accepted practice to calibrate instruments by statistical methods such as multiple regression analysis using large numbers of samples. One of the pitfalls in this approach is that of using too many terms in the calibration equation.

The most common methods are the stepwise regression procedures, the steps being taken either forwards or backwards. A forward stepwise regression begins by selecting the single wavelength at which the spectral data correlate best with the reference data, then a second wavelength is added and so on, building up the equation step by step and maximizing the correlation at each step. With log $(1/R)$ data, this procedure is highly unsatisfactory because of the high intercorrelations between such data (Table I) and the very poor correlation of log $(1/R)$ at any single wavelength with the analyte concentration. For example, moisture in wheat is usually measured at 1,940 nm, but a plot of moisture against log $(1/R)$ at 1,940 nm shows little relationship between these readings and the moisture content of the samples (Fig. 3). However, if the difference between log $(1/R)$ at 1,940 nm and at a reference wavelength, say 2,310 nm, is plotted against moisture content, there is now a strong relationship since differencing takes out most of the base-level variability (Fig. 4). The situation is not usually as simple as this because of overlapping bands and the fact that the shifts in the spectra caused by variations in scattering between

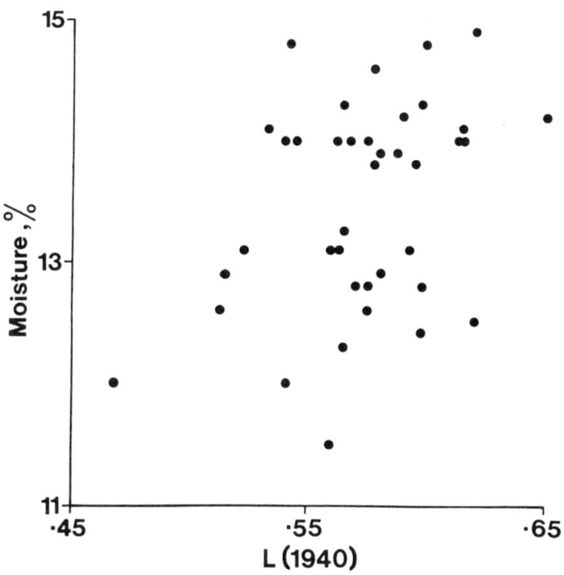

Fig. 3. Relationship between percentage moisture content of 40 flour samples and log reflectance at 1,940 nm.

TABLE I
Typical Independent Variable Correlation Matrix for log $(1/R)$ Data at Six Wavelengths

	Nanometers					
Nanometers	**2,310**	**2,230**	**2,180**	**2,100**	**1,940**	**1,680**
2,310	1.00000					
2,230	0.99721	1.00000				
2,180	0.99858	0.99959	1.00000			
2,100	0.99886	0.99340	0.99558	1.00000		
1,940	0.97624	0.99610	0.97086	0.98231	1.00000	
1,680	0.99418	0.99750	0.99625	0.99162	0.96368	1.00000

individual samples are not constant but vary in a complex fashion with the wavelength and intensity of individual absorption bands. This is why a $\Delta \log (1/R)$ algorithm has not been found to give rise to sufficiently accurate and stable calibrations.

The backwards selection procedure begins with an equation utilizing all the predictors and deletes terms one at a time, maintaining the maximum correlation at each step. Once again, there is no guarantee that this procedure will give, for example, the best two-term equation, because one of the pair may be deleted at an early stage. Also, great care must be taken in selecting the significance level used as the criterion for inclusion of terms; achieving significance greatly depends on the number of samples used, and it is therefore easy to include more terms simply because more samples are used.

Although calibration equations may be produced by a purely mathematical procedure, they should utilize wavelengths known to be relevant and be capable of interpretation in conventional spectroscopic terms. It is possible to use multiple regression with care to carry out calibrations, and this is still desirable on the grounds of simplicity and ease of understanding the calibration. Several mathematical procedures have been proposed as alternatives to stepwise multiple linear regression (Hunston, 1980; Hruschka and Norris, 1982; Martens and Jensen, 1982).

An alternative approach to stepwise regression is that of studying all the subsets. The aim of this procedure is for the computer to present the operator with a list of possible equations, all of which are acceptable from a statistical point of view in terms of statistical parameters. The operator then makes the final selection on chemical grounds.

As an example, an equation (Osborne et al, 1982a) used in the U.K. and Eire for the measurement of protein in flour on an InfraAlyzer 300B was derived by the minimal adequate subset procedure as follows:

$$\% \text{ Protein} = 12.68 + (493.7 \times L(2{,}180)) - (323.1 \times L(2{,}100)) - (243.4 \times L(1{,}680)),$$

where $L(N)$ are the $\log (1/R)$ values at wavelengths N nm.

This may be rewritten as:

$$\% \text{ Protein} = 12.68 + 493.7 \times [(L(2{,}180) - C1 \times L(1{,}680)) - 0.65 \times (L(2{,}100) - C2 \times L(1{,}680))],$$

where $C1$ and $C2$ are unknown constants.

The term in brackets is made up of the absorbance of the protein peak at 2,180 nm (referred to the level of the spectrum at 1,680 nm) minus 65% of the absorbance of the starch peak at 2,100 nm (referred to the level of the spectrum at 1,680 nm). The standardized, corrected protein peak is then multiplied by a calibration factor (493.7) to give percentage protein, the F_0 (12.68) being necessary because of the arbitrary scaling of the log values and arbitrary baseline correction. This equation has given excellent results, without further adjustment, over a long period and has been successfully transferred to other instruments of the same model (Osborne and Fearn, 1983a).

The only limitation to the use of minimal adequate subsets is one of computing power: there are $2^K - 2$ possible subsets from K independent variables. There are, for example, 62 subsets from 6 variables, 1,022 from 10, and about 10^6 from 19 variables, which is about the maximum practical number that can be handled. The technique is not practicable for scanning spectrophotometers. With the 700 or more wavelengths available on a scanning instrument, the number of possible subsets is 10^{485} and there are 10^{10} with only 4 or fewer wavelengths. The only way that the procedure can be used practically is to select a handful of likely wavelengths, either on chemical or statistical grounds, and thus reduce the array to a manageable size (Osborne et al, 1984a).

A strong reason for seeking chemical explanations for statistically acceptable equations is the possibility of one or more secondary correlations with hidden variables. If, for example, in order to calibrate for a particular component in flour a set of samples such as flour mill streams from the same grist is used for the purpose of achieving a suitably wide range, it is possible to overlook the effect of the high intercorrelations between the sample constituents (Table II) when only one is under investigation. The intercorrelations arise in this case because the concentrations of each of these constituents tend to vary in a very similar fashion through layers of the wheat kernel, and such relationships may not be maintained with samples from different grists or different mill settings. Therefore, it would be unwise to use a calibration for ash, say, based on such samples as part of a mill control system that aimed to minimize ash content for a given protein content: the control actions taken may very well destroy the secondary correlations on which the ash measurement is based. In the case of processed food products such as biscuits, it is possible to devise a calibration sample set such that the levels of ingredients are randomized (Osborne et al, 1984a), but this is not possible in the case of natural materials such as wheat.

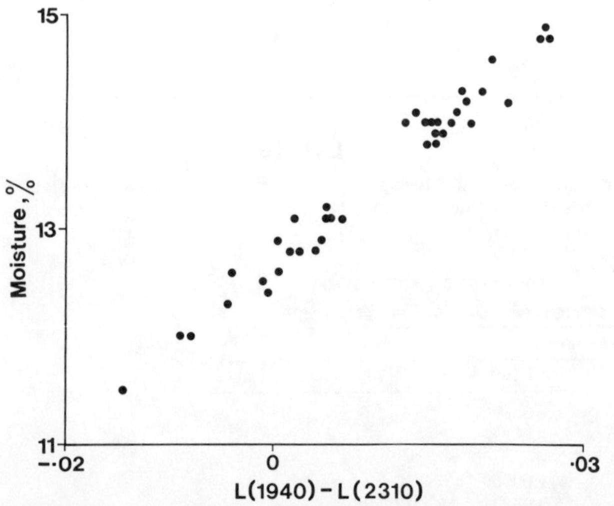

Fig. 4. Relationship between percentage moisture content of 40 flour samples and difference in log reflectance at 1,940 and 2,310 nm.

II. WHEAT-BREEDING PROGRAMS AND NEW VARIETY TRIALS

A. Criteria of Milling and Baking Quality of Wheat

Quality of wheat in the general sense means suitability for a particular purpose, and criteria of quality are ease of processing and yield and quality of the end product (flour, bread, or other baked goods). For example, the FMBRA classification of U.K. homegrown wheat varieties according to their value for milling and baking purposes is based on the following quality criteria:

1. Endosperm texture: wheats are classified as hard or soft according to an air-jet sieve measurement.

TABLE II
Dependent Variable Intercorrelations for Flour Mill Stream Samples

	Ash	Fiber	Oil	Protein	Color
Ash	···	0.96	0.92	0.89	0.92
Fiber		···	0.95	0.86	0.74
Oil			···	0.91	0.84
Protein				···	0.63
Color					···

2. Milling quality assessment is based on flour yield and on freedom of flour from bran specks as measured by grade color figure (GCF).
3. Flour water absorption is determined following a standard milling procedure and is calculated on a 10% protein basis.
4. Bread baking quality is assessed by a baking test.
5. Biscuit baking quality is based on extensometer testing of unyeasted dough and on endosperm texture.

Some of these criteria are dependent largely on environmental factors, but within given limits of such factors quality can be influenced by characteristics that can be varied by breeding.

The quality of wheat on the mill is measured by the yield and purity of the flour obtained from it; good milling wheats, when properly conditioned and milled under standard conditions, yield relatively more flour of lower mineral content and low grade color than poor milling wheats. Ease of processing is strongly influenced by grain texture: in hard wheats the endosperm cells come away more cleanly and remain more intact, whereas in soft wheats the subaleurone endosperm cells tend to fragment, part coming away while part is left attached to the bran. Also, the milling of hard wheats tends to result in a higher proportion of damaged starch granules in the flour, which has a strong influence on its water absorption characteristics.

Protein content is a very important aspect of wheat quality, as it determines to a large extent the end use of the wheat. For the production of yeast-leavened bread, flour of a minimum 11% protein content is preferred, and to produce such flour the wheat must have a protein content of about 12%. Such wheat protein contents have been uncommon under northern European climatic conditions, and consequently countries in these areas have traditionally imported wheat of high-protein content to blend with locally grown wheat. The economic importance of breeding wheat varieties with high average protein content is therefore great. Gluten quality is also important in determining the bread baking quality of wheat. This is largely a varietal characteristic and therefore it should be possible to breed varieties of good quality, but it is a rather elusive characteristic to define. A rapid method for the assessment of breadmaking quality would be of invaluable assistance to wheat breeders. Finally, rainy weather after maturation but before harvesting may cause sprouting of wheat kernels with a concomitant rise in alpha-amylase activity. If such wheat is milled into flour, the optimum level of alpha-amylase activity for bread dough fermentation will be greatly exceeded and the flour will become unsuitable for breadmaking.

B. Role of NIR

For successful selection in the early stages of wheat-breeding programs, methods are needed that require only small quantities of sample and are suitable for examining large numbers of samples in a reasonable time. The small amount of sample and degree of accuracy and precision of testing required make the NIR method useful for screening samples as early as the F generation (Starr et al, 1983). In later stages of breeding the numbers of samples decrease and the amounts available increase, but rapidity is still important. By the time a genotype reaches the stage of new variety trials for potential recommendation for particular purposes, experimental milling is performed so that in addition to the wheat samples a number of flour samples are also available for examination.

The potential of NIR has been examined with respect to the quality criteria in wheat and flour (Table III).

NIR has successfully been applied to the determination of protein and moisture contents of wheat during plant-breeding programs (Saurer, 1978; Starr et al, 1981), and it has also been useful in determining the milling quality of wheat with respect to grain texture. Saurer (1978) was able to predict hardness on the basis of NIR measurement of particle size of wholemeals, and similarly Svensson (1981) obtained a satisfactory level of precision for early-generation tests by calibrating an NIR instrument against pearling resistance. Starr et al (1981) assessed grinding resistance by calibrating against the method of Stenvert (1974), and Osborne et al (1981) reported the NIR measurement of air-jet sieve characteristics of flour Bühler-milled from wheats involved in new variety trials. This work was later extended to the direct measurement of air-jet sieve data on wholemeals that would be more useful for early-generation breeding or grain trading. A single log reflectance reading at 2,230 nm using an InfraAlyzer 300B instrument enabled air-jet sieve values to be measured to an SEP of 3% (SD of differences between NIR and air-jet sieve) in a range of from 45 to 75%, which covered a large number of varieties and growing locations. Bolling and Zwingelberg (1982) calibrated an InfraAlyzer 400 for a similar sieving test with an SEP of 4% in the range 31–66%.

Bran cleanliness is another important aspect of wheat kernel structure and is related to the ease with which the starchy endosperm separates from the bran during milling. In selecting for this character, samples were coarsely milled and evaluated subjectively on a 1–5 scale; NIR analysis was judged suitable to screen for bran cleanliness (Starr et al, 1981).

Bertrand et al (1985) reported that NIR could be a rapid, useful tool for variety identification of wheat. These researchers studied 202 wheat flour samples consisting of 23 new trial varieties and 6 varieties of proven baking quality, all grown in six regions differing in climate and soil conditions. An InfraAlyzer 500 was used for NIR measurements and an IBM-PC microcomputer for mathematical investigation. Variations in particle size were reduced by mathematical manipulation. Principal component analysis (PCA) was applied to reduce the number of variables to be studied to 10 and to enable spectral data to be represented in graphic form. Using PCA, three varietal groups were identified. Multiple discriminant analysis (MDA) was then applied to identify wheat varieties. Samples were randomly divided into two sets—a calibration set of 50 samples, including the six control varieties, and a prediction set of 16 other samples. With MDA, four varieties were consistently correctly identified. The percentage of correctly identified varieties was 84% for the calibration set and 87% for the analysis set. Although calculation of corrected values and discriminant factors required about 2 hr, computing time for variety identification of an unknown sample was reported to be less than 20 sec. The authors (Bertrand et al, 1985) felt that the ease and rapidity of the use of NIR would make it a useful technique for varietal screening.

The ultimate measurement for the quality assessment of wheat would be an assessment of the overall bread-baking quality of the milled product. Since the baking quality of a flour depends on so many different factors, some of which are

TABLE III
Quality Criteria of Wheat and Flour Determined by Near-Infrared Reflectance

Wheat	Flour
Protein content	Protein content
Moisture content	Moisture content
Grain texture	Ash content
Bran cleanliness	Sieving tests
Zeleny sedimentation	Grade color
SDS sedimentation	Water absorption
Loaf volume	Loaf score (volume)
Alpha-amylase	Alpha-amylase
	Starch damage

dependent on the milling and baking techniques themselves, rather than on the intrinsic properties of the wheat, it is difficult to envisage a single method such as NIR successfully measuring such a parameter. However, as NIR can measure protein, moisture, and hardness of the wheat, it may be that it can contribute sufficient measurements for a bread-baking quality assessment to be possible. For example, Saurer (1978) proposed a "quality mark" based on NIR determination of protein and hardness together with a Zeleny sedimentation value (ICC Standard Method 116). The estimation of the Zeleny values of Italian wheat samples was attempted by Cubadda and Douglas (1982). Using a Neotec 6350 they investigated 129 samples (15 varieties, 41 locations), with 63 for calibration and 66 for analysis; the SEP was ±4 ml. The wavelengths selected using second-derivative transformation were 1,992, 2,072, and 1,726 nm, but an equal result was obtained using untransformed log (1/*R*) data at 2,180, 2,100, and 1,680 nm. Since these latter wavelengths have already been established for the measurement of protein in wheat (Osborne et al, 1982d), it seems reasonable to suppose that a secondary correlation between protein quantity and Zeleny value was operating. In fact the correlation between these two parameters was 0.76, which would be sufficient to explain the NIR calibration bearing in mind the accuracy of NIR protein measurement. This view was further confirmed by an unsuccessful calibration attempt for Zeleny values on U.K. wheats (SEP was 7 ml) on which the correlation between Zeleny and protein content was 0.19. Bolling and Zwingelberg (1982) reported an SEP of 9 ml for sedimentation volume of German wheats using an InfraAlyzer 400. More recently, Höhne (1983) reported that 7 out of 10 samples of wheat tested with an InfraAlyzer 400 gave agreement within five points on the sedimentation value.

Starr et al (1981) investigated the possible analysis of another protein quality index, the SDS sedimentation volume. However, independent samples gave poor results (SEP was 19 ml), and they concluded that the calibration was not suitable for the estimation of SDS sedimentation volume. They also investigated wholemeal loaf volume and alpha-amylase by Falling Number, also with unsuccessful results (the SEPs were 98 ml and 62 sec respectively). Bolling and Zwingelberg (1982) reported an SEP of 40 ml for loaf volume and no success in attempted calibration of Falling Number.

As far as plant breeding is concerned, NIR analysis allows rapid screening for several characters simultaneously, but the advantages of speed must be weighed against the error that can be tolerated. Once a genotype has become sufficiently advanced, it is usually tested more extensively before being recommended to farmers; such testing operations are termed new variety trials. Wheats are then grown in sufficient quantities for experimental milling to be performed, and at that stage it is beneficial to examine the flour samples by NIR, as well as the wheats. Accordingly, a feasibility study was undertaken in the U.K. to assess the suitability of NIR for screening a number of parameters on Bühler-milled flour (Table IV). Protein, moisture, and hardness (air-jet sieve) all gave good results as expected, and GCF is probably acceptable for screening purposes. Results for these calibrations when used over a period of time were satisfactory (Osborne et al, 1982a). The calibrations for loaf score, water absorption, ash, and Falling Number were all considered too poor to justify attempting determination.

In summary, NIR analysis is a useful aid to the quality assessment of wheat for plant-breeding programs and new variety trials in that some of the labor involved in manual analyses of very large numbers of samples can be avoided. However, not all quality characteristics are capable of measurement to sufficient accuracy.

III. MILLING INDUSTRY APPLICATIONS

A. Grain Trading

As already stated, the determination of protein in wheat for grain trading purposes is by far the most important application of NIR analysis in the flour milling and baking industries in Europe. The reported accuracy of this determination relative to Kjeldahl has varied from 0.2 to 0.5% standard deviation (Popineau and Godon, 1976; Schafer and Weinard, 1978; Bolling, 1979; Bolling and Zwingelberg, 1979, 1982; Allvin, 1980; Ringlund, 1980; Weinard, 1981; Downey et al, 1982; Osborne et al, 1982d; Waltl, 1982; Lasztity and Varga, 1984).

The potential of NIR analysis for very rapid testing of protein, moisture, and hardness enables decisions to be made as to the suitability of samples for purchase and the composition of grists. Samples may be accepted or rejected, the price agreed, and the wheat appropriately binned on the basis of NIR readings. An essential prerequisite for this is that the industry must be satisfied that the level of accuracy is good enough for these decisions to be made with confidence. A thorough evaluation of the measurement of protein in wheat by NIR was

TABLE IV
Results of Feasibility Study into Use of Near-Infrared Reflectance to Analyze Buhler-Milled Flours (InfraAlyzer 300B)

Calibration				
Parameter	Number of Samples	Range	Residual Standard Deviation	Multiple Correlation
Protein	379	6.1–12.7%	0.25%	0.98
Moisture	57	12.2–13.6%	0.09%	0.96
Air-jet sieve	119	34–78%	2%	0.99
GCF[a]	379	0.2–5.4	0.3	0.94
Ash	50	0.36–0.71%	0.05%	0.76
Loaf score	379	0–35	6	0.70
Water absorption	379	44–58%	2%	0.60
Falling number	379	61–254 sec	46 sec	0.51

Prediction			
Parameter	Number of Samples	Range	Standard Deviation of Differences
Protein	357	6.9–14.5%	0.22%
Moisture	97	12.3–15.5%	0.17%
Air-jet sieve	72	31–66%	2%
GCF	209	0.1–7.4	0.4

[a] Grade color figure.

carried out in Eire in 1981 as a result of the introduction of a quality standard for milling wheat based on grain protein content (Downey et al, 1982). This evaluation showed that unskilled operators using NIR equipment were able to analyze wheat for protein content as precisely as, and much more rapidly than, skilled laboratory personnel using the Kjeldahl method. As a result, NIR was accepted as a criterion of protein content of wheat, and by 1982 all Irish wheat was traded using NIR. Collaborative testing of protein, moisture, and hardness on ground wheat and moisture on whole grain samples was also carried out in the U.K. (Osborne and Fearn, 1983b). The results showed that the precision and accuracy of all these determinations were excellent and the hard/soft classifications produced only 10 significant errors out of 160 results (20 samples in eight laboratories).

The Irish and U.K. experiences proved that NIR is eminently suitable for trading of wheat grain according to protein, moisture, and hardness specifications. These findings have been borne out by the experiences of other European countries, such as West Germany and Sweden.

B. Control of the Milling Process

Just as millers need to monitor the protein and other quality characteristics of the intake wheat, so they also wish to monitor the flour produced. In the U.K. it is usually the millers who provide certification of the agreed flour specification to their customers, and this is increasingly being performed on the basis of NIR analysis. Again, as with wheat, a collaborative experiment (Osborne and Fearn, 1983a) involving both milling and baking laboratories was carried out to ensure that consistent results for protein and moisture contents of flour could be obtained. Nine InfraAlyzer 300B instruments were set up in separate laboratories with the same calibration constants, adjusted for bias using one set of samples, and the reproducibility and repeatability determined several months later using a second set of samples. The results were that for protein the repeatability was 0.07% and the reproducibility 0.17%, compared with 0.04% and 0.14% respectively for Kjeldahl determinations on the same samples in the same laboratories. For moisture the NIR figures were 0.05% and 0.16%.

NIR has been used for internal control of flour protein content in Europe for a number of years (Bolling, 1979; Bolling and Zwingelberg, 1979, 1982; Clover, 1979; Colas, 1979; Schorch, 1983). It can be used to check that the flour from a given grist conforms to specifications, and also for quality control of the final product. For more sophisticated mill control, it is necessary to make measurements on individual mill streams so that the results can be used to make decisions as to which streams to blend for production of the required flour. It may be necessary to carry out separate calibrations for streams with greatly differing particle size in order to achieve this (Colas, 1979). Bolling and Zwingelberg (1979) extended the application of NIR to mill process control even further. Samples were taken every 30 min prior to the first break and every 10 min from the flour worm-conveyor, and protein readings were made by NIR. The high number of individual samples was intended to determine the trend of possible fluctuations in flour protein as a result of grist variations. In addition, this type of control provided information as to the possibility of continuous measurements. There was good agreement between the protein content of wheat before first break and the theoretical protein content of the flour. The effect of addition to, or withholding of break flours from, the final product was also examined, but again difficulties were encountered as a result of the wide variations in particle size of the products of different breaks. Schorch (1983) reported similar findings with good correlations ($r = 0.96$) between NIR and Kjeldahl protein results for patent and bakers' flours, but poor results for production control of protein content in milling fraction flours. Control actions on the basis of NIR readings are difficult because changes in mill setting cause changes to the particle size of intermediate products, which in turn have an adverse effect on the NIR calibrations.

There are measurements other than protein and moisture that would be of use in mill control, and one of these is the purity of the flour as determined by ash or GCF. Ash measurements by conventional incineration procedures are widely used in Europe as flour quality criteria, except in the U.K. where mineral additions are made to flour by government regulation. Attempts at NIR measurement of ash by means of an InfraAlyzer 400 fitted with cellulose absorption filters were made by Bolling (1979), who obtained an SEP of 0.015% and correlation coefficient of 0.80 in the range 0.41–0.60%, and Bolling and Zwingelberg (1982), with an SEP of 0.021% and correlation coefficient of 0.99 in the range 0.37–0.80%. In the U.K., ash measurements are only used for flour destined for export since no mineral additions are made in this case. Calibration of an InfraAlyzer 300B for ash in 56 export flour samples resulted in an SEP of 0.017% in a range 0.43–0.60%, and calibration for 70 Greek flour samples gave an SEP of 0.025% in the range 0.50–0.75%. All these results appear to be acceptable compared with the precision of the standard method, which is between 0.01 and 0.03% (Bolling, 1979), but the possibility of secondary correlation (Table II) reduces confidence in the measurement. For example, it was possible to measure the ash content of the U.K. export samples with an SEP of 0.019% by means of Kjeldahl protein figures. This does not necessarily mean that NIR ash measurements should not be used but that this limitation should be borne in mind. This limitation also applies to fiber measurement, but GCF, which is designed to measure the extent of contamination of flour with bran particles regardless of background color, can be analyzed to a limited degree of accuracy using an InfraAlyzer fitted with a 540-nm filter (Osborne et al, 1982a).

The other important milling parameter that can be analyzed by NIR is starch damage. This measurement was first developed in the U.K. using a Neotec 6350 (Osborne and Douglas, 1981), and then appropriate filters were obtained for use in an InfraAlyzer 300B that was calibrated and extensively tested (Osborne et al, 1982a). The SEP between NIR and the Farrand method was 3 Farrand units (FU) in a range of 5–35. For the control of starch damage of bread flours where values lay in the range of about 20–35, the range of Farrand units may not be adequate. Nevertheless the NIR measurement of starch damage, protein, and moisture together is satisfactory for the estimation of water absorption (see Section IV.A). This SEP figure represents results obtained by several operators over a period of time. Further work on possible environmental influences on the calibration were carried out with the following results. The published data (Osborne et al, 1982a) represented five separate analytical sets examined over a period of six months. There were separate biases on each set of up to 2.1 FU, but no evidence of a trend. These biases were not corrected since they were considered part of the general scatter of the calibration. Had the bias been adjusted for each sample set, then the SEP would only have improved from 3.5 to 3.3 FU. The effect of different operators on the determination was difficult to assess because the apparent biases between operators observed when the same samples were read on different days were not confirmed if those operators read the samples on the same day.

This observation indicates an environmental effect, the most likely being the result of temperature changes. The magnitude of the calibration constants for starch damage measurements was at the limit of instrumental precision. The wavelengths used corresponded to bands due to O-H bonds in various states of hydrogen bonding. Since the degrees of hydrogen bonding and hence band intensities are strongly affected by temperature, fluctuation of temperature could have affected the results.

There was no effect over and above the general scatter in the calibration because of up to ±5°C change in ambient temperature, but there was an effect of +0.4 FU per °C in sample temperature. Although sample temperature variations most likely accounted for the fairly high SEP, quite large fluctuations in sample temperature (say ±10°C or more) would be necessary in order to disturb the calibration to greater than ±6 FU.

As a prerequisite to automated mill control using NIR, it is necessary to consider the degree of variation presently occurring in commercial mill production. Although it is not enormous, there is scope for improvement through the use of NIR to control protein, moisture, GCF, and water absorption with concomitant financial benefit to the miller in more efficient usage of expensive raw materials whilst maintaining the consistent quality required by the baker. Areas where NIR could be of great benefit to mill control either on-line or off-line are (i) as part of wheat conditioning control equipment; (ii) in the control of grists by protein content; (iii) in the control of flour protein and moisture (this would probably follow from [i] and [ii], since there is evidence that in some mills at least the flour protein and moisture values can be accurately estimated from those of the corresponding grist); (iv) in the blending of flours or supplementation of flour with gluten to achieve a desired protein level; and (v) in the control of flour grade color and water absorption, since although the NIR results are subject to more variability than those for protein and moisture, so is the sensitivity of those factors to mill settings.

IV. BAKING INDUSTRY APPLICATIONS

A. Flour Intake Quality Control

As previously mentioned, flour specifications in the U.K. are usually provided by the miller and therefore use of NIR for quality control of intake flour at bakeries is not widespread. However, NIR together with independent alpha-amylase measurements has the potential to provide an indication of bread baking quality. As indicated in Section III.B, the contribution of NIR to the estimation of loaf volume lies in measuring protein quantity and possibly starch damage. To confirm this, 49 commercial flour samples were examined for protein, moisture, damaged starch, water absorption, total alpha-amylase (Farrand method), and Falling Number (5 g). The samples were then baked using the Chorleywood Bread Process after the potassium bromate level had been adjusted to a constant figure of 45 ppm. The replicate standard deviation of this test baking procedure was 23 ml in a range of 1,170–1,600 ml. NIR measurements on the flour samples were made using an InfraAlyzer 300B, and the best calibration between NIR readings and loaf volume gave an SEP of 50 ml using the protein subset 2,180, 2,100, and 1,680 nm (Osborne et al, 1982a). An equation derived from protein, damaged starch, Falling Number, and total alpha-amylase data had an SEP of 36 ml, whereas one using protein and damaged starch measurements only had an SEP of 46 ml. Therefore, NIR alone performs neither better nor worse than protein and damaged starch figures in the estimation of loaf volume. This finding was further supported by studies reported by Osborne (1984) in which 52 samples of U.K. wheat were used to calibrate a Neotec 6350 Mk1 Research Composition Analyzer for protein, moisture, alpha-amylase, and SDS sedimentation volume (protein quality).

Water absorption is probably the most critical factor as far as performance on automatic dough-handling equipment is concerned, and this can be estimated by NIR determination of protein, moisture, and starch damage and by application of the Farrand equation. The standard deviation of differences between the calculated and actual water absorption (Simon extrusion meter, 10-min unyeasted) of the same flour samples used for the baking experiment described above was 2% in the range 52–67% (Osborne et al, 1982b), which is the same figure as found for Bühler-milled flours by direct NIR measurement (Osborne et al, 1982b, Table IV, Section II). The result for Bühler-milled flour was carried out using an InfraAlyzer 300B without the special starch-damage filters (1,445, 2,050, and 2,139 nm), and a direct NIR measurement using the commercial flours on the modified 300B gave an SEP of 1.4%, which on analysis of 83 further samples resulted in an SEP of 1.5% in the range 50–64%. The SEP with the calibration obtained using the InfraAlyzer with the extra starch-damage filters was significantly better (at the 5% level) than that obtained without them. This indicated that the NIR starch damage measurement was satisfactory using the extra filters, but the result was not conclusive because different sample sets were used for the two experiments.

Practical use of NIR for flour evaluation at the bakery lies in the estimation of loaf volume. The evaluation of loaf volume estimation was carried out using a carefully controlled procedure for mixing and processing with a constant recipe (except for water addition). Other recipes and baking methods, especially those using brown or wholemeal flours, are likely to give different calibrations, although there is good reason to expect that the same flour characteristics would be important. Finally, the effect of changes in the level and composition of improvers used in the Chorleywood Bread Process should also be considered. The levels of additives permitted in the improver mix must be carefully controlled to maintain the correct amounts in the flour or dough after supplementation. In (1983c), Osborne reported results of studies designed to investigate the use of NIR as a means of rapid quality control of bread improvers. Results of analysis of 51 samples of mixtures of ascorbic acid, potassium bromate, azodicarbonamide, and L-cysteine with wheat starch using a Neotec 6350 RCA Mk1 showed that although the accuracy of this method for determining levels of the improvers was satisfactory in bulk concentrates, it was not sufficiently sensitive for measurement of improvers in diluted mixes or in flour. Furthermore, calibration for potassium bromate estimation was unsuccessful, as the best SEP that could be achieved was ±4%.

B. Monitoring of Product Composition

Very little work has been carried out on the possible application of NIR to the analysis of baked products (Osborne, 1983a). Osborne et al (1984a) have evaluated the use of NIR for the monitoring of short dough biscuits and biscuit doughs for the major ingredients fat, dry flour, sucrose, and water. SEP values on dough pieces, for example, with writing and docker holes facing the light beam, were 0.2, 0.9, 1.1, and 0.5% for fat, dry flour, sucrose, and water respectively. On this basis, only the results for fat may be considered satisfactory for the detection of recipe errors of ±5% relative. The NIR method would be suitable for screening for gross errors in sucrose, dry flour, and water in biscuit doughs, especially in view of the speed with which results can be obtained. As a production control measure, NIR has an important place in applications such as biscuit baking. The determinations carried out on dough or dough pieces would allow feedback control if levels of ingredients were beyond control limits and the baking could be aborted.

An example of a control chart for fat in dough is shown in Fig. 5. This chart was obtained by simulation of a quality control operation by test baking of biscuits in which most recipes were close in composition to the "normal" and a few "errors" of magnitude 5% relative were deliberately included. The chart shows results of NIR fat measurement on a run of 20 samples. The triangles inside circles represent ±2.5% relative (±25 g in average fat weight of 1,082 g ±0.4% fat absolute), and the triangles inside squares represent ±5% relative (±50 g in average fat weight of 1,082 g ±0.8% fat absolute). Control of moisture content of the final biscuit is already a practical reality using NIR (McFarlane, 1981; Verbiese, 1981).

Another similar application of NIR for the analysis of baked products is the determination of fat and sucrose in dry cake mixes (Osborne et al, 1983). An InfraAlyzer 400R was calibrated in a research laboratory and then assessed under quality control conditions in a food processing factory. This assessment showed that it was possible to measure fat with an accuracy ($\pm 2\sigma$) of ±3.4% compared with the Soxhlet method for products with a fat content of 8–25%. In the case of sucrose, the accuracy was ±5.4% compared with a high-pressure liquid chromatography method for products with a sucrose content of 10–40%.

Osborne et al (1984b) reported the results of studies to determine the protein, fat, and moisture content of bread by NIR. "Unknown" samples of fresh bread slices gave SEP values of 0.20% for protein, 0.18% for fat, and 0.51% for moisture. Calibrations derived from samples of the same bread that had been air-dried and ground showed similar SEP values for protein and fat.

NIR has great value to the flour milling and baking industries, and the factors that influence the accuracy and precision of the technique have now been thoroughly studied (Osborne, 1981a, 1981b). It finds application for the rapid analysis of wheat and its products from breeding of new varieties through new variety trials, grain trading, mill control, and flour quality control to compositional analysis of the baked products. The range of applications reported so far is summarized in Table V.

Further applications of NIR include automatic on-line continuous measurements, for which it is ideal. This development has now become established (Bolling and Zwingelberg, 1984; Klein Schiphorst, 1986; Osborne, 1986).

V. ADDITIONAL CEREALS AND OILSEEDS

In addition to its demonstrated usefulness in the milling and grain trade, the technique of NIR has been applied to the analysis of cereals other than wheat as well as to oilseeds for quality assessment and control purposes.

NIR spectroscopy has been used successfully to monitor barley protein and moisture at point of delivery in much the same way as it is used for wheat. Studies of 326 samples of malting barley analyzed with a Percon Inframatic showed that NIR protein results varied by <0.3% from Kjeldahl results for 80.1% of the samples on a dry matter basis (Höhne, 1983). Similar results were reported by the same author when an InfraAlyzer 400 was used. Schmutz and Do (1982) analyzed a total of 1,079 barley samples from different areas and crop years, ranging in protein content from 8.1 to 16.3%. Using an InfraAlyzer 400, they found that the correlation between NIR and Kjeldahl results was 0.961–0.995, with standard deviations ranging from ±0.21 to ±0.25%. Lasztity and Varga (1984) reported correlations of 0.94 and 0.99 with standard deviations of 0.31 and 0.17 between NIR and standard methods for barley protein and moisture respectively. Lipid content was also reported to give favorable results; however, the statistics for this component were not reported. Oil content of oilseeds, meals, and various cereals by NIR determination has been reported in many European countries by numerous researchers, including Calandra (1977), Fossati and Perrotta (1977), Allvin (1980), Cardinale (1980), Morgan and Starr (1983), and Davies et al (1985). Good correlations between the NIR results and standard methods of solvent extraction were reported.

In recent years, NIR has been used by a number of researchers as a rapid screening method in barley breeding programs. Gill et al (1979) reported that NIR was sensitive to changes in lysine and protein nitrogen to a high degree and thus was a suitable method for rapid screening of materials in breeding programs designed to improve the lysine content of barley proteins. It was all the more so since the gross morphological and biochemical abnormalities of high-lysine mutants did not affect the NIR measurements.

TABLE V
Summary of Applications of Near-Infrared Reflectance in Flour Milling and Baking Industries in Europe

Parameter	Wheat	Flour	Baked Products
Protein	+[a]	+	+
Moisture	+	+	+
Hardness	+	0	0
Sieving test	+	+	0
Bran cleanliness	±	0	0
Ash	–	±s	0
Grade color	0	±	0
Alpha-amylase	–	–	0
Starch damage	0	±	0
Water absorption	0	±s	0
Loaf volume	±s	±s	0
Fat	0	0	+
Sucrose	0	0	±

[a] 0 = not applicable or not attempted; – = unacceptable; ± = marginally acceptable; + = acceptable; and s = secondary measurement arising from ability of near-infrared reflectance to measure other parameters.

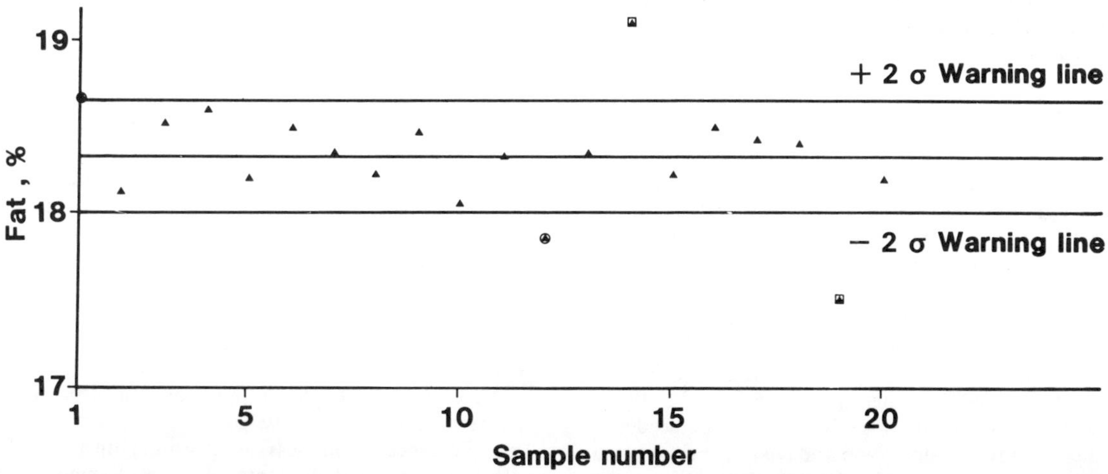

Fig. 5. Control chart showing successful detection of recipe errors of ±50 g of fat in biscuit dough by near-infrared reflectance measurements; triangle within circle = ±25 g, triangle within square = ±50 g.

A major concern in barley breeding programs is the assessment of malting quality in early generations. A number of tests have been proposed; however, the more reliable of these, such as small-scale full malting, require larger amounts of grain than are available in early-generation trials. In (1979), Morgan and Gothard reported the use of an InfraAlyzer to provide rapid analysis of malt hot water extract (HWE). This extract, which is a function of the amount of sugar and low-molecular-weight dextrans produced during mashing by the action of α-amylase on the barley starch, is one of the most important variables used to determine malting quality. In preliminary studies, the authors found that the best results were obtained by using separate calibrations for spring and winter barleys. Calibrations for each type were made using 60 samples representing an even distribution over a range of HWE. In all, 168 winter barleys and 134 spring barleys were used and correlations between NIR and HWE were 0.70 and 0.82 respectively. When NIR estimates of HWE were plotted against micromalting data, good linear fit was obtained for both types of barley although the near-infrared method appeared to be more accurate for the spring barleys based on the standard errors. In both cases, the authors believed that the accuracy of the NIR method was adequate for such rapid screening as was necessary in a breeding program. The NIR method required the use of only 3–4 g of barley flour per sample, which would make it pertinent to early-generation screening, and samples could be tested at the rate of 30 per hour. Morgan and Gothard (1979) also pointed out that the NIR instrument could be calibrated to measure protein, moisture, β-glucan, and lysine content of the barley sample at the same time as the HWE was being determined.

Further studies by these authors (1981) involved the assessment of HWE, and hence malting quality, based on NIR measurement of nitrogen content, viscosity, and glucose content of an acid extract of barley flour. Again, the accuracy of analysis was greater with spring barleys than with winter ones. This technique required only 2 g of seed, and up to 70 samples per person per day could be accommodated.

Not only has NIR been used in plant-breeding programs for barley, it has also been used to measure protein (Davies and Wright, 1984), starch and lipid content (Davies et al, 1985) in pea flour, as well as oil and nitrogen content of oilseeds (Morgan and Starr, 1983) in various breeding programs. Davies and Wright (1984) found the accuracy of the NIR technique for determining protein in pea flour to be better than ±1.5% ($P = 0.95$). Coefficients of variation for analyses of protein, starch, and lipid were respectively 1.6%, 2.0%, and 3.4% for chemical determinations and 3.0%, 2.0%, and 5.6% for NIR (Davies and Wright, 1984; Davies et al, 1985). Scans of pea flour, starch, and lipid made using a Neotec 6350 Research Composition Analyzer indicated that although pea starch made a significant contribution to the spectrum of pea flour, the lipid component did not. An InfraAlyzer 400 was used for calibration and testing purposes. Calibrations for pea starch and lipid contents were found to be bimodal in nature, and precision varied with the level of the component present. For low-starch samples, precision was ±2.63 ($P = 0.95$), whereas high-starch samples showed precision of ±1.81. The reverse was observed for lipids where calibration with low-lipid samples exhibited a precision of ±0.16 and high-lipid calibration precision was ±0.31 (Davies et al, 1985). These authors reported that the starch:lipid ratio appeared to be a genotype characteristic associated with the morphology of the pea. Attempts were also made to calibrate for a number of individual fatty acids present in the lipid component, but no significant correlations were found.

Measurement of oil content in oilseeds is another application of NIR in plant-breeding programs (Davies et al, 1985), although with high-oil-content seeds such as rapeseed, oil extrusion during sample preparation and analysis may prove a problem. Morgan and Starr (1983) examined the use of samples frozen in liquid nitrogen before milling and the use of whole seed samples as alternative methods of analyzing rapeseed by NIR. The results from both these treatments compared favorably with the original calibration using seed milled at room temperature. The frozen seed did not lose excessive amounts of oil during milling, but problems were still encountered with oil contamination of the sample cell and window; hence, thorough cleaning with alcohol between samples was still necessary. The standard error of difference between NIR and reference values was slightly higher for frozen samples, suggesting that the problem of oil extrusion encountered during cell loading caused greater errors than did the loss during milling. For this reason, the authors believed that freezing samples before milling offered no advantage over milling at room temperature. Although reproducibility of duplicate packings was poorest using whole seeds and the standard errors of prediction, especially for oil content, were the highest, the accuracy using whole seeds was still seen as adequate for the requirements of a screening program for plant breeding. The fact that the seed is not ground increases the speed of sample handling and allows the plant breeder the option of growing the seed if desired following analysis.

Screening for disease resistance is another important aspect of plant-breeding programs. In (1982), Asher et al reported the use of NIR as a technique for determining the susceptibility of barley to powdery mildew, yellow rust, and brown rust and of wheat to brown rust. These researchers used a Neotec 6350 Research Composition Analyzer to scan spore samples on glass-fiber filter disks. Use of three wavelengths—1,900, 2,252, and 2,308 nm—showed the best multiple correlation against spore populations counted on a hemacytometer slide. Derivation of partial regression coefficients for each species was accomplished by relating reflectance energy values to direct spore counts. This method provides a rapid, high-precision ($r = 0.9$) tool for counting of spore samples and can be adapted to the smaller, fixed-filter instruments for routine testing.

NIR has been used as a technique for quality control in the malting and brewing industry in Europe for a number of years. Studies by Moll et al (1976) examined the feasibility of using an InfraAlyzer and a GQA 31 to determine protein, moisture, and extract contents in 56–77 malt samples. The NIR results for both instruments were compared with the standard methods of Kjeldahl for protein and European Brewery Convention (EBC) methods for moisture and extracts. Correlations between the standard methods and the InfraAlyzer and GQA 31 respectively were 0.96 and 0.92 for moisture, 0.96 and 0.97 for protein, and 0.29 and 0.14 for extract. Results for the last component were believed to be poor partly because of the low precision of the reference method itself. Repeatability, as indicated by the coefficient of variability (CV) for multiple tests on a single malt sample, was better with the GQA 31. The CVs for GQA 31 moisture, protein, and extract were 0.65, 1.21, and 0.21% respectively, whereas corresponding CVs for the InfraAlyzer were 2.07, 1.56, and 0.52%. The authors concluded that both the GQA 31 and the InfraAlyzer were of similar quality and well suited to large-scale brewery control. Similar findings were reported in studies by Donhauser et al (1983), who used an InfraAlyzer 400 to measure protein and moisture in barley, wheat and barley, and wheat malt, as well as to determine original gravity and nitrogen content of worts. Results for both solid and liquid samples demonstrated reproducibility comparable to that of standard methods, as well as good correlation.

Carnielo (1984) reported correlations similar to those found by Moll et al (1976) for protein and moisture. Correlations for extract content in this case ranged from 0.85 to 0.92.

NIR has found further application in measuring the alcohol content of beer ($r = 0.98$, $SD = 0.26$, $n = 108$) using the transflectance technique (Coventry and Hunston, 1984) and the alcohol and sugar contents of wine (Kemeny et al, 1983; Altmeyer, 1984).

VI. MIXED FEEDS AND FORAGES

The introduction of scanning monochromators such as the Neotec 6350 and 6250 Research Composition Analyzers, the InfraAlyzer 500, and Trebor and LT Industries instruments (respectively Model 77 and Quantum 1200) has enabled the research scientist to apply NIR techniques to the analysis of complex mixtures. One of the major areas of application has been in the formulation of feed mixes, analysis of forages, and the prediction of their digestibility for livestock nutrition.

The potential for use of NIR at a feed mill is in many ways similar to that at a flour mill as far as monitoring raw materials is concerned. In this case, though, the range of raw materials to be examined is much broader, demanding equipment capable of accommodating a wider variety of calibrations. Earlier work in Canada had revealed significant interactions between ingredients (Williams and Starkey, 1980). Traditional proximate analysis of animal feeds has consisted of determining moisture, protein, oil, ash, fiber, and gross energy using both wet chemical and in vivo or in vitro analyses. As has been discussed previously in this chapter, the NIR technique can provide accurate determinations of protein, moisture, oil, and starch in wheat, barley, and many of the other cereals used in the production of mixed feeds. Current research is also proceeding into assessment of this technique to determine crude fiber (Gerstenkorn, 1985) in barley as well as digestibility and metabolizable energy (ME) in various grass and forage crops (Murray and Hall, 1983; Bengtsson and Larsson, 1984).

A number of studies have been reported recently concerning NIR analysis of forages. In most cases, the need for specific calibrations for different types of forages as well as for different methods of forage preparation has been recommended. The determination of a single "universal" calibration for forages has been hampered by a number of things, including the wide variety of forages, growing environments, sampling problems, and difficulties with sample preparation because particle-size distribution on grinding varies so greatly. Despite these drawbacks, numerous researchers in several European countries have been able to derive reasonable calibrations for a number of forage crops in various forms. In his summary of papers presented at the 5th International NIR Symposium in Paris, Linsenmann (1984) reported on four studies involving the use of NIR spectroscopy in areas of feed and forage analysis. Two of these papers represented work done in the Netherlands; work for one was done in France, and one in Finland. In each case, NIR was felt to be a useful technique for analysis purposes; however, further work was recommended to improve the accuracy of the results.

Biston (1979) used an FQA 51 and a Neotec 6350 Research Composition Analyzer to study the nutritional parameters of a number of products used for animal feeds. He divided these products into two broad classes of homogeneous and heterogeneous. The former class included cereals such as oats, spelt, barley, and wheat as well as noncereal products such as *Vicia faba* and potatoes. The heterogeneous group, which represented analysis of 8,000 samples over two years, included such substances as feed corn, grass, legumes, and legume-grass mixtures. These substances were further classified as fresh, dried, hay, or ensiled. The homogeneous products required a maximum of 50 samples for calibration and gave results on NIR analysis that were in close agreement with laboratory results for crude protein, starch, and amino acids. The heterogeneous products, on the other hand, required 100–200 samples for an adequate calibration. In addition, separate calibrations were required for tropical and temperate grasses and for the different types of forage preservations—i.e., fresh, hay, or ensiled. Once these adjustments were made, the correlations between laboratory and NIR result for crude protein, crude fiber, ash, and in vivo digestibility were good.

Further studies by Biston and Dardene (1984) showed similar results for products of both homogeneous and heterogeneous nature. They reported that in vivo digestibility of organic matter could be analyzed with better precision using NIR than by use of the normal laboratory in vitro method using rumen liquor from fistulated sheep. These authors also described a number of research projects employing NIR analysis currently under consideration in their center. They include measurement of amino acids for screening of new cereal varieties; determination of optimal forage-harvesting time based on predicting the best production/quality ratio; and prediction of fitness for ensiling and necessity for addition of preservatives, based on the determination of sugar/protein ratio and estimation of the intake of grazed forage based on NIR fecal analysis for protein and lignin. Brief reference is also made to a study based on the computation of the projection matrix to differentiate between products and possibly detect defective spectra.

Bengtsson and Larsson (1984) used an InfraAlyzer 400R to estimate crude protein, digestibility of organic matter, and ME in different species of forage and in grass forages preserved by different methods. Samples of pure timothy (*Phleum pratense* L.), meadow fescue (*Festuca pratensis* Huds.), alfalfa (*Medicago sativa* L.), and red clover (*Trifolium pratense* L.) grown in different localities and harvested at different growth stages were used to calibrate the instrument for crude protein (Kjeldahl method) and in vitro digestibility (rumen digestible organic matter—RDOM). Known mixtures of these grass and legume species were used as test samples. Repeated analysis of one coarse and one fine sample was done to estimate the precision of the NIR method in the presence of widely varying sample particle size. Fine samples showed lower SD and CV for both crude and RDOM than coarse samples. Overall results were, for the calibration samples ($n = 185$), SD = 0.33, CV = 2.0 for crude protein, SD = 0.52, CV = 0.66 for RDOM; and, for the mixture test samples ($n = 103$), SD = 0.35, CV = 2.12 for crude protein, SD = 0.59, CV = 0.76 for RDOM. In addition to analyzing samples with the standard filters used for grain analysis, combinations of six filters providing the smallest residual standard deviation (RSD) and highest correlation between the NIR and standard methods were selected for each of the four pure sample calibrations. When this was done, it was apparent that the combinations of filters selected were species specific.

The study was repeated using a series of grass samples, mainly timothy and meadow fescue, that had been preserved either as frozen green chop, hay, or silage. The frozen samples were used for calibration purposes and samples were analyzed for in vitro digestibility as well as for crude protein and in vivo digestibility. A mixture of samples preserved by all three techniques was used as test samples. Again, filter selection was shown to be dependent on forage treatment and was different for each treatment. The accuracy in analyzing standard results by NIR was tested using both the filter combination selected for each forage type or preservation method and those used for standard grain analysis. In vivo ME of samples of frozen chop grass could be fairly accurately analyzed using the calibration based on in vitro values for grass/legume samples. In this case, standard filters gave better results than the selected ones, whereas selected filters were better suited for analysis of in vitro ME in grass/legume samples. Results indicated that filters selected for in vitro calibrations were better suited for in vitro rather than for in vivo predictions. In addition, when standard filters were used, in vivo ME in the grass samples was analyzed with the same accuracy as the in vitro ME, even though the calibration was performed with in vitro data.

The authors (Bengtsson and Larsson, 1984) concluded that the use of NIR to assess forage quality was restricted by the differences in chemical composition and reflectance properties among common forage species and that its use was inferior to the RDOM method for estimating in vivo ME in forages preserved by different techniques. However, they also concluded that NIR results could be improved significantly with proper calibration and use of a system based on spectral recognition that would allow selection of calibrations from data stored in a computer.

Barnes (1984) used a Neotec 6350 Research Composition Analyzer to measure the nitrogen content of cocksfoot, ryegrass, and timothy taken at weekly intervals over a period of nine weeks. The samples were dried and ground in a Christy Norris lab mill to pass through a 1-mm sieve. The best correlation between NIR and Kjeldahl results ($r = 0.998$, SEP = 0.07) was obtained by using the second derivative of log $(1/R)$. This calibration was then tested against pure and mixed grass samples grown in the same location as the calibration samples and at other locations as permanent pasture and grazed areas. Samples from areas that had been grazed, whether they were pure grasses or mixed, showed the greatest variance between NIR and lab results ($r = 0.907$, SEP = 0.34 for pure grasses grazed; $r = 0.952$, SEP = 0.07 for permanent pasture grazed). The calibration was felt to be adequate to analyze nitrogen content of ungrazed reseeded or permanent pastures during one season of growth. Data on regrowth samples and effect of crop year were not available for their report.

The same set of samples was used to establish a calibration for in vitro digestibility of organic matter in the dry matter (DOMD). In this case, a two-term equation using the second derivative of log $(1/R)$ gave a standard error of calibration of 0.87 and a correlation between NIR and lab results of 0.99. The SEP on a second set of samples was 1.17 with a correlation of 0.98.

Barnes (1984) also investigated the in vivo DOMD of six varieties of grasses collected from eight sites over four growing seasons. Fifty of these samples were used for calibration and the remaining 75 were used as unknowns to test the calibration, with a resulting r of 0.85 and SEP of 2.5.

Studies by Murray and his co-workers (Murray et al, 1982; Murray, 1983; Murray and Hall, 1983) have centered around evaluation of grass silages and compound feeds for DOMD, ME, and crude protein using a spectrocomputer. The standard chemical method for crude protein measurement in these studies was Kjeldahl, whereas DOMD was estimated from a regression of in vivo DOMD on lignin determined spectrophotometrically in duplicate using acetyl bromide. ME calculations were estimated from the DOMD, assuming a linear relationship. Tests on 24 grass silages with a wide range of ME values showed good correlation: $r = 0.90$, SEE = 2.59; and $r = 0.96$, SEE = 0.40 between NIR analyzed values and percentage DOMD and ME respectively (Murray et al, 1982). Corresponding values for the standard method (lignin prediction) were $r = -0.92$, SEE = 2.41 and $r = 0.79$, SEE = 0.81. As can be seen, the NIR method was considerably better at estimating ME than the method based on lignin. The authors suggested that this may result from the fact that NIR measures the -CH stretch first-overtone region at 1.7–1.8 μm, a region highly correlated to both DOMD and ME.

Further studies on grass silage by Murray (1983) showed that NIR was capable of measuring crude protein in silage with sufficient accuracy ($r = 0.92$) to allow the silages to be grouped into five or seven protein quality groups. This protein calibration was less precise than that obtained with grass samples ($r = 0.98$) because of higher sampling error with wet silage and loss of ammonia nitrogen during sample drying. Other factors that affect silage sample variability include the complex range of products included in the silage, stage of maturity at cutting, postharvest treatment, and type of silo used. One of the benefits of using the spectrocomputer, as shown in this study, was that grossly abnormal silages, such as those containing potatoes or grain, or silage carried over in storage from another year, were obvious to the operator because of abnormal changes in the spectral patterns. An examination of six different predictors of in vivo DOMD showed the NIR method to be equivalent to that of pepsin-cellulase solubility and slightly inferior to the two-stage rumen in vitro and acetyl bromide lignin methods. When a similar examination of analyzers of in vivo ME was conducted, the NIR method emerged as the best analyzer.

Murray and Hall (1983) also attempted NIR calibrations on 24 compound feeds containing up to 35 ingredients and analyzed for 15 chemical-biochemical parameters. With the exception of sugar, ash, and gross energy on a dry-matter basis, the NIR showed good prediction correlations and standard errors. Since no test set of similar samples was available to test the calibration equations, it was not possible to separate chemical and NIR precision. These authors suggest that NIR may be capable of better precision than some of the methods used for its calibration and that it offers a method for standardizing the measurement of tests used to predict animal performance. In many cases, wet chemical methods used for animal feed evaluation vary from one laboratory to the next, as do the regression equations used to relate one method to another. For advisory and comparison purposes, these laboratories should agree upon some common form of ranking of sample quality regardless of test methods used, and NIR may provide the solution to this problem.

To date, most of the research involving mixed feeds and forages and their analysis by NIR has involved the use of a computerized spectrophotometer, a type of equipment not generally available in a feed mill or for forage analysis on a routine basis. Research has also been conducted to determine whether calibrations for animal feedstuffs conducted on a spectrocomputer can be transferred to the smaller instruments and whether the filters in these instruments normally used for grain analysis are adequate for feed and forage analysis or whether a different combination of filters would provide more accurate results. Some of the findings in this field have been alluded to in considering the work of Bengtsson and Larsson (1984), who reported that selection of filters other than the standard ones only gave accurate results if the composition of the test samples was very similar to that of the calibration samples, but that on the whole the standard filters for grain analysis gave reasonable results. Similar conclusions were reached by van Lonkhuysen (1984), who studied maize gluten protein and fat and wheat middlings protein, fat, and fiber using both an InfraAlyzer 400 and 500. He reported that the standard set of filters for the InfraAlyzer 400 gave good results for all the constituents measured and that additional filters did not markedly improve instrument performance. Further studies on transferability of calibrations indicated that for the materials under investigation, calibrations obtained on the InfraAlyzer 500 could be successfully transferred to the InfraAlyzer 400.

VII. AGRICULTURAL PRODUCTS FOR HUMAN CONSUMPTION

A survey of the literature dealing with the practical applications of NIR in Europe would not be complete without some mention of the wide range of information dealing with its use in food analysis. It appears to have found practical application in the fields of dairy products, meat, fruit and vegetables, pasta, starch, vegetable oils, tobacco, and even rapid analysis of packaging laminates used in the food processing industry.

Applications of NIR in the dairy industry have been reported for fluid and dry milk powders as well as dairy products such as yogurt, butter, cream, and *Quark* in Germany (Thomasow and Paschke, 1981), Switzerland (Dillier-Zulauf and Egli, 1982; Egli, 1984), France (Verbiese, 1981; Leuckx, 1984), Italy (Cardinale, 1980), Hungary (Lasztity et al, 1981), Sweden (Allvin, 1980; Larsson, 1980), and Great Britain (McGann, 1978). The most commonly analyzed constituents are protein, moisture, fat, and lactose, although such elements as total solids, solids-not-fat, and ash content may also be measured with good accuracy. The equipment may be used for on-line process control, for batch testing of bulk milk coming into the dairy, or for quality control of the finished product.

NIR spectroscopy has been used in the meat industry as a rapid method for measurement of fat, moisture, and protein in meat and meat products. Hauser and Weber (1980) described the use of an InfraAlyzer Plus to determine the main meat parameters of water, fat, crude protein, connective-tissue-protein-free meat protein, and ash. They obtained good agreement and high correlations between results of NIR and standard methods and reported that the results were useful for production control although not good enough for control laboratories. Trials with a temperature-controlled cell ranging from 5° to 20° C, as opposed to the usual operating temperature of 40° C, showed improved reproducibility with use of constant temperature. The smallest deviations were observed at 10° C.

Applications of NIR for quality assessment of a number of fruits and vegetables and their products, including stone-fruit pulp, tomato paste, whole tomatoes, tomato juice and puree, carrots, and red paprika, have been described by Kaffka (1981) and Horvath and Horvathova-Mosonyova (1981).

Davies (1983) described the use of a Neotec 6350 Research Composition Analyzer for the analysis of egg and oil in salad cream. The levels of these substances are subject to legislation in the U.K., hence a rapid method for their determination would have direct industrial application. The calibrations obtained were subsequently adapted for use in a Neotec 102, InfraAlyzer 400, and Instalab 800 with the expectation that they could be used on-line in the factory to monitor the level of these two constituents during salad cream production.

Experience with NIR in routine analysis of starches, their derivatives, and saccharification products was outlined by Marihart (1984). Analyses for moisture, protein, starch, degree of substitution, and degree of degradation for such starches as corn, potato, wheat, and waxy maize as well as dextrose and malt dextrose were carried out with an InfraAlyzer 400 and 400D. In addition to the acceptable results obtained by this technique, its flexibility, ease of analysis by untrained personnel, lack of need for sample pretreatment, and economy were cited by the author.

Near-infrared reflectance analysis has also been applied to the pasta industry with much success (Cardinale, 1980; Lasztity and Varga, 1984). The technique has been used to determine levels of moisture, protein, and fat in various types of pasta. From the fat content, it is then possible to predict the egg content (Lasztity and Varga, 1984), an important consideration in countries where food legislation exists concerning minimum standards for addition of this component.

Another application of NIR to the analysis of agricultural products for human consumption concerns its use in the European tobacco industry. Gough (1983) described some of the problems encountered with the use of NIR for measuring the composition of tobacco. These included moisture content of the sample, effect of particle size, and contending with rogue or spurious data. These problems are not exclusive to work with tobacco, and methods for dealing with them are common to NIR research with most agricultural products. They include avoiding selection of filters that are sensitive to moisture, normalizing optical data to eliminate particle-size effects, and eliminating rogue samples from the calibration set prior to instrument calibration. Despite these problems, the author allows that NIR analysis is potentially a very powerful tool for measuring tobacco composition.

Finally, mention should be made of a recent description of the application of NIR to the analysis of peat (Downey and Byrne, 1986). NIR was used to determine moisture over a range of 34–71% with an SEP of 1.1 and a precision (SET) of 0.7. Bulk density of the peat was analyzed with an SEP of 15.1 over a range of 193–402 g/liter. Respective coefficients of correlation were 0.99 and 0.94 between NIR and standard determinations of moisture and bulk density.

LITERATURE CITED

ALLVIN, B. 1980. New analytical technique in the food area—near infrared reflectance (NIR). (In Swedish) Kem. Tidskr. 92:24-27.

ALTMEYER, F. 1984. Rapid and practical alcohol measurement in wine with the Technicon InfraAlyzer 400L. (In German) Monatsschr. Brau. 37:88-91.

ANONYMOUS. 1984. Near infrared reflectance analysis—the first ten years. Milling, December, pp. 24-26.

ASHER, M. J. C., COWE, I. A., THOMAS, C. E., and CUTHBERTSON, D. C. 1982. A rapid method of counting spores of fungal pathogens by infrared reflectance analysis. Plant Pathol. 31:363-372.

BARNES, R. J. 1984. Forage quality analysis by N.I.R. reflectance. Proc. Annu. Users Conf. NIR Res., 3rd. Pacific Scientific, Rockville, MD.

BENGTSSON, S., and LARSSON, K. 1984. Prediction of the nutritive value of forages by near infrared reflectance photometry. J. Sci. Food Agric. 35:951-958.

BERTRAND, D., ROBERT, P., and LOISEL, W. 1985. Identification of some wheat varieties by near infrared reflectance spectroscopy. J. Sci. Food Agric. 36:1120-1124.

BEUCH, H. G. 1983. Classification of wheat at storage with the aid of infrared spectroscopy. (In German) Getreide Mehl Brot 37:21-23.

BISTON, R. 1979. Potential uses of N.I.R.S. for the analysis of agricultural productions. Pac. Sci. Res. Notes NIR 2016.

BISTON, R., and DARDENE, P. 1984. Applications of near infrared spectroscopy in the field of agronomy. Proc. Annu. Users Conf. NIR Res., 3rd. Pacific Scientific, Rockville, MD.

BOLLING, H. 1979. The use of NIR spectroscopy in grain analysis. (In German) Getreide Mehl Brot 33:264-268.

BOLLING, H., and ZWINGELBERG, H. 1979. The use of infrared spectroscopy in the milling industry. (In German) Muehle Mischfuttertechnik 116:53-55.

BOLLING, H., and ZWINGELBERG, H. 1982. Infrared spectroscopy—Applications in grain intake and end product control in milled products. (In German) Getreide Mehl Brot 36:197-201.

BOLLING, H., and ZWINGELBERG, H. 1984. Continuous determination of flour components by NIR. (In German) Getreide Mehl Brot 38:3-5.

CALANDRA, F. 1977. Use of infrared reflection for immediate assessment of residual oil in sunflower meal. (In Italian) Riv. Ital. Sostanze Grasse 54:171-176.

CARDINALE, E. 1980. Near infrared reflectance instrumentation. (In Italian) Tec. Molitoria 31:910-916.

CARNIELO, M. 1984. The use of NIRA in malting and brewing industries. (In German) Monatsschr. Brau. 37:88-91.

CLOVER, M. B. 1979. The use of the InfraAlyzer in a small/medium sized flour mill. Autoanalyst, November, pp. 3-4.

COLAS, A. 1979. The InfraAlyzer at the Grands Moulins de Paris. Proc. Technicon Int. Symp., 9th, 14–16 November, Paris.

COVENTRY, A. G., and HUNSTON, M. J. 1984. Application of near-infrared spectroscopy to the analysis of beer samples. Cereal Foods World 29:715-718.

CUBADDA, R., and DOUGLAS, S. 1982. Investigations on the near infrared reflectance techniques for measuring the properties of wheat and its products. Proc. World Cereal Bread Congr., 7th, 24 June–2 July, Prague. Elsevier, in press.

DAVIES, A. M. C. 1983. Human food analysis by NIR—still a promising technique. Paper 3 in: NIR Analysis—How Near Infrared Reflects Composition. Chemistry Division, School of Agriculture, Aberdeen, Scotland.

DAVIES, A. M. C., COXON, D. T., GAVREL, G. M., and WRIGHT, D. J. 1985. Determination of starch and lipid in pea flour by near infrared reflectance analysis. The effect of pea genotype on starch and lipid content. J. Sci. Food Agric. 36:49-54.

DAVIES, A. M. C., and WRIGHT, D. J. 1984. Determination of protein in pea flour by near infrared analysis. J. Sci. Food Agric. 35:1034-1039.

DAY, M. S., and FEARN, F. R. B. 1982a. Near infrared reflectance as an analytical technique. Part I. History and development. Lab. Pract. 31:328-330.

DAY, M. S., and FEARN, F. R. B. 1982b. Near infrared reflectance as an analytical technique. Part II. Design and development of practical NIR instruments. Lab. Pract. 31:439-443.

DILLIER-ZULAUF, A., and EGLI, H. R. 1982. Quantitative rapid determination of fat, nitrogen and dried substances in whey products with the InfraAlyzer. (In German) Dtsch. Molk. Ztg. 103:820-822, 824-825.

DONHAUSER, E. G., LINSENMANN, O., and FALTERMEIER, E. 1983. Evaluation of near infrared spectrophotometry in the malting and brewery laboratory. (In German) Monatsschr. Brau. 36:474-477.

DOWNEY, G., and BYRNE, P. 1986. Prediction of moisture and bulk density in milled peat by near-infrared reflectance. J. Sci. Food Agric. 37:231-238.

DOWNEY, G., MORRISEY, A., and FLYNN, S. 1982. Protein testing of wheat by near infrared reflectance. Farm Food Res. 13:14-16.

EGLI, H. R. 1984. Standardization, quality insurance and control of milk products with the Technicon InfraAlyzer 400D. (In German) Monatsschr. Brau. 37:88-91.

FEARN, F. 1981. Sampling best grinders for British market. Milling Feed Fert. 164:18.

FOSSATI, M., and PERROTTA, F. 1977. Rapid protein, oil and moisture determination with the Technicon InfraAlyzer system. (In Italian) Riv. Ital. Sostanze Grasse 54:159-160.

GERSTENKORN, P. 1985. NIR-spectroscopy as a practical method for measurement of crude fiber in barley. (In German) Getreide Mehl Brot 39:167-170.

GILL, A. A., STARR, C., and SMITH, D. B. 1979. Lysine and nitrogen measurement by infrared reflectance analysis as an aid to barley breeding. J. Agric. Sci. 93:727-733.

GOUGH, K. 1983. Statistical considerations of near-infrared analysis applied to tobacco. Paper 5 in: NIR Analysis—How Near Infrared Reflects Composition. Chemistry Division, School of Agriculture, Aberdeen, Scotland.

HAUSER, E., and WEBER, U. 1980. Application of IR-reflection analysis for quick determination of the valuable portions in meat and meat products. II. Plotting curves for contents of the main constituents of meat and meat products (as in I, but with a greater number of samples), using existing contents curves. Experiments with a refrigerated cell. (In German) Fleischwirtschaft 60:482-485.

HOFFMANN, K. 1963. Moisture measurement through infrared reflection. Chem. Ing. Tech. 35:55-62.

HÖHNE, A. 1983. Use of infrared spectroscopy at cereal grain reception (wheat and barley). (In German) Getreide Mehl Brot 37:20-21.

HORVATH, L., and HORVATHOVA-MOSONYOVA, M. 1981. Measurement of quality of vegetables. (In Hungarian) Prum. Potravin 32:647-648.

HRUSCHKA, W. R., and NORRIS, K. H. 1982. Least-squares curve fitting of near infrared spectra predicts protein and moisture content of ground wheat. Appl. Spectrosc. 36:261-265.

HUNSTON, M. J. 1980. Infra-red analysis: Quality control of the future. Food Flavor Ingred. Process. 2:24-27.

KAFFKA, K. 1981. Non-destructive methods of analysis in the food industry. (In Hungarian) Prum. Potravin 32:634-639.

KEMENY, G., POKORNY, T., FORIZS, K., and LEKO, L. 1983. Use of the near infrared measurement technique in the wine industry. (In Hungarian) Borgazdasag 31:127-132.

KLEIN SCHIPHORST, H. A. J. 1986. Determination of flour components using the NIROS system. (In German) Getreide Mehl Brot 40:78-81.

LANDA, I. 1979. High-energy spectrophotometer for rapid constituent analysis in the range of 0.25–2.4 μm. Rev. Sci. Instrum. 50:34-40.

LARSSON, A. 1980. Analytical work for milk quality assessment. (In Swedish) Nord. Mejeriindustri 7:598, 600.

LASZTITY, R., MERÉSZ, P., BÉKÉS, F., SMIED, I., and ZSIGMOND, A. 1981. Rapid methods for the determination of fat content in certain foods. (In Hungarian) Elelmiszervizsgalati Kozl. 27:113-122.

LASZTITY, R., and VARGA, J. 1984. Determination of some cereal constituents by NIR technique in Hungary—experience and future trends. Pages 106-109 in: Proc. World Cereal Bread Congr., 11th, Vienna.

LEUCKX, M. 1984. Process monitoring with the Technicon InfraAlyzer 400D. (In German) Monatsschr. Brau. 37:88-91.

LINSENMANN, O. 1984. The 5th International NIRA-Symposium in Paris. (In German) Monatsschr. Brau. 37:88-91.

MADDIX, C. 1976. Survey of five different grinders in the United Kingdom for potential application to infrared reflectance analysis. Technicon Ind. Syst. News 2:7-9, 12.

MARIHART, J. 1984. Experiences with NIR-reflectance spectroscopy in routine analysis of starches, starch derivatives and starch saccharification products. (In German) Staerke 36:51-55.

MARTENS, H., and JENSEN, S. A. 1982. Partial least squares regression: A new two-stage NIR calibration method. Pages 607-647 in: Proc. World Cereal Bread Congr., 7th. J. Holas and J. Kratochvil, eds. Elsevier, Amsterdam.

MASSIE, D. R., and NORRIS, K. H. 1965. Spectral reflectance and transmittance properties of grain in the visible and near infrared. Trans. ASAE 8:598-600.

McFARLANE, I. D. 1981. Measurement of moisture content. U.K. Patent Application GB 2 078 938 A. Filed 13 April.

McGANN, T. C. A. 1978. Automated physicochemical methods for the analysis of milk—a review of major advances (1960–1978). Ir. J. Food Sci. Technol. 2:141-156.

MOLL, M., FLAYEUX, R., and LEHUEDÉ, J. M. 1976. Rapid determination of moisture and proteins in malt by infrared reflectance spectroscopy. (In French) Bios (Nancy) 7:3-6.

MORGAN, A. G., and GOTHARD, P. G. 1979. Rapid prediction of malt hot water extract by near infrared reflectance spectroscopy studies on barley. J. Inst. Brew. (London) 85:339-341.

MORGAN, A. G., and GOTHARD, P. G. 1981. A rapid test for identifying potential malting quality in the early generations of a barley breeding program. J. Sci. Food Agric. 32:333-338.

MORGAN, A. G., and STARR, C. 1983. A comparison of the use of milled and whole rapeseeds for the measurement of nitrogen and oil content by near infrared spectroscopy. Paper 6 in: NIR Analysis—How Near Infrared Reflects Composition. Chemistry Division, School of Agriculture, Aberdeen, Scotland.

MURRAY, I. 1983. Rapid silage evaluation by near infrared reflectance (NIRR) spectroscopy. Pages 83-87 in: Research Investigations and Field Trials 1982–83. School of Agriculture, Aberdeen, Scotland.

MURRAY, I., GELMAN, A. L., and TOPPS, J. H. 1982. Evaluation of grass silage by near infrared reflectance or chemical analysis. Pages 391-394 in: Proc. Int. Network of Feed Information Centres Symp., Sydney, Australia.

MURRAY, I., and HALL, P. A. 1983. Animal feed evaluation by use of near infrared reflectance (NIR) spectrocomputer. Anal. Proc. (London) 20:75-79.

NORRIS, K. H. 1964. Reports on design and development of a new moisture meter. Agric. Eng. 45:370-372.

NORRIS, K. H., and BUTLER, W. L. 1961. Techniques for obtaining absorption spectra on intact biological samples. I.R.E. Trans. Bio-Med. Electron. 8:153-157.

NORRIS, K. H., and HART, J. R. 1965. Direct spectrophotometric determination of moisture content of grain and seeds. Pages 19-25 in: Principles and Methods of Measuring Moisture in Liquids and Solids. Vol. 4. A. Wexler, ed. Reinhold, New York.

OSBORNE, B. G. 1981a. Application of near infrared reflectance spectroscopy to the analysis of food. Anal. Proc. (London) 18:488-489.

OSBORNE, B. G. 1981b. Principles and practice of near infra-red (NIR) reflectance analysis. J. Food Technol. 16:13-19.

OSBORNE, B. G. 1983a. Applications of NIR in the baking industry. Anal. Proc. (London) 20:79-83.

OSBORNE, B. G. 1983b. Investigation of the performance of an improved calibration for the determination of protein in UK home-grown wheat by near infrared reflectance analysis. J. Sci. Food Agric. 34:1441-1443.

OSBORNE, B. G. 1983c. Measurement of levels of bread improvers in concentrates by means of near infrared reflectance spectroscopy. J. Sci. Food Agric. 34:1297-1301.

OSBORNE, B. G. 1984. Investigations into the use of near infrared reflectance spectroscopy for the quality assessment of wheat with respect to its potential for bread baking. J. Sci. Food Agric. 35:106-110.

OSBORNE, B. G. 1986. The use of near infrared reflectance spectroscopy in the control of flour protein content. J. Chem. Technol. Biotechnol. 36:364-366.

OSBORNE, B. G., BARRETT, G. M., CAUVAIN, S. P., and FEARN, T. 1984b. The determination of protein, fat and moisture in bread by near infrared reflectance spectroscopy. J. Sci. Food Agric. 35:940-945.

OSBORNE, B. G., and DOUGLAS, S. 1981. Measurement of the degree of starch damage in flour by near infrared reflectance analysis. J. Sci. Food Agric. 32:328-332.

OSBORNE, B. G., DOUGLAS, S. G., and FEARN, T. 1981. Assessment of wheat grain texture by near infrared reflectance measurements on Buhler-milled flour. J. Sci. Food Agric. 32:200-202.

OSBORNE, B. G., DOUGLAS, S., and FEARN, T. 1982a. The application of near infrared reflectance analysis to rapid flour testing. J. Food Technol. 17:355-363.

OSBORNE, B. G., DOUGLAS, S. G., and FEARN, T. 1982b. Recent progress in the application of NIR to the measurement of quality parameters in flour. Proc. World Cereal Bread Congr., 7th, 24 June–4 July, Prague.

OSBORNE, B. G., DOUGLAS, S., FEARN, T., and WILLIS, K. H. 1982d. The development of universal calibrations for measurement of protein and moisture in UK home-grown wheat by near-infrared reflectance analysis. J. Sci. Food Agric. 33:736-740.

OSBORNE, B. G., and FEARN, T. 1983a. Collaborative evaluation of universal calibrations for the measurement of protein and moisture in flour by near infrared reflectance. J. Food Technol. 18:453-460.

OSBORNE, B. G., and FEARN, T. 1983b. Collaborative evaluation of near infrared reflectance analysis for the determination of protein, moisture and hardness in wheat. J. Sci. Food Agric. 34:1011-1017.

OSBORNE, B. G., FEARN, T., and BARRETT, G. M. 1982c. FMBRA takes the grind out of measuring moisture. Milling Feed Fert. 165:34-35.

OSBORNE, B. G., FEARN, T., MILLER, A. R., and DOUGLAS, S. 1984a. Application of near infrared reflectance spectroscopy to the compositional analysis of biscuits and biscuit doughs. J. Sci. Food Agric. 35:99-105.

OSBORNE, B. G., FEARN, T., and RANDALL, P. G. 1983. Measurement of fat and sucrose in dry cake mixes by near infrared reflectance spectroscopy. J. Food Technol. 18:651-656.

POPINEAU, Y., and GODON, B. 1976. Testing of the Technicon InfraAlyzer. (In French) Tech. Ind. Cerealieres 152:7-12.

RINGLUND, K. 1980. Use of near-infrared reflectance spectroscopy for the determination of total nitrogen. (In Norwegian) Nord. Jordbrugsforsk. 62:273-275.

SAURER, W. 1978. Use of infra-red reflectance measurement for the determination of protein and moisture content and kernel hardness in wheat. (In German) Getreide Mehl Brot 32:272-276.

SCHAFER, K., and WEINARD, W. 1978. Infrared reflectance spectroscopy: A new method for simultaneous rapid determination of protein, fat and moisture. (In German) Muehle Mischfuttertechnik 115:240-243.

SCHMUTZ, W., and DO, Q. N. 1982. Improved fast protein determination in barley by IR reflectance spectroscopy. (In German) Muehle Mischfuttertechnik 119:258-261.

SCHORCH, K. H. 1983. Potential for production in wheat flour mills using infrared spectroscopy. (In German) Getreide Mehl Brot 37:23-25.

STARR, C., MORGAN, A. G., and SMITH, D. B. 1981. An evaluation of near infrared reflectance analysis in some plant breeding programmes. J. Agric. Sci. 97:107-118.

STARR, C., SMITH, D. B., BLACKMAN, J. A., and GILL, A. A. 1983. Applications of near infrared reflectance analysis in breeding wheats for bread-making quality. Anal. Proc. (London) 20:72-74.

STENVERT, N. L. 1974. Grinding resistance. A simple measure of wheat hardness. Flour Anim. Feed Milling, July, pp. 24-27.

SVENSSON, G. 1981. Varietal and environmental effects on wheat milling quality. Agric. Hort. Genet. 39:1-103.

THOMASOW, J., and PASCHKE, M. 1981. Determination of fat, protein and lactose in milk by infrared absorption using the Multispec instrument. (In German) Milchwissenschaft 36:65-68.

VAN LONKHUYSEN, H. J. 1984. Some experience with the InfraAlyzer-500 on feeding stuffs. Pages 103-105 in: Proc. World Cereal Bread Congress, 11th, Vienna.

VERBIESE, Y. 1981. Continuous moisture determination by infrared absorption in the agricultural and food industries. (In French) Ind. Agric. Aliment. 98:759-768.

WALTL, K. 1982. The determination of protein content in grain and mill products. (In German) Allg. Muehlen-Markt 83:264-267.

WEINARD, W. 1981. New opportunities for protein determination in small and medium-sized mills. (In German) Muehle Mischfuttertechnik 118:482-483.

WILLIAMS, P. C. 1975. Application of near infrared reflectance spectroscopy to analysis of cereal grains and oilseeds. Cereal Chem. 52:561-576.

WILLIAMS, P. C. 1981. Choosing grinders for sample analysis. Milling Feed Fert. 164:30-31, 33.

WILLIAMS, P. C., NORRIS, K. H., and ZAROWSKI, W. S. 1982. Influence of temperature on estimation of protein and moisture in wheat by near-infrared reflectance. Cereal Chem. 59:473-477.

WILLIAMS, P. C., and STARKEY, P. M. 1980. Influence of feed ingredients upon the prediction of protein in animal feed mixes by near-infrared reflectance spectroscopy. J. Sci. Food Agric. 31:1201-1213.

WILLIAMS, P. C., and THOMPSON, B. N. 1978. Influence of whole meal granularity on analysis of HRS wheat for protein and moisture by near infrared reflectance spectroscopy (NRS). Cereal Chem. 55:1014-1037.

ZWINGELBERG, H. 1983. Infrared spectroscopy—significance and measurement of grain hardness. (In German) Getreide Mehl Brot 37:25-28.

CHAPTER 10

Application of Near-Infrared Reflectance Spectroscopy in North America

JANET A. PANFORD
Grain Research Laboratory
Canadian Grain Commission
Winnipeg, Manitoba, Canada

I. INTRODUCTION

Most tests used for grading grains under the present grain standards in Canada and the United States are subjective. They are based on a determination of physical characteristics, and the relationship of these characteristics to the end use of the grain is often difficult to identify. The tests that are objective tend to provide data relative only to the purity of the grain, except in the case of moisture and test weight. Although some reliable objective tests (e.g., protein) are available, they are usually too time-consuming to perform, require considerable capital investment, use small sample size, require considerable operator technique and lengthy calibration procedures, and generally do not satisfy today's trade needs. For example, it takes about 2 hr to complete a Kjeldahl test for protein in wheat, one of the most important factors in predicting the performance of flour made from that wheat, whereas an extraction test for oil takes about 18 hr and a lysine determination about two days. Yet it takes only about 5 min to unload a truck or railcar at an elevator. Any quality tests to be performed at receiving points in the marketing chain must be both rapid and as accurate as possible. The advent of near-infrared reflectance (NIR) spectroscopy as a high-throughput and accurate method for the analysis of grains and oilseeds has been well timed to fill a burgeoning need in agriculture and in the food and feed processing industries in North America. NIR has the potential to furnish a series of objective tests for the determination of end-use processing potential of cereal and pulse grains and of oilseeds.

The use of NIR for analysis of grains and seeds was first introduced by Norris and Hart in 1965. The technique was originally developed for the measurement of moisture content (Norris, 1964). These workers measured the moisture content of whole peanuts, lima bean cotyledons, and intact wheat kernels by transmittance spectroscopy, using a monochromator that incorporated a wedge interference filter. The instrument had a scanning range of 0.7–2.4 μm, a tungsten lamp (150 W, 115 V) as the light (energy) source, and a lead-sulfide detector. They found that three wavelengths—1.8, 1.94, and 2.08 μm—were applicable to moisture determination, with 1.94 μm being the most sensitive, and they also found that sample thickness affected the absorbance. They measured moisture in wheat and soybeans over a range of 0–20% and obtained coefficients of correlation of 0.97 for wheat and 0.90 for soybeans, with respective standard deviations of 0.28 and 0.37. They concluded that moisture content of a range of material could be measured at 1.94 μm if the sample was uniform and thin (or a fine powder) with no interfering absorption bands. This work became the basis of modern NIR technology in the food, feed, and grain industries. The first technical presentations on the application of a commercial NIR instrument to analysis of cereals and oilseeds were made in 1973 (Williams, 1973a, 1973b). Since that time, NIR technology has become widely used in food analysis. Instruments capable of measuring starch, oil, sugar, fiber, moisture, and protein are commercially available, and recent research on five continents has shown that many other applications are possible, such as the determination of amino acids, fiber components, and mineral constituents. This chapter summarizes the work carried out in North America.

II. APPLICATION TO PLANT AND ANIMAL PRODUCTS

A. Wheat

There are several functions served by NIR technology in industry, especially in those areas where analytical results are "time-critical." This applies, for example, to the measurement of protein, fat/oil, carbohydrate, and water in grain handling and food processing operations. It often involves the monitoring of incoming, in-process, and outgoing products (Watson, 1977).

NIR technology is used at the receiving stage at mills, processing plants, and elevators to examine incoming grains before storage or processing. For example, NIR is used in the segregation of wheat on the basis of protein content (Williams, 1977). Western Canada produces, on the average, about 20 million tonnes of hard red spring wheat annually, the bulk of which is exported on the basis of guaranteed minimum protein contents of 11.5, 12.5, 13.5, and 14.5%. Accurate protein measurement is essential to the binning or segregation of the wheat into the various protein levels. NIR has been used for this purpose in terminal (port) and country elevators in the United States and Canada since 1976 (Hunt et al, 1977; Miller et al, 1978). To improve the efficiency of binning on the basis of protein content and to prevent delays in unloading producer trucks, a protein result must be available in about 1.5 min from the time the truck or railcar arrives at the elevator until the grain is directed into receiving bins. NIR instruments have made this possible (Williams and Panford, 1980).

Since the original work of Norris and Hart, the analysis of ground or whole wheat for protein and moisture has been reported by several other workers in North America. Williams (1973a, 1973b) reported the application of NIR to the analysis of wheat, flour, barley, oats, rapeseed, and soybeans. This work was later reported in more detail (Williams, 1975). Law and Tkachuk (1977a) reported similar work on wheat. These workers compared NIR with oven moisture results for three different varieties of wheat and found very high correlation coefficients (0.98–1.00) between the two methods. They concluded that NIR was an accurate and reliable method for the estimation of moisture in different wheat types with a wide range of moisture contents (2.8–16.5%). These measurements were again made at 1.93 and 2.10 μm in the NIR region of the electromagnetic spectrum. Later, Stermer and co-workers (1977a) discussed the application of NIR to the direct determination of moisture in whole wheat kernels. Tkachuk (1981a) reported a multiple correlation coefficient of 0.988 and standard error of estimate of 0.30 for the prediction of protein content in 45 whole wheat samples using six different wavelength points between 1.18 and 1.23 μm, 1.63 and 1.69 μm, and 1.73 and 1.79 μm. More recently, Williams et al (1985b) demonstrated that near-infrared transmittance (NIT) could provide accurate analysis of wheat and barley for protein and moisture. These workers used a Trebor GT-90 NIT analyzer and showed that small weed seeds interfered with NIT analysis. They also showed that the NIT instrument was capable of accurate moisture determination on recently tempered wheat.

Hardness is one of the most important quality parameters in the marketing of wheat because it affects milling and baking behavior. In North America, kernel hardness and vitreousness are associated with high protein, and both are used as a means of segregating hard wheat types by visual appearance. Opaqueness is traditionally associated with softness and low protein content. Kernel hardness and vitreousness do not result from the same fundamental cause (e.g., genetics, climate, etc.), although it is true that hard wheats are usually vitreous and soft wheats are rarely so. As a result, the degree of hardness, as distinct from vitreousness, could be established to enable segregation of various wheat classes relative to their potential end use. The most satisfactory standard method for the determination of wheat hardness is the particle-size index (PSI) test (Symes, 1965). This is very time-consuming and not applicable to elevator operations. The application of NIR technology to the determination of this parameter has been investigated by researchers. For example, Williams (1979) reported on the differentiation between wheat types on the basis of their hardness and protein content using an NIR instrument. He used a total of 154 wheat samples from seven regions (including Canada, the United States, South America, South Africa, Europe, India, and Australia), ranging from 9.8% to 17.7% in protein content, to calibrate a Neotec Model 31 Grain Quality Analyzer against Kjeldahl protein and PSI (as determined by Ro-tap sieving method). Correlation coefficients and standard error of estimates acceptable for screening purposes were obtained, and the conclusion was that, for best accuracy in protein testing, NIR instruments should be calibrated separately for hard and soft wheats.

In another study, Sampson and co-workers (1983) used an NIR instrument and grinding time (GT) to differentiate wheat classes based on their hardness. These workers segregated soft from hard wheat cultivars by both methods (NIR and GT) using a Technicon InfraAlyzer Model 2.5A (with a wavelength filter of 1.68 μm) and a Brabender burr-type mill. Here again, acceptable correlations and standard errors of estimate were obtained. More recently, the validity of the NIR method for hardness testing has been confirmed by an international collaborative study and a proposed near-infrared index that ranges from 0 to 100, with durum wheats (the hardest) having the highest values (Williams and Sobering, 1986). Williams and others (1986, unpubl. data) have also reported very good correlations between NIR and wheat strength. These workers developed calibration sets with minimum ranges in hardness and protein content and were able to demonstrate high correlations (0.85–0.92) between NIR optical data and such parameters as farinograph stability and loaf volume (remix method). Oil bands, as well as protein bands, were prominent in the wavelength selections for strength parameters. The correlations were generally higher than those between, e.g., Pelshenke and SDS sedimentation tests, and the recognized quality parameters and NIR measurements based on such factors as farinograph stability may be superior to other tests for early-generation screening of wheat on the basis of quality.

The application of NIR in the flour mill and for bakery flour quality control has also been reported. Williams and co-workers (1981) applied NIR to the determination of ash content of flour. A Technicon InfraAlyzer 400 (I 400) was used, and the results indicated that ash content of different commercial flour streams could be measured at 2,345 μm using a single MULTICAL calibration based on all flour streams. The ash test, which is a criterion of flour purity and efficiency of milling, is a lengthy process, and the successful use of NIR could be beneficial to the flour mill laboratory in quality control. Similarly, application of NIR in bakery control of flour for the analysis of protein, moisture, and ash has proved reliable enough for it to become the standard procedure used in the Quality Bakers of America (QBA) Laboratory on all bakers' hard wheat patent flour shipments to member bakeries from major commercial flour mills (Diachuk et al, 1981). A recent publication by Posner and Wetzel (1986) describes the continuous on-line monitoring of flour mill streams for flour ash using a Technicon 400 remotely located and with a specially designed sampling device.

B. Barley

The protein content of barley and barley products is important to the breeder as well as the brewing industry. In programs designed to improve quality characters, there is always the requirement that large numbers of samples must be evaluated and the methods used must be rapid as well as accurate. In breeding, to improve protein quality in barley, the lysine content is important because lysine is the first limiting amino acid for monogastric animals (Carpenter and Ellinger, 1955). The estimation of protein quality involving the determination of the amino acid content by conventional methods such as chromatography is accurate but too slow and expensive for routine assays of the large numbers of samples generated by breeding programs. Hence, the use of NIR for the determination of lysine content and lysine as a proportion of the protein content has enabled renewed research on the breeding of barley genotypes with enriched lysine content (Gill et al, 1979). The technique has applications both in early-generation screening and in the evaluation of advanced and parental lines. NIR researchers have expanded the technology to measure protein content of malt. Pomeranz and co-workers (1977a) compared NIR with Kjeldahl protein (and with four other conventional methods) and found that NIR was highly correlated (0.97) to Kjeldahl results, proving again to be a viable alternative. Rubenthaler and Bruinsma (1978) calibrated a Technicon InfraAlyzer Plus (interfaced with a microprocessor) and estimated the lysine content of barley. A comparison of the chemical and NIR lysine results showed a high correlation coefficient of 0.975 between the two methods. These workers concluded that the NIR method is economical, simple, and rapid and the best method for screening large populations of early generations in breeding programs.

Williams and co-workers (1984) demonstrated that NIR spectroscopy could be used to predict the amino acid composition in wheat and barley in breeding programs. Their first study involved the Beltsville Universal Computerized Spectrophotometer to generate spectral information for the wheat, whereas data for barley were obtained from a Neotec Model 6350 Research Composition Analyzer and a commercial bench-type instrument, the Pacific Scientific Feed Quality Analyzer Model 51A. The concentrations of lysine, threonine, tryptophan, and methionine in wheat were determined with accuracies acceptable for screening purposes. For the barley samples, low standard errors of performance were obtained for 14 amino acids. First-derivative mathematical treatments of spectral data gave the best accuracy for the barley samples for both instruments (i.e., the RCA 6350 and the bench-type unit). Different wavelength sets were reported for all individual amino acids and for protein. The interpretation was that the NIR system was actually measuring the individual amino acid residues, distinct from each other and from protein.

The malting quality of barley is related to the amount of β-glucan dissolved at low pH under defined conditions (Greenberg, 1974). Accurate and rapid analytical methods for the estimation of β-glucan would be beneficial to the brewing industry. Malt extract is another parameter used in the assessment of the malting quality of barley. The conventional methods used in this assessment are tedious and poorly adapted to rapid screening of early-generation breeding lines. McGuire (1982) examined NIR technology as a potential analytical tool in this area. He calibrated a Technicon 400-R InfraAlyzer with 54 barley genotypes, then estimated the malt extract of 133 F_2 and 199 selected F_3 lines. The NIR values were highly correlated with the laboratory malt extract values. Coefficients of correlation of 0.95 and 0.98 were respectively obtained for F_2 and F_3 series.

C. Maize (Corn)

Moisture content of corn is of direct economic importance in corn marketing. Dry matter per unit weight is inversely related to the percentage of moisture in corn. Low moisture content, which implies high dry-matter content, indicates that a high yield of milling corn products can be obtained. Moisture content also affects keeping quality. High moisture content provides ideal conditions for mold growth and for heating and spoilage during storage. These undesirable factors lead to losses in both nutritive and economic value. A rapid and accurate method would be very beneficial to this industry. NIR has been used in the analysis of the constituents of corn and corn products. Massie and Norris (1965) published the first NIR spectra for corn, wheat, oats, soybean, and milled rice. Since then, several other researchers have applied the technique to the analysis of corn composition. These include the simultaneous estimations of protein and oil content in corn, soybean, and oats by Hymowitz and co-workers (1974). In this study, a computerized NIR grain analyzer (DICKEY-john Corporation) was compared with protein and oil determinations by Kjeldahl and nuclear magnetic resonance (NMR), respectively. The accuracy of NIR as compared with Kjeldahl and NMR is reflected in the correlation coefficients and the standard errors of difference (Table I) for independent samples. These workers concluded that NIR technology was satisfactory for use in routine analyses on a large scale. Stermer and colleagues (1977a) reported on the estimation of moisture content of whole grain corn and sorghum by use of NIR. These scientists reported correlation coefficients of 0.959 for sorghum grain and 0.993 for corn (for samples ranging in moisture content from 13% to 55%) between oven moisture and a Neotec Grain Quality Analyzer Model 41 (GQA 41). A year later, the estimation of moisture content of corn by use of NIT was reported by Finney and Norris (1978). These workers measured moisture in individual kernels of corn using NIT over the spectral region of 700–1,100 nm. A correlation coefficient of 0.96 between oven moisture and NIT results and a standard error of 2% for corn with moisture ranging from 8% to 35% were reported. These workers indicated that NIT measurements of moisture content of individual corn kernels might incur relatively large errors for high-moisture corn.

NIR has been utilized in the determination of the free and bound starch present in the fiber portion during corn wet milling (Bell, 1983) and also of the protein, moisture, and oil contents of feed and prime gluten fractions. The corn wet-milling process consists of a series of washings and separations. The efficiency of these separation processes can be monitored with NIR instruments by measuring the drop in protein content between separations. Bell (1983) reported correlation coefficients of 0.987 for protein, 0.984 for moisture, and 0.942 for oil in gluten feed and correlations of 0.992 for protein, 0.992 for moisture, and 0.963 for oil in prime gluten. The same worker also utilized NIR technology to estimate dextrose equivalent, total solids, and percentage of fructose in corn syrup (Bell, 1982). High-fructose corn syrup (HFCS) has also been analyzed for fructose and solids using NIR. Here, the analysis measured the conversion of dextrose to fructose in the process and allowed the HFCS plant to monitor the efficiency of this conversion. In a similar study, Psotka and Bahl (1982) used a Technicon InfraAlyzer 400L to determine the concentrations of lower saccharides in HFCS by correlation of instrument response to saccharide profiles that had been determined by high-pressure liquid chromatography. Monitoring saccharide concentration during production of HFCS permits the plant operator to control the efficiency of the conversion system and thereby improve the productivity of the manufacturing facility.

NIR has also been used in the determination of fructose, glucose, and sucrose in model systems intended to represent the major constituents of dried apple tissue (Giangiacomo et al, 1979). Giangiacomo and others recorded spectra (950–1,850 nm) of a series of standard samples. Correlation coefficients for standard vs. observed values were 0.995 for fructose, 0.994 for glucose, and 0.986 for sucrose with respective standard errors of 1.48%, 1.39%, and 1.44%. These results led to the conclusion that NIR can be used for nondestructive determination of fructose, glucose, and sucrose in multicomponent dry mixtures. Current procedures for analysis of starch ethers involve acid hydrolysis followed by alkyl derivatization and measurement by titration or gas chromatography. Mauro and Owen (1982) used NIR in the quantitative analysis of ether and ester substituent groups in starch derivatives. Excellent regression statistics were obtained, with a correlation coefficient of 0.97 and standard error of estimate of less than 7% for modification levels of 0.5% when they determined propylene oxide and ethylene oxide ether modifications of starch products and succinic anhydride and acetic anhydride ester modifications of starch products by means of NIR.

D. Other Cereal Grains

The analysis of other cereal grains by use of NIR includes measurement of the degree of milling of rice. Stermer and co-workers (1977b) predicted the degree of milling rice from the estimation of total lipids in the rice as an index of milling.

TABLE I
Accuracy of Near-Infrared Reflectance Determinations of Protein and Oil Percentages in Corn, Soybean, and Oat Meals[a]

		Percentage Protein[b]		Percentage Oil	
Meal	**N**	***r***	**Sd**	***r***	**Sd**
Corn					
Calibration	44	0.994	0.06	0.993	0.07
Independent	30	0.910	0.10	0.913	0.09
Soybean					
Calibration	60	0.996	0.09	0.992	0.06
Independent	30	0.943	0.08	0.969	0.04
Oat					
Calibration	69	0.982	0.06	0.990	0.03
Independent	30	0.945	0.12	0.978	0.04

[a] Adapted from Hymowitz et al (1974); used by permission of the Crop Society of America.

[b] Standard methods were Kjeldahl (total nitrogen × 6.25) for protein and nuclear magnetic resonance for oil; *r* = correlation coefficient and Sd = standard error of difference between means using paired *t*-test.

Reflectance at three principal oil absorption bands (0.928, 1.215, and 1.725 μm) was found to be highly correlated with surface ($R = 0.764$) and total ($R = 0.856$) lipid contents of milled rice. Also reported in the literature is the simultaneous determination of protein, moisture, and oil in oats by use of NIR (Massie and Norris, 1965; Hymowitz et al, 1974; Williams 1975). Pomeranz et al (1977b) reported on the determination of protein and ash in grain sorghum. The following year, Stevenson and Williams (1978) reported on the determination of protein and moisture in sorghum from different sources and of protein, moisture, and oil in four different types of millet including pearl, foxtail, proso, and finger millets. They observed a significant effect of growing location on sorghum analysis and of species on millet analysis, and they also reported successful application of NIR to the determination of protein in Ethiopian teff (*Eragrostis abyssinica*).

E. Oilseeds

In the fats and oils industry, estimation of fat/oil content is normally carried out by time-consuming solvent extraction methods. Most of the official extraction methods require 3–16 hr of extraction (e.g., AOCS, 1973). In recent years, wide-line NMR has been applied to the estimation of oil in oilseeds and legumes. For example, NMR was used for the estimation of oil content of intact, partially dehydrated seeds from plants including castor beans, corn, crambe, isano, jojoba, linseed, oiticica, soybean, tung, and other commodities (Conway et al, 1963). The estimations by NMR of oil content of maize (Alexander et al, 1967), soybeans (Collins et al, 1967), and sunflower seeds (Robertson and Morrison, 1979) have also been reported.

NMR is a form of radio frequency spectroscopy. The basic principle of its application to oil analysis is that it provides accurate counts of hydrogen nuclei in liquid oils, even in a surrounding matrix consisting of starch, protein, and other components. The hydrogen population density is then related to the mass of the sample, and the sample in turn may be analyzed for oil by classical gravimetric methods. Hydrogen in liquid water has resonance qualities similar to those in oil, and accurate estimates of total hydrogen in the liquid oil component of a sample can be made only if water is reduced to 4.5% or lower (Alexander et al, 1967). Samples must be dried by forced air or vacuum oven at temperatures ranging from 50° C to 130° C for from 3 hr to five days. The NMR method is generally too time-consuming for industrial analysis. Present-day instruments take at least 4 hr (and up to five days) for a single test. In contrast, the application of NIR to the analysis of oilseeds, other oleaceous seeds, and pulses has been under study in recent years. The technique has been applied to the determination of oil content of corn, soybean, and oats (Hymowitz et al, 1974) and of rapeseed and soybean meals (Williams, 1975). For the meals, excellent correlation was obtained between NIR and conventional extraction results for oil and protein contents. Williams (1975) drew attention to the fact that the accuracy and precision of NIR analysis were influenced by factors associated with calibration (including sample preparation and presentation), by factors that influence the nature of grain, and by factors associated with the instrument itself.

In subsequent work (J. A. Panford and P. C. Williams, 1986, unpubl. data), sample preparation was found to be critical in NIR analysis of oilseeds. A high-speed (30,000 rpm) impeller mill was found suitable for most types of oilseed. A compromise was needed between particle size and extrusion of oil. Prolonged storage after grinding was found to increase the standard error of prediction and decrease the r^2, possibly due to agglomeration of the ground material. NIR analysis performed immediately after grinding improved the accuracy. Iwamoto and colleagues (1981) tested 54 soybean samples for protein and oil using two NIR instruments: a DICKEY-john GAC III and a Technicon InfraAlyzer 400. Accuracy and precision were good and were comparable for both instruments and similar to results reported by other workers. Tkachuk (1981b) investigated the analysis of whole rapeseed kernels for oil and protein by NIR. He obtained good accuracy for the prediction of oil content in 89 rapeseed samples using eight wavelengths in the 1,100- to 2,400-nm region. Two of the wavelengths, 1,734 and 1,210 nm, were close to the first and second C-H overtones (1,720/1,760 nm and 1,210 nm) of the CH group in fatty acids. A later study (Tkachuk and Kuzina, 1982) showed that chlorophyll could be determined in whole rapeseed by NIR. Analysis of sunflower seeds by NIR was reported by Robertson and Barton (1984) and by Kaffka and co-workers (1982). Both reports verified the feasibility of NIR as an analytical tool for sunflower seeds, although there was more variability in the results than had been encountered in studies of other oilseeds. This might have been caused by the fibrous sunflower seed hulls affecting the mean particle size and shape, factors that have been found to influence NIR accuracy of cereal grain analysis significantly (Williams and Thompson, 1978).

Spectrophotometric studies of the structure of oils have been conducted for many years by use of midinfrared technology. The infrared spectra for mono-, di-, and tri-glycerides were described by O'Connor and co-workers (1955). Infrared determination of fats and oils has also been reported by Arnold and Hartung (1971). Infrared spectrometry is used as the official rapid method of distinguishing *cis*- and *trans*- forms of monoglycerides (AOCS, 1979) and for the determination of *cis*, *cis*-polyunsaturated fatty acid content of oils (AOCS, 1983). It is only in recent years that NIR has also found application in this area. Holman and Edmondson (1956) published NIR spectra for several fatty acids and related substances. They showed that the first overtone of the C-H stretching vibration at 1.7 μm and the C-H combination band near 2.2 and 2.5 μm were exhibited with differing intensities for all the fatty acids. Holman and Edmondson found that short-chain fatty acids in which methyl groups are dominant showed a discrete band at 2.28 μm, whereas in the long-chain molecules where methylene groups predominate the bands at 2.30 and 2.34 μm were more clearly expressed. They concluded that fatty acids containing *cis*-double bonds have moderate combination absorption bands at 2.19 and 2.15 μm, a weak first overtone at 1.68 μm, and a second overtone at 1.18 μm. In addition, the absorption at 2.15 and 2.19 μm increased with increasing unsaturation, and *trans*- isomers of these acids did not exhibit any strong absorption in this region or elsewhere in the NIR range.

Holman et al (1959) also reported the determination of *cis*-unsaturation in lipid components and hydrogenated oils by NIR. They measured the absorbance of the band at 2.15 μm and related this to the iodine value of the sample corrected for *trans*-content. They reported that *trans*- unsaturation did not interfere at the analytical wavelength. Following this report, Fenton and Crisler (1959) developed the analytical method for the determination of *cis*- unsaturation present in refined oils, hydrogenated oils, and finished shortenings. Their measurements were made in the NIR region at 2.14 μm, a band caused by *cis*- unsaturation (Holman and Edmondson, 1956). Holman and co-workers (1958) used NIR to detect and quantitatively measure hydroperoxides in fatty acids. Absorption maxima at 2.07 and 1.46 μm were found to be specific for hydroperoxides by these researchers. Their results have permitted studies of the oxidation of unsaturated fatty acids by distinguishing between hydroperoxides and other substances that also oxidize KI (and are therefore measured chemically as peroxides). They attributed the absorption maxima at 2.07 and 1.46 μm to hydroxyl groups in these compounds.

Goddu (1957) also reported on the determination of unsaturation in organic compounds. He found that bands at 1.62 and 2.10 μm could be used for determination of terminal unsaturation and a band at 2.14 μm for *cis*- unsaturation in aliphatic hydrocarbons. His results confirmed earlier work reported by Holman and Edmondson (1956). Goddu (1958) determined phenolic hydroxyl and terminal epoxides qualitatively and quantitatively by near-infrared spectrophotometry. He found that phenolic antioxidants could be identified and determined between 2.7 and 3.0 μm, whereas terminal epoxides exhibited sharp absorption bands at 1.65 and 2.20 μm. Until this time, all structural studies of organic compounds including fatty acids and proteins had been carried out using slightly modified infrared spectrophotometers (e.g., Beckman Models DK-1 and DK-2). Samples had to be liquefied and reduced to small quantities and concentrations before analysis. Holman, Edmondson, Goddu and the earlier NIR researchers realized that sample preparation was less elaborate and analysis faster than for infrared spectrophotometry. However, dedicated NIR instruments were not available until 1963 when Norris designed and introduced the first NIR spectroscope for laboratory use.

Using such an NIR instrument, Wetzel (1983) investigated the relationship between the structure of oil and NIR response. He reported a correlation coefficient of 0.983 and standard error of calibration of 0.163 between structural information (e.g., degree of unsaturation, solid fat index, and mean carbon number in triglycerides) and NIR response. He also reported three major bands at 1.680, 2.139, and 2.208 μm for the oils.

F. Forages

Forage quality is best defined in terms of the value of the forage in the production of milk, meat, and wool when fed to animals, but the time and cost of large animal feeding trials limit their use in routine analysis of forage samples. Laboratory methods that predict feeding value on the basis of chemical composition and in vitro digestibility are also expensive and time-consuming. Agronomists, plant breeders, and ruminant nutritionalists require a rapid, accurate technique for determining the chemical composition and nutritive value of forages. Norris and co-workers (1976) tested 87 samples of ground dry forages including the temperate forage species alfalfa, tall fescue, and alfalfa-bromegrass mixtures such as hay, silage, and frozen forages. Tropical forages included bermudagrass and pangola digitgrass. The samples were chemically analyzed for crude protein, acid detergent fiber (ADF), neutral detergent fiber (NDF), lignin, and in vitro and in vivo digestibility. They recorded reflectance spectra as log $(1/R)$ vs. wavelength for all samples. The log $(1/R)$ data were transformed to second derivative of log $(1/R)$ for correlation with chemical data. Multiple linear regression techniques were used to determine the optimum wavelengths for predicting the forage composition. The data are summarized in Table II. They concluded that NIR had the potential for use in rapid evaluation of forage quality. Shenk and colleagues (1976) reported results for a similar study of forages fed to small herbivores (e.g., sheep). They drew similar conclusions of the potential for NIR to predict chemical composition and nutritional value of forage and feedstuff. Templeton and others (1978) carried out a collaborative study of 835 forage samples involving six different laboratories and for five parameters. A summary of their chemical and NIR data is presented in Table III. The errors associated with NIR analysis compared favorably with those obtained with routine chemical procedures.

Other reports on similar work include those by Shenk and co-workers (1981), Park et al (1980), and Ward et al (1982). In these studies, scanning spectrophotometers with grating monochromators were used. However, the optical energy output of the monochromator used was relatively low, and its use for generating quantitative data in the NIR region may be limited for certain forage constituents (Burdick et al, 1981). The high cost of monochromators also precludes their use for routine analyses. Hence, multiple interference filter NIR instruments have been applied to the analysis of feeds. Barton and Burdick (1979) evaluated a Neotec FQA 51 for the analysis of forages including such grasses as bermudagrass, bromegrass, orchardgrass, tall fescue, and Kentucky bluegrass. Burdick and others (1981) and Barton and Burdick (1982) have also applied the FQA 51 to the analysis of forage composition. The spectral range of the FQA 51 extends from 1.50 to 2.36 μm, and six sections of that range are scanned. The initial spectral data, log $(1/R)$, were transformed into second-derivative data that were regressed against the chemical data. From two to six wavelength points were identified for each constituent or quality parameter. Correlation coefficients ranged from 0.54 for crude fiber to 0.96 for crude protein.

G. Fruits and Vegetables

Near-infrared reflectance and transmittance technology has been applied to the analysis of internal quality including the determination of the stage of maturity of fruits (Bittner and

TABLE III
Standard Error (%) of 30 Forage Samples, Five Quality Parameters, and Six Laboratories[a]

Methods of Analysis and Effects	Parameter[b]				
	CP	NDF	ADF	ADL	IVDMD
Wet chemistry					
Between labs	0.75	1.82	2.24	1.01	2.80
Error	0.43	1.70	1.40	0.90	3.67
Near-infrared reflectance					
Between labs	0.83	2.45	1.81	1.07	1.33
Error	0.56	1.12	0.60	0.63	0.83

[a] Source: Templeton et al (1978); used by permission. Statistical models are apparently not the same for the two analytical methods; hence, direct comparisons between them should be made with caution.
[b] CP = crude protein, NDF = neutral detergent fiber, ADF = acid detergent fiber, ADL = acid detergent lignin, and IVDMD = in vitro dry matter disappearance.

TABLE II
Summary of Results with log $(1/R)$ Reflectance Data to Predict Forage Components[a]

Parameter[b]	Component[c]						
	CP	NDF	ADF	L	IVDMD	DMD	Intake
N	87	87	87	87	75	75	75
R	0.98	0.92	0.90	0.72	0.90	0.81	0.79
SE	1.07	5.3	2.5	2.1	2.5	4.4	7.8
τ	5.98	12.80	5.60	2.88	7.75	6.90	13.26

[a] Source: Norris et al (1976); used by permission.
[b] N = sample population, R = correlation coefficient, SE = standard error of prediction, and τ = standard deviation.
[c] CP = crude protein, NDF = neutral detergent fiber, ADF = acid detergent fiber, L = lignin, IVDMD = in vitro dry matter disappearance, DMD = in vivo digestibility, and Intake = dry matter intake.

Norris, 1968) in pears, apples, and peaches. Reflectance spectra were recorded for these fruits in the region of 1.0–2.1 μm using a Cary Model 14R spectrophotometer. This instrument was fitted with an integrating sphere for reflectance measurements and a specially designed end-window multiple-type phototube that permitted measurements in the visible and ultraviolet regions while a large-area lead-sulfide detector was installed for the NIR region. These workers concluded that reflectance response in the visible region gives more reliable results in the evaluation of the stage of maturation of fruits. Near-infrared instruments have also been applied to the detection of lettuce maturity in greenhouses during winter months and in the field during the summer season by Brach and co-workers (1982). These workers obtained a correlation coefficient of 0.96 between change in the reflectance pattern of the lettuce and the physiological and anatomical changes as the lettuce head matured. The changes were noted to be more prominent when measured over the wavelength range of 1.10–1.30 μm. Here again a scanning NIR spectrophotometer was utilized. Birth and others (1982) applied a scanning NIR spectrophotometer to the estimation of soluble solids in onions. The soluble solids content of the bulbs is an important quality factor in the marketing of onions. These workers found a correlation coefficient of 0.95 between optical measurements and soluble solids. They selected a wavelength range of 0.7–1.0 μm because the absorption by water is sufficiently low in that spectral region to allow transmittance through a substantial portion of the bulb to be measured. The data were then recorded as relative absorbance, or log (1/*R*), and processed into the second derivative.

H. Tobacco

In the tobacco industry, knowledge of the sugar levels in tobacco is important to all segments. For example, nutritional deficiencies, such as low boron levels, can be detected by measuring sugar variations, and Fusarium wilt results in a significant decrease in sugars (McClure et al, 1977). In addition, the quantity of sugars remaining in tobacco at the end of curing is often used as an indication of proper curing practices as well as an indication of the mildness and aromaticity of smoke. Before the advent of NIR technology, tedious chemical extraction procedures were used in this industry. McClure and his group (1977) used a scanning NIR spectrophotometer to determine the total reducing sugar in tobacco at different curing stages. Reflectance spectra in the 1.4- to 2.4-μm region were recorded for 152 samples. The second-derivative transformation of the log (1/*R*) reflectance signals was correlated to chemical analyses data. A correlation coefficient of 0.99 and standard error of 1.15% were obtained between reducing sugar content and NIR instrument response, using seven wavelengths. Hamid and co-workers (1978) reported on the analysis of flue-cured tobacco for total alkaloid content by use of NIR spectroscopy. The estimation of alkaloid by wet chemistry also involves cumbersome procedures. Using second-derivative treatment of the log (1/*R*) data, the coefficient of determination was found to be 0.975, with a 10-wavelength calibration, and a standard error of 0.438% was obtained. These workers concluded that NIR had good potential for measuring total alkaloids in tobacco. Using identical NIR techniques, McClure and Williamson (1982) measured the polyphenol content of tobacco, adding to the versatility of NIR in the tobacco industry.

I. Pulses

Grain or food legumes, the dry seeds of which are known as pulses, provide an important contribution to the protein intake of the population of the world's developing countries. Pulses offer great potential for the economic and nutritional improvement of both human and livestock diets that depend largely upon vegetable sources of protein. In North America, pulses, especially soybean meal, account for about 60% of the protein source in livestock feeds (FAO, 1984). The chemical composition of grains and seeds, including the percentages of moisture, protein, oil, and fiber, is important for both commercial and nutritional quality evaluation of these commodities. A rapid and accurate procedure for the determination of these constituents in the commodities is valuable in their marketing and handling.

Most of the work reported in the literature has been carried out on soybeans; however, continuing research indicates a broadening interest in other major pulses. Ben-Gera and Norris (1968a) reported the determination of moisture content of ground soybeans by direct NIR transmittance spectrophotometry. They found that instrument response at 1,940 nm was highly correlated to moisture contents of up to 20%, and they obtained a standard deviation of 0.10. Jasanski and Bilanski (1973) also evaluated the spectral absorbance of whole and ground soybeans in the region of 750–2,300 nm. The spectra obtained in this study were used in the development of a compact, efficient, and simple heat-treating unit for raw legumes. Soybean meal protein and oil percentages were determined by an NIR instrument and compared with standards for protein analysis by Kjeldahl and oil analysis by Soxhlet extraction method (Rinne et al, 1975). Tests of independent soybean samples gave correlation coefficients of 0.971 between NIR and Kjeldahl protein and 0.977 between NIR and Soxhlet oil results. Similar studies showing similar results have also been reported by Wilson and others (1973) and Williams and co-workers (1978). The measurement of protein and oil content of winged bean seeds by NIR has also been reported by Hildebrand and co-workers (1981). These scientists utilized a computerized DICKEY-john NIR instrument to compare results for protein and oil content of winged beans with those obtained from conventional methods (Kjeldahl and Soxhlet). Excellent correlations of 0.998 for protein and 0.995 for oil were reported. A feasibility study of the use of NIR for quality control in soybean wet processing has also been conducted (Moen, 1979). In this study, NIR was applied to monitor the protein concentration of fractions at various stages of processing. More recently, Williams and co-workers (1985a) tested numerous types of peas for methionine, using NIR instruments. Three different NIR techniques were employed to measure protein and methionine directly in ground peas without hydrolysis or any chemical treatment. Accuracy of predictions of methionine was ±0.011% and of protein ±0.76% for the whole peas with a commercially available NIR instrument.

J. Meat and Meat Products

The first recorded work to utilize NIR spectroscopy for the analysis of a meat product (gelatin) was carried out by Ellis and Bath in 1938. These workers reported spectra for dry and moist gelatin using polarized infrared waves in the 1.0- to 2.5-μm region. They also concluded that the absorption bands at 2.18, 2.05, and 1.72 μm were the result of hydrogenic vibrational overtones of C = O...H-N, H-N, and C-H groups in the protein molecule, whereas bands at 1.44 and 1.93 μm were attributed to H_2O absorption. Ben-Gera and Norris (1968b) determined fat and moisture content of meat products (including bacon, ham, all-meat frankfurters, and bologna) by use of the wavelengths in the NIR region (1.0–2.5 μm) of the spectrum. They found that the difference in optical density between 1.80 and 1.725 μm gave a high correlation (0.974) with moisture content and that the difference between 1.725 and 1.65 μm gave a high correlation (0.974) with fat content. They attributed the resulting spectra to the absorptions from O-H and C-H stretching vibrations. Direct spectrophotometric analysis predicted fat content within a standard error of ±2.1% and moisture content within ±1.4%.

Using similar techniques, Massie (1974) analyzed ground beef for fat content. Kruggel and co-workers (1981) reported the estimation of fat, protein, and moisture in fresh ground beef and lamb, and Iwamoto and colleagues (1981) also predicted the chemical composition of pork by NIR spectroscopy. Other reports in the literature include the estimation of fat, protein, carbohydrate, and water in meat pie ingredients in the 2.5- to 10-μm region (Steel and Bell, 1983) and raw pork and beef (Lanza, 1983) in the 1.1- to 2.5-μm region. In these studies, spectral data were correlated with moisture, protein, fat, and calorie data determined by standard methods. Optimum wavelengths reported were 1.446 μm for moisture, 1.726 μm for protein, 1.718 μm for fat, and 1.726 μm for carbohydrates. Begley and others (1984) determined the quantity of sodium chloride (NaCl) in cured ham, salted fresh ham (uncured), and ground beef by use of NIR spectroscopy. These workers found that the second derivative treatment of the log ($1/R$) data gave the best results. Optimum wavelength selected for the estimation of NaCl in meats was 1.806 μm with a correlation coefficient of 0.96 and standard error of prediction of 0.17%.

K. Dairy Products

Infrared transmission spectrometry has been applied to the analysis of milk and milk products since the early 1960s. The analysis of milk for fat, protein, and lactose content has been reported by Goulden (1964) and Biggs (1967). Biggs and Stanley (1972) measured quantitatively the fat, protein, lactose, and solids-not-fat content of milk and milk products, including skim milk, evaporated milk, heavy cream, and ice cream mixes. Their measurements were made in the 5.0- to 10.5-μm range of the midinfrared region. They reported coefficients of determinations for each sample to be in excess of 0.999. They concluded that infrared technology was a useful method for the analysis of milk and milk products. Since that time, infrared instruments have been automated and computerized and are used for on-line quality control evaluation of milk and milk products in the dairy industry.

Ahmed and Helal (1977) demonstrated the direct relationship between the degree of unsaturation of milk fat and the absorbance at 3.3 μm with a high correlation coefficient of 0.97. They observed that as the iodine value increased (from 32.4 to 67.2), the infrared absorbance at 3.3 μm increased from 0.5839 to 1.0453. Arnold and Hartung (1971) in a previous study have also indicated the positive relation between iodine value and ratio of absorbance at 3.3 μm to the absorbance at 3.5 μm with a correlation coefficient of 0.98. Infrared spectrophotometry has also been used to measure the degree of unsaturation of fats and oils, serving as a monitor of unsaturation in process control application (Ahmed and Helal, 1977).

Ben-Gera and Norris (1968c) reported the first work involving the application of NIR spectroscopy to the analysis of milk. These workers analyzed raw milk for fat and solids-not-fat by diffuse-transmittance spectrophotometry in the NIR region. They found that milk, with its water content of more than 80%, has a near-infrared absorption spectrum similar to that of water and that the water absorption band at 0.97 μm is a function of the fat content of milk; they suggested necessary corrections for future work to establish the degree of correlation between fat content and the 0.97-μm water band. Giangiacomo and colleagues (1979) applied NIR to the analysis of blue cheese. They compared NIR with classical methods for the estimation of protein, fat, and water in blue cheese and concluded that NIR results were comparable to those obtained from classical methods. Frank and Birth (1982) also obtained very high correlations (0.942 for fat, 0.938 for protein, and 0.965 for moisture) between NIR and standard laboratory results for the constituents of semihard cheeses, including cheddar, Colby, Parmesan, Romano, and Gruyère. Wavelengths selected for predicting fat, protein, moisture, and moisture in nonfatty substance were 2.31 μm, 1.76 μm, 1.455 μm, and 1.157 μm respectively.

L. Ready-to-Eat Foods

Processors of industrially prepared foods must ensure that their finished products conform to specifications of the purchaser/consumer and meet ingredient labeling laws prior to shipping. Rapid and accurate analytical methods are needed to ensure this and to remove the risk and cost of shipping the wrong products or products of incorrect composition. NIR has been applied to the analyses of several types of foods, including the estimation of crude fiber in processed cereal foods (breakfast cereals) by Baker and co-workers (1979) and Baker (1983). The relationship of NIR measurement to NDF values of breakfast cereals was determined (Baker et al, 1979). Reflectance spectra were recorded from the cyclone-ground cereals. The wavelength range from 1.0 to 2.6 μm was scanned using the Beltsville computerized spectrophotometer. Readings were recorded every 1.6 nm. Second-derivative transformation of the log ($1/R$) data gave high correlations between NIR measurements at three wavelengths, 2.26, 2.23, and 1.97 μm, and NDF values for 213 samples. The correlation coefficient was 0.985 with a standard error of 1.36%. Typical data are presented in Table IV. The study was extended in 1983 when the analysis of more processed cereal foods for fiber was reported. Forty cereal samples were used to calibrate the NIR instrument. Determination of fiber in 50 samples gave a coefficient of correlation of 0.955 and a standard error of 1.8%. Although interference was found by high sugar content in some samples, it was concluded that NIR was a suitable rapid method for the determination of the fiber content of processed cereal foods.

Williams and co-workers (1981) reported on the determination of ash content (bran level) of wheat flour by use of NIR technology. NIR spectroscopy has also been applied to the analysis of prepared snack foods. The use of an NIR instrument has been reported for the determination of the fiber, starch, and total carbohydrate content of snack foods including potato chips, corn chips, popcorn, crackers, and pretzels (Baker, 1985). The Kellogg Corporation also uses the NIR technique to determine moisture, protein, and sugar levels in a variety of food products, including frozen and fresh baked pies, frozen waffles, desserts, teas, soups, soup bases, and breakfast cereals (McKay and Dorsheimer, 1982). NIR spectroscopy is also used by Kellogg for the analysis of moisture in finished products, analysis of sugar in cereal products, monitoring incoming raw materials, and measurement of protein in Kellogg's Special K cereal.

M. Other Products

NIR spectroscopy has been used for the determination of moisture in walnut seeds (Campbell and Martin, 1976). Sample preparation procedure in this study was different from that normally used for grains and most other seeds in that NIR measurement was not made directly on the ground nuts. Instead, these workers measured the moisture content of the nuts in methanol by grinding the nuts in a 1:5 (nut:methanol) ratio. The slurry was centrifuged and the clarified supernatant was then scanned in the NIR region from 1.3 to 2.1 μm by transmittance. The instrument was first calibrated using samples containing from 3% to 30% moisture. It was found that samples with low (percentage) water content exhibited strong absorbance at 1.94 μm, the wavelength at which moisture measurement is made directly in grains and most seeds, whereas high (percentage) moisture samples showed strong absorbance at 1.44 μm, which corresponds to the second-overtone band for water. Although the correlation coefficient of 0.999 obtained between absorbance and (percentage) water in this study was excellent, this procedure has not been widely adopted because direct and accurate NIR measurement can be made of ground grains and seeds without having to use the extract in a solvent. Cocoa bean is another economically important commodity that requires chemical analysis. Because supply of cocoa beans and cost of chocolate products in the producing countries are constantly fluctuating, the pattern of routine laboratory analyses has been altered. The cocoa bean is being replaced by chocolate liquor in the form of blocks, cakes, drops, and flakes. This change to a semifinished product has increased the demand for rapid analytical methods. Miner and co-workers (1979) applied a Neotec Research Composition Analyzer 6350 to the estimation of oil and moisture in cocoa powder, raw cocoa beans, and cocoa liquor. Excellent correlations were obtained.

Spectrophotometric analysis of wool for the amount of dye on the fibers has also been carried out by Larsen (1979). This area of wool research has been extensively investigated by scientists in Australia.

III. RESEARCH APPLICATIONS

Research applications have been directed mainly at wavelength assignment and the development of mathematical treatments, sample selection, calibration methods, and discriminant analysis. Law and Tkachuk (1977b) reported on the assignment of absorption bands for gluten protein at 2.18 and 1.76 μm. These they attributed to a -CONH (amide III) and N-H combination and C-H first-overtone vibrations respectively. The starch bands at 1.78 and 2.09 μm were attributable to a C-H first overtone and an O-H combination. Other assignments were described for gluten protein and for the starch, pentosan, oil, and water in wheat. The authors also commented on wavelength shifts caused by changes in hydrogen bonding in samples with low moisture content. Hecht and Wood (1956) published the NIR absorption spectra of protein molecules in silkworm gut, collagen, elephant hair, eagle feather, quill, gelatin, horn (bovine), porcupine quill, and mono-substituted amides. The band at 2.17 μm was assigned to the peptide group and a band at 1.949 μm was attributed to H_2O associated with compounds containing the peptide group. The origin of the -NH_2 overtone and combination vibrations in the NIR region for simplified primary and secondary amides has been identified (Wheeler, 1959; Goddu and Delker, 1960; Krikorian and Mahpour, 1973). These workers identified and assigned three bands in the 1.428- to 1.492-μm region to the first and second overtone and combination vibrations of the -NH group in primary amides, whereas bands involving combination vibrations of amide II and amide III were identified in the 1.960- to 2.127-μm region. Other investigations carried out in the NIR region are studies of the interamide hydrogen bonds in hydrogen-bonding solvents (Susi and Ard, 1966), structural studies of ribonuclease by examination of the hydrogen-bonded peptide-N-H group (Hermans and Scheraga, 1960), and the development of correlation charts for major functional groups (Kaye, 1954; Goddu and Delker, 1960). Other research in this area (e.g., Hruschka and Norris, 1982; Williams and Norris, 1983; Norris and Williams, 1984) has involved the development of mathematical treatments of the log $(1/R)$ data and is reported in more detail in earlier chapters in this monograph. Subsequently, Williams and Cordeiro (1985) described the application of a two-sample calibration based on quotient first-derivative mathematics to the analysis for protein of wheat with a very wide range of seasonal and environmental variance. Mark and Tunnell (1985) have described the application of discriminant analysis to the characterization of pharmaceuticals using an ingenious NIR procedure. Rose and co-workers (1982) described the application of NIR spectroscopy in the analysis of parenteral products.

IV. BREEDING PROGRAMS

Plant-breeding programs require rapid analytical methods in the screening of very large numbers of small samples (2–20 g) for protein, amino acids, hardness, and other parameters. In this area, NIR has been applied to protein testing in pulse-breeding programs in the Middle East (Williams et al, 1978). These scientists screened early-generation pulses including lentils, chickpeas, and faba beans on the basis of the protein content using a Neotec GQA Model 31. More recently, Williams and co-workers (1985) were able to determine the levels of methionine in peas. Lysine determination is of

TABLE IV
Neutral Detergent Fiber of Some Breakfast Cereals[a]

Cereal Type	Description	No. of Samples	NDF[b] Average (%)	NIR GQA 41[c] Average (%)
Wheat	Nabisco 100% Bran	9	30.8	32.4
	Kellogg All Bran	9	29.0	33.5
	Post 40% Bran	9	14.8	14.5
	Post Raisin Bran	9	11.5	···
	Nabisco Shredded Wheat	12	10.0	11.7
	General Mills Total	12	7.4	10.1
	Post Grape Nuts Flakes	9	6.9	9.1
	Post Grape Nuts	9	5.1	5.4
	Post Sugar Crisp	9	1.5	···
Corn	Kellogg Corn Flakes	12	1.2	4.6
	Post Country Crisp	9	1.1	···
	Post Toasties	9	1.9	3.1
	Post Honey Combs	9	1.3	···
Oat	Quaker Life	12	3.2	5.3
	Post Oat Flakes	9	2.8	4.3
	Post Alpha Bits	9	1.2	···
Rice	Kellogg Special K	12	0.9	1.3
	Post Rice Krinkles	9	0.1	···
	Post Cocoa Pebbles	9	0.3	···
	Post Fruity Pebbles	9	0	···
Granola	C.W. Post	9	2.5	···
	C.W. Post—Raisins	9	1.9	···

[a] Source: Baker et al (1979); used by permission.
[b] NDF = neutral detergent fiber.
[c] Results obtained by near-infrared reflectance with Grain Quality Analyzer 41.

particular interest in breeding programs for barley and maize, and methionine is the most frequently limiting amino acid in pulses.

Near-infrared spectroscopy has also been utilized by remote sensing in the investigation of crop discriminability of cultivars of wheat, oats, barley, faba beans, soybeans, rapeseed (Rao et al, 1978), and wheat (Mack et al, 1978). These researchers scanned field crops over a range of 0.95 to 1.84 μm using a spectroradiometer. They obtained unique differences among cultivars at certain narrow bands in relation to the overall mean radiance, based on coefficients of variation. An index of discriminability, determined to assess separability of crops throughout the spectrum, was used to distinguish between two wheat cultivars at 0.95 and 1.40 μm, which corresponds to the water absorption region for leaves. Similarly, NIR monitoring of the protein (or nitrogen) level in the vegetative plant as an index of potential protein content of the mature seed at harvest has been investigated by several workers. These include the measurement by NIR of protein differences within wheat plants (Klepper and Wilhelmi, 1979; Löffler and Busch, 1983), and protein level in young corn leaves has been studied by Dorsheimer and Isaac (1982). In these reports, commercial bench-type, fixed-wavelength-filter NIR instruments were utilized. Correlation coefficients between NIR response and Kjeldahl protein results ranged from 0.82 for wheat head protein to 0.97 for wheat vegetative protein. The general conclusion reached was that NIR spectroscopy was sufficiently accurate to be utilized in crop protein translocation studies.

Spectrophotometric determination of soil water content has also been investigated. Bowers and Smith (1972) used a Beckman DK-2A spectrophotometer with NIR analysis capability to measure the moisture content of soil samples. Methanol extracts of soil water were tested using the water absorption band at 1.94 μm. A linear relation between absorbance and soil water content was adequate for moisture determinations ranging from air dry to the equilibrium moisture content.

LITERATURE CITED

AHMED, N. S., and HELAL, F. R. 1977. The infrared absorption spectra for measuring the unsaturation in milk fat and oils. Milchwissenschaft 32:272-273.

ALEXANDER, E., L. SILVELA, S., COLLINS, F. I., and RODGERS, R. C. 1967. Analysis of oil content of maize by wide-line NMR. J. Am. Oil Chem. Soc. 44:555-558.

AOCS. 1973. Official Methods of Analysis Aa, 4-38. Assoc. Off. Anal. Chem., Washington, DC.

AOCS. 1979. Official Methods of Analysis Cd, 14-16. Assoc. Off. Anal. Chem., Washington, DC.

AOCS. 1983. Official Methods of Analysis Cd, 15-78. Assoc. Off. Anal. Chem., Washington, DC.

ARNOLD, R. C., and HARTUNG, T. E. 1971. Infrared spectroscopic determination of degree of unsaturation of fats and oils. J. Food Sci. 36:166-168.

BAKER, D. 1983. The determination of fiber in processed cereal foods by near-infrared reflectance spectroscopy. Cereal Chem. 60:217-219.

BAKER, D. 1985. The determination of fiber, starch, and total carbohydrate in snack foods by near-infrared reflectance spectroscopy. Cereal Foods World 30:389-392.

BAKER, D., NORRIS, K. H., and LI, B. W. 1979. Food fiber analysis: Advances in methodology. Pages 67-78 in: Dietary Fibers: Chemistry and Nutrition. G. E. Inglett and S. I. Falkenhag, eds. Academic Press, New York.

BARTON, F. E., II, and BURDICK, D. 1979. Preliminary study on the analysis of forages with a filter-type near-infrared reflectance spectrometer. J. Agric. Food Chem. 27:1248-1252.

BARTON, F. E., II, and BURDICK, D. 1982. The prediction of forage quality with NIR reflectance spectroscopy. Pacific Scientific, Gardner/Neotec Instrument Div., Tech. Pap. 2012.

BEGLEY, T. H., LANZA, E., NORRIS, K. H., and HRUSCHKA, W. R. 1984. Determination of sodium chloride in meat by near-infrared diffuse reflectance spectroscopy. J. Agric. Food Chem. 32:984-987.

BELL, W. 1982. The analysis of fructose concentration in high fructose corn syrups using near infrared reflectance analysis (NIRA). (Abstr.) Cereal Foods World 27:468.

BELL, W. L. 1983. The use of NIR as a process control tool for wet corn milling. Cereal Foods World 28:249-251.

BEN-GERA, I., and NORRIS, K. H. 1968a. Determination of moisture content in soybeans by direct spectrophotometry. Isr. J. Agric. Res. 18:125-132.

BEN-GERA, I., and NORRIS, K. H. 1968b. Direct spectrophotometric determination of fat and moisture in meat products. J. Food Sci. 33:64-67.

BEN-GERA, I., and NORRIS, K. H. 1968c. Influence of fat concentration on the absorption spectrum of milk in the near-infrared region. Isr. J. Agric. Res. 18:117-124.

BIGGS, D. A. 1967. Milk analysis with the infrared milk analyser. J. Dairy Sci. 50:799-803.

BIGGS, D. A., and STANLEY, D. W. 1972. Measurement of total solids in foods by transmission infrared spectroscopy. Can. Inst. Food Sci. Technol. J. 5:203-206.

BIRTH, G. S., TURLEY, W. C., and KAYS, S. 1982. Non-destructive determination of soluble solids in onions. Pacific Scientific, Gardner/Neotec Instruments Div., Tech. Pap. 4036.

BITTNER, D. R., and NORRIS, K. H. 1968. Optical properties of selected fruits vs. maturity. Trans. ASAE 11:534-536.

BOWERS, S. A., and SMITH, S. J. 1972. Spectrophotometric determination of soil water content. Soil Sci. Soc. Am. Proc. 36:978-980.

BRACH, E. J., PHAN, C. T., POUSHINSKY, G., JASMIN, J. J., and AUBÉ, C. B. 1982. Lettuce maturity detection in the visible (380–720 nm) far red (680–750 nm) and near infrared (800–1850 nm) wavelength band. Agronomie 2:685-694.

BURDICK, D., BARTON, F. E., II, and NELSON, B. D. 1981. Prediction of bermudagrass composition and digestibility with a near-infrared multiple filter spectrophotometer. Agron. J. 73:399-403.

CAMPBELL, R. C., and MARTIN, G. C. 1976. Determination of moisture in walnut seeds by near-infrared spectrophotometry. HortScience 11:494-496.

CARPENTER, K. J., and ELLINGER, G. M. 1955. Protein quality and 'available lysine' in animal products. Poult. Sci. 34:1451-1454.

COLLINS, F. I., ALEXANDER, D. E., RODGERS, R. C., and L. SILVELA, S. 1967. Analysis of oil content of soybeans by wide-line NMR. J. Am. Oil Chem. Soc. 44:708-710.

CONWAY, T. F. 1963. Nuclear magnetic resonance for determining oil content of seeds. J. Am. Oil Chem. Soc. 40:265-268.

DIACHUK, V. R., HAMILTON, E., SAVCHUK, N., and JACKEL, S. S. 1981. Bakery flour control using near infrared reflectance analysis. Baker's Dig. 55:72-75, 102.

DORSHEIMER, W. T., and ISAAC, R. A. 1982. Application of NIR analysis. Am. Lab. 14:58-63.

ELLIS, J. W., and BATH, J. 1938. Modification in the near infrared absorption spectra of protein and of light and heavy water molecules where water is bound to gelatin. J. Chem. Phys. 6:723-729.

FAO. 1984. Commodity Review and Outlook—1983–84. Food and Agricultural Organization, United Nations, Rome.

FENTON, A. J., Jr., and CRISLER, R. O. 1959. Determination of *cis* unsaturation in oils by near infrared spectroscopy. J. Am. Oil Chem. Soc. 36:620-623.

FINNEY, E. E., Jr., and NORRIS, K. H. 1978. Determination of moisture in corn kernels by near infrared transmittance measurement. Trans. ASAE 21:581-584.

FRANK, J. F., and BIRTH, G. S. 1982. Application of near infrared reflectance spectroscopy to cheese analysis. J. Dairy Sci. 65:1110-1116.

GIANGIACOMO, R., TORREGGIANI, D., FRANK, J. F., LOEWENSTEIN, M., and BIRTH, G. S. 1979. Analysis of blue cheese using infrared reflectance spectroscopy. J. Dairy Sci. 62:39-42.

GILL, A. A., STARR, C., and SMITH, D. B. 1979. Lysine and nitrogen measurement by infrared reflectance analysis as an aid to barley breeding. J. Agric. Sci. 93:727-733.

GODDU, R. F. 1957. Determination of unsaturation by near-infrared spectrophotometry. Anal. Chem. 29:1790-1794.

GODDU, R. F. 1958. Determination of phenolic hydroxyl by near infrared spectrophotometry. Anal. Chem. 30:2009-2013.

GODDU, R. F., and DELKER, D. A. 1960. Spectra-structure correlations for the near infrared region. Anal. Chem. 32:140-141.

GOULDEN, J. D. S. 1964. Analysis of milk by infrared absorption.

J. Dairy Res. 31:273-284.

GREENBERG, D. C. 1974. β-Glucan and extract viscosity. J. Inst. Brew. London 80:435.

HAMID, A., McCLURE, W. F., and WEEKS, W. W. 1978. Rapid spectrophotometric analysis of the chemical composition of tobacco. Part 2. Total alkaloids. Beitr. Tabakforsch. Int. 9:267-274.

HECHT, K. T., and WOOD, D. L. 1956. The near infra-red spectrum of the peptide group. Proc. R. Soc. London 235:174-188.

HERMANS, J., Jr., and SCHERAGA, H. A. 1960. Structural studies of ribonuclease. IV. The near infrared absorption of the hydrogen-bonded peptide NH group. J. Am. Chem. Soc. 82:5156-5160.

HILDEBRAND, D. F., CHAVEN, C., HYMOWITZ, T., BRYAN, H. H., and DUNCAN, A. A. 1981. Protein and oil content of winged bean seeds as measured by near-infrared light reflectance. Agron. J. 73:623-625.

HOLMAN, R. T., and EDMONDSON, P. R. 1956. Near-infrared spectra of fatty acid and some related substances. Anal. Chem. 28:1533-1538.

HOLMAN, R. T., ENER, S., and EDMONDSON, P. R. 1959. Detection and measurement of *cis* unsaturation in fatty acids. Arch. Biochem. Biophys. 80:72-79.

HOLMAN, R. T., NICKELL, C., and PRIVETT, O. S. 1958. Detection and measurement of hydroperoxides by near infrared spectrophotometry. J. Am. Oil Chem. Soc. 35:422-425.

HRUSCHKA, W. R., and NORRIS, K. H. 1982. Least-squares curve-fitting of near-infrared spectra predicts protein and moisture content of ground wheat. Applied Spectrosc. 36:261-265.

HUNT, W. H., FULK, D. W., NOLAN, T., and ABBOTT, D. 1977. Evaluation of the InfraAlyzer as a potential grain inspection tool for the USDA. I. Collaborative study on hard red winter wheat. Pages 114-117 in: Advances in Automated Analysis. Vol. 2. Industrial Symposia. Mediad, Inc., Tarrytown, NY.

HYMOWITZ, T., DUDLEY, J. W., COLLINS, F. I., and BROWN, C. M. 1974. Estimation of protein and oil concentration in corn, soybean, and oat seed by near-infrared light reflectance. Crop Sci. 14:713-715.

IWAMOTO, M., NORRIS, K. H., and KIMURA, S. 1981. Rapid prediction of chemical compositions of wheat, soybean, pork and fresh potatoes by near infrared spectrophotometric analysis. Nippon Shokuhin Kogyo Gakkai-Shi 28:85-90.

JASANSKI, A., and BILANSKI, W. K. 1973. Spectral absorbance of soybeans and its importance for efficient heat processing. Can. Inst. Food Sci. Technol. J. 6:151-155.

KAFFKA, K. J., NORRIS, K. H., PEREDI, J., and BALOGH, A. 1982. Attempts to determine oil, protein, water and fibre content in sunflower seeds by the NIR technique. Acta Aliment. 11:253-269.

KAYE, W. 1954. Near-infrared spectroscopy. I. Spectral identification and analytical applications. Spectrochim. Acta 6:257-287.

KLEPPER, L., and WILHELMI, K. 1979. Comparison between Kjeldahl and near infrared protein analyses on vegetative and head samples of wheat. Crop Sci. 19:923-925.

KRIKORIAN, E., and MAHPOUR, M. 1973. The identification and origin of N-H overtone and combination bands in the near-infrared spectra of simple primary and secondary amides. Spectrochim. Acta 29A:1233-1246.

KRUGGEL, W. G., FIELD, R. A., RILEY, M. L., RADLOFF, H. D., and HORTON, K. M. 1981. Near-infrared reflectance determination of fat, protein and moisture in fresh meat. J. Assoc. Off. Anal. Chem. 64:692-696.

LANZA, E. 1983. Determination of moisture, protein, fat and calories in raw pork and beef by near infrared spectroscopy. J. Food Sci. 48:471-474.

LARSEN, S. A. 1979. Examination of wool top by infrared reflectance: A note on the influence of dyeing. J. Text. Inst. 70:526-528.

LAW, D. P., and TKACHUK, R. 1977a. Determination of moisture content in wheat by near-infrared diffuse reflectance spectrophotometry. Cereal Chem. 54:874-881.

LAW, D. P., and TKACHUK, R. 1977b. Near infrared diffuse reflectance spectra of wheat and wheat components. Cereal Chem. 54:256-265.

LÖFFLER, C. M., and BUSCH, R. H. 1983. Determination of protein percentage of wheat vegetation at two growth stages by near infrared reflectance. Crop Sci. 23:167-168.

MACK, A. R., BRACH, E. J., and RAO, V. R. 1978. Investigation of crop spectra and selection of optimal spectral channels. Can. J. Spectrosc. 23:42-51.

MARK, H. L., and TUNNELL, D. 1985. Quantitative near-infrared reflectance analysis using Mahalanobis distances. Anal. Chem. 57:1449-1456.

MASSIE, D. R. 1974. Fat measurement of ground beef with a gallium arsenide infrared emitter. Pages 24-26 in: Quality Detection in Foods. J. J. Gaffney, ed. Am. Soc. Agric. Eng., St. Joseph, MI.

MASSIE, D. R., and NORRIS, K. H. 1965. Spectral reflectance and transmittance properties of grain in the visible and near infrared. Trans. ASAE 8:598-600.

MAURO, D. J., and OWEN, R. L. 1982. Rapid quantitation of ether and ester substituent groups in starch derivatives by NIR. (Abstr.) Cereal Foods World 27:469.

McCLURE, W. F., NORRIS, K. H., and WEEKS, W. W. 1977. Rapid spectrophotometric analysis of the chemical composition of tobacco. Part I. Total reducing sugars. Beitr. Tabakforsch. Int. 9:13-18.

McCLURE, W. F., and WILLIAMSON, R. E. 1982. Rapid spectrophotometric analysis of the chemical composition of tobacco. Part 3. Polyphenols. Beitr. Tabakforsch. Int. 11:219-227.

McGUIRE, C. F. 1982. Near-infrared reflectance estimates of malt extract. Cereal Chem. 59:510-511.

McKAY, D., and DORSHEIMER, W. 1982. Rapid analysis aids quality assurance. Food Eng. 54:78-79.

MILLER, B. S., POMERANZ, Y., THOMPSON, W. O., NOLAN, T. W., HUGHES, J. W., DAVIS, G., JACKSON, N. G., and FULK, D. W. 1978. Interlaboratory and intralaboratory reproducibility of protein determination in hard red winter wheat by Kjeldahl and near infrared procedures. Cereal Foods World 23:198-201.

MINER, D. C., ZIOMEK, J. V., and LANDA, I. J. 1979. Near-infrared instrumentation for controlling the quality of chocolate products. Pacific Scientific Tech. Pap. 4008.

MOEN, R. 1979. Soybean processing and NIR measurement. Pacific Scientific Tech. Pap. 1015.

NORRIS, K. H. 1964. Simple spectroradiometer for 0.4 to 1.2 micron region. Trans. ASAE 7:240-242.

NORRIS, K. H., BARNES, R. F., MOORE, J. E., and SHENK, J. S. 1976. Predicting forage quality by infrared reflectance spectroscopy. J. Anim. Sci. 43:889-897.

NORRIS, K. H., and HART, J. R. 1965. Direct spectrophotometric determination of moisture content of grain and seeds. Pages 19-25 in: Principles and Methods of Measuring Moisture in Liquids and Solids. Vol. 4. A. Wexler, ed. Reinhold, New York.

NORRIS, K. H., and WILLIAMS, P. C. 1984. Optimization of mathematical treatments of raw near-infrared signal in the measurement of protein in hard red spring wheat. I. Influence of particle size. Cereal Chem. 61:158-165.

O'CONNOR, R. T., DUPRE, E. F., and FEUGE, R. O. 1955. The infrared spectra of mono-, di- and triglycerides. J. Am. Oil Chem. Soc. 32:88-93.

PARK, Y. W., ANDERSON, M. J., ASAY, K. H., and MAHONEY, A. W. 1980. Predicting soluble nitrogen and fibrous fractions in crested wheatgrass with near-infrared spectroscopy. Pacific Scientific. Tech. Pap. 2008.

POMERANZ, Y., DAVIS, G. D., STOOPS, J. L., and HUBBARD, J. D. 1977b. Estimation of protein content and quality in grain sorghum. (Abstr.) Cereal Foods World 22:472.

POMERANZ, Y., MOORE, R. B., and LAI, F. S. 1977a. Reliability of five methods for protein determination in barley and malt. Am. Soc. Brew. Chem. J. 35:86-93.

POSNER, E. S., and WETZEL, D. L. 1986. Control of flour mills by NIR on-line monitoring. Assoc. Oper. Millers Bull., April, pp. 4711-4722.

PSOTKA, J. J., and BAHL, J. 1982. Measurement of saccharide distribution in high fructose corn syrup by near-infrared spectroscopy. (Abstr.) Cereal Foods World 27:469.

RAO, V. R., BRACH, E. J., and MACK, A. R. 1978. Crop discriminability in the visible and NIR regions. Photogramm. Eng. Remote Sensing 44:1179-1184.

RINNE, R. W., GIBBONS, S., BRADLEY, J., SEIF, R., and BRIM, C. A. 1975. Soybean protein and oil percentages determined by infrared analysis. U.S. Dep. Agric. Agric. Res. Publ. ARS-NC-26.

ROBERTSON, J. A., and BARTON, F. E., II. 1984. Oil and water analysis of sunflower seed by near-infrared reflectance spectroscopy. J. Am. Oil Chem. Soc. 61:543-547.

ROBERTSON, A., and MORRISON, W. H., III. 1979. Analysis of oil content of sunflower seed by wide line NMR. J. Am. Oil Chem. Soc. 56:961-964.

ROSE, J. J., PRUSIK, T., and MARDEKIAN, J. 1982. Near infrared multi-component analysis of parenteral products using the InfraAlyzer 400. J. Parenter. Sci. Technol. 36:71-78.

RUBENTHALER, G. L., and BRUINSMA, B. L. 1978. Lysine estimation in cereals by near-infrared reflectance. Crop Sci. 18:1039-1042.

SAMPSON, D. R., FLYNN, D. W., and JUI, P. 1983. Genetic studies on kernel hardness in wheat using grinding time and near infrared reflectance spectroscopy. Can. J. Plant Sci. 63:825-832.

SHENK, J. S., LANDA, I., HOOVER, M. R., and WESTERHAUS, M. O. 1981. Description and evaluation of a near infrared reflectance spectrocomputer for forage and grain analysis. Crop Sci. 21:355-358.

SHENK, J. S., MASON, W. N., RISIUS, M. L., NORRIS, K. H., and BARNES, R. F. 1976. Application of infrared reflectance analysis to feedstuff evaluation. Pages 242-248 in: Proc. Int. Symp. Feed Composition Anim. Nutr. Require. Comput. Diets, 1st. P. V. Fonnesbeck, L. E. Harris, and L. C. Kearl, eds. Utah Agric. Exp. Stn., Logan.

STEEL, P., and BELL, C. 1983. Multicomponent meat measurement. Food Flavourings Ingreds. Process. 5:45-56.

STERMER, R. A., POMERANZ, Y., and McGINTY, R. J. 1977a. Infrared reflectance spectroscopy for estimation of moisture of whole grain. Cereal Chem. 54:345-351.

STERMER, R. A., WATSON, C. A., and DIKEMAN, E. 1977b. Infrared spectra of milled rice. Trans. ASAE Pap. 76-3030:547-549, 557.

STEVENSON, S. G., and WILLIAMS, P. C. 1978. The determination of protein, moisture and oil in diverse cereals, cereal products and oilseeds by NIRS. (Abstr.) Cereal Foods World 23:462.

SUSI, H., and ARD, J. S. 1966. Near infrared investigation of interamide hydrogen bonds in hydrogen-bonding solvents. Arch. Biochem. Biophys. 117:147-153.

SYMES, K. J. 1965. The inheritance of grain hardness in wheat as measured by the particle size index. Austr. J. Agric. Res. 16:113-116.

TEMPLETON, W. C., Jr., SHENK, J. S., NORRIS, K. H., FISSEL, G. W., MARTEN, G. C., ELGIN, J. H., Jr., and WESTERHAUS, M. O. 1978. Forage analysis with near-infrared reflectance spectroscopy—current status and outline of national research project. Pages 528-531 in: Proc. Int. Grassl. Congr., 14th. J. A. Smith and V. W. Hays, eds. Westview Press, Boulder, CO.

TKACHUK, R. 1981a. Protein analysis of whole wheat kernels by near infrared reflectance. Cereal Foods World 26:584-587.

TKACHUK, R. 1981b. Oil and protein analysis of whole rapeseed kernels by near infrared reflectance spectroscopy. J. Am. Oil Chem. Soc. 58:819-822.

TKACHUK, R., and KUZINA, F. D. 1982. Chlorophyll analysis of whole rapeseed kernels by near infrared reflectance. Can. J. Plant Sci. 62:875-884.

WARD, R. G., SMITH, G. S., WALLACE, J. D., URQUHART, N. S., and SHENK, J. S. 1982. Estimates of intake and quality of grazed range forage by near infrared reflectance spectroscopy. J. Anim. Sci. 54:399-402.

WATSON, C. A. 1977. Near infrared reflectance spectrophotometric analysis of agricultural products. Anal. Chem. 49(a):835a-840a.

WETZEL, D. L. 1983. Lipid structures and NIRA. Proc. Int. Symp. Near Infrared Reflectance Anal., 4th. Technicon Industrial Systems. Tarrytown, NY.

WHEELER, O. H. 1959. Near infrared spectra of organic compounds. Chem. Rev. 39:629-666.

WILLIAMS, P. C. 1973a. The application of the Neotec Grain Quality Analyzer to the analyses of cereal grains and oilseeds. I. Preliminary observations. (Abstr.) Cereal Sci. Today 18:272.

WILLIAMS, P. C. 1973b. The application of the Neotec Grain Quality Analyzer to the analyses of cereal grains and oilseeds. II. Technical aspects of operation. (Abstr.) Cereal Sci. Today 18:272.

WILLIAMS, P. C. 1975. Application of near infrared reflectance spectroscopy to analysis of cereal grains and oilseeds. Cereal Chem. 52:561-576.

WILLIAMS, P. C. 1977. Technical aspects of instituting a protein grading system at a grain elevator, using near infrared reflection spectroscopy (NIRS). Pages 118-121 in: Advances in Automated Analysis. Techicon Int. Congr., 1976. Vol. 2, Industrial Symposia. Mediad, Inc., Tarrytown, NY.

WILLIAMS, P. C. 1979. Screening wheat for protein and hardness by near infrared reflectance spectroscopy. Cereal Chem. 56:169-172.

WILLIAMS, P. C., and CORDEIRO, H. M. 1985. Effect of calibration practice on correction of errors induced in near-infrared protein testing of hard red spring wheat by growing location and season. J. Agric. Sci. 104:113-123.

WILLIAMS, P. C., MacKENZIE, S. L., and STARKEY, P. M. 1985. Determination of methionine in peas by near-infrared reflectance spectroscopy (NIRS). J. Agric. Food Chem. 33:811-815.

WILLIAMS, P. C., and NORRIS, K. H. 1983. Effect of mutual interactions on the estimation of protein and moisture in wheat. Cereal Chem. 60:202-207.

WILLIAMS, P. C., NORRIS, K. H., and SOBERING, D. C. 1985b. Determination of protein and moisture in wheat and barley by near-infrared transmission. J. Agric. Food Chem. 33:239-244.

WILLIAMS, P. C., and PANFORD, J. A. 1980. The development and use of universal calibrations for the measurement of protein and moisture in HRS wheat by near-infrared reflectance spectroscopy (NIR). (Abstr.) Cereal Foods World 25:530.

WILLIAMS, P. C., PRESTON, K. R., NORRIS, K. H., and STARKEY, P. M. 1984. Determination of amino acids in wheat and barley by near-infrared reflectance spectroscopy. J. Food Sci. 49:17-20.

WILLIAMS, P. C., and SOBERING, D. C. 1986. Attempts at standardization of hardness testing of wheat. II. The near-infrared reflectance method. Cereal Foods World 31:417-420.

WILLIAMS, P. C., STEVENSON, S. G., STARKEY, P. M., and HAWTIN, G. C. 1978. The application of near infrared reflectance spectroscopy to protein testing in pulse breeding programmes. J. Sci. Food Agric. 29:285-292.

WILLIAMS, P. C., and THOMPSON, B. N. 1978. Influence of whole meal granularity on analysis of HRS wheat for protein and moisture by near infrared reflectance spectroscopy (NRS). Cereal Chem. 55:1014-1037.

WILLIAMS, P. C., THOMPSON, B. N., WETZEL, D., McLAY, G. W., and LOEWEN, D. 1981. Near-infrared instruments in flour mill quality control. Cereal Foods World 26:234-237.

WILSON, J. M., KRAMER, A., and BEN-GERA, I. 1973. Quantitative determination of fat, protein, and carbohydrate of soy products with infrared attenuated total reflectance. J. Food Sci. 38:14-17.

CHAPTER 11

Use of Near-Infrared Reflectance Spectrophotometry in Investigations in the USSR

V. P. KRISCHENKO
Central Institute for Agrochemical Services
Moscow, USSR

I. INTRODUCTION

The use of near-infrared reflectance (NIR) spectrophotometry in research has long been known in the USSR. Investigations of food products (Rogov and Zhukov, 1971; Avraamenko et al, 1974) and studies concerned with the processes occurring in food products under the effect of infrared irradiation (Ginzburg, 1966; Ilyasov and Krasnikov, 1972; Krischenko and Bolling, 1982) have gained popularity. Advances made in the past decade in the technique have made it possible to determine quantitatively various quality characteristics of crop products without subjecting them to chemical treatment. The technique is considered a valuable tool for the routine analyses of materials of plant origin in both research and quality control laboratories. At present the technique is being used in several countries as a standard method for protein determination in wheat. It has also been found suitable for the determination of gluten, oil, crude fiber content, and some other properties (Krischenko et al, 1980a, 1980b).

Plant material analysis by measuring the intensity of near-infrared radiation reflected from the surface of ground samples provides the possibility of determining several parameters simultaneously during a period of 30–60 sec, thus allowing NIR to increase the efficiency of analytical technologists by 1.5–2 orders of magnitude, depending on the number of characteristics to be determined.

In view of the high efficiency of the NIR technique, studies have been carried out at the Central Institute for Agrochemical Services, Moscow, on the potential usefulness of various models of NIR analyzers for the determination of protein, oil, crude fiber, starch, and other properties of plant materials. The instruments tested included the InfraAlyzer 2.5A and InfraAlyzer 400, manufactured by the Technicon Company (U.S.A.). Reflectance measurements are conducted at six or more wavelengths isolated by interference filters, provision being made in some instrument models for as many as 19 filters. The Neotec 21 and Neotec 41 NIR analyzers (Pacific Scientific Corporation, U.S.A.) and the Infrapid 31 instrument (Labormim Amalgamation, Hungary) have only three filters, in which the angle of inclination is varied in the course of measurement relative to the direction of the path of the light beam. This arrangement allows the scanning of a greater range of wavelengths. Individual wavelengths for making measurements are identified to the instrument detection circuitry via a computerized encoder. In these instruments, multiple linear regression is used to relate the magnitude of the parameter being measured to the intensity of NIR radiation reflected from the surface of the ground material. A rapid infrared moisture meter (type EVI) has been developed in the USSR at the Research Institute of Analytical Instrumentation Engineering (Krischenko et al, 1980c). The instrument uses two interference filters.

Sample preparation (comminution) for subsequent analysis was shown to be of great importance in the NIR method. In the course of using wheat grain samples ground on a cyclone mill (Tekator Company, Sweden) for the calibration of the InfraAlyzer 2.5A, researchers found that the analyses of wheat samples ground on other mill types differed markedly as to the apparent content of protein. They concluded that the degree of meal fineness affected the results of protein content determination by NIR. Studies were carried out using four different types of grinders (Table I).

A critical aspect of NIR analysis resides not only in the selection of samples for instrument calibration but in the calibration itself, insofar as these factors govern the accuracy of subsequent analyses and the coincidence of the data obtained with the results of determinations by standard methods.

II. INSTRUMENT EVALUATION

The NIR instruments evaluated included the Neotec models 21 and 41, the Labormim Infrapid 31, and the Technicon InfraAlyzer models 2.5A and 400. Evaluation of the Neotec 41 and Infrapid 31 analyzers involved instrument calibration, check analyses of the statistically representative number of samples, and mathematical treatment of the results obtained (Krischenko et al, 1982). Primary emphasis was given to selecting samples for instrument calibration that would adequately represent the diversity of varieties and plant growth ecological conditions. To calibrate the instruments against barley, samples of 32 barley varieties were compiled from 25 regions of the USSR. Calibration of wheat was carried out separately for Krasnozernaya spring wheat (type I) and Krasnozernaya winter wheat (type IY). The selected samples included 56 lines from 42 regions. The samples were materially dissimilar with regard to baking properties. Dough elastic resilience varied from 140 to 580 J, bread volume from 780 to 1,500 cm^3, and valorimeter figure from 38 to 75. The content of the components being analyzed varied as follows in the samples used for instrument calibration: Krasnozernaya spring wheat: protein, 11.1–18.0% (graduation increment, 0.2%); Krasnozernaya winter wheat: protein, 10.1–21.5% (graduation increment, 0.2%); soybean: protein, 30.8–46.0% (graduation increment, 0.5%), fat, 16.8–24.1% (graduation increment, 0.3%); sunflower: fat, 40.3–59.3% (graduation increment, 0.5%), protein, 13.4–25.0% (graduation increment, 0.6%); flaxseed: oil, 41.0–49.5% (graduation increment, 0.5%), protein, 21.0–29.0% (graduation increment, 0.5%); poppyseed: oil, 43.4–49.4% (graduation increment, 0.4%), protein, 21.1–25.0% (graduation increment, 0.3%). *Protein* in wheat and flour refers to total nitrogen × 5.7, dry basis. For all other commodities, the nitrogen/protein factor was 6.25. In the samples used for instrument calibration, protein was determined by the Kjeldahl method as the standard technique (USSR State Standard GOST 10846-74), oil was determined by the Soxhlet method (GOST 10857-64), and starch was determined by the polarimetric technique of Evers (GOST 10845-76). The samples intended for instrument calibration were analyzed twice by different analytical chemists in different laboratories. The results were used for calculating regression equations, standard deviations, and also correlation coefficients for the data obtained by the standard method and the NIR method.

For Krasnozernaya spring wheat, 65% of the samples had deviations of less than 0.5% in the results of protein determinations by NIR on the Neotec 41 analyzer, whereas for Krasnozernaya winter wheat the figure rose to 75% of the samples. In spring wheat, 6% of the results were biased downward in the 0.6–0.8% range, and 4% of the winter wheat results were biased downward. For upwardly biased deviations, the respective figures were 24% and 21%. Some substantial deviations (0.9–1.0%) were noted for Krasnozernaya spring wheat, but not for the winter wheat. In general, deviations in protein content were more frequently biased upward than downward, but the opposite occurred in oil content. The fraction of samples for which the results of protein determination by NIR were within the acceptable limits (deviation from the results according to the standard method) was found to be different for various wheat types. The correlation coefficients between protein content by NIR and standard methods for Krasnozernaya spring and winter wheat were found to be 0.97 and 0.98 respectively. In the case of barley, soybean, flaxseed, and poppyseed, the correlation coefficients were respectively 0.94, 0.96, 0.92, and 0.71.

Winter wheat and soybean differed considerably in protein content, in ranges of protein content, and in distribution of acceptable deviations. In wheat, samples having a protein content of from 14.1% to 16.0% gave closest agreement to the Kjeldahl results. For barley, even though a high coefficient of correlation occurred between NIR and Kjeldahl data, only 34% of the 284 samples studied gave acceptable deviations. In the remaining 66% the results of protein determinations differed significantly from the results obtained by the standard method, with deviations ranging from 0.6% to 1.9%. Determinations of oil and protein in sunflower yielded similar results, and the percentages of large deviations from standard method data were high. The correlation coefficients for protein and oil determinations in sunflower seeds were respectively 0.92 and 0.73. In research that determined the content of starch in 303 barley samples, the percentage of acceptable deviations was 85% and $r = 0.83$. Experience gained in the present study pointed to the feasibility of improving the accuracy of analyses significantly by refining instrument calibration, sample preparation procedure, and the technique of loading the sample cell in the NIR analyzer.

In oil content determinations, the highest percentage of acceptable deviations (70%) was observed in flax and poppyseed, whereas in soybean the best data were obtained in the range from 16.1% to 19.9%. The standard deviation (SD) figures indicated that the Neotec 41 and Infrapid 31 analyzers were practically identical in terms of the reproducibility of protein content analyses in wheat and barley, but they were somewhat dissimilar with regard to starch determination in barley.

In a comparative test, 31 samples of type I wheat and 40 samples of type IY wheat were used to calibrate the Neotec 21, and 43 and 46 samples respectively were used to calibrate the Neotec 41 (Table II). All samples were ground on a Udy cyclone sample mill fitted with a 1.0-mm sieve. Protein content in the samples selected for calibration varied over a wide range. Samples of type I wheat varied from 11.98% to 17.97% protein, and samples of type IY wheat varied from 9.80% to 17.22% protein (Table II). The standard errors of calibration were high, particularly so for Neotec 21 (0.64 and 1.20), and the correlation coefficients obtained on calibration were relatively low (0.78 and 0.80). Results from the Neotec 41 calibration were better, although the standard deviation for instrument calibration with type I wheat was still quite high (0.75).

To evaluate the accuracy of the calibrations of the Neotec 21 analyzer, 30 samples of type I wheat and 40 samples of type IY wheat were used, whereas for the Neotec 41 analyzer 23 samples of type I wheat and 22 samples of type IY wheat were used. In these samples, protein content was in the same range as in those used for calibration.

Deviations between the results of measuring the protein content by the standard method and on the different types of infrared analyzers were dissimilar. For the Neotec 21, deviations were large, particularly in the analysis of type I wheat, with only 43% of the results falling within the deviation range of up to 0.3%. Better data were obtained using the Neotec

TABLE I
Effect of Fineness of Grind on Results of Protein Determination in Wheat by Near-Infrared Reflectance Technique Using InfraAlyzer 400[a]

Mill Type	Specific Surface Area of Cake Meal (cm²/g)	Protein (%)
Cyclone	2,145	14.7
Ball	5,269	14.3
3M	2,675	15.5
MY	1,640	17.5

[a] $r = -0.69$ between specific surface area and protein.

TABLE II
Protein Content in Wheat Samples Used in Instrument Calibration and for Check Measurements

	Neotec 21				Neotec 41			
Wheat Type	*n*	Protein Content Range (%)[a]	SEC[b]	*r*[c]	*n*	Protein Content Range (%)	SEC	*r*
			Calibration Measurements					
I	31	12.66–17.97	0.64	0.78	43	11.98–17.97	0.75	0.92
IY	40	9.80–17.22	1.20	0.80	46	9.90–17.22	0.27	0.99
			Check Measurements					
I	30	12.8–18.1	···	···	23	10.4–18.9	···	···
IY	40	9.9–17.9	···	···	22	9.5–18.2	···	···

[a] Total N × 5.7, dry basis.
[b] Standard error of calibration.
[c] Coefficient of multiple correlation for calibration.

41 analyzer, especially for type I wheat, with 82% of samples falling within the deviation range of up to 0.3% (Table III).

To check the reproducibility of analysis, the protein content was determined in each sample 10 times using the same or different sample cells. The results obtained demonstrated good reproducibility of analysis on both instruments. Maximum discrepancy for the Neotec 21 was 0.2%, whereas for the Neotec 41 the standard deviation of a single test was 0.06 (cell I) and 0.09 (cell 2).

III. ANALYSIS OF FORAGES

Crude protein, oil, and crude fiber analyses are included in the evaluation of fodder grass. Chemical methods for analyzing these substances have been refined, but they remain laborious and time-consuming and they fail to provide the prompt results needed for the balanced feeding of farm animals. These analyses are carried out routinely in the course of fodder crop research, in breeding new varieties of forage plants, and in other applications. The NIR method was evaluated for the determination of forage grass quality.

Forage grass samples were taken from experimental plots of the Forage Research Institute (Table IV). Calibration and verification sample series consisted of plants grown under identical conditions, and both series included samples of different varieties taken during hay cuttings at different grass vegetative phases. The grass mixture samples included cock'sfoot, awnless bromegrass, meadow fescue, timothy grass, and a variety of herbs. The proportions of these components differed from sample to sample. The legume-grass samples, in addition to the listed grasses, included various proportions of red and white clover; consequently, the samples tested differed widely with respect to varietal composition, and the grass mixture samples differed further in terms of the species composition. The samples also differed as to the vegetative phases and time of cutting. Use of such widely diversified samples for examining the potentialities both of the standard method and the NIR analyses presented particular interest, since in the Soviet Union feedstuffs consist for the most part of grass mixtures.

Twenty-five to 30 samples analyzed by conventional chemical methods were used for calibration, each sample being analyzed in duplicate. The verification series also comprised 20–30 samples. In both series, the dried samples were ground on a cyclone-type mill. Correlations between the results of NIR spectrophotometry and conventional methods are illustrated in Table V. Statistical treatment verified that deviations between the mean results obtained by both methods were not significant.

A comparison of two series of mean values, x^{chem} and x^{NIR}, made it possible to assess the difference of parallel results. For tentative estimates, it is reasonable to assume that SD_{chem} was approximately equal to SD_{NIR}.

The NIR method was shown to be suitable for the analysis of plant samples with an accuracy comparable to that of standard methods (Krischenko and Bolling, 1982). The method is relatively free of systematic errors. Student's t criterion was employed as that of mean difference. In the analyses for protein, oil, starch, and crude fiber, the observed error was found to be lower than the theoretical error. The high correlations (0.72–0.96) between the results of chemical and NIR analyses corroborated these observations.

On the basis of these results, NIR appeared to be practicable for the analysis of both forage grass and grain. However, the experiments that provided the data for this conclusion were carried out using material with a high degree of uniformity with regard to varietal characteristics and also (to a certain degree) in terms of growing location.

Of particular interest was the possibility of using a universal instrument calibration for related materials, such as different

TABLE III
Deviations in Results of Protein Content in Two Wheat Types As Determined by Standard Method and Infrared Analysis

Deviations (% Protein)	Percentage of Samples			
	Neotec 21		Neotec 41	
	I[a]	IY	I	IY
0.3	43	60	32	65
0.4–0.5	20	15	0	22
0.6–1.0	30	23	18	13
1.0	7	2	0	0

[a]Wheat type.

TABLE IV
Test Component Concentration Ranges in Calibration and Check Sample Series of Forage Grasses

Sample Type	Total Nitrogen (%)		Crude Fat (%)		Crude Cellulose (%)	
	1[a]	2[b]	1	2	1	2
Alfalfa	2.52–4.02	2.62–3.73	2.50–5.88	2.50–5.88	19.5–28.1	19.7–26.6
Cock'sfoot grass	1.15–4.20	1.29–4.13	2.70–6.80	2.86–6.54	18.4–29.5	18.5–28.7
Legume-grass mixture	1.55–3.30	1.61–3.46	3.40–6.55	3.42–6.48	16.0–26.8	16.1–25.9
Grass mixture	0.90–4.19	1.40–4.21	1.72–5.45	1.64–5.40	20.1–30.3	20.2–30.3

[a]Calibration series.
[b]Check series.

TABLE V
Evaluation of Results of Nitrogen (Protein), Oil, and Crude Fiber Determination in Forage Grasses by Near-Infrared Reflectance (NIR) Spectroscopy[a]

Sample Type	Total Nitrogen				Crude Fat				Crude Cellulose			
	t	r[b]	$\bar{x}$[c]	SD[d]	t	r	$\bar{x}$	SD	t	r	$\bar{x}$	SD
Alfalfa	0.25	0.88	3.19	0.19	1.58	0.84	4.2	0.29	1.32	0.88	24.9	0.05
Cock'sfoot grass	1.32	0.87	2.66	0.21	0.60	0.72	3.7	0.50	2.10	0.86	23.5	1.62
Legume-grass mixture	1.32	0.95	2.52	0.17	1.55	0.93	4.5	0.67	0.10	0.94	21.4	0.61
Grass mixture	0.77	0.96	2.32	0.24	2.25	0.93	3.5	0.35	1.63	0.95	25.0	2.05

[a]For all samples and analytical techniques, $t_{0.95} = 2.1$.
[b]Coefficient of correlation between standard and NIR results.
[c]Mean of respective analytical results.
[d]Standard deviation of differences between standard and NIR results (SEP).

varieties or species of crops. Universal calibration equations were considered as possibilities for extending the usefulness of NIR spectrophotometry.

IV. UNIVERSAL CONSTANTS

As a further development of the studies reported above, researchers investigated the possibility of applying universal regression equations to instrument calibration for the determination of crude protein (total nitrogen content multiplied by an appropriate conversion factor). Also studied were crude fiber, starch, and moisture in the grain and plant material irrespective of growing location, variety, and crop type.

The NIR analyzer was calibrated to read total nitrogen content separately and the content of nitrogen was determined in the test samples. The method is known to determine the content of nitrogen indirectly, and the data are based on the determination of the actual amount of protein substances and free amino acids in test samples. To perform relevant investigations, samples of the grains of a large number of varieties of wheat, oats, rice, maize, and barley were selected from the World Collections at the USSR Research Institute of Plant Growth, care being taken to include samples having the maximum range of composition. The contents of crude protein, crude fiber, oil, starch, and moisture were determined by standard methods. The investigation also encompassed plant material samples tested for crude protein, crude fiber, and oil by the agrochemical service laboratories.

In these investigations, the InfraAlyzer 2.5A and 400 instruments were used. Coefficients of correlation and linear regression were obtained from the results of grain analysis of each separate crop and of plant material. In the crops listed, crude protein was determined by means of the calibrated NIR analyzer using individual regression constants for each crop. To develop a universal calibration for the determination of crude protein (nitrogen), oil, crude fiber, starch, and moisture in grain, batches of 30 samples of different crops characterized by the maximum available range of constitution were selected. The ranges in composition were as follows: crude protein, 6–31%; crude fiber, 1–10%; oil, 1–5%; starch, 30–90%; and moisture, 2–11%. Correlation coefficients and universal regression constants were obtained for all constituents in all crops.

The agreement between results obtained by standard methods and NIR was established by correlation and null hypothesis check with Student's t test. The actual value of the Student criterion (t_{act}) was calculated for two biased samples using the formula:

$$t_{act} = \frac{\bar{d}}{SD} \quad \text{where } S\bar{d} = S_{D^2}/n .$$

The correlation coefficients between standard and NIR data were as follows: for crude protein, 0.99; for crude fiber, 0.93; for oil, 0.97; for starch, 0.99; and for moisture, 0.95.

Among the factors responsible for high correlation coefficients is the selection of samples characterized by a broad range of composition. In the case of starch, the correlation coefficients were generally appreciably lower than those for protein when universal or individual calibration was used in analysis of individual crops. The correlation coefficients for starch determination in the case of individual and universal calibration of the instrument had the following respective values: for rice, 0.80 and 0.52; for barley, 0.90 and 0.79; and for maize, 0.91 and 0.96. Analogous results were obtained for other sample characters. The experiments verified that r depends markedly on the range of test parameters.

Accuracy of starch analysis by the two InfraAlyzer models was compared. For starch content determination, the standard deviation and the coefficients of correlation obtained as a result of analysis were as follows: for the InfraAlyzer 2.5, SD = 3.25 and $r = 0.74$; for the InfraAlyzer 400, SD = 1.14 and $r = 0.86$. The InfraAlyzer 400 displayed substantially greater accuracy, probably because of its more advanced optical system with integrating sphere. However, during the starch and moisture content determinations, a time-dependent drift of instrument readings was observed (1–2 absolute percentages), which was ascribed to slight test sample heating in the measuring cell.

With a view to deriving a universal regression equation for plant material, researchers used a further series of samples, each containing 30 samples. The range of composition was as follows: crude protein, 4–22%; crude fiber, 21–45%; and oil, 1–4%. The samples were used for calibration. Results of analysis of verification samples were subjected to correlation analysis. The r values for crude protein were 0.99; for crude fiber, 0.85; and for oil, 0.77. The results indicated that satisfactory agreement between standard NIR analyses was achieved using universal calibration constants. The correlation coefficients were somewhat lower than those found for grain. Cereal crops were analyzed by compiling several series, each consisting of 30 samples of grain or plant material. The range of composition was as follows: crude protein, 4–22%; crude fiber, 2–45%; and oil, 1–5%. Relevant regression equations were obtained, and check samples were analyzed by standard and NIR methods. The following correlation coefficients were established: crude protein, 0.99%; crude fiber, 0.88%; and oil, 0.88%.

In these experiments, satisfactory values for the coefficients of correlation were found, particularly for crude protein. In some instances, the correlation coefficients obtained using the universal calibrations were higher than those using individual calibration. It was concluded that the universal calibration constants can be used to advantage, particularly for the determination of quality characteristics of plant material.

V. DETERMINATION OF LYSINE IN MAIZE BY NIR SPECTROSCOPY

The InfraAlyzer 400 was used in an attempt to determine the content of lysine in ground maize grain by NIR spectrophotometry. Samples of different maize varieties were selected having a high and a low lysine content in the 2.0–4.8% range (in the protein). The results of lysine determination on a Durrum automatic amino acid analyzer were used for InfraAlyzer 400 calibration. The calibration equations were used in the analysis of check samples of maize for lysine content on the NIR analyzer. Correlation analysis of the results of analyses by the standard and NIR methods gave a coefficient of correlation of 0.93, showing that the NIR method was capable of accurate determination of lysine in maize and was much more rapid than the automatic amino acid analyzer. It may be argued that an indirect correlation was observed for lysine because of a relationship between variations in the protein content and the amino acid, but in these experiments no such relationship was detected.

VI. DETERMINATION OF PRECISION AND ANALYTICAL ERROR

To evaluate the stability of NIR analyzer operation, two sample cells were loaded with finely ground whole meal from wheat grain of known crude protein content, and the sample was analyzed five times in succession for the content of crude protein. The data showed a fairly high precision when the InfraAlyzer 2.5A was used. The average standard deviation for two cells was 0.12%, and the coefficient of variability (CV) was 0.8% at a protein content of 15.46%.

The total error of analysis incorporates the error associated with the measurement itself and with sample preparation for analysis. Sample preparation for an NIR analyzer, or for other instruments for nondestructive determination of chemical properties, generally consists of two operations: grinding the material to be tested and loading it into the sample cell. Experiments were carried out to determine the magnitude of probable errors arising during these operations. To determine the error associated with loading the sample cell, a sample of ground wheat grain was repeatedly loaded into the same cell, the content of crude protein being determined for each loading. The results of this study showed the error due to this operation to be very small. According to the results of these experiments, the total measurement error, including both instrument error and the error due to sample cell loading, totaled 0.24% (CV = 1.54%). The error due to sample cell loading was calculated from the total error (SD_{total}) minus the error due to the instrument itself ($SD_{instrument}$) using the formula:

$$\sqrt{[(SD_{total})^2 - (SD_{instrument})^2]} = \sqrt{SD_{loading}}\ .$$

Substituting the values of 0.24 for SD_{total} and 0.12 for $SD_{instrument}$, the value of 0.21 was obtained for $SD_{loading}$. The error due to sample cell loading was nearly twice as high as the error due to the instrument itself. Fineness of sample grinding was also found to be associated with a significant error.

The next stage of the work involved determining the influence of the number of samples used for calibration on the accuracy of the NIR analyzer. Protein content was determined by the Kjeldahl method in 50 samples of ground Saratovskaya 29 wheat grain, and the instrument was calibrated using the results of chemical analysis of 20 and 40 samples. The instrument calibrated with 20 samples was used for determining protein in 10 samples. The same procedure was repeated for a further 10 samples, but after the analyzer had been calibrated with 40 samples.

Mathematical treatment of the results showed a probability (P) of 0.95 that discrepancies between the results obtained on the analyzer calibrated with 20 and 40 samples were randomly distributed and statistically not significant. In the case of calibration using the results of analysis of 20 samples, $t_{act} = 1.31$ and $t_{theor} = 2.30$; for calibration using 40 samples, $t_{act} = 0.95$ and $t_{theor} = 2.30$. In our experiment, the average measurement error in crude protein analysis on the analyzer calibrated using 20 and 40 samples was 4.9% (relative).

The absolute error in the analysis of crude protein and moisture on the InfraAlyzer was determined using a test material noted for its uniformity as regards variety composition, namely, Mironovskaya 808 winter wheat grown in a single multivariant experiment in the Moskva region. The differences between the samples lay in the variance in conditions of nitrogen, phosphorus, and potassium nutrition. Researchers analyzed 72 samples for protein and moisture by standard methods; 40 of the samples were for InfraAlyzer calibration and the remaining 32 were analyzed as verification samples on the instrument (Table VI). The InfraAlyzer error was found to be 6.9% and 7 2% (relative) for crude protein and moisture respectively. The NIR method did not give evidence of systematic errors according to statistical treatment of the results. Student's t test was used to assess mean difference validity and to determine both crude protein and moisture, and actual error was found to be below the theoretical value: for protein, $t_{act} = 1.00$, $t_{theor} = 2.00$; for water, $t_{act} = 1.83$, $t_{theor} = 2.00$.

The usefulness of NIR for the determination of starch and oil has been investigated to a lesser degree, particularly where the concentration of these substances is low. Starch and oil were determined by NIR in ground wheat grain having known contents of these substances. The InfraAlyzer was calibrated using 40 samples of Mironovskaya 808 winter wheat that had been analyzed by standard methods for starch and oil, and 32 samples were analyzed as check samples (Table VI).

In starch and oil analysis on the InfraAlyzer, the measurement error did not exceed 6% for starch, but it was as high as 24–26% for oil in some cases. The relatively high error in oil determination resulted not so much from the inherent variability of the NIR method as from the standard method for oil determination, which gives large relative errors in samples of low oil content. This affects the efficiency of instrument

TABLE VI
Crude Protein, Moisture, Starch, and Oil Contents of Randomly Selected Samples of Mironovskaya 808 Wheat Flour As Determined by Standard and Near-Infrared Reflectance (NIR) Methods

Crude Protein (%)			Moisture (%)			Starch (%)			Oil (%)		
Standard Method	NIR	Measurement Error[a] (%)	Standard Method	NIR	Measurement Error[a] (%)	Standard Method	NIR	Measurement Error[a] (%)	Standard Method	NIR	Measurement Error[a] (%)
12.54	13.09	4.39	9.61	9.55	−0.62	64.6	66.0	2.1	1.48	1.57	6.08
12.26	12.39	1.06	9.04	9.36	3.54	59.1	61.2	3.4	1.63	1.38	−15.34
12.03	12.50	3.91	9.15	9.18	0.33	63.3	63.0	−0.5	1.32	1.37	3.79
13.57	13.91	2.51	9.65	9.59	−0.62	61.0	61.2	0.3	1.40	1.29	−7.86
13.05	12.61	−3.37	10.38	10.32	−0.58	59.5	59.6	0.2	1.29	1.31	1.55
12.43	12.59	1.29	9.34	9.38	0.43	60.0	60.6	1.0	1.16	1.33	14.66
13.00	12.76	−1.85	9.19	9.85	7.18	67.0	66.1	−1.3	1.17	1.38	17.95
13.96	13.71	−1.79	9.98	9.44	−0.64	65.4	67.1	2.6	1.21	1.30	7.44
12.20	12.45	2.05	9.61	9.65	0.42	65.4	64.3	−1.7	1.54	1.34	−12.99
12.82	12.81	−0.08	9.41	9.41	0.00	64.7	68.6	6.0	1.23	1.40	13.82

Statistical Analysis

	Crude Protein	Moisture	Starch	Oil
r	0.727	0.678	0.849	0.295
byx	0.8515	0.889	0.9136	0.5476
ayx	1.95	0.973	5.15	0.554
bxy	0.620	0.5163	0.7881	0.1589
axy	4.80	4.69	13.67	1.17
$\bar{x}$	12.80	9.52	63.00	1.31
SDx	0.52	0.42	2.94	0.13
$\bar{y}$	12.74	9.61	63.33	1.38
SDy	0.45	0.32	2.73	0.07
SDD (SEP)	0:366	0.313	1.570	0.131
N	32	32	32	32

[a] Relative error.

calibration and reduces the correlation coefficient in the calibration. The range of oil content was very small, which also precludes good correlation and regression statistics.

VII. DETERMINATION OF FIBER

The applicability of the NIR method for crude fiber analysis has also been studied, although to a lesser extent than analysis of crude protein, moisture, or even starch and oil. To study the usefulness of NIR for crude fiber analysis, wheat and rice plant material samples (50 of each crop) were selected. The samples, which were selected from single varieties of rice or wheat, were analyzed using a standard method to determine the content of crude fiber. Forty samples were used for InfraAlyzer calibration, and the remaining 10 were used as verification samples (Table VII). The analytical errors in crude fiber determination for wheat and rice on the InfraAlyzer were respectively 11.2% and 11.6% (relative).

Student's t test verified the absence of systematic errors from wheat and rice at a confidence level of 0.95. For crude fiber analysis in wheat plant material, $t_{act} = 1.14$ and $t_{theor} = 2.30$; in rice, $t_{act} = 0.31$ and $t_{theor} = 2.30$. These data verified the possibility of analyzing plant material for crude fiber with an accuracy equivalent to the currently accepted standard method. The material used in this experiment possessed high uniformity in varietal composition and growth location. A further experiment was aimed at determining the quality of a crop by NIR using the InfraAlyzer with a universal calibration—i.e., against samples with wide variance in growth zones, nutrition conditions, and genotype. The investigation involved 58 samples of spring wheat grain of different varieties and from various zones of the USSR submitted by the Research Institute of Grain Products. All samples were analyzed by standard methods for crude protein and starch. InfraAlyzer calibration was performed using 50 samples, the remaining 8 being used for verification.

The results (Table VIII) showed the analytical error for crude protein and starch to be respectively 6.5% and 4.2% (relative).

TABLE VII
Percentage of Crude Fiber in Vegetable Matter of Winter Wheat and Rice As Determined by Standard and Near-Infrared Reflectance (NIR) Methods

	Winter Wheat			Rice	
Standard Method	NIR	Measurement Error[a] (%)	Standard Method	NIR	Measurement Error[a] (%)
28.5	29.3	2.8	32.3	32.8	1.5
28.5	32.1	12.6	32.8	34.8	6.1
35.8	36.4	1.7	29.6	33.0	11.5
31.2	31.4	0.6	29.3	30.0	2.4
35.2	35.5	0.9	31.4	31.0	−1.3
30.1	31.7	5.3	27.6	26.6	−3.6
33.2	32.7	−1.5	36.2	32.0	−11.6
33.3	33.2	−0.3	32.9	32.8	−0.3
37.2	34.7	−6.7	19.1	20.9	9.4
27.7	29.4	6.1	33.2	32.5	−2.1
Statistical Analysis					
r	0.902			0.897	
byx	1.273			1.026	
ayx	−0.948			−0.999	
bxy	0.639			0.783	
axy	12.14			8.79	
$\bar{x}$	32.07			30.44	
SDx	3.37			4.66	
$\bar{y}$	32.84			30.64	
SDy	2.39			4.07	
SDD (SEP)	1.59			2.07	
N	10			10	

[a] Relative error.

In the crude protein analyses, $t_{act} = 1.46$ and $t_{theor} = 2.30$; in the starch analyses, $t_{act} = 0.21$ and $t_{theor} = 2.30$. For routine analysis, therefore, a single comprehensive instrument calibration was found to be sufficient for the proximate analysis of different varieties of a given crop grown in various soil and climatic zones and under dissimilar nutrition conditions.

VIII. RAPID MOISTURE ANALYSIS

A rapid NIR moisture meter (EVI) was developed at the USSR Research Institute of Analytical Instrumentation Engineering (Larchenko et al, 1980) using the principle of diffuse reflectance at two wavelengths in the near-infrared region. The operating wavelengths were selected in the absorption maxima of the OH and CH groups. The ratio of two output signals of the photoelectric receiver, corresponding to the sample and the reference beams (1,950- and 1,730-nm regions) reflected from the test sample, is correlated with the moisture content of the sample being analyzed.

The EVI near-infrared moisture meter incorporates an infrared radiation source, a reflectometer, and a stabilizer. Test sample irradiation by two monochromatic beams is affected by the radiation source via the reflectometer block. Conversion of the energy reflected from the test sample into electric energy, amplification, and detection of signals takes place in the reflectometer. Infrared double-wavelength moisture meters ensure that the moisture content of finely dispersed powders is accurately monitored, provided the following conditions are observed: (i) uniform moisture distribution over the powder bulk and individual particles; (ii) adequate transparency of the sample material to infrared radiation; (iii) consistency of transparency; (iv) constant macrodispersity and microdispersity of the powder; and (v) instrument calibration for a given material.

When these conditions were met, the accuracy of measurement was fairly good. The standard deviation was 0.05% (absolute) and the systematic component of the basic error was 0.3% (absolute) at a confidence level of 0.90. The EVI instrument was designed to monitor the moisture content of dry dairy products, and the object of the investigation was to evaluate it for determining the moisture of ground wheat and other cereals.

Wheat flour, barley, and oatmeal samples with moisture contents of 1–10% were prepared by a special procedure. The moisture contents as determined on the instrument were compared statistically with those by oven drying to constant weight. The procedure for calibrating the meter was modified by the use of two samples with maximum and minimum moisture contents. After placing a beaker containing the sample of lowest moisture content onto the sample port, the instrument was zeroed using the SCALE ORIGIN knob. Next a beaker containing the sample with the highest moisture content was placed onto the sample port, and by using the SCALE END knob the instrument was set to show the difference between the maximum and minimum moisture contents. The SCALE ORIGIN knob was used to set the instrument to indicate the moisture content of each sample as compared with the standard procedure. The sample with the lowest moisture content was used to check the instrument, after which the moisture content of all samples of the same product was measured. The instrument must be calibrated separately for each test material. For example, if the instrument is calibrated for determining the moisture content of barley, it cannot be used for moisture measurements in other cereals without slope/bias adjustment.

Determination of the moisture content should be accompanied by calibration of three reflecting plates. In subsequent moisture testing, the calibrated plates (reflectors) may be used to simulate the moisture content of a given test material, as well as to check calibration stability and introduce relevant corrections.

Table IX lists comparative data on the determination of moisture content in wheat, flour, barley, and oatmeal by the instrument and by oven drying. Statistical treatment showed that these methods of moisture content determination agreed closely. Student's t test results were as follows: $t_{act} = 0.12$ and $t_{theor} = 2.0$. The currently available EVI model is not adapted for measuring moisture contents above 10%, but the instrument yields good results when analyzing samples with moisture contents below that level.

IX. WHEAT GLUTEN DETERMINATION

Since gluten contains 75% protein, an investigation was initiated to determine the suitability of the NIR method for determining wet and dry gluten content. The problem of gluten analysis in the USSR is acute. Manual washing of gluten (according to GOST 13861-1-68 with amendment 1 of 1978) is a very laborious process involving several major sources of error. The development of the MOK-I and UOK-2 gluten-washing instruments failed to solve the problem because these analyzers are not adapted for carrying out complete analysis and require manual afterwashing of the gluten.

In the investigation, 120 samples of Mironovskaya 808 winter wheat grown in a field multivariant experiment in the Moskva region were used. The variables were the levels of nitrogen, phosphorus, and potassium nutrition. The performance of MOK-I instruments was evaluated on 30 samples of winter wheat at the Alma-Ata branch of the Central Institute for Agrochemical Services (TSINAO). The samples were selected during field experiments with mineral fertilizers.

Gluten was determined using the MOK-I and UOK-2 instruments and the InfraAlyzer. For InfraAlyzer calibration, crude protein in the grain and protein content of the gluten were determined by the Kjeldahl method. Before the NIR study, the MOK-I instrument was compared with manual washing. The MOK-I operates on the same principle as the UOK-2 instrument, which was the one used for determination of gluten content for InfraAlyzer calibration. From comparative analysis of the results of gluten determination by the manual washing technique and by the MOK-I instrument (30 samples), a coefficient of correlation of 0.80 was found. Student's t data were $t_{act} = 0.77$ and $t_{theor} = 2.00$ ($P = 0.95$).

The NIR method of gluten determination was compared with the gluten washing technique using the UOK-2 instrument. Gluten was determined in samples of ground grain of Mironovskaya 808 wheat using the standard method, and the results were used for InfraAlyzer calibration. The InfraAlyzer was then used to determine gluten content in 120 samples of

TABLE IX
Comparative Measurements of Moisture Content in Cereal Products by Standard Method and EVI Instrument[a]

Sample	Moisture Content by Standard Method (%)	EVI Moisture (%)	Reflector Calibration (*mcA* II)	$(x_1 - x_2)$ d
Wheat	3.4	3.6		−0.2
	5.4	5.2	26	0.2
	5.7	5.7		0.0
	2.8	2.8		0.0
Barley	4.1	4.1		0.0
	9.6	9.6		0.0
	5.7	5.4	76	0.3
	6.8	6.4		0.4
	7.0	7.0		0.0
	8.4	9.1		−0.7
	6.9	6.4		0.5
Oats	2.8	2.8		0.0
	4.8	5.1		−0.3
	9.8	9.8	41	0.0
	9.7	9.4		0.3
	5.7	5.3		0.4
$\bar{d}$				0.056
SDD[b]				0.30
t_{act}				0.12
t_{theor} at $n - 1$; $P = 0.95$				2.00

[a] EVI instrument = near-infrared moisture meter.
[b] SDD = standard deviation of differences between standard and EVI readings.

TABLE VIII
Content of Crude Protein and Starch in Flour from Spring Wheat Grown Under Diverse Soil and Climatic Conditions As Determined by Standard and Near-Infrared Reflectance (NIR) Methods

		Crude Protein (%)			Starch (%)		
Region	Spring Wheat Variety	Standard Method	NIR	Measurement Error[a] (%)	Standard Method	NIR	Measurement Error[a] (%)
Kuibyshev	Bezenchukskaya 98	17.12	16.08	−6.07	64.0	61.3	−4.2
Saratov	Saratovskaya 29	12.94	12.24	−5.41	66.7	66.8	0.1
Saratov	Albidum 24	15.08	14.39	−4.58	62.7	61.1	−2.5
Leningrad	Leningradka	13.30	13.05	−1.88	64.5	62.3	−3.4
Leningrad	Otechestvenaya	14.44	13.64	−5.54	64.1	62.3	−2.8
Minsk	Minskaya	15.60	16.13	3.40	62.7	61.6	−1.7
Zelinograd	Bezenchukskaya 98	13.86	13.55	−2.24	58.3	58.7	0.7
Saratov	Saratovskaya 29	14.08	14.79	5.04	60.7	61.2	−0.8
		Statistical Analysis					
r		0.89			0.87		
byx		0.871			0.984		
ayx		2.16			2.06		
bxy		0.917			0.778		
axy		0.89			12.93		
$\bar{x}$		14.55			62.96		
SDx		1.35			2.56		
$\bar{y}$		14.23			61.91		
SDy		1.39			2.27		
SDD (SEP)		0.63			1.24		
N		8			8		

[a] Relative error.

ground grain of the same variety that had been analyzed for gluten by the standard method (GOST 13861-1-68). The correlation coefficient between the results of gluten content determinations by these methods was 0.87. Student's t data were $t_{act} = 1.43$ and $t_{theor} = 2.00$; it was concluded that the NIR method was capable of determining gluten content with satisfactory accuracy and more effectively than the manual washing technique.

To study the determination of protein content in dried ground gluten of 70% protein and upward by the NIR method, gluten was isolated from winter wheat grains, dried, and ground. In 40 samples of gluten, protein content was determined by the Kjeldahl method. Twenty samples were used for InfraAlyzer calibration, and the remaining 20 samples were analyzed for protein content by the NIR method. The statistical treatment of the results of protein determination in gluten samples by Kjeldahl and NIR yielded a correlation coefficient of 0.97. Student's t test at a confidence level of 0.95 gave data of $t_{act} = 0.16$ and $t_{theor} = 2.00$. It was concluded that NIR spectrophotometry could be useful for protein analysis in high-protein samples of isolated, dried, and ground gluten (Krischenko et al, 1980c).

An analysis of data reported in the literature and results obtained in these experiments indicates that the NIR method is potentially capable of providing accuracy superior to that of conventional methods and that the accuracy of sample analysis by standard methods for NIR calibration is the most important factor limiting the accuracy of NIR analysis. The strong linear relationship between the content of protein and that of gluten in wheat grain was used to confirm the validity of this conclusion. In carrying out the experiment, it was assumed that the correlation coefficient between the content of protein and of gluten would be higher as the accuracy of analyses was greater.

The experiment involved two stages. First, after analyzing 120 samples of Mironovskaya 808 winter wheat, the coefficients of correlation were determined between (a) the results of gluten analysis on the UOK-2 instrument and by the NIR method using the analyzer calibrated against the data obtained from analysis of 40 samples using the UOK-2; and (b) the results of protein content determination by the standard method and by NIR on the analyzer calibrated against 40 samples that had been analyzed by the standard method. The coefficient of correlation between protein content determinations by these two methods was 0.91, and, according to Student's t test, $t_{act} = 1.33$ and $t_{theor} = 2.00$.

Second, from the results of analysis of 120 wheat samples, the regression equation and coefficient of correlation were calculated between protein and gluten content as determined (a) by the standard method and on the UOK-2 instrument; (b) on the NIR analyzer calibrated against 40 samples analyzed for protein by the standard methods and for gluten on UOK-2; and (c) with different sample sets on the NIR analyzer calibrated using the results of protein and gluten content determinations on the same analyzer, which had been calibrated against the results of analysis for protein by the standard method and for gluten on the UOK-2 instrument.

The coefficients of correlation between protein and gluten content as determined by techniques (a), (b), and (c) were respectively 0.83, 0.95, and 0.97. These results led to the conclusion that NIR analyzer calibration using the results of analyses on the same analyzer (preliminarily calibrated against the results obtained by the standard methods) made it possible to improve the accuracy of analysis. The implication was that the NIR method was potentially more accurate than protein analysis by the Kjeldahl method and gluten analysis by either the manual washing technique or the UOK-2 instrument. It was further concluded that the NIR method was potentially more accurate, not only in the analysis of protein and gluten, but also for the determination of starch, moisture, oil, and crude fiber. There does not appear to be any other documented demonstration of NIR analysis using calibrations based on NIR rather than standard analysis.

Further investigations concerned the application of the NIR method (InfraAlyzer 400) to determine gluten in patent (70%) flour from winter and spring wheat. The InfraAlyzer 400 was calibrated against the results of manual gluten analysis (GOST 13586-1-68) in different varieties of spring and winter wheats (30 samples of each wheat type). Calibration was performed separately on each wheat type (individual calibration) and also by using a mixed set of the same samples (universal calibration). Based on three sets of regression constants, the analyses of check samples of diverse spring and winter varieties were performed on the analyzer with respective individual calibrations and the universal calibration. The results of these analyses are listed in Tables X and XI. Statistical treatment yielded the following correlation coefficients:

1. $r = 0.89$ for gluten determination in patent (70%) spring wheat flour on the NIR analyzer calibrated against samples of different spring wheat varieties analyzed by the standard method.
2. $r = 0.97$ for gluten determination in patent (70%) flour of winter wheat on the NIR analyzer calibrated against samples of different winter wheat varieties analyzed by the standard method.
3. $r = 0.97$ for gluten determination in patent (70%) flour of spring and winter wheats on the NIR analyzer calibrated against samples of spring and winter wheats analyzed by the standard method.

Student's t test for a confidence level of 0.95 gave $t_{act} = 1.2$, i.e., smaller than t_{theor} (2.0).

The results obtained demonstrate that the content of gluten in patent (70%) wheat flour can be determined with adequate

TABLE X

Typical Results of Gluten Analysis in Patent (70%) Flour from Randomly Selected Samples of Spring and Winter Wheats by Standard (GOST) and Near-Infrared Reflectance (NIR) Methods (Individual Calibrations)[a]

	Spring Wheat			Winter Wheat		
	Gluten (%)		Measurement Error[b] (%)	Gluten (%)		Measurement Error[b] (%)
	GOST	NIR		GOST	NIR	
	25.0	25.9	0.9	40.5	40.4	0.1
	32.4	32.3	−0.1	48.6	48.0	−0.6
	36.6	37.4	0.8	31.3	31.7	0.4
	30.1	30.3	0.2	22.8	23.9	1.1
	31.9	31.7	−0.2	31.2	32.0	0.8
	24.1	24.2	0.1	23.8	23.3	−0.5
	27.8	27.7	−0.1	26.9	26.4	−0.5
	37.7	33.8	0.1	43.1	43.2	0.1
	36.9	36.2	−0.7	28.9	28.9	0.0
	37.1	37.8	0.7	30.5	31.1	0.6
	31.3[c]	30.6[c]	±0.6[c]	33.3[c]	33.3[c]	±0.6[c]
Statistical Analysis						
r			0.983			0.995
byx			0.951			1.011
ayx			1.55			−0.222
bxy			1.015			0.980
axy			−0.49			0.517
$\bar{x}$			31.66			32.54
SDx			4.17			6.74
$\bar{y}$			31.65			32.41
SDy			4.31			6.63
SDD (SEP)			0.80			0.65
N			24			24

[a] NIR analyzer calibration was carried out separately for spring and winter wheat varieties.

[b] Mean value.

[c] Relative error.

accuracy by NIR. The content of gluten in different samples of various wheat types and varieties can be achieved by using an NIR analyzer calibrated with a universal calibration developed by the procedure described. Results based on individual calibrations for spring and winter wheats were slightly better for the individual series than those obtained when a single calibration, based on spring and winter samples, was used for analysis of both types.

NIR has been applied to the determination of several baking quality parameters (Krischenko et al, 1986a). These include vitreous kernel percentage, protein and wet gluten content, alveograph W value, valorimetric index, loaf volume, and total bread score. Samples were accumulated from six of the republics of the USSR—Estonia (40), Lithuania (36), Byelorussia (30), Ukraine (19), Latvia (9), and Moldavia (3), all from the western USSR. Figures in parentheses indicate the number of samples. The variability of the samples is summarized in Table XII.

If A is less than 0.25, asymmetry is considered to be insignificant; between 0.25 and 0.50, moderate; and above 0.5, significant (Voznesenskii, 1969). The samples contained a relatively small range of protein. All samples represented different bread wheat varieties, and the objective was to predict the value of the varieties for individual republics and for the whole region. Growing conditions in the season preceding the growth of varieties included peas and fallow (Ukraine), annual grasses (Lithuania), raw crops (Latvia), potato and winter rye (Byelorussia), and occupied fallow (Moldavia and Estonia). Using the NIR method to determine individual baking quality parameters, researchers developed a mathematical model to enable discriminant analysis of wheat varieties from the aspect of bread-baking quality (Krischenko et al, 1980a). As a result, the number of wheat varieties recommended for growing in the western USSR was restricted to those most suitable for baking quality.

Kremer and co-workers (1986) studied the effect of moisture content on the magnitude of calibration constants. Thirty-three samples of wheat were ground in a Cyclotec grinder and divided into six groups. One of these was dried in an oven at 105° C to dryness. The remaining five were humidified to varying degrees ranging from 5.3% to 16.2% mean moisture content. The protein range was from about 11% to 17%. An Inframatic 8100 was used for the study. Although there was a tendency for the first three constants to decrease as the moisture content rose, the sizes of the constants were not consistent. For example, constants A_1 and A_3 increased from the 0 to 5.3% moisture level and then decreased. The influence of moisture on the remaining constants, including the important protein-testing constants at 2,180, 2,100, and 1,680 nm, was not significant.

Efremov et al (1986) studied the determination of moisture in fertilizers by means of NIR. Using an Infrapid 61, they noted that water is present in several states in fertilizers and recognized the need for careful sample preparation; they also noted that the errors of determination of moisture in mineral fertilizers varied with water content in a nonlinear fashion. They observed that losses of volatile constituents other than water occurred when water was determined by an oven-drying method. This factor constitutes a source of error in preparing and analyzing samples to be used for calibration of an NIR instrument. Organic fertilizers, such as manures, composts, sewage residues, and peat, presented further problems resulting from loss of volatiles and nonhomogeneity. Errors of from 0.5% to 1.2% moisture were obtained from the NIR determination of water in organic fertilizers with moisture ranging from 5% to 25%.

Korovin and co-workers (1986b) studied the application of NIR to determine the effect of growth regulators on vegetables. An Infrapid 61 NIR instrument was used, calibrated directly to nitrogen, oil, and cellulose contents and to the ratios of each of these to the other. Plant height, assimilation surface area, seedling weight, and dry weight were among the plant parameters studied. The Lastochka variety of tomato and Moskovski teplichni cucumber were the species involved. Most of the plant parameters were highly correlated with each other, particularly in the tomato; however, relatively few of these were correlated to the chemical constituents (nitrogen, oil, and cellulose), although the mineral content of tomatoes appeared to be correlated to plant height and assimilatory surface area, as well as to dry weight. The authors concluded that NIR determinations of chemical composition, coupled with statistical regression, would enable the prediction of the action of growth regulators such as benzimidazole phosphorylated derivatives on growing vegetables.

TABLE XI
Typical Results of Gluten Analysis in Patent (70%) Flour from Randomly Selected Spring and Winter Wheats by Standard (GOST) and Near-Infrared Reflectance (NIR) Methods (Universal Calibrations)[a]

	Spring Wheat			Winter Wheat	
Gluten (%)		Measurement Error[b]	Gluten (%)		Measurement Error[b]
GOST	NIR	(%)	GOST	NIR	(%)
31.6	30.7	−0.9	36.8	34.7	−2.1
31.5	32.8	1.3	35.1	34.0	−1.1
35.6	34.8	−0.8	41.0	43.2	2.2
32.6	31.1	−1.5	33.4	33.6	0.2
24.6	25.0	1.0	40.5	41.2	0.7
25.7	23.4	−2.3	37.8	37.7	−0.1
36.9	35.2	−1.7	31.2	31.6	0.4
29.0	30.2	1.2	37.8	39.5	1.7
22.8	22.4	−0.4	22.8	25.0	2.2
36.6	37.3	0.7	29.3	29.9	0.6
31.7[c]	31.7[c]	±0.6[c]	32.9[c]	32.8[c]	−0.5[c]

Statistical Analysis

	Spring Wheat	Winter Wheat
r	0.948	0.983
byx	1.019	0.941
ayx	0.048	1.91
bxy	0.882	1.027
axy	3.06	−0.84
$\bar{x}$	31.26	33.26
SDx	4.69	7.29
$\bar{y}$	30.63	33.32
SDy	4.37	7.62
SDD (SEP)	1.49	1.41
N	30	30

[a] NIR analyzer calibration was carried out against mixed set of samples selected from different spring and winter wheat varieties.
[b] Mean value.
[c] Relative error.

TABLE XII
Overall Statistics for Wheat from Western USSR

Parameter	$\bar{x}$[a]	CV[b]	A[c]	E[d]	95% Confidence Level Lower	Upper
Vitreous kernel (%)	60.8	31.5	−0.8	0.3	57.5	64.2
Protein (%)	13.1	10.0	−0.6	0.9	12.8	13.3
Wet gluten (%)	27.9	14.5	−1.0	2.8	27.2	28.7
Alveograph W	198	38.9	0.3	−0.05	184	211
Valorimeter index (%)	51.2	19.4	0.1	−0.2	49.5	53.0
Loaf volume, cc	1,068	15.2	−0.2	−0.2	1,039	1,097
Bread score[e]	3.8	17.0	−0.7	−0.07	3.7	3.9

[a] Mean result.
[b] Coefficient of variability.
[c] Asymmetry coefficient.
[d] Kurtosis coefficient.
[e] Arbitrary scoring units.

Gorpinchenko and Osanova (1986) have studied the prediction of malting quality in barley using NIR instruments. Chemical parameters determined included protein and starch contents. Protein was determined by Kjelfoss for calibration samples and starch by a polarimetric method. Very wide ranges of chemical composition and growing location were incorporated into the samples used for calibration. Protein ranged from 8.7% to 20.7% and starch from 50.3% to 65.6%. The analytical determinations were performed on a separate series drawn from the same wide region (from Kirgiz in the east, south to Georgia, and west to the Polish People's Republic) over a three-year period. The authors were able to determine protein and starch within the 95% confidence limits using a single calibration. Respective coefficients of correlation for protein and starch were 0.99 and 0.96.

Krischenko and Sazonov (1986) studied the diagnosis of the status of nitrogen nutrition of cereals using a Technicon InfraAlyzer 400. A standard error of 0.124 and coefficient of correlation of 0.99 were obtained for the analysis of wheat vegetative material ranging from 1.03% to 4.3% total nitrogen. These workers experienced difficulties in preparation of samples of high moisture content. Both the moisture content itself and the method of sample presentation added extra variables to the determination. Krischenko and co-workers (1986b) next considered the possibility of studying the evolution and breeding of plants via the medium of NIR spectroscopy. Using the apple as a test subject, they studied the series from wild species to breeding varieties to modern advanced varieties. Species included the oriental wild apple, crab apple, and the Sievers apple. X-ray fluorescence analysis (XRFA) was used as a standard nondestructive instrumental method for comparison, and all mineral constituents were determined by XRFA. Total nitrogen and fat (dry matter basis) were determined by NIR. Seeds and leaves were analyzed. A Retsch centrifugal grinder was used to prepare the samples for analysis. The authors were able, by using complementary NIR and XRFA analysis, to identify promising high-nutrient varieties. They were also, by determination of the differences in variability in certain constituents in pure species and advanced varieties, able to judge the degree to which the evolutionary process had occurred.

A group of workers headed by Korovin (1986a) studied the error of NIR testing between and within laboratories. They found that the error of testing within a laboratory increased from 2 to 14 times when the sample cells were reloaded before retesting and that the magnitude of the error was related both to the constituent and the commodity being tested. The same criteria applied to interlaboratory testing, where the magnitude of errors also appeared to be related to the concentration of the constituent, being generally lower at higher concentrations. The interlaboratory error of testing wheat rose almost by a factor of 10 from protein-gluten-oil-cellulose. In plant vegetation, the largest errors were associated with oil determination. These workers concluded that, in general, NIR analysis was quite satisfactory compared with standard chemical methods.

LITERATURE CITED

AVRAAMENKO, V. N., ESELSON, M. P., and ZAIKA, A. A. 1974. Infrared spectra of food products. Moscow.

EFREMOV, E. N., KRISCHENKO, V. P., NOSIKOV, V. V., and SAZONOV, Yu. G. 1986. Determination of water content in fertilizers with near-infrared spectroscopy. Pages 73-83 in: Agricultural Applications of Near-Infrared Spectroscopy. Central Inst. for Agrochemical Services, Moscow.

GINZBURG, A. S. 1966. Infrared technique in food industry. Moscow.

GORPINCHENKO, T. V., and OSANOVA, M. A. 1986. Near-infrared (NIR) spectroscopy applications for evaluation of brewing quality of barley. Pages 64-72 in: Application of Near-Infrared (NIR) Spectroscopy in the Analysis of Grain and Other Agricultural Products. Central Inst. for Agrochemical Services, Moscow.

ILYASOV, S. I., and KRASNIKOV, V. V. 1972. Methods of determining the optical and thermal radiation characteristics of food products. Moscow.

KOROVIN, A. I., KRISCHENKO, V. P., and SAMOKHVALOV, S. G. 1986a. Metrological characteristics of the near-infrared spectroscopy method used in the analysis of crop products. Pages 30-39 in: Application of Near-Infrared (NIR) Spectroscopy in the Analysis of Grain and Other Agricultural Products. Central Inst. for Agrochemical Services, Moscow.

KOROVIN, A. I., KRISCHENKO, V. P., SAMOKHVALOV, S. G., KOZLOVA, G. A., SHEVELEVA, I. V., and RADOSHKEVITCH, Ye. O. 1986b. Application of near-infrared (NIR) spectroscopy in the determination of the effect of growth regulators on vegetables. Pages 55-71 in: Agricultural Applications of Near-Infrared Spectroscopy. Central Inst. for Agrochemical Services, Moscow.

KREMER, A. A., ANCHURINA, S. P., BER, A. Yu., and GORPINCHENKO, T. V. 1986. Effect of moisture content of samples on calibration constants in infrared (IR) analysis of grain. Pages 99-102 in: Application of Near-Infrared (NIR) Spectroscopy in the Analysis of Grain and Other Agricultural Products. Central Inst. for Agrochemical Services, Moscow.

KRISCHENKO, V. P., and BOLLING, G. 1982. Employment of near infrared reflectance spectroscopy for determining the content of protein, gluten and ash in grain and grain processing products. Agrokhimiya 7:124-129.

KRISCHENKO, V. P., KOROVIN, A. I., and GORPINCHENKO, T. V. 1986a. Application of infrared spectroscopy as a means of analysis of some parameters and prediction of processing qualities of wheat. Pages 73-87 in: Application of Near-Infrared (NIR) Spectroscopy in the Analysis of Grain and Other Agricultural Products. Central Inst. for Agrochemical Services, Moscow.

KRISCHENKO, V. P., PASHCHINSKY, V. P., LARCHENKO, V. I., KRUCHININA, L. K., and ZHIGULEV, K. K. 1980b. Determination of plant material humidity content by an infrared analyzer. Paper presented at USSR–USA Symposium on Employment of Infrared Spectroscopy for Determining the Quality of Grain and Other Produce, Moscow.

KRISCHENKO, V. P., SAMOKHVALOV, S. G., GORPINCHENKO, T. V., MARTYANOVA, A. I., KHUDYAKOVA, F. K., ZHIGULEV, K. K., SOROCHINSKAYA, O. F., EVDOKIMOV, T. N., ZVYAGINTSEV, A. I., and GORSHKOVA, G. I. 1982. The employment of near infrared spectrophotometry for determining the quality characteristics of forage grasses, cereal grain and oil plant seeds. Agrokhimiya 6:112-124.

KRISCHENKO, V. P., SAMOKHVALOV, S. G., KOROVIN, A. I., BUKHARIN, P. B., and CHUGUNOVA, N. P. 1986b. Possible applications of near-infrared (NIR) reflectance spectroscopy in studies of evolution and breeding in plants. Pages 23-36 in: Agricultural Applications of Near-Infrared Spectroscopy. Central Inst. for Agrochemical Services, Moscow.

KRISCHENKO, V. P., and SAZONOV, Yu. G. 1986. Applications of infrared (IR) spectroscopy in the diagnosis of the nitrogenous nutrition of cereals. Pages 95-98 in: Application of Near-Infrared (NIR) Spectroscopy in the Analysis of Grain and Other Agricultural Products. Central Inst. for Agrochemical Services, Moscow.

KRISCHENKO, V. P., SAZONOV, Yu. G., CHUIKOVA, L. A., VALITOVA, E. G., RUSAKOVA, M. P., and LITVINOVA, S. I. 1980c. Gluten analysis by infrared reflectance measurements. Agrokhimiya 7:103-107.

KRISCHENKO, V. P., SAZONOV, Yu. G., GORSHKOVA, G. I., and VESELITSKAYA, T. V. 1980a. Determination of protein, starch, crude fiber, oil and moisture in vegetable matter and grain by infrared reflectance measurements. Agrokhimiya 6:128-133.

LARCHENKO, V. I., PASHCHINSKY, V. P., and KRUCHININA, L. K. 1980. Pages 99-102 in: Development of Methods, Instrumentation and Equipment for the Agrochemical Analyses of Soil, Plants, Fodder and Fertilizers. Central Inst. for Agrochemical Services, Moscow.

ROGOV, I. A., and ZHUKOV, N. N. 1971. Employment of infrared radiation in various food industries. Moscow, unpublished.

VOZNESENSKII, V. L. 1969. The Primary Treatment of Experimental Data. Practical Receivings and Examples. Leningrad Science. 82 pp.

CHAPTER 12

Near-Infrared Research in Asia

MUTSUO IWAMOTO
National Food Research Institute
Ministry of Agriculture, Forestry and Fisheries
Yatabe, Ibaraki, Japan

I. INTRODUCTION

Statistics from the U.N. Food and Agricultural Organization in 1981 report that 43 Asian countries have a total population of about 2.63 billion and that 57% of the economically active population is engaged in agriculture. Although the agricultural population in Asia is gradually decreasing year by year, it is still larger than the world average (44.7%). Agriculture, therefore, is the most important economic activity in Asian countries at the present time.

The potential of agricultural production in several Asian countries, mostly in eastern Asia, is shown in Table I. Agricultural production in Asia accounts for nearly 40–50% of the total world production, with the exception of meat and dairy products. These data indicate that agriculture in Asia occupies an important position in world food production. When compared with agricultural production during 1974–76, indices of production in 1981 increased more rapidly than the world average in most Asian countries, except Japan and Korea.

Rice is the principal crop in the countries listed in Table I. Rice production is about 25% of all cereal production in the world, occupying third place following wheat and maize. However, rice production in Asian countries reaches an average of 57% of cereal production, and more than 90% in Malaysia and Japan.

In addition, the food processing industry is growing rapidly, paralleling the "modernization" of eating habits in the region. As a major industry, it plays a key role in providing readily processed foods, creating markets for raw materials from the primary product sector, and generating employment opportunities. In Japan alone, the proportion of processed foods in the daily diet increased dramatically during the past decade, reaching 60% by the early 1980s. This figure is expected to increase on a parallel with that of other highly developed countries. The food manufacturing industry in Japan numbers over 83,000 firms, with a total economic value of $132 billion. More than 1.2 million employees are involved in the industry that produced 11.1% of the total output of Japanese manufacturing during 1983, ranking third (next to transportation and electronics). The output of the processed foods industry was more than double that of the primary production sector.

There is great potential for the introduction of near-infrared reflectance (NIR) technology for the determination of composition of raw materials in primary production and assurance of quality in both raw materials and final products in the food industry. The technique was first applied in agriculture, and its acceptance was originally delayed because of lack of expertise and information, particularly in the less developed countries of Asia. This chapter describes NIR applications in Asia outside of the USSR and discusses applications and research in Japan in more detail.

II. DISTRIBUTION OF NIR INSTRUMENTS IN ASIA

NIR instrumentation was first introduced to southeast Asia in the early 1970s by the Hokkaido Agriculture Experiment Station, Japan, for use in soybean breeding. At that time, the technique did not take root because the technical information was not comprehensive and the instruments were less reliable than present ones. With improved accuracy and simplified operation, NIR was reintroduced in 1977–78 by the Neotec Instrument Company. However, two to three years elapsed before the technology was accorded recognition by customers. Thus, the NIR technique was actually adopted in Japan nearly five years after its adoption in the United States, Canada, and the United Kingdom.

In recent years, NIR instruments have been brought to Japan by the commercial firms of DICKEY-john, Percon, Pacific Scientific, Technicon, and Trebor. The number of NIR instruments in use in Japan can be roughly estimated at 250–300. Considering the scale of industrial activity in Japan, it is likely that many more instruments will be required in the future.

About half (52%) of the instruments in Japan are used in the flour mill and feed industries; oil processors use 12%, brewers 3%, and other applications account for the remaining 33%. Several flour companies intend to use the equipment in quality control by means of universal calibrations, and one company has succeeded in developing a satisfactory set of constants. The formulas of wheat flour are very complicated in Japan, to meet the diverse requirements of customers, and more stable calibrations will be necessary in the future. More than half of the instruments used in feed quality control are found in prefectural animal experiment stations. The expense of these instruments has been subsidized by the Ministry of Agriculture, Forestry and Fisheries (MAFF) since 1982. They are mainly used for forage analysis. Recently the Food Agency, under the auspices of MAFF, began an investigation of the possibility of adopting NIR for compositional analysis of imported grains such as wheat and corn (maize).

The introduction of the technique in other Asian countries is indicated by the estimates in Table II. NIR use in China has expanded rapidly in recent years. Most of the instruments are used in quality evaluation and control in wheat and the food industry. A totally new approach is attributed to Zhong (1982), who was the first to describe the application of NIR to soil analysis. He was able to determine total nitrogen in soils.

Most of the instruments in the Asian region covered by this chapter are used in the flour milling, food production, and feed manufacturing industries. Some instruments, such as those at the International Center for Agricultural Research in the Dry Areas (Aleppo, Syria) and the Arab Center for the Studies of Arid Zones and Dry Lands (Douma, Syria), are used in breeding programs for selection of improved lines of cereal grains, pulses, and forages. Similarly, an instrument at the Rice Research Institute of Thailand is used in rice-breeding programs, particularly in connection with identification of rice genotypes with improved lysine content. This instrument was supported financially by the International Atomic Energy Agency. Researchers at the institute report that it would be useful for NIR instruments to have a self-diagnosis feature to help customers in making repairs. Service is generally poor in developing countries, a factor that has inhibited more general acceptance of the technique. Some of the instruments in the Philippines are being used in the coconut and copra industries. In Taiwan, the Food Industry Research and Development Institute is using NIR to determine routine compositional analysis of mixed feeds. Another instrument is used industrially for quality control in soybean processing. Instruments in Korea are used for product formulation and quality control in the food industry, particularly with products such as flour and coffee. In Turkey, NIR is being used in the tobacco industry as well as in the wheat-breeding program centered in Arikan.

As a general rule, instruments are used for specific applications and the information is generated for internal purposes, with no publication of results. Consequently, it is difficult to obtain detailed, reliable information about the numbers, types, and applications of NIR instruments that are in operation.

III. TRENDS OF NIR RESEARCH IN JAPAN

Several papers on the NIR technique itself have reported on the application of NIR to both quantitative and qualitative analyses in the same manner as other methods of spectrophotometric analyses (Kato et al, 1964; Takeuchi et al, 1969). However, these reports contain no description of the concept of nondestructive application of the NIR technique. NIR research in Japan is growing more active in various fields of industrial application and basic research, despite a lag time before the consistent introduction of the technique. Although work conducted in private companies is generally not published, the research has certainly become more active and progressive. Research in specific areas is summarized in the following subsections.

A. Rice

Rice is the most important crop in Japan. The Hokkaido Agriculture Experiment Station has been using NIR technology in its breeding program for determination of protein, moisture, amylose, and ash content in rice grain. More than 10,000 samples have been analyzed during each of the last three years. Although fairly good results have been reported, reproducibility for amylose and ash measurements needs to be improved (O. Inatsu, 1983, pers. comm.). According to the results of spectral analysis using the second derivative of the log $(1/R)$ in connection with amylose content, one of the key wavelengths selected is located around the 1,800-nm region. Absorption at this wavelength was assigned to the C-O-H group in the molecule. This wavelength is also a key band for measurement of the degree of gelatinization of rice starch.

High correlation coefficients have been obtained in the calibration between NIR optical data and contents of several amino acids, including lysine, glycine, alanine, and others. NIR was found to be as efficient in the determination of protein in rice as in wheat. The question arises as to whether it is possible

TABLE II
Estimated Number of Near-Infrared Reflectance Instruments in Asian Region

Country	Estimated Number of Instruments	Principal Uses
Burma	5	Feed, agriculture
China	200–250	Food, feed, agriculture
India	50	Flour, feed, agriculture
Indonesia	12	Feed, agriculture
Israel	8	Food, feed, agriculture
Japan	250–300	Food, feed, industry
Korea	7–10	Food
Malaysia	4	Food, agriculture, flour
Pakistan	10	Flour, food
Philippines	20	Feed, rice, copra
Singapore	12	Flour, food
Sri Lanka	1	Flour
Syria	6	Grain, forage, agriculture
Taiwan	9	Food, feed, flour
Thailand	3	Food, agriculture
Turkey	4	Tobacco, food, agriculture
Approximate total	600	

TABLE I
Potential of Agriculture in Some Asian Countries[a]

Country	Agricultural Products (1,000 t): Cereals	Root Crops	Fruits, Vegetables, and Tree Nuts	Pulses	Oil Crops	Meat	Milk	Eggs	Indices of Production: Agricultural Total (1974–76 = 100)
China	286,104	145,275	90,888	6,856	26,835	23,481	7,975	4,902	121
India	149,720	16,916	58,733	11,207	18,856	916	31,940	776	121
Indonesia	36,991	16,329	5,243	320	13,259	454	79	96	135
Japan	13,886	5,252	21,257	95	277	3,022	6,610	1,999	94
Korea	8,026	1,666	10,257	58	331	490	513	280	116
Malaysia	2,155	507	1,379	···	4,279	235	33	126	130
Philippines	10,896	3,612	8,259	50	9,632	772	28	210	127
Thailand	23,084	18,255	7,214	324	1,321	715	12	108	129
Asian total	661,439	225,271	269,971	22,241	84,206	35,991	72,948	9,716	118
	(39.8%)[b]	(40.1%)	(42.3%)	(52.5%)	(37.5%)	(25.0%)	(15.4%)	(33.3%)	
World total	1,663,828	561,567	638,486	42,403	224,315	143,459	472,955	29,210	114

[a]Source: FAO (1981).
[b]Figures in parentheses refer to share of regional production in world production.

to determine individual amino acids directly or whether this would be jeopardized by the high correlations between protein and the amino acids. This aspect is still being researched. In NIR analysis for protein, use of a derivative of the log $(1/R)$ signal led to a reduction in the standard error of prediction of protein, but the bias increased in the analysis of unknown samples (Iwamoto et al, 1986b).

Rice is also important as a raw material for brewing Japanese sake. The technique was studied for the development of a discriminant analysis technique to select desirable genotypes for brewing, but this research is not yet complete (Kusama et al, 1984).

B. Wheat

At the beginning of the 1970s, the production of homegrown wheat had decreased in Japan. In recent years, however, wheat production has increased as a substitute crop for surplus rice production, and it is expected to reach 1.2 million tons in the near future. Since homegrown wheat is inferior to imported wheat as a raw material for processing, it is necessary to pay more attention to quality control.

NIR instruments could be used successfully to determine moisture and protein in homegrown wheat flour, but not ash (Iwamoto et al, 1984a). Ash content, as well as protein and moisture, is one of the more important factors involved in quality control of wheat flour. Generally speaking, ash does not produce significant absorption in the NIR spectrum. There is really no "ash" in flour, only minerals. Another difficulty in the determination of ash in wheat flour with NIR is that flour containing more ash tends to have a smaller particle size, and vice versa. Since diffuse reflectance spectroscopy is strongly influenced by particle size, this characteristic can cause a serious bias in NIR analysis. It is thus necessary to find a suitable method to compensate for the bias.

As reported by Williams et al (1981), ash can be determined because of its association with other components, such as oil and cellulose. The accuracy of prediction for ash in homegrown wheat flour was improved by adding a 2,344-nm wavelength to the calibration obtained by multiple linear regression analysis using second-derivative mathematics. Because the absorption at 2,344 nm disappeared with defatted flour as well as defatted bran or short bran, the absorber that produces this absorption can be assumed to be a solvent-extractable component (Iwamoto et al, 1986a).

C. Buckwheat

Buckwheat is an important cereal in Japan as raw material for the popular *soba* noodles. Feasibility of determination of protein and ash in buckwheat flour was reported by Suzuki et al (1984).

D. Soybean

More than 95% of the soybeans consumed in Japan are imported and used for oil extraction and for preparation of soybean protein isolate and traditional foods such as soy sauce, fermented soybean (*natto*), fermented soybean paste (*miso*), or soybean curd (*tofu*), among others. In addition to total contents of oil and protein, an acceptable result was reported concerning a classifying analysis for the content of main protein components such as 7S or 11S by means of second-derivative mathematics. The data also showed the potential to identify soybeans wherein the protein had been denatured, raising the possibility that NIR could replace the time-consuming ultracentrifuge analysis as a simple method to estimate protein components (Cho et al, 1987).

Study of the influence of grain color based on soybean variety on NIR accuracy showed the NIR technique to be independent of color (Nakamura et al, 1978).

E. Red Pepper

Capsaicin, one of the pungent components in red pepper, is an important factor in the quality of red pepper. Because it takes a long time to analyze capsaicin content by the chemical method, it would be advantageous to develop a convenient method using NIR. According to spectral analysis, capsaicin was found to be sensitive to the NIR spectrum by reason of its characteristic molecular structure. A fairly satisfactory result was obtained using second-derivative mathematics (Iwamoto et al, 1984b).

F. Laver

Laver is a seaweed popularly consumed by Japanese people. An objective method for grading the quality of laver is needed to replace the laborious grading work that is now performed. Generally, laver containing more protein (and, conversely, fewer carbohydrates) is classified to be of higher quality. Since NIR has the capability of determining protein and carbohydrate, this technique is a possible method to be adopted. Although further study is needed to develop a universal calibration that is independent of laver variety, this technique certainly has a high capability for performing the on-line qualitative measurement expected (Iwamoto et al, 1983a, 1983b).

G. Feed

About one fourth of the NIR instruments in Japan are used for feed analysis. During the last three years, MAFF has adopted NIR for forage control in the prefectural animal experiment stations. The method is reported to have great possibilities for determining moisture, protein, and other nutrient components, even in undried herbaceous material (T. Ota, 1985, pers. comm.).

There are many difficulties in the analysis of mixed feed, mainly resulting from the complicated formulas of the different feeds. Appropriate methods need to be developed for data processing of the NIR signal to compensate for the effects of nonhomogeneity in the mixed feed. A preliminary study suggested that the 2,180-nm wavelength that is ordinarily used for protein analysis with NIR should be reconsidered. Because the intensity of absorption at this wavelength was found to be significantly dependent on the different types of protein present and the mixing ratio of raw materials, other protein-related wavelengths at 2,140, 2,056, and 1,978 nm, in combination with 2,180 nm, may provide a more stable calibration for mixed feed analysis.

H. Green Tea

Green tea is the most popular beverage in Japan. Quality control is presently based on experience and perception, and researchers explored the possibility of developing a standardized method by means of NIR.

The NIR technique has been used to determine total nitrogen content of green tea, which is highly correlated to quality (higher nitrogen content means better quality). The major sources of nitrogen are protein (75–85%), caffeine (15–20%), and free amino acids (3–5%). Contents of these components have been determined with acceptable accuracy (Ikegaya et al, 1985, 1987). NIR can also be used to detect adulteration of the green tea with monosodium glutamate.

I. Other Foods

The NIR technique is applied as a screening test for malting quality of barley, particularly in connection with protein, hot water extract, and other components. About 120 samples can be analyzed every day with NIR, improving the efficiency of analysis as much as four times over the Kjeldahl method, for example. The technique is used for determination of components in hops. NIR is also used in quality control in Japanese sake, with good results being reported on the determination of such parameters as alcohol, acid, amino acids, sugar, and others (Wakai et al, 1984).

Several edible-oil refineries are using NIR in quality control, and investigations are being carried out on the development of an on-line monitoring system. Similarly, a factory manufacturing fructose syrups is using NIR in quality control, and this factory also intends to develop an on-line system pending satisfactory results with conventional instruments.

Studies carried out on minced ham have demonstrated good results for the determination of protein and moisture (S. Ozawa, 1984, pers. comm.). Satisfactory data have also been obtained for the analysis of fat in ground pork, beef, and chicken (Iwamoto et al, 1981; Nagao et al, 1985). A linear regression analysis using quotient mathematics of log $(1/R)$ absorbance has the potential of providing high correlations on a log $(1/R)$ (nonderivative) instrument.

NIR analysis of raw milk gave results for fat, protein, lactose, and total milk solids that were equivalent to those from the infrared instruments recommended by the AOAC (Sato et al, 1985). Water structure in food systems has been studied by the NIR method (Niwa et al, 1980). This study concerned the state of water in fish flesh. Later studies by Iwamoto et al (1986c) extended the work to high-moisture foods such as milk, sliced potato and pear, and tofu, and foods of lower moisture content including cereal flours were also studied. The absorption band of water near 1,400 nm that was observed in the high-moisture materials virtually disappeared and was replaced by a fairly strong band near 1,430 nm in the lower-moisture foods. These workers concluded that NIR using the second derivative of the log $(1/R)$ may be a valuable tool for the study of the state of water in foods. An important implication of their work is that different wavelengths are almost certainly necessary for the determination of water in foods and raw materials varying widely in moisture content.

ACKNOWLEDGMENTS

I wish to thank T. Toda of Technicon Japan Company and K. Tatsubayashi of Nippon Regulator Company for their courtesies in exchanging information.

LITERATURE CITED

CHO, R. K., IWAMOTO, M., and SAIO, K. 1987. Determination of 7S and 11S globulins in soybean seeds by near-infrared spectroscopic analysis. In press.

FAO. 1981. World Agricultural Statistics. Food and Agriculture Organization, Rome.

IKEGAYA, K., HINO, A., UOZUMI, J., TAKAYANAGI, H., ANAN, T., and IWAMOTO, M. 1985. Near-infrared spectrophotometric analysis of total nitrogen content in green tea. (In Japanese with English summary) Nippon Shokuhin Kogyo Gakkai-Shi 32:553-559.

IKEGAYA, K., IWAMOTO, M., UOZUMI, H., and NISHINARI, K. 1987. Near infrared spectra of caffeine and its related compounds and their application to determination of caffeine content in green tea. Nippon Shokuhin Kogyo Gakkai-Shi 34:254-258.

IWAMOTO, M., CHO, R. K., SUZUKI, T., and UOZUMI, J. 1984a. Near infrared reflectance analysis for determining moisture, protein and ash contents in home-grown wheat flours. Nippon Shokuhin Kogyo Gakkai-Shi 31:50-53.

IWAMOTO, M., CHO, R. K., UOZUMI, J., and IINO, K. 1984b. Near-infrared reflectance spectrum of red pepper and its applicability to determination of capsaicin content. Nippon Shokuhin Kogyo Gakkai-Shi 31:120-125.

IWAMOTO, M., HIRATA, T., SUZUKI, T., UOZUMI, J., and ISHITANI, T. 1983a. Feasibility of near-infrared diffuse reflectance for determination of the quality of dried laver (*Porphyra yezoensis*). (In Japanese with English summary) Nippon Shokuhin Kogyo Gakkai-Shi 30:397-403.

IWAMOTO, M., KONGSEREE, N., UOZUMI, J., and SUZUKI, T. 1986a. Determination of ash content in Japanese home-grown wheat flour by near-infrared diffuse reflectance analysis. Nippon Shokuhin Kogyo Gakkai-Shi 33:842-847.

IWAMOTO, M., NORRIS, K. H., and KIMURA, S. 1981. Rapid prediction of chemical compositions of wheat, soybean, pork and fresh potatoes by near infrared spectrophotometric analysis. Nippon Shokuhin Kogyo Gakkai-Shi 28:85-90.

IWAMOTO, M., SUZUKI, T., HIRATA, T., NODA, H., UOZUMI, J., and ISHITANI, T. 1983b. Some characteristics of infrared and near-infrared spectra of dried laver (*Porphyra yezoensis*) in connection with the quality. (In Japanese with English summary) Nippon Shokuhin Kogyo Gakkai-Shi 30:544-551.

IWAMOTO, M., SUZUKI, T., and UOZUMI, J. 1986b. Analysis of protein and amino acid contents in rice flour by near-infrared spectroscopy. Nippon Shokuhin Kogyo Gakkai-Shi 33:848-853.

IWAMOTO, M., UOZUMI, J., and NISHINARI, K. 1986c. Preliminary investigation of the state of water in foods by near-infrared spectroscopy. Pages 1-9 in: Proc. Int. NIR Symp., Budapest.

KATO, A., OGURO, M., and TSUTSUI, Y. 1964. Near-infrared spectra of oleamide and saturated fatty acid amides (C_8-C_{18}). (In Japanese with English summary) Yukagaku 13:426-431.

KUSAMA, T., OCHIAI-YANAGI, S., and IWAMOTO, M. 1984. Comparison of rice (*Oryza sativa* L.) seed proteins from several varieties. Agric. Biol. Chem. 48:1649-1651.

NAGAO, A., UOZUMI, J., IWAMOTO, M., and YAMAZAKI, M. 1985. Determination of fat content in meats by near-infrared reflectance spectroscopy. Yukagaku 34:257-261.

NAKAMURA, S., WATANABE, I., and KOYAMA, T. 1978. Some inquiries on usage of grain quality analyzer (GQA). (In Japanese) Rep. Tohoku Bur. Crop Sci. Jpn. 21:67-68.

NIWA, E., NAKAYAMA, T., and HAMADA, I. 1980. State of water in fish flesh determined by near IR techniques. Nippon Suisan Gakkai-Shi 46:1147-1150.

SATO, T., IWAMOTO, M., HASHIZUME, K., YOSHINO, M., FURUKAWA, S., SOMEYA, Y., and YANO, N. 1985. Near infrared spectrophotometric analysis for measuring major constituents of raw milk. Nippon Chikusan Gakkai-Ho 56:878-882.

SUZUKI, T., IWAMOTO, M., CHO, R. K., UOZUMI, J., and IMAI, T. 1984. Applicability of near infrared reflectance method to moisture, protein and ash measurements of buckwheat flours. Nippon Shokuhin Kogyo Gakkai-Shi 31:200-202.

TAKEUCHI, T., TSUGE, S., and SUGIYAMA, Y. 1969. Near-infrared spectrophotometric analysis of ethylene-propylene copolymers. Anal. Chem. 41:184-186.

WAKAI, Y., INOUE, Y., NISHIKAWA, Y., MURATA, J., and MIURA, T. 1984. Determination of various properties of sake and rice using the near-infrared reflectance analyzer. (In Japanese) J. Brew. Soc. Jpn. 79:445-446.

WILLIAMS, P. C., THOMPSON, B. N., WETZEL, D., McLAY, G. W., and LOEWEN, D. 1981. Near-infrared instruments in flour mill quality control. Cereal Foods World 26:234-237.

ZHONG, S. 1982. Direct spectrophotometric determination of total nitrogen in soils. (In Chinese) Fenxi Huaxue 10:191-192.

CHAPTER 13

Near-Infrared Research in Australia and New Zealand

S. G. STEVENSON
Grain Research Laboratory
Canadian Grain Commission
Winnipeg, Manitoba, Canada

I. INTRODUCTION

Near-infrared reflectance (NIR) technology has been in use in Australia since 1974 and was used as a routine method for screening barley for protein by the Queensland Barley Board by the 1975 harvest season. As is the case in many countries, the need for a rapid, nondestructive, and accurate method for quality evaluation and control of agricultural products was responsible for the consideration of NIR instrumentation in Australia and New Zealand. In both countries, the technique was first examined for its suitability for use in products other than cereal grains. In Australia, NIR was initially considered for use in the wool trade rather than in the grain industry.

II. WOOL AND SYNTHETICS

Australia is a major wool-exporting country, and the introduction of sale by sample resulted in an increase in the volume of tests required to the point that the economic and time limitations of the traditional test methods became obvious. The standard test methods for determination of "wool base" of greasy (raw) wool require washing, drying, and weighing of replicate subsamples of a greasy-wool core sample. Gravimetric determinations are then made to determine residual grease, mineral, and vegetable matter. Tests include ethanol extraction of residual grease, ash content of washed wool, and digestion of wool with hot sodium hydroxide, resulting in a residue that is then dried. This dried residue is manually separated into five groups: three types of vegetable matter (VM), skin pieces, and other material, each of which is weighed and used to calculate the content of its group in the washed wool. The wool content (WC) of the washed wool is calculated by difference and these figures are then used to calculate the wool base of the greasy wool.

The standard method requires about one and one-half operator hours per sample and includes errors inherent in the measurement of ethanol extract, ash, and total alkali in solubles, since WC is calculated by difference.

Because wool is almost entirely fibrous protein and the VM contaminants are mainly polysaccharide material, diffuse reflectance spectroscopy at appropriate wavelengths in the near-infrared region should provide a useful tool for the analysis of wool. In preliminary studies it was found that NIR alone did not give very precise results because of the variability of the vegetable matter present. Connell and Brown (1978) calibrated a Technicon InfraAlyzer with 71 greasy-wool core samples for which wool base (WB) and vegetable matter base (VMB) had been determined by standard methods. The calibration was then tested in a validation experiment using synthetic mixtures of wool and vegetable matter made by combining clean wool and VM in various proportions by weight to provide samples whose composition was accurately known. Wools with a wide range of diameter, from 18 to 37 μm, were used to determine the effect of variations in mean fiber diameter. The samples were also tested with the Macbeth Corporation Color Eye, which makes measurements of diffuse reflectance in the visible spectrum at 19 wavelengths ranging from 380 to 760 nm at intervals of 20 nm.

The results of these studies indicated that the reflectance measurements at the 6 infrared and 19 visible wavelengths were highly correlated. Selection of a combination of more than seven wavelengths showed no improvement in terms of the reduction of residual variance of a linear calibration equation. If only two wavelengths were used for calibration, the near-infrared wavelengths 2.10 and 2.18 μm, which are respectively associated with polysaccharide and protein, explained 93.4% of the total variation. When more wavelengths were selected, the aforementioned wavelengths were included, but the rest of the wavelengths came from the visible region. The addition of the 460-nm wavelength in the visible spectrum reduced the residual variance markedly. Use of more than three wavelengths improved the precision but reduced the accuracy and therefore was not recommended. Mean fiber diameter was found to have no influence on prediction of wool content, although the nature of the size and shape of VM particles was felt to have an influence on the results. Connell and Brown (1978) concluded that measurements obtained using the InfraAlyzer and Color Eye were adequate for situations where rapid measurements of moderate precision were acceptable. In addition, they suggested that the wavelengths used in the InfraAlyzer, since they had been chosen for measurements of cereals, were perhaps not optimum for measurement of wool components.

Slack-Smith et al (1979) reported that they had calibrated both a Neotec GQA 31 and a Technicon InfraAlyzer with a series of 32 scoured wool cores varying in alcohol-extractable matter from 0.75% to 4.14% and in VM from 0.2% to 25.2%. These calibrated instruments were then used to test over 900 "unknown" wool samples under various conditions. They reported no bias between the NIR values and the Soxhlet-extractable-matter values and found between-test-specimen variances for NIR measurements similar to those for Soxhlet values on the same samples.

Connell and Norris (1980) reported the results of further studies in which NIR was used to predict the yield of greasy wool from the estimated wool content of washed wool. Using a Cary Model 14 monochromator in single-beam mode, they determined the reflectance spectra of washed wool and its natural contaminants. The wavelength range scanned was 1.0–2.6 μm and the spectrum of each sample was recorded as log $(1/R)$. The washed wool was ground in a Wiley-type mill with attached cyclone and 0.5-mm screen before being loaded into the sample cell. During sample preparation, VM was hand harvested from the wool samples and chopped before testing. Wool grease (petroleum spirit extract), residual grease (ethanol extract), and suint (the water-soluble salts) were also obtained. The spectra of oils, liquids, and pastes of the contaminants were obtained from thin films of these samples applied to a ceramic surface.

The instrument was calibrated with 126 core samples of greasy wool obtained from the Australian Wool Testing Authority, a commercial test house. The samples represented an even distribution of VM from 0 to 35%. These samples were also representative of the types of VM, ash range, ethanol extracts, and mean wool fiber diameter generally found in commercially tested Australian wool. Reflectance spectra were measured in the visible (0.5–1.0 μm) and near-infrared (1.0–2.6 μm) regions. Forward stepwise regression was used to select the optimum combination of wavelengths for prediction, and analyses were conducted on the log $(1/R)$ and first and second derivatives. An examination of the reflectance spectra of the wool and vegetable matter showed water absorption at 1.94 μm, protein at 2.06 μm and 2.18 μm, and possibly chlorophyll at 0.67 μm. Scans of the lipid contaminants showed sharp absorption peaks at 2.32 and 2.37 μm. The cellulose in subterranean clover, a contaminant often present in wool, caused absorption in the 2.10-μm region. Three wavelengths were sufficient for analysis. Variability in the ash content of washed wool resulted in changes in the 1.0- to 1.2-μm region, and variation in mean fiber diameter had a negligible effect on prediction of wool content.

Connell and Norris (1980) estimated the improvement possible in commercially available NIR instruments when wavelengths specific for wool and vegetable matter were used rather than using the wavelengths chosen for cereals and currently installed in most NIR instruments. This improvement, which was estimated by simulating the instruments with the spectrophotometer, suggested that optimizing the instruments for wool would improve calibration results. Comparison of results from NIR analysis and those obtained by standard test methods was obscured by the variability in subsampling and in the standard methods. The authors concluded that wool base, calculated from wool content predicted by NIR, was generally free from systematic error and approached the precision of the standard yield test and that the error of the spectroscopic method was low, especially for wools of low vegetable matter content (i.e., 0–5%), which in Australia represent about 80% of the wools tested commercially.

The same authors subsequently (1981) examined the feasibility of predicting the wool base of greasy wool from spectral measurements made directly on greasy wool. This study involved 47 samples of greasy wool that were obtained from a commercial Australian test house. The samples were held and tested under controlled atmospheric conditions of 20°C and 65% relative humidity to reduce variations resulting from moisture content of the wool. The best combinations of up to eight wavelengths were selected from the total of eight wavelengths used for routine analysis of agricultural products (1.68, 1.94, 2.10, 2.12, 2.18, 2.23, 2.28, and 2.31 μm). Wavelengths at 1.68, 2.12, 2.18, and 2.23 μm gave a prediction of wool base with a residual mean square error of about 3. Addition of more wavelengths resulted in no significant improvement. These four wavelengths include 2.18 μm, at which protein absorbs, and 2.12 μm, at which polysaccharide absorbs.

When the wavelengths were chosen on the basis of the statistical properties of the spectral data instead of those used for agricultural products, the results were not as good. A much lower residual mean square error was achieved by reducing the number of wavelengths considered for analysis to 60 by using a wavelength interval of 32 nm from 1.2 to 1.68 μm and an interval of 16 nm from 1.68 to 2.4 μm. By employing this method to determine a starting combination for further optimization by successive replacement using the forward stepwise addition method, the researchers selected the best wavelength for one term, keeping all other terms fixed. After this procedure was performed for each term, the best combination of n terms from this group was selected and used as the starting combination for the next cycle. The procedure was repeated until there was no change in wavelength for each term. The six wavelengths selected for prediction of wool base by this method were 1.70, 1.74, 1.77, 1.79, 2.05, and 2.22 μm. Calibrations using this combination of wavelengths produced a residual mean square error of 0.64 and an analytical error similar to that of the standard methods.

Marler (1980) reported studies to determine the efficacy of using NIR to measure the amount of wool in blends with nylon or terylene (polyethylene terephthalate). A Neotec GQA 31 was used to record the spectral characteristics for 50-mm disks of dyed wool/terylene blend fabrics ranging from 40% to 90% wool. The wool content of each sample was determined with standard wet-chemical procedures, and then these data were correlated with the spectral data using multiple linear regression. No significant difference was found for calibration purposes between samples that had been Hercosett treated and untreated samples.

The protein channel of the GQA 31 is set at 2,180 nm. The absorption peak in this wavelength region is caused by the presence of amide groups and as such is related to C-H stretching and N-H in-plane bending. Terylene has no amide bonds, so a second trial was performed using nylon as the synthetic fiber. In this case the samples used were undyed wool/terylene and wool/nylon blend fabrics, and their spectral responses were measured on both a GQA 31 and a GQA 31EL. The results showed that the protein channel was sensitive to the differences in synthetic used, suggesting that a separate calibration would be required for each different synthetic blended with wool if the NIR method were to be used for routine analysis. If the responses of all three channels were used in a multiple regression equation, it was possible to derive a common prediction equation for wool/terylene and wool/nylon blends. The GQA 31EL produced a significantly better fit for the data than the GQA 31. This was attributed to the particular wavelengths used in the instrument.

The third experiment consisted of testing 11 blends of wool and terylene tops ranging in wool content from 0 to 100% in increments of approximately 10% on both the GQA 31 and GQA 31EL. The results for all possible one-, two-, and three-variable combinations of the three spectral responses were correlated against standard test data. In both instruments, the overall response of the protein channel was curvilinear, although it was linear over the 40–90% range. The best single variable for the GQA 31 was the protein channel (C2) and for the GQA 31EL was the oil channel (C1). Addition of a second variable, C2 for the GQA 31EL and C3 for the GQA 31, resulted in a significant improvement, although the addition of a third variable did not. Statistical analysis of calibration performance showed the quadratic fit of the protein channel data to be best for the GQA 31 and the multiple linear regression fit best for the GQA 31EL.

III. THE DAIRY INDUSTRY

The applicability of NIR to quality measurements in dairy products has been examined in New Zealand. The use of NIR was first permitted in the late 1970s by the Dairy Division of the Ministry of Agriculture and Fisheries for monitoring moisture, fat, and protein levels in milk powder and casein products (Woollard, 1985). Stannard and Scotter (1975) reported studies on whole and skim milk powders using a Neotec GQA. These authors calibrated the GQA for protein in skim milk powder and for protein, fat, and moisture in whole milk powder using 25–27 samples obtained from several factories and representing as wide a range as possible in composition. Following calibration, unknown samples from a number of factories were tested and the NIR results were compared with laboratory analyses by standard methods. Correlations between NIR and chemical data ranged from 0.992 for protein in whole milk powder to 0.998 for protein in skim milk powder and moisture in whole milk powder. Precision of measurements determined by sampling and reading a whole-milk powder sample 30 times by three different operators was 0.082 for fat, 0.094 for protein, and 0.057 for moisture.

Roberts (1977) reported that the Neotec GQA was capable of giving results as accurate as standard wet-chemistry methods and with the precision required by the dairy industry for fat, protein, moisture, lactose, and ash when used for analysis of milk powders and casein. Seasonal fluctuations in milk and the resulting powder do not appear to affect a calibration, but changes in processing specifications for powders call for recalibration or necessitate an intercept adjustment. The GQA 31 was reported to be sensitive to different types of dryers, with separate calibrations required for each kind (nozzle and atomizer). The GQA 41R could accommodate influences caused by dryer differences. The age and temperature of the samples used in calibration were also reported to have an effect on the results. If the instrument was used on-line or for control of the drying process, then fresh samples were required for calibration. In addition, since the powder was warm when it came out of the dryer, it was concluded that warm powders should be used for calibration purposes. Roberts (1977) recommended the GQA 31 as being suitable for the average dairy company and the GQA 41 for more sophisticated products and the measurement of parameters such as sucrose and ash.

NIR has largely replaced the traditional wet-chemistry procedures for quality control in factories producing milk powder and casein. Woollard (1985) reported a decreased reaction time to processing problems and an increased use of NIR for classification of finished products into their appropriate quality codes. His paper covers the history of introduction of NIR into the dairy industry, the rationalization of the present situation where these instruments have now become an integral part of factory hardware, the role of the Ministry of Agriculture and Fisheries in supervising the instrumentation program, pitfalls of NIR use, and some new applications that will likely affect the dairy industry. At the time of writing there were 42 NIR instruments in use in the New Zealand dairy industry, representing the three major NIR instrument manufacturers (DICKEY-john, Pacific Scientific, and Technicon). NIR instruments are currently authorized for official use on whole milk powder, buttermilk powder, skim milk powder, casein, caseinates, and whey protein concentrates. These products can be analyzed for fat, moisture, protein, lactose, acidity, and ash using NIR (Woollard, 1985).

Toohey (1985) described the use of NIR over a four-year period to analyze whole milk powder for fat and moisture as a routine control in a spray-drying operation at Toora, Victoria (Australia). Calibration of the equipment is updated on an ongoing basis by adding the wet chemistry results from all outlier samples to the calibration. Check testing of samples is carried out on a regular basis to assess the reliability of the NIR results as compared with standard methods. Analysis of the differences between these results is used to fine-tune the instrument.

IV. FORAGE APPLICATIONS

Another area in which NIR has been applied to agricultural products is the analysis of forages. Minson et al (1983) analyzed forages for crude protein (CP), dry matter digestibility (DMD), and voluntary intake (VI) using 19 wavelengths with a Technicon InfraAlyzer Model 400. They used 97 samples including six tropical grass species, separated leaf and stem fractions, pelleted and chopped forms, and regrowths varying from 28 to 150 days. The samples were ground in a 20-cm (8-in.) Christy and Norris lab mill fitted with a 1-mm screen. No significant differences in CP were attributable to differences in physical form (pelleted or chopped), but strong differences were caused by differences in species and plant parts (leaf or stem). There was also a significant interaction between the latter two variables. Addition of species and plant part information to the regression equation resulted in a decrease in the residual standard deviation. The time of regrowth (late summer vs. midsummer) also affected the prediction equation. The accuracy of DMD values was not affected by differences relating to plant parts, but species, form, and age of material introduced serious biases into NIR predictions of DMD if the method was used to analyze forage types that had not been included in the original calibration. The negative relationship between accuracy of analysis and maturity of the samples indicated that the reflectances did not fully distinguish the higher level of organization of the cell walls in the mature forage. NIR determination of voluntary intake was affected by species, form, and plant part as well as their two-factor interactions. There were also significant differences between plant age within each study.

Although the differences in reflectances could not completely account for differences in VI associated with pelleting, the NIR technique responded well to differences in VI associated with leaf and stem fractions. Although all 19 wavelengths were used in this study, the authors point out that increasing the number used would not automatically reduce the type of bias reported in the study because much of the residual error in the regression relating nutritional value to reflectance was the result of differences in forage species, plant parts, and physical form. By taking these differences into account, they could still predict nutritional value with smaller errors than had been quoted by other authors working with spectrocomputers that select wavelengths specific for each type of forage. Although the bias patterns may be complicated, Minson et al (1983) recommended that they should be taken into account when doing NIR forage analysis.

Minson et al (1985) subsequently compared three different methods for improving the precision of NIR for estimating CP, DMD, and VI in forages. The three methods, which were compared with coarse grinding using a 1-mm screen, were fine grinding in a Spex mill and adjusting the reflectance data by centering it around the mean reflectance spectrum using two procedures. As in the previous study (Minson et al, 1983), the reflectances were measured at 19 wavelengths and the regressions obtained contained reflectances measured at different numbers of wavelengths. The samples included six tropical grass species, separated leaf and stem fractions, and a range of maturity. The authors found that fine grinding improved the CP regressions but not those for DMD or VI. The best procedure for improving the regressions used to estimate VI appeared to be that of centering the reflectance data. Minson et al (1985) concluded that in NIR studies dealing with tropical forages, fine grinding of samples should not be advocated as it simply required extra work that did not contribute to any improvement in the accuracy of predicting nutritive value. In the same way, centering and similar mathematical manipulations of spectral data showed no advantage over use of crude reflectance data. The authors believed that such manipulations were unnecessary in forage analysis. Use of all 19 wavelengths with no mathematical manipulation provided the best regressions for coarse-ground tropical forages.

In a joint study involving the Pastoral Research Institute in Hamilton, Victoria, and the School of Agriculture in Aberdeen, Scotland (Flinn and Murray, 1985), 195 samples of pasture hay harvested in Hamilton from 1979 to 1981 were analyzed for crude protein, neutral detergent fiber (NDF), and in vitro dry matter digestibility (IVDMD). Analysis by conventional methods was conducted in Hamilton and the NIR scanning with a Neotec 6100 monochromator was conducted in Aberdeen. Half of the samples (98) were used for calibration of the instrument and the other half for analysis. Between-year differences were also investigated. The equations chosen for prediction of CP, NDF, and IVDMD contained two, three, and four wavelength terms respectively and the optimum wavelengths chosen for each constituent were consistent with those selected on other sets of similar forages grown in the United Kingdom and the United States. The SEPs for NIR determination of CP, NDF, and IVDMD were ±1%, ±2%, and ±2.5% respectively despite the wide range of seasonal conditions, composition, and haymaking practices used.

V. 1984 NIR INTERNATIONAL SYMPOSIUM

In October of 1984, the Royal Australian Chemical Institute (RACI), Cereal Chemistry Division, sponsored an international symposium on near-infrared reflectance spectroscopy. Of the 23 papers presented as part of this symposium, 12 were concerned with NIR research currently being undertaken in Australia and New Zealand. In addition, eight poster papers were presented, all of which dealt with NIR research in these two countries. The proceedings of this symposium have added considerably to the literature on Australian and New Zealand NIR research.

Marler (1985a) reported the significance of NIR instrument, samples, standard test procedures, and computing software and statistics in the calibration of an NIR spectrocomputer. The instrument used was a Neotec 6100 monochromator coupled to a DEC LSI 11/23 computer. Wavelength selection, stability, and analytical performances of calibrations based on various derivative transformations were examined using the protein content of washed wool as the sample constituent. Some of the limitations of multiple linear regression as applied to NIR instrument calibration were also considered.

The application of NIR to measurements on washed wool was further examined by Marler (1985b) in light of the savings in labor costs and testing time. Two sets of wool samples were used in this study. One set consisted of pure wool samples and vegetable matter combined gravimetrically (artificial samples) whereas the second set represented samples of naturally occurring wool. The Neotec 6100 scanning monochromator was calibrated to determine wool content, moisture content, and vegetable matter type and content. For the artificial samples, the first derivative of the log $(1/R)$ signal gave consistently lower SEPs for wool content than did log $(1/R)$ and second-derivative transformations. The largest SEPs and biases occurred with attempts at estimating the type and amount of vegetable matter present in the samples. This was attributed to the lack of information in the NIR spectra needed to permit discrimination between the plant species. The best equation for the natural samples was found to be a second-derivative calibration of three or five terms using samples covering as wide a range of characteristics as possible. Again, comparison between NIR and standard methods showed large variations, especially for vegetable matter by species estimation. The problems involved in determination of wool content were reported to be related to the protein content of the contaminants, particularly the vegetable matter, rather than to the analytical ability of the NIR equipment.

The use of NIR for analysis of wool in New Zealand was reported by Hammersley et al (1985). The increasing emphasis on certification of nonwool components, especially moisture and solvent-extractable material, in shipments has necessitated consideration of rapid test methods such as NIR. Determination of moisture content by this method was satisfactory (on prediction tests SEP = 0.21, $n = 200$), with NIR results being more precise and reliable than those obtained by manual methods. Measurement of solvent extractables, however, necessitated refinement of the chemical method (Soxhlet) and careful sample selection. When this was done, the SEP on 193 prediction samples was 0.049. New sampling methods and handling procedures were also necessary. Because of the variability of the product, throughput was reduced since multiple readings were necessary. NIR analysis of clean wool has now advanced to the stage of in-plant trials, and future work will be concerned with estimating the content of wool and other constituents in greasy wool.

The use of NIR in cereal grains analysis was also discussed in a number of presentations. Tattersall (1985) reported on the development of wheat protein calibrations for two InfraAlyzer Model 300 instruments, one of which was fitted with the six standard wavelength filters and the other with additional filters. The calibration statistics for the two instruments were similar, with correlations of 0.996 and SEPs of 0.15. The calibration data were transferred to 10 other InfraAlyzer 300 instruments and initial bias adjustments were made. The Australian Wheat Board then installed seven of these instruments in country seaboard terminals in South Australia and placed the remaining five in their main Adelaide laboratory. These instruments tested up to 200 wheat samples daily for protein during the harvest season. One in 50 of each of these samples was also tested by the Kjeldahl method and the data accumulated to provide an overall evaluation of calibration performance by instrument. Results indicated that the protein calibrations were transferable between instruments providing initial bias adjustments were made.

Further information concerning the use of NIR for protein segregation of Australian hard wheat (AHW) was provided by Small and Harry (1985). The introduction of protein testing for segregation of AHW in South Australia is a recent innovation. Before now, segregation for this wheat class was made on the basis of visual assessment of grain mottling with subsequent protein testing on postharvest samples. The introduction of NIR protein analysis in the 1983–84 harvest season was found to be cost-effective, improved the uniformity and marketability of the AHW class as well as Australian standard white class, and resulted in equitable premium payments to growers on the basis of individual deliveries. Miskelly and Murray (1985) summarized the use of NIR in wheat protein testing in Australia by indicating that its availability has made possible a detailed and objective assessment of the Australian wheat crop and has enabled the Australian Wheat Board to tailor supplies of wheats of all classes to individual customer requirements.

Panozzo and O'Brien (1985) reported studies using an InfraAlyzer 400 to measure wheat protein, hardness (particle-size index), and flour-milling yield for early-generation selection of wheat quality. They obtained correlations of 0.996 for protein, 0.972 for particle-size index, and 0.841 for flour-milling yield when NIR values were compared with those obtained by standard methods. SEPs for these parameters were acceptable for a wheat-breeding program. In addition to determining protein in wheat, NIR has been used to determine α-amylase in wheat samples of feed and milling quality. Stopford et al (1985) indicated that preliminary regression analysis between scanning NIR and the standard test methods of Phadebas method, Brabender amylograph, and Falling Number test produced high enough multiple correlation coefficients to justify further examination of this area.

Hillier and Clucas (1985a) used scanning NIR to determine the extent of wheat starch contamination in tapioca starch produced by the batch process. In this study, levels of up to 5% wheat starch were added to tapioca samples having different pH and moisture levels. The multiple correlations for the initial regression analysis were high ($r = 0.995$, SEE = 0.19, $n = 11$), and addition of a wider variety of samples improved the overall calibration statistics (SEE = 0.87) but reduced the multiple correlation coefficients (r = 0.82, n = 78). The authors concluded that, in this study, NIR analysis using these calibrations was not sufficiently accurate to determine batch cutoffs between batches of starch mixes when compared with the current laboratory method.

The ability to transfer calibrations from one instrument to another is an important consideration. Hillier and Clucas (1985b) reported studies conducted in their facility using InfraAlyzer models 300 and 500. Samples of dry gluten were used to calibrate the Model 300 for protein and moisture. The same set of samples was used to generate a calibration for the Model 500 that was then transferred to the 300. Multiple-wavelength calibration equations were developed on both instruments. These calibrations were statistically similar, and the long-term accuracy and precision are currently being assessed.

Protein content is an important factor in segregation of malting grades of barley (Saint, 1985). In 1975, the Australian Barley Board began to use NIR (Neotec GQA 31) for testing the protein content of malting varieties. Further improvement to this program was effected, starting in 1982, by installing NIR equipment at major delivery centers. As of 1984, 11 instruments (InfraAlyzer 300) were in use in South Australia and 3 (2 InfraAlyzer 300 instruments and 1 Neotec 102) in Victoria. These provide benefits to the Australian Barley Board similar to those noted by the Wheat Board, namely improved quality as a result of better segregation of deliveries, better marketing strategies, and better relations with growers and buyers.

NIR has been used in the screening and selection of barley for malting quality in breeding programs. Law (1985) reviewed the use of NIR for estimation of barley quality parameters (protein, hot water extract, and β-glucan), with respect to malting, in breeding programs. Stable barley calibrations with widespread application could be established provided suitable techniques were used to overcome the multicollinear nature of the data. The same caveat applied to the prediction of hot water extract. Law also reported that total β-glucan concentration of barley could be determined by NIR analysis. This was based on studies with an InfraAlyzer 400 using 64 samples for calibration and 47 other samples for analysis. The main difficulty in studying the use of NIR for β-glucan measurement was that of obtaining sufficient samples of known β-glucan content to establish the calibration. To this end, the Cereal Division of RACI has recently started a worldwide collaborative survey of β-glucan methods. Preliminary studies on soluble β-glucans suggest that NIR may be able to differentiate them from insoluble β-glucans. Law (1985) concluded that NIR could be a useful technique for the prediction of malting quality in barley-breeding programs as long as those using this method possessed a full understanding of the statistics involved, its deficiencies and capabilities.

Using a combinations search as well as a step-up search method, Henry (1985a, 1985b, 1985c) selected wavelengths to calibrate a scanning NIR spectrophotometer for barley protein, moisture, aleurone color, total β-glucan, and predicted malt extract. (The term *malt extract* was preferred to *hot water extract* by this author.) Fifty samples covering the range of these constituents normally encountered in barley-breeding selection were used. All five characteristics could be assessed in approximately 25 sec and were sufficiently precise for routine screening of breeding material. In studying aleurone color, the author used the wavelength range from 550 nm to 1,950 nm to account for the fact that this characteristic is of a visible nature. The best calibration obtained contained two wavelengths (690 and 802 nm) below the normal range and had a correlation coefficient of 0.945 (Henry, 1985b). Endosperm modification was also predicted most successfully by NIR when the malt was produced without additives. Henry (1985b) indicated that the success of malt extract predictions from barley is determined by the malting procedure the NIR method is required to predict. Calibrations were also determined for prediction of malt extract and endosperm modification from barley and for hot water extract and endosperm modification from malt (Henry, 1985a, 1985b). Malt hot-water extract values resulting from malting without addition of gibberellic acid and potassium bromate gave the best prediction of this constituent. NIR is now being used for the measurement of barley moisture, protein, β-glucan, malt extract, and aleurone color in the preliminary evaluation of all barley samples from the Queensland barley-breeding program as well as for malt moisture determinations.

NIR spectra of carbohydrates can be used to develop calibrations for the measurement of specific carbohydrate fractions in mixtures. Henry (1984) used mixtures of barley β-glucan and starch to select wavelengths useful for the determination of (1→3) and (1→4) β-glucans in barley. The wavelengths selected were then used to calibrate an NIR reflectance spectrophotometer to predict the β-glucan content of barley.

NIR has also found a niche in the confectionery industry. Products including chocolate crumb, milk chocolate, and cocoa powder have been tested for fat, sucrose, and moisture using NIR. The compositions of milk chocolate and chocolate crumb are summarized in Table I. Chocolate crumb is a precursor of the familiar block chocolate and the chocolate used in coating. It is prepared from milk, cocoa liquor, and sugar. Milk chocolate is a high-energy food, with a significant amount of high-quality (milk-derived) protein. Gardam (1985) described the use of the Neotec GQA 41, FQA 51 (with computer terminal), and GQA 102 to generate calibrations for various quality attributes against the standard test methods of Soxhlet and Mojonnier methods for fat, polarimetry for sucrose, and oven-drying for moisture. NIR analysis was cited as reducing laboratory worker hours and increasing the level of process control. Techniques for sample preparation and presentation included the development of a special two-piece sample cup by Pacific Scientific/Neotec. The glass cell cover was cemented in place and the cap carried a stainless steel plate that produced a layer of chocolate. The chocolate was presented to the instrument at 45°C in the semimolten state. The high fat content (30%) of chocolate caused saturation at 1.76 and 2.36 μm, and the instrument gain had to be adjusted. Instruments used included the GQA 41 and 102 and the FQA 51. Wavelength selection, instrument modification, selection of sample sets for calibration, and evaluation of calibrations were also discussed.

TABLE I
Composition (%) of Some Chocolate Products

Product	Fat	Sucrose	Lactose	Protein	Moisture
Chocolate crumb	13	60	···	···	1
Milk chocolate	30	46	9	9	1
Cocoa powder	16	···	···	···	3.5

LITERATURE CITED

CONNELL, J. P., and BROWN, G. H. 1978. The yield-testing of wool by reflectance spectroscopy. J. Text. Inst. 69:357-363.

CONNELL, J. P., and NORRIS, K. H. 1980. The prediction of the yield of greasy wool by near-infrared reflectance spectroscopy. I: Principles and washed wool. Text. Res. Inst. J. 50:371-380.

CONNELL, J. P., and NORRIS, K. H. 1981. The prediction of the yield of greasy wool by near-infrared reflectance spectroscopy. II: Greasy wool. Text. Res. Inst. J. 51:339-344.

FLINN, P. C., and MURRAY, I. 1985. Evaluation of hay in Western Victoria using NIR. Pages 180-182 in: NIR 84 Proc. Int. Symp. Near Infrared Reflectance Spectrosc., 1st. RACI, Cereal Chem. Div., Melbourne, Australia.

GARDAM, C. M. 1985. Applications of Neotec NIRA instruments to chocolate processing. Pages 119-125 in: NIR 84 Proc. Int. Symp.

Near Infrared Reflectance Spectrosc., Ist. RACI, Cereal Chem. Div., Melbourne, Australia.

HAMMERSLEY, M. J., RANFORD, S. L., and PATRICK, V. C. 1985. Wool testing in New Zealand with the InfraAlyzer 400R. Pages 78-83 in: NIR 84 Proc. Int. Symp. Near Infrared Reflectance Spectrosc., Ist. RACI, Cereal Chem. Div., Melbourne, Australia.

HENRY, R. J. 1984. Analysis of carbohydrates by near infrared reflectance techniques. (Abstr.) Chem. Aust. 51:240.

HENRY, R. J. 1985a. Calibration and use of a scanning near infrared reflectance spectrophotometer for assessment of malting quality in barley. Pages 93-94 in: NIR 84 Proc. Int. Symp. Near Infrared Reflectance Spectrosc., Ist. RACI, Cereal Chem. Div., Melbourne, Australia.

HENRY, R. J. 1985b. Evaluation of barley and malt quality using near-infrared reflectance techniques. J. Inst. Brew. London 91:393-396.

HENRY, R. J. 1985c. Use of a scanning near-infrared reflectance spectrophotometer for assessment of the malting potential of barley. J. Sci. Food Agric. 36:249-254.

HILLIER, G., and CLUCAS, A. 1985a. Scanning NIR-contamination of tapioca starch by wheat starch. Pages 175-176 in: NIR 84 Proc. Int. Symp. Near Infrared Reflectance Spectrosc., Ist. RACI, Cereal Chem. Div., Melbourne, Australia.

HILLIER, G., and CLUCAS, A. 1985b. I/A 500 to I/A 300—transferability of calibrations. Pages 177-179 in: NIR 84 Proc. Int. Symp. Near Infrared Reflectance Spectrosc., Ist. RACI, Cereal Chem. Div., Melbourne, Australia.

LAW, D. P. 1985. Selection for malting quality in barley breeding using near infrared reflectance. Pages 84-92 in: NIR 84 Proc. Int. Symp. Near Infrared Reflectance Spectrosc., Ist. RACI, Cereal Chem. Div., Melbourne, Australia.

MARLER, J. W. 1980. Quantitative analysis of wool/synthetic blends by near infrared reflectance spectroscopy. Pacific Scientific Tech. Pap. NIR 5000.

MARLER, J. W. 1985a. Calibrating a spectrocomputer. Pages 31-41 in: NIR 84 Proc. Int. Symp. Near Infrared Reflectance Spectrosc., Ist. RACI, Cereal Chem. Div., Melbourne, Australia.

MARLER, J. W. 1985b. The application of NIR to measurements on washed wool. Pages 65-77 in: NIR 84 Proc. Int. Symp. Near Infrared Reflectance Spectrosc., Ist. RACI, Cereal Chem. Div., Melbourne, Australia.

MINSON, D. J., BUTLER, K. L., GRUMMITT, N., and LAW, D. P. 1983. Bias when predicting crude protein, dry matter digestibility and voluntary intake of tropical grasses by near-infrared reflectance. Anim. Feed Sci. Technol. 9:221-237.

MINSON, D. J., TUCKETT, P. G., and LAW, D. P. 1985. A comparison of three methods of increasing the precision of regressions used for estimating nutritive value of forages by near infrared reflectance. Pages 138-148 in: NIR 84 Proc. Int. Symp. Near Infrared Reflectance Spectrosc., Ist. RACI, Cereal Chem. Div., Melbourne, Australia.

MISKELLY, D. M., and MURRAY, I. C. 1985. The use of NIR in wheat protein testing in Australia. Pages 169-170 in: NIR 84 Proc. Int. Symp. Near Infrared Reflectance Spectrosc., Ist. RACI, Cereal Chem. Div., Melbourne, Australia.

PANOZZO, J., and O'BRIEN, L. 1985. Application of near infrared reflectance spectroscopy to early generation selection for wheat quality. Pages 183-186 in: NIR 84 Proc. Int. Symp. Near Infrared Reflectance Spectrosc., Ist. RACI, Cereal Chem. Div., Melbourne, Australia.

ROBERTS, G. M. 1977. Infra-red analysis of milk powders and casein products. Pacific Scientific Tech. Pap. NIR 4077.

SAINT, K. J. 1985. Use of NIR equipment by the Australian Barley Board. Pages 171-172 in: NIR 84 Proc. Int. Symp. Near Infrared Reflectance Spectrosc., Ist. RACI, Cereal Chem. Div., Melbourne, Australia.

SLACK-SMITH, T., FONG, D., and DOUGLAS, S. A. S. 1979. The potential application of near-infrared reflectance to estimate the alcohol-extractable-matter content of scoured wool. J. Text. Inst. 70:33.

SMALL, D. M., and HARRY, S. P. 1985. Introduction of protein testing by NIR for the segregation of Australian hard wheat in South Australia. Page 126 in: NIR 84 Proc. Int. Symp. Near Infrared Reflectance Spectrosc., Ist. RACI, Cereal Chem. Div., Melbourne, Australia.

STANNARD, D. J., and SCOTTER, A. 1975. Evaluation study of the GQA Neotec IR analyzer. Appended to: Infra-red analysis of milk powders and casein products, by G. M. Roberts. Pacific Scientific Tech. Pap. NIR 4077.

STOPFORD, S., HILLIER, G., and CLUCAS, A. 1985. Scanning NIR-prediction of α-amylase in wheat samples. Pages 173-174 in: NIR 84 Proc. Int. Symp. Near Infrared Reflectance Spectrosc., Ist. RACI, Cereal Chem. Div., Melbourne, Australia.

TATTERSALL, A. K. 1985. Transferring near infrared reflectance calibrations: Evaluation of protein in wheat. Pages 28-29 in: NIR 84 Proc. Int. Symp. Near Infrared Reflectance Spectrosc., Ist. RACI, Cereal Chem. Div., Melbourne, Australia.

TOOHEY, F. S. 1985. Testing of dried milk products by NIR. Pages 114-118 in: NIR 84 Proc. Int. Symp. Near Infrared Reflectance Spectrosc., Ist. RACI, Cereal Chem. Div., Melbourne, Australia.

WOOLLARD, D. C. 1985. NIR technology within the New Zealand dairy industry—its dramatic impact. Pages 95-104 in: NIR 84 Proc. Int. Symp. Near Infrared Reflectance Spectrosc., Ist. RACI, Cereal Chem. Div., Melbourne, Australia.

CHAPTER 14

Analysis of Whole Grains by Near-Infrared Reflectance

RUSSELL TKACHUK
Grain Research Laboratory
Canadian Grain Commission
Winnipeg, Manitoba, Canada

I. INTRODUCTION

Reflectance properties of grain in whole seed form are of much interest to cereal chemists because they offer the possibility of providing the most convenient testing procedure. Such a procedure has ultimate benefits in terms of convenience, speed, and cost. This chapter deals with some studies using reflectance in the visible and near-infrared wavelength regions for analyzing several constituents in whole kernel samples of wheat, field peas, and rapeseed. For comparative purposes, some results are also presented for ground samples.

II. DEFINITIONS

Reflectance = R.

Percentage reflectance = % R.

$R = I/I_0$, where I = intensity of radiation reflected from medium.

I_0 = intensity of incident radiation.

Absolute reflectance, the quantitative estimate of the reflected energy from a solid sample, is difficult to measure. More commonly, relative reflectance, R, is measured, i.e., relative to the reflectance of a standard substance:

$$R' = \frac{R \text{ sample}}{R \text{ standard}}$$

or apparent absorbance, A', where $A' = \log (1/R')$. If the absolute reflectance of the standard is equal to unity, then the absolute and relative values are identical, but there are no known standard materials that are perfect reflectors. MgO and $BaSO_4$ are satisfactory standards for the visible region. Compressed sulfur or polyperfluoroethylene powders are satisfactory for the near-infrared region, and many other custom-built or commercially available near-infrared reflectance (NIR) systems use a ceramic or sand-blasted gold standard.

III. INSTRUMENTATION

A Cary 17DI spectrophotometer was used in all the studies. This double monochromator, which is designed for optimum performance in the near-infrared region, is equipped with a 30° prism, 750-nm blazed echelette-grating dispersing elements, and a curved slit optimal for the near-infrared region. The transmission optics were fabricated from Suprasil-W fused silica to minimize absorption due to -OH groups that are found in ordinary silica. To prevent spurious radiation from reaching the detector, the instrument was used in the double-beam mode. This mode of operation has an electronic null balance system where the sample and reference signals are separated in time periods by a rotating mirror splitter and where each signal is further separated by equivalent dark periods by a blocking chopper blade. All periods are of 5-millisecond duration.

For diffuse reflectance, a type I integrating sphere allows monochromator sample and reference illumination (Fig. 1). For increased response in the longer wavelength region, a gold-coated sphere was used. The gold coating was achieved by evaporating gold from a tungsten filament at 5×10^{-5} torr onto a surface roughened by emery paper to produce a matte finish. For improved response in the middle and shorter wavelength regions, a $BaSO_4$-coated sphere was used. It was made by spraying a $BaSO_4/H_2O$ slurry into the sphere. Use of the $BaSO_4$ sphere resulted in a twofold increase in signal strength over much of the wavelength region (Fig. 2). To maximize signal/noise ratios, PbS detectors can be used at lower temperatures; for example, lowering the temperature from 25° to 2° C causes about a twofold increase in signal output (Fig. 2).

While obtaining reflectance data, it is necessary that the temperature of the PbS be controlled to within ±0.01° C. Such control is needed because of the sensitivity that the detector has to temperature (Fig. 2), particularly when single-beam recordings are carried out, since in this instance sample and reference recordings are obtained at different times. Temperature control is also needed with double-beam spectrophotometers. The latter need was clearly indicated when reference baselines were being recorded at the time when the PbS detector was directly connected to a circulating bath, an arrangement that did not provide much temperature buffering between detector and bath. This arrangement resulted in a baseline that had superimposed upon it a regular sine wave variation of ±0.004 A, the variation being directly correlated to the bath normal temperature cycle of ±0.08° C—i.e., about 0.5 A per 1° C.

The temperature of the samples being analyzed should also be controlled at some convenient level, otherwise spectral differences are obtained (Fig. 3). The effect of sample temperature on analytical results is difficult to predict because it is a function not only of the sample temperature but also of the measuring wavelengths and the algorithm being used. For example, it is easily seen from Fig. 3 that temperature changes will show a greater effect when a wavelength of 1,420 nm is used rather than 1,260 nm. The type of algorithm and data transformation used each has an effect since the incorporation of a superior normalizing wavelength will obviously result in lower analytical deviations; for example, a quotient consisting of a suitably spaced wavelength will have a better normalizing effect than a closely spaced first-derivative variable.

To reduce noise further, a number of other modifications were also carried out to the spectrophotometer. Quartz-halogen coiled tungsten lamps rated at 500 W and 115 V were selected for minimum noise characteristics; all mechanical switches in the PbS amplifier were replaced with soldered connections; the 10-Mohm load resistor in the PbS detector amplifier was selected for its low noise and minimum temperature-variability characteristics; and the instrument was purged with dry nitrogen to avoid moisture interferences and to keep the optics clean. Under these conditions, and if the instrument was kept on at all times, stray light at 2,300 nm was 0.008% *R* and 1-sec readings were precise to ±0.0008 *A* (at absorptions of 0.5 *A*, 1-nm bandpass, 1-sec 0–100% *R* response times, signal/noise ratios of 1,000, lamp voltage of 90 V, and amplifer gain of 1.5). Further details about achieving high-accuracy reflectance spectrophotometry with Cary 14 spectrophotometers are given in an excellent paper by Clarke (1972).

Using only a PbS detector, workers could operate the spectrophotometer over a 500- to 2,600-nm wavelength range. Measurements at lower wavelengths required the use of a photomultiplier detector.

IV. COMPUTER PROGRAMS

Several computer programs were written in Macro assembly or FORTRAN languages to control the spectrophotometer and to record and manipulate data (Table I). In some cases, the programs were adapted from published algorithms (Savitzky and Golay, 1964) or from programs written for handling infrared data (Jones et al, 1976). There are a number of advantages to user-assembled apparatus and user-written computer programs. Firstly, most experimental procedures, conditions, and variables are under user control and one is not dependent on the limitations of an equipment manufacturer's program. Secondly, it is simple to alter a user-written program because the source code is available. Thirdly, having written the program, the user is clear as to instrument operation and how the data are being handled. On the other hand, there is a cost to the user approach described above, as it takes much time and effort to assemble the equipment and programming expertise to write, debug, and test programs. Also, the software packages provided by manufacturers of computerized spectrophotometers, have been compiled on the basis of the experiences and input of many users with a wide range of needs, and they may be correspondingly comprehensive. Table I, which illustrates the type of computer programs used, is comparable with Table V of Chapter 5.

V. DATA ACQUISITION

The spectrophotometer is controlled by a Digital Equipment Corporation (DEC) PDP 11/84 minicomputer. The mini-

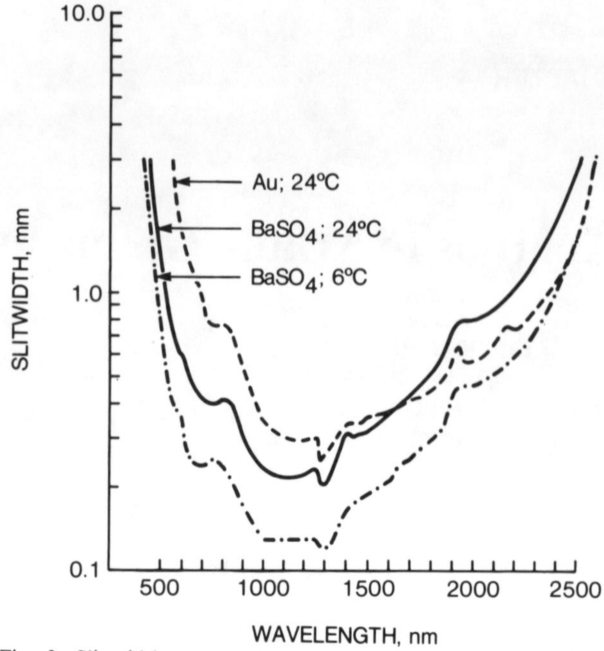

Fig. 2. Slit-width response curves for various integrating-sphere coatings at PbS detector temperatures of 6° and 24° C.

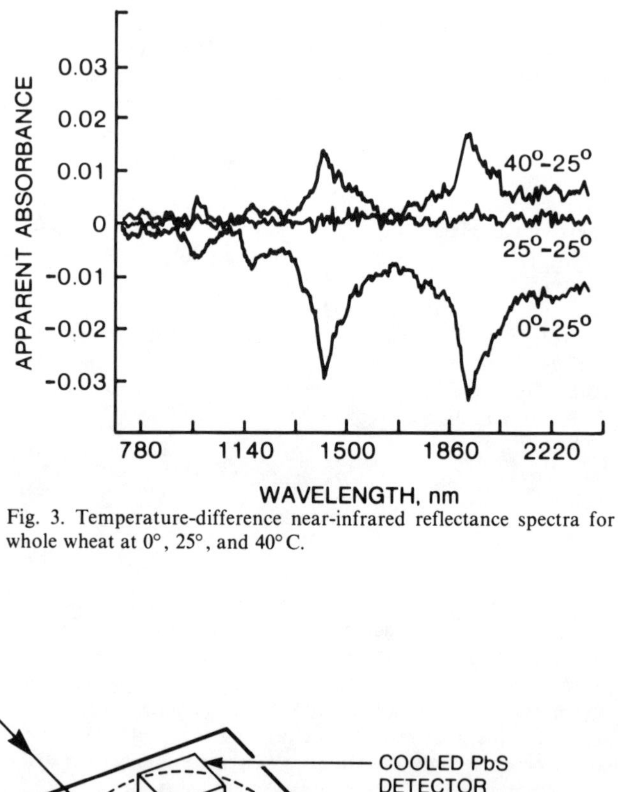

Fig. 3. Temperature-difference near-infrared reflectance spectra for whole wheat at 0°, 25°, and 40° C.

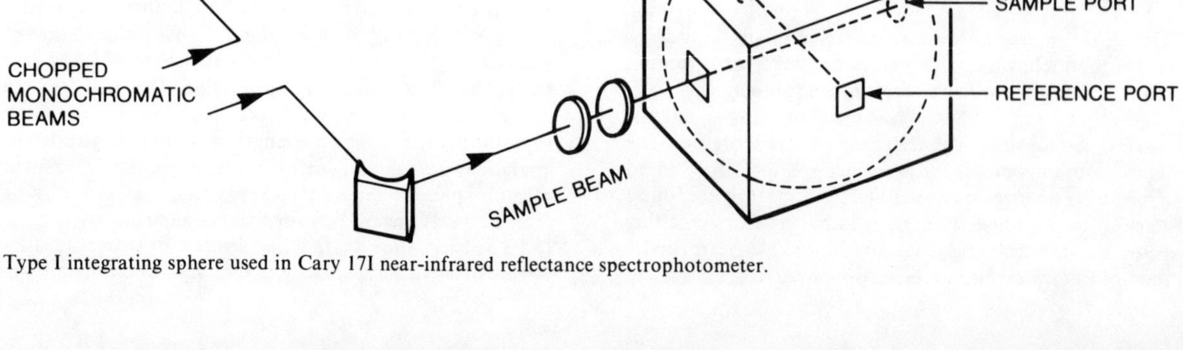

Fig. 1. Type I integrating sphere used in Cary 17I near-infrared reflectance spectrophotometer.

computer is equipped with 1 megabyte of core memory, two 10.4-megabyte and 205-megabyte disks, and connected to DEC hard-copy and high-resolution Tektronix 4010 and 4027 CRT terminals and a nine-track TE 16 Magtape. The RSX-11M+ operating system was used with a FORTRAN 77 compiler and a Macro assembler. A DEC DR11L/M interface, which has the capability of two parallel 16-bit words each for I/O, was used to connect the spectrophotometer to the computer.

Diffuse reflectance measurements were made by placing the analytical substrate in a holder at the sample port and the reference material in the reference port (Fig. 1). A compressed pellet of sulfur powder was used as the reference material (Tkachuk and Kuzina, 1978). The sample and reference materials were contained in cylindrical Bakelite holders, 1 cm deep and 6.5 cm in diameter, faced with matched, 2-mm-thick Infrasil windows. Transmittance measurements were carried out by placing samples in Infrasil cells at the sample entrance port of the integrating sphere.

The wavelength region scanned, scanning speed, wavelength interval, number of readings per interval, and noise rejection limits were all variables under the control of the operator. Readings were collected for the wavelength region at 600–2,400 nm. Although all studies did not need such an extended wavelength scan, experience showed that it was more convenient to record extended data files. In this manner, data for other wavelength regions would be available in the future, often for samples that would no longer have been available or that would have deteriorated during storage. Data were recorded at set wavelength readings, unlike most other NIR instruments, which record data continuously while wavelength scanning is in progress.

Once a scan is completed, the file is examined for noise spikes by examining a first-derivative plot of the spectrum, on which spurious noise will appear as very sharp spikes. Usually the problem is solved by substituting for the noise spike the average value of the adjacent reflectance readings on both sides of the

TABLE I
Computer Programs Used in Near-Infrared Reflectance Analysis

Program	Function
CARYTEST	Tests operation of Cary 17 spectrophotometer
CARYSCAN	Controls Cary 17 scanning operation and averages and records reflectance data
RDDPM	Checks for panel meter error
PACK	Checks that scanning range is correct
CHECK	Checks that starting wavelength, step interval, and number of steps per reading are correct
MOVE	Controls stepwise movement of wavelength selector on Cary 17
RDIN	Collects reflectance data from Cary 17
FIXUP	Checks for and rejects noisy readings
CRTPLT	Outputs data to CRT
TRANSFRM	Transforms row spectral data; permits subtraction of baseline, converts transmittance data to log $(1/R')$; performs data smoothing and/or first and second derivatization
SPECTRA	Plots spectral data (raw or transformed) on Tektronix CRT 4027
READDATA	Outputs selected unformatted reflectance (raw or transformed) data files
SPECWAVE	Outputs specified wavelength data
CORRECT	Permits correction or addition to dependent variable data in existing files
MERGE	Merges NIR and visible reflectance data files
CONVERT	Converts formatted to unformatted data files
FILES95	Creates single data file from up to 95 files for various correlation programs
FILES180	Same as FILES95 but handles 180 files
CORRELAT	Computes correlation between variable and smoothed or derivatized reflectance data and creates correlation data file for CORPLOT
CORPLOT	Plots correlations between variable and smoothed or derivatized reflectance data on Tektronix CRT 4027
CORQUO	Computes correlation between variable and quotient of reflectance data at two wavelengths and creates correlation data file for CORGRAM
CORDIF	Computes correlation between variable and difference between reflectance data of two wavelengths and creates correlation data file for CORGRAM
CORGRAM	Plots quotient or difference correlation data as colored correlogram on Tektronix CRT 4027
WAVESET	Creates data file of 180 wavelengths selected for regression analysis
WAVEPAIR	Creates data file of 180 pairs of wavelengths selected for regression analysis
REGFILE1	Creates binary data file consisting of smoothed or derivatized reflectance data for maximum of selected wavelengths and dependent variable for regression analysis
REGFILE2	Reflectance data for up to 80 selected pairs of wavelengths and dependent variable for regression analysis
P1R	Multiple linear regression: BMDP program that gives estimates of multiple linear regression equation on all (or subsets) of data
P2R	Stepwise regression: BMDP program that estimates multiple linear regression equations in forward or backward stepping manner; four criteria available for influencing stepping are F-to-enter, F-to-remove, Tolerance, and number and sequence of variable entry
P9R	All-possible subsets regression: BMDP (Dixon et al, 1983) program that provides regression equation estimates for "best" subsets of predictor variables and carries out extensive residue analyses; three criteria defining "best" are sample R^2, adjusted R^2, and Mallows' Cp.
PnR.CTL	Provides control instructions for P1R, P2R, P9R; these programs describe data input, name variables, state missing value codes and upper and lower limits for variables, specify analysis type, request optional results to be printed, and specify variables to be plotted

spike. However, if an extensive noise peak is present, or if there are many noise spikes, it is best to rerun the sample on the spectrophotometer. The presence of even one small noise spike in 1 out of 90 samples can have a drastic effect on optimum wavelength selection. Accordingly, it is essential that all known spurious noise be removed from the reflectance scans. The manual technique for noise removal has been found to be necessary, since simple computer algorithms do not remove all the noise and all-encompassing algorithms are very difficult to prepare and can eliminate essential experimental data.

VI. DATA STORAGE AND TRANSFORMATIONS

Raw reflectance data from each sample are stored as data files (records) on a 10.4-megabyte RL02 storage disk. Current data files consist of 900 averaged five-digit readings for each 2-nm interval in a wavelength range of 600–2,400 nm. Stored as unformatted files, each data file occupies 4,096 bytes on the disk. Taking into account the disk storage overhead space that is needed to keep track of disk files, approximately 2,275 individual data files can be stored on the disk.

Raw data are smoothed with a nine-point quartic convolution function (Savitzky and Golay, 1964) using a modified program that had originally been written for an IBM mainframe (Jones et al, 1976). Next, a number of transformations are carried out:

1. Conversion from R' to A' values, or vice versa.
2. Computation of first-derivative values using a nine-point quartic function (Jones et al, 1976).
3. Computation of second-derivative values using a quadratic convolution function (Jones et al, 1976).
4. Calculation of $(A_{\lambda x'} - A_{\lambda y'})/A_{\lambda y'}$ quotient ratios where x and y can assume all wavelengths scanned except $x = y$.

VII. CORRELATION ANALYSIS

Once the data transformations are carried out, the resulting data are correlated with the constituent component being studied—e.g., protein content. It is useful to plot the correlations vs. wavelengths to determine the highest values and the correlation variation patterns. This procedure is carried out for each transformation mode. In the instance of the quotient transformation, $R_{\lambda x'}/R_{\lambda y'}$, the correlations that can number in the hundreds of thousands are plotted in color on a high-resolution CRT to obtain three-dimensional plots termed *correlograms* (Tkachuk, 1981a). Correlograms provide interesting summaries of each datum. A similar procedure has been used by Davies et al (1984) to summarize and find best wavelength pairs of reflectance data.

VIII. REGRESSION ANALYSIS AND OPTIMUM WAVELENGTH SELECTION

Samples are analyzed by conventional laboratory techniques and also scanned by the NIR spectrophotometer. A regression equation is then computed to relate sample reflectance properties to a sample component—hitherto assumed to be (for example) total protein content. To obtain the most accurate regression equation, it is necessary to find the optimum wavelength reflectances by applying standard multiple regression analyses.

The multiple linear regression computations used in this study were carried out by using programs supplied by the BMDP statistical software package developed by the UCLA Health Sciences Computing Facility (Dixon et al, 1983). In current use is the 1983 BMDP version, as converted for PDP-11 computers (Software Development Inc., Middlebury, VT). The three main programs used in these computations, P1R, P2R, and P9R, are described in Table I.

On a PDP 11/84 computer, a maximum of up to about 90, 92, and 20 variables can be run with the P1R, P2R, and P9R programs respectively. The number of variables chosen for the regression equations by these programs is controlled by limits set for the parameters F and tolerance (T), which are defined (UCLA Health Sciences Computing Facility) as:

$$F\text{ (to enter or remove)} = \frac{\text{SS (residuals)} - \text{SS (next)}}{\text{SS (next)}/N - p - 1)}$$

where SS = sum of squares, N = number of cases, p = number of parameters, $p = p + 1$, and $T = 1 - R^2$ where R = multiple correlation coefficient. The values of F are used to test the coefficient of each variable as if it were entered (removed) separately into the equation at each step.

If R^2 of any independent variable with the other independent variables is high, that variable will not be entered into the regression equation if it does not pass the tolerance limit. Unless stated otherwise, in these studies limits of 0.01 and 4.0 were used for T and F.

Experience has shown that three properly chosen wavelength variables can provide sufficiently accurate regression equations in many instances. These three variables appear to be equivalent to reflectance parameters representing the top of a component absorption peak, a nearby wavelength serving as a baseline for the peak, and a third wavelength that acts as a normalizing variable. The normalizing variable seems to take care of general reflectance curve differences caused by different seed (particle) sizes or by weathered seed surfaces. The two other reasons for desiring only two or three variables per chemical component being measured are that one wants to avoid overfitting reflectance data to the sample set size (Hill, 1979) and that for economical reasons it is best for only three interference filters, per component to be analyzed, to be installed in simpler NIR instruments designed for commercial usage.

The first variable entered into the regression equation is the one that correlates most highly with the constituent to be determined (e.g., protein). The wavelength varies somewhat with the grain being analyzed (Table II). Computation is then used to find the remaining best two or three variables. In this iterative process, subsets of up to 90 variables are entered and analyzed until the best overall subset is found. It is an aid in the selection process to view simple correlation plots, or when using the wavelength quotient transformations to view three-dimensional correlograms.

Once the optimum wavelengths are determined and the corresponding regression equation computed, the equation is tested by determination of protein levels for an analytical set of samples. The statistical parameters obtained for the analytical set, R and Sy (sometimes referred to as the standard error of performance, SEP), are compared with the corresponding values for the sample test to evaluate the accuracy of the NIR test procedure and to see that there is no undue influence to sample sets (Table II). Finally, to use all the available data, the regression equation is recomputed for all the samples so that the data to be reported make use of all the samples analyzed.

IX. REFLECTANCE AND TRANSFLECTANCE PROPERTIES OF WHEAT

During the initial phases of the near-infrared research program in this laboratory, some experiments were carried out in attempts to decide which of the possible two analytical modes, reflectance or transflectance, was the best one to use in analyzing whole-kernel cereal and oilseed samples. "Transflectance" is a term introduced by the Technicon Corporation to describe a method for the reflectance spectroscopy of liquid. The principle is illustrated in Chapter 6. When used in connection with reflectance spectroscopy of wheat, the term implies that the lower layers of the wheat sample serve as a reflecting surface. Some of the signal used in the analysis is reflectance of this type, and some of it is undoubtedly true diffuse reflectance.

Typical reflectance spectra for whole and ground wheat are shown in Fig. 4. It is clear that both spectra probably have similar analytical information incorporated in them, as the number of visible peaks and valleys is approximately the same in both. Accordingly, the analysis of whole wheat kernels for at least some of the constituent chemical and physical parameters should be theoretically feasible and this has been shown to be so (Tkachuk, 1981a).

One significant difference in the two spectra in Fig. 4 is the amount of reflected energy. Whole wheat kernels reflect only 40–50% of the energy reflected by ground wheat from 800 to 2,400 nm, and only 22% at 500 nm. The low reflectance in the visible region confirms the visible perception of whole wheat (hard red spring wheat is dark red in appearance). Instruments designed to analyze whole wheat must be able to compensate for the decreased available reflectance energy, either by additional electronic amplification, using more intense light sources or different detector/light source geometries, or by using wider bandwidth interference filters.

Transflectance spectra of whole wheat layers are shown in Fig. 5. Relatively flat spectra were obtained for single wheat layers (3 mm thick) because of the large amounts of incident light passing through the kernel spacings. To reduce incident light to negligible levels, it is necessary to analyze wheat thicknesses of at least 9 mm. Unfortunately, the analysis of thicker wheat layers results in very high absorptions in the longer wavelength region, effectively limiting the working region to about 800–1,100 nm. (Commercial transmittance instruments available for routine testing use a sample thickness of approximately 2 cm and wavelengths in the 900- to 1,050-nm region).

Consideration of the above evidence led to the decision to emphasize near-infrared studies using the reflectance mode in this laboratory. As studies progressed, additional evidence appeared that buttressed this decision. The evidence favoring the use of the reflectance mode of analysis of whole cereal and oilseed samples is summarized below:

1. The reflectance technique offers the use of a much wider wavelength working range (500–2,500 nm with a PbS detector).
2. The availability of a wide wavelength region permits the analysis of both whole and ground wheat samples, allowing the direct comparison of results from these two techniques. It is very difficult to carry out transmittance studies on ground samples because of the extremely low

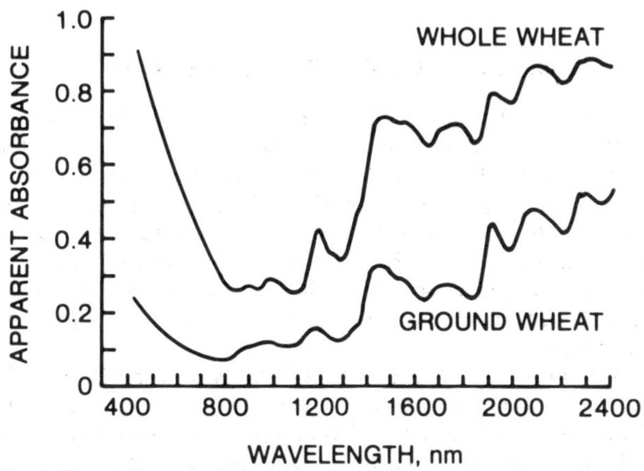

Fig. 4. Near-infrared reflectance spectra for whole and ground wheat recorded vs. a sulfur-disk reflectance standard.

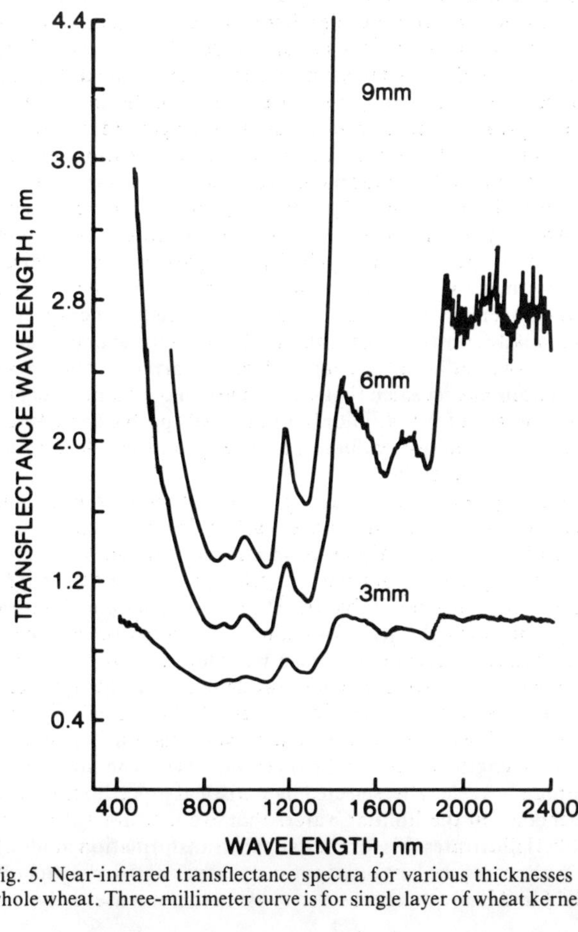

Fig. 5. Near-infrared transflectance spectra for various thicknesses of whole wheat. Three-millimeter curve is for single layer of wheat kernels.

TABLE II
Optimum Wavelength Selection for Protein Prediction of Whole Wheat Kernels[a]

Transformation	Step	Wavelength Selected (nm) λ_1	λ_2	Calibration R	Sy
Quotients	1	1,214	1,190	0.9569	0.544
	2	1,788	1,726	0.9765	0.406
	3	1,680	1,644	0.9828	0.350
Differences	1	1,216	1,188	0.9548	0.557
	2	1,680	1,642	0.9758	0.412
	3	1,788	1,728	0.9811	0.367
First derivative	1	1,196	···	0.7621	1.214
	2	1,206	···	0.9499	0.589
	3	1,006	···	0.9641	0.504
Second derivative	1	1,144	···	0.7735	1.188
	2	1,186	···	0.9288	0.698
	3	924	···	0.9503	0.590
Apparent absorbance I[b]	1	836	···	0.5366	1.582
	2	2,096	···	0.6646	1.409
	3	2,002	···	0.8078	1.118
	4[c]	978	···	0.8258	1.076
Apparent absorbance II[d]	1	1,214	···	0.2146	1.831
	2	1,190	···	0.9548	0.561
	3	1,788	···	0.9554	0.560
	4	1,726	···	0.9759	0.416
	5	1,680	···	0.9766	0.412
	6	1,644	···	0.9823	0.361

[a]Regression parameters, multiple correlation coefficient (R), and standard error of estimate (Sy) were computed for 90 samples using the BMDP-77 programs.
[b]Tolerance = 0.01.
[c]No further stepping because tolerance below limit of 0.01.
[d]Tolerance = 0.0001.

amounts of radiant energy that emerge from the transverse side of the sample.

3. Wavelengths that correlate the best with protein and moisture levels lie in the region from 1,200 to 2,096 nm, which are outside the transflectance region (Fig. 5).
4. The higher-energy reflectance spectra in the longer wavelength regions offer significant signal/noise advantages.
5. Reflectance spectra are not as markedly affected by the presence of small seeds and dust as are transflectance spectra, probably because the transflectance mode is more dependent on kernel void spaces to impart energy to the detector.
6. It is relatively easy to adapt existing commercial reflectance instruments designed to analyze ground samples to analysis of whole kernel samples.

X. ANALYSIS FOR WHEAT PROTEIN

The application of an NIR technique for the analysis of whole wheat kernels for protein content was first described by the author in 1979 (Tkachuk, 1979). Additional details were subsequently published in 1981 (Tkachuk, 1981a). This research showed that whole wheat kernel samples could be analyzed with acceptable accuracy on a research spectrophotometer. Subsequent research (unpublished) in this area has centered on noise reduction to improve the accuracy of the technique.

A detailed examination of the data used in the 1981 study revealed that all extraneous noise had not been removed from the reflectance data files. It has become clear that noise is very difficult to remove from reflectance spectra through the sole application of computer algorithms, because simple algorithms do not remove all of the noise and complex algorithms can remove useful signal information. Smoothing by using various averaging and digital least-squares techniques is useful in removing low-magnitude, high-frequency components, but it is not useful in removing larger magnitude effects. In addition, excessive use of the latter Savitzky and Golay (1964) technique removes data points from the extreme ends of the spectra and merely spreads the noise over a wider wavelength range. Besides using simpler algorithms and moderate application of the Savitzky and Golay smoothing technique, manual techniques have also been adopted for removing noise. The technique involves examining first-derivative transformations of the reflectance spectra, on which (by comparison with similar spectra) noise usually shows up as very sharp spikes. Such spikes are removed either by obtaining a new reflectance spectra or manually substituting for the noise region the average value of the two adjacent "clean values." Carrying out these corrections has revealed that even the presence of a noise peak, with a height of 0.1 *A*, in only 1 out of 90 spectra has a large effect on correlation values and thereby on the selection of optimum wavelengths.

Carrying out the regression and optimum wavelength selection procedures described earlier in this chapter on manually corrected reflectance spectra has resulted in several minor changes in the optimum wavelengths for the determination of wheat protein (Table II). The best results were obtained using the variable transformation of quotients, although reflectance differences at two wavelengths were almost identical. The two principal wavelengths in the quotient transformation are identical to those previously found (Tkachuk, 1981a), but there are some very slight differences in the wavelength values for the other variables. The number of optimum wavelengths selected by statistical procedures is very dependent on the limiting values that are used for tolerance. Table II illustrates this for the variable transformation mode of *A*. In choosing the best set of optimum wavelengths, researchers would find it advantageous also to use commonsense analytical principles in addition to estimates provided by statistical procedures. For example, all regression procedures will use as their first variable the one with the highest *R* value. This *R* could occur, however, as a result of a noise spike or could be the maximum value in a narrow peak of correlation values; in this instance, it is probably best to use a slightly lower *R* value if such exists as part of a broader peak of *R* values. For this and other reasons, it is useful to examine correlation values as simple plots or as correlograms.

A plot for the calibration NIR reflectance protein vs. Kjeldahl is shown in Fig. 6. The regression line has a slope of 45° and an intercept of zero. These factors, combined with a relatively high value of 0.9830 for *R*, indicate that the NIR method for determination of protein in whole wheat kernel samples can have an acceptable accuracy for many commercial needs.

If additional wavelengths are used in quotient data transformation, the prediction improves only slightly. Use of a fourth quotient variable (Table II) results in *R* and *Sy* values changing to 0.984 and 0.34, and one risks the danger of overfitting by using wavelengths that are highly intercorrelated.

The only other published report of optimum wavelengths for whole wheat kernel analysis describes the use of primary and denominator wavelengths of 1,018 and 970 nm in a study over a 900- to 1,050-nm wavelength region involving a near-infrared transmittance (NIT) technique to determine protein and moisture in whole wheat and barley (Williams et al, 1985). These workers used a computerized spectrophotometer in transmittance mode and a Trebor GT 90 whole grain analyzer, which also operates in transmittance mode. They reported SEP values of 0.26 and 0.20 respectively for hard red spring (HRS) wheat protein and moisture and 0.33 and 0.24 respectively for barley protein and moisture. When the study was extended to all types of wheat, including soft and durum wheat, the mean SEP was 0.36 for protein and 0.28 for moisture, using the second derivative of the log $(1/R)$ divided by a second D2OD term. In testing HRS wheat by the GT 90, they found the most important variables to be small seeds (although "dockage" of up to 7.5% led to only small increases in protein content) and temperature. The presence of 17–20% of small weed seeds could reduce the apparent protein content of wheat by over 50% (from 14% to 6.5%), presumably because of interference with light transmittance through the sample. The instrument temperature-correction procedure was fairly effective but was greatly improved if samples of different temperature were used in the

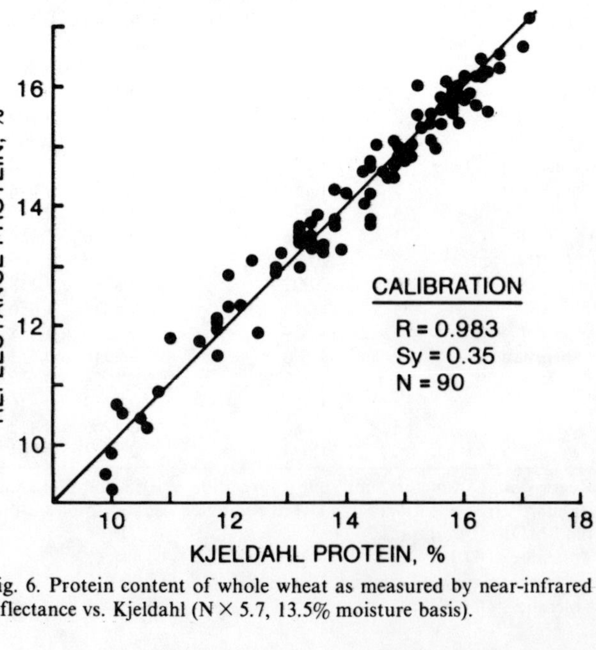

Fig. 6. Protein content of whole wheat as measured by near-infrared reflectance vs. Kjeldahl (N × 5.7, 13.5% moisture basis).

TABLE III
Calibration Coefficients for Estimating Moisture in Wheat and Barley

Material	Moisture Range	Wavelength (nm) 2,310	2,230	2,180	2,100	1,940	1,680	Development n[a]	R[b]	RSD[c]	Evaluation n	RSD
Whole grain barley	13.6–20.0	−107.4	47.15	122.4	−102.6	82.6	−39.1	134	0.97	0.38	66	0.50
Whole grain wheat	13.6–20.0	−76.27	−98.68	275.2	−208.2	105.7	−0.76	37	0.99	0.27	40	0.36
Coarsely ground barley	13.4–25.9	−19.72	−100.7	0.17	−28.12	95.95	62.22	90	0.99	0.38	115	0.46
	13.4–20.0	−40.73	−52.61	−4.68	−11.84	83.55	34.26	46	0.99	0.25	88	0.33
	20.0–25.9	−18.97	−64.46	−103.5	30.11	93.97	76.19	44	0.97	0.39	30	0.64
Coarsely ground wheat	13.1–26.1	−276.0	−5.46	19.18	169.5	75.5	38.9	117	0.99	0.43	64	0.46
	13.1–20.0	−231.4	−14.13	14.74	130.3	74.4	50.4	70	0.99	0.33	33	0.34
	20.0–26.1	−216.5	2.16	121.8	−15.4	104.9	−1.1	47	0.99	0.31	29	0.36

[a]Number of samples.
[b]Multiple correlation coefficient.
[c]Residual standard deviation.

calibration. Kernel size had no significant effect on the testing of wheat for protein, but it had a significant effect on barley: larger kernels gave an apparent increase in protein content. Growing season also had a significant effect on the accuracy of protein testing, although the visually determined grade was not a significant factor provided all grades were included in the calibration.

Stermer et al (1977) studied the determination of moisture in high-moisture corn (maize) and sorghum using a conventional tilting-filter NIR instrument. Both series of grains ranged up to nearly 50% moisture. Two series were studied, native wet grains and grains that had been tempered to a very high moisture range. Using three wavelengths, 1,940, 2,190, and 2,330 nm, they obtained a standard error of about 2.0 with tempered corn. With native wet grains, wavelengths ranging from 970 to 1,500, or in a different range, from 1,890 to 2,240 nm, gave an average SEP of about 1.2% for corn and 3.0% for sorghum. Coefficients of correlation ranged from 0.99 for corn to 0.96 for sorghum. In a parallel study using a series of capitance moisture meters, the respective SEPs were 2.43 for corn and 3.1 for sorghum. The authors concluded that the NIR instrument was satisfactory for the determination of moisture in high-moisture grains.

XI. WHEAT AND BARLEY MOISTURE

The application of the NIR technique for the analysis of whole-grain wheat and barley samples for moisture content has recently been reported by Downey (1985). A Technicon InfraAlyzer 300B instrument was used in this study, with samples being presented to the instrument with the supplied moisture-free quartz covered sample cup. In this instrument, the sample cup does not rotate and compresses the sample into a layer approximately 6 mm in thickness. The study gives thorough details for results obtained for analysis of both whole grain and ground samples. For moisture contents of up to 20% for whole-grain wheat and barley samples, an accuracy was obtained of ±0.72% and ±1.0% respectively, and the respective precisions (standard deviation between replicates) were 0.4 and 0.3 (Table III). These precision values, which are much poorer than those usually obtained for ground samples, were attributed to difficulties in maintaining constant packing densities of the whole grain samples. The author suggested that the NIR method will not replace the standard air-oven technique for moisture determination but is of sufficient accuracy and convenience to be useful in commerce, since on coarsely ground wheat and barley samples the NIR method performed better than the moisture meters usually used in Ireland. Turning to the analysis of unground samples, the author could not provide a direct comparison between NIR and moisture meter results (i.e., for the same sample set), but using residual standard deviation as an indicator of accuracy, the study produced strong evidence that NIR determination of the moisture content of unground samples was also the more accurate method.

XII. RAPESEED, PROTEIN, OIL, AND CHLOROPHYLL

Whole rapeseed samples have been analyzed by NIR methods for the determination of several constituents. In one study, determination of oil, protein, chlorophyll, and glucosinolate contents was carried out using data in the 1,000- to 2,400-nm wavelength region (Tkachuk, 1981b), and a further study reexamined the determination of chlorophyll by using an extended wavelength region from 600 to 2,400 nm (Tkachuk and Kuzina, 1982). These results showed that standard errors of estimate could be obtained of approximately 0.83, 0.88, 0.99, and 1.5 for oil, protein, chlorophyll, and glucosinolate contents. Analytical errors of the above magnitude may be satisfactory for the analysis of rapeseed samples for plant-breeding purposes. However, ordinary analytical requirements of commerce require that the errors be smaller by a factor of about two for oil, protein, and chlorophyll contents and by a factor of four for glucosinolate content. Accordingly, further improvements in accuracy are needed before the NIR technique can be used for whole rapeseed analysis.

ACKNOWLEDGMENTS

I would like to thank V. J. Mellish for assistance in programming and computing, J. C. Babb for statistical analysis, G. H. Rogers for help in systems operations, and F. D. Kuzina and S. Lowe for technical assistance.

LITERATURE CITED

CLARKE, F. J. J. 1972. High accuracy spectrophotometry at the National Physical Laboratory. J. Res. Natl. Bur. Stand. 76A:375-403.

DAVIES, A. M. C., GEE, M. G., and FOSTER, P. W. 1984. A colour graphics display system to aid the selection of "best-pair" wavelengths for regression analysis of near infrared data. Lab. Pract. 33:78-80.

DIXON, W. J., BROWN, M. B., ENGELMAN, L., FRANE, J. W., HILL, M. A., JENNRICH, R. I., and TOPOREK, J. D. 1983. BMDP Statistical Software. Printing with additions. University of California Press, Berkeley. (PDP-11 version, Software Development Inc., Middlebury, VT.)

DOWNEY, G. 1985. Estimation of moisture in undried wheat and barley by near infrared reflectance. J. Sci. Food Agric. 36:951-958.

HILL, M. A. 1979. Annotated computer output for regression analysis. Tech. Rep. 48. Health Sciences Computing Facility, University of California, Los Angeles.

JONES, R. N., BACH, T. E., FUHRER, H., KARTHA, V. B., PITHA, J., SASHADRIE, K. S., VENKATARAGHAVAN, R., and YOUNG, R. P. 1976. Computer Programs for Infrared Spectrophotometry. 2nd ed., Vol. 1. Bull. 11, pp. 65-97. Natl. Res. Council Can., Ottawa.

SAVITZKY, A., and GOLAY, M. J. E. 1964. Smoothing and differentiation of data by simplified least squares procedures. Anal. Chem. 36:1627-1638.

STERMER, R. A., POMERANZ, Y., and McGINTY, R. J. 1977.

Infrared reflectance spectroscopy for estimation of moisture of whole grain. Cereal Chem. 54:345-351.

TKACHUK, R. 1979. Whole kernel analysis with a computerized double beam Cary 17I spectrophotometer. (Abstr.) Cereal Foods World 24:456.

TKACHUK, R. 1981a. Protein analysis of whole wheat kernels by near infrared reflectance. Cereal Foods World 26:584-587.

TKACHUK, R. 1981b. Oil and protein analysis of whole rapeseed kernels by near infrared reflectance spectroscopy. J. Am. Oil Chem. Soc. 58:819-822.

TKACHUK, R., and KUZINA, F. D. 1978. Sulfur as a proposed near infrared reflectance standard. Appl. Opt. 17:2817-2820.

TKACHUK, R., and KUZINA, F. D. 1982. Chlorophyll analysis of whole rapeseed kernels by near infrared reflectance. Can. J. Plant Sci. 62:875-884.

WILLIAMS, P. C., NORRIS, K. H., and SOBERING, D. C. 1985. Determination of protein and moisture in wheat and barley by near-infrared transmission. J. Agric. Food Chem. 33:239-244.

CHAPTER 15

Qualitative Applications of Near-Infrared Reflectance Spectroscopy

PHILIP C. WILLIAMS
Grain Research Laboratory
Canadian Grain Commission
Winnipeg, Manitoba, Canada

KARL H. NORRIS
Instrumentation Research Laboratory
Agricultural Research Service
U.S. Department of Agriculture
Beltsville, Maryland, U.S.A.

I. INTRODUCTION

There are four main types of qualitative application of near-infrared reflectance (NIR) spectroscopy. These are (i) determination of the general composition of a substance; (ii) detection of impurities; (iii) identification of areas of maximum difference in composition for given constituents; and (iv) identification or classification of unknown substances. The first three applications are limited to the computerized spectrophotometer because it is necessary to display the whole spectrum from 1,000 to 2,500 nm. The advantages of such qualitative examinations with NIR are speed, applicability to small amounts of sample, and nondestructiveness. The second and third applications are particularly important, for example, in the field of preparative biochemistry, where it is simple to test a prepared enzyme, protein fraction, or other experimental material for the presence of impurities without redissolving the material. As little as 10 mg of sample has been successfully studied in this manner.

In comparing spectra between computerized spectrophotometers, we may notice small differences as to the wavelengths at which bands occur. For example, a band in a spectrum recorded on one instrument may appear as a shoulder in the spectrum of the same sample recorded on another instrument. One reason is a difference in the bandwidth over which computerized spectrophotometers work. Some have a narrower bandwidth and accordingly are able to identify more bands, and with generally better resolution. Bands that appear to be slightly displaced in one instrument relative to another are still there and available to the computer for wavelength selection. It is also important to use exactly the same algorithm in comparing spectra—e.g., segment and derivative (gap) size. In using a first-derivative algorithm, the "bands" are "valleys" in the same way as with the second derivative, but they will be offset to one side or the other. The wavelength corresponding to the band displayed in a second-derivative trace is a zero point in a first-derivative trace.

II. DETERMINATION OF GROSS COMPOSITION

The purpose of displaying the spectrum of an unknown substance is to detect rapidly the major constituents present, such as high protein and fiber contents or the presence of significant quantities of sugars. Figure 1 illustrates the spectrum of feathermeal. The prominent bands for protein at 2,054 and 2,172 nm and oil at 2,308 and 2,347 nm are clear. Feathermeal is practically all protein, and water and fat are the only other significant constituents. Figure 2 gives the spectrum of fishmeal, where the protein bands are again prominent, and also the oil bands at 2,308 and 2,347 nm. Figure 3 gives the spectrum of dried molasses. The angular spectrum with a strong band in the 1,425-nm area is typical of substances high in crystalline sugar. The fiber bands at 2,270 and 2,485 nm can be detected, as well as water in the 1,940-nm area. Displaying the spectrum enables the operator to locate the most important constituents for analysis and those most likely to cause interferences. Figure 4 illustrates the areas where bands have been found for certain constituents. It differs from Fig. 21 in Chap. 2, which illustrates areas where bands occur for known absorbers, and indicates areas where the computer is likely to select wavelengths for the determination of constituents normally determined using wavelengths that may be affected by interferences from major constituents in certain substances. The 1,300- to 1,750-nm range is a very fruitful area for absorbers for use in the determination of protein and oil, other than the familiar bands at 2,172 and 2,308 nm. The alternate bands often become more prominent in wavelength selection when the constituent to be determined is in low concentration: for example, protein content in straws, where the protein is present in much lower concentration than cellulose-type fiber components.

Changes in physical characteristics can also be detected by displaying the NIR log $1/R$ spectrum. The most familiar example is the effect on mean particle size of grinding to different degrees (Fig. 4, Chap. 7). Changes in crystalline form can also be detected.

III. DETECTION OF IMPURITIES

The detection of impurities is another useful application of qualitative analysis. The most important areas in this field concern verification of purity in biological and other substances and the detection of water. Figure 5 is the spectrum of a freeze-dried, dialyzed 70% ethanol extract of wheat flour. The objective of the extraction was to isolate a fraction of the protein. The derivatized spectrum shows bands in the area where starch is known to have absorbers, and subsequent analysis verified that the freeze-dried material was far from pure. Figure 6 is of a freeze-dried preparation of wheat gluten that contained 95% protein (N × 5.7 moisture-free basis). The preparation was practically free of starch and oil, and the only constituent present in significant quantities other than protein was moisture. Another application is the detection of traces of salts, such as tris-(hydroxymethyl)-amino methane. Many of the chemicals used in the preparation of buffered extracting solutions have characteristic bands in the near infrared. Inspection of the spectra of freeze-dried preparations indicates the efficiency of the dialysis and other methods used to remove the salts. Figures 7 and 8 give the spectra of tris-(hydroxymethyl)-amino methane and sodium bicarbonate. Presence of bands characteristic of extracting salts in the spectrum of a biological preparation signifies that the preparation is not pure and that further dialysis or other treatment is necessary.

Moisture is the most common contaminant in organic substances, and a great many ostensibly pure substances contain traces of moisture. Water has a very strong characteristic absorbance from the combination band at 1,925–1,945 nm, and the presence of this band is evidence that the substance is not completely dry. Moreover, the presence of moisture changes the band pattern of the spectra of some substances because of the disappearance after exhaustive drying of bands caused by hydrogen bonding. In some cases the moisture has the effect of masking certain bands. Furthermore, water causes hydration of many constituents and bond formations within constituents, and the degree of hydration may change the frequency of molecular vibration and, consequently, the position of some absorption bands. Figures 9, 10A, and 10B illustrate starch before and after drying. The band at 1,934 nm was considerably reduced, whereas those in the 2,100-, 2,270-, and 1,630-nm areas were more prominent as a result of the drying. This suggests that the band at 1,430 nm may not be caused by water alone.

A common source of error in analysis can occur when substances that are believed to be dry contain small amounts (1–2%) of moisture. When the results of analysis for other constituents are reported on a dry basis, this moisture can have the effect of causing significant errors in concentration because its presence is unnoticed. For example, the presence of 1% moisture in soybean meal that is believed to be dry can introduce an error of over 0.4% in a protein analysis. Most commodities, even materials with a high oil content, become very hygroscopic when dried, and unless they are carefully protected their fibrous components will rapidly pick up small amounts of water. The most effective method of sample preparation and chemical analysis of oil-bearing commodities is to grind them as received, then to dry the well-mixed meal overnight in a vacuum oven at 100° C. They should be cooled in a desiccator and then weighed for analysis immediately to avoid errors induced by traces of moisture. Figures 11A and 11B illustrate ground soybeans dried before grinding and after grinding. The prominent moisture band at 1,930 nm in the sample dried as whole seeds is practically absent from the spectrum of the dried meal. The sample "dried" before grinding contained nearly 4% moisture, equivalent to an error of over 1.5% protein if protein is reported on a dry matter basis.

IV. DETECTION OF AREAS OF MAXIMUM DIFFERENCES IN CONCENTRATION

Areas in the spectrum where maximum differences occur in concentration of substances to be determined can indicate the wavelengths where accurate analysis is most likely. This is done by using as a standard a sample low in concentration of, for example, cellulosic fiber and subtracting the spectrum from that of a sample high in that constituent. The low-concentration

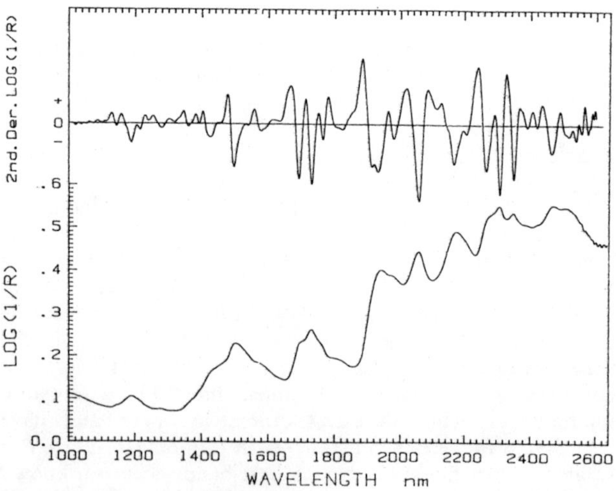

Fig. 1. Spectrum of feather meal (shown as spectrum 227 in illustrations of spectra given in the following section).

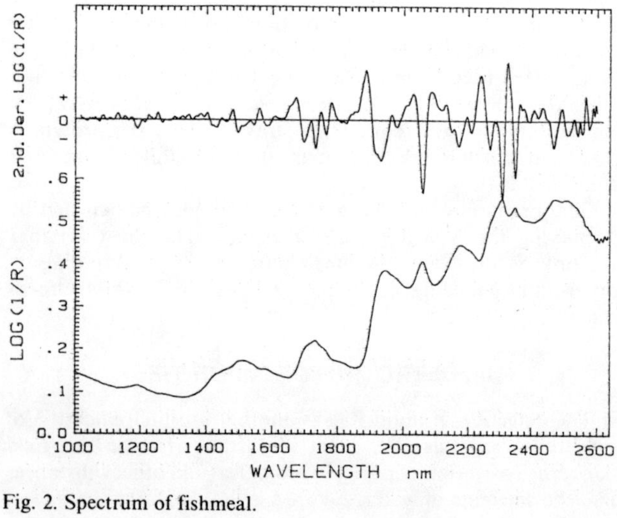

Fig. 2. Spectrum of fishmeal.

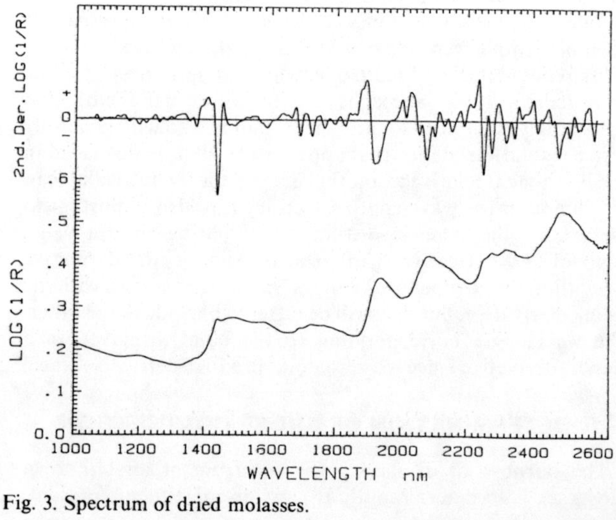

Fig. 3. Spectrum of dried molasses.

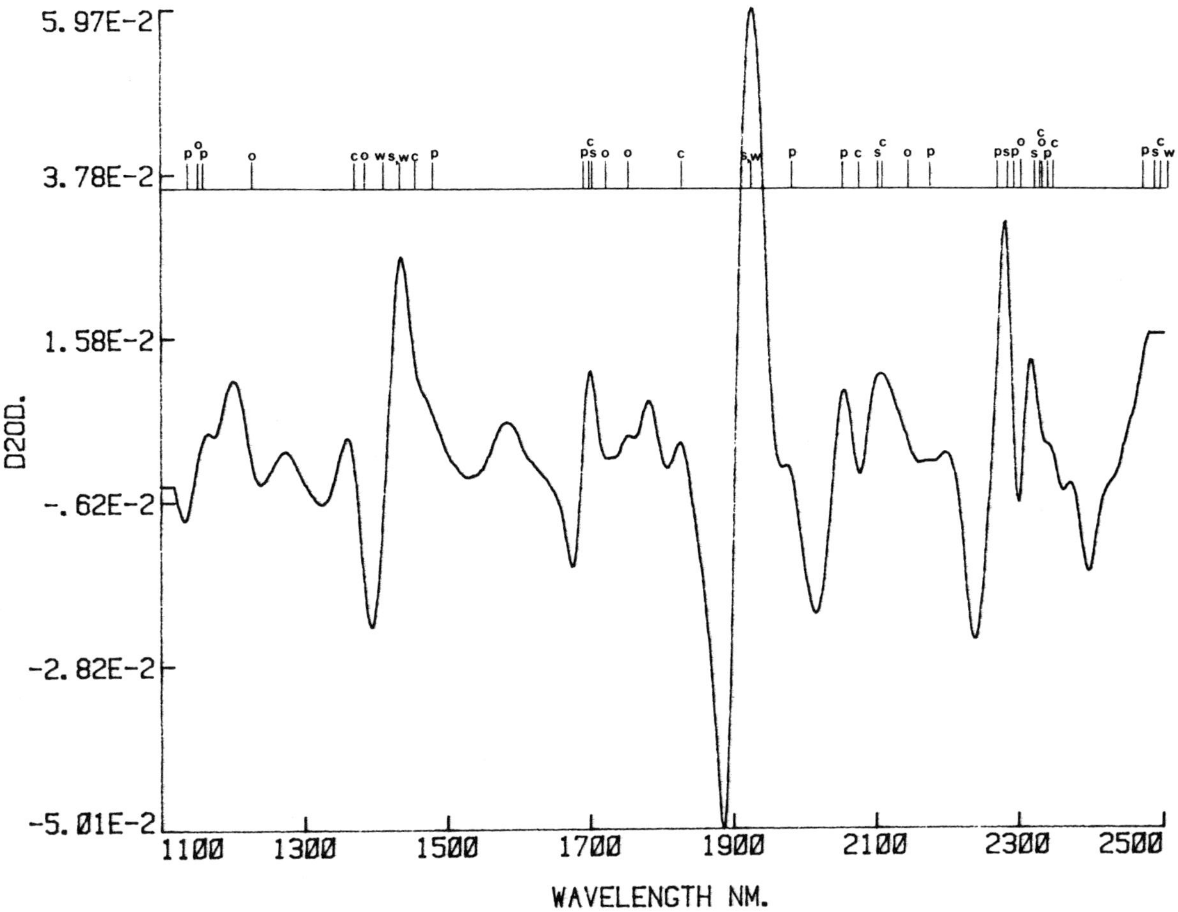

Fig. 4. Common absorbers in the near-infrared area; c = cellulose, o = oil, p = protein, s = starch, and w = water.

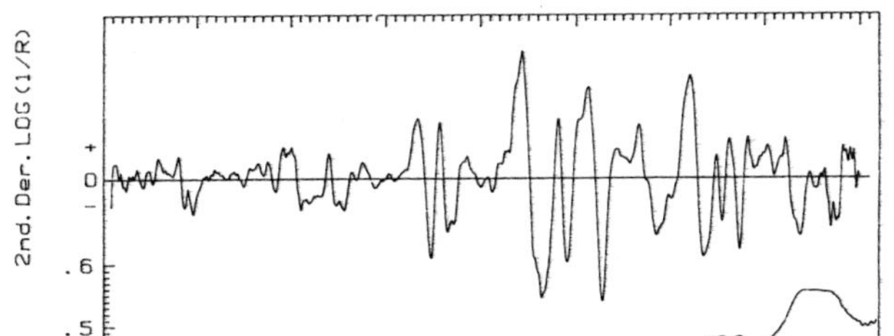

Fig. 5. Spectrum of freeze-dried, dialyzed 70% ethanol extract of wheat flour.

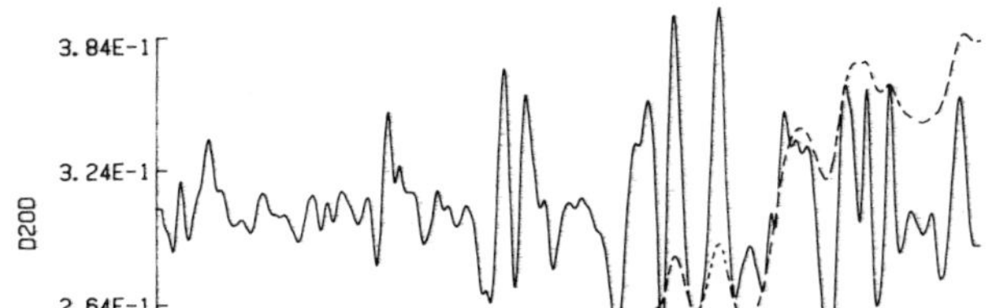

Fig. 6. Spectrum of freeze-dried preparation of wheat gluten containing 95% protein (N × 5.7 moisture-free basis). Broken line = log $1/R$, continuous line = D2OD.

sample is recorded as a standard. The high-concentration sample is then read in the normal way and recorded as a sample. "Bands" in the spectrum will indicate areas where maximum differences occur between the two samples. Usually, five or six of these areas can be identified and the wavelengths more closely specified by displaying the first- or second-derivative "spectrum" of the differences. The difference "spectrum" can be used to generate a calibration stipulating these wavelengths. Typical results of this type of exercise are given in Chap. 6. This technique can also be used to study spectra of, for example, "dry" substances by subtracting the water spectrum from the "as is" substances.

V. CLASSIFICATION OF UNKNOWN SUBSTANCES

The classification of unknown substances is of particular value in operations in which it is necessary to identify chemical compounds, both simple and complex. The petrochemical and pharmaceutical industries are typical areas where such an operation is of value. It is often necessary to ensure that one or more compounds of known composition are being produced in a form of sufficient purity to conform to specifications. This can be done by monitoring the product over a large wavelength area, using a computerized spectrophotometer, or over a more

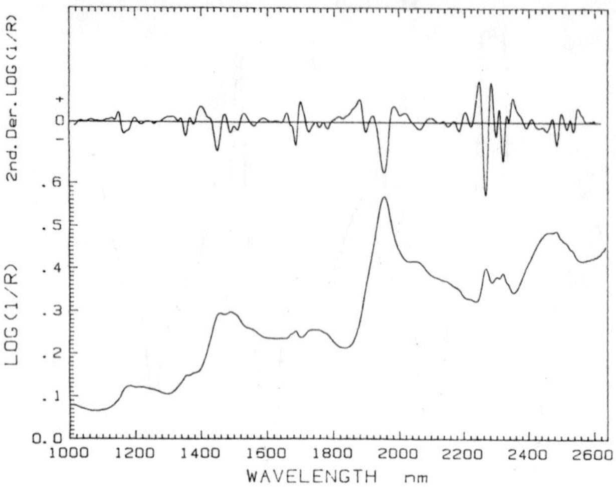

Fig. 7. Spectrum of tris-(hydroxymethyl)-amino methane.

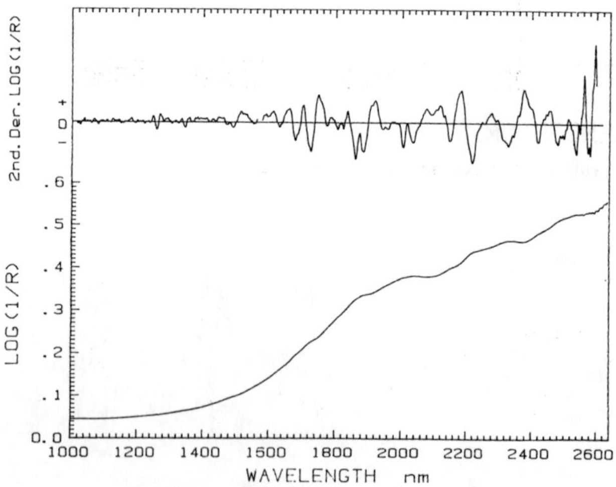

Fig. 8. Spectrum of sodium bicarbonate.

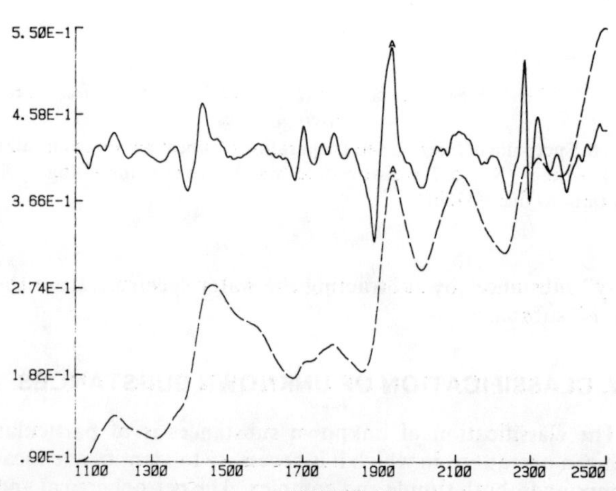

Fig. 9. Spectrum of wheat starch before drying. Broken line = $1/R$, continuous line = D20D 5/2. Main moisture band is at A.

restricted wavelength area for compounds with known absorbance in a particular region. A recent communication by Mark and Tunnell (1985) introduced the concept of discriminant analysis for the classification of chemical compounds by calibration over the near-infrared spectrum from 1,100 to 2,500 nm using a large number of different compounds as "calibration" samples. This calibration set would contain the compounds essential to the operation. If the process was producing substances other than those for which the operation was intended, the spectrophotometer could indicate this to the quality control laboratory. This could be achieved by programming the system to compute the deviations in absorbance from standard absorbances at a large number of wavelengths. "Foreign" substances would differ significantly from the desired materials at certain wavelengths. The distance in absorbance from the "normal" is referred to as the Mahalanobis distance (Mahalanobis, 1936). This system of compound identification is of particular value in situations where the substances under investigation do not display excessive variability in particle size or shape.

More recently, discriminant analysis has been applied to the identification of wheat varieties of superior baking quality (Bertrand et al, 1985). These workers developed a calibration based on groups of samples of each variety. Varieties and genotypes that did not conform to the calibration would be identified as being unacceptable. Discriminant analysis of this type is likely to become a very important aspect of NIR technology in future years.

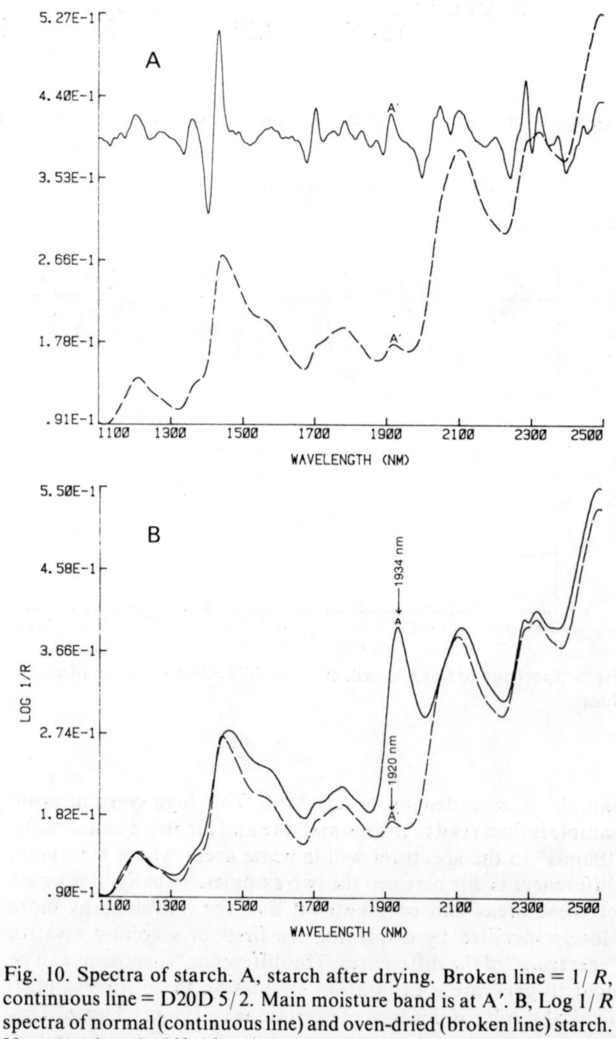

Fig. 10. Spectra of starch. A, starch after drying. Broken line = $1/R$, continuous line = D20D 5/2. Main moisture band is at A′. B, Log $1/R$ spectra of normal (continuous line) and oven-dried (broken line) starch. Note wavelength shift of moisture band A, A′.

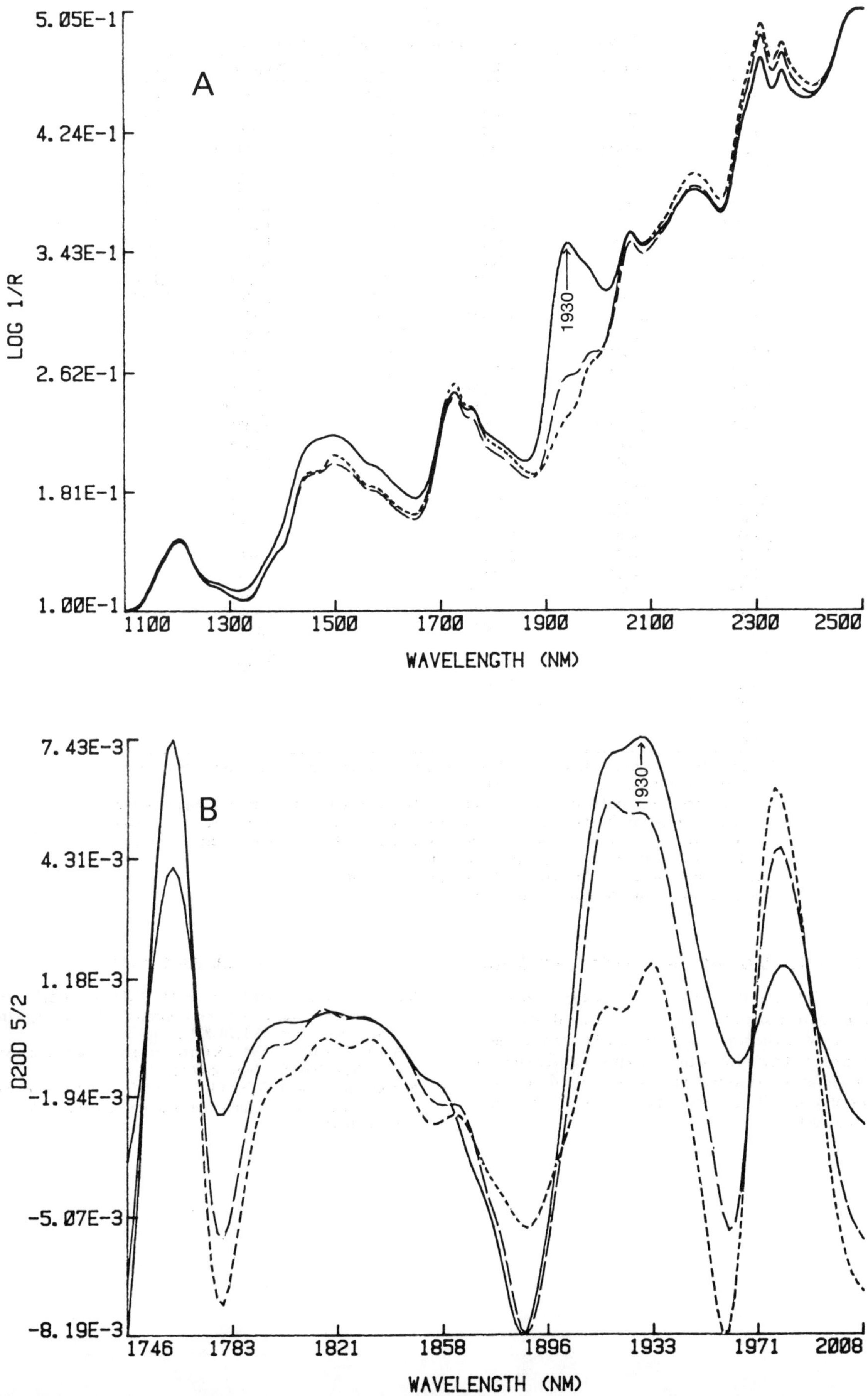

Fig. 11. Spectra of soybean. Log 1/*R* spectra (A) and D20D 5/2 spectra (B) of undried soybean (continuous line) and of soybean dried before (dashed line) and after (dotted line) grinding. B is the second derivative (segment 10, derivative 4 nm) of A in main moisture area.

TABLE I
Identification of Wavelengths (in nm) of Near-Infrared Absorbers[a,b]

Protein	Starch	Oil	Water	Cellulose	Sugar
874*[c]	878*	891*	834*	860	838
909**	901*	913	938	905*	888
979	918**	931*	958**	920*	913
1,018**	979***	965	978	978***	978*
1,051*	1,030	979	986	1,058	1,005
1,143**	1,053	998	994	1,160	1,380*
1,171*	1,068	1,018*	1,010	1,190*	1,437*
1,187***	1,088	1,054*	1,030	1,220*	1,687
1,240*	1,160	1,161**	1,099	1,275*	2,080***
1,276*	1,198	1,188*	1,153*	1,363**	2,202
1,357*	1,254	1,212**	1,409**	1,425**?[d]	2,275*
1,365*	1,360*	1,360	1,460**	1,460**	2,320*
1,385*	1,430***?	1,387***	1,780*	1,520	
1,422*	1,580*	1,410*	1,910**	1,585*	
1,458*	1,700**	1,433	2,305	1,702**	
1,485***	1,745*	1,703**	2,345	1,825**	
1,570*	1,780*	1,722***	2,510**	2,050*	
1,690***	1,825*	1,760**		2,079**	
1,735*	1,928***?	1,820		2,103**	
1,757	2,100***	1,900		2,145*	
1,928	2,180	1,930		2,172	
1,972***	2,200	2,005		2,268***	
2,055***	2,282***	2,120*		2,335**	
2,140	2,320**	2,142**		2,355**	
2,162**	2,370	2,179*		2,370	
2,182*	2,445*	2,265*		2,390	
2,203*	2,485**	2,306***		2,410	
2,265**	2,540*	2,342***		2,445	
2,300**		2,380*		2,480***	
2,345**		2,460*		2,530	
2,390		2,480		2,560	
2,420*		2,515		2,582*	
2,462**		2,535			
2,520					
2,545					
2,560					

[a] Source: Instrumentation Research Laboratory, Beltsville Agricultural Research Center, U.S. Department of Agriculture, Beltsville, MD.

[b] Wavelengths determined from second derivative log $(1/T)$ spectra for oil, water, and sugar and log $(1/R)$ spectra for protein, starch, and cellulose. Data recorded with a spectral bandpass of 7 nm and derivatives computed with a spacing of 10 nm between points. The sugar spectra were obtained by subtracting the water from a 50% sugar-water solution.

[c] Asterisks indicate relative strength with *** for strongest bands.

[d] Question marks indicate possible water bands.

VI. ILLUSTRATIONS OF NEAR-INFRARED SPECTRA

In the section following this chapter, the NIR spectra of many substances of pure chemicals and of agricultural and food products are shown. The reader is encouraged to refer to these spectra for further illustration of the concepts discussed herein. Table I illustrates principal bands identified in some constituents of foods. The constituents were isolated in as pure a form as practical.

LITERATURE CITED

BERTRAND, D., ROBERT, P., and LOISEL, W. 1985. Identification of some wheat varieties by near-infrared reflectance spectroscopy. J. Sci. Food Agric. 36:1120-1124.

MAHALANOBIS, P. C. 1936. On the generalized distance in statistics. Proc. Natl. Inst. Sci. India 2:49-55.

MARK, H. L., and TUNNELL, D. 1985. Qualitative near-infrared reflectance analysis using Mahalanobis distances. Anal. Chem. 57:1449-1456.

Near-Infrared Spectra

The figures presented on the following pages illustrate the near-infrared reflectance spectra of a large number of substances of pure chemicals and agricultural and food products. The second-derivative spectra are displayed above the log $(1/R)$ spectra to accentuate the absorbance peaks. For the second-derivative traces, segment size was one wavelength point (about 1.6 nm) and derivative (gap) was six points (about 10 nm). A key to spectra is on pages 289–290.

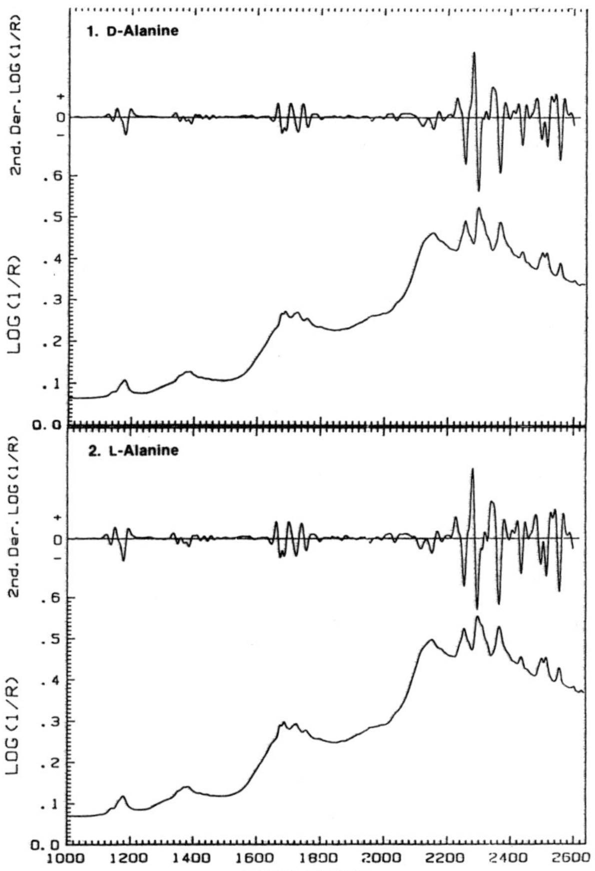

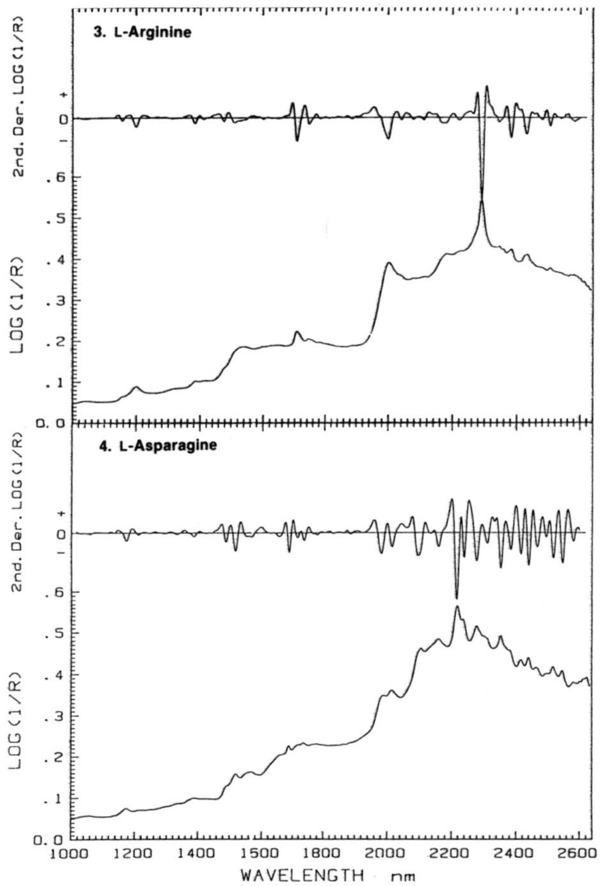

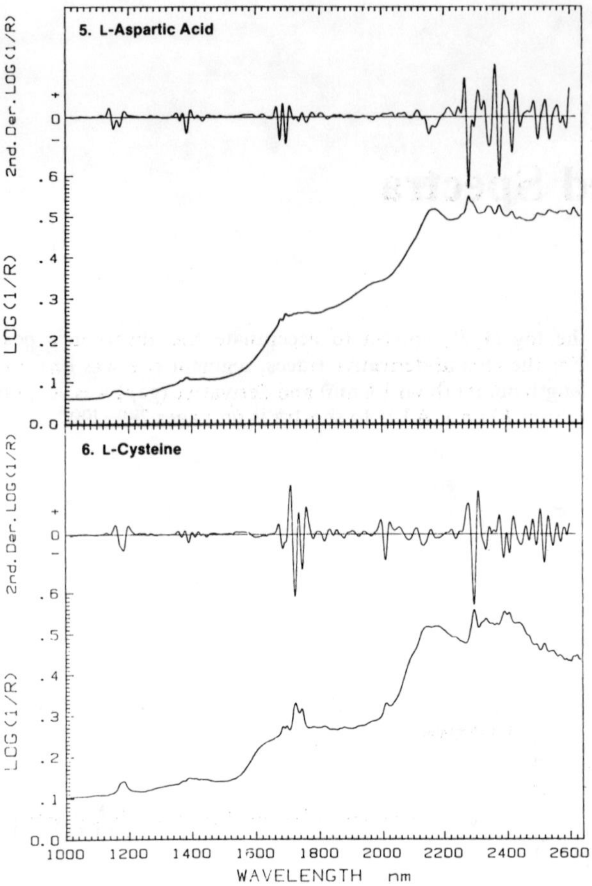
5. L-Aspartic Acid
6. L-Cysteine
2nd. Der. LOG (1/R)
LOG (1/R)
WAVELENGTH nm

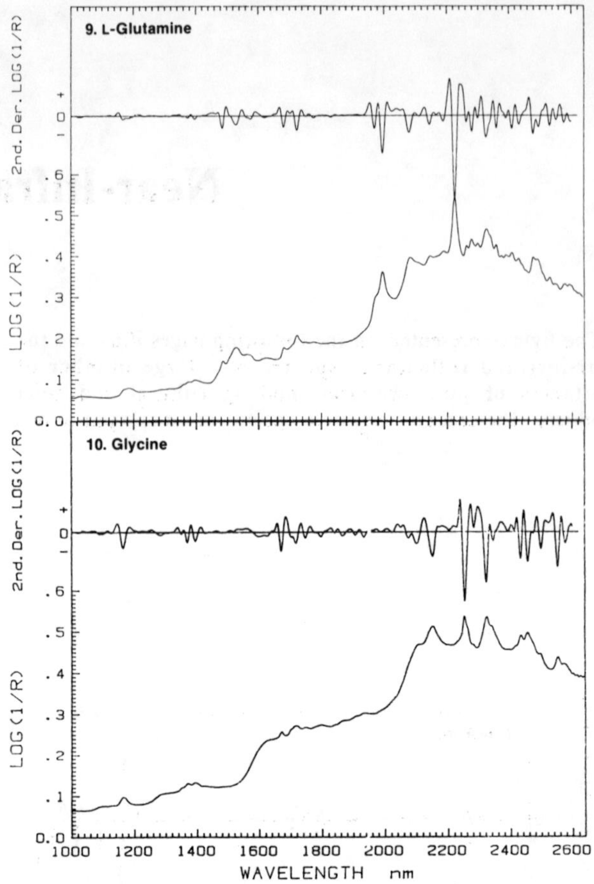
9. L-Glutamine
10. Glycine
2nd. Der. LOG (1/R)
LOG (1/R)
WAVELENGTH nm

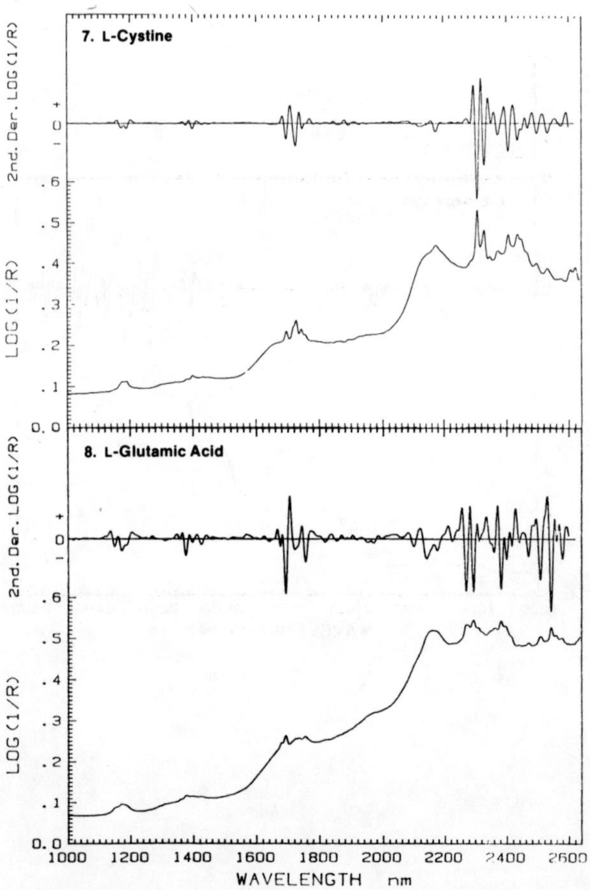
7. L-Cystine
8. L-Glutamic Acid
2nd. Der. LOG (1/R)
LOG (1/R)
WAVELENGTH nm

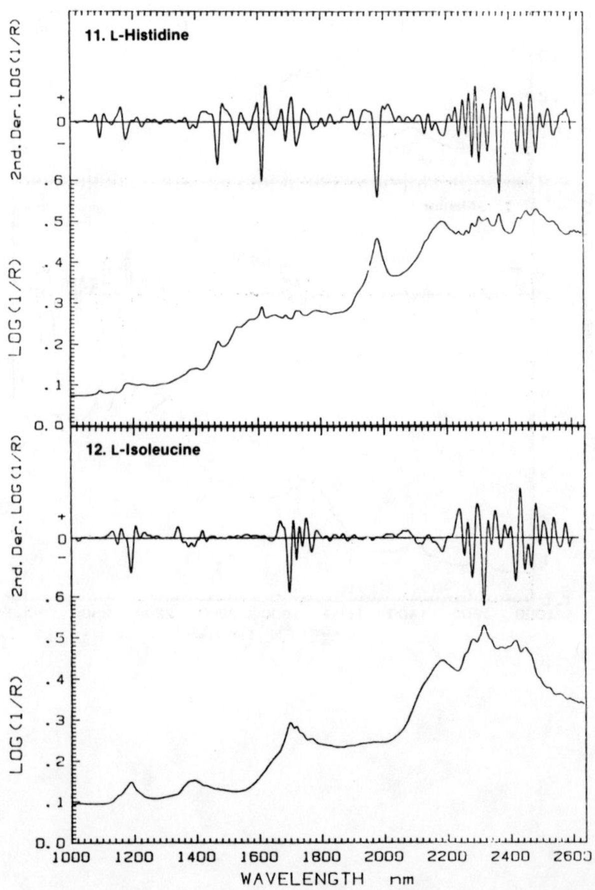
11. L-Histidine
12. L-Isoleucine
2nd. Der. LOG (1/R)
LOG (1/R)
WAVELENGTH nm

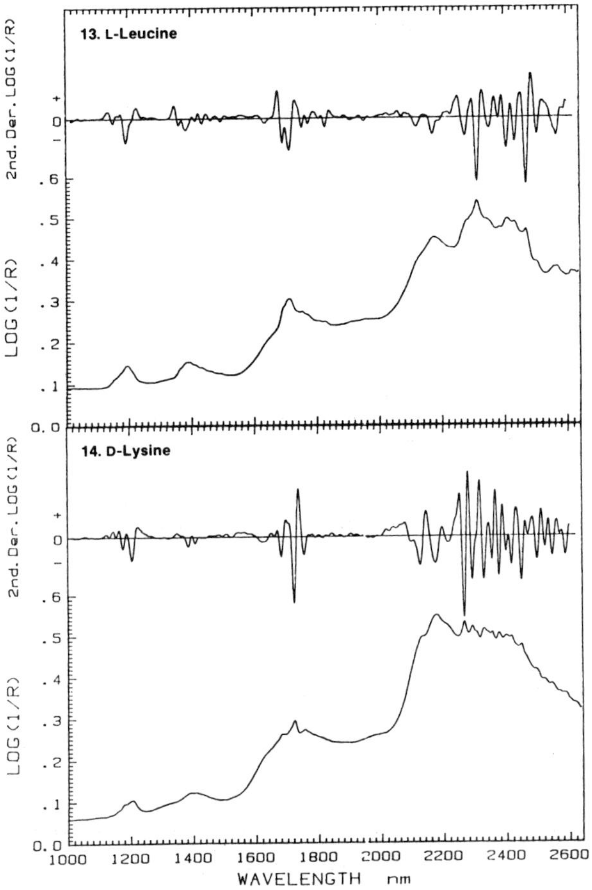
13. L-Leucine
14. D-Lysine
2nd. Der. LOG (1/R)
LOG (1/R)
WAVELENGTH nm

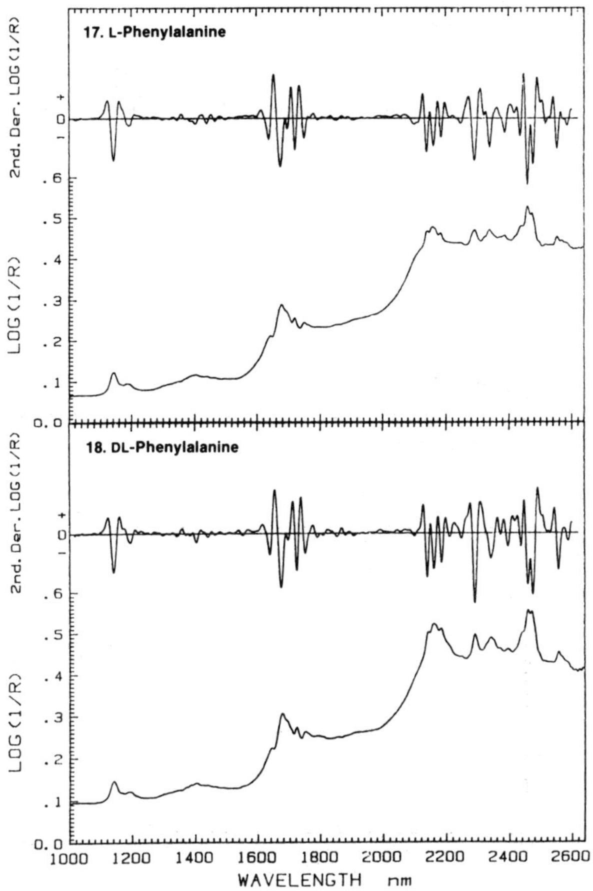
17. L-Phenylalanine
18. DL-Phenylalanine
2nd. Der. LOG (1/R)
LOG (1/R)
WAVELENGTH nm

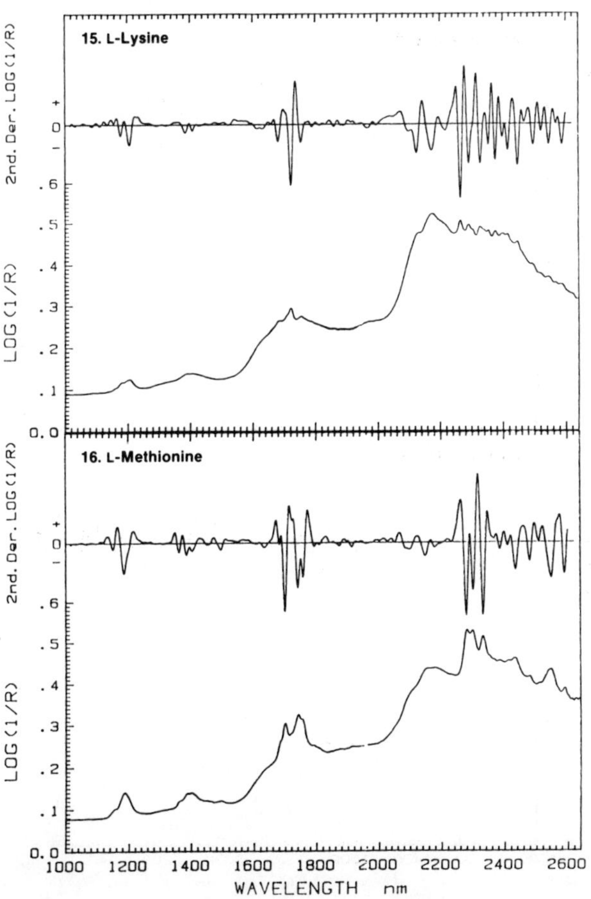
15. L-Lysine
16. L-Methionine
2nd. Der. LOG (1/R)
LOG (1/R)
WAVELENGTH nm

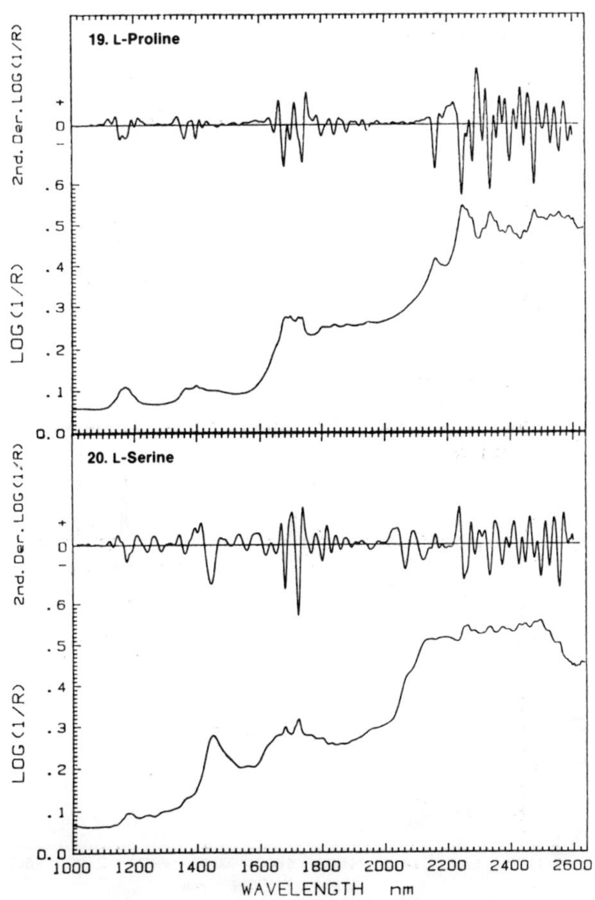
19. L-Proline
20. L-Serine
2nd. Der. LOG (1/R)
LOG (1/R)
WAVELENGTH nm

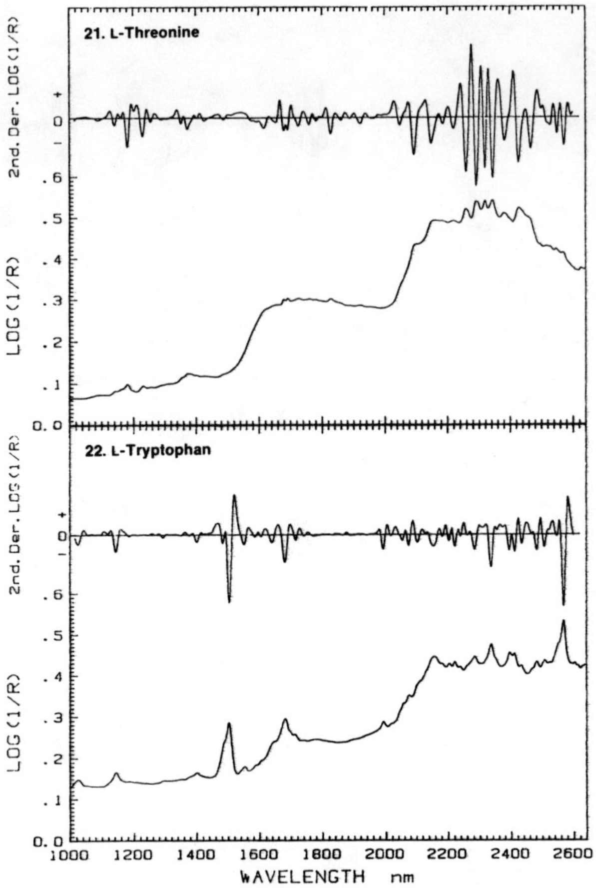
21. L-Threonine
22. L-Tryptophan
2nd. Der. LOG (1/R)
LOG (1/R)
WAVELENGTH nm

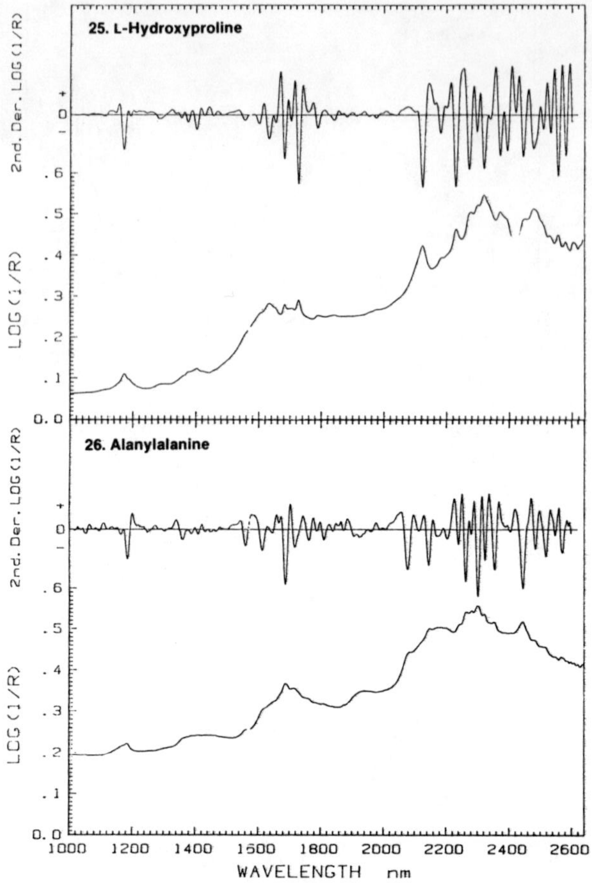
25. L-Hydroxyproline
26. Alanylalanine
2nd. Der. LOG (1/R)
LOG (1/R)
WAVELENGTH nm

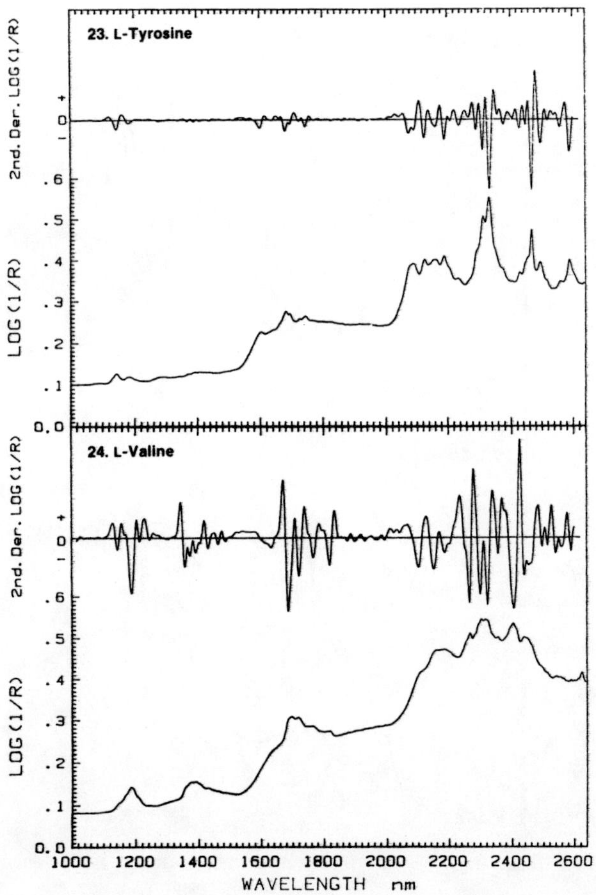
23. L-Tyrosine
24. L-Valine
2nd. Der. LOG (1/R)
LOG (1/R)
WAVELENGTH nm

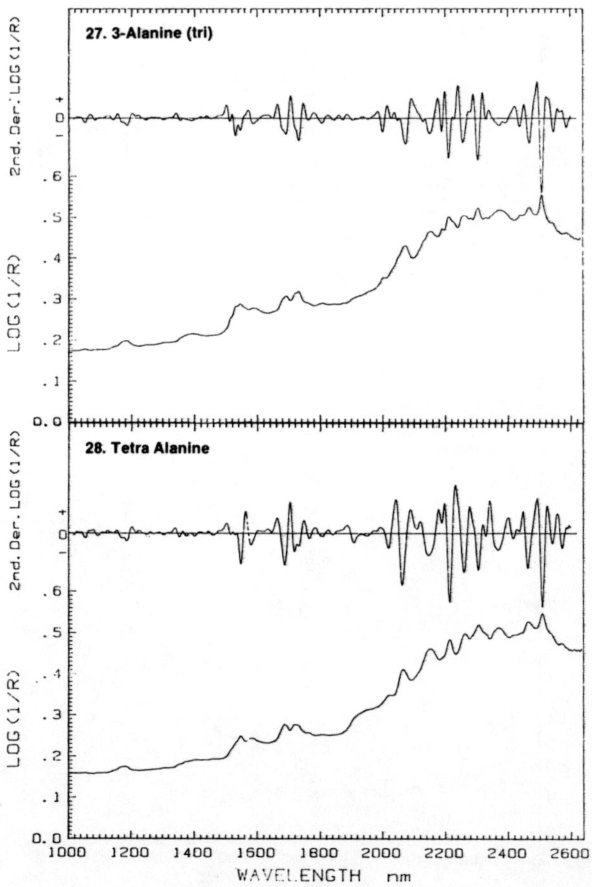
27. 3-Alanine (tri)
28. Tetra Alanine
2nd. Der. LOG (1/R)
LOG (1/R)
WAVELENGTH nm

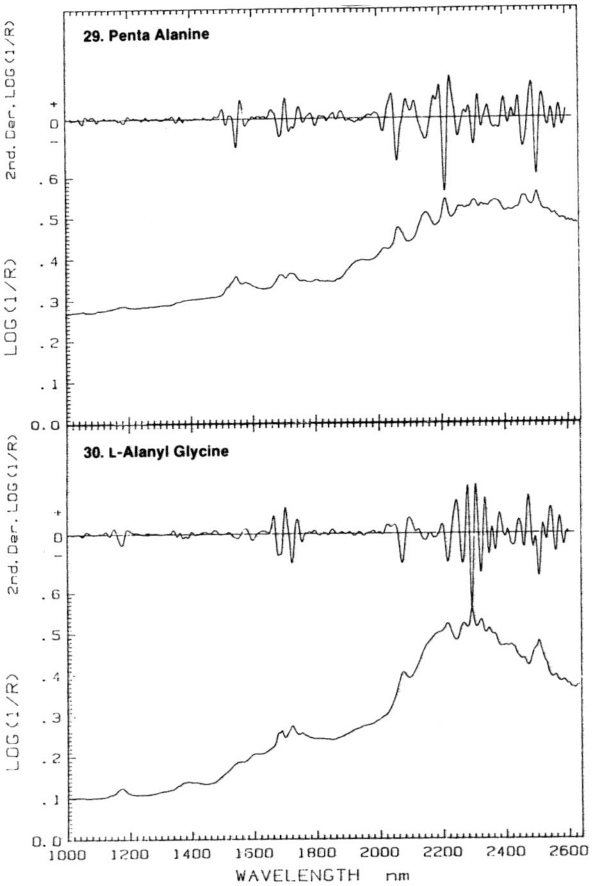
29. Penta Alanine
30. L-Alanyl Glycine
2nd. Der. LOG(1/R)
LOG(1/R)
WAVELENGTH nm

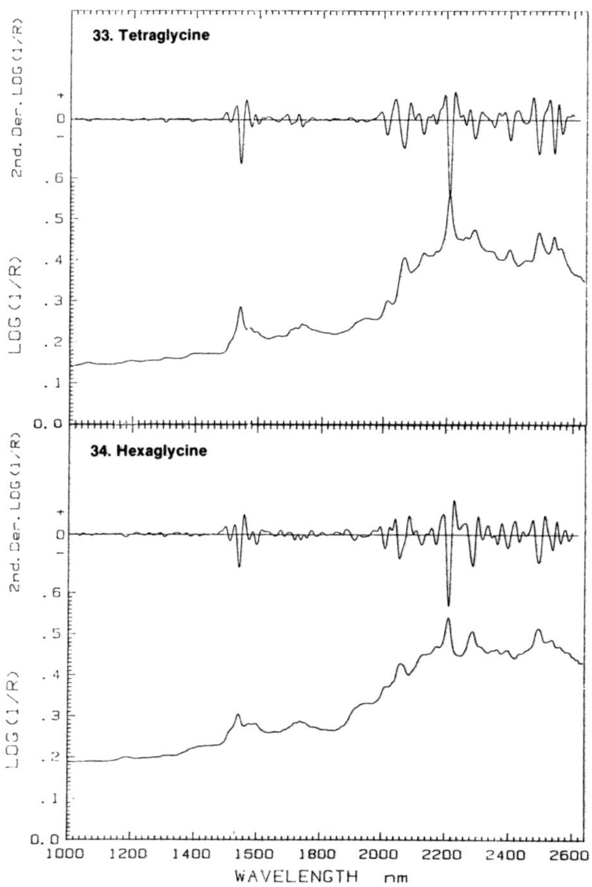
33. Tetraglycine
34. Hexaglycine
2nd. Der. LOG(1/R)
LOG(1/R)
WAVELENGTH nm

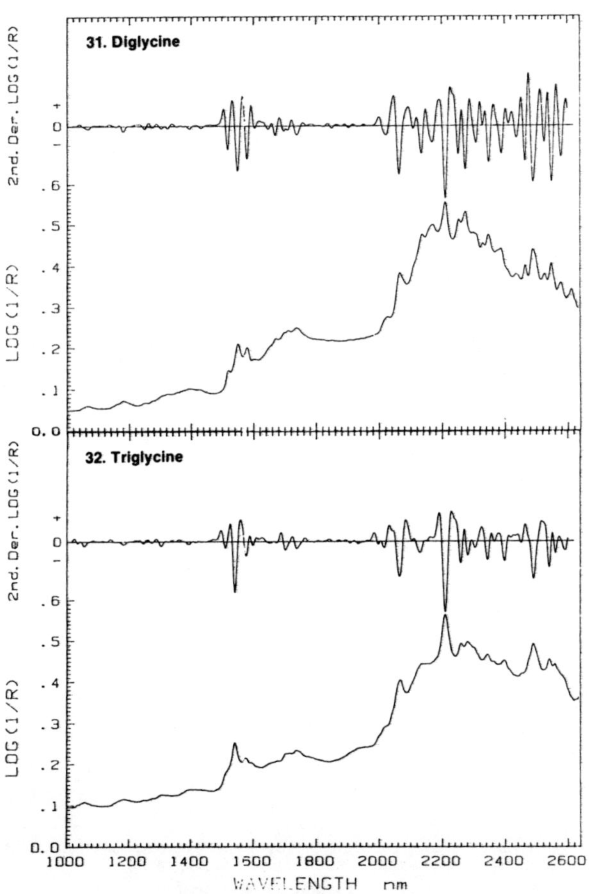
31. Diglycine
32. Triglycine
2nd. Der. LOG(1/R)
LOG(1/R)
WAVELENGTH nm

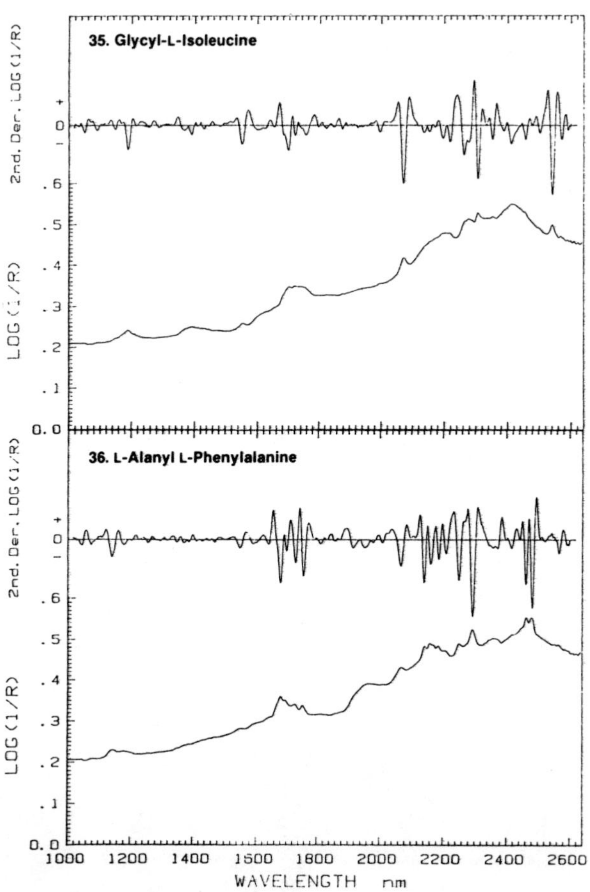
35. Glycyl-L-Isoleucine
36. L-Alanyl L-Phenylalanine
2nd. Der. LOG(1/R)
LOG(1/R)
WAVELENGTH nm

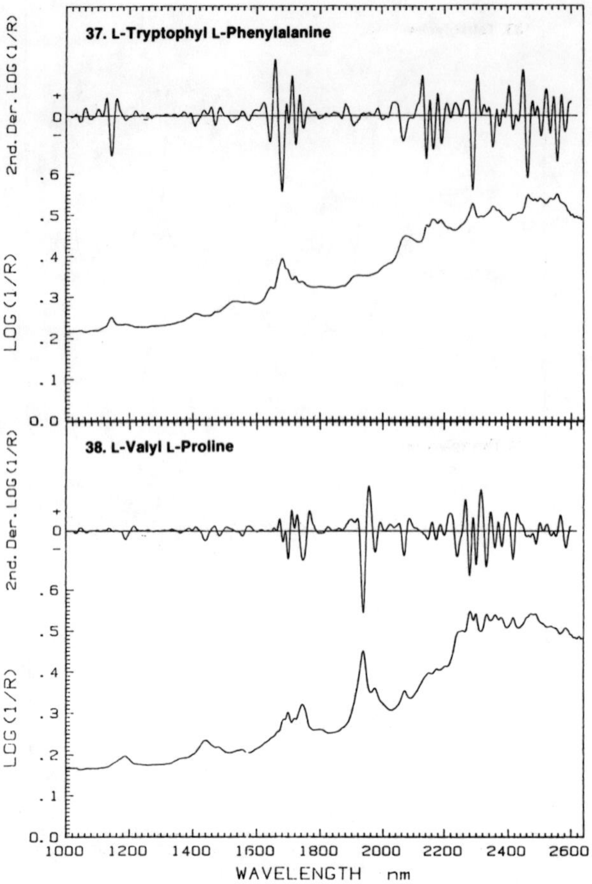
37. L-Tryptophyl L-Phenylalanine
38. L-Valyl L-Proline
2nd. Der. LOG (1/R)
LOG (1/R)
WAVELENGTH nm

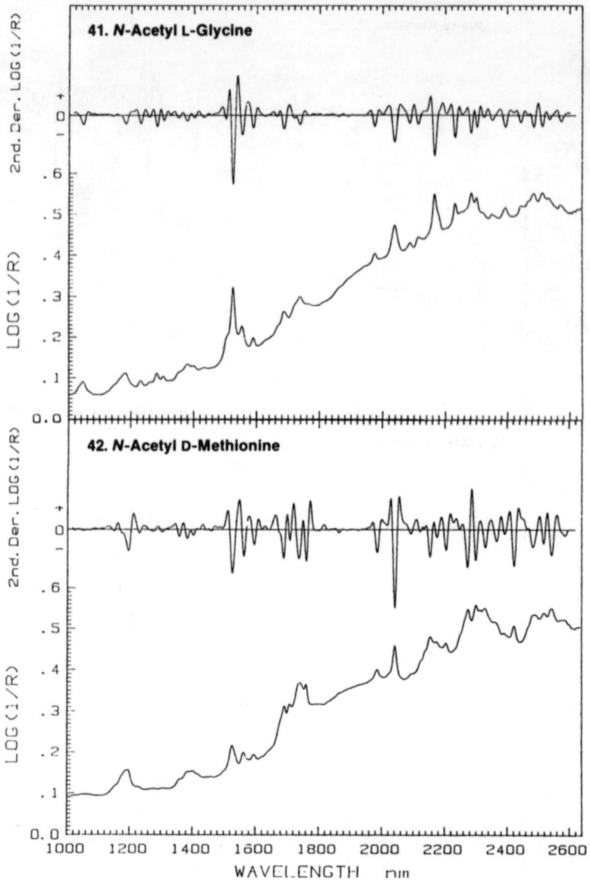
41. N-Acetyl L-Glycine
42. N-Acetyl D-Methionine
2nd. Der. LOG (1/R)
LOG (1/R)
WAVELENGTH nm

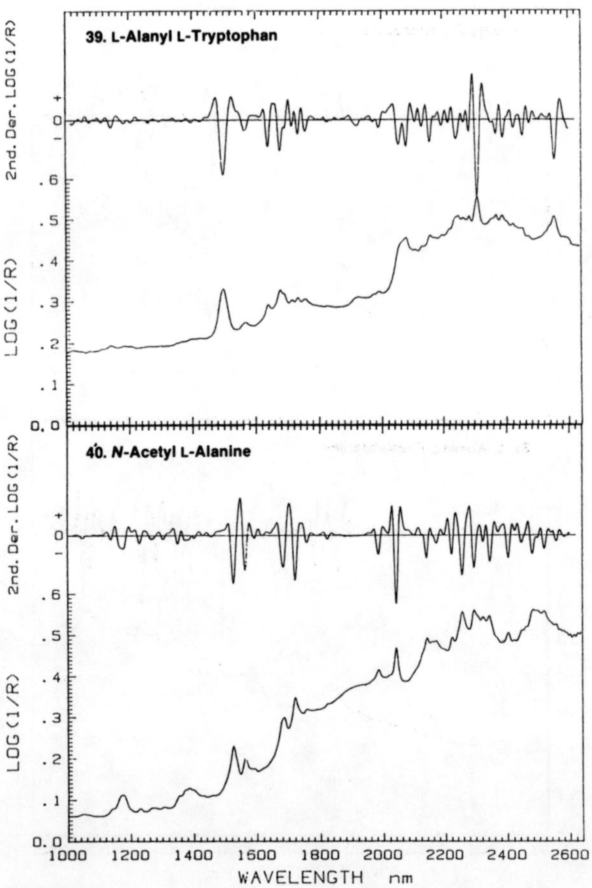
39. L-Alanyl L-Tryptophan
40. N-Acetyl L-Alanine
2nd. Der. LOG (1/R)
LOG (1/R)
WAVELENGTH nm

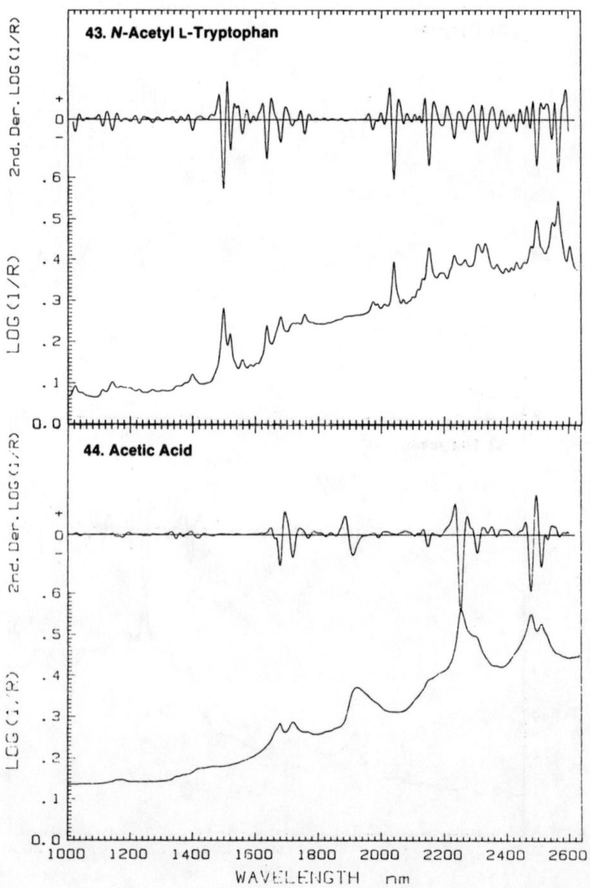
43. N-Acetyl L-Tryptophan
44. Acetic Acid
2nd. Der. LOG (1/R)
LOG (1/R)
WAVELENGTH nm

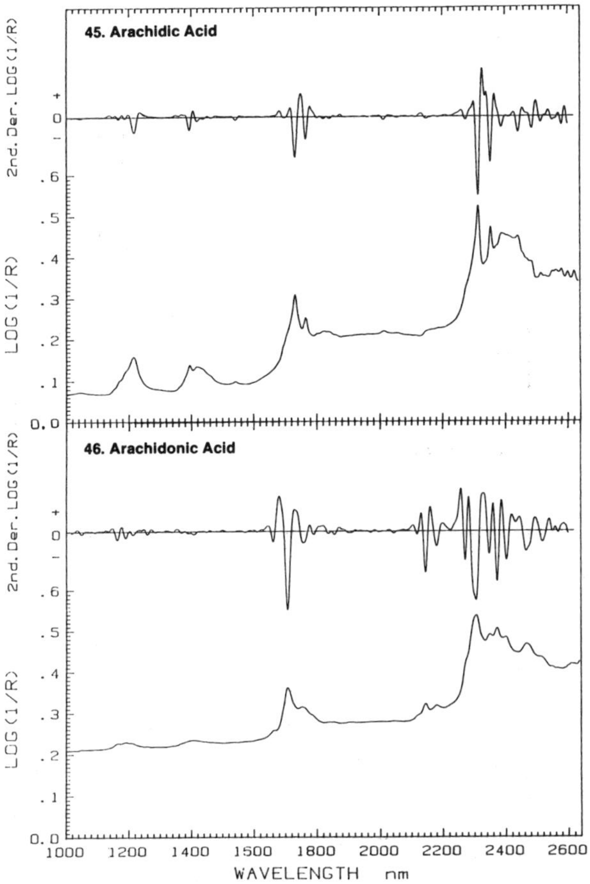
45. Arachidic Acid
46. Arachidonic Acid
2nd. Der. LOG (1/R)
LOG (1/R)
WAVELENGTH nm

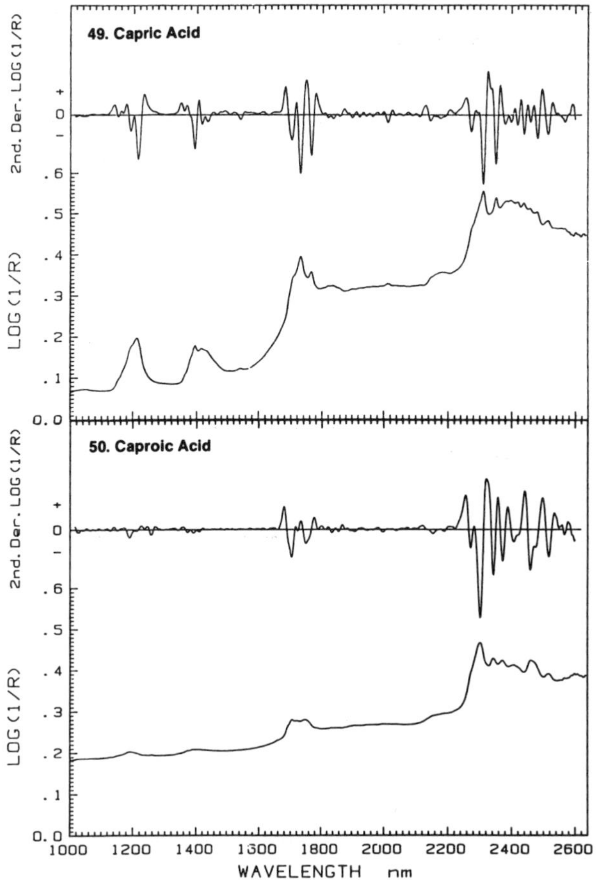
49. Capric Acid
50. Caproic Acid
2nd. Der. LOG (1/R)
LOG (1/R)
WAVELENGTH nm

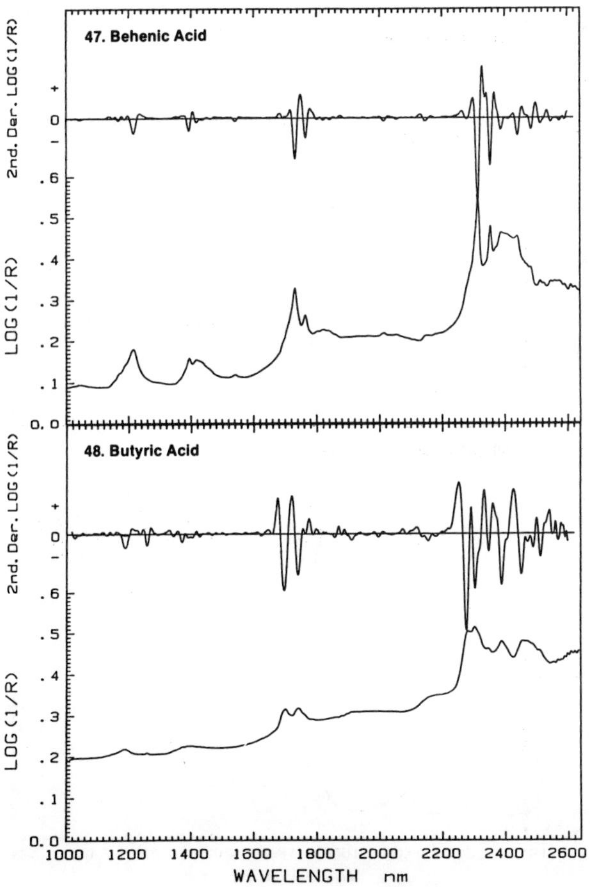
47. Behenic Acid
48. Butyric Acid
2nd. Der. LOG (1/R)
LOG (1/R)
WAVELENGTH nm

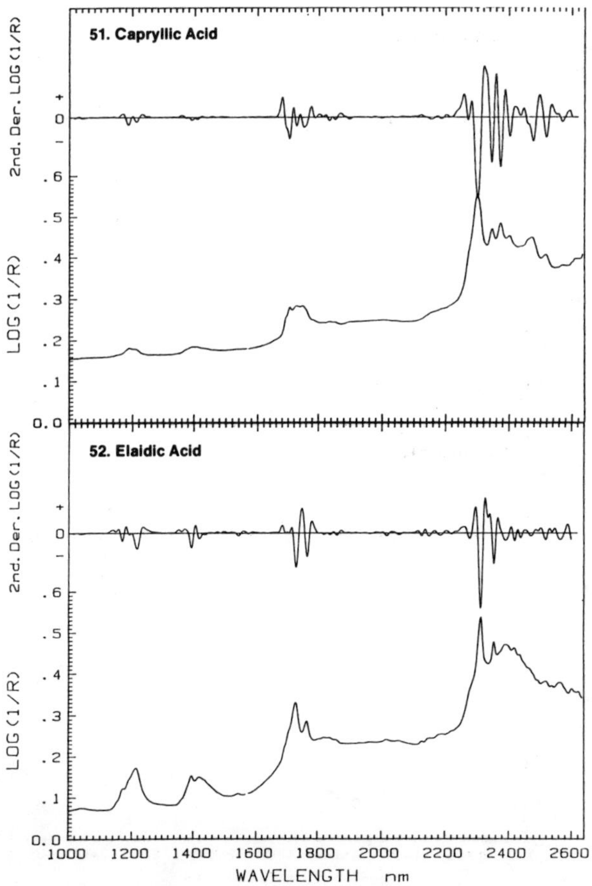
51. Caprylic Acid
52. Elaidic Acid
2nd. Der. LOG (1/R)
LOG (1/R)
WAVELENGTH nm

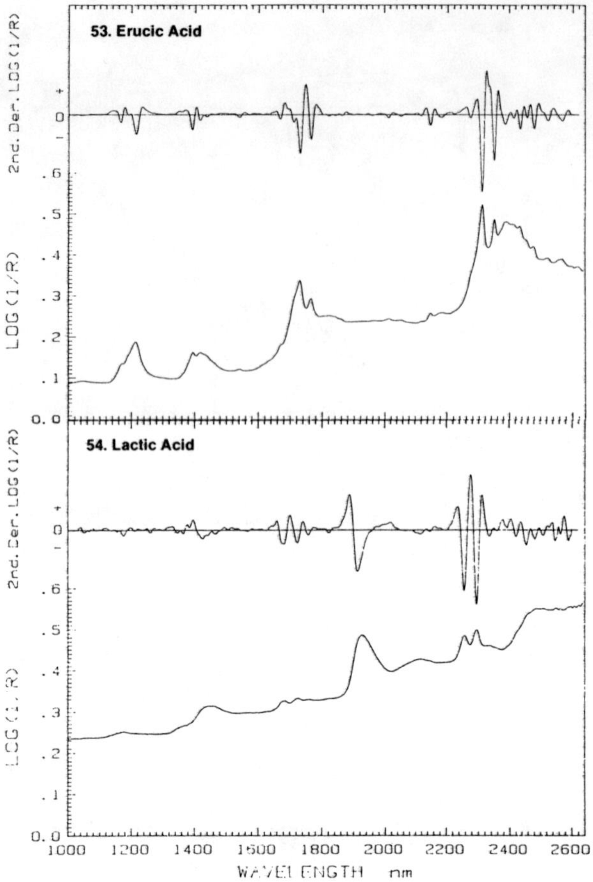
53. Erucic Acid
54. Lactic Acid
2nd. Der. LOG (1/R)
LOG (1/R)
WAVELENGTH nm

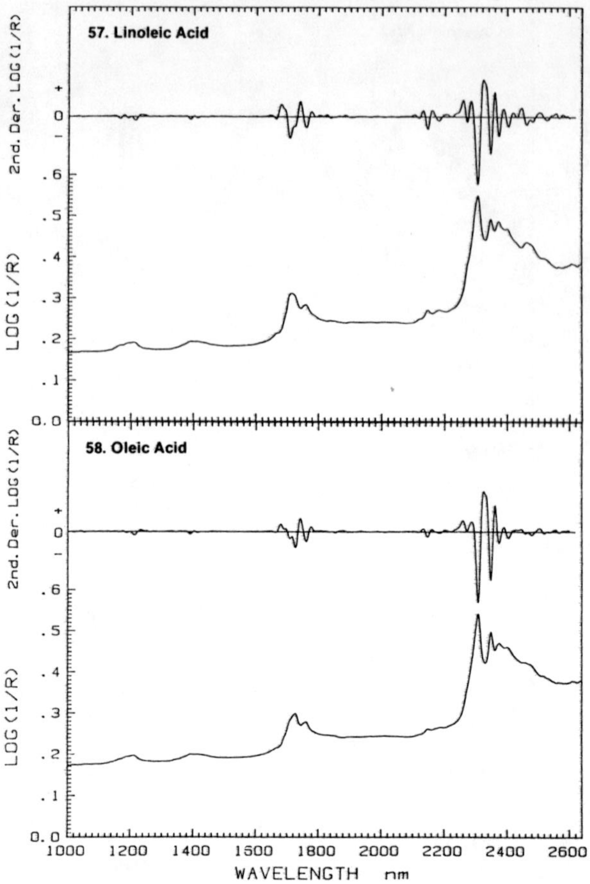
57. Linoleic Acid
58. Oleic Acid
2nd. Der. LOG (1/R)
LOG (1/R)
WAVELENGTH nm

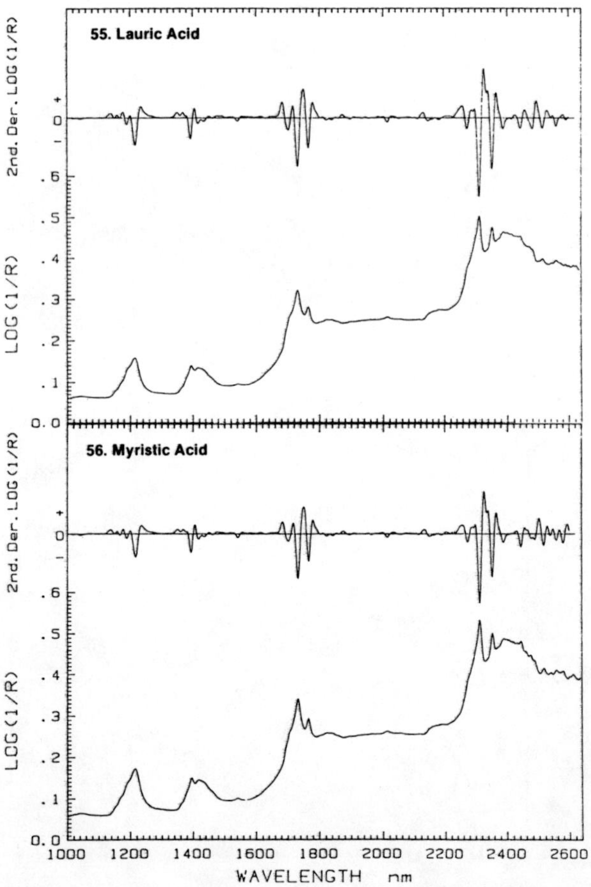
55. Lauric Acid
56. Myristic Acid
2nd. Der. LOG (1/R)
LOG (1/R)
WAVELENGTH nm

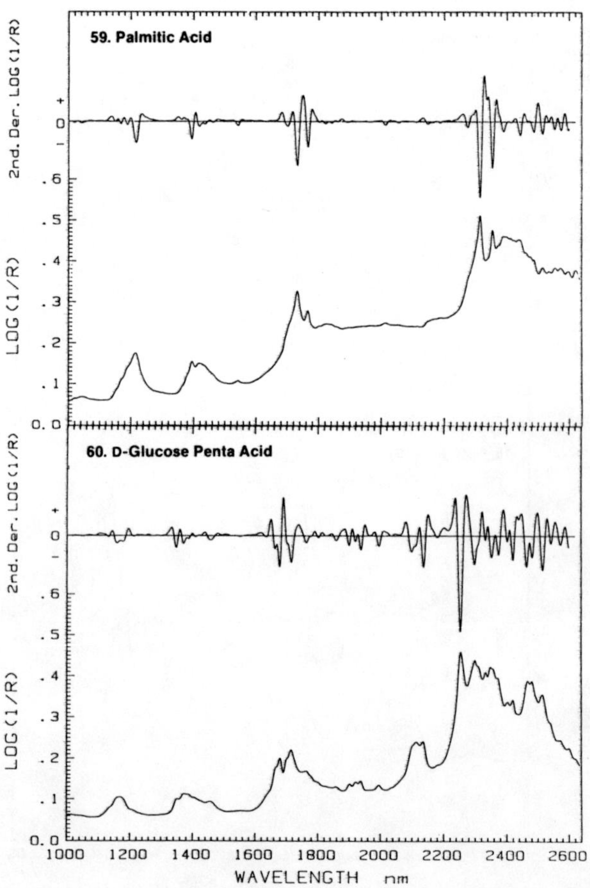
59. Palmitic Acid
60. D-Glucose Penta Acid
2nd. Der. LOG (1/R)
LOG (1/R)
WAVELENGTH nm

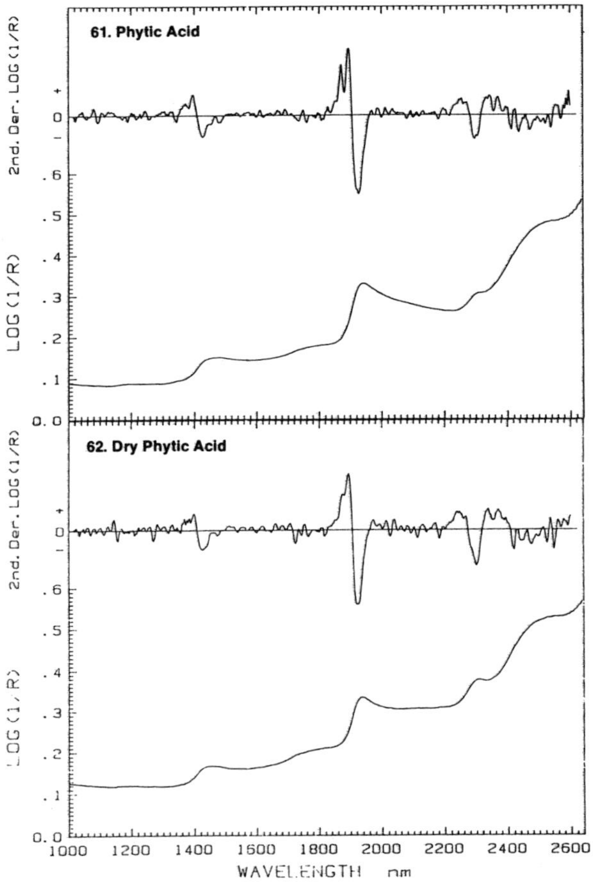
61. Phytic Acid
62. Dry Phytic Acid
2nd. Der. LOG (1/R)
LOG (1/R)
WAVELENGTH nm

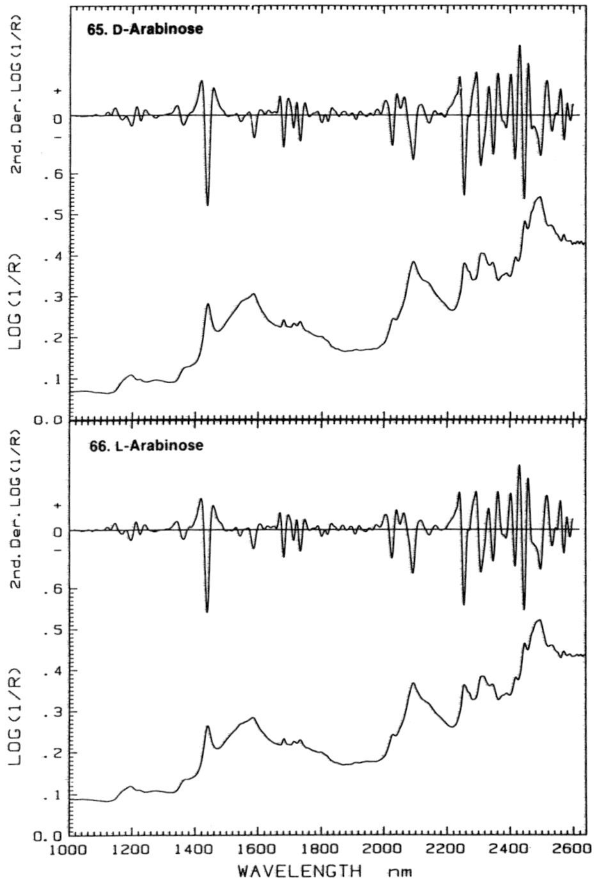
65. D-Arabinose
66. L-Arabinose
2nd. Der. LOG (1/R)
LOG (1/R)
WAVELENGTH nm

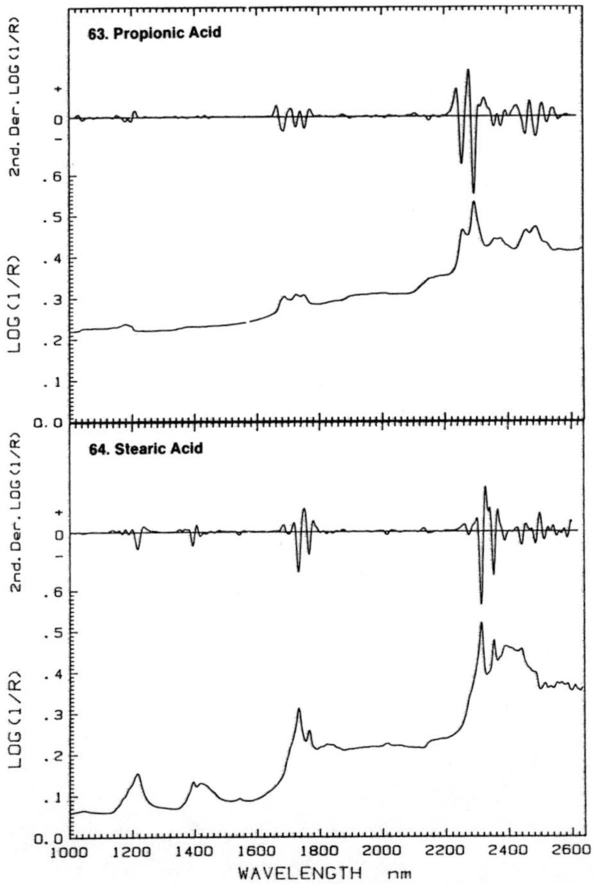
63. Propionic Acid
64. Stearic Acid
2nd. Der. LOG (1/R)
LOG (1/R)
WAVELENGTH nm

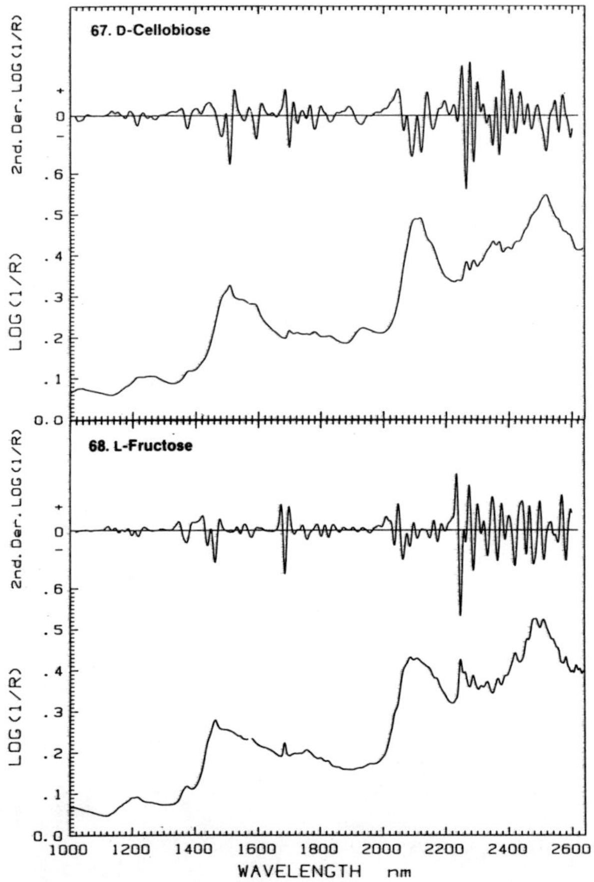
67. D-Cellobiose
68. L-Fructose
2nd. Der. LOG (1/R)
LOG (1/R)
WAVELENGTH nm

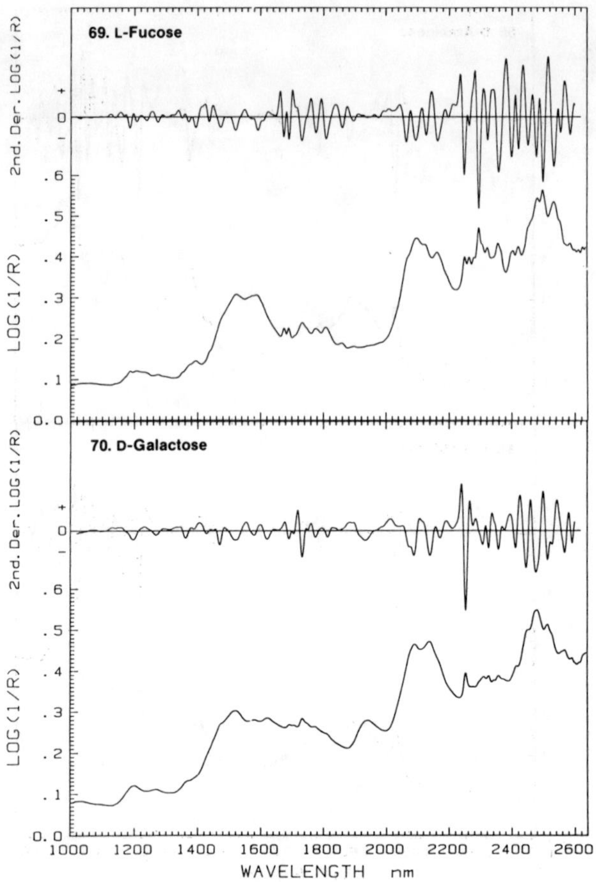
69. L-Fucose
70. D-Galactose
2nd. Der. LOG (1/R)
LOG (1/R)
WAVELENGTH nm

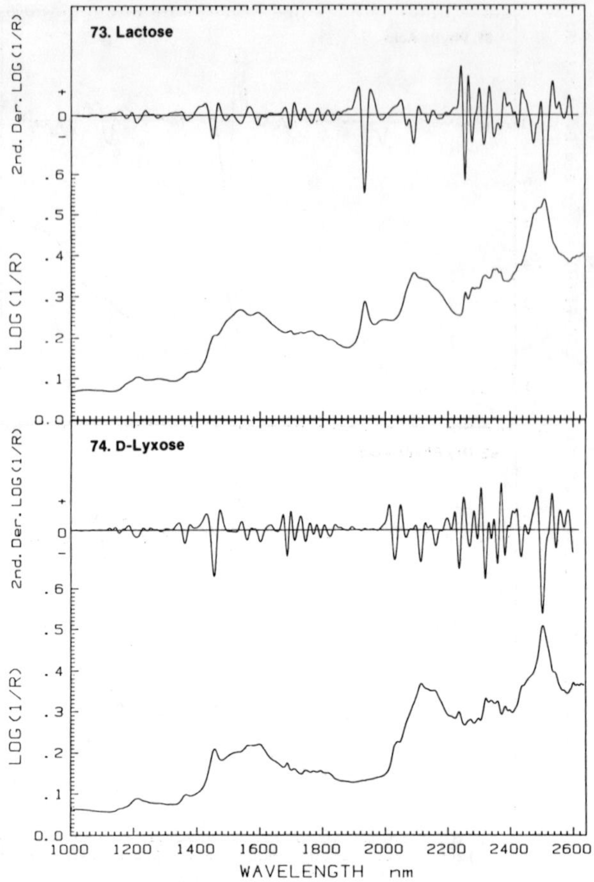
73. Lactose
74. D-Lyxose
2nd. Der. LOG (1/R)
LOG (1/R)
WAVELENGTH nm

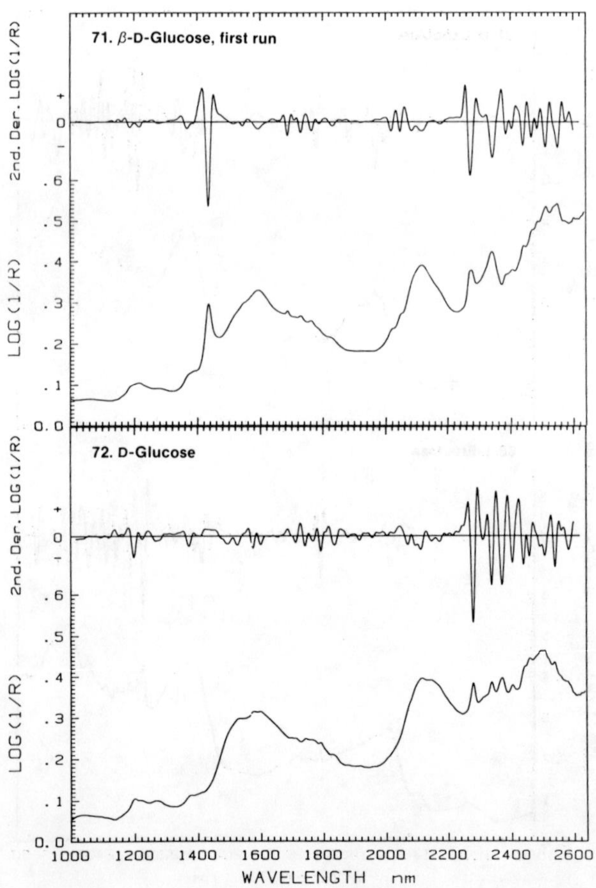
71. β-D-Glucose, first run
72. D-Glucose
2nd. Der. LOG (1/R)
LOG (1/R)
WAVELENGTH nm

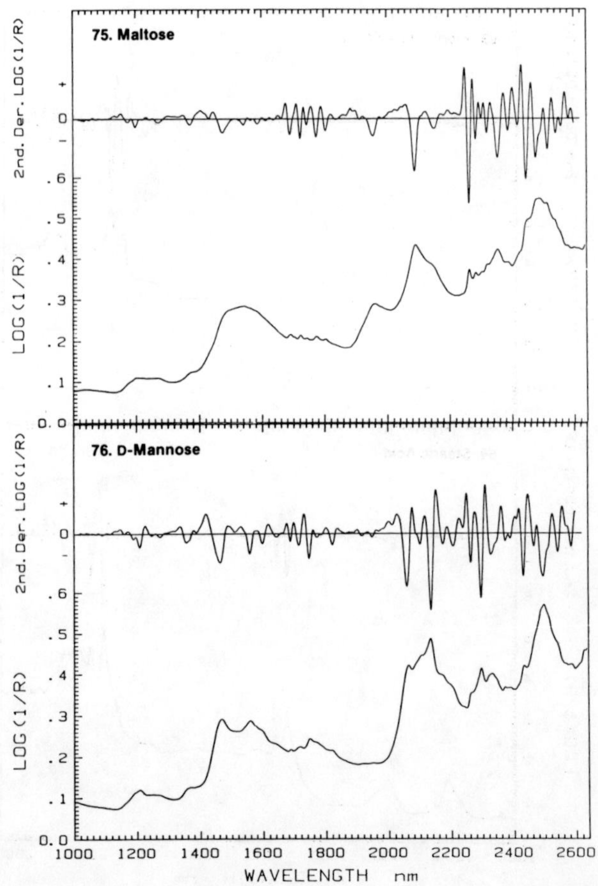
75. Maltose
76. D-Mannose
2nd. Der. LOG (1/R)
LOG (1/R)
WAVELENGTH nm

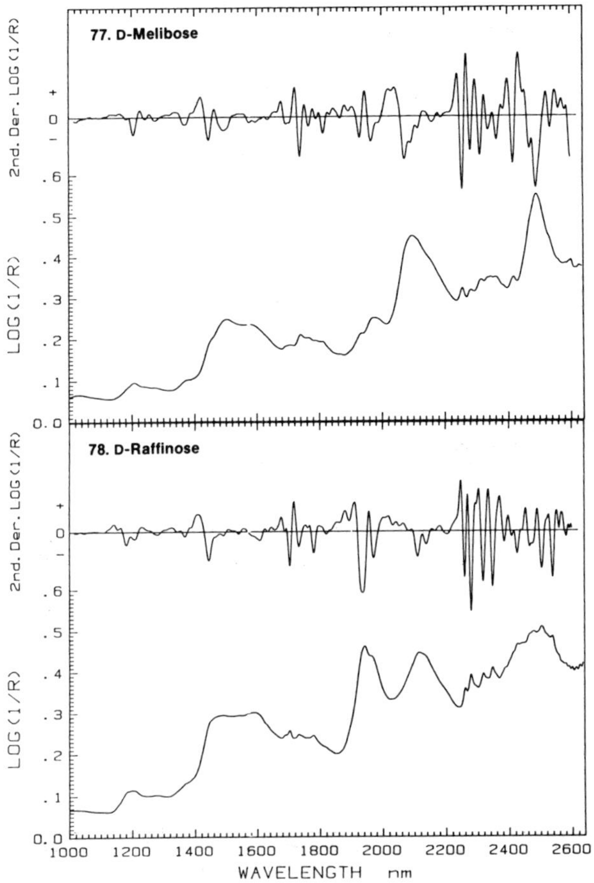
77. D-Melibose
78. D-Raffinose
2nd. Der. LOG (1/R)
LOG (1/R)
WAVELENGTH nm

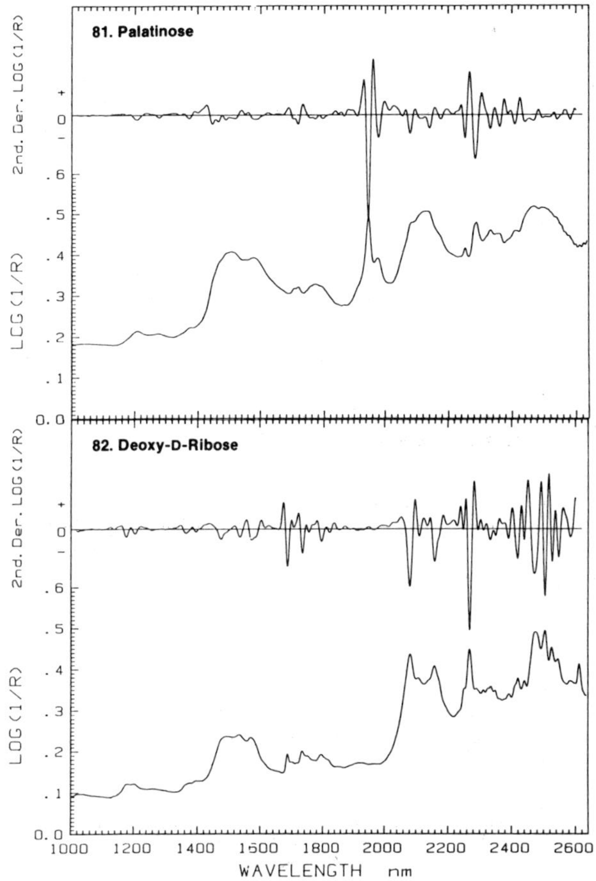
81. Palatinose
82. Deoxy-D-Ribose
2nd. Der. LOG (1/R)
LOG (1/R)
WAVELENGTH nm

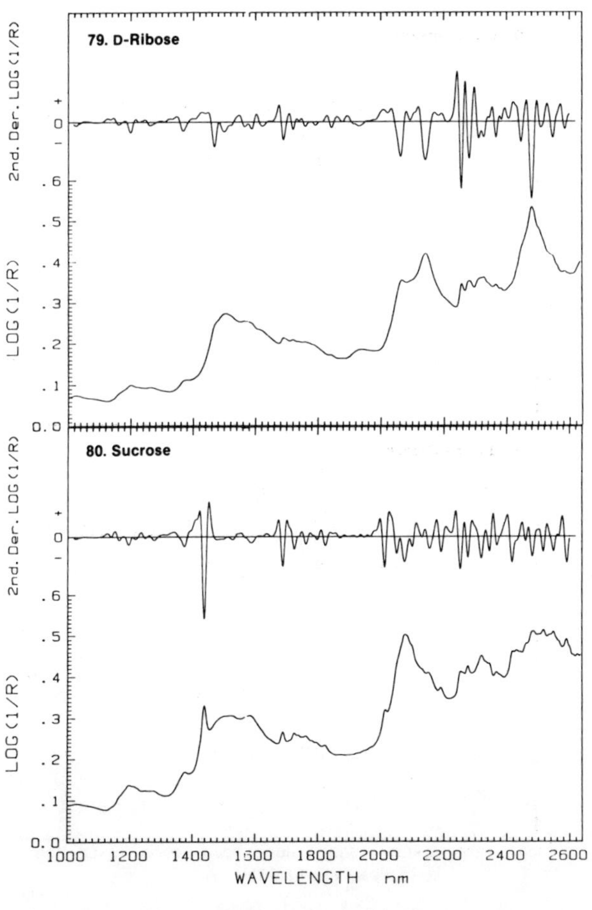
79. D-Ribose
80. Sucrose
2nd. Der. LOG (1/R)
LOG (1/R)
WAVELENGTH nm

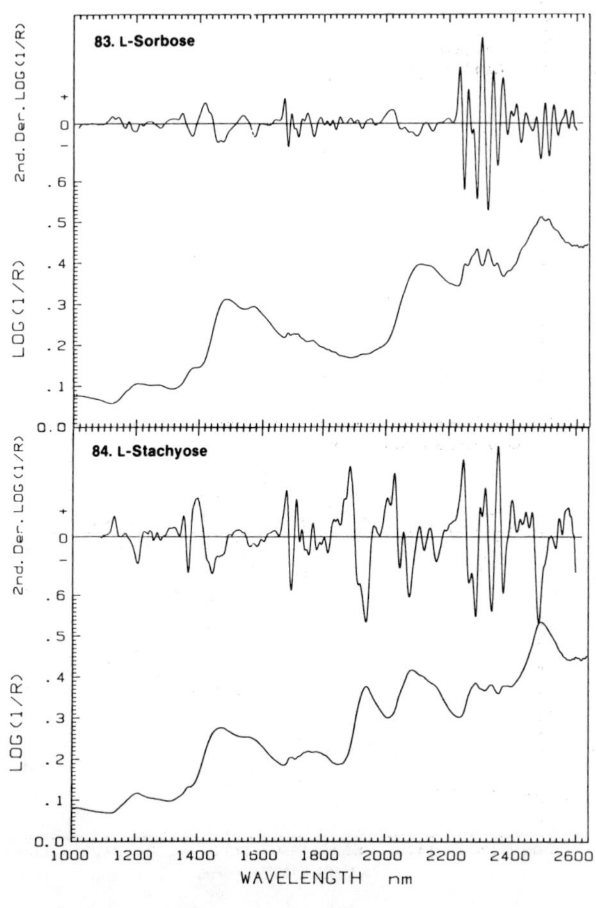
83. L-Sorbose
84. L-Stachyose
2nd. Der. LOG (1/R)
LOG (1/R)
WAVELENGTH nm

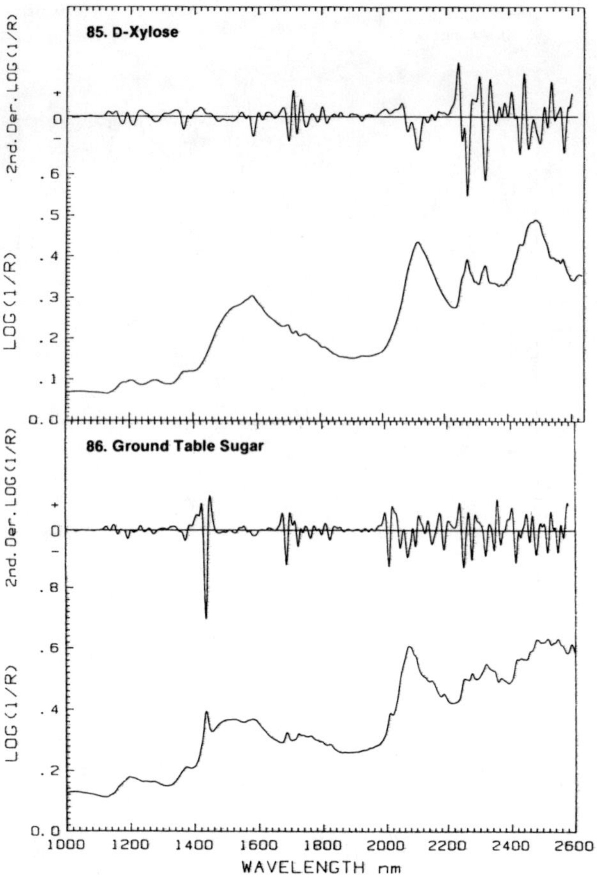
85. D-Xylose
86. Ground Table Sugar
2nd. Der. LOG(1/R)
LOG(1/R)
WAVELENGTH nm

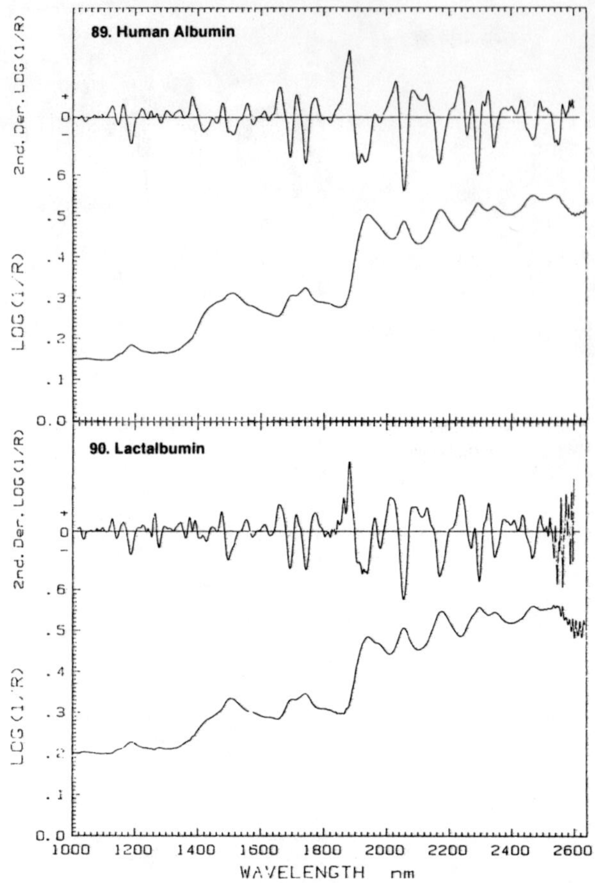
89. Human Albumin
90. Lactalbumin
2nd. Der. LOG(1/R)
LOG(1/R)
WAVELENGTH nm

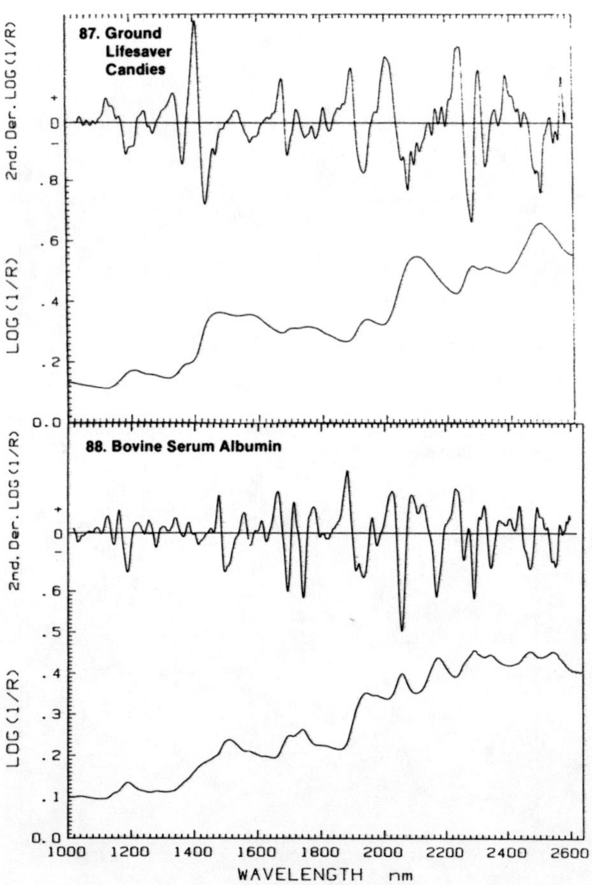
87. Ground Lifesaver Candies
88. Bovine Serum Albumin
2nd. Der. LOG(1/R)
LOG(1/R)
WAVELENGTH nm

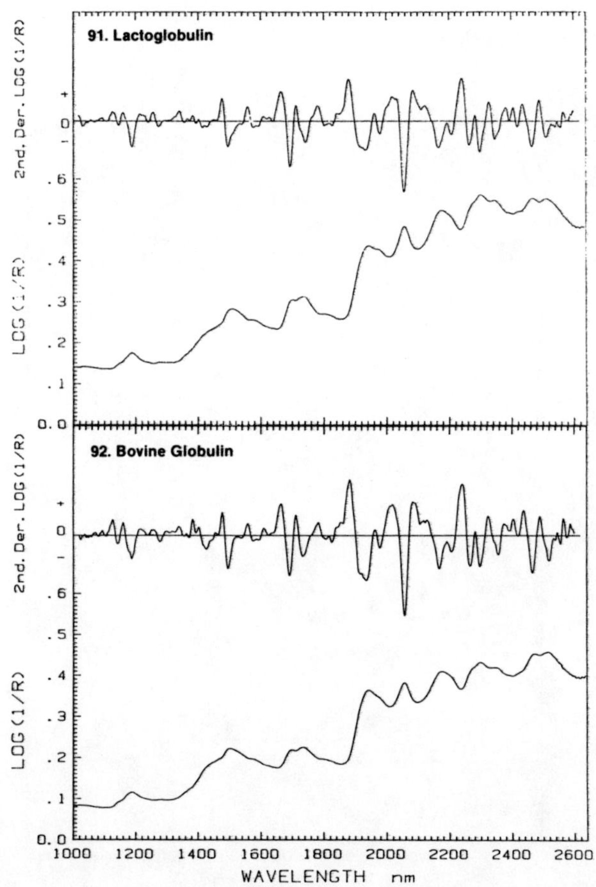
91. Lactoglobulin
92. Bovine Globulin
2nd. Der. LOG(1/R)
LOG(1/R)
WAVELENGTH nm

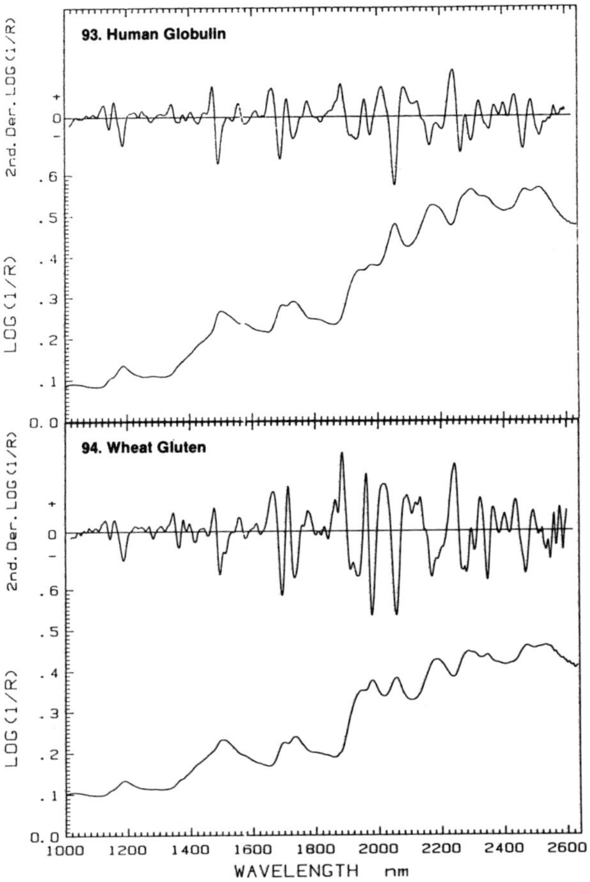
93. Human Globulin
94. Wheat Gluten
2nd. Der. LOG (1/R)
LOG (1/R)
WAVELENGTH nm

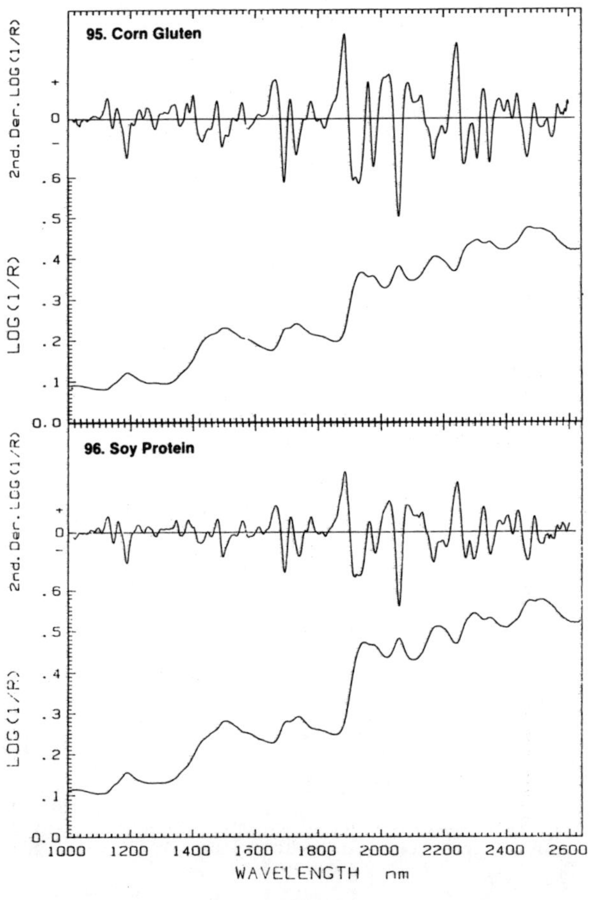
95. Corn Gluten
96. Soy Protein
2nd. Der. LOG (1/R)
LOG (1/R)
WAVELENGTH nm

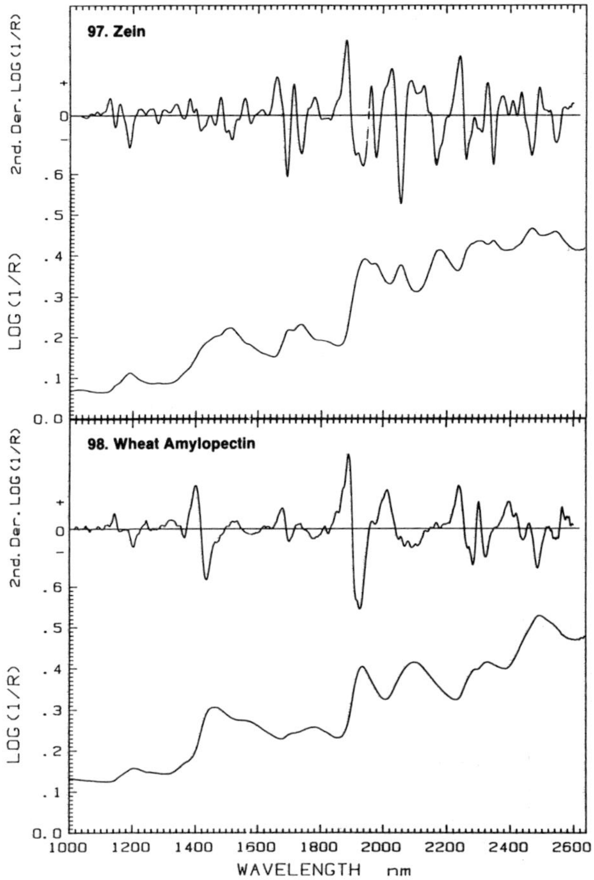
97. Zein
98. Wheat Amylopectin
2nd. Der. LOG (1/R)
LOG (1/R)
WAVELENGTH nm

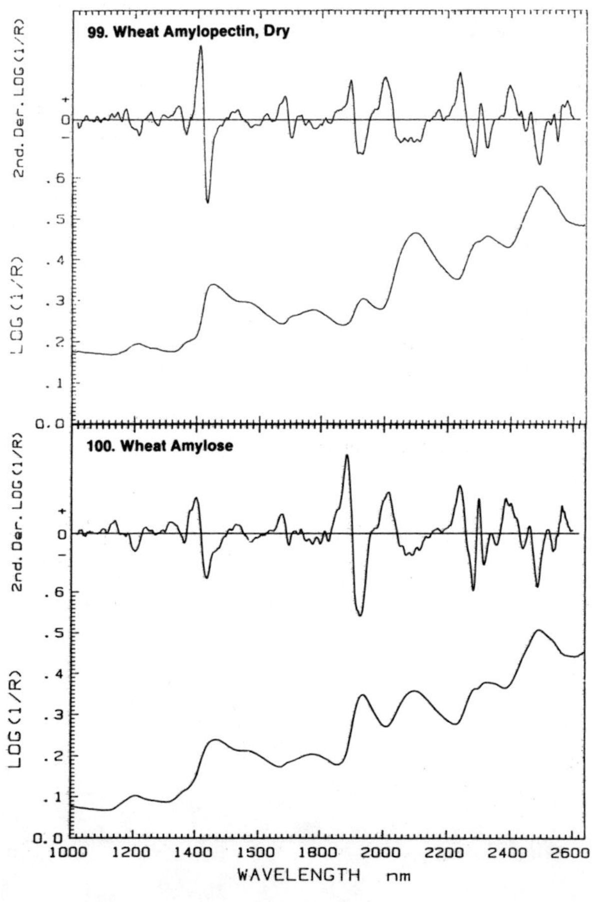
99. Wheat Amylopectin, Dry
100. Wheat Amylose
2nd. Der. LOG (1/R)
LOG (1/R)
WAVELENGTH nm

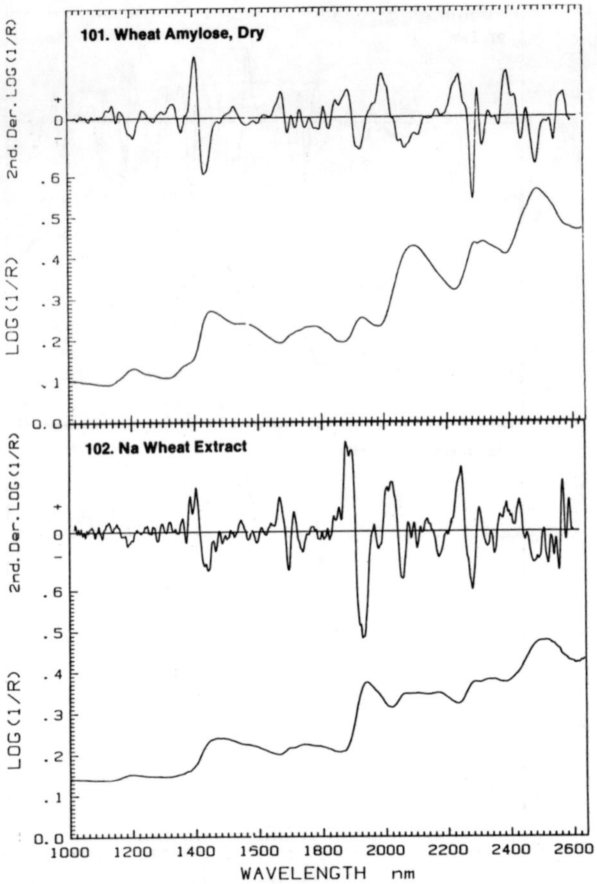
101. Wheat Amylose, Dry
102. Na Wheat Extract
2nd. Der. LOG (1/R)
LOG (1/R)
WAVELENGTH nm

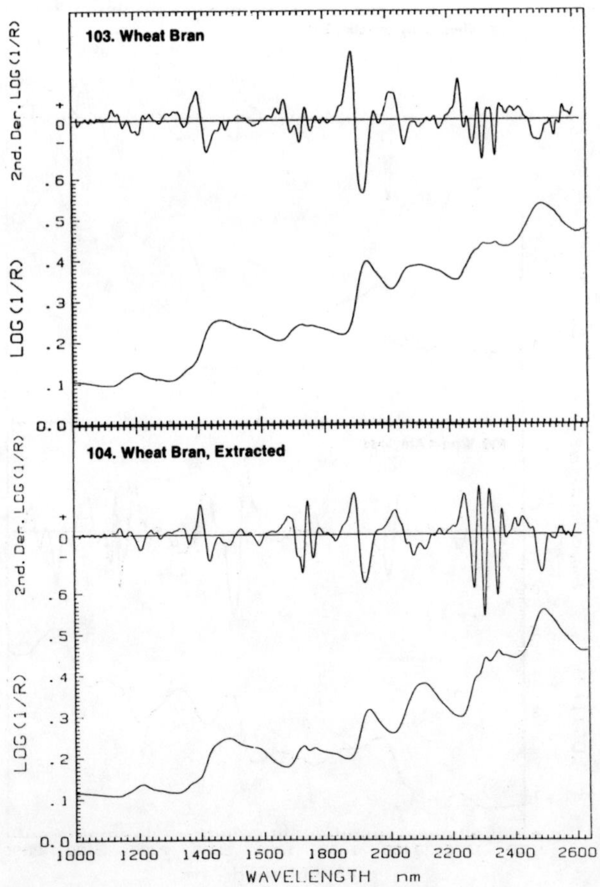
103. Wheat Bran
104. Wheat Bran, Extracted
2nd. Der. LOG (1/R)
LOG (1/R)
WAVELENGTH nm

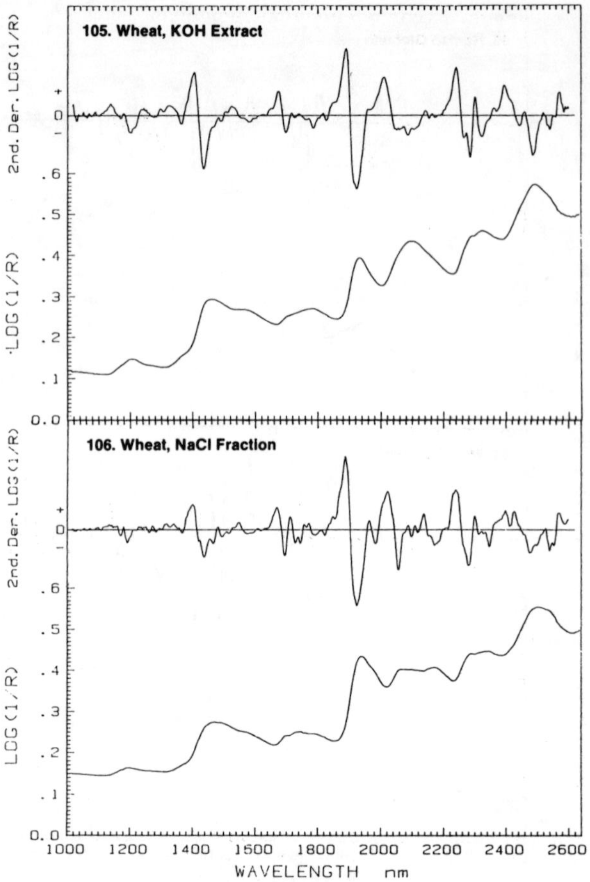
105. Wheat, KOH Extract
106. Wheat, NaCl Fraction
2nd. Der. LOG (1/R)
LOG (1/R)
WAVELENGTH nm

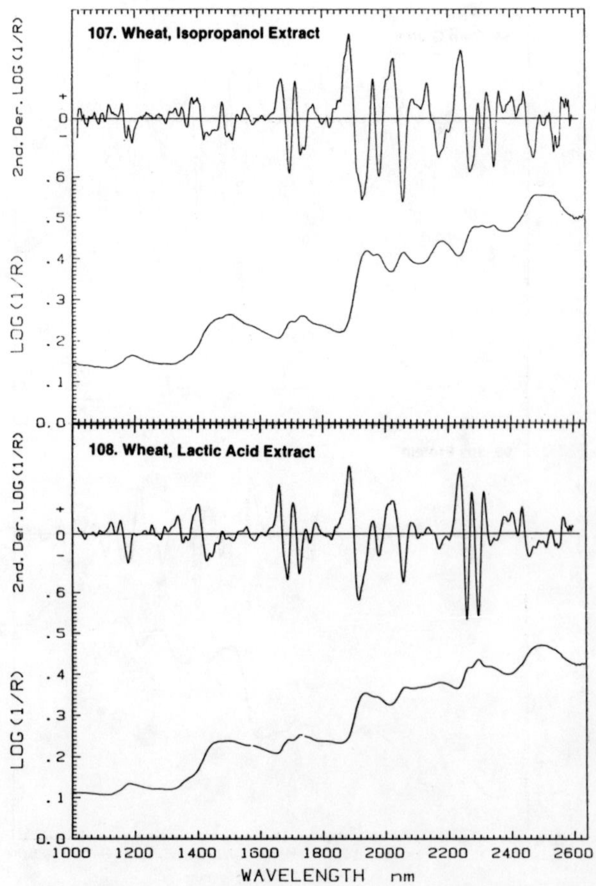
107. Wheat, Isopropanol Extract
108. Wheat, Lactic Acid Extract
2nd. Der. LOG (1/R)
LOG (1/R)
WAVELENGTH nm

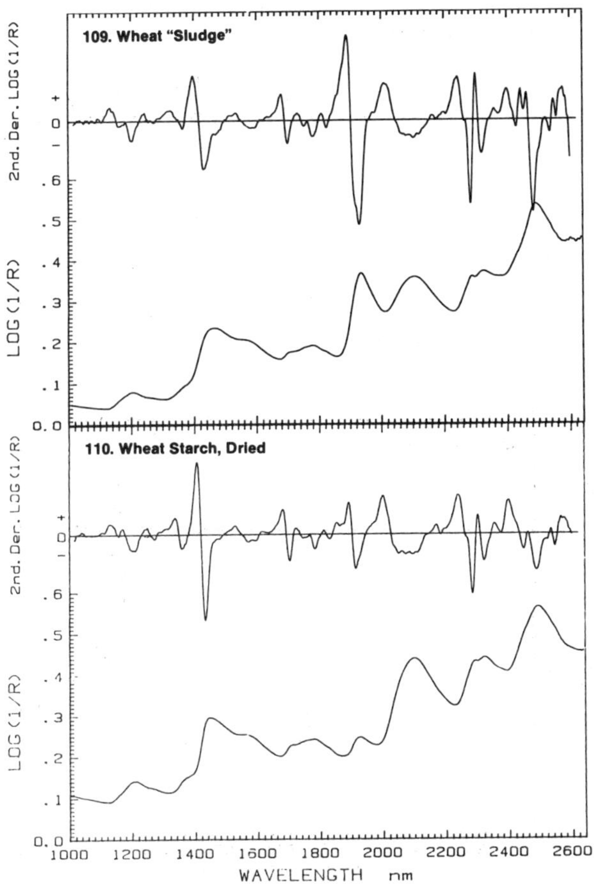
109. Wheat "Sludge"
110. Wheat Starch, Dried
2nd. Der. LOG (1/R)
LOG (1/R)
WAVELENGTH nm

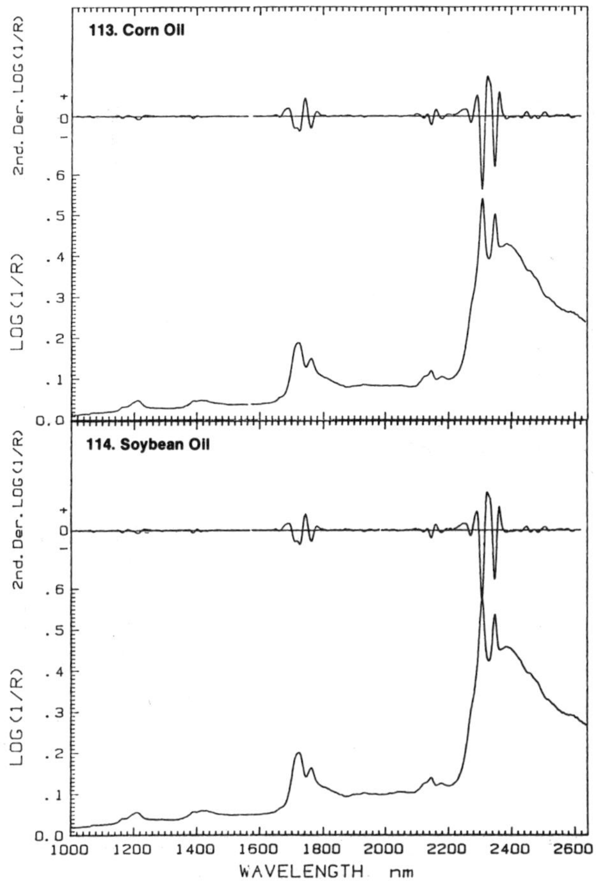
113. Corn Oil
114. Soybean Oil
2nd. Der. LOG (1/R)
LOG (1/R)
WAVELENGTH nm

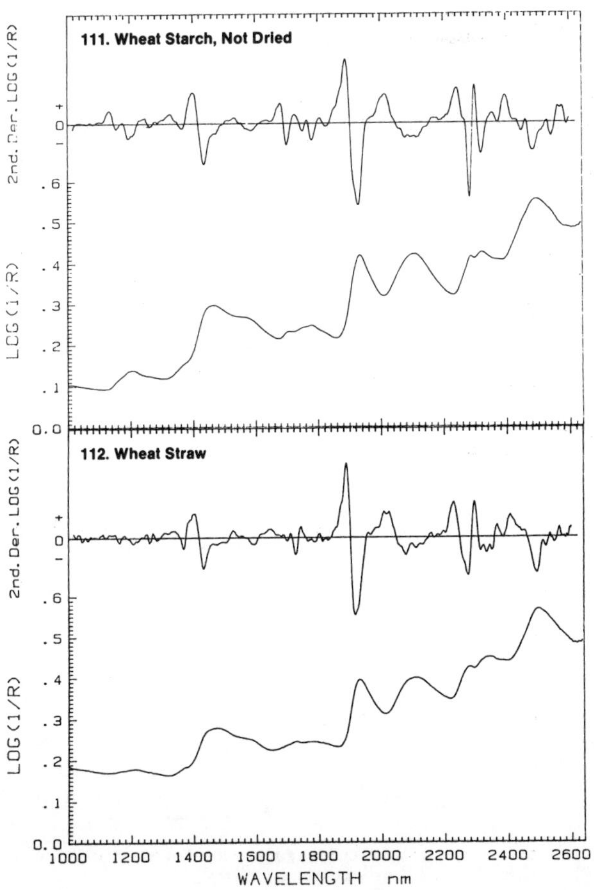
111. Wheat Starch, Not Dried
112. Wheat Straw
2nd. Der. LOG (1/R)
LOG (1/R)
WAVELENGTH nm

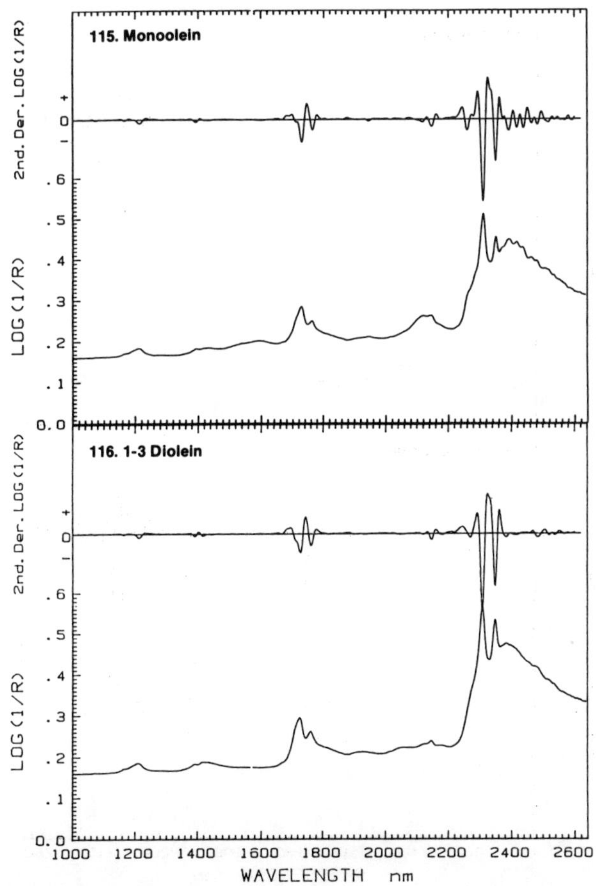
115. Monoolein
116. 1-3 Diolein
2nd. Der. LOG (1/R)
LOG (1/R)
WAVELENGTH nm

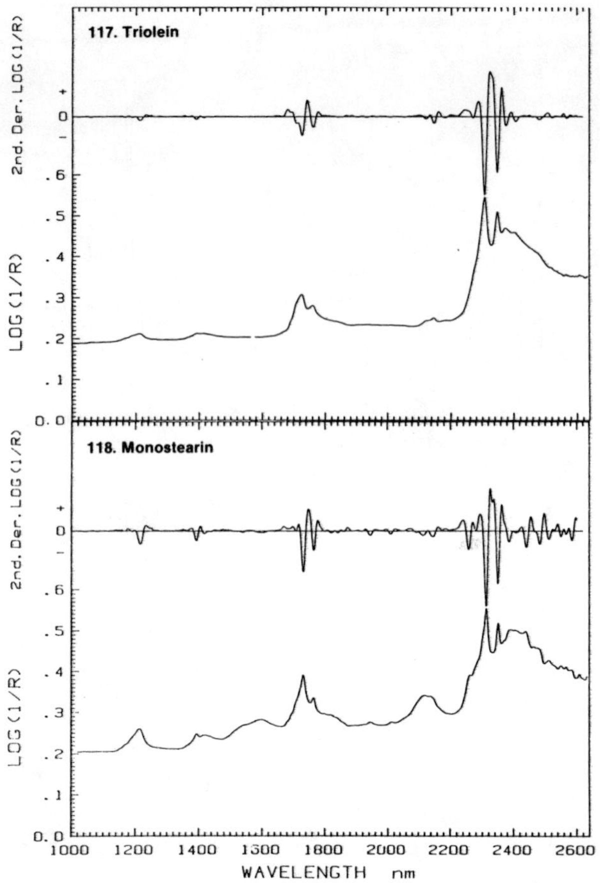
117. Triolein
118. Monostearin
2nd. Der. LOG(1/R)
LOG(1/R)
WAVELENGTH nm

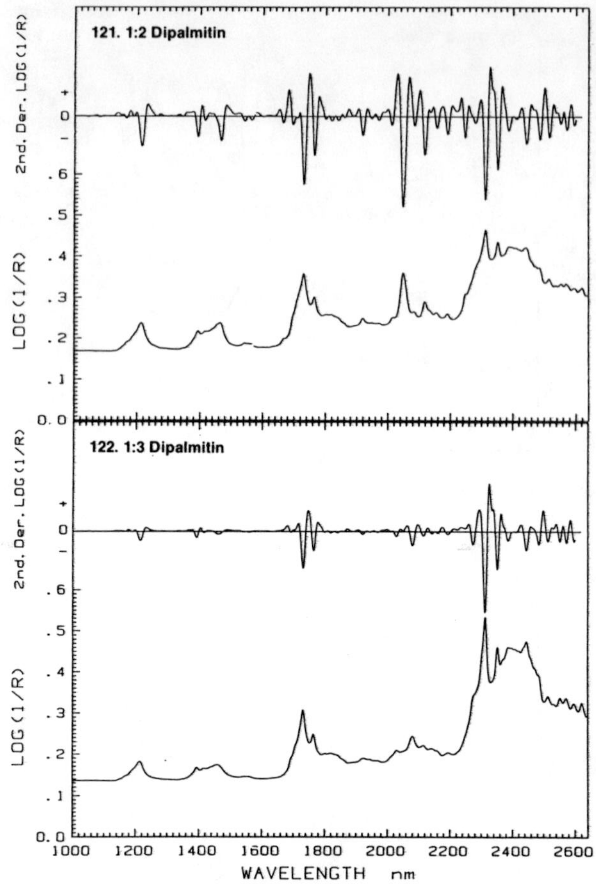
121. 1:2 Dipalmitin
122. 1:3 Dipalmitin
2nd. Der. LOG(1/R)
LOG(1/R)
WAVELENGTH nm

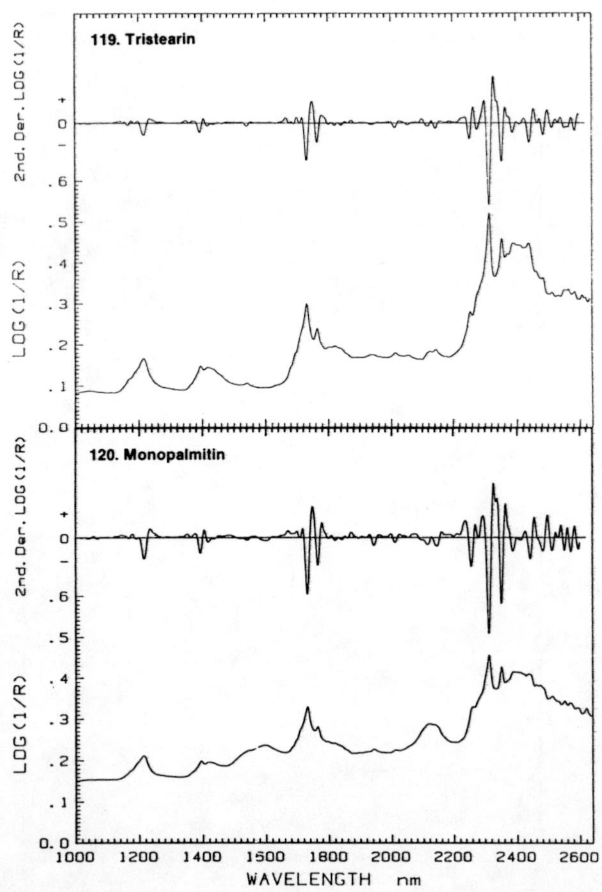
119. Tristearin
120. Monopalmitin
2nd. Der. LOG(1/R)
LOG(1/R)
WAVELENGTH nm

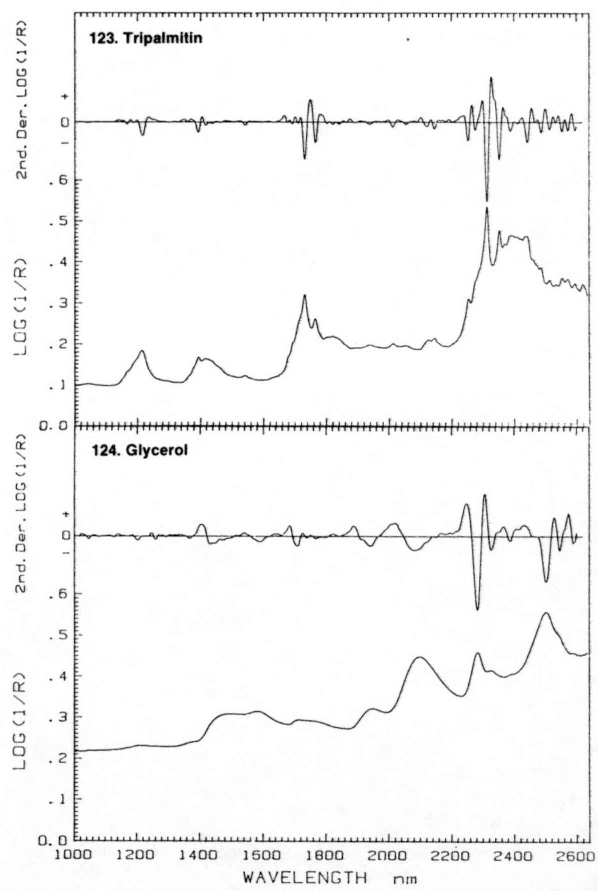
123. Tripalmitin
124. Glycerol
2nd. Der. LOG(1/R)
LOG(1/R)
WAVELENGTH nm

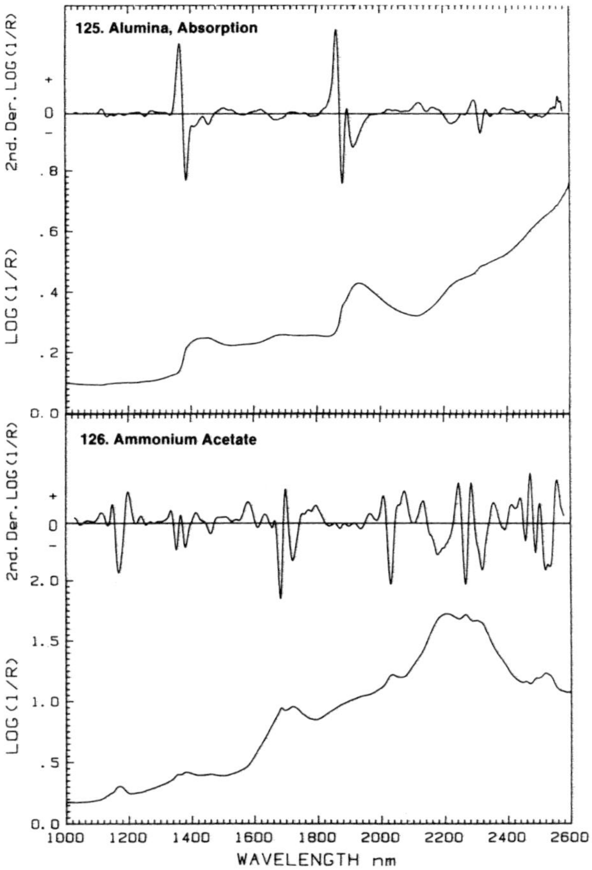
125. Alumina, Absorption
126. Ammonium Acetate
2nd. Der. LOG (1/R)
LOG (1/R)
WAVELENGTH nm

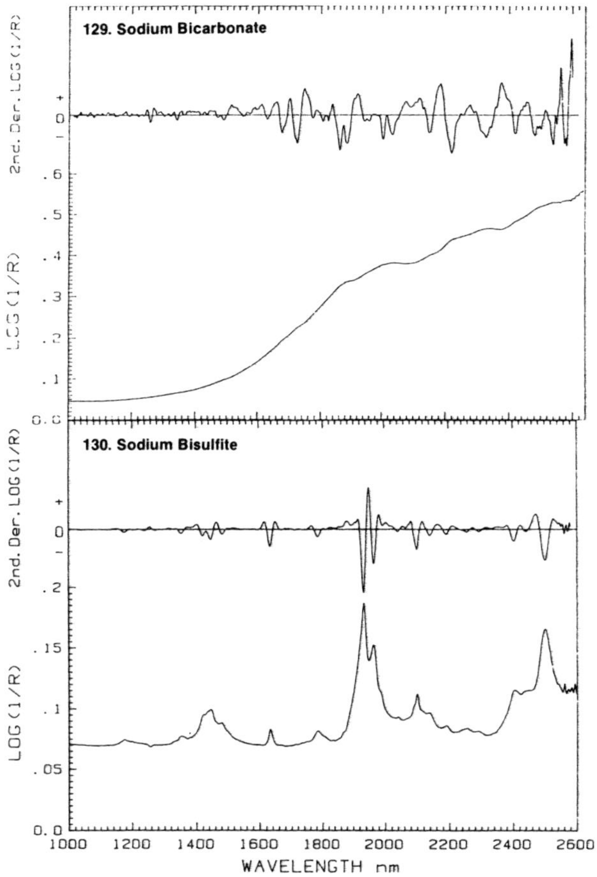
129. Sodium Bicarbonate
130. Sodium Bisulfite
2nd. Der. LOG (1/R)
LOG (1/R)
WAVELENGTH nm

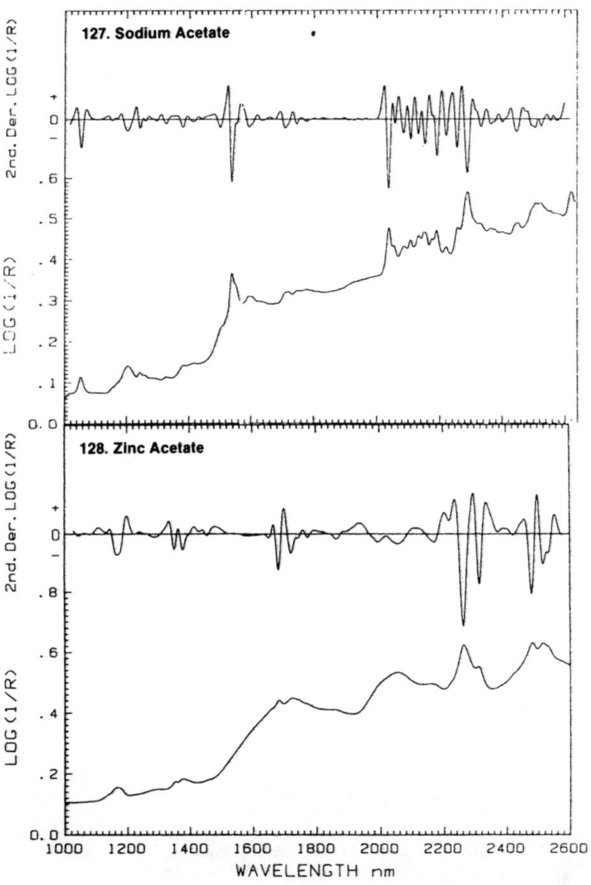
127. Sodium Acetate
128. Zinc Acetate
2nd. Der. LOG (1/R)
LOG (1/R)
WAVELENGTH nm

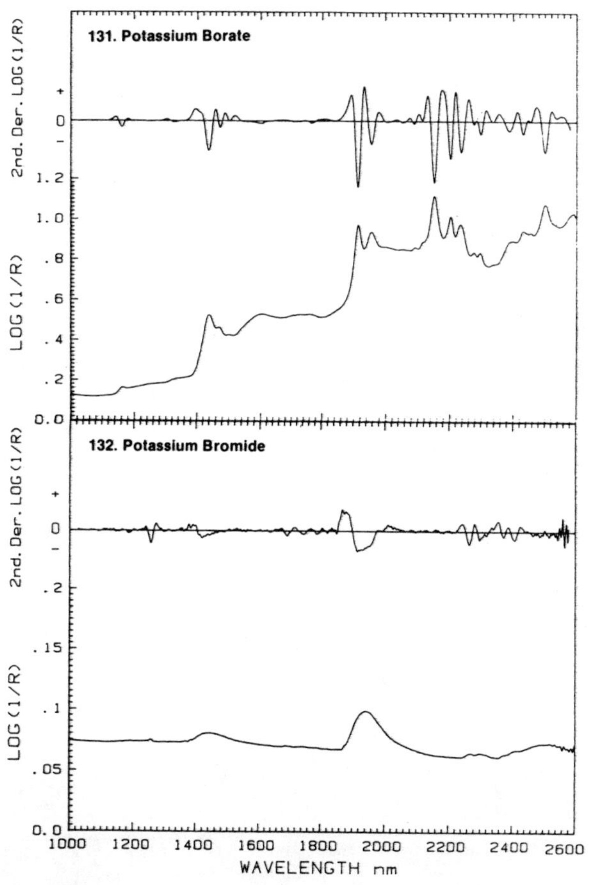
131. Potassium Borate
132. Potassium Bromide
2nd. Der. LOG (1/R)
LOG (1/R)
WAVELENGTH nm

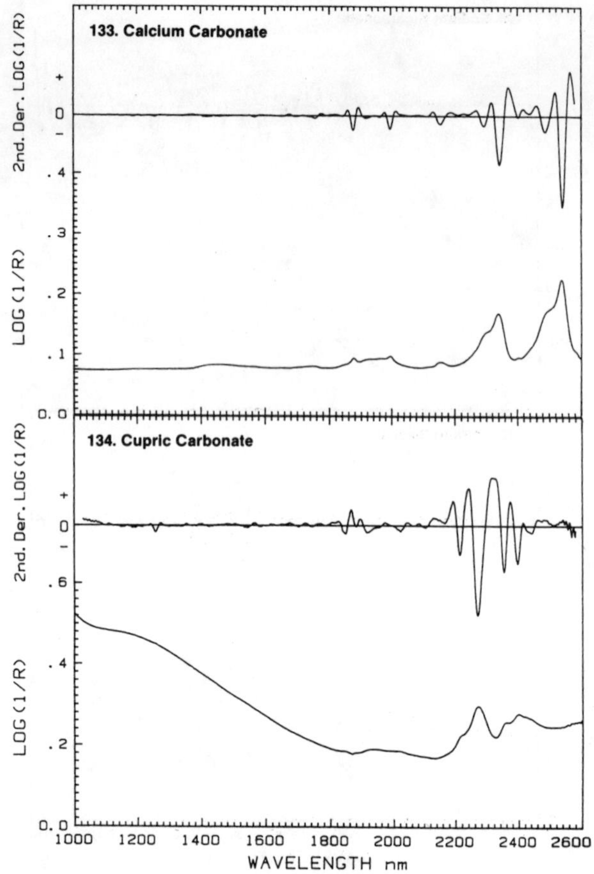
133. Calcium Carbonate
134. Cupric Carbonate
2nd. Der. LOG (1/R)
LOG (1/R)
WAVELENGTH nm

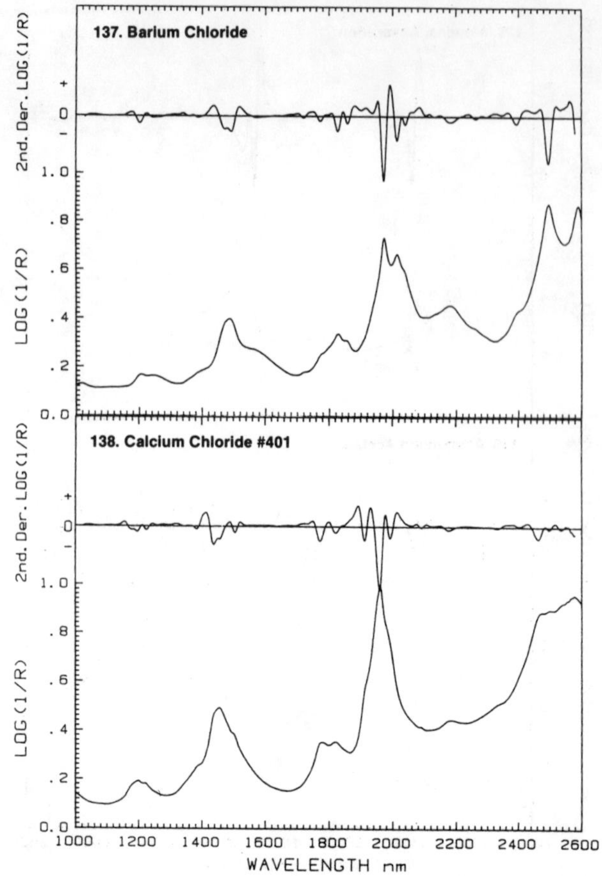
137. Barium Chloride
138. Calcium Chloride #401
2nd. Der. LOG (1/R)
LOG (1/R)
WAVELENGTH nm

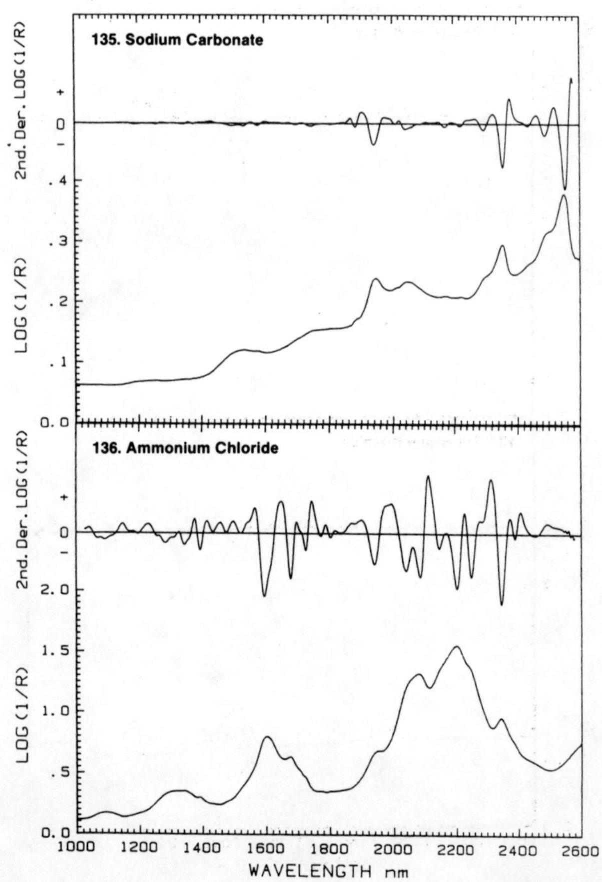
135. Sodium Carbonate
136. Ammonium Chloride
2nd. Der. LOG (1/R)
LOG (1/R)
WAVELENGTH nm

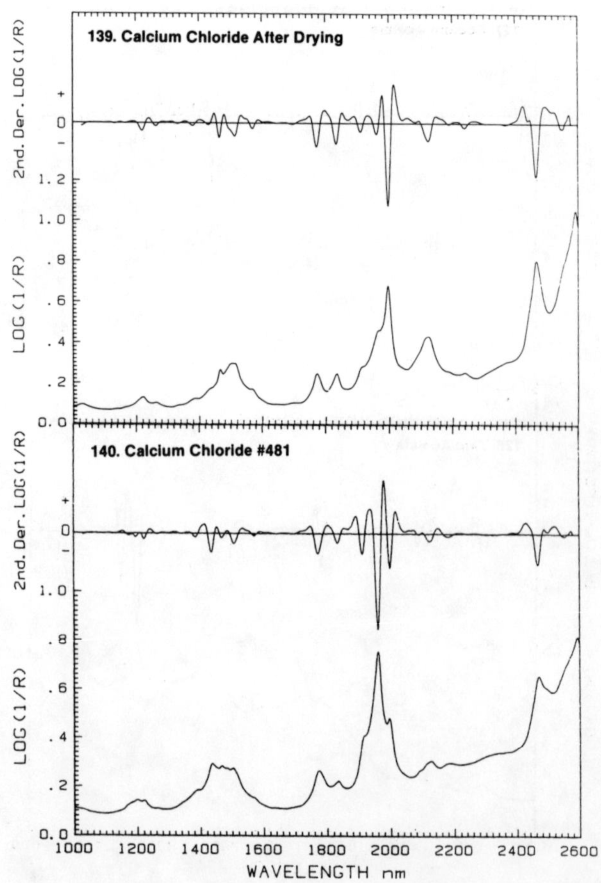
139. Calcium Chloride After Drying
140. Calcium Chloride #481
2nd. Der. LOG (1/R)
LOG (1/R)
WAVELENGTH nm

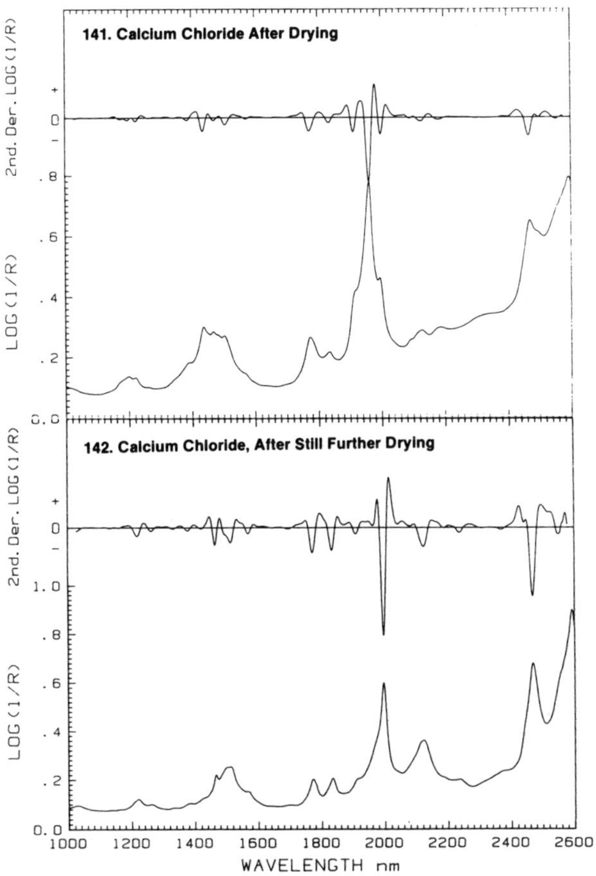
141. Calcium Chloride After Drying
142. Calcium Chloride, After Still Further Drying
2nd. Der. LOG (1/R)
LOG (1/R)
WAVELENGTH nm

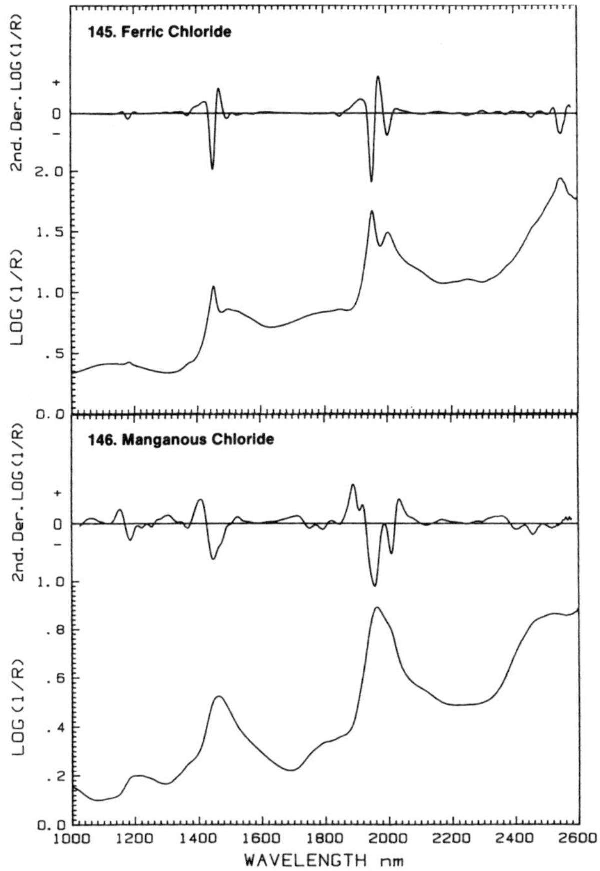
145. Ferric Chloride
146. Manganous Chloride
2nd. Der. LOG (1/R)
LOG (1/R)
WAVELENGTH nm

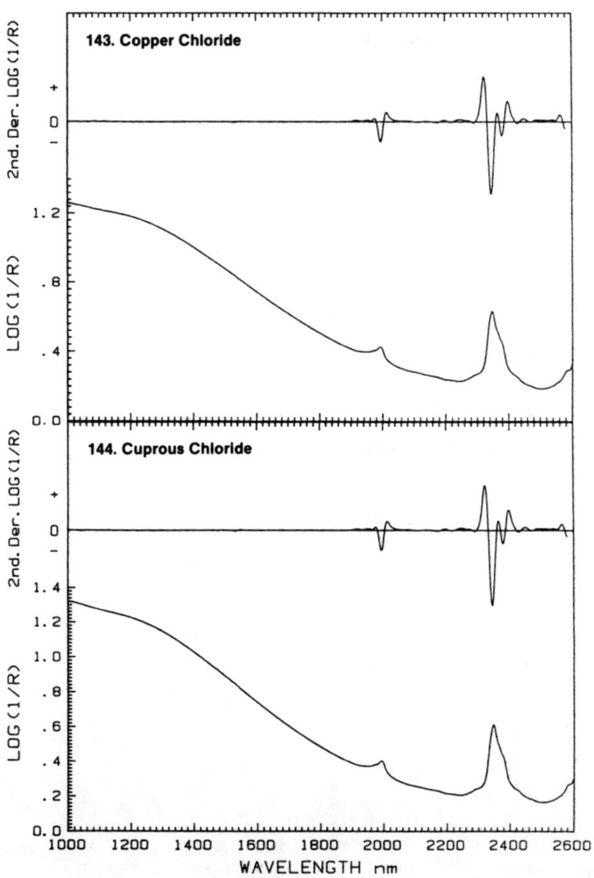
143. Copper Chloride
144. Cuprous Chloride
2nd. Der. LOG (1/R)
LOG (1/R)
WAVELENGTH nm

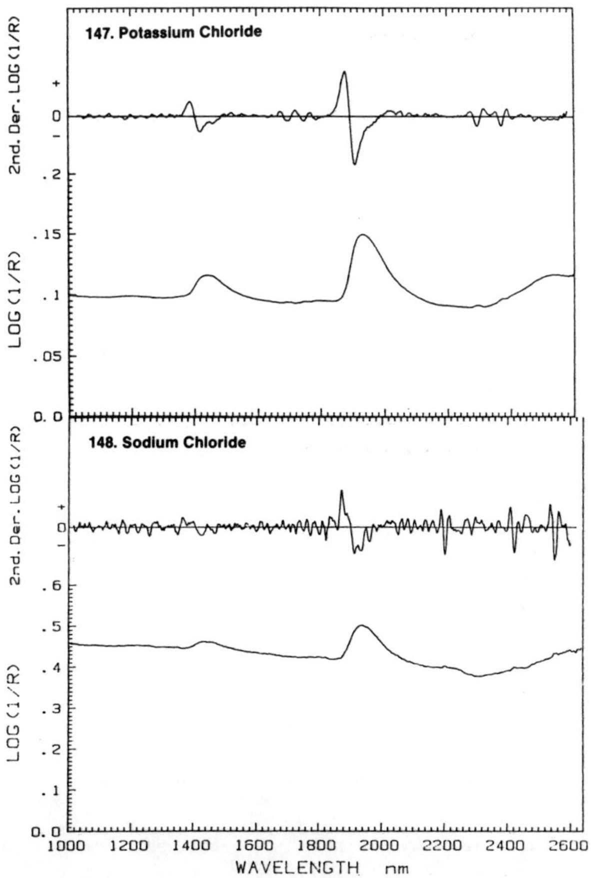
147. Potassium Chloride
148. Sodium Chloride
2nd. Der. LOG (1/R)
LOG (1/R)
WAVELENGTH nm

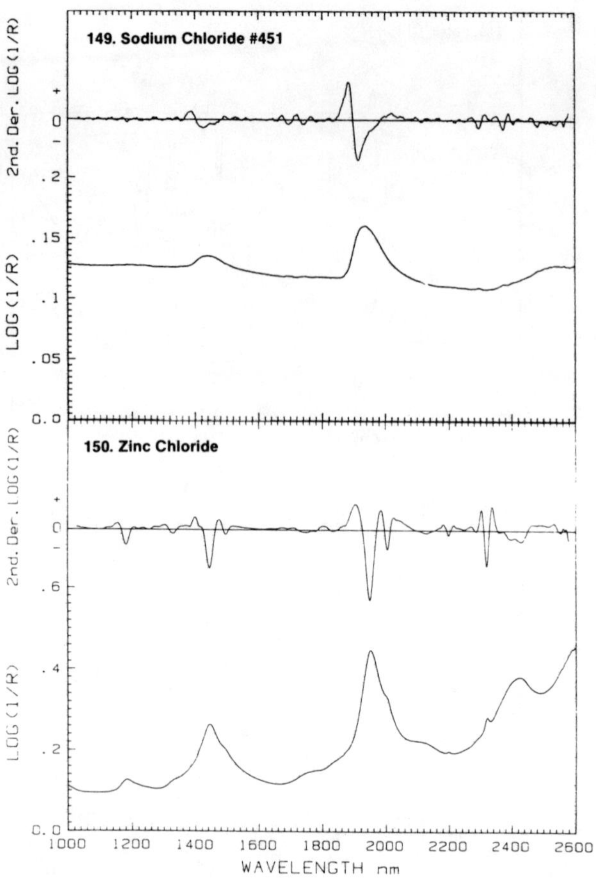
149. Sodium Chloride #451
150. Zinc Chloride
2nd. Der. LOG(1/R)
LOG(1/R)
WAVELENGTH nm

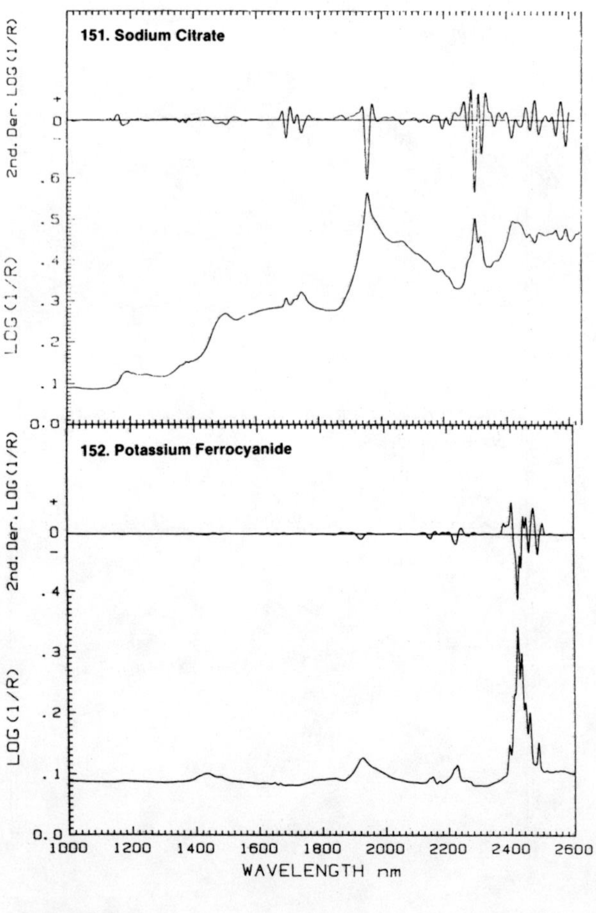
151. Sodium Citrate
152. Potassium Ferrocyanide
2nd. Der. LOG(1/R)
LOG(1/R)
WAVELENGTH nm

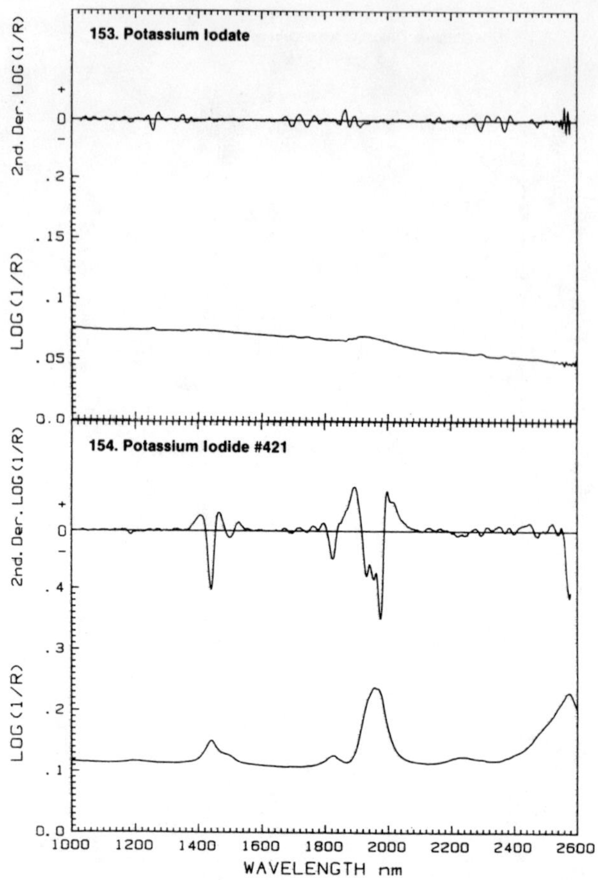
153. Potassium Iodate
154. Potassium Iodide #421
2nd. Der. LOG(1/R)
LOG(1/R)
WAVELENGTH nm

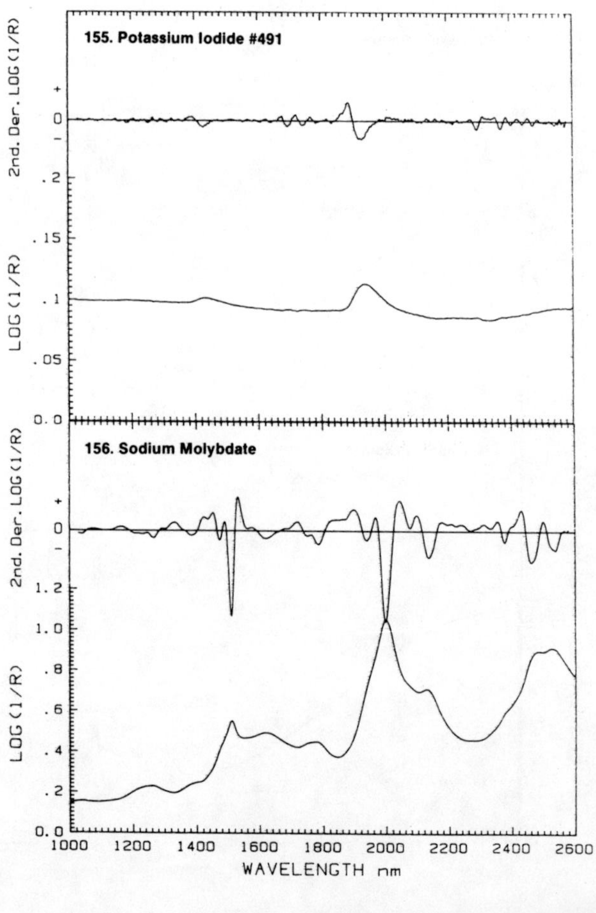
155. Potassium Iodide #491
156. Sodium Molybdate
2nd. Der. LOG(1/R)
LOG(1/R)
WAVELENGTH nm

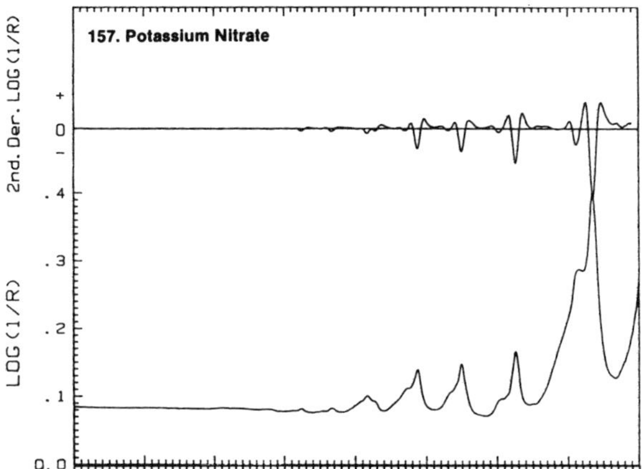
157. Potassium Nitrate
2nd. Der. LOG(1/R)
LOG(1/R)

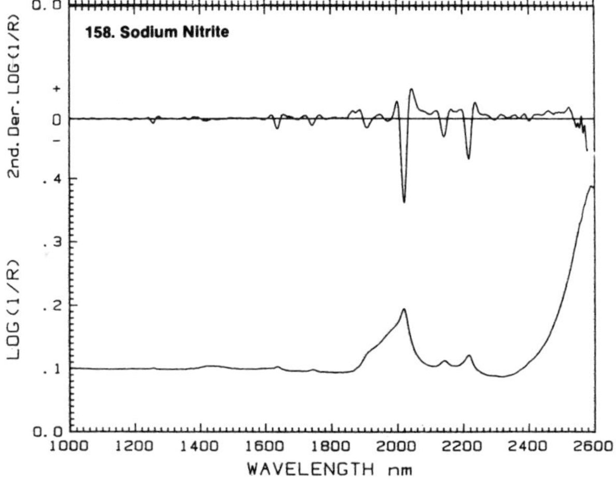
158. Sodium Nitrite
2nd. Der. LOG(1/R)
LOG(1/R)
WAVELENGTH nm

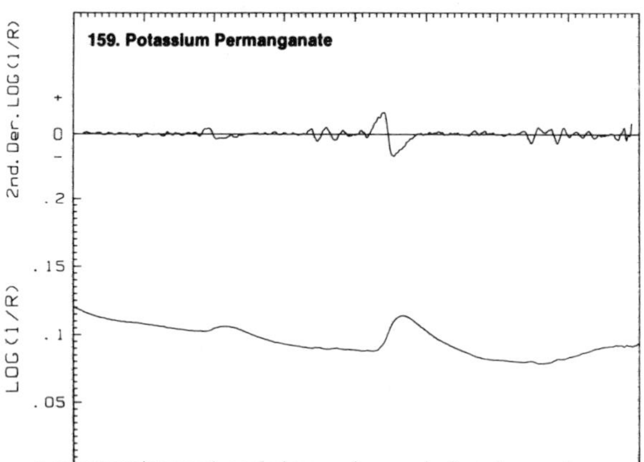
159. Potassium Permanganate
2nd. Der. LOG(1/R)
LOG(1/R)

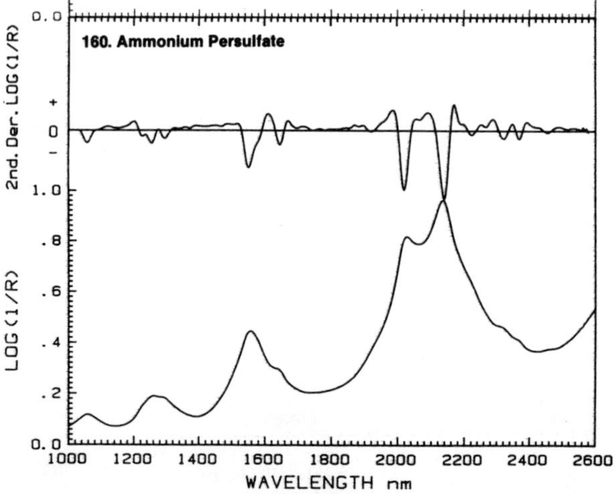
160. Ammonium Persulfate
2nd. Der. LOG(1/R)
LOG(1/R)
WAVELENGTH nm

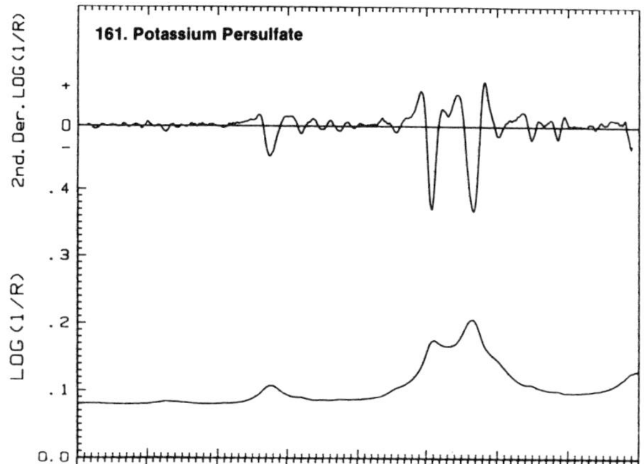
161. Potassium Persulfate
2nd. Der. LOG(1/R)
LOG(1/R)

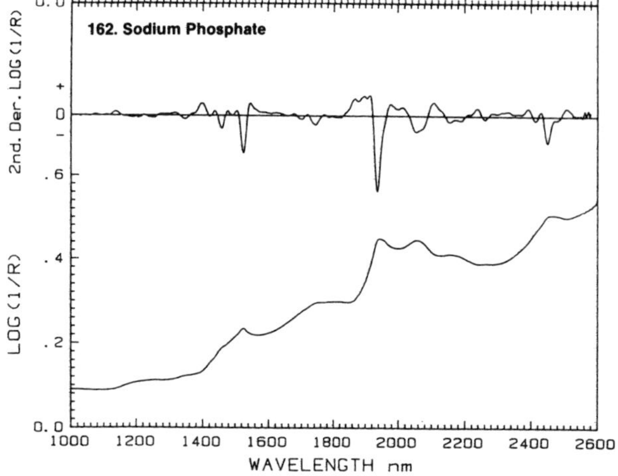
162. Sodium Phosphate
2nd. Der. LOG(1/R)
LOG(1/R)
WAVELENGTH nm

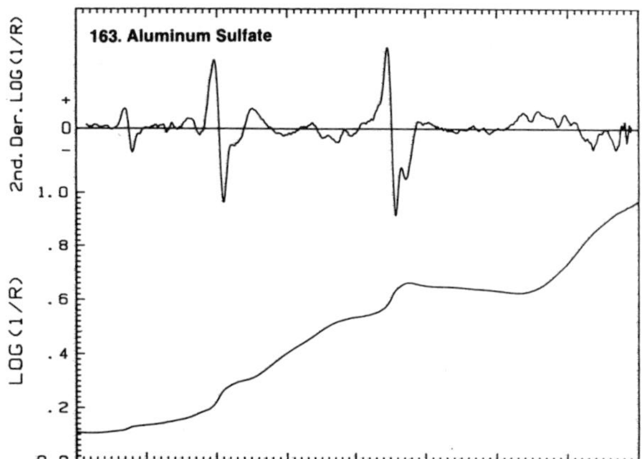
163. Aluminum Sulfate
2nd. Der. LOG(1/R)
LOG(1/R)

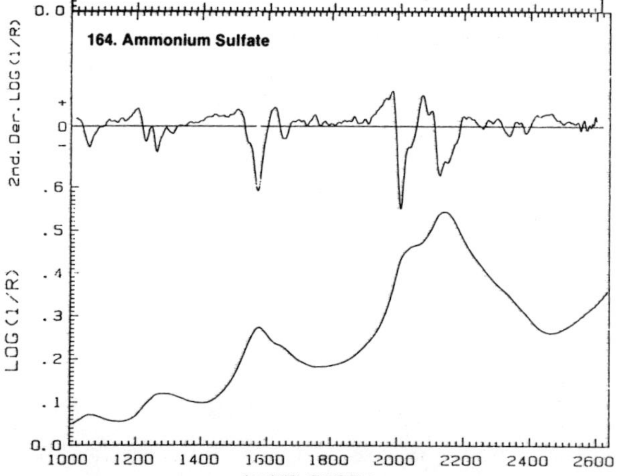
164. Ammonium Sulfate
2nd. Der. LOG(1/R)
LOG(1/R)
WAVELENGTH nm

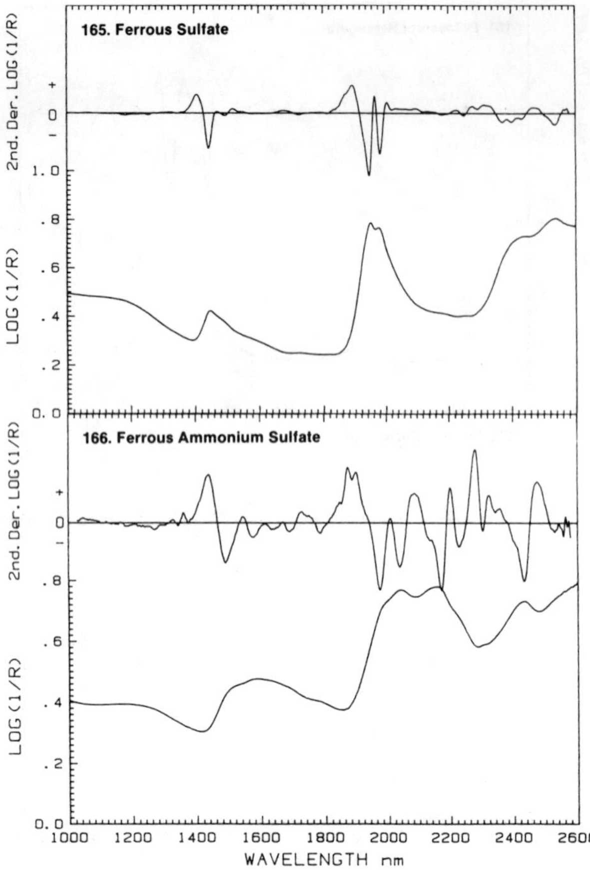
165. Ferrous Sulfate
166. Ferrous Ammonium Sulfate
2nd. Der. LOG (1/R)
LOG (1/R)
WAVELENGTH nm

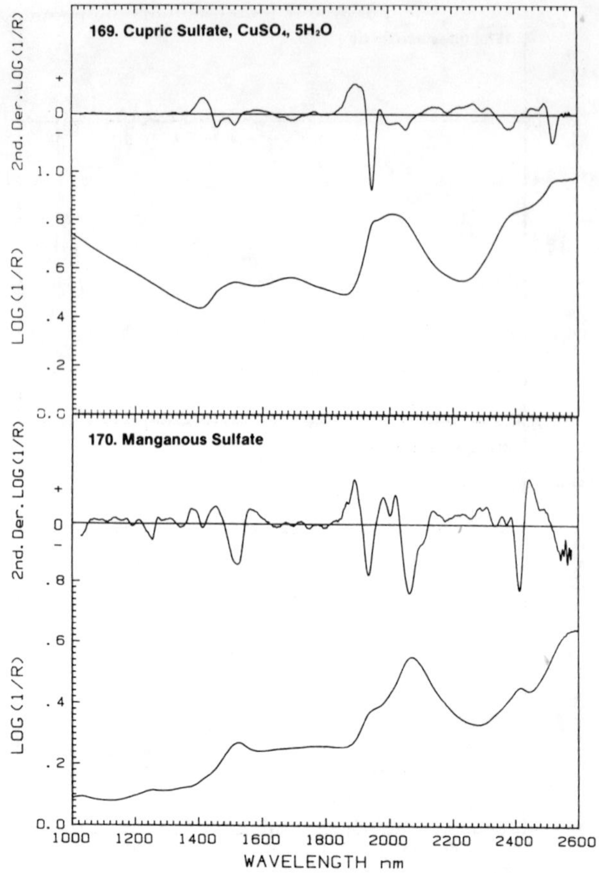
169. Cupric Sulfate, CuSO₄, 5H₂O
170. Manganous Sulfate
2nd. Der. LOG (1/R)
LOG (1/R)
WAVELENGTH nm

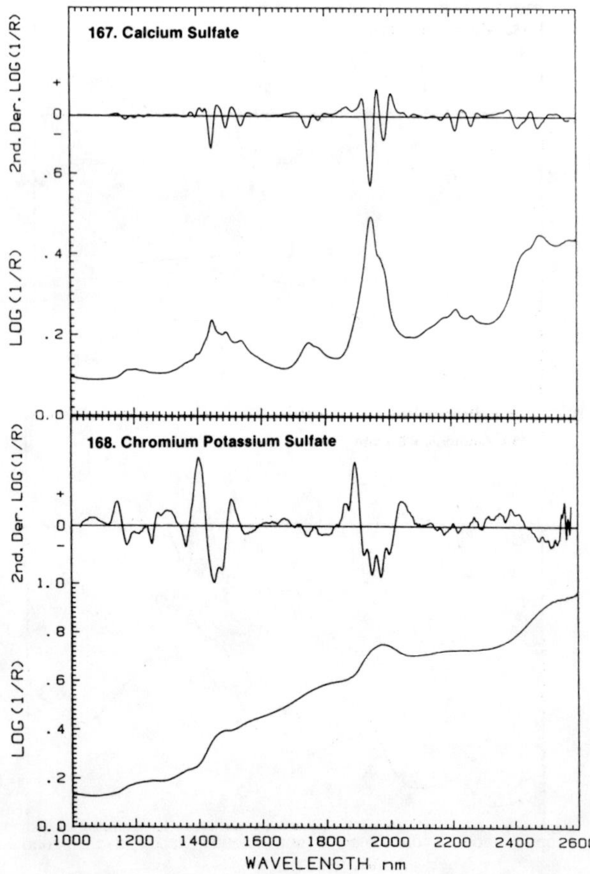
167. Calcium Sulfate
168. Chromium Potassium Sulfate
2nd. Der. LOG (1/R)
LOG (1/R)
WAVELENGTH nm

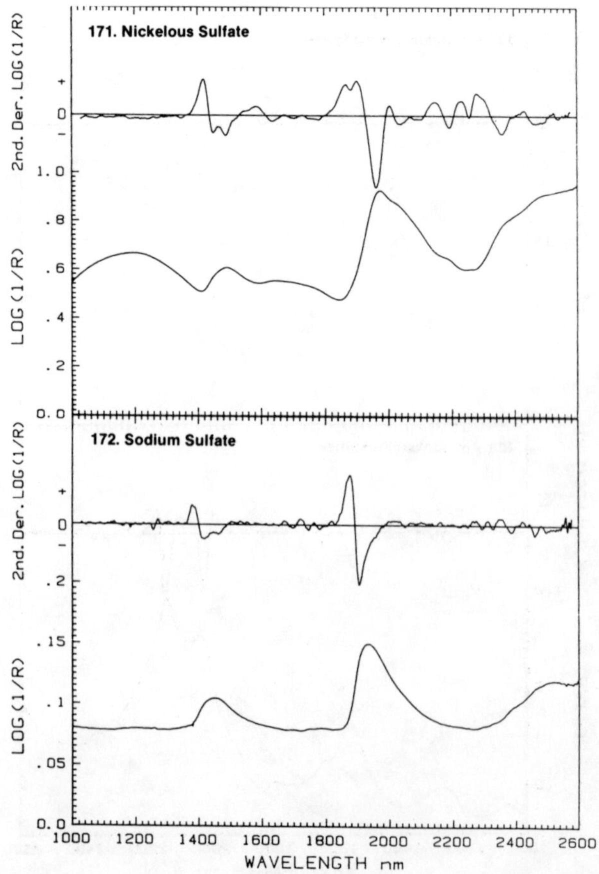
171. Nickelous Sulfate
172. Sodium Sulfate
2nd. Der. LOG (1/R)
LOG (1/R)
WAVELENGTH nm

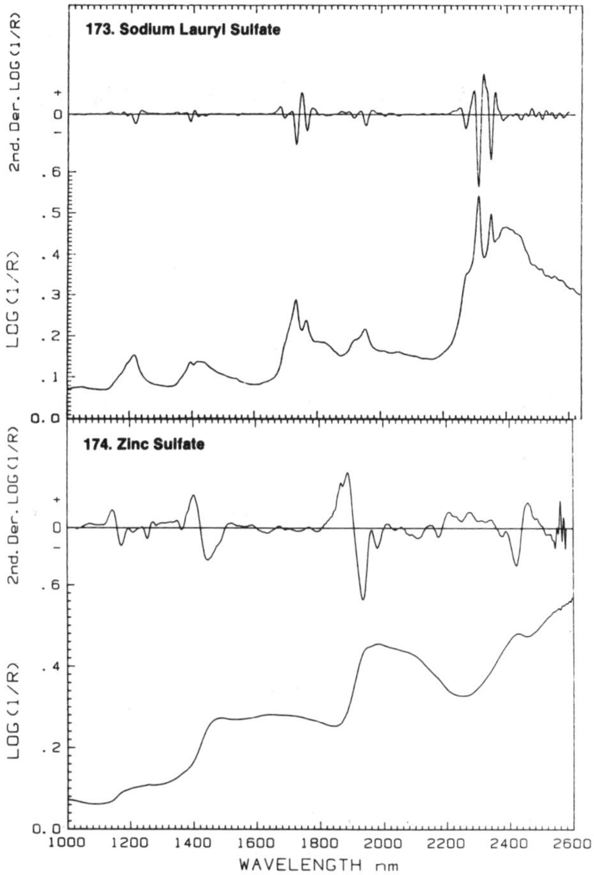
173. Sodium Lauryl Sulfate
174. Zinc Sulfate
2nd. Der. LOG(1/R)
LOG(1/R)
1000 1200 1400 1600 1800 2000 2200 2400 2600
WAVELENGTH nm

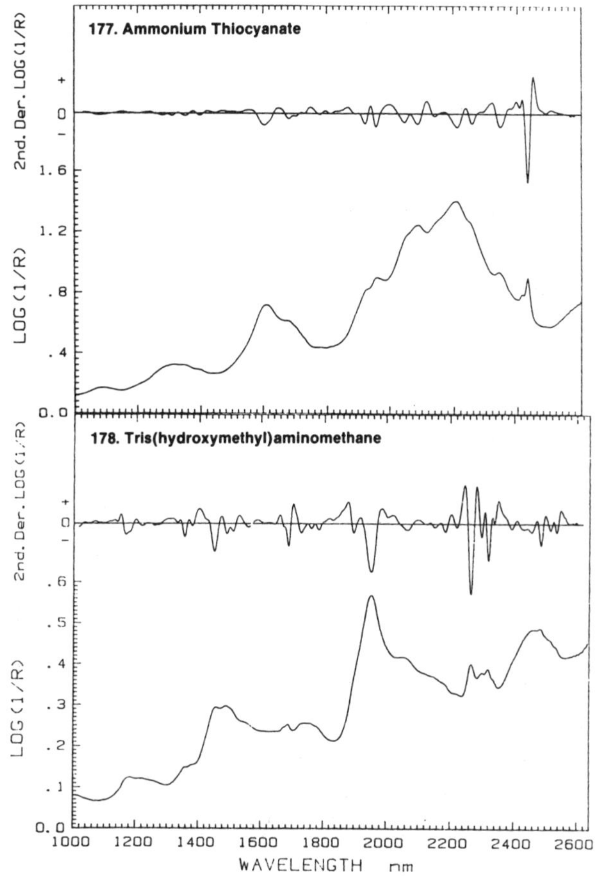
177. Ammonium Thiocyanate
178. Tris(hydroxymethyl)aminomethane
2nd. Der. LOG(1/R)
LOG(1/R)
1000 1200 1400 1600 1800 2000 2200 2400 2600
WAVELENGTH nm

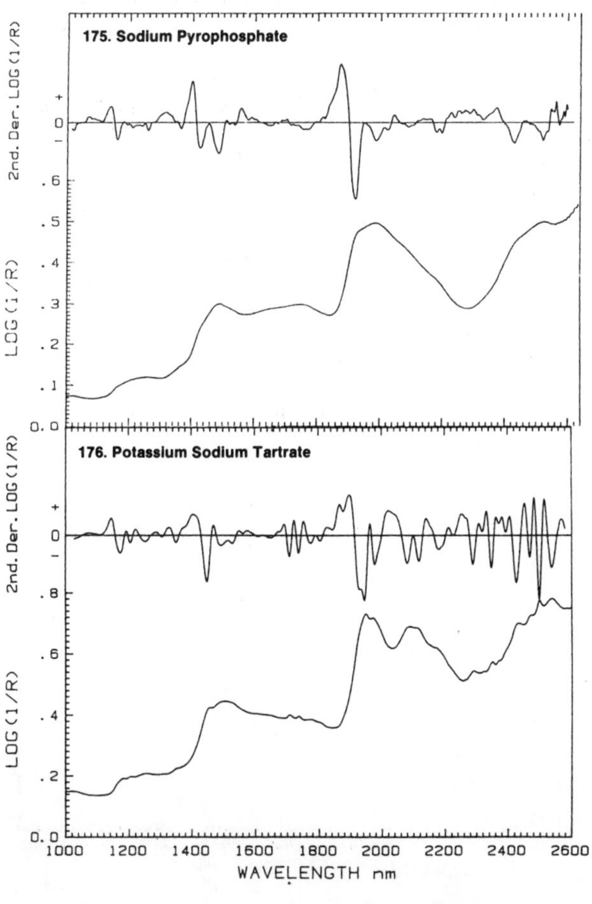
175. Sodium Pyrophosphate
176. Potassium Sodium Tartrate
2nd. Der. LOG(1/R)
LOG(1/R)
1000 1200 1400 1600 1800 2000 2200 2400 2600
WAVELENGTH nm

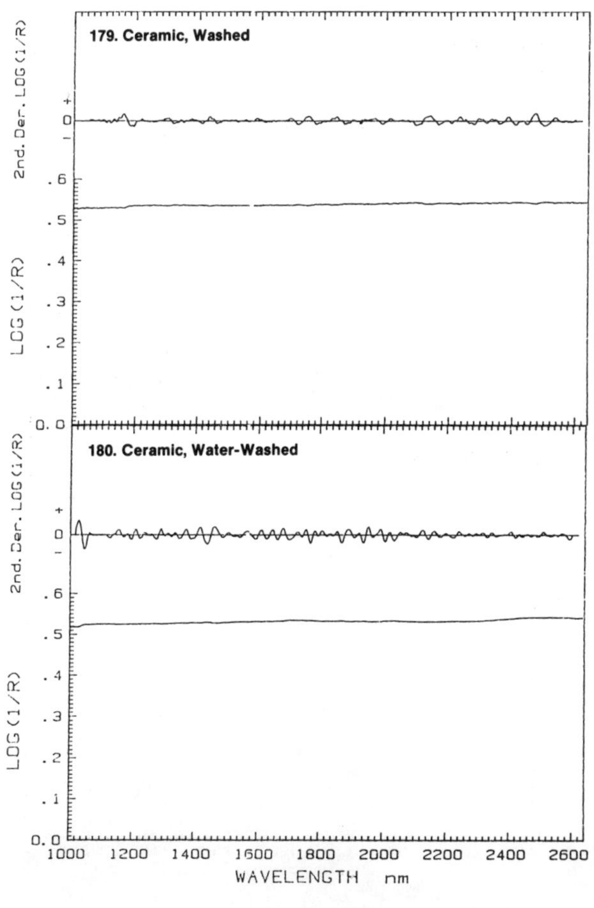
179. Ceramic, Washed
180. Ceramic, Water-Washed
2nd. Der. LOG(1/R)
LOG(1/R)
1000 1200 1400 1600 1800 2000 2200 2400 2600
WAVELENGTH nm

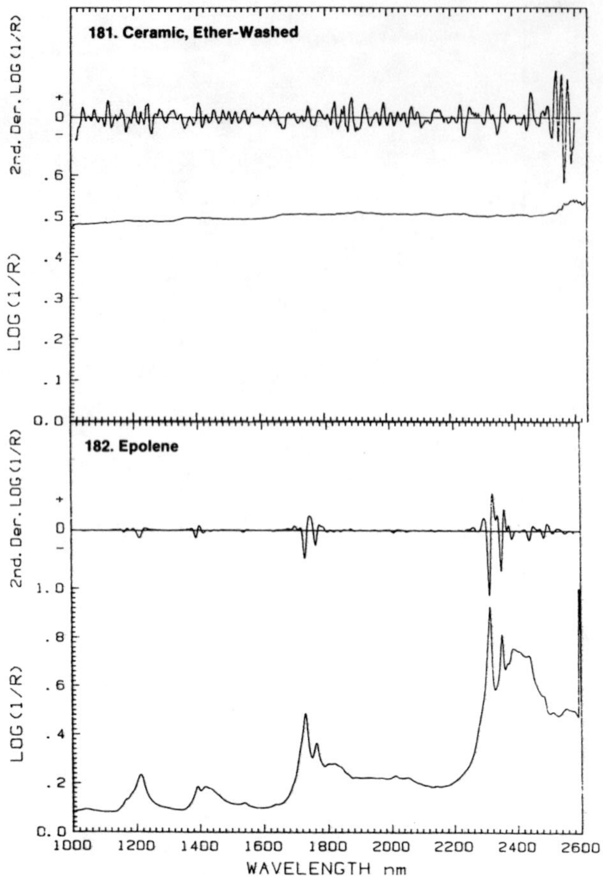
181. Ceramic, Ether-Washed
182. Epolene
2nd. Der. LOG(1/R)
LOG(1/R)
WAVELENGTH nm

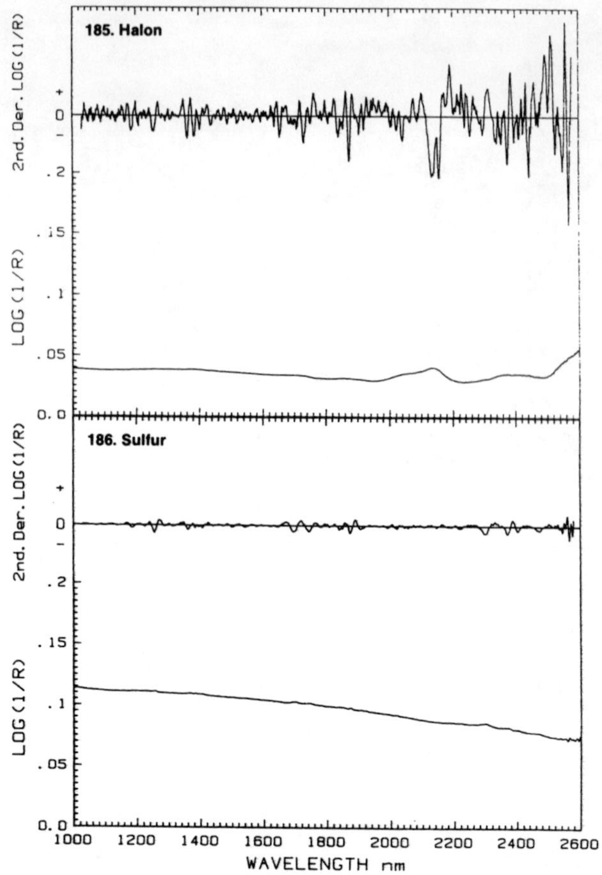
185. Halon
186. Sulfur
2nd. Der. LOG(1/R)
LOG(1/R)
WAVELENGTH nm

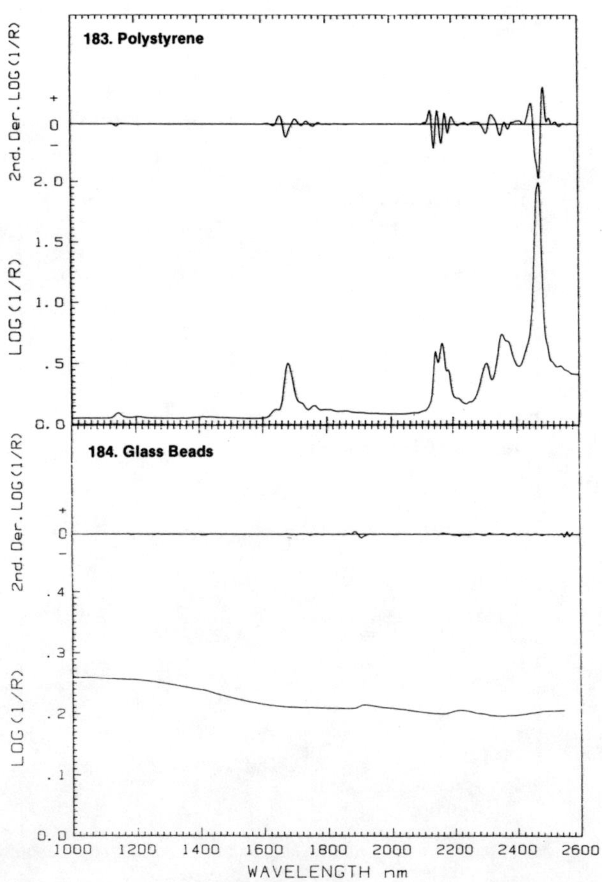
183. Polystyrene
184. Glass Beads
2nd. Der. LOG(1/R)
LOG(1/R)
WAVELENGTH nm

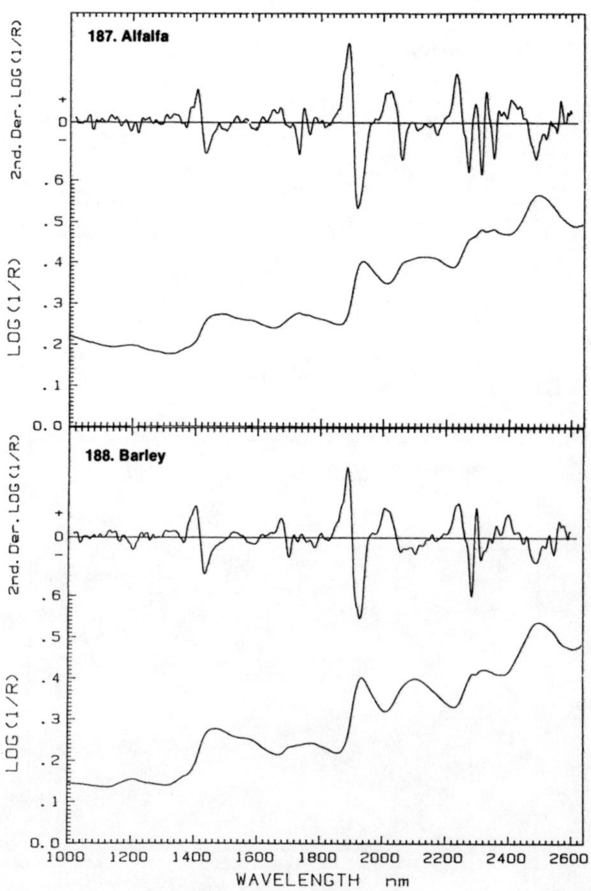
187. Alfalfa
188. Barley
2nd. Der. LOG(1/R)
LOG(1/R)
WAVELENGTH nm

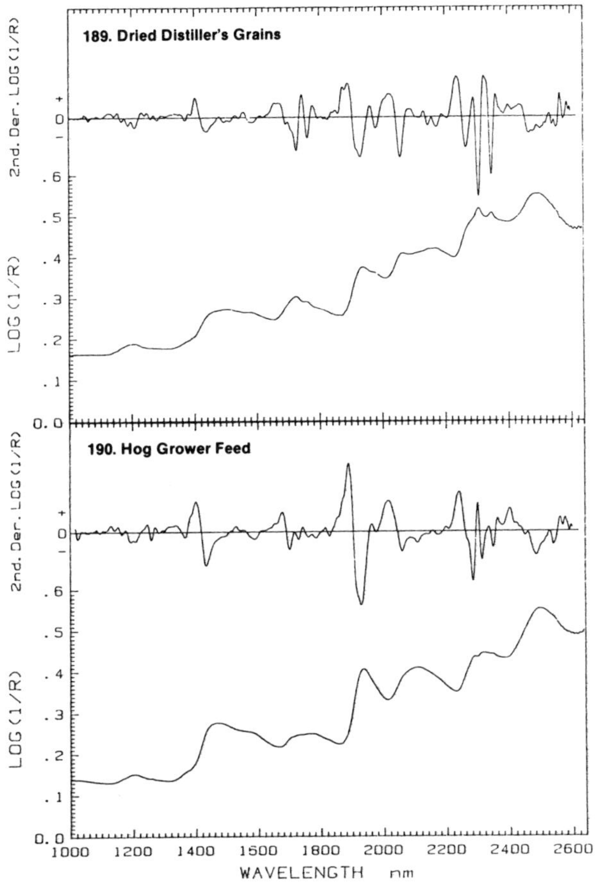
189. Dried Distiller's Grains
190. Hog Grower Feed
2nd. Der. LOG (1/R)
LOG (1/R)
WAVELENGTH nm

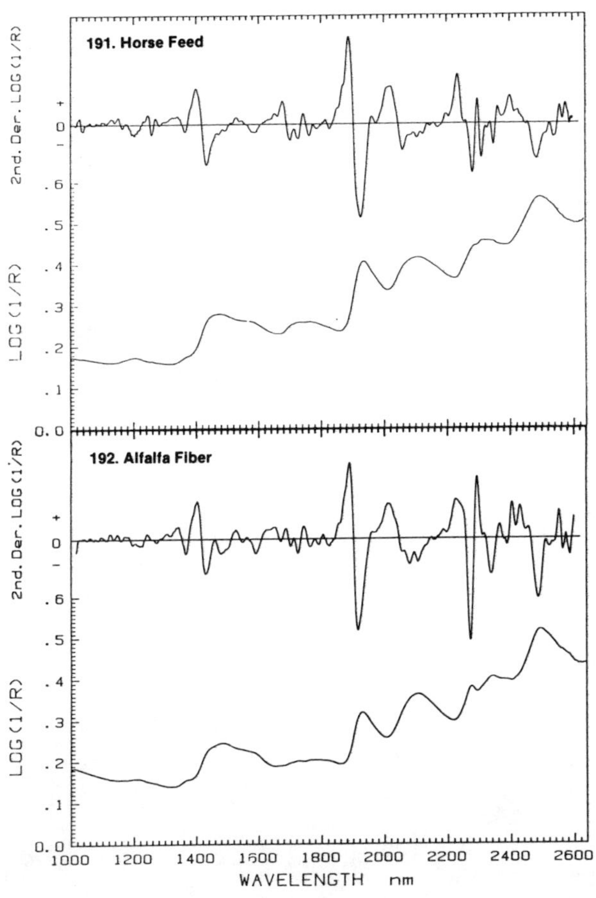
191. Horse Feed
192. Alfalfa Fiber
2nd. Der. LOG (1/R)
LOG (1/R)
WAVELENGTH nm

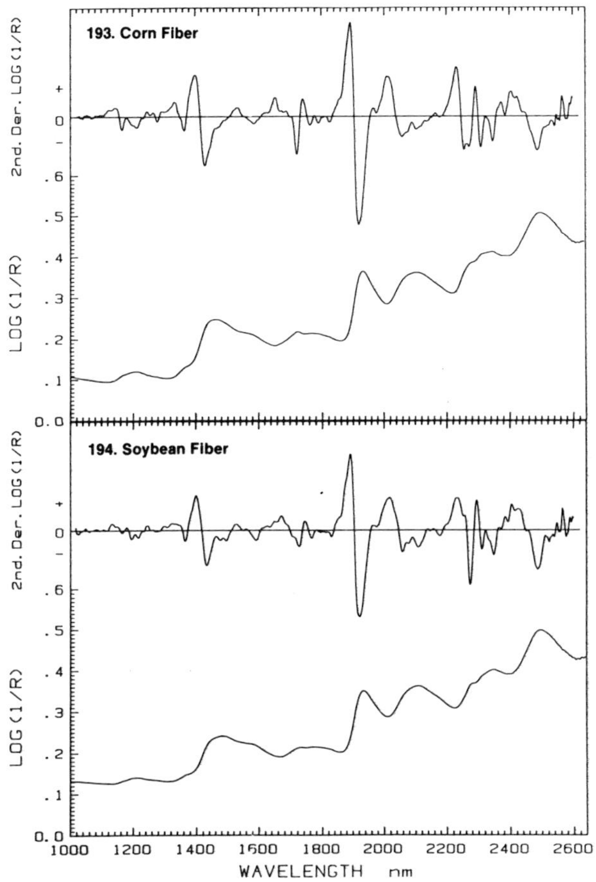
193. Corn Fiber
194. Soybean Fiber
2nd. Der. LOG (1/R)
LOG (1/R)
WAVELENGTH nm

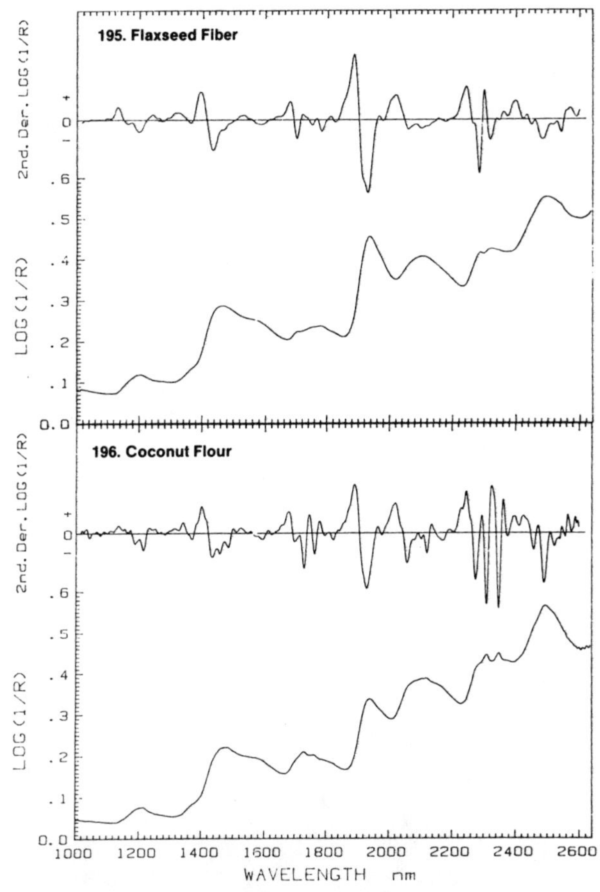
195. Flaxseed Fiber
196. Coconut Flour
2nd. Der. LOG (1/R)
LOG (1/R)
WAVELENGTH nm

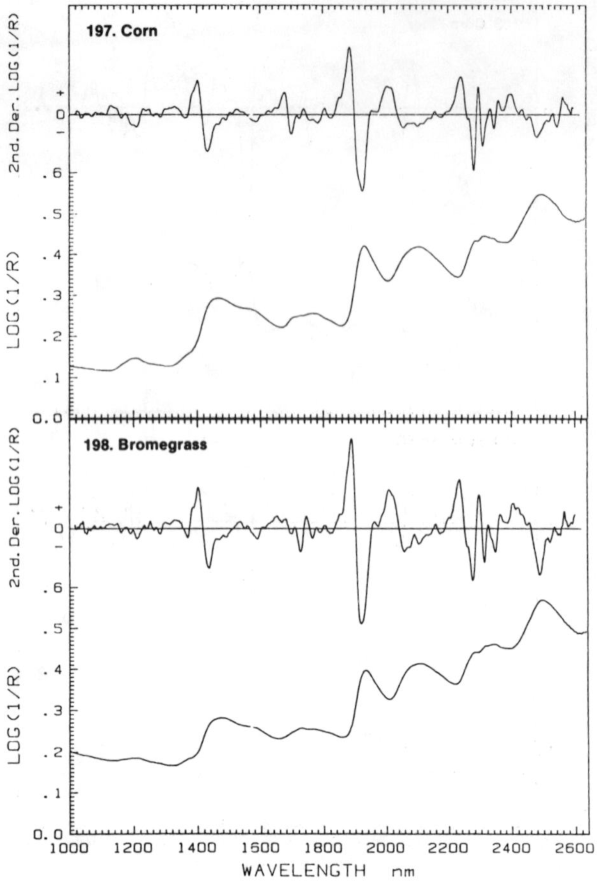
197. Corn
198. Bromegrass
2nd. Der. LOG(1/R)
LOG(1/R)
WAVELENGTH nm

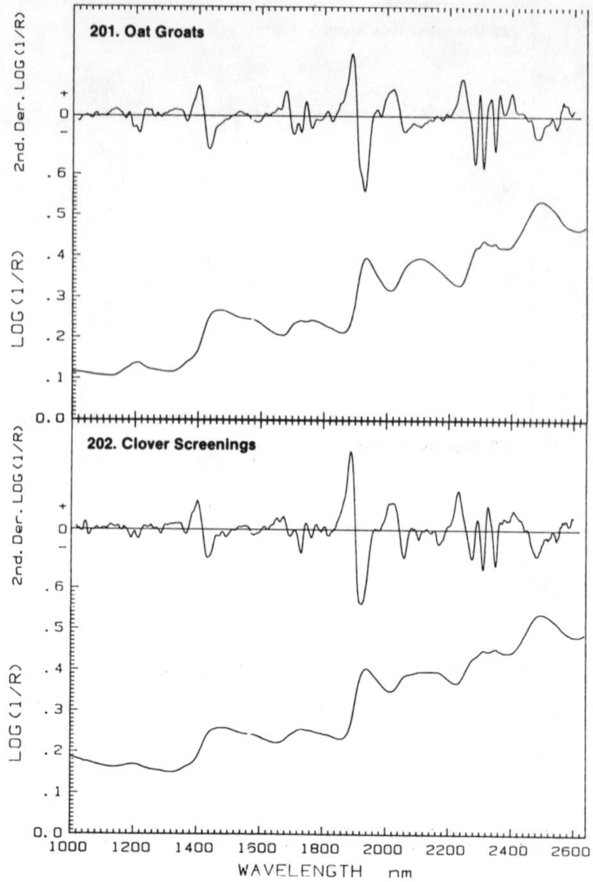
201. Oat Groats
202. Clover Screenings
2nd. Der. LOG(1/R)
LOG(1/R)
WAVELENGTH nm

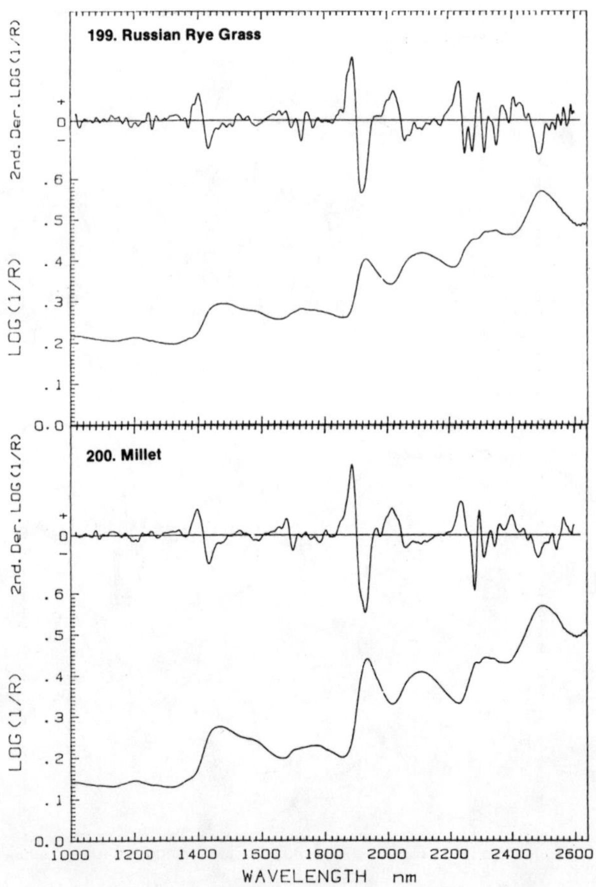
199. Russian Rye Grass
200. Millet
2nd. Der. LOG(1/R)
LOG(1/R)
WAVELENGTH nm

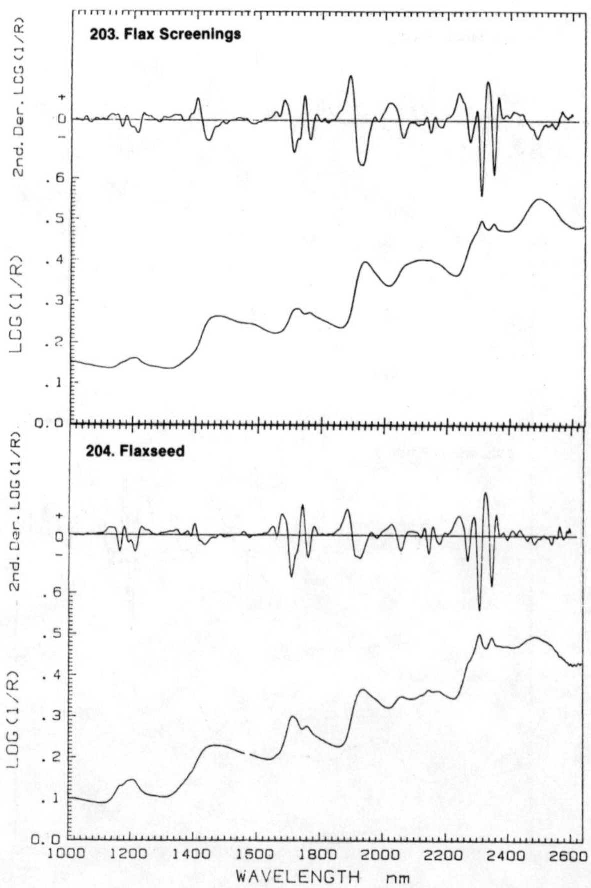
203. Flax Screenings
204. Flaxseed
2nd. Der. LOG(1/R)
LOG(1/R)
WAVELENGTH nm

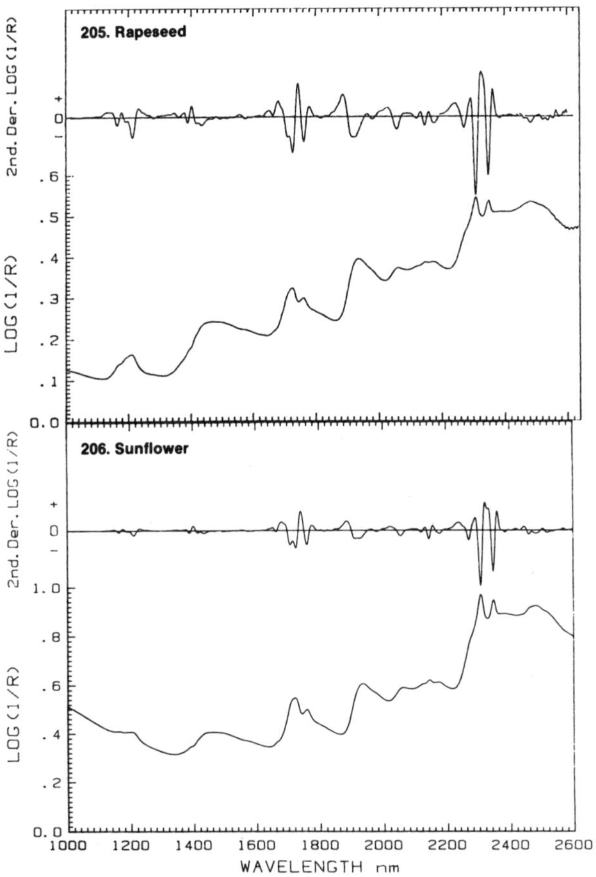
205. Rapeseed
206. Sunflower
2nd. Der. LOG (1/R)
LOG (1/R)
WAVELENGTH nm

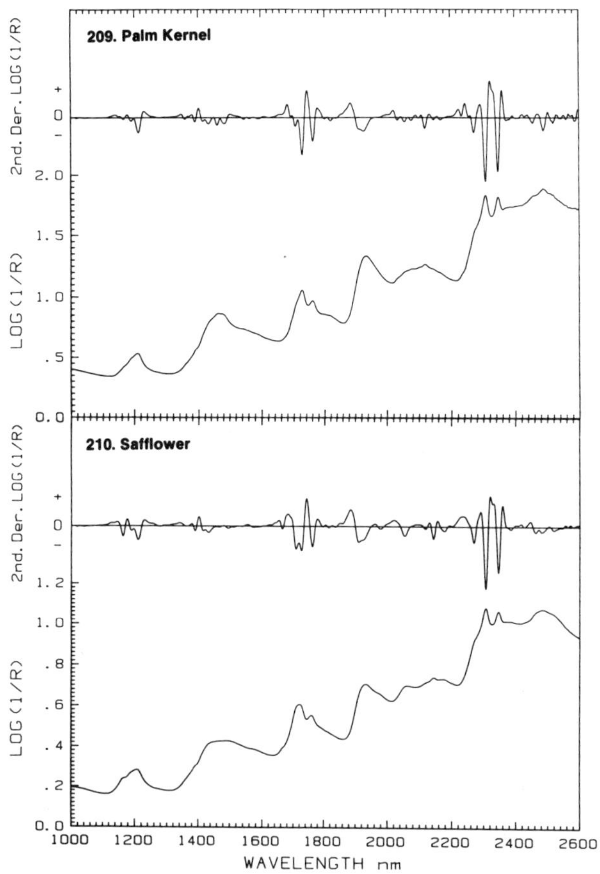
209. Palm Kernel
210. Safflower
2nd. Der. LOG (1/R)
LOG (1/R)
WAVELENGTH nm

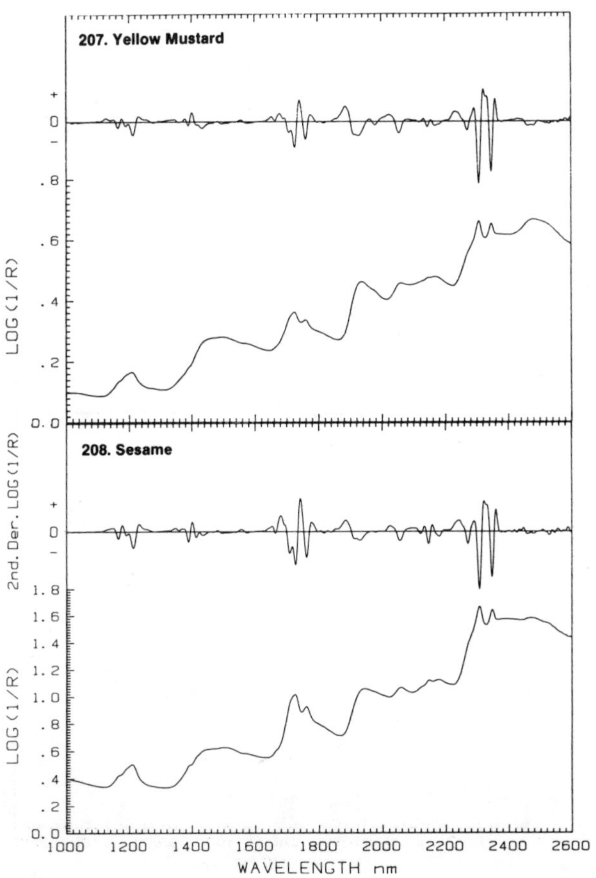
207. Yellow Mustard
208. Sesame
2nd. Der. LOG (1/R)
LOG (1/R)
WAVELENGTH nm

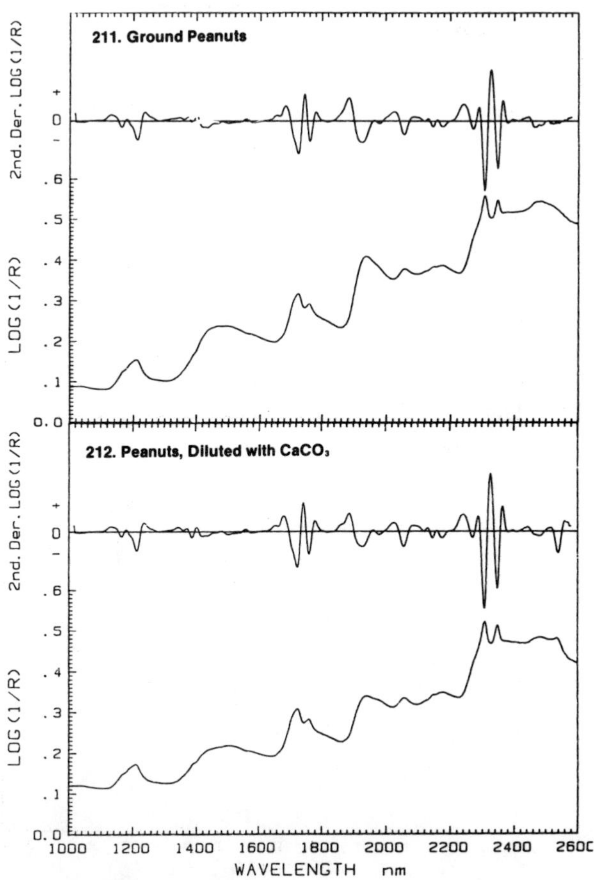
211. Ground Peanuts
212. Peanuts, Diluted with CaCO₃
2nd. Der. LOG (1/R)
LOG (1/R)
WAVELENGTH nm

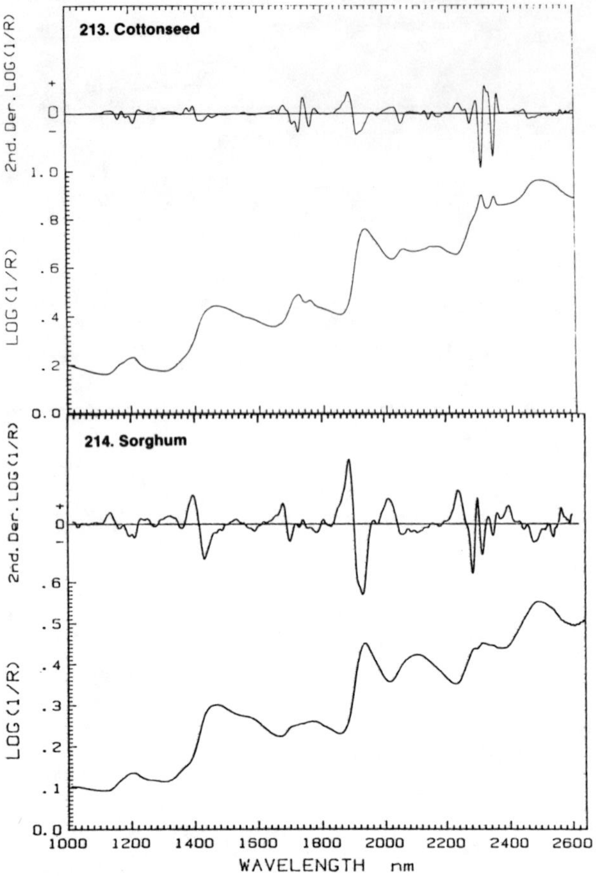
213. Cottonseed
214. Sorghum
2nd. Der. LOG (1/R)
LOG (1/R)
WAVELENGTH nm

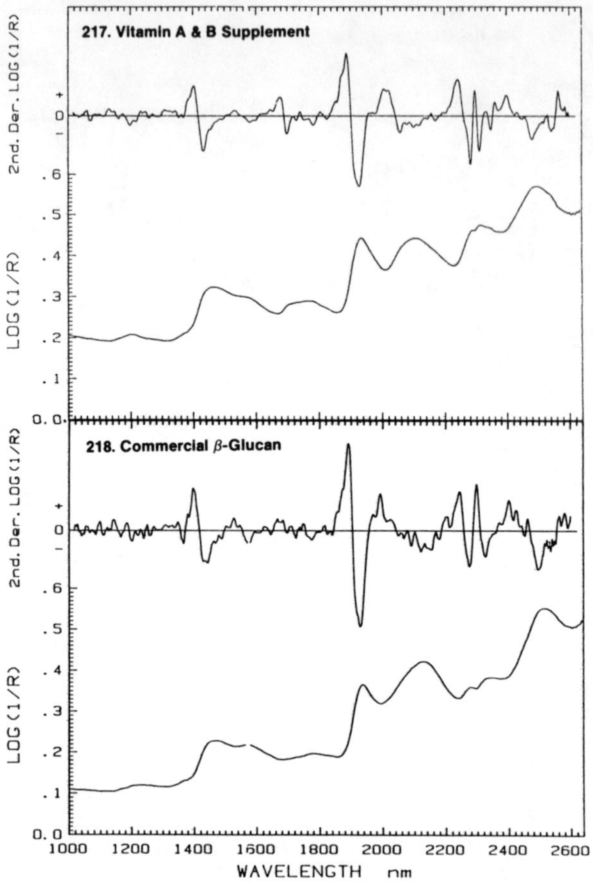
217. Vitamin A & B Supplement
218. Commercial β-Glucan
2nd. Der. LOG (1/R)
LOG (1/R)
WAVELENGTH nm

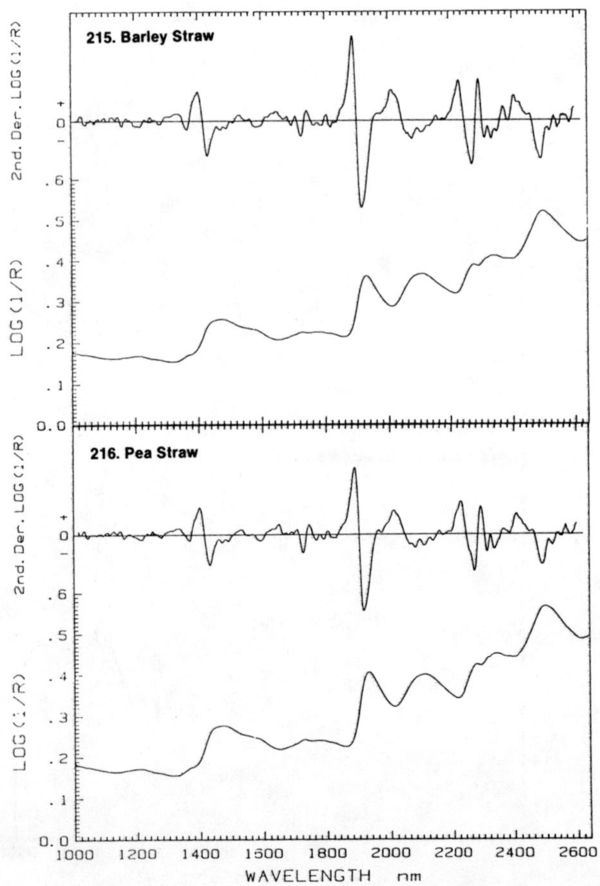
215. Barley Straw
216. Pea Straw
2nd. Der. LOG (1/R)
LOG (1/R)
WAVELENGTH nm

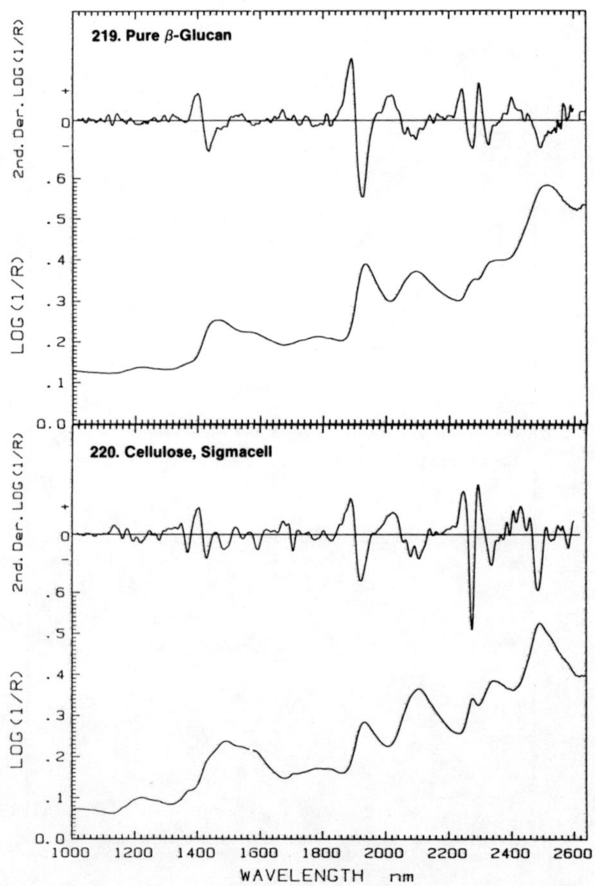
219. Pure β-Glucan
220. Cellulose, Sigmacell
2nd. Der. LOG (1/R)
LOG (1/R)
WAVELENGTH nm

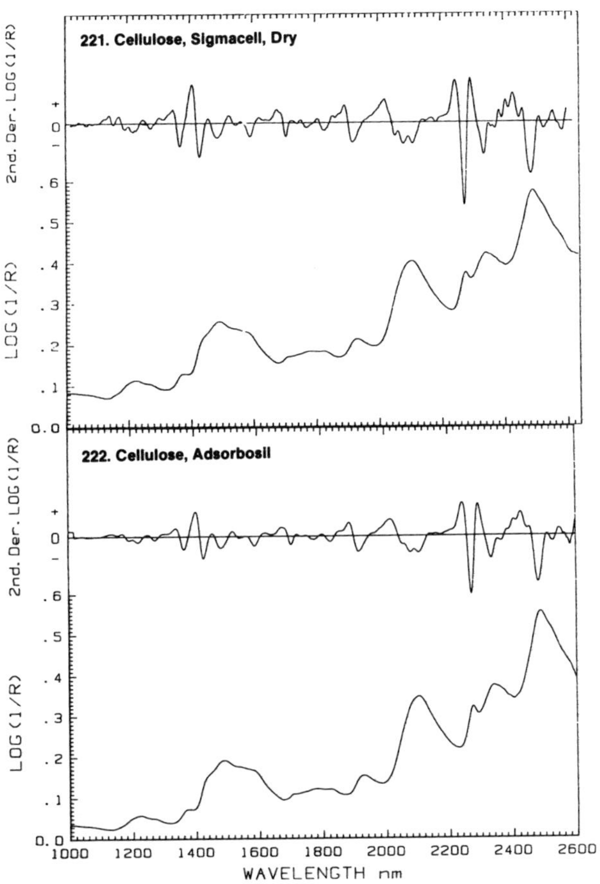
221. Cellulose, Sigmacell, Dry
222. Cellulose, Adsorbosil
2nd. Der. LOG(1/R)
LOG(1/R)
WAVELENGTH nm

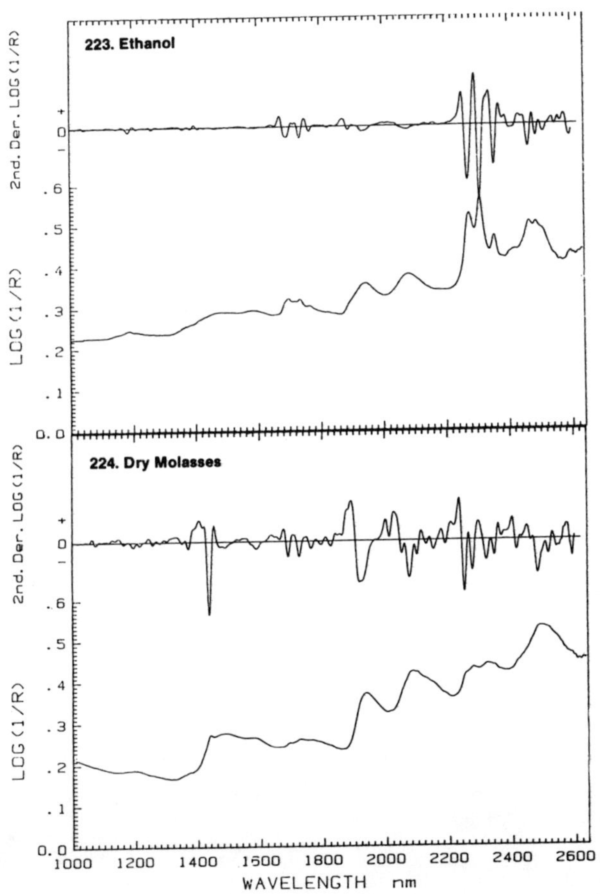
223. Ethanol
224. Dry Molasses
2nd. Der. LOG(1/R)
LOG(1/R)
WAVELENGTH nm

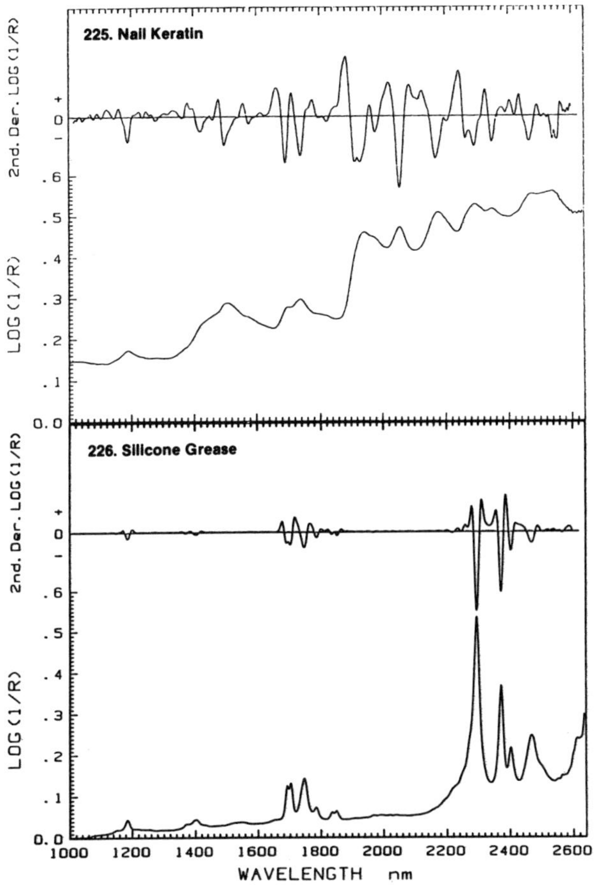
225. Nail Keratin
226. Silicone Grease
2nd. Der. LOG(1/R)
LOG(1/R)
WAVELENGTH nm

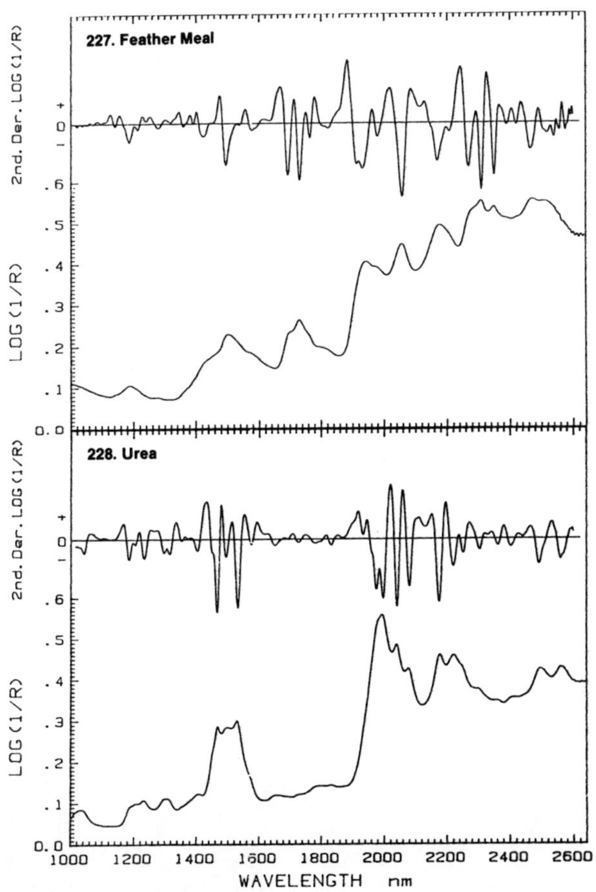
227. Feather Meal
228. Urea
2nd. Der. LOG(1/R)
LOG(1/R)
WAVELENGTH nm

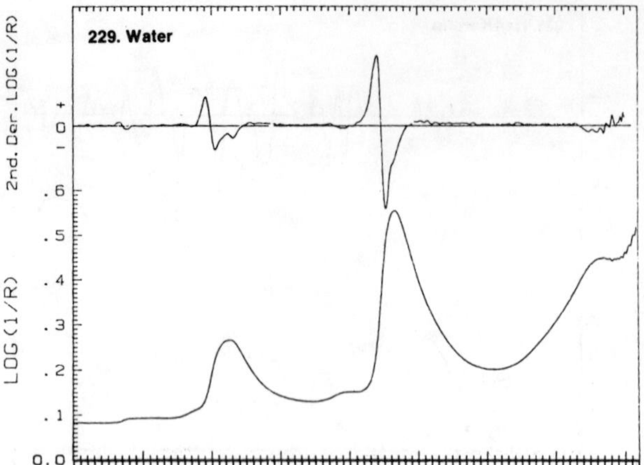

229. Water

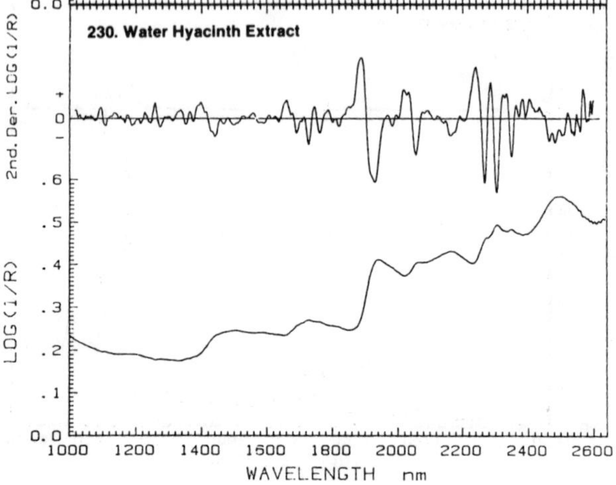

230. Water Hyacinth Extract

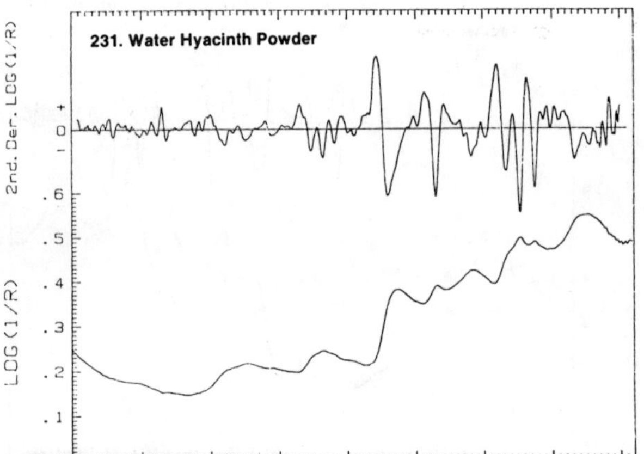

231. Water Hyacinth Powder

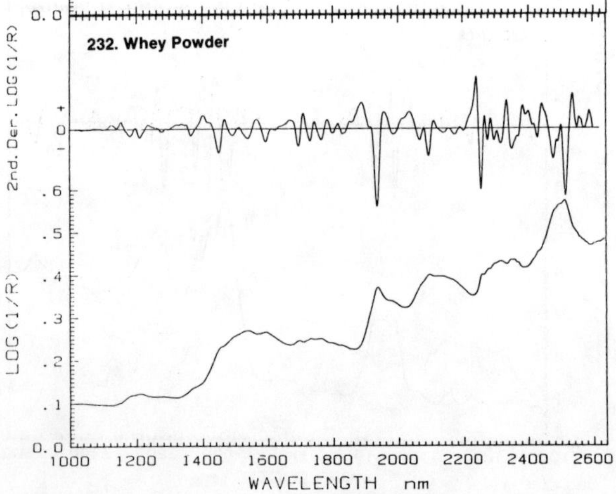

232. Whey Powder

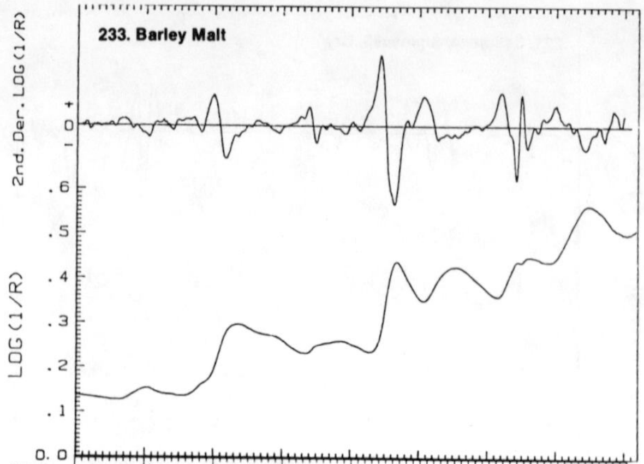

233. Barley Malt

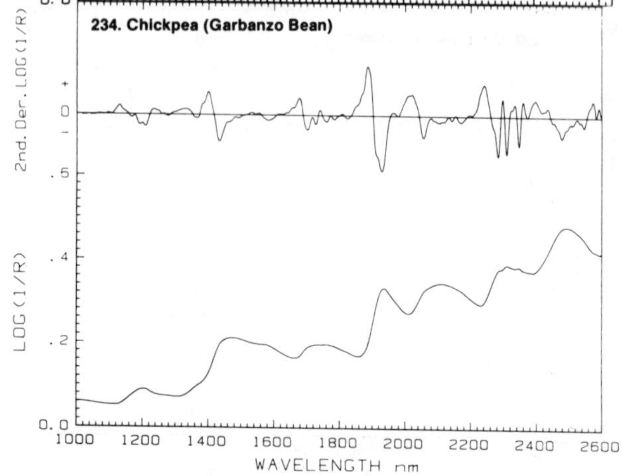

234. Chickpea (Garbanzo Bean)

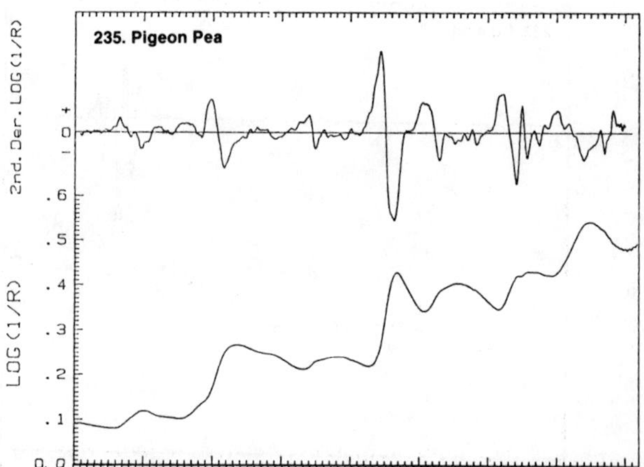

235. Pigeon Pea

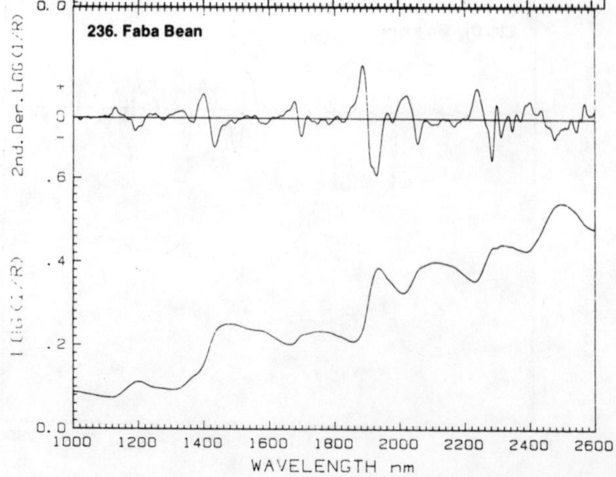

236. Faba Bean

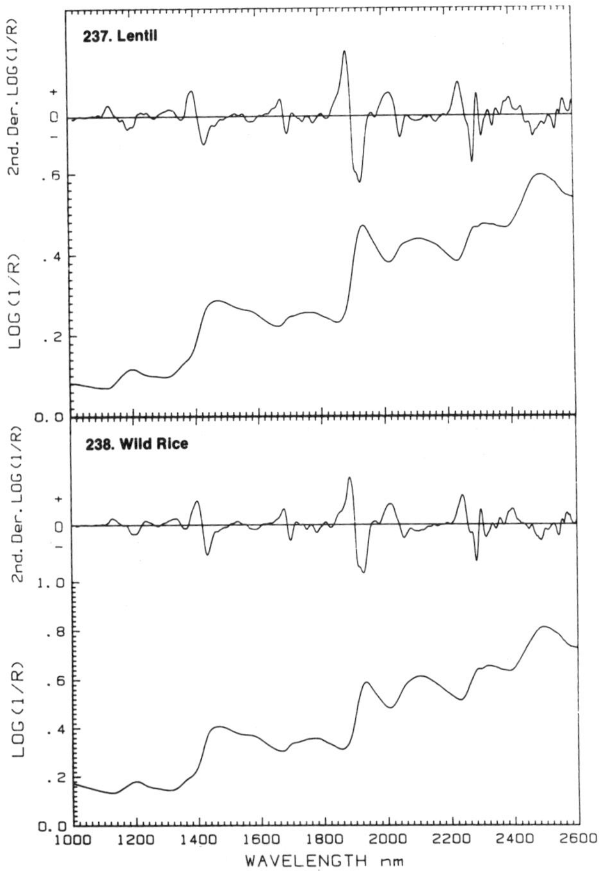
237. Lentil
238. Wild Rice
2nd. Der. LOG (1/R)
LOG (1/R)
WAVELENGTH nm
1000 1200 1400 1600 1800 2000 2200 2400 2600

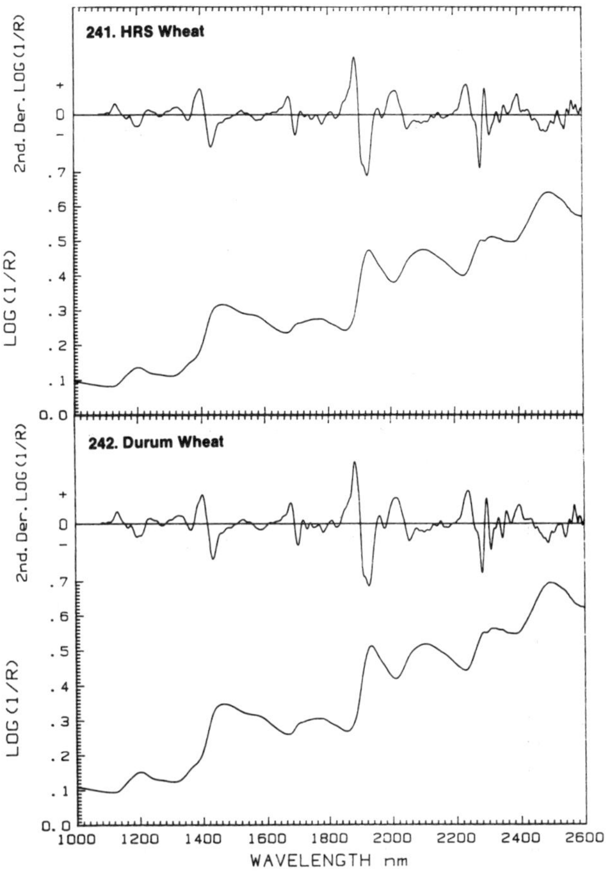
241. HRS Wheat
242. Durum Wheat
2nd. Der. LOG (1/R)
LOG (1/R)
WAVELENGTH nm
1000 1200 1400 1600 1800 2000 2200 2400 2600

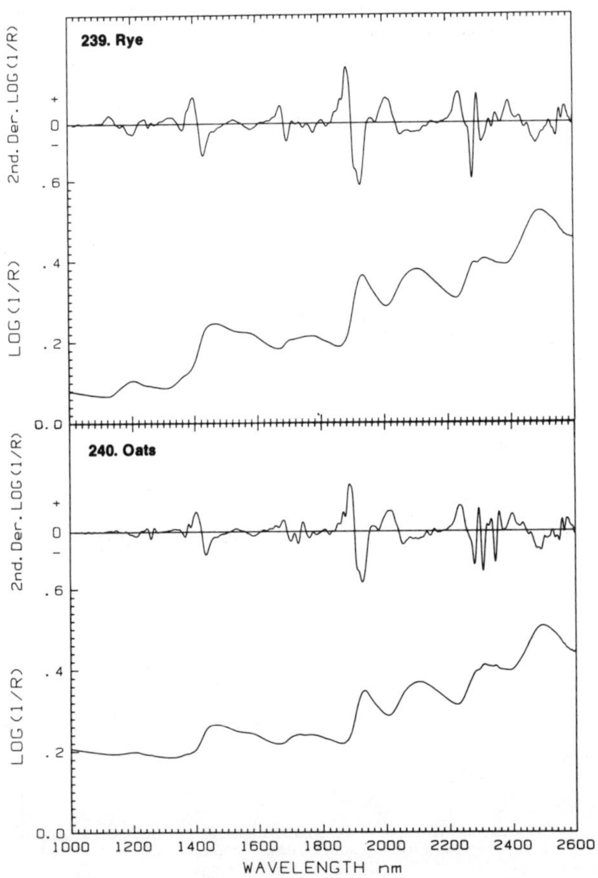
239. Rye
240. Oats
2nd. Der. LOG (1/R)
LOG (1/R)
WAVELENGTH nm
1000 1200 1400 1600 1800 2000 2200 2400 2600

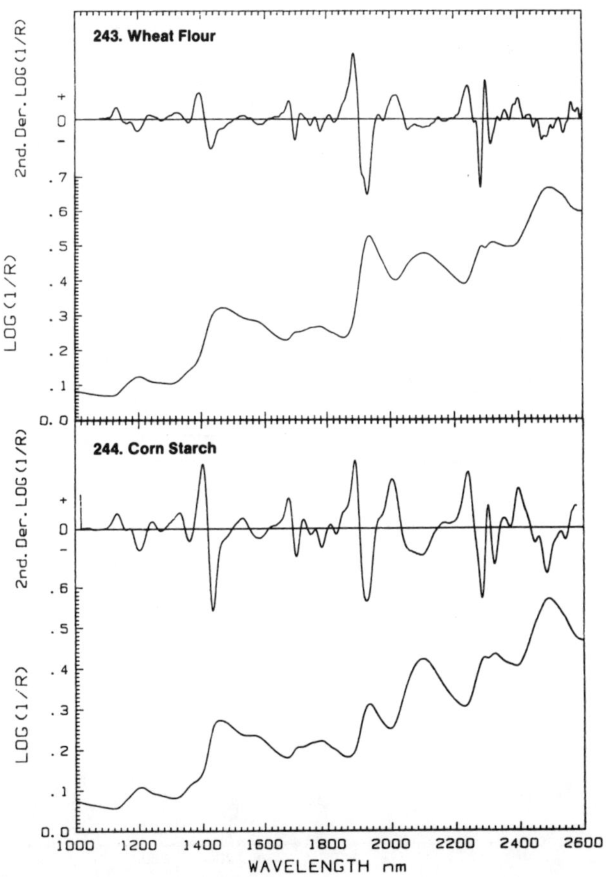
243. Wheat Flour
244. Corn Starch
2nd. Der. LOG (1/R)
LOG (1/R)
WAVELENGTH nm
1000 1200 1400 1600 1800 2000 2200 2400 2600

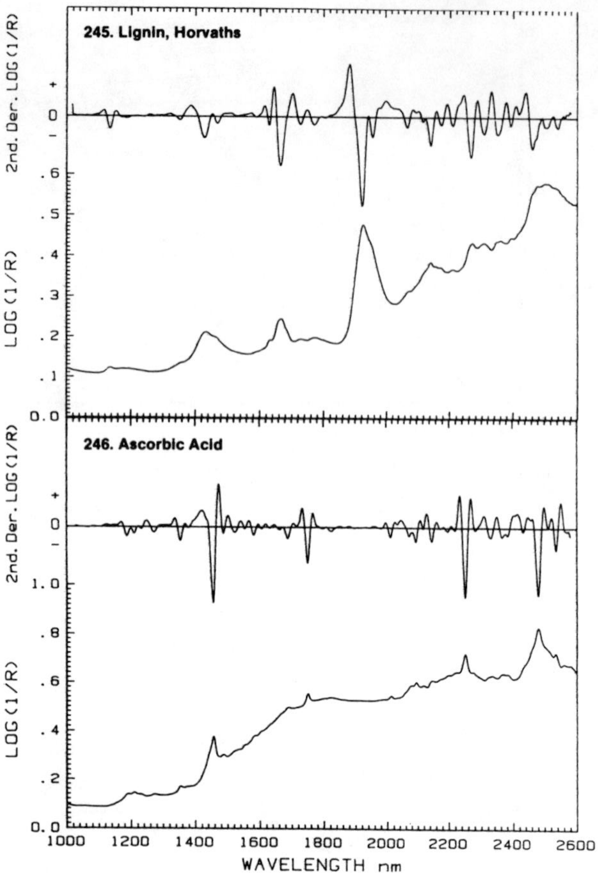
245. Lignin, Horvaths
246. Ascorbic Acid
2nd. Der. LOG (1/R)
LOG (1/R)
WAVELENGTH nm

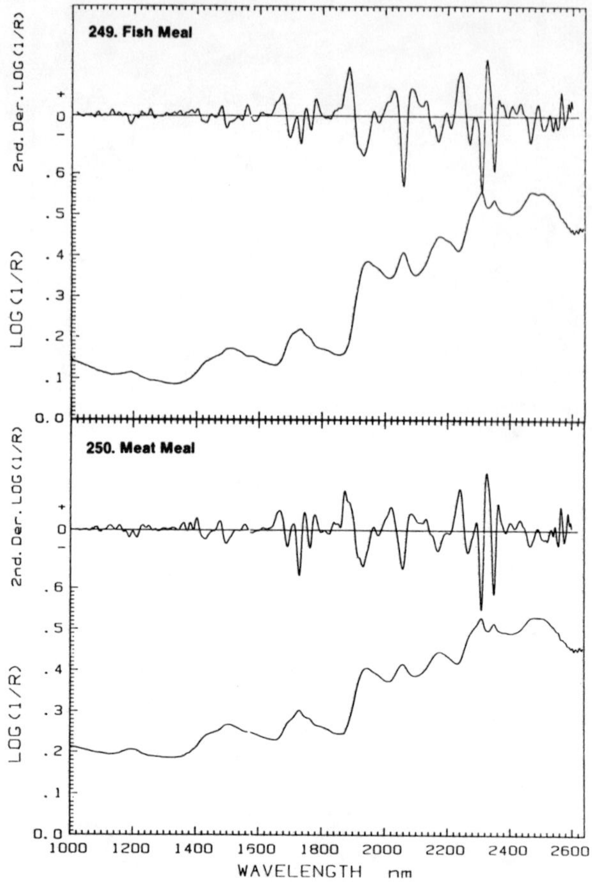
249. Fish Meal
250. Meat Meal
2nd. Der. LOG (1/R)
LOG (1/R)
WAVELENGTH nm

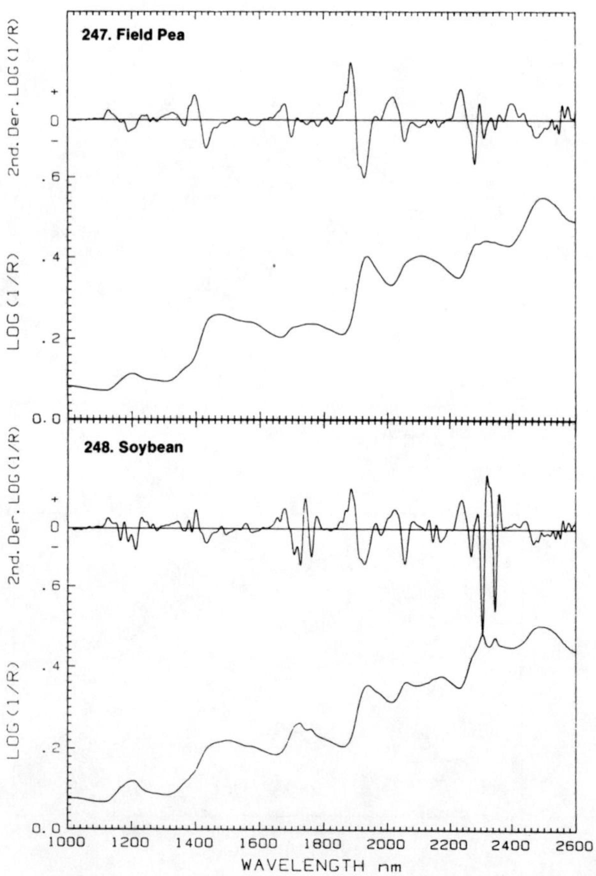
247. Field Pea
248. Soybean
2nd. Der. LOG (1/R)
LOG (1/R)
WAVELENGTH nm

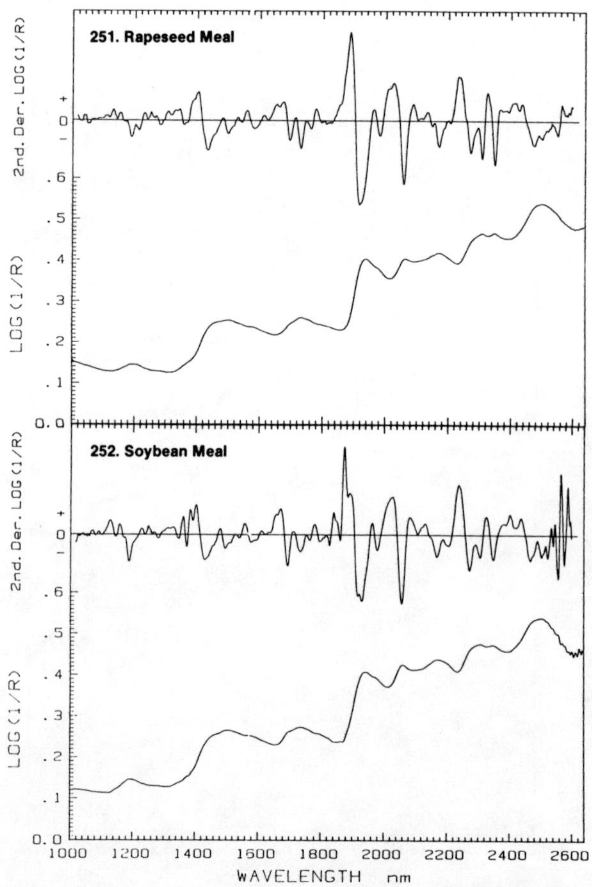
251. Rapeseed Meal
252. Soybean Meal
2nd. Der. LOG (1/R)
LOG (1/R)
WAVELENGTH nm

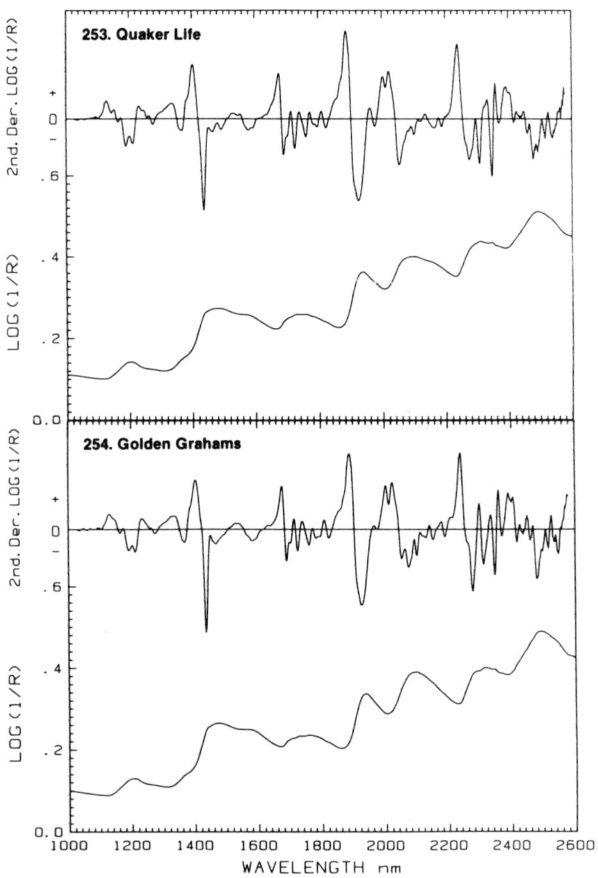
253. Quaker Life
254. Golden Grahams
2nd. Der. LOG (1/R)
LOG (1/R)
WAVELENGTH nm

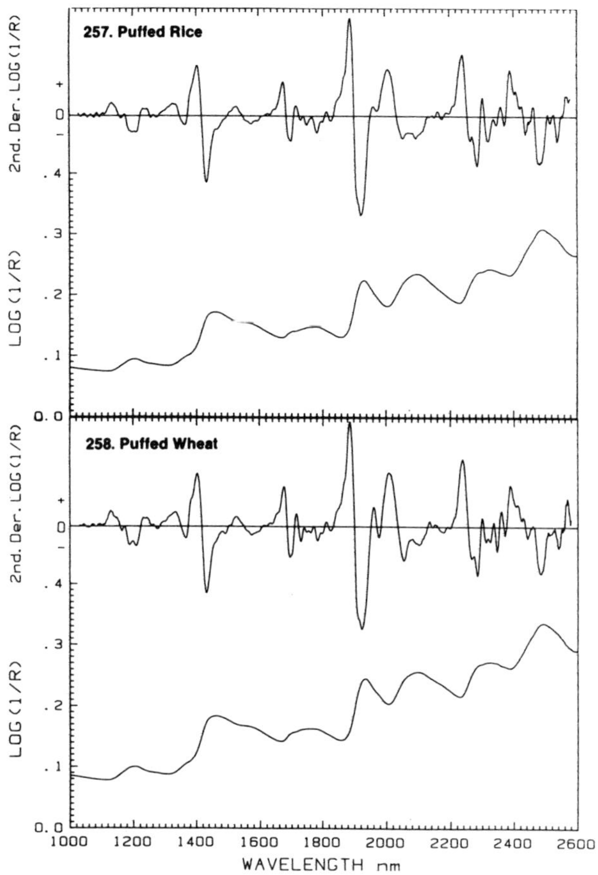
257. Puffed Rice
258. Puffed Wheat
2nd. Der. LOG (1/R)
LOG (1/R)
WAVELENGTH nm

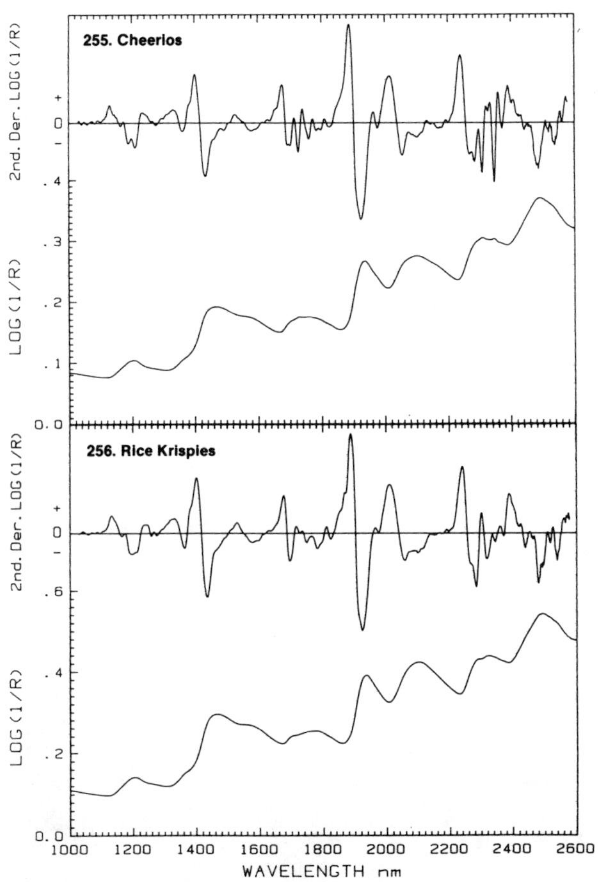
255. Cheerios
256. Rice Krispies
2nd. Der. LOG (1/R)
LOG (1/R)
WAVELENGTH nm

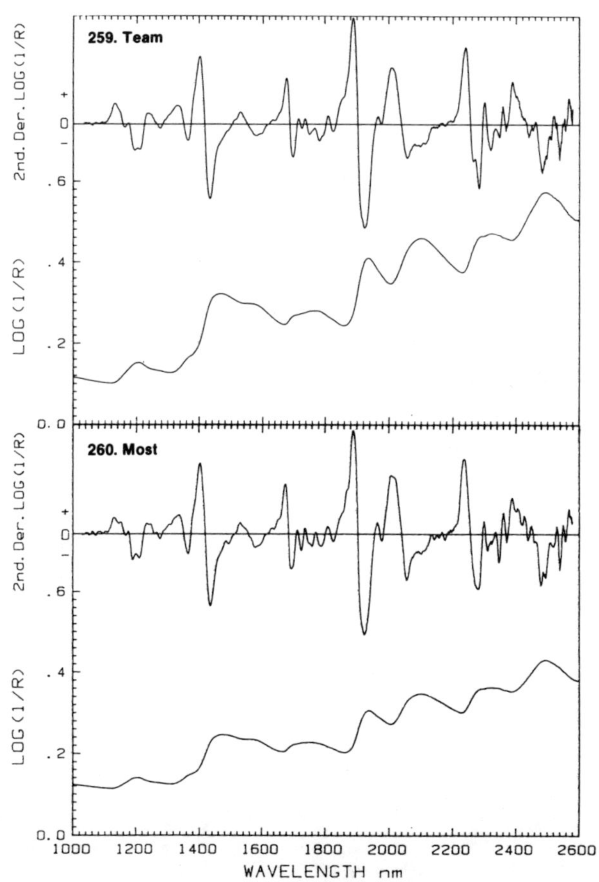
259. Team
260. Most
2nd. Der. LOG (1/R)
LOG (1/R)
WAVELENGTH nm

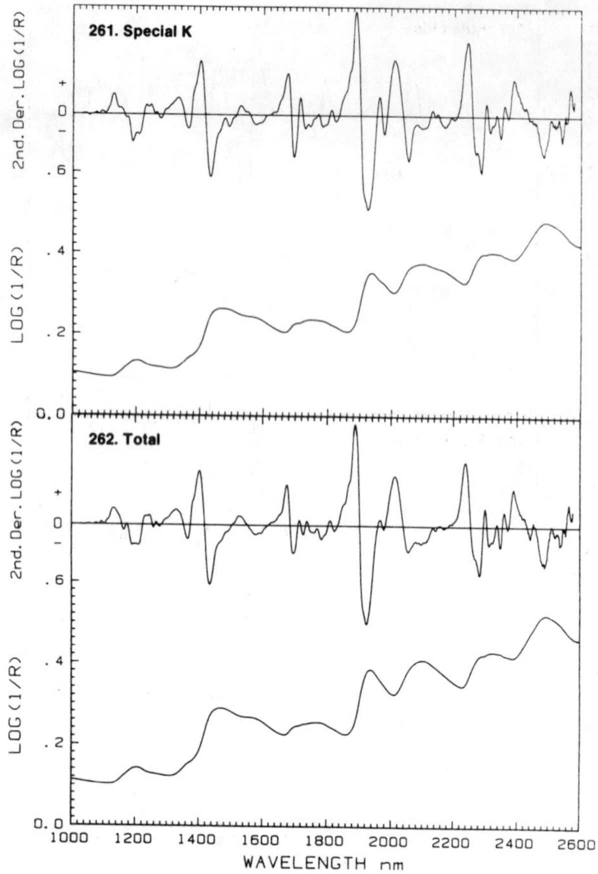
261. Special K
262. Total
2nd. Der. LOG (1/R)
LOG (1/R)
WAVELENGTH nm

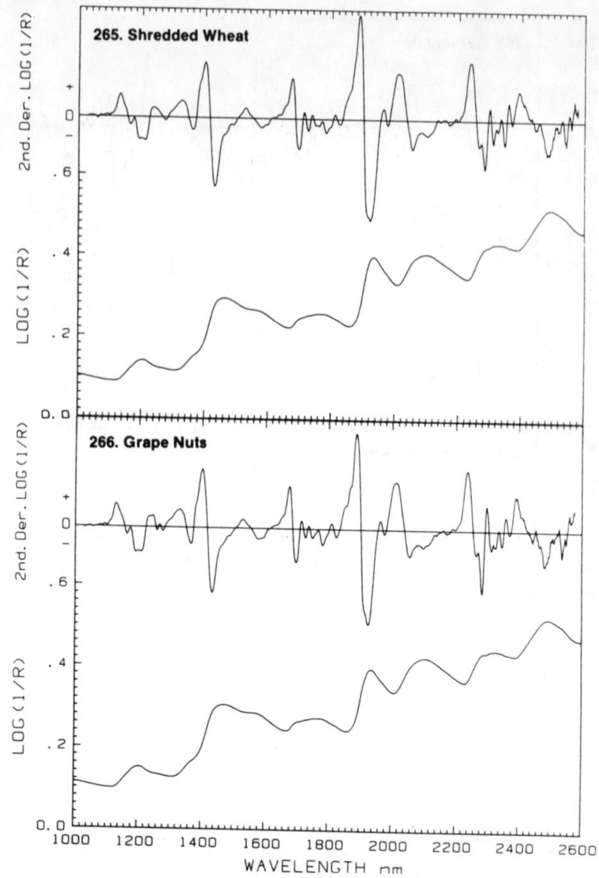
265. Shredded Wheat
266. Grape Nuts
2nd. Der. LOG (1/R)
LOG (1/R)
WAVELENGTH nm

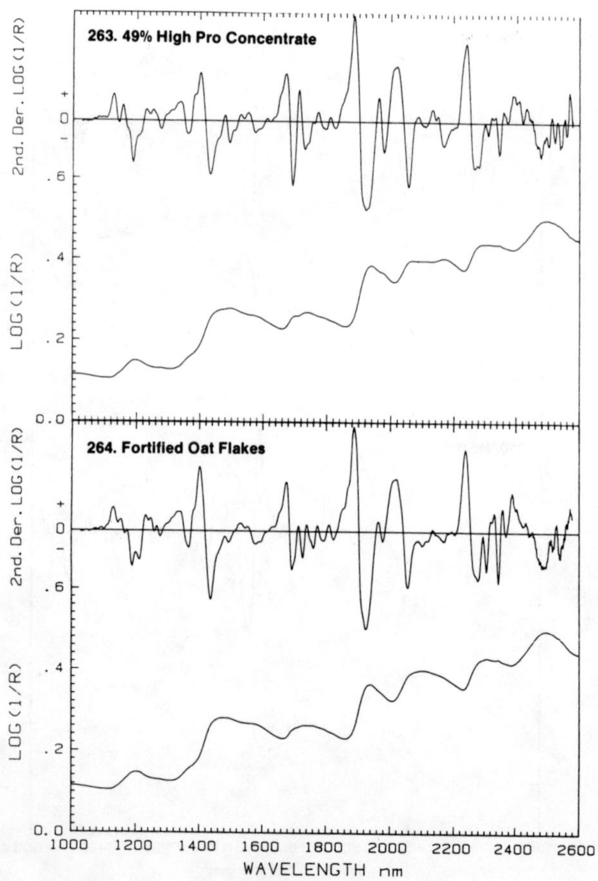
263. 49% High Pro Concentrate
264. Fortified Oat Flakes
2nd. Der. LOG (1/R)
LOG (1/R)
WAVELENGTH nm

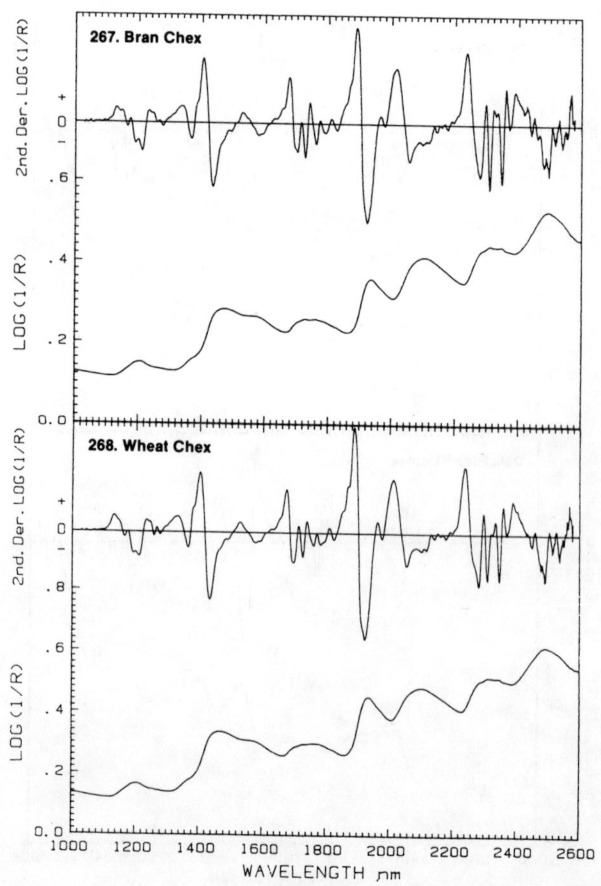
267. Bran Chex
268. Wheat Chex
2nd. Der. LOG (1/R)
LOG (1/R)
WAVELENGTH nm

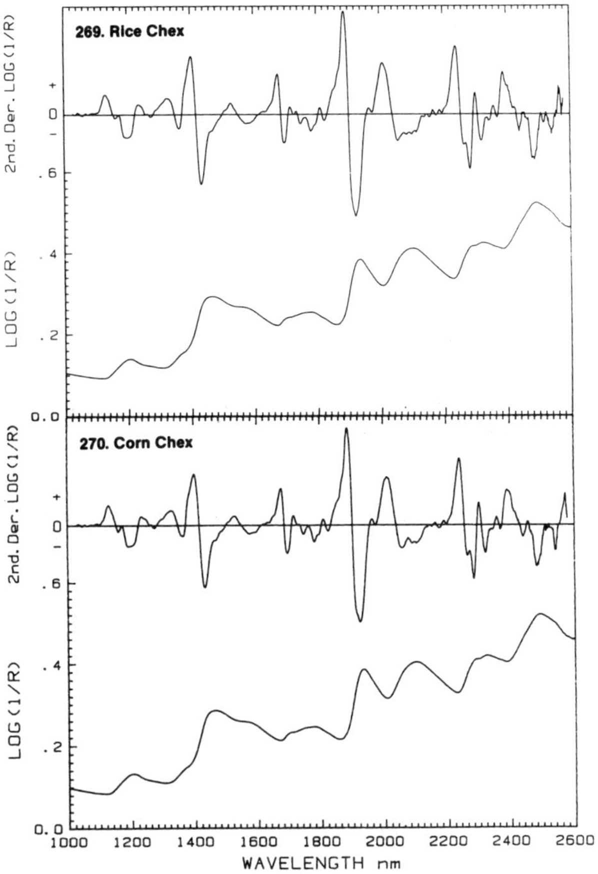
269. Rice Chex
270. Corn Chex
2nd. Der. LOG(1/R)
LOG(1/R)
WAVELENGTH nm

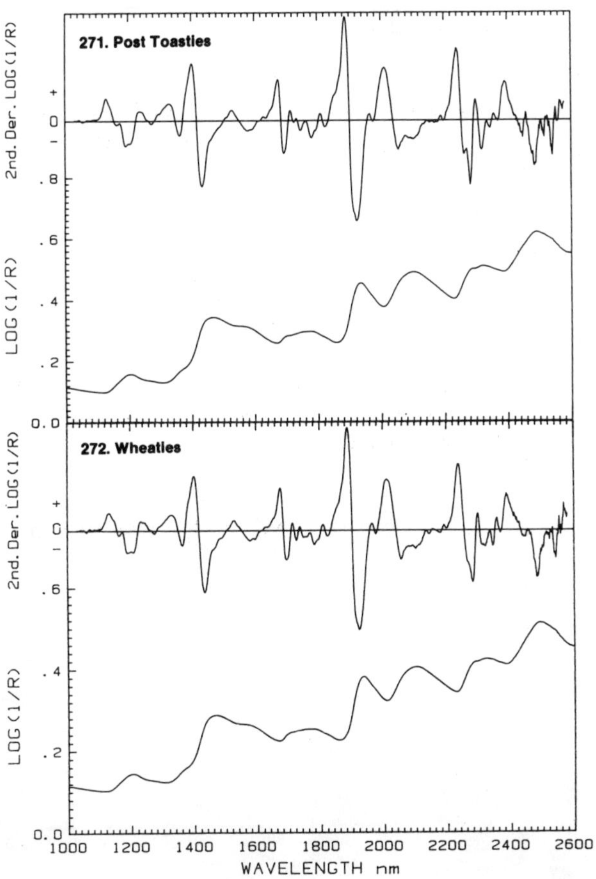
271. Post Toasties
272. Wheaties
2nd. Der. LOG(1/R)
LOG(1/R)
WAVELENGTH nm

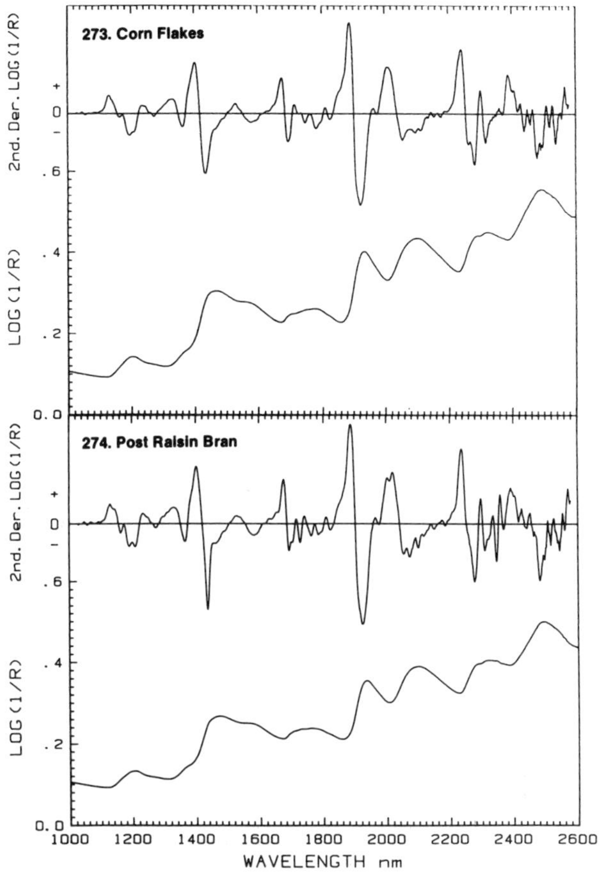
273. Corn Flakes
274. Post Raisin Bran
2nd. Der. LOG(1/R)
LOG(1/R)
WAVELENGTH nm

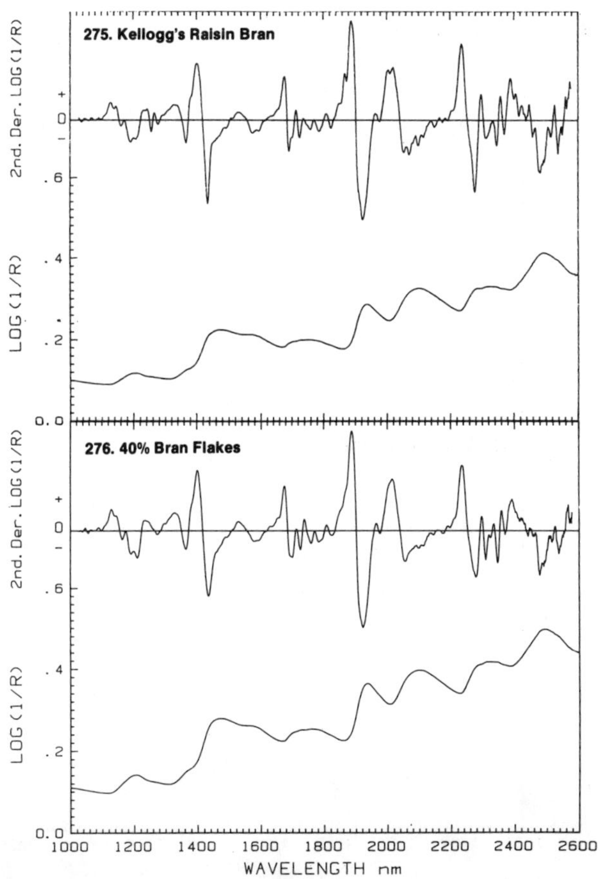
275. Kellogg's Raisin Bran
276. 40% Bran Flakes
2nd. Der. LOG(1/R)
LOG(1/R)
WAVELENGTH nm

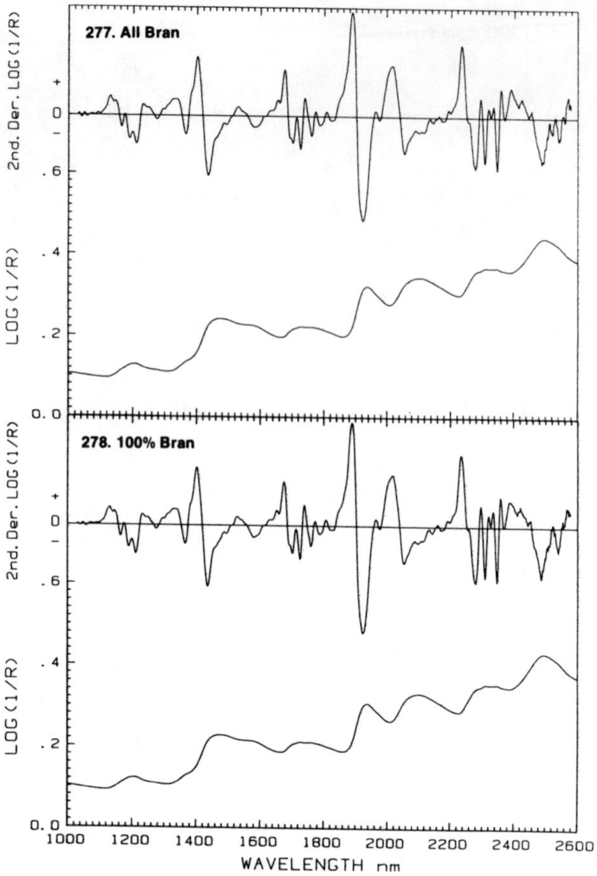
277. All Bran
278. 100% Bran
2nd. Der. LOG (1/R)
LOG (1/R)
WAVELENGTH nm

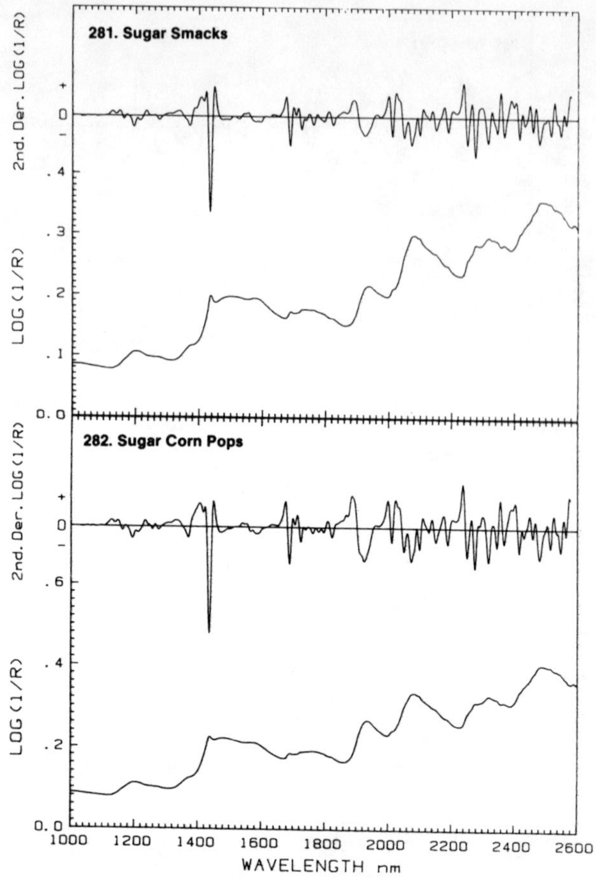
281. Sugar Smacks
282. Sugar Corn Pops
2nd. Der. LOG (1/R)
LOG (1/R)
WAVELENGTH nm

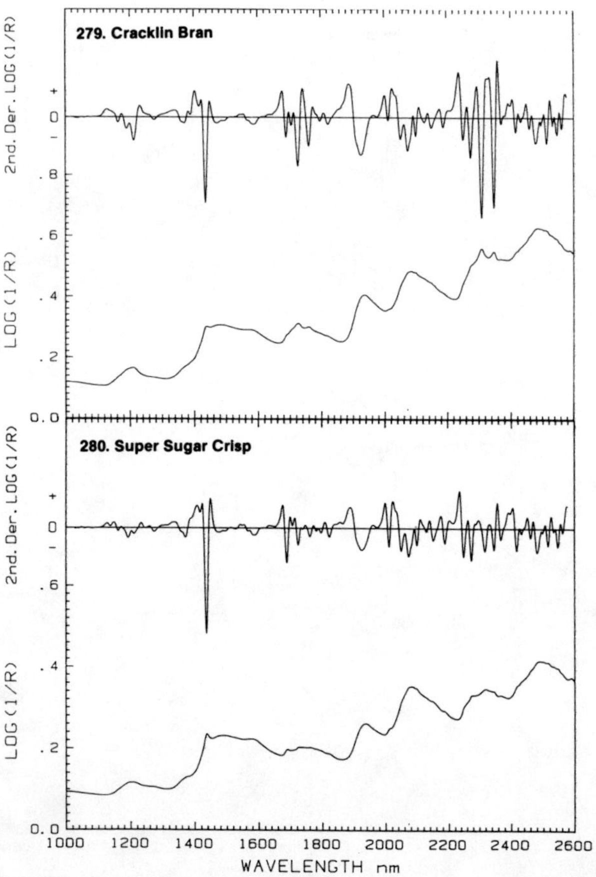
279. Cracklin Bran
280. Super Sugar Crisp
2nd. Der. LOG (1/R)
LOG (1/R)
WAVELENGTH nm

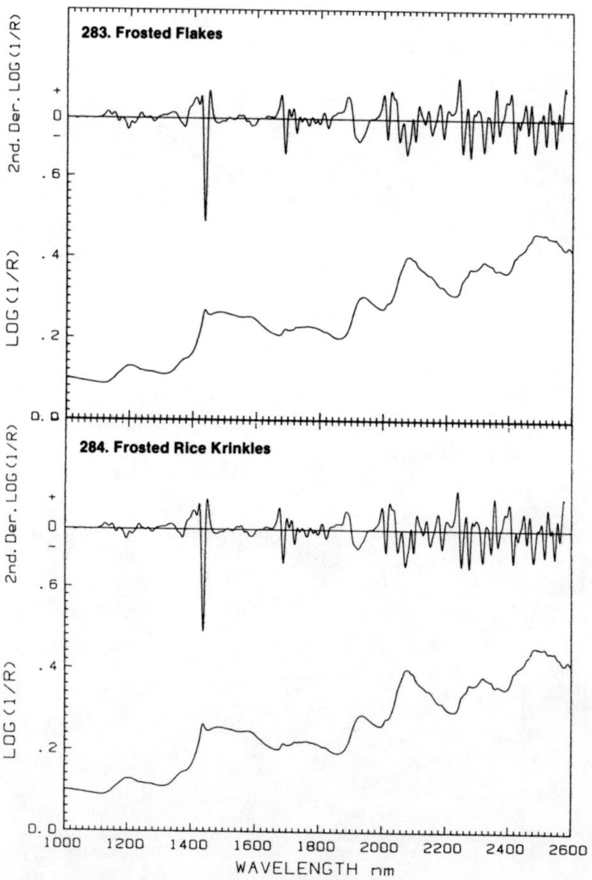
283. Frosted Flakes
284. Frosted Rice Krinkles
2nd. Der. LOG (1/R)
LOG (1/R)
WAVELENGTH nm

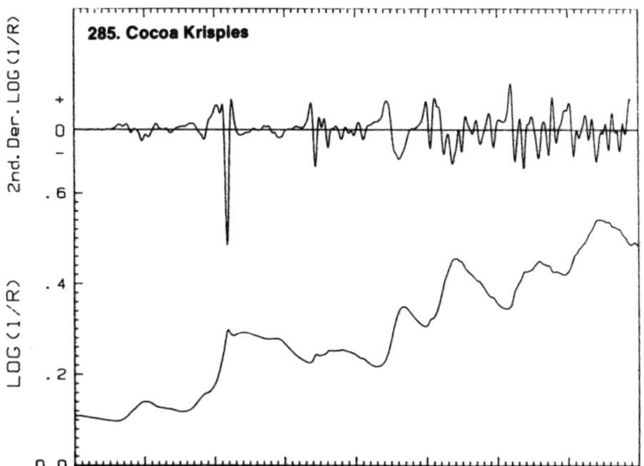
285. Cocoa Krispies
2nd. Der. LOG(1/R)
LOG(1/R)

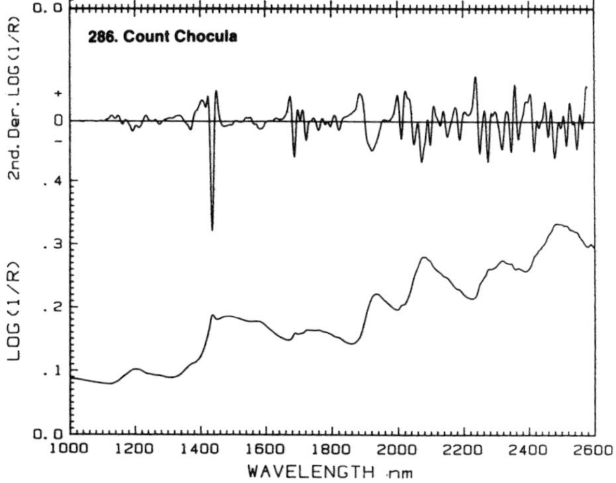
286. Count Chocula
2nd. Der. LOG(1/R)
LOG(1/R)
WAVELENGTH nm

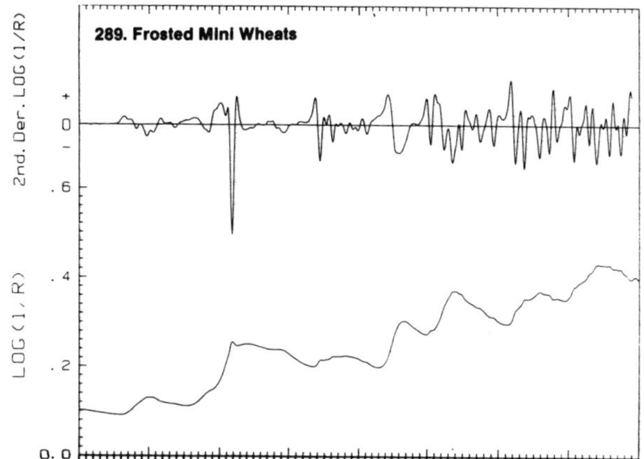
289. Frosted Mini Wheats
2nd. Der. LOG(1/R)
LOG(1/R)

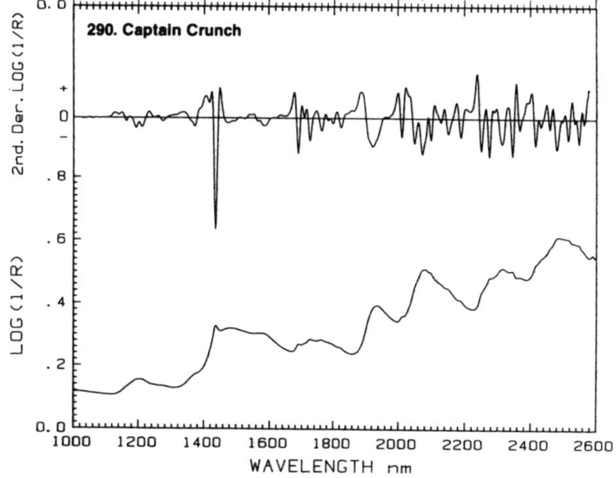
290. Captain Crunch
2nd. Der. LOG(1/R)
LOG(1/R)
WAVELENGTH nm

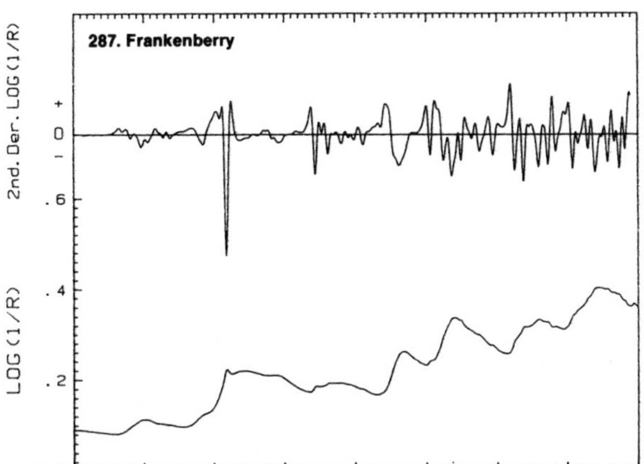
287. Frankenberry
2nd. Der. LOG(1/R)
LOG(1/R)

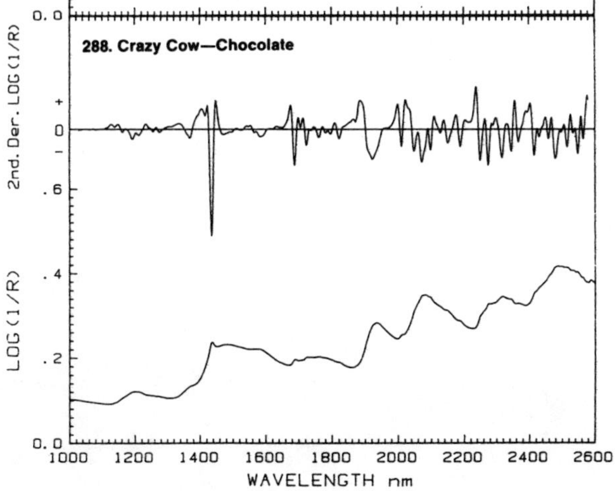
288. Crazy Cow—Chocolate
2nd. Der. LOG(1/R)
LOG(1/R)
WAVELENGTH nm

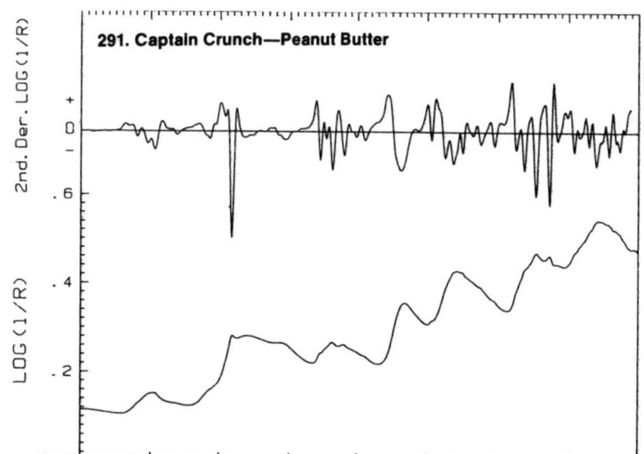
291. Captain Crunch—Peanut Butter
2nd. Der. LOG(1/R)
LOG(1/R)

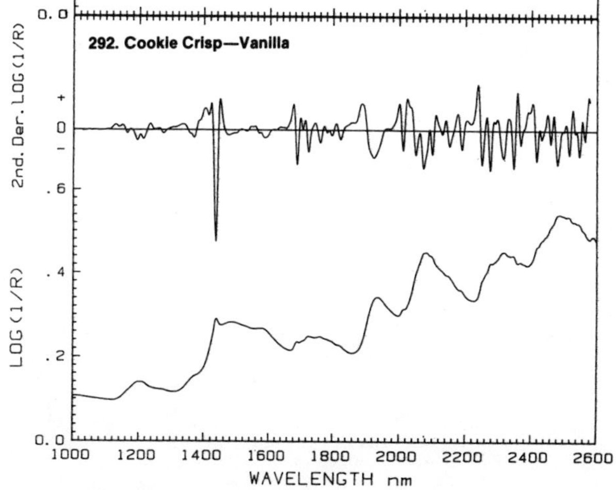
292. Cookie Crisp—Vanilla
2nd. Der. LOG(1/R)
LOG(1/R)
WAVELENGTH nm

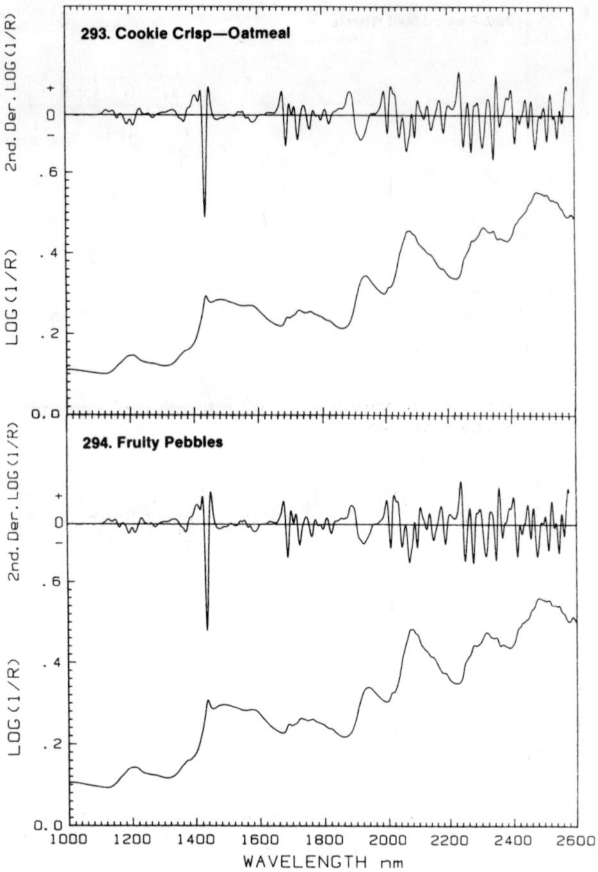

293. Cookie Crisp—Oatmeal

294. Fruity Pebbles

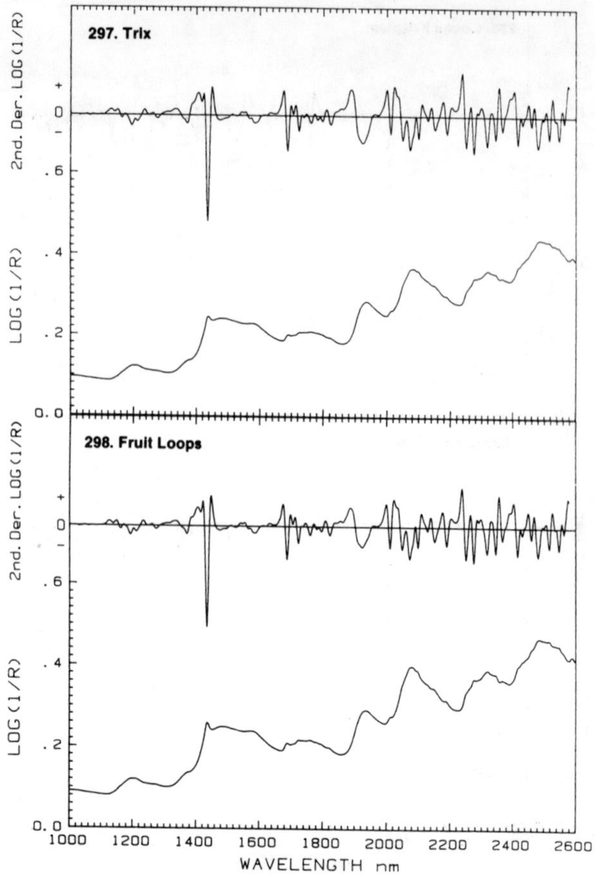

297. Trix

298. Fruit Loops

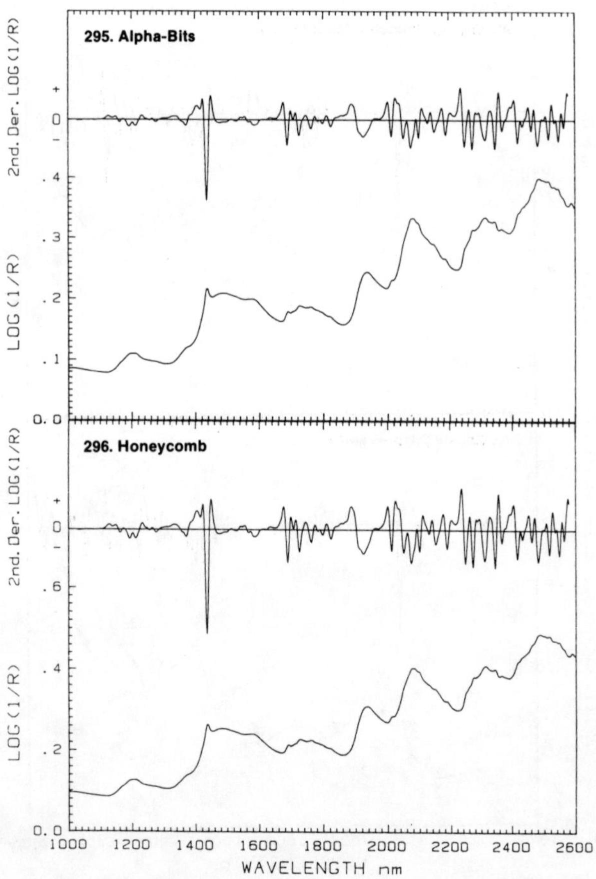

295. Alpha-Bits

296. Honeycomb

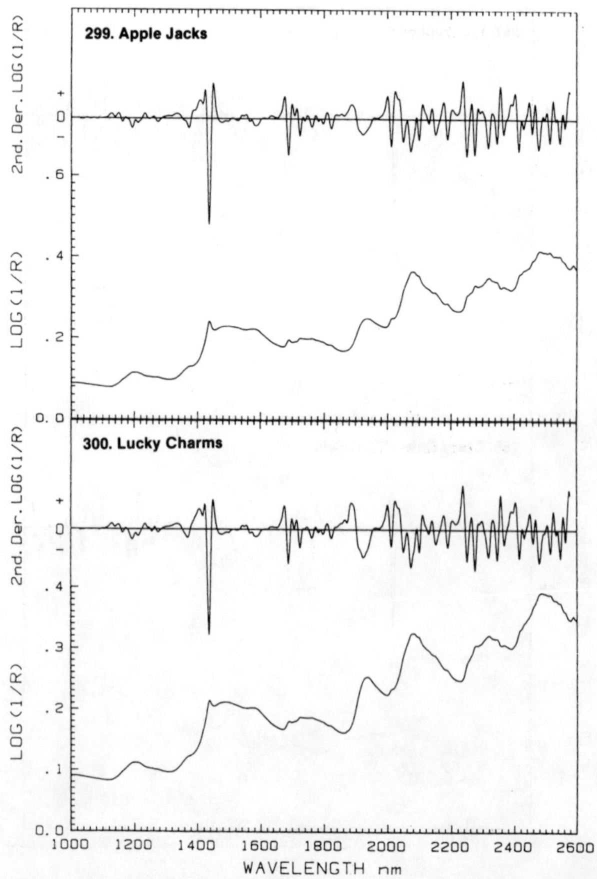

299. Apple Jacks

300. Lucky Charms

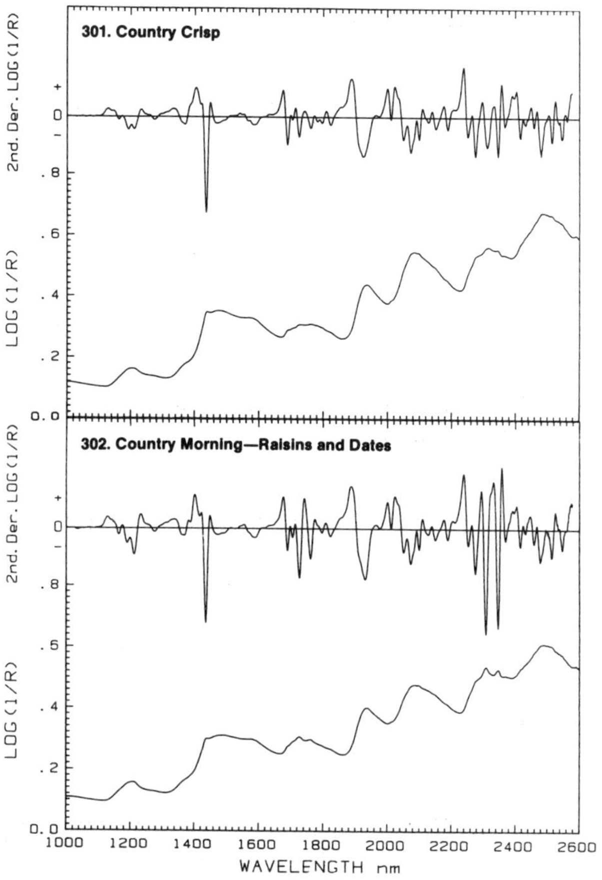
301. Country Crisp
302. Country Morning—Raisins and Dates
2nd. Der. LOG (1/R)
LOG (1/R)
1000 1200 1400 1600 1800 2000 2200 2400 2600
WAVELENGTH nm

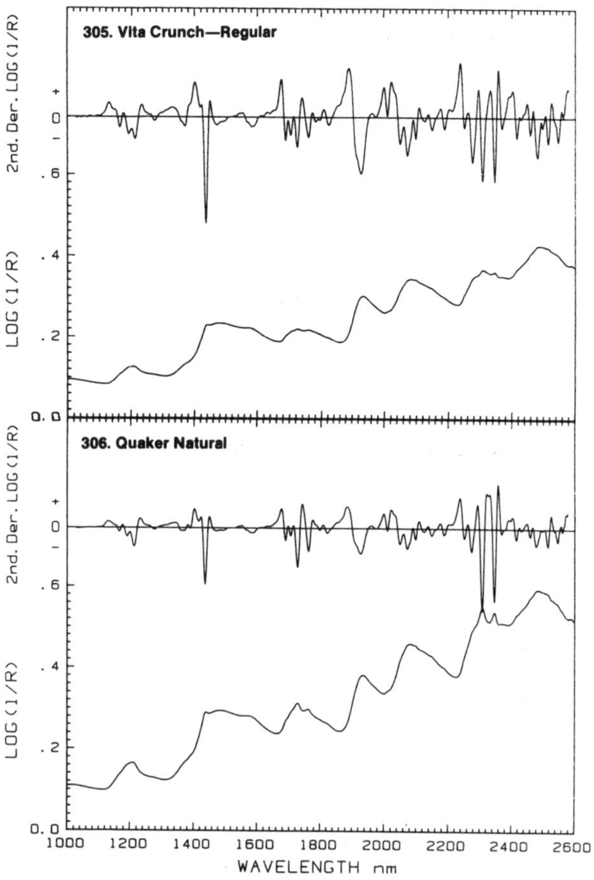
305. Vita Crunch—Regular
306. Quaker Natural
2nd. Der. LOG (1/R)
LOG (1/R)
1000 1200 1400 1600 1800 2000 2200 2400 2600
WAVELENGTH nm

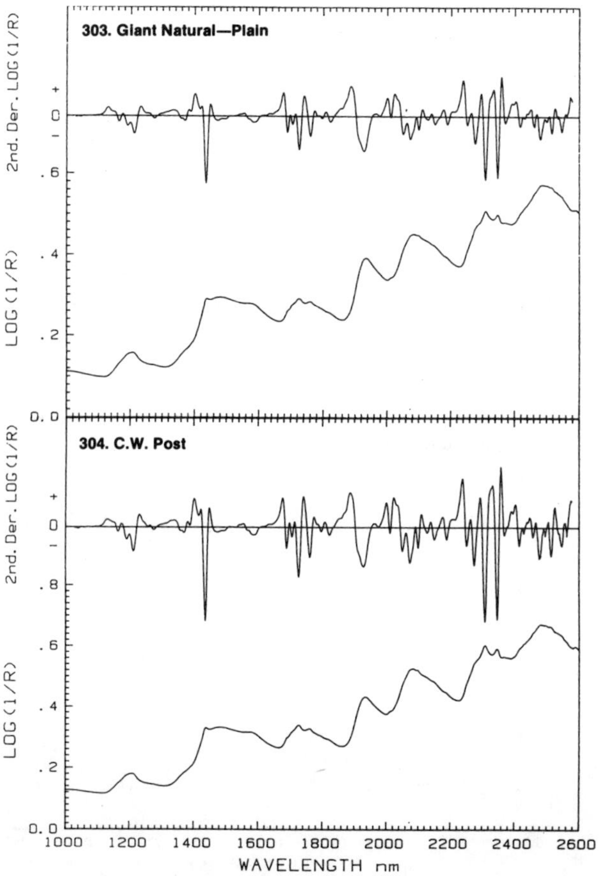
303. Giant Natural—Plain
304. C.W. Post
2nd. Der. LOG (1/R)
LOG (1/R)
1000 1200 1400 1600 1800 2000 2200 2400 2600
WAVELENGTH nm

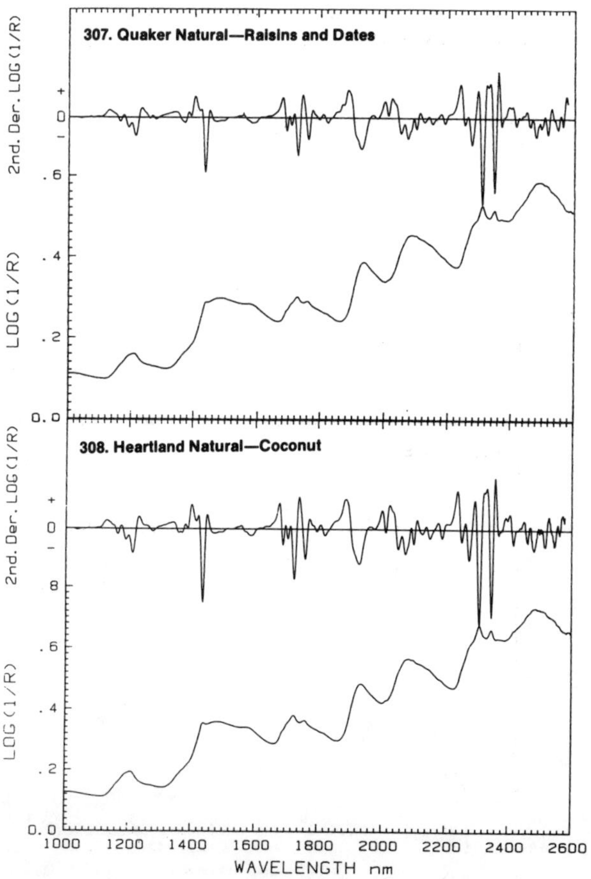
307. Quaker Natural—Raisins and Dates
308. Heartland Natural—Coconut
2nd. Der. LOG (1/R)
LOG (1/R)
1000 1200 1400 1600 1800 2000 2200 2400 2600
WAVELENGTH nm

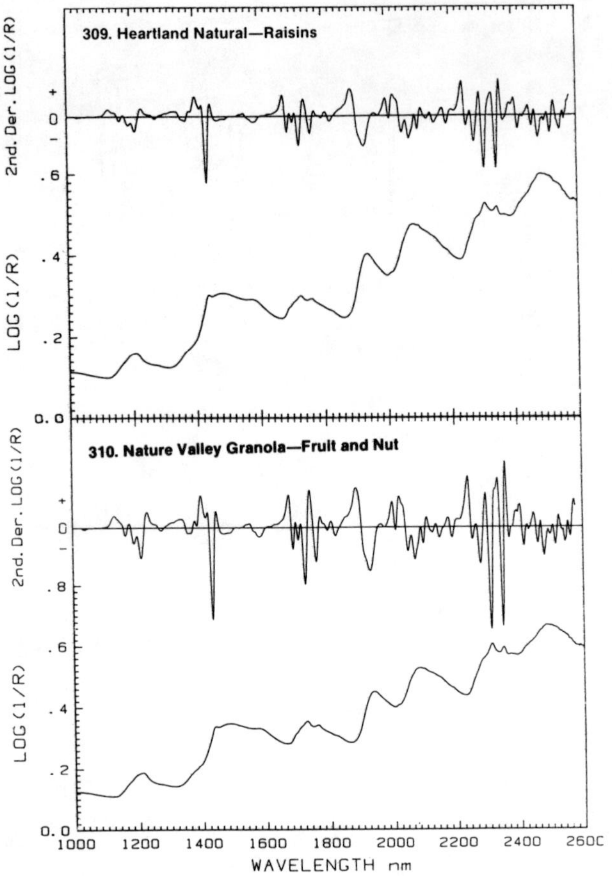
309. Heartland Natural—Raisins
310. Nature Valley Granola—Fruit and Nut
2nd. Der. LOG (1/R)
LOG (1/R)
WAVELENGTH nm

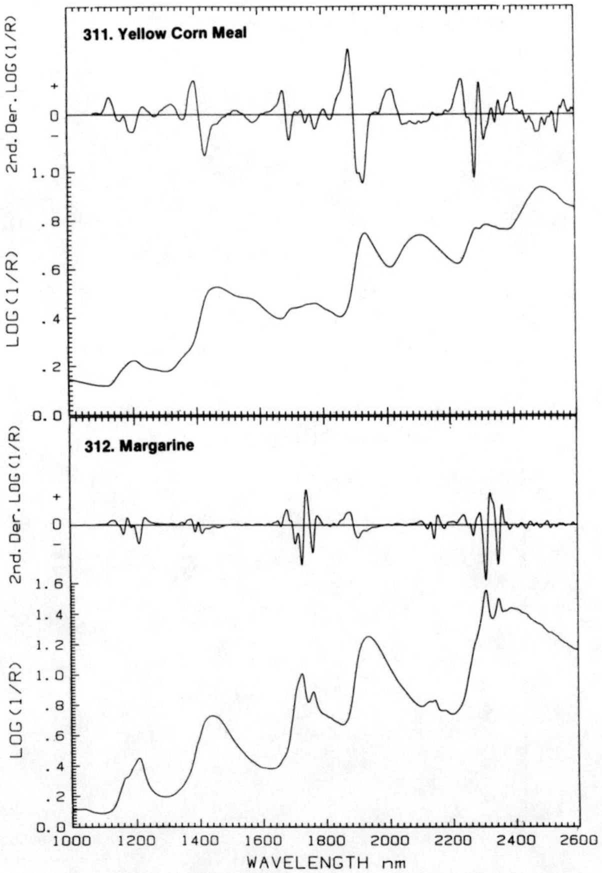
311. Yellow Corn Meal
312. Margarine
2nd. Der. LOG (1/R)
LOG (1/R)
WAVELENGTH nm

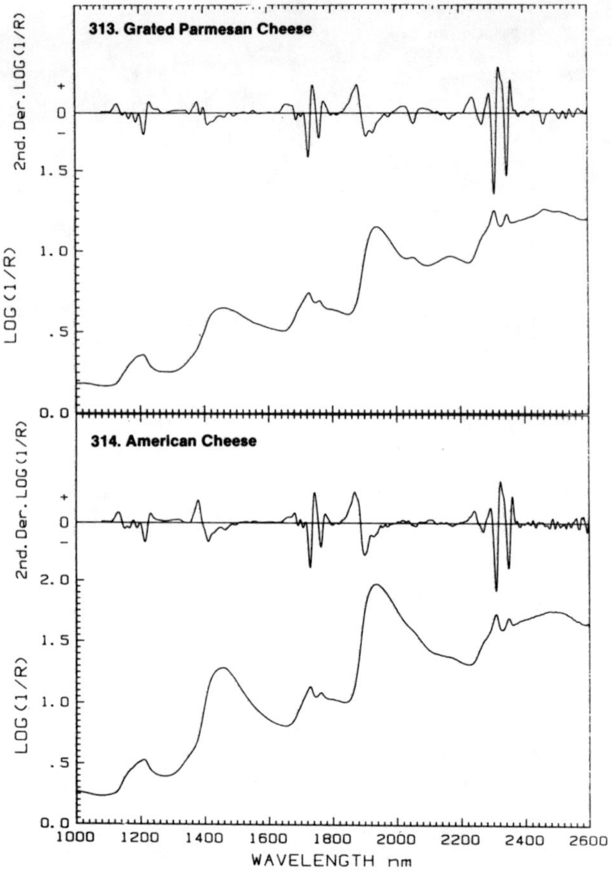
313. Grated Parmesan Cheese
314. American Cheese
2nd. Der. LOG (1/R)
LOG (1/R)
WAVELENGTH nm

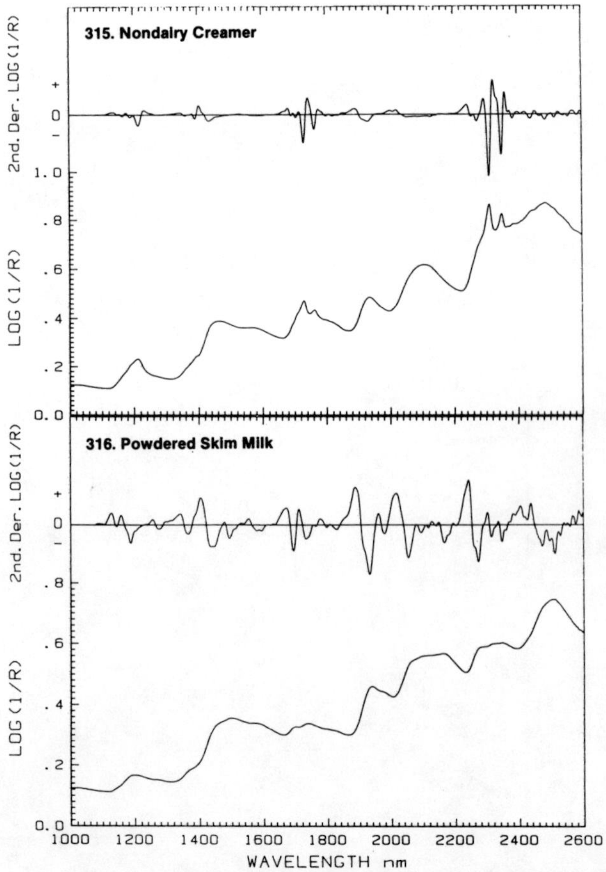
315. Nondairy Creamer
316. Powdered Skim Milk
2nd. Der. LOG (1/R)
LOG (1/R)
WAVELENGTH nm

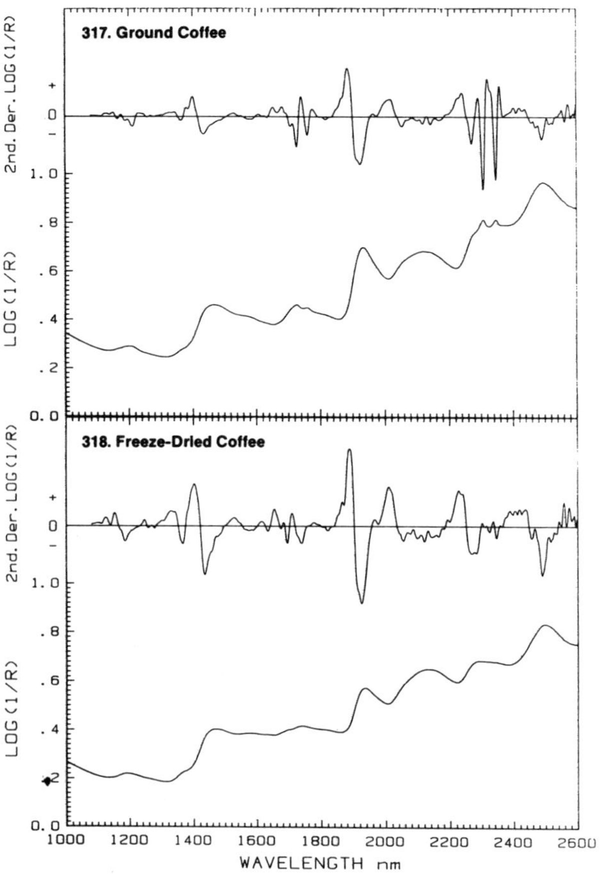
317. Ground Coffee
318. Freeze-Dried Coffee
2nd. Der. LOG (1/R)
LOG (1/R)
WAVELENGTH nm

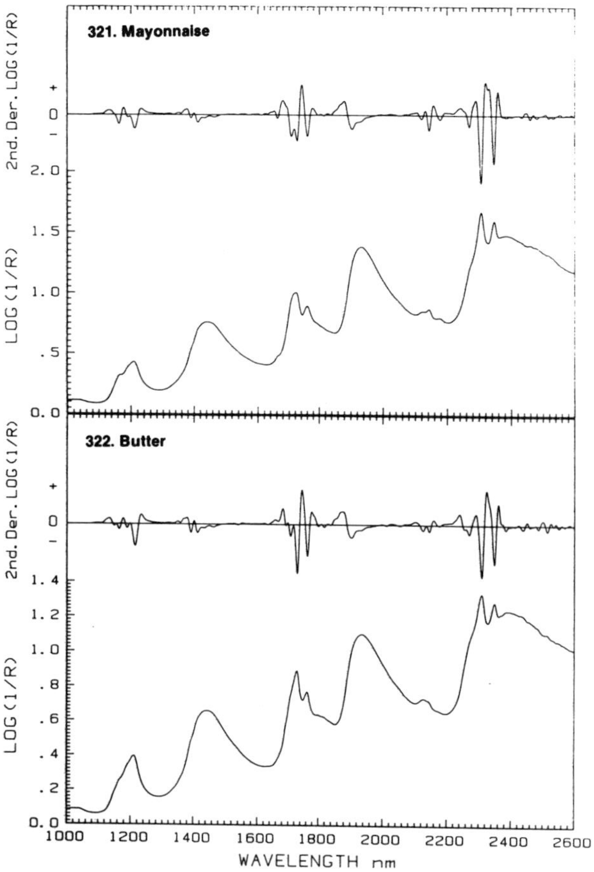
321. Mayonnaise
322. Butter
2nd. Der. LOG (1/R)
LOG (1/R)
WAVELENGTH nm

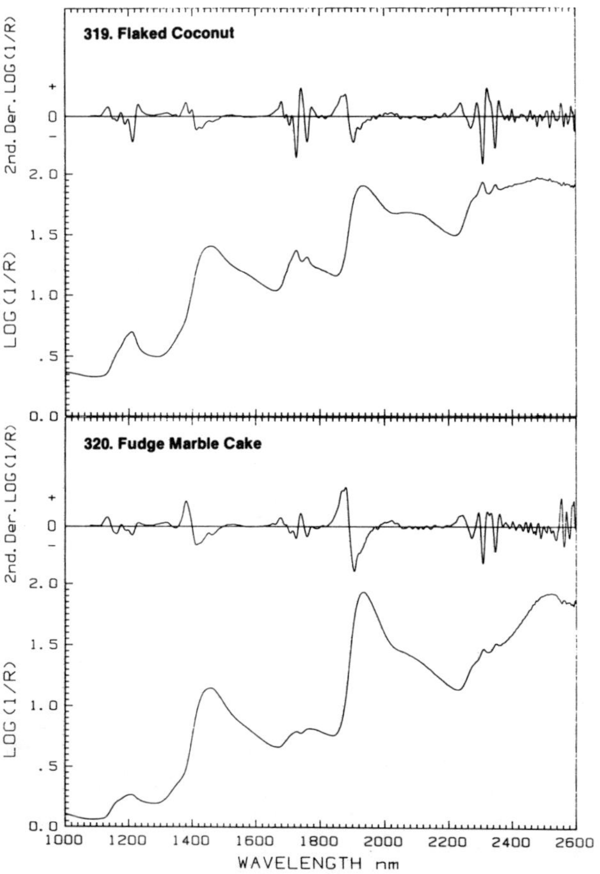
319. Flaked Coconut
320. Fudge Marble Cake
2nd. Der. LOG (1/R)
LOG (1/R)
WAVELENGTH nm

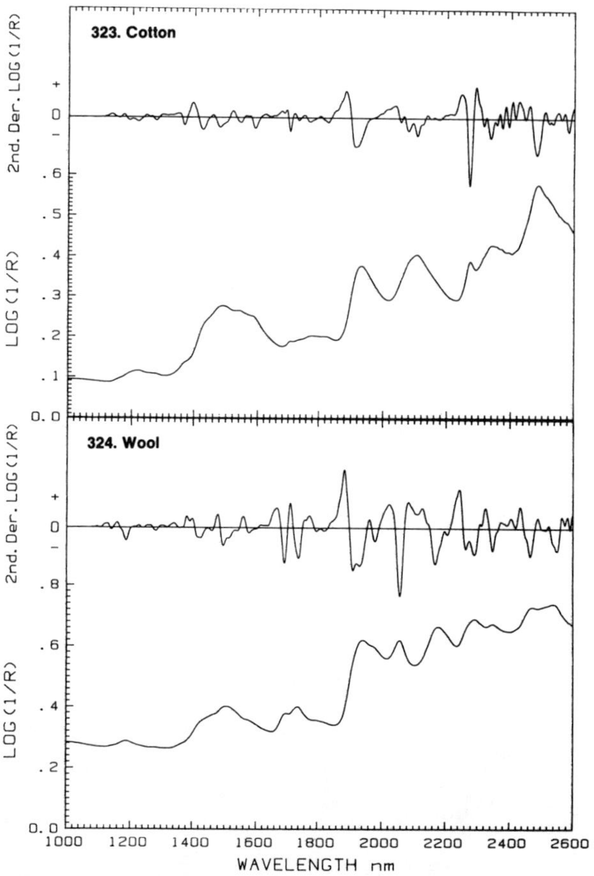
323. Cotton
324. Wool
2nd. Der. LOG (1/R)
LOG (1/R)
WAVELENGTH nm

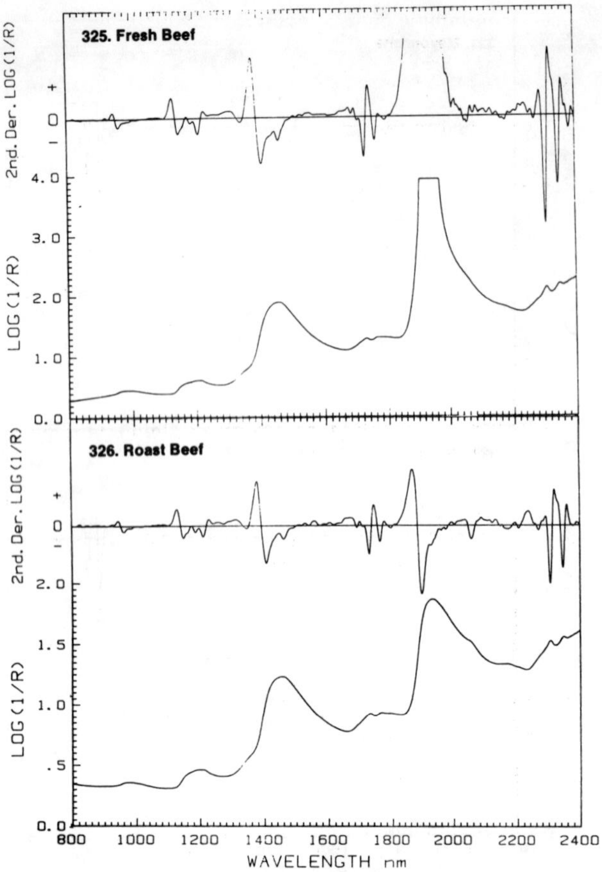
325. Fresh Beef
326. Roast Beef
2nd. Der. LOG (1/R)
LOG (1/R)
WAVELENGTH nm

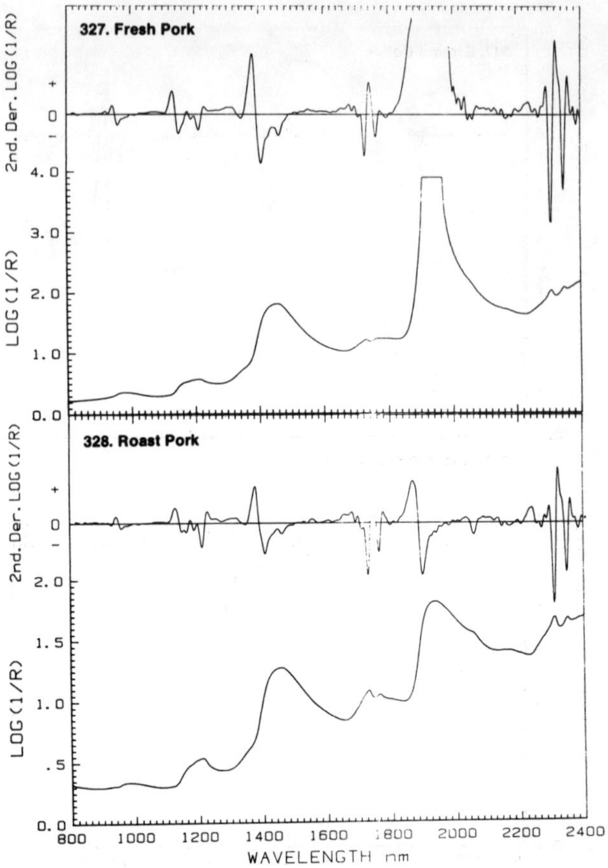
327. Fresh Pork
328. Roast Pork
2nd. Der. LOG (1/R)
LOG (1/R)
WAVELENGTH nm

Key to Near-Infrared Spectra

Numbers refer to spectrum numbers in the section beginning on page 247.

AMINO ACIDS

D-Alanine, 1
L-Alanine, 2
L-Arginine, 3
L-Asparagine, 4
L-Aspartic acid, 5
L-Cysteine, 6
L-Cystine, 7
L-Glutamic acid, 8
L-Glutamine, 9
Glycine, 10
L-Histidine, 11
L-Hydroxyproline, 25
L-Isoleucine, 12
L-Leucine, 13
D-Lysine, 14
L-Lysine, 15
L-Methionine, 16
DL-Phenylalanine, 18
L-Phenylalanine, 17
L-Proline, 19
L-Serine, 20
L-Threonine, 21
L-Tryptophan, 22
L-Tyrosine, 23
L-Valine, 24

BREAKFAST CEREALS

All Bran, 277
Alpha-Bits, 295
Apple Jacks, 299
Bran, 100%, 278
Bran Chex, 267
Bran Flakes, 40%, 276
Captain Crunch, 290
Captain Crunch—peanut butter, 291
Cheerios, 255
Cocoa Krispies, 285
Cookie Crisp—oatmeal, 293
Cookie Crisp—vanilla, 292
Corn Chex, 270
Corn Flakes, 273
Count Chocula, 286
Country Crisp, 301
Country Morning—raisins and dates, 302
Cracklin Bran, 279
Crazy Cow—chocolate, 288
C.W. Post, 304
Fortified Oat Flakes, 264
Frankenberry, 287
Frosted Flakes, 283
Frosted Mini Wheats, 289
Frosted Rice Krinkles, 284
Fruit Loops, 298
Fruity Pebbles, 294
Giant Natural—plain, 303
Golden Grahams, 254
Grape Nuts, 266
Heartland Natural—coconut, 308
Heartland natural—raisins, 309
High Pro Concentrate, 40%, 263
Honeycomb, 296
Kellogg's Raisin Bran, 275
Lucky Charms, 300
Most, 260
Nature Valley Granola—fruit and nut, 310
Post Raisin Bran, 274
Pot Toasties, 271
Puffed Rice, 257
Puffed Wheat, 258
Quaker Life, 253
Quaker Natural, 306
Quaker Natural—raisins and dates, 307
Rice Chex, 269
Rice Krispies, 256
Shredded Wheat, 265
Special K, 261
Sugar Corn Pops, 282
Sugar Smacks, 281
Super Sugar Crisp, 280
Team, 259
Total, 262
Trix, 297
Vita Crunch—regular, 305
Wheat Chex, 268
Wheaties, 272

CHEMICALS

Alumina, absorption, 125
Aluminum sulfate, 163
Ammonium acetate, 126
Ammonium chloride, 136
Ammonium persulfate, 160
Ammonium sulfate, 164
Ammonium thiocyanate, 177
Barium chloride, 137
Calcium carbonate, 133
Calcium chloride #401, 138
Calcium chloride #401, after drying, 139
Calcium chloride #481, 140
Calcium chloride #481, after drying, 141
Calcium chloride #481, after still further drying, 142
Calcium sulfate, 167
Ceramic, ether-washed, 181
Ceramic, washed, 179
Ceramic, water-washed, 180
Chromium potassium sulfate, 168
Copper chloride, 143
Cupric carbonate, 134
Cupric sulfate, $CuSO_4$, $5H_2O$, 169
Cuprous chloride, 144
Epolene, 182
Ferric chloride, 145
Ferrous ammonium sulfate, 166
Ferrous sulfate, 165
Glass beads, 184
Halon, 185
Manganous chloride, 146
Manganous sulfate, 170
Nickelous sulfate, 171
Polystyrene, 183
Potassium borate, 131
Potassium bromide, 132
Potassium chloride, 147
Potassium ferrocyanide, 152
Potassium iodate, 153
Potassium iodide #421, 154
Potassium iodide #491, 155
Potassium nitrate, 157
Potassium permanganate, 159
Potassium persulfate, 161
Potassium sodium tartrate, 176
Sodium acetate, 127
Sodium bicarbonate, 129
Sodium bisulfite, 130
Sodium carbonate, 135
Sodium chloride, 148
Sodium chloride #451, 149
Sodium citrate, 151
Sodium lauryl sulfate, 173
Sodium molybdate, 156
Sodium nitrite, 158
Sodium phosphate, 162
Sodium pyrophosphate, 175
Sodium sulfate, 172
Sulfur, 186
Tris(hydroxymethyl)aminomethane, 178
Zinc acetate, 128
Zinc chloride, 150
Zinc sulfate, 174

MISCELLANEOUS COMMODITIES

Alfalfa, 187
Alfalfa fiber, 192
American cheese, 314
Ascorbic acid, 246
Barley, 188
Barley malt, 233
Barley straw, 215
Beef, fresh, 325
Beef, roast, 326
Bromegrass, 198
Butter, 322
Cake, fudge marble, 320
Cellulose, Adsorbosil, 222
Cellulose, Sigmacell, 220
Cellulose, Sigmacell, dry, 221
Chickpea (garbanzo bean), 234
Clover screenings, 202
Coconut, flaked, 319
Coconut flour, 196
Coffee, freeze-dried, 318
Coffee, ground, 317
Corn, 197
Corn fiber, 193
Corn meal, yellow, 311
Corn starch, 244
Cotton, 323
Cottonseed, 213
Creamer, nondairy, 315
Distiller's grains, dried, 189
Durum wheat, 242

Ethanol, 223
Faba bean, 236
Feather meal, 227
Field pea, 247
Fish meal, 249
Flax screenings, 203
Flaxseed, 204
Flaxseed fiber, 195
β-Glucan, commercial, 218
β-Glucan, pure, 219
Hog grower feed, 190
Horse feed, 191
Lentil, 237
Lignin, Horvath's, 245
Margarine, 312
Mayonnaise, 321
Meat meal, 250
Millet, 200
Molasses, dry, 224
Mustard, yellow, 207
Nail keratin, 225
Oat groats, 201
Oats, 240
Palm kernel, 209
Parmesan cheese, grated, 313
Pea straw, 216
Peanuts, diluted with $CaCo_3$, 212
Peanuts, ground, 211
Pigeon pea, 235
Pork, fresh, 327
Pork, roast, 328
Rapeseed, 205
Rapeseed meal, 251
Russian rye grass, 199
Rye, 239
Safflower, 210
Sesame, 208
Silicone grease, 226
Skim milk, powdered, 316
Sorghum, 214
Soybean, 248
Soybean fiber, 194
Soybean meal, 252
Sunflower, 206
Urea, 228
Vitamin A & B supplement, 217
Water, 229
Water hyacinth extract, 230
Water hyacinth powder, 231
Wheat, common (hard red spring), 241
Wheat flour, 243
Whey powder, 232
Wild rice, 238
Wool, 324

OILS
Corn oil, 113
1:3-Diolein, 116
1:2 Dipalmitin, 121
1:3 Dipalmitin, 122
Glycerol, 124
Monoolein, 115
Monopalmitin, 120
Monostearin, 118
Soybean oil, 114
Triolein, 117
Tripalmitin, 123
Tristearin, 119

ORGANIC ACIDS
Acetic acid, 44
Arachidic acid, 45
Arachidonic acid, 46
Behenic acid, 47
Butyric acid, 48
Capric acid, 49
Caproic acid, 50
Capryllic acid, 51
Elaidic acid, 52
Erucic acid, 53
D-Glucose penta acid, 60
Lactic acid, 54
Lauric acid, 55
Linoleic acid, 57
Myristic acid, 56
Oleic acid, 58
Palmitic acid, 59
Phytic acid, 61
Phytic acid, dry, 62
Propionic acid, 63
Stearic acid, 64

PEPTIDES
N-Acetyl L-alanine, 40
N-Acetyl L-glycine, 41
N-Acetyl D-methionine, 42
N-Acetyl L-tryptophan, 43
3-Alanine (Tri), 27
L-Alanyl glycine, 30
L-Alanyl L-phenylalanine, 36
L-Alanyl L-tryptophan, 39
Alanylalanine, 26
Diglycine, 31
Glycyl-L-isoleucine, 35
Hexaglycine, 34
Penta alanine, 29
Tetra alanine, 28
Tetraglycine, 33
Triglycine, 32
L-Tryptophyl L-phenylalanine, 37
L-Valyl L-proline, 38

PROTEINS
Bovine globulin, 92
Bovine serum albumin, 88
Corn gluten, 95
Human albumin, 89
Human globulin, 93
Lactalbumin, 90
Lactoglobulin, 91
Soy protein, 96
Wheat gluten, 94
Zein, 97

SUGARS
D-Arabinose, 65
L-Arabinose, 66
D-Cellobiose, 67
Deoxy-D-ribose, 82
L-Fructose, 68
L-Fucose, 69
D-Galactose, 70
β-D-Glucose, first run, 71
D-Glucose, 72
Lactose, 73
Lifesaver candies, ground, 87
D-Lyxose, 74
Maltose, 75
D-Mannose, 76
D-Melibose, 77
Palatinose, 81
D-Raffinose, 78
D-Ribose, 79
L-Sorbose, 83
L-Stachyose, 84
Sucrose, 80
Table sugar, ground, 86
D-Xylose, 85

WHEAT COMPONENTS
Wheat amylopectin, 98
Wheat amylopectin, dry, 99
Wheat amylose, 100
Wheat amylose, dry, 101
Wheat bran, 103
Wheat bran, extracted, 104
Wheat, isopropanol extract, 107
Wheat, KOH Extract, 105
Wheat, lactic acid extract, 108
Wheat, Na extract, 102
Wheat, NaCl fraction, 106
Wheat "sludge", 109
Wheat starch, dried, 110
Wheat starch, not dried, 111
Wheat straw, 112

Near-Infrared Method for Protein Determination AACC Approved Method 39-10

SCOPE

Wheat of all classes.

APPARATUS

1. Infrared reflectance spectrophotometer such as those manufactured by DICKEY-john Corporation, Pacific Scientific Company, Percon Prufgerate, Technicon Industrial Systems, or equivalent. Operator should become familiar with instruction manual before calibrating instrument or running any samples.
2. Mill capable of grinding samples to less than 1 mm.
   a. Udy Corporation grinder with 1.0-mm screen, Falling Number Model KT3300 with 0.8-mm screen, or equivalent is satisfactory.
   b. Different types of grinders give different values on instruments because of variations in particle size and/or configuration. Therefore, always use same type of grinder for routine samples as was used for calibration of instrument.

PROCEDURE

1. Calibration.
   a. Follow manufacturer's daily check procedure before starting tests.
   b. Select at least 40 samples of commodity to be tested, covering full range of protein and moisture that can be expected to be encountered in future testing. It is imperative that there be uniform distribution of samples throughout entire range of protein and moisture.
   c. Grind each sample in turn, mix thoroughly, place in test cell, and record raw optical data as per manufacturer's instructions. Record readings at 0 position and then after 120° rotations of test cell. Thorough mixing is essential for precise readings.
   d. Make replicate standard laboratory tests on all samples (at least in duplicate).
   e. Calculate calibration constants and enter into instrument by technique recommended in instruction manual.
   f. Select 20 additional samples covering full range of protein and test to determine bias.
   g. If necessary, correct for slope and bias according to instruction manual.
2. Routine tests.
   a. Before tests each day, follow manufacturer's recommended daily check procedures.
   b. Grind each sample in the same way as calibration samples were ground, mix thoroughly, place in test cell, and take readings at 0 position and after 120° rotation.

NOTE

It is essential that near-infrared reflectance instruments be calibrated on as-is basis with respect to moisture status of samples used in calibration. Protein results can be reported on constant moisture basis using moisture figure given by instrument. Major sources of error in near-infrared reflectance measurement are covered in the following references.

REFERENCES

1. HUNT, W. H., FULK, D. W., ELDER, B., and NORRIS, K. H. 1977. Collaborative study on infrared reflectance devices for determination of protein in hard red winter wheat, and for protein and oil in soybeans. Cereal Foods World 22:534-536.
2. HYMOWITZ, T., DUDLEY, J. W., COLLINS, F. I., and BROWN, C. M. 1974. Estimations of protein and oil concentration in corn, soybean, and oat seed by near infrared light reflectance. Crop Sci. 14:713-715.
3. WATSON, C. A. 1977. Near infrared reflectance spectrophotometric analysis of agricultural products. Anal. Chem. 49:835a-840a.
4. WATSON, C. A., CARVILLE, D., DIKEMEN, E., DIAGGER, G., and BOOTH, G. D. 1976. Evaluation of two infrared instruments for determining protein content of hard red winter wheat. Cereal Chem. 53:214-222.
5. WILLIAMS, P. C. 1975. Application of near infrared reflectance spectroscopy to analysis of cereal grains and oilseeds. Cereal Chem. 52:561-576.
6. WILLIAMS, P. C., and THOMPSON, B. N. 1978. Influence of whole meal granularity on analysis of HRS wheat for protein and moisture by near infrared reflectance spectroscopy (NRS). Cereal Chem. 55:1014-1037.

Bibliography

Abstracts are included in this bibliography where they indicate to the reader the source of the work and the workers.

1. AACC. 1983. Approved Methods of the AACC. Method 39-10, first approval October 1979; final approval October 1982. Am. Assoc. Cereal Chem., St. Paul, MN.
2. ABU-SHUMAYS, A. 1982. Near infrared spectroscopy. VIA (Varian Newsl.) 16:14-15.
3. AHMED, N. S., and HELAL, F. R. 1977. The infrared absorption spectra for measuring the unsaturation in milk fat and oils. Milchwissenschaft 32:272-273.
4. AHMED, W., and GOLDSTEIN, M. 1976. Analytical applications of reflectance spectroscopy in the near infrared region: A method for quality control of fat in dried milk. Lab. Pract. 25:385-386.
5. AITKIN, M. A. 1974. Simultaneous inference and the choice of variable subsets in multiple regression. Technometrics 16:221-227.
6. ALEXANDER, E. L., SILVELA, S., COLLINS, F. I., and RODGERS, R. C. 1967. Analysis of oil content of maize by wide-line NMR. J. Am. Oil Chem. Soc. 44:555-558.
7. ALLISON, M. J. 1983. A rapid screening method for determining the digestibility of kale using near infrared reflectance. J. Sci. Food Agric. 34:175-180.
8. ALLISON, M. J., BORZUCKI, R., COWE, I. A., and McHALE, R. 1979. Variation in a barley collection for endosperm attributes that relate to malting quality. J. Inst. Brew. London 85:86-88.
9. ALLISON, M. J., COWE, I. A., and McHALE, R. 1978. The use of infrared reflectance for the rapid estimation of the soluble β-glucan content of barley. J. Inst. Brew. London 84:153-155.
10. ALLVIN, B. 1980. New analytical technique in the food area—near infrared reflectance (NIR). (In Swedish) Kem. Tidskr. 92:24-27.
11. ALTMEYER, F. 1984. Rapid and practical alcohol measurement in wine with the Technicon InfraAlyzer 400L. (In German) Monatsschr. Brau. 37:88-91.
12. ANDREWS, M. W., CONNELL, J. P., DAVID, H. G., MARLER, J. W., and O'DONNELL, R. E. 1978. Predicting wool yield by diffuse reflectance spectroscopy. Part 2: Some results on greasy wool. Proc. I.W.T.O. Tech. Committee.
13. ANONYMOUS. Undated. Application note on the determination of iodine value, peroxide value and free fatty acids in edible oils. Technicon Instruments Co., Basingstoke, U.K.
14. ANONYMOUS. 1978a. Application note on measurement of fat and total solids in milk samples using a GQA-31EL. Pacific Scientific Tech. Pap. NIR 4000.
15. ANONYMOUS. 1978b. New test for lysine. Agric. Res. 26:15.
16. ANONYMOUS. 1981. The Atlas of Near Infrared Spectra. Sadtler Research Laboratories, Philadelphia.
17. ANONYMOUS. 1982a. Investigations into the use of near infrared (NIR) reflectance spectroscopy for the quality assessment of wheat. Min. Agric. Fisheries Food Commission M15 to FMBRA, Chorleywood, England. Annu. Rep. for year ending 31 March.
18. ANONYMOUS. 1982b. Rapid analysis for food fiber. Agric. Res. 30:13.
19. ANONYMOUS. 1984. Near infrared reflectance analysis—the first ten years. Milling, December, pp. 24-26.
20. ARNOLD, R. G., and HARTUNG, T. E. 1971. Infrared spectroscopic determination of degree of unsaturation of fats and oils. J. Food Sci. 36:166-168.
21. ASHER, M. J. C., COWE, I. A., THOMAS, C. E., and CUTHBERTSON, D. C. 1982. A rapid method of counting spores of fungal pathogens by infrared reflectance analysis. Plant Pathol. 31:363-372.
22. AVRAMENKO, V. N., and SABOLEV, V. A. 1978. Spectral analysis of powdered foods using reflectance spectroscopy. (In Russian) Pishch. Prom. 24:72-75.
23. AXCELL, B., TULEJ, R., and MURRAY, J. 1981a. An ultra-fast system for hop analysis. I. The determination of alpha acids and moisture by near infrared reflectance spectroscopy. Brew. Dig. 56:18-19, 41.
24. AXCELL, B., TULEJ, R., and MURRAY, J. 1981b. An ultra-fast system for hop analysis. II. The determination of beta-acids and prediction of hop storage index by near infrared reflectance spectroscopy. Brew. Dig. 56:32-33.
25. BAER, R. J., FRANK, J. F., and LOEWENSTEIN, M. 1982. Application of NIRA to dry dairy product analysis. Proc. NIRA Symp., 2nd. Tarrytown, NY, July.
26. BAHL, J. 1982a. Analysis of hydrogenated soybean oil by NIRA. Proc. NIRA Symp., 2nd. Tarrytown, NY, July.
27. BAHL, J. 1982b. Use of near infrared reflectance analysis for process control in the manufacture of ready-to-eat breakfast cereals. (Abstr.) Cereal Foods World 27:462.
28. BAKER, C. D. 1985. Automatic measurement of original gravity, color and pH using a near infrared analyzer. Pages 531-538 in: Eur. Brew. Conv., Proc. Congr. 20th, Helsinki.
29. BAKER, D. A. 1980. Report of the technical sub-committee on barley analysis. J. Am. Soc. Brew. Chem. 38:94-96.
30. BAKER, D. 1983. The determination of fiber in processed cereal foods by near-infrared reflectance spectroscopy. Cereal Chem. 60:217-219.
31. BAKER, D. 1985. The determination of fiber, starch, and total carbohydrate in snack foods by near-infrared reflectance spectroscopy. Cereal Foods World 30:389-392.
32. BAKER, D., and NORRIS, K. H. 1985. Near-infrared reflectance measurement of total sugar content of breakfast cereals. Appl. Spectrosc. 39:618-621.
33. BAKER, D., NORRIS, K. H., and LI, B. W. 1979. Food fiber analysis: Advances in methodology. Pages 67-78 in: Dietary Fibers: Chemistry and Nutrition. G. E. Inglett and S. I. Falkenhag, eds. Academic Press, New York.
34. BARNES, R. J. 1984. Forage quality analysis by N.I.R. reflectance. Proc. Annu. Users Conf. NIR Res., 3rd. Pacific Scientific, Rockville, MD.
35. BARR, I. S., and HARP, W. R. 1943. The near infrared absorption spectra of some vegetable oils. Phys. Rev. 63:457.
36. BARTON, F. E., II. 1984. Advantage of the chemometric method. Proc. Annu. Users Conf. NIR Res., 3rd. Pacific Scientific, Rockville, MD.
37. BARTON, F. E., II. 1985. Near-infrared reflectance spectroscopy (NIRS): Analysis of forage quality. Page 96 in: U.S. Dep. Agric. Handb. 643. G. C. J. Marten, J. S. Shenk, and F. E. Barton II, eds.
38. BARTON, F. E., II, and BURDICK, D. 1979. Preliminary study on the analysis of forages with a filter-type near infrared reflectance spectrometer. J. Agric. Food Chem. 27:1248-1252.
39. BARTON, F. E., II, and BURDICK, D. 1980. Prediction of crude protein in dehydrated coastal bermudagrass by NIR reflectance. Pages 103-106 in: Proc. Res. Ind. Conf. Coastal Bermudagrass Processors Assoc., 10th. Athens, GA.
40. BARTON, F. E., II, and BURDICK, D. 1982. The prediction of forage quality with NIR reflectance spectroscopy. Pacific Scientific, Gardner/Neotec Instrument Div., Tech. Pap. NIR 2012.
41. BARTON, F. E., II, and BURDICK, D. 1983. Prediction of forage quality with NIR reflectance spectroscopy. Pages 532-534 in: Proc. Int. Grassl. Congr., 14th. Lexington, KY.
42. BAUMAN, R. P. 1962. Absorption Spectroscopy. John Wiley, New York.
43. BEGLEY, T. 1984a. Derivative spectroscopy. Proc. Annu. Users

Conf. NIR Res., 3rd. Pacific Scientific, Rockville, MD.

44. BEGLEY, T. 1984b. Determination of NaCl in meat by near infrared diffuse reflectance spectroscopy. Proc. Annu. Users Conf. NIR Res., 3rd. Pacific Scientific, Rockville, MD.
45. BEGLEY, T. H., LANZA, E., NORRIS, K. H., and HRUSCHKA, W. R. 1984. Determination of sodium chloride in meat by near-infrared diffuse reflectance spectroscopy. J. Agric. Food Chem. 32:984-987.
46. BELL, W. L. 1980. A universal calibration for U.S. white pan bread flours. (Abstr.) Cereal Foods World 25:531.
47. BELL, W. L. 1982a. Near-infrared reflectance comes of age. Grain Age 23(10):98-99.
48. BELL, W. L. 1982b. The use of a liquid near infrared reflectance analyzer to analyze liquid baking ingredients. (Abstr.) Cereal Foods World 27:462.
49. BELL, W. L. 1982c. The analysis of fructose concentration in high fructose corn syrup using near infrared reflectance analysis (NIRA). (Abstr.) Cereal Foods World 27:468.
50. BELL, W. L. 1983. The use of NIR as a process control tool for wet corn milling. Cereal Foods World 28:249-251.
51. BELL, W. L., and MELAND, P. 1983. Improved accuracy of near infrared reflectance measurement of "ash" in flour. (Abstr.) Cereal Foods World 28:575.
52. BELLAMY, L. J., and WILLIAMS, R. L. 1957. The NH stretching frequencies of primary amines. Spectrochim. Acta 9:341-345.
53. BELTON, P. S., and TANNER, S. F. 1983. Determination of the moisture content of starch using near infrared photoacoustic spectroscopy. Analyst (London) 108:591-596.
54. BEN-GERA, I., and NORRIS, K. H. 1968a. Influence of fat concentration on the absorption spectrum of milk in the near infrared region. Isr. J. Agric. Res. 18:117-124.
55. BEN-GERA, I., and NORRIS, K. H. 1968b. Determination of moisture content in soybeans by direct spectrophotometry. Isr. J. Agric. Res. 18:125-132.
56. BEN-GERA, I., and NORRIS, K. H. 1968c. Direct spectrophotometric determination of fat and moisture in meat products. J. Food Sci. 33:64-67.
57. BENGTSSON, S., and LARSSON, K. 1984. Prediction of the nutritive value of forages by near infrared reflectance photometry. J. Sci. Food Agric. 35:951-958.
58. BENSE, T., and GASTELU, C. 1983. Estimation of total volatile bases in tobacco by near infrared (NIR) reflectance spectrophotometry. Tobacco Int. 185(15):183-185.
59. BERNARDINI, M., FEDELI, E., and BARONI, D. 1985. Near-infrared reflectance analysis. Riv. Ital. Sostanze Grasse 62:357-364.
60. BERTRAND, D., and ROBERT, P. 1983. Les techniques d'analyse rapide par spectrométrie infrarouge. Bios (Nancy) 14:31-35.
61. BERTRAND, D., ROBERT, P., and LOISEL, W. 1985. Identification of some wheat varieties by near infrared reflectance spectroscopy. J. Sci. Food Agric. 36:1120-1124.
62. BERTRAND, D., ROBERT, P., and TRAN, V. 1984. Mathematical treatments of NIR spectra of mixtures. (In French) Pages 93-97 in: Proc. World Cereal Bread Congr., 11th. Vienna.
63. BEUCH, H. G. 1983. Classification of wheat at storage with the aid of infrared spectroscopy. (In German) Getreide Mehl Brot 37:21-23.
64. BIGGS, D. A. 1972. Precision and accuracy of infrared milk analysis. J. Assoc. Off. Anal. Chem. 55:488-497.
65. BIGGS, D. A. 1978. Instrumental infrared estimation of fat, protein, and lactose in milk: Collaborative study. J. Assoc. Off. Anal. Chem. 61:1015-1034.
66. BIGGS, D. A. 1979a. Applications of infrared instrumentation in cheese analysis. Page 409 in: Proc. Bienn. Marschall Inst. Cheese Conf., 1st. Madison, WI, Miles Labs.
67. BIGGS, D. A. 1979b. Infrared estimation of fat, protein, and lactose in milk: Evaluation of multispec instrument. J. Assoc. Off. Anal. Chem. 62:1202-1207.
68. BIGGS, D. A. 1979c. Performance specifications for infrared milk analysis. J. Assoc. Off. Anal. Chem. 62:1211-1214.
69. BIGGS, D. A., and STANLEY, D. W. 1972. Measurement of total solids in foods by transmission infrared spectroscopy. Can. Inst. Food Sci. Technol. J. 5:203-206.
70. BIRTH, G. S. 1960a. Agricultural applications of the dual-monochromator spectrophotometer. Agric. Eng. 41:432-435, 452.
71. BIRTH, G. S. 1960b. Measuring the smut content of wheat. Trans. ASAE 3:19-21.
72. BIRTH, G. S. 1960c. A nondestructive technique for detecting internal discolourations in potatoes. Am. Potato J. 37:53-60.
73. BIRTH, G. S. 1967. A fiber optic reflectance attachment. Agric. Eng. 48:448-449.
74. BIRTH, G. S. 1975. Interpretation of the optical scattering coefficient. Am. Soc. Agric. Eng. Natl. Meet., Davis, CA, June.
75. BIRTH, G. S. 1976. How light interacts with foods. Pages 6-11 in: Quality Detection in Foods. J. J. Gaffney, ed. ASAE Publ. 1-76.
76. BIRTH, G. S. 1979a. Radiometric measurement of food quality—a review. J. Food Sci. 44:949-953, 957.
77. BIRTH, G. S. 1979b. The physics of near infrared reflectance of foods. (Abstr.) Cereal Foods World 24:455.
78. BIRTH, G. S. 1980. Diffuse reflectance as a measure of light scattering. Fed. Anal. Chem. Spectrosc. Soc. Meet., Philadelphia, September.
79. BIRTH, G. S., NORRIS, K. H., and YEATMAN, J. N. 1957. Non-destructive measurement of internal color of tomatoes by spectral transmission. Food Technol. 11:552-557.
80. BIRTH, G. S., and RAMEY, H. H., Jr. 1982. Near-infrared reflectance for analysis of cottonseed for gossypol. Cereal Chem. 59:516-519.
81. BIRTH, G. S., TURLEY, W. C., and KAYS, S. 1982. Non-destructive determination of soluble solids in onions. Pacific Scientific, Gardner/Neotec Instruments Div. Tech. Pap. 4036.
82. BIRTH, G. S., and ZACHARIAH, G. L. 1976. Spectrophotometry of agricultural products. Pages 1-5 in: Quality Detection in Foods. J. J. Gaffney, ed. ASAE Publ. 1-76.
83. BISTON, R. 1979. Potential uses of N.I.R.S. for the analysis of agricultural productions. Pacific Scientific Res. Notes NIR 2016.
84. BISTON, R., and CLAMOT, G. 1982. Use of near-infrared reflectance spectroscopy and dye-binding techniques for estimating protein in oat groats. Cereal Chem. 59:333-335.
85. BISTON, R., and DARDENE, P. 1984. Applications of near infrared spectroscopy in the field of agronomy. Proc. Annu. Users Conf. NIR Res., 3rd. Pacific Scientific, Rockville, MD.
86. BISTON, R., DARDENNE, G., SINDIC, M., DARDENNE, P., WATHELET, B., and DE SOUSA, B. 1983. Near infrared spectroscopy as a rapid and accurate technique for estimating *Lupinus* seed protein and *Vicia faba* seeds amino acids and protein. Pages 247-255 in: Perspectives for Peas and *Lupinus* as Protein Crops. R. Thompson and R. Casey, eds. Martinus Nijhoff, The Hague. (Pacific Scientific Tech. Pap. NIR 1009.)
87. BITTNER, D. R., and NORRIS, K. H. 1968. Optical properties of selected fruits vs. maturity. Trans. ASAE 11:534-536.
88. BJARNO, O. C. 1981. Multicomponent analysis of meat products. J. Assoc. Off. Anal. Chem. 64:1392-1396.
89. BJARNO, O. C. 1982. Multicomponent analysis of meat products by infrared spectrophotometry: Collaborative study. J. Assoc. Off. Anal. Chem. 65:696-700.
90. BLACK, L. T., ELDRIDGE, A. C., HOCKRIDGE, M. E., and KWOLEK, W. F. 1985. Determination of texturized soybean flour in ground beef by near infrared reflectance spectroscopy. J. Agric. Food Chem. 33:823-826.
91. BLACK, L. T., HAMERSTRAND, G. E., and KWOLEK, W. F. 1985. Analysis of rubber, resin and moisture content of guayule by near infrared reflectance spectroscopy. Rubber Chem. Technol. 58:304-313.
92. BLACK, R. G. 1985. Analysis of raw milk by NIR. Pages 105-113 in: NIR 84 Proc. Int. Symp. Near Infrared Reflectance Spectrosc., 1st. RACI, Cereal Chem. Div., Melbourne, Australia.
93. BLAKE, C. J. 1981. Infrared reflectance techniques in food analysis—a literature survey. Leatherhead Food RA Sci. Tech. Surv. 126. 22 pp.
94. BLANCO, C., CASADO, P., and POZA, A. 1980. Application of diffuse reflectance spectroscopy. Its application to the determination of powder milk composition. (In Spanish) Quim. Ind. (Madrid) 26:361-363, 366-367.
95. BLANK, R. E., and WAKEFIELD, T., II. 1979. Double-beam photoacoustic spectrometer for use in the ultraviolet, visible and near infrared spectral regions. Anal. Chem. 51:50-54.
96. BLEVIN, W. R., and BROWN, W. J. 1965. An infrared reflectometer with a spheroidal mirror. J. Sci. Instrum. 42:385-389.
97. BOLLING, H. 1979. The use of NIR spectroscopy in grain analysis. (In German) Getreide Mehl Brot 33:264-268.
98. BOLLING, H. 1984. NIR studies—application to production control in milling. (In German) Pages 98-102 in: Proc. World

Cereal Bread Congr., 11th. Vienna.

99. BOLLING, H., and ZWINGELBERG, H. 1979. The use of infrared spectroscopy in the milling industry. (In German) Muehle Mischfuttertechnik 116:53-55.

100. BOLLING, H., and ZWINGELBERG, H. 1982. Infrared spectroscopy—applications in grain intake and end product control in milled products. (In German) Getreide Mehl Brot 36:197-201.

101. BOLLING, H., and ZWINGELBERG, H. 1984. Continuous analysis of flour by NIR. Getreide Mehl Brot 38:3-5.

102. BORMAN, S. 1982. Math is cheaper than physics. Anal. Chem. 54:1379A.

103. BRACH, E. J., PHAN, C. T., POUSHINSKY, G., JASMIN, J. J., and AUBÉ, C. B. 1982. Lettuce maturity detection in the visible (380–720 nm) far red (680–740 nm) and near infrared (800–1850 nm) wavelength band. Agronomie 2:685-694.

104. BRINK, G. E., and MARTEN, G. C. 1986. Analysis of alfalfa root carbohydrate concentration by near infrared reflectance spectroscopy. Crop Sci. 26:159-161.

105. BROWN, G. K., SERGERLINE, L. J., and SUMMITT, R. 1976. Near infrared reflectance of bruised apples. Pages 68-70 in: Quality Detection in Foods. J. J. Gaffney, ed. ASAE Publ. 1-76.

106. BRUINSMA, B. L., and RUBENTHALER, G. L. 1978. Lysine estimation in cereals by near-infrared reflectance. Crop Sci. 18:1039.

107. BUCHANAN, B. R., and HONIGS, D. E. 1986a. Trends in near infrared analysis. Trends Anal. Chem. 5:154-157.

108. BUCHANAN, B. R., and HONIGS, D. E. 1986b. Advances in near infrared spectroscopy. J. Spectrosc. 1:40-42.

109. BURDICK, D., BARTON, F. E., II, and NELSON, B. D. 1981. Prediction of bermudagrass composition and digestibility with a near-infrared multiple filter spectrophotometer. Agron. J. 73:399-403.

110. BURKHARDT, T. H., and MROZEK, R. F. 1976. Light reflectance as a criterion for sorting dried prunes. Pages 80-82 in: Quality Detection in Foods. J. J. Gaffney, ed. ASAE Publ. 1-76.

111. BUTLER, L. A. 1983. The history and background of NIR. Cereal Foods World 28:238-240.

112. CAHN, L., and HENDERSON, B. D. 1954. Design of a laboratory spectrophotometer for the near infrared. Abstr. Pittsburgh Conf. Anal. Chem. Appl. Spectrosc. Exposition Modern Lab. Equip. Anal. Chem. 26:428.

113. CALANDRA, F. 1977a. Effect of soybean meal particle size on the spectroscopic response in infrared refractometry. (In Italian) Riv. Ital. Sostanze Grasse 54:161-170.

114. CALANDRA, F. 1977b. The use of infrared reflection for immediate assessment of residual oil contained in sunflower meal. (In Italian) Riv. Ital. Sostanze Grasse 54:171-176.

115. CAMPBELL, R. C., and MARTIN, G. C. 1976. Determination of moisture in walnut seeds by near-infrared spectrophotometry. HortScience 11:494-496.

116. CARBALLIDO, A., OLMEDO, R. G., and DEL ARCO, M. P. S. 1975. Applications of infrared spectrometry in the analysis of foods. (In Spanish) An. Bromatol. 26:293-336.

117. CARDINALE, E. 1980. Near infrared reflectance instrumentation. (In Italian) Tec. Molitoria 31:910-916.

118. CARNIELO, M. 1984. The use of NIRA in malting and brewing industries. (In German) Monatsschr. Brau. 37:88-91.

119. CARNIELO, M., GRANDCLERC, J., MULLER, P., MOLL, M., GRANDVOINET, P., BERGER, M., SIMIAND, J. P., LEMAITRE, M., MARY, D., L'HOMMEL, F., LEBOEUF, J. P., ROUILLER, M., and MABILLE, M. 1985. Evaluation of the InfraAlyzer 400 for the analysis of barley and malt. J. Inst. Brew. London 91:174-179.

120. CARPENTER, K. J., and ELLINGER, G. M. 1955. Protein quality and 'available lysine' in animal products. Poult. Sci. 34:1451-1454.

121. CASCIERCO, J. 1983. Evaluation of new NIR calibration in bench model instruments: Prediction of "known/unknowns." (Abstr.) Cereal Foods World 28:571.

122. CASTLEDEN, S. L., KIRKBRIGHT, G. F., and MENON, K. R. 1980. Determination of moisture in single-cell protein utilising photo-acoustic spectroscopy in the near infrared region. Analyst (London) 105:1076-1081.

123. CAUVAIN, S. P., DOUGLAS, S., FEARN, T., and OSBORNE, B. G. 1982. Relationship between bread baking quality of flour and analytical data including NIR. Flour Milling Baking Res. Assoc. Bull. 1:23-28.

124. CHAMBERLAIN, N. 1982. The addition of wheat gluten to bread flour. Flour Milling Baking Res. Assoc. Bull. 6:243-252.

125. CHAPMAN, D., and NACEY, J. F. 1958. A rapid spectroscopic method for the determination of water in glycerol. Analyst (London) 83:377-379.

126. CHASSEVENT, F., and DALGER, G. 1973. Determination of the water content of soluble powdered coffee extracts by near infrared spectrophotometry, the Karl Fischer method, and drying oven methods. (In French) Colloq. Int. Chim. Cafes (C.R.) 5:162-174.

127. CHO, R. K., IWAMOTO, M., and SAIO, K. 1987. Determination of 7S and 11S globulins in soybean seeds by near-infrared spectroscopic analysis. In press.

128. CLANCY, P. J. 1985. Near infrared transmission analysis 700 to 1100 nm—practical applications. Pages 62-64 in: NIR 84 Proc. Int. Symp. Near Infrared Reflectance Spectrosc., 1st. RACI, Cereal Chem. Div., Melbourne, Australia.

129. CLANCY, P. J., and STREAMER, R. 1985. Forage analysis—using near infrared reflectance. Pages 134-137 in: NIR 84 Proc. Int. Symp. Near Infrared Reflectance Spectrosc., 1st. RACI, Cereal Chem. Div., Melbourne, Australia.

130. CLARK, D. H., ADAMS, R. P., LAMB, R. C., and ANDERSON, M. J. 1985. Near-infrared analysis of hydrocarbon producing plant species. Biomass 8:1-11.

131. CLARK, M. J., and HICKIE, T. S. 1975. The quantitative analysis of some binary mixtures of fibers by methods of infra-red spectroscopy. J. Text. Inst. 66:243-248.

132. CLAY, A. F. 1973. Aspects of the theory and practice of color measurement. Scan 2:7-14.

133. CLOVER, M. B. 1979. The use of the InfraAlyzer in a small/medium sized flour mill. Autoanalyst, November, pp. 3-4.

134. COBLENTZ, W. K. 1982. A study in the near infrared reflectance spectra of protein and related compounds. M.Sc. thesis, Penn. State Univ., University Park.

135. COBLENZ, W. W. 1905. Investigations of IR spectra. Part I: Absorption spectra. Part II: Emission spectra. Carnegie Inst. Washington Publ. 35, Washington, DC.

136. COLAS, A. 1979. The InfraAlyzer at the Grands Moulins de Paris. Proc. Technicon Int. Symp., 9th. 14–16 November, Paris.

137. CONNELL, J. P., and BROWN, G. H. 1978. The yield-testing of wool by reflectance spectroscopy. J. Text. Inst. 69:357-363.

138. CONNELL, J. P., and NORRIS, K. H. 1980. The prediction of the yield of greasy wool by near infrared reflectance spectroscopy. I. Principles and washed wool. Text. Res. J. 50:371-380.

139. CONNELL, J. P., and NORRIS, K. H. 1981. The prediction of the yield of wool by near-infrared reflectance spectroscopy. II. Greasy wool. Text. Res. J. 51:339-344.

140. COOKE, M. V. 1979. Evaluation of near infrared techniques for the analysis of oilseeds. Text of a lecture given at a meeting of the American Soybean Association, 12 September, Brussels.

141. COOPER, P. J. 1983a. NIR analysis for process control. Cereal Foods World 28:241-245.

142. COOPER, P. J. 1983b. Normalization of optical responses of NIR spectrophotometers for calibration transfers. (Abstr.) Cereal Foods World 28:571.

143. COOPER, P. J. 1984. Special considerations for the calibration of on-line NIR instrumentation. (Abstr.) Cereal Foods World 29:511.

144. COOPER, P. J., WEBSTER, D. R., and CASCIERO, J. L. 1985. Developing analytical NIR calibrations. (Abstr.) Cereal Foods World 30:567.

145. COVENTRY, A. G., and HUNSTON, M. J. 1984. Application of near-infrared spectroscopy to the analysis of beer samples. Cereal Foods World 29:715-718.

146. COWE, I. A. 1983a. Factors affecting the complexity of NIRR calibrations. Pap. 2 in: NIR Analysis—How Near Infrared Reflects Composition. Chemistry Division, School of Agriculture, Aberdeen, Scotland.

147. COWE, I. A. 1983b. NIR: Past, present and future. Anal. Proc. (London) 20:65-68.

148. COWE, I. A., and McNICOL, J. W. 1985a. Some techniques for the interpretation of near infrared reflectance data. Pages 13-26 in: NIR 84 Proc. Int. Symp. Near Infrared Reflectance Spectrosc., 1st. RACI, Cereal Chem. Div., Melbourne, Australia.

149. COWE, I. A., and McNICOL, J. W. 1985b. The use of principal components in the analysis of near-infrared spectra. Appl. Spectrosc. 39:257-266.

150. COWE, I. A., McNICOL, J. W., and CUTHBERTSON, D. C.

1985a. A designed experiment for the examination of techniques used in the analysis of near-infrared spectra. Part I. Analysis of spectral structure. Analyst (London) 110:1227-1232.

151. COWE, I. A., McNICOL, J. W., and CUTHBERTSON, D. C. 1985b. A designed experiment for the examination of techniques used in the analysis of near-infrared spectra. Part II. Derivatization and testing of regression models. Analyst (London) 110:1233-1240.

152. CRIADA, B. G., MORAN, L. L., and CIUDAD, A. G. 1978. Automatic analysis and evaluation of forage by infrared reflectance spectroscopy. Optimization of wavelengths. (In Spanish) Rev. Pastas 8:311-323.

153. CUBADDA, R., and DOUGLAS, S. 1982. Investigations on the near infrared reflectance techniques for measuring the properties of wheat and its products. Proc. World Cereal Bread Congr., 24 June–2 July, Prague.

154. DALAL, R. C., and HENRY, R. J. 1986. Simultaneous determination of moisture, organic carbon and total nitrogen in soil by near infrared reflectance spectrophotometry. Soil Sci. Soc. Am. J. 50:120-123.

155. DANTOIN, V. L. 1984. NIR application in public forage testing. Proc. Annu. Users Conf. NIR Res., 3rd. Pacific Scientific, Rockville, MD.

156. DAVIES, A. M. C. 1983. Human food analysis by NIR—still a promising technique. Pap. 3 in: NIR Analysis—How Near Infrared Reflects Composition. Chemistry Division, School of Agriculture, Aberdeen, Scotland.

157. DAVIES, A. M. C. 1984. Progress in human food analysis by near-infrared. Anal. Proc. (London) 21:488-491.

158. DAVIES, A. M. C. 1985. NIR spectroscopy for fast food analysis. Lab. Pract. 39:32-34.

159. DAVIES, A. M. C., and BROCKLEHURST, T. F. 1986. Near-infrared reflectance analysis of oil concentration in an emulsion: A study of oil and water migrations in a mayonnaise-based salad. J. Sci. Food Agric. 37:310-316.

160. DAVIES, A. M. C., COXON, D. T., GAVREL, G. M., and WRIGHT, D. J. 1985. Determination of starch and lipid in pea flour by near infrared reflectance analysis. The effect of pea genotype on starch and lipid content. J. Sci. Food Agric. 36:49-54.

161. DAVIES, A. M. C., GEE, M. G., and FOSTER, P. W. 1984. A colour graphics display system to aid the selection of "best-pair" wavelengths for regression analysis of near infrared data. Lab. Pract. 38:78-80.

162. DAVIES, A. M. C., GRANT, A., GAVREL, G. M., and STEEPER, R. V. 1985. Rapid analysis of packaging laminates by near-infrared spectroscopy. Analyst (London) 110:643-647.

163. DAVIES, A. M. C., and McCLURE, W. F. 1985. Near infrared analysis in the Fourier domain with special reference to process control. Anal. Proc. (London) 22:321-322.

164. DAVIES, A. M. C., and WRIGHT, D. J. 1984. Determination of protein in pea flour by near infrared analysis. J. Sci. Food Agric. 35:1034-1039.

165. DAY, M. S., and FEARN, F. R. B. 1982a. Near infrared reflectance as an analytical technique. Part I. History and development. Lab. Pract. 31:328-330.

166. DAY, M. S., and FEARN, F. R. B. 1982b. Near infrared reflectance as an analytical technique. Part II. Design and development of practical NIR instruments. Lab. Pract. 31:439-443.

167. DE GROEN, A. 1980. Rapid determination of protein, water and β-glucan in barley and malt. (In German) Monatsschr. Brau. 33:131-135.

168. DERKSEN, W. L., MONAHAN, T. J., and LAWES, A. J. 1957. Automatic recording reflectometer for measuring diffuse reflectance in the visible and infrared regions. J. Opt. Soc. Am. 47:995-999.

169. DEVAUX, M. F., BERTRAND, D., and MARTIN, G. 1986. Discrimination of bread-baking quality of wheats according to their variety by near-infrared reflectance spectroscopy. Cereal Chem. 63:151-154.

170. DE VILDER, J., and BASSUYT, R. 1983. Practical experiences with an InfraAlyzer 400 in determining the water, protein and fat content of milk powder. Milchwissenschaft 38:65-69.

171. DIACHUK, V. R., HAMILTON, E., SAVCHUK, N., and JACKEL, S. S. 1981. Bakery flour control using near infrared reflectance analysis. Baker's Dig. 55:72-75, 102.

172. DILLIER-ZULAUF, A., and EGLI, H. R. 1982. Quantitative rapid determination of fat, nitrogen and dried substances in whey products with the InfraAlyzer. (In German) Dtsch. Molk. Ztg. 103:820-822, 824-825.

173. DONHAUSER, E. G., LINSENMANN, O., and FALTERMEIER, E. 1983. Evaluation of near infrared spectrophotometry in the malting and brewery laboratory. (In German) Monatsschr. Brau. 36:474-477.

174. DONHAUSER, S., and LINSENMANN, O. 1983. Experiences with NIR (near-infrared reflectance) spectroscopy for the analysis of brewery raw materials and intermediate products. (In German) GIT Fachz. Lab. 27:1063-1067.

175. DORSHEIMER, W. T., and ISAAC, R. A. 1982. Application of NIR analysis. Am. Lab. 14:58-63.

176. DOUGLAS, S. A. S., and MARLER, J. W. 1983. Near-infrared-reflectance estimation of alcohol-extractable-matter content of scoured wool. J. Text. Inst. 74:292-293.

177. DOUGLAS, S. G., FEARN, T., MILLER, A. R., and OSBORNE, B. G. 1983. The determination of the principal constituents of short dough and biscuit by NIR. Flour Milling Baking Res. Assoc. Bull. 2:92-101.

178. DOUGLAS, S. G., FEARN, R., and OSBORNE, B. G. 1981. Determination of starch damage using near infra-red (NIR) reflectance. Flour Milling Baking Res. Assoc. Bull. 1:31-33.

179. DOUGLAS, S., FEARN, T., and OSBORNE, B. G. 1982. A new low cost NIR instrument for grain trading. Flour Milling Baking Res. Assoc. Bull. 1:33-34.

180. DOWNEY, G. 1981. Grain analysis by near infrared reflectance. An Foras Taluntais Res. Rep., pp. 35-36.

181. DOWNEY, G. 1985. Estimation of moisture in undried wheat and barley by near-infrared reflectance. J. Sci. Food Agric. 36:951-958.

182. DOWNEY, G., and BYRNE, P. 1986. Prediction of moisture and bulk density in milled peat by near infrared reflectance. J. Sci. Food Agric. 37:231-238.

183. DOWNEY, G., MORRISEY, A., and FLYNN, S. 1982a. Poor harvest forces Irish to solve testy problems. Milling Feed Fert. 165:18-19.

184. DOWNEY, G., MORRISEY, A., and FLYNN, S. 1982b. Protein testing of wheat by near infrared reflectance. Farm Food Res. 13:14-16.

185. DOWNEY, G., MORRISEY, A., and BYRNE, S. 1982. Protein determination of wheat in trade by near infrared reflectance spectroscopy—the Irish experience. Proc. NIRA Symp., 2nd. Tarrytown, NY, July.

186. DULL, G. G., BIRTH, G. S., and MAGEE, J. B. 1980. Nondestructive evaluation of internal quality. HortScience 15:60-63.

187. DZWINCZYK, M., and MARK, H. 1986. Method for daily check of wavelength accuracy in near-infrared reflectance spectrophotometers. Spectroscopy 1:51.

188. EGLI, H. R. 1984. Standardization, quality insurance and control of milk products with the Technicon InfraAlyzer 400D. Page 90 in: The 5th International NIRA Symposium in Paris. O. Linsenmann, ed. (In German) Monatsschr. Brau. 37:88-91.

189. EISENBRAND, J., and BAUMANN, K. 1969. Light absorption measurement in clear liquid foods in the near infrared. II. Determination of water in beer. (In German) Dtsch. Lebensm. Rundsch. 65:169-172.

190. ELLIOT, A., HANBY, W. E., and MALCOLM, B. R. 1954. The near infrared absorption spectra of natural and synthetic fibres. Br. J. Appl. Phys. 5:377-381.

191. EREIFEJ, K. I., MARKAKIS, P., CHARALAMBOUS, G. (ed.), and INGLETT, G. (ed.). 1983. Analysis of dried milk and cheese powders by near infrared reflectance spectroscopy. Pages 237-242 in: Instrumental Analysis of Foods: Recent Progress. Vol. 2. Proc. Symp. Int. Flavor Conf., 3rd.

192. EVERS, A. D. 1982. The production and measurement of starch damage. Flour Milling Baking Res. Assoc. Bull. 6:256-268.

193. FAO. 1984. Commodity Review and Outlook—1983–84. Food and Agricultural Organization, United Nations, Rome.

194. FARNELL, P. J. 1975. An infrared spectroscopic method for the rapid simultaneous determination of fat and moisture in meat and meat products. Analyst (London) 100:661-667.

195. FEARN, F. 1980. The feed package and further developments. Text of lecture given at Symp. Devel. Automated Anal. Tech. Equip, 28–30 January. Society of Feed Technologists, York, England.

196. FEARN, F. 1981. Sampling best grinders for British market. Milling Feed Fert. 164:18.

197. FEARN, F. R. B. 1982. Near infrared reflectance as an analytical technique. Part 3. New advances. Lab. Pract. 31:658-660.
198. FEARN, T. 1983. A review of calibration methods for NIRR. Pap. 7 in: NIR Analysis—How Near Infrared Reflects Composition. Chemistry Division, School of Agriculture, Aberdeen, Scotland.
199. FEARN, T. 1986. Some statistical comments on the errors in NIR calibrations. Anal. Proc. (London) 23:123-125.
200. FEARN, T., and OSBORNE, B. G. 1982a. The use and abuse of statistics in NIR. Lecture given at Joint Symp. Near Infrared Analysis—Today or Tomorrow? Royal Society of Chemistry Automatic Methods and Special Techniques Group. 23 March, University of East Anglia, Norwich, U.K.
201. FEARN, T., and OSBORNE, B. G. 1982b. The use and abuse of statistics in NIR. Flour Milling Baking Res. Assoc. Bull. 5:222-232.
202. FENTON, A. J., Jr., and CRISLER, R. O. 1959. Determination of *cis* unsaturation in oils by near infrared spectroscopy. J. Am. Oil Chem. Soc. 36:620-623.
203. FINNEY, E. E., Jr., and NORRIS, K. H. 1978. Determination of moisture in corn kernels by near infrared transmittance measurement. Trans. ASAE 21:581-584.
204. FLINN, P. C., and MURRAY, I. 1985. Evaluation of hay in Western Victoria using NIR. Pages 180-182 in: NIR 84 Proc. Int. Symp. Near Infrared Reflectance Spectrosc., 1st. RACI, Cereal Chem. Div., Melbourne, Australia.
205. FORD, M. A., PRICE, W. C., and WILKINSON, T. R. 1958. A high resolution grating spectrometer for the infrared region. J. Sci. Instrum. 35:55-58.
206. FOSSATI, M., and PERROTTA, F. 1977. Rapid protein, oil and moisture determination with the Technicon InfraAlyzer system. (In Italian) Riv. Ital. Sostanze Grasse 54:159-160.
207. FRANK, J. F., and BIRTH, G. S. 1982. Application of near infrared reflectance spectroscopy to cheese analysis. J. Dairy Sci. 65:1110-1116.
208. FRANKHUIZEN, R., and VAN DER VEEN, N. G. 1985. Determination of major and minor constituents in milk powders and cheese by near infra-red reflectance spectroscopy. Ned. Melk Zuiveltijdschr. 39:191-207.
209. FREI, R. W. 1976. Diffuse reflectance spectroscopy: Applications, standards and calibration (with special reference to chromatography). J. Res. Natl. Bur. Stand. Sect. A. 80A:551-565.
210. FREI, R. W., and RYAN, D. E. 1966. Simultaneous analysis of two-component systems by diffuse reflectance spectroscopy. Can. J. Chem. 44:1945-1950.
211. FULLER, M. P., and GRIFFITHS, P. R. 1978a. Infrared analysis by diffuse reflectance spectrometry. Am. Lab. 10:69-80.
212. FULLER, M. P., and GRIFFITHS, P. R. 1978b. Diffuse reflectance measurements by infrared Fourier transform spectrometry. Anal. Chem. 50:1906-1910.
213. FYSH, S. A., SWINKELS, D. A. J., and FREDERICKS, P. M. 1985. Near-infrared diffuse reflectance spectroscopy of coal. Appl. Spectrosc. 39:354-357.
214. GAFFNEY, J. J. 1976a. Reflectance properties of citrus fruits. Pages 60-64 in: Quality Detection in Foods. J. J. Gaffney, ed. ASAE Publ. 1-76.
215. GAFFNEY, J. J. 1976b. Light reflectance of radishes as a basis for automatic grading. Pages 75-79 in: Quality Detection in Foods. J. J. Gaffney, ed. ASAE Publ. 1-76.
216. GARCIA-CIUDAD, A., GARCIA-CRIADO, B., and GARCIA BELLIDO, I. 1983. Rapid analysis of herbaceous species by near-infrared reflectance spectroscopy. (In Spanish) An. Edafol. Agrobiol. 42:1643-1653.
217. GARCIA-CIUDAD, A., GARCIA-CRIADO, B., and PONTON-SAN EMETERIO, C. 1985. Determination of fluoride in plant samples by a potentiometric method and near-infrared reflectance spectroscopy. Commun. Soil Sci. Plant Anal. 16:1107-1122.
218. GARCIA-CRIADO, B., NORAY, L. L., and GARCIA-CIUDAD, A. 1977. Determinacion directa de proteina, NDF, AD, lignina DNDF y DMD en plantas herbaceas mediante reflectancia de infrarrojos. Natl. Meet. Span. Soc. Grass Study, 17th. Cordoba, Spain, May.
219. GARDAM, C. M. 1985. Applications of Neotec NIRA instruments to chocolate processing. Pages 119-125 in: NIR 84 Proc. Int. Symp. Near Infrared Reflectance Spectrosc., 1st. RACI, Cereal Chem. Div., Melbourne, Australia.
220. GERSTENKORN, P. 1985. NIR-spectroscopy as a practical method for measurement of crude fiber in barley. (In German) Getreide Mehl Brot 39:167-170.
221. GHOSH, S., and RODGERS, J. E. 1985. Determining heat set temperature by near-infrared reflectance spectroscopy. Text. Res. J. 55:556-560.
222. GIANGIACOMO, R., MAGEE, J. B., BIRTH, G. S., and DULL, G. G. 1981. Predicting concentrations of individual sugars in dry mixtures by near infrared reflectance spectroscopy. J. Food Sci. 46:531-534.
223. GIANGIACOMO, R., TORREGGIANI, D., FRANK, J. F., LOEWENSTEIN, M., and BIRTH, G. S. 1979. Analysis of blue cheese using infrared reflectance spectroscopy. J. Dairy Sci. 62:39-42.
224. GIER, J. T., DUNKLE, R. V., and BEVANS, J. T. 1954. Measurement of absolute spectral reflectivity from 1.0 to 15 microns. J. Opt. Soc. Am. 44:558-562.
225. GIESBRECHT, F. G., McCLURE, W. F., and HAMID, A. 1981. The use of trigonometric polynomials to approximate visible and near-infrared spectra of agricultural products. Appl. Spectrosc. 35:210-214.
226. GIESE, A. T., and FRENCH, C. S. 1955. The analysis of overlapping spectral absorption bands by derivative spectrophotometry. Appl. Spectrosc. 9:78-96.
227. GILL, A. A., STARR, C., and SMITH, D. B. 1979. Lysine and nitrogen measurement by infrared reflectance analysis as an aid to barley breeding. J. Agric. Sci. 93:727-733.
228. GODDU, R. F. 1957. Determination of unsaturation by near infrared spectrophotometry. Anal. Chem. 29:1790-1794.
229. GODDU, R. F. 1958. Determination of phenolic hydroxyl by near infrared spectrophotometry. Anal. Chem. 30:2009-2013.
230. GODDU, R. F. 1960. NIR spectrophotometry. Adv. Anal. Chem. Instrum. 1:347-417.
231. GODDU, R. F., and DELKER, D. A. 1958. Determination of terminal epoxides by near infrared spectrophotometry. Anal. Chem. 30:2009-2013.
232. GODDU, R. F., and DELKER, D. A. 1960. Spectra-structure correlations for the near-infrared region. Anal. Chem. 32:140-141.
233. GOUGH, K. 1983. Statistical considerations of near-infrared analysis applied to tobacco. Pap. 5 in: NIR Analysis—How Near Infrared Reflects Composition. Chemistry Division, School of Agriculture, Aberdeen, Scotland.
234. GOULDEN, J. D. S. 1964. Analysis of milk by infrared absorption. J. Dairy Res. 31:273-284.
235. GOULDEN, J. D. S., and MANNING, D. J. 1970. Determination of moisture in dairy products by near infrared absorption of methanol extracts. J. Dairy Res. 37:107-112.
236. GOULDEN, J. D. S., SHIELDS, J., and HASWELL, R. 1964. The infrared milk analyser. J. Soc. Dairy Technol. 17:28-33.
237. GRACIAN, J., and MARTEL, J. 1968. Determination of the hydroxyl number of the insaponifiable fraction of olive oil and olive oil cake by near infrared spectrophotometry. (In Spanish) Grasas Aceites (Seville) 19:99-109.
238. GRAHAM, T. F., and COLLINS, F. I. 1976. Optimum sample preparation: The key to accurate analysis by infrared reflectance. Technicon Ind. Syst. News 2:1-3.
239. GRIFFITHS, P. R. 1982. Mid-infrared and near infrared reflectance: Advantages and disadvantages. Proc. NIRA Symp., 2nd. Tarrytown, NY, July.
240. HALL, P. A., JESSIMAN, C. S., and MURRAY, I. 1982. A bibliography of near infrared reflectance (NIR) spectroscopy. Unpublished list of 198 references. School of Agriculture, Aberdeen, Scotland.
241. HALSEY, S. A. 1985. The use of transmission and transflectance near infrared spectroscopy for the analysis of beer. J. Inst. Brew. (London) 91:306-312.
242. HALSEY, S. A. 1986. Application of near infrared analysis in brewing. Anal. Proc. (London) 23:126-127.
243. HALSEY, S. A., and BUCKEE, G. K. 1985. Near infrared analysis applied to malts, worts and beers. Pages 523-530 in: Proc. Congr., Eur. Brew. Conv.
244. HAMID, A., McCLURE, W. F., and WEEKS, W. W. 1978. Rapid spectrophotometric analysis of the chemical composition of tobacco. Part 2. Total alkaloids. Beitr. Tabakforsch. Int. 9:267-274.
245. HAMID, A., McCLURE, W. F., and WHITTAKER, T. B. 1981. Rapid NIR measurement of food and food products. Part II: Stepwise linear regression in memory-limited environments. Am. Lab. 13:108-121.

246. HAMMERSLEY, M. J., RANFORD, S. L., and PATRICK, V. C. 1985. Wool testing in New Zealand with the InfraAlyzer 400R. Pages 78-83 in: NIR 84 Proc. Int. Symp. Near Infrared Reflectance Spectrosc., 1st. RACI, Cereal Chem. Div., Melbourne, Australia.

247. HARBOURNE, D. 1980. Near infrared analysis of barley. Neotec Int. News Items 16.

248. HARBUTT, R. P. 1980. Evaluation of the Neotec 6350 Research Composition Analyzer for potential use of near infrared quantitative analysis in the packaging industry. Pira Rep. PK5(R)-July. Abstracted in Pira Packaging Abstr. 37:423.

249. HARDING, L., and MORRIS, J. L. 1973. Reconstitution of churned or "oiled" milk samples for instrumental determination of fat content. J. Soc. Dairy Technol. 26:91.

250. HARRICK, N. J. 1967. Internal Reflection Spectroscopy. Interscience, New York. 327 pp.

251. HARRIS, W. M. 1986. Automated determination of fat, crude protein and lactose in ewe milk by infrared spectrometry. Analyst (London) 3:37-39.

252. HART, H. V. 1976. Protein determination in (home grown) wheat by the InfraAlyzer. Flour Milling Baking Res. Assoc. Bull. 6:192-204.

253. HART, H. V. 1978. The InfraAlyzer—further information for determination of protein and moisture in wheat. Flour Milling Baking Res. Assoc. Bull. 5:173-175.

254. HART, J. R., NORRIS, K. H., and GOLUMBIC, C. 1962. Determination of the moisture content of seeds by near-infrared spectrophotometry of their methanol extracts. Cereal Chem. 39:94-99.

255. HATTORI, T., SHIRAI, K., NIWA, M., and MURAKAMI, Y. 1981. Quantitative interpretation of infrared diffuse reflectance spectra over whole concentration range. Kagaku To Kogyo (Tokyo) 54:1964-1967.

256. HAUSER, E., and WEBER, U. 1980. Application of IR-reflection analysis for quick determination of the valuable portions in meat and meat products. II. Plotting curves for contents of the main constituents of meat and meat products (as in I, but with a greater number of samples), using existing contents curves. Experiments with a refrigerated cell. (In German) Fleischwirtschaft 60:482-485.

257. HAYASHI, H., and MORITA, S. 1980. Near-infrared absorption spectra of light harvesting bacteriochlorophyll protein complexes from *Chromatium vinosum*. J. Biochem. (Tokyo) 88:1251-1258.

258. HEAVENS, O. S., KING, J., and SMITH, S. D. 1957. Interference filters for the infrared. Spectrochim. Acta 10:179-194.

259. HECHT, H. G. 1968. The present status of diffuse reflectance theory. Pages 1-26 in: Mod. Aspects Reflectance Spectrosc., Proc. Symp. W. W. Werdlandt, ed. Plenum Press, New York.

260. HECHT, K. T., and WOOD, D. L. 1956. The near infra-red spectrum of the peptide group. Proc. R. Soc. London 235:174-188.

261. HELLÄMÄKI, M., and MOISIO, T. 1983. Prediction of protein and fiber contents in silage by near infrared reflectance analysis. Milchwissenschaft 38:14-15.

262. HENDERSON, G., and BRYANT, M. F. 1980. Comparison of near-infrared reflectance and photoacoustic spectra of grain samples. Anal. Chem. 52:1787-1790.

263. HENRY, R. J. 1984. Analysis of carbohydrates by near infrared reflectance techniques. (Abstr.) Chem. Aust. 51:240.

264. HENRY, R. J. 1985a. Calibration and use of a scanning near infrared reflectance spectrophotometer for assessment of malting quality in barley. Pages 93-94 in: NIR 84 Proc. Int. Symp. Near Infrared Reflectance Spectrosc., 1st. RACI, Cereal Chem. Div., Melbourne, Australia.

265. HENRY, R. J. 1985b. Evaluation of barley and malt quality using near-infrared reflectance techniques. J. Inst. Brew. (London) 91:393-396.

266. HENRY, R. J. 1985c. Use of a scanning near-infrared reflectance spectrophotometer for assessment of the malting potential of barley. J. Sci. Food Agric. 36:249-254.

267. HENRY, R. J. 1985d. Near-infrared reflectance analysis of carbohydrates and its application to the determination of (1.FWDARW.3), (1.FWDARW.4)-.Beta.-D-Glucan in barley. Carbohydr. Res. 141:13-19.

268. HILDEBRAND, D. F., CHAVEN, C., HYMOWITZ, T., BRYAN, H. H., and DUNCAN, A. A. 1981. Protein and oil content of winged bean seeds as measured by near-infrared light reflectance. Agron. J. 73:623-625.

269. HILLIARD, J. H., and DAYNARD, T. B. 1976. Measurement of protein and oil in grains and soybeans with reflected near infra-red light. Can. Inst. Food Sci. Technol. J. 9:11-14.

270. HILLIER, G., and CLUCAS, A. 1985a. Scanning NIR-contamination of tapioca starch by wheat starch. Pages 175-176 in: NIR 84 Proc. Int. Symp. Near Infrared Reflectance Spectrosc., 1st. RACI, Cereal Chem. Div., Melbourne, Australia.

271. HILLIER, G., and CLUCAS, A. 1985b. I/A 500 to I/A 300—transferability of calibrations. Pages 177-179 in: NIR 84 Proc. Int. Symp. Near Infrared Reflectance Spectrosc., 1st. RACI, Cereal Chem. Div., Melbourne, Australia.

272. HIRSCHFELD, T. 1979. Near infrared reflectance analysis: Theory and applications. Fed. Anal. Chem. Spectrosc. Soc. Meet., Philadelphia, September.

273. HIRSCHFELD, T. 1980. The present and future of near infrared spectroscopy. Proc. Int. Technicon Symp., 9th. Paris, November.

274. HIRSCHFELD, T. 1982a. Fast correlation transform. Eur. Technicon NIRA Symp., London, October.

275. HIRSCHFELD, T. 1982b. NIRA. The technique, its advantages and disadvantages. Eur. Technicon NIRA Symp., London, October.

276. HIRSCHFELD, T. 1982c. NIRA—Where do we grow from here? Proc. NIRA Symp., 2nd. Tarrytown, NY, July.

277. HIRSCHFELD, T. 1982d. Principal components analysis and other methods applied to NIRA. Eur. Technicon NIRA Symp., London, October.

278. HIRSCHFELD, T. 1982e. Principles of regression transforms: Pulling the rabbit from the hat. Proc. NIRA Symp., 2nd. Tarrytown, NY, July.

279. HIRSCHFELD, T. 1983. Near infrared bibliography. Unpublished list of 346 references. Page 18 in: A Bibliography of Applications of Near Infrared Reflectance Spectroscopy to Food Analysis. B. C. Osborne, ed. Flour Milling Baking Res. Assoc., U.K.

280. HIRSCHFELD, T. 1985. Sample area optimization in a diffuse reflectance near-infrared spectrophotometer. Appl. Spectrosc. 39:1085-1086.

281. HIRSCHFELD, T. B., HONIGS, D. E., and HIEFTJE, G. M. 1985. Optimization of sample absorbance for quantitative analysis in the presence of pathlength error in the IR and NIR regions. Appl. Spectrosc. 39:430-433.

282. HIRSCHFELD, T., and STARK, E. 1980. A comparison of NIRA and classical IR analysis. Fed. Anal. Chem. Spectrosc. Soc. Meet., Philadelphia, September.

283. HIRSCHFELD, T., and STARK, E. W. 1984. Near-infrared reflectance analysis of foodstuffs. Pages 505-551 in: Analysis of Foods and Beverages: Modern Techniques. G. Charalambous, ed. Academic Press, Orlando.

284. HIRSCHFELD, T., STARK, E., and JUDGE, J. 1981. Near infrared reflectance analysis (NIRA). Instrumentation and performance. Pittsburgh Conf., Atlantic City, March.

285. HIRSCHFELD, T., STARK, E., and LUCHTER, K. 1981. Near infrared reflectance in the laboratory. Fed. Anal. Chem. Spectrosc. Soc. Meet., Philadelphia, September.

286. HITACHI LTD. 1985. Method and apparatus for obtaining IR absorption spectra using reflectance technique. Jpn. Kokai Tokkyo Koho Patent JP 85117119 A2, JP 60117119, 24 June.

287. HOAGLIN, D. C., and WELSCH, R. E. 1978. The hat matrix in regression and ANOVA. Am. Stat. 32:17-22.

288. HODGE, T. P., MORGAN, A. G., and SMITH, D. B. 1979/80. Rapid identification of potential malting quality. HGCA Prog. Rep. 34-38.

289. HOFFMANN, K. 1963. Moisture measurement through infrared reflection. (In German) Chem. Ing. Tech. 35:55-62.

290. HÖHNE, A. 1983. Use of infrared spectroscopy at cereal grain reception (wheat and barley). (In German) Getreide Mehl Brot 37:20-21.

291. HOLMAN, R. T., and EDMONSON, P. R. 1956. Near-infrared spectra of fatty acids and some related substances. Anal. Chem. 28:1533-1538.

292. HONIGS, D. E. 1984. Improvements of the near-infrared diffuse-reflectance technique. Thesis, Indiana Univ. (UM DA8426654).

293. HONIGS, D. E. 1985. Near-infrared analysis. Anal. Instrum. 16:1-62.

294. HONIGS, D. E. 1986. Near-infrared analysis. Eur. Spectrosc. News 65:32-34.

295. HONIGS, D. E., FREELIN, J. M., HIEFTJE, G. M., and HIRSCHFELD, T. B. 1983. Near-infrared reflectance analysis by Gauss-Jordan linear algebra. Appl. Spectrosc. 37:491-497.

296. HONIGS, D. E., HIEFTJE, G. M., and HIRSCHFELD, T. 1984a. A new method for obtaining individual component spectra from those of complex mixtures. Appl. Spectrosc. 38:317-322.

297. HONIGS, D. E., HIEFTJE, G. M., and HIRSCHFELD, T. 1984b. Number of samples and wavelengths required for the training set in near-infrared reflectance spectroscopy. Appl. Spectrosc. 38:844-847.

298. HONIGS, D. E., HIEFTJE, G. M., MARK, H. L., and HIRSCHFELD, T. B. 1985a. Unique sample selection via near-infrared spectral subtraction. Anal. Chem. 57:2299-2303.

299. HONIGS, D. E., HIRSCHFELD, T. B., and HIEFTJE, G. M. 1985b. Near infrared spectrophotometric methods development with a limited number of samples: Application to carbonate in geological samples. Appl. Spectrosc. 39:1062-1065.

300. HOOK, S. C. W. 1982. Determination of wheat hardness—an evaluation of this aspect of wheat specification. Flour Milling Baking Res. Assoc. Bull. 1:12-23.

301. HOOPER, A. W., SMITH, R. A., and BOWMAN, G. E. 1979. A near-infrared diffuse reflectance spectrophotometer. J. Agric. Eng. Res. 24:79.

302. HOOTON, D. E. 1978. The versatility of near infrared reflectance devices. Cereal Foods World 23:176-179.

303. HOPKINS, D. W. 1983. Calibrating NIR reflectance spectrometers: Factors affecting accuracy of predictions of the composition of cereal and oilseed samples. (Abstr.) Cereal Foods World 28:571.

304. HOPKINS, D. W., and KENEASTER, K. K. 1980. Measurement of surface oil in milled rice by near infrared reflectance. (Abstr.) Cereal Foods World 25:518.

305. HORVATH, L., and HORVATHOVA-MOSONYOVA, M. 1981a. Determination of the quality of raw tomato juice with an optical research composition analyser. Acta Aliment. Acad. Sci. Hung. 10:87-96.

306. HORVATH, L., and HORVATHOVA-MOSONYOVA, M. 1981b. Measurement of quality of vegetables. (In Hungarian) Prum. Potravin 32:647-648.

307. HORVATH, L., NORRIS, K. H., and HORVATH-MOSONYI, M. 1984. Determining dietary fibre of wheat bran and carrots using NIR technique. Proc. Annu. Users Conf. NIR Res., 3rd. Pacific Scientific, Rockville, MD.

308. HORVATH, L., NORRIS, K. H., and HORVATH-MOSONYI, M. 1985. Comparative investigations into the determination of protein by the Kjeldahl method and NIR (near infrared reflectance) technique. Acta Aliment. Acad. Sci. Hung. 14:113-124.

309. HOWARD, J. N., and CHAPMAN, R. M. 1956. Near-infrared spectra of fatty acids and some related substances. Anal. Chem. 28:1533-1538.

310. HRUSCHKA, W. R., and MARTENS, H. 1982. Application of factor analysis to NIRA. Proc. NIRA Symp., 2nd. Tarrytown, NY, July.

311. HRUSCHKA, W. R., and NORRIS, K. H. 1980. Fourier transform near-infrared spectra of individual soybean seeds. (Abstr.) Cereal Foods World 24:460.

312. HRUSCHKA, W. R., and NORRIS, K. H. 1982. Least-squares curve-fitting of near-infrared spectra predicts protein and moisture content of ground wheat. Appl. Spectrosc. 36:261-265.

313. HSU, J. F., and YONG, S. S. 1982. Potential use of near infrared reflectance spectroscopy in Taiwan. (In Chinese) Shih P'in Kung Yeh (Taiwan) 14:46-49.

314. HUBER, H., and BACKMANN, F. 1976. Infrared reflection spectroscopy (IRS), a new analytical measurement of simultaneous estimation of water, crude protein and crude fat in plant material. (In German) Schweiz. Landwirtsch. Forsch. 15:497-502.

315. HUGHES, R. H., MARTIN, R. J., and COGGESHALL, M. D. 1956. Temperature dependence of infrared absorptions. J. Chem. Phys. 24:489-490.

316. HUNSTON, M. J. 1980. Infra-red analysis: Quality control of the future. Food Flavourings Ingred. Process. 2:24-27.

317. HUNSTON, M. J. 1983. Application of NIR spectroscopy to the analysis of beer samples. (Abstr.) Cereal Foods World 28:575.

318. HUNT, W. H., FULK, D. W., ELDER, B., and NORRIS, K. H. 1977. Collaborative study on infrared reflectance devices for determination of protein in hard red winter wheat, and for protein and oil in soybeans. Cereal Foods World 22:534-536.

319. HUNT, W. H., FULK, D., MOEN, R., NORRIS, K., THOMAS, T., and LIEBE, E. B. 1976. Collaborative study on infrared reflectance devices for determination of moisture, protein and oil in soybeans. (Abstr.) Cereal Foods World 21:438.

320. HUNT, W. H., FULK, D. W., NOLAN, T., and ABBOTT, D. 1977. Evaluation of the InfraAlyzer as a potential grain inspection tool for the USDA. I. Collaborative study on hard red winter wheat. Pages 114-117 in: Advances in Automated Analysis. Vol. 2. Industrial Symposia. Mediad Inc., Tarrytown, NY.

321. HUNT, W. H., FULK, D. W., THOMAS, T., and NOLAN, T. 1978. Effect of type of grinder on protein values of hard red winter wheat when analyzed by infrared reflectance devices. Cereal Foods World 23:143-144.

322. HUNT, H., NOLAN, T., and HAWK, A. 1978. Report on the 1978 collaborative study on the NIR technique. (Abstr.) Cereal Foods World 23:469.

323. HYMOWITZ, T., DUDLEY, J. W., COLLINS, F. I., and BROWN, C. M. 1974. Estimations of protein and oil concentration in corn, soybean, and oat seed by near-infrared light reflectance. Crop Sci. 14:713-715.

324. IKEGAYA, K., HINO, A., UOZOMI, J., TAKAYANAGI, H., ANAN, T., and IWAMOTO, M. 1985. Near-infrared spectrophotometric analysis of total nitrogen content in green tea. (In Japanese) Nippon Shokuhin Kogyo Gakkai-Shi 32:553-559.

325. ISAAC, R. A. 1982. Field monitoring of nitrogen in plant tissue by NIRA. Proc. NIRA Symp., 2nd. Tarrytown, NY, July.

326. ISAAC, R. A., and JOHNSON, W. C. 1980. Determination of nitrogen in plant tissue using near infrared reflectance spectroscopy. Fed. Anal. Chem. Spectrosc. Soc. Meet., Philadelphia, September.

327. ISAAC, R. A., and JOHNSON, W. C. 1984. Near infrared reflectance spectroscopic determination of protein nitrogen in plant tissue. J. Assoc. Off. Anal. Chem. 67:506-509.

328. IWAMOTO, M. 1980a. Analysis of food components by near infrared irradiation without decomposing the food. (In Japanese) Kagaku To Seibutsu 18:369-370.

329. IWAMOTO, M. 1980b. Determination of food components by near infrared analysis. (In Japanese) Jpn. Food Sci. 19:65-72.

330. IWAMOTO, M. 1980c. Nondestructive quality measurement of food by near infrared spectroscopic technique. (In Japanese) Nippon Shokuhin Kogyo Gakkai-Shi 27:467-472.

331. IWAMOTO, M. 1981a. Evaluation of food components by near infrared spectrophotometry. (In Japanese) New Food Ind. 23:57-70.

332. IWAMOTO, M. 1981b. Nondestructive evaluation of food composition by near infrared radiation. (In Japanese) Shokuhin No Bussei 7:79-92.

333. IWAMOTO, M. 1982. Near infrared spectrophotometric analysis of fats in foods. (In Japanese) Yukagaku 31:199-204.

334. IWAMOTO, M., CHO, R. K., SUZUKI, T., and UOZUMI, J. 1984. Near infrared reflectance analysis for determining moisture, protein and ash contents in Japanese wheat flours. Nippon Shokuhin Kogyo Gakkai-Shi 31:50-53.

335. IWAMOTO, M., CHO, R. K., UOZUMI, J., and IINO, K. 1984. Near-infrared reflectance spectrum of red pepper and its applicability to determination of capsaicin content. Nippon Shokuhin Kogyo Gakkai-Shi 31:120-125.

336. IWAMOTO, M., HIRATA, T., SUZUKI, T., UOZUMI, J., and ISHITANI, T. 1983. Feasibility of near-infrared diffuse reflectance for determination of the quality of dried laver (*Porphyra yezoensis*). (In Japanese with English summary) Nippon Shokuhin Kogyo Gakkai-Shi 30:397-403.

337. IWAMOTO, M., KONGSEREE, N., UOZUMI, J., and SUZUKI, T. 1986a. Determination of ash content in Japanese home-grown wheat flour by near-infrared diffuse reflectance analysis. Nippon Shokuhin Kogyo Gakkai-Shi 33:842-847.

338. IWAMOTO, M., NORRIS, K. H. and KIMURA, S. 1981. Rapid prediction of chemical compositions of wheat, soybean, pork and fresh potatoes by near infrared spectrophotometric analysis. Nippon Shokuhin Kogyo Gakkai-Shi 28:85-90.

339. IWAMOTO, M., SUZUKI, T., HIRATA, T., NODA, H., UOZUMI, J., and ISHITANI, T. 1983. Some characteristics of infrared and near-infrared spectra of dried laver (*Porphyra yezoensis*) in connection with the quality. (In Japanese with English summary) Nippon Shokuhin Kogyo Gakkai-Shi 30:544-551.

340. IWAMOTO, M., SUZUKI, T., and UOZUMI, J. 1986b. Analysis of protein and amino acid contents in rice by near-infrared spectroscopy. Nippon Shokuhin Kogyo Gakkai-Shi 33:848-853.

341. IWAMOTO, M., UOZUMI, J., and NISHINARI, K. 1986. Preliminary investigation of the state of water in foods by near-

infrared spectroscopy. Pages 1-9 in: Proc. Int. NIR Symp., Budapest.

342. JACQUEZ, J. A., McKEEHAN, W., HUSS, J., DIMITROFF, J. M., and KUPPENHEIM, H. F. 1955. An integrating sphere for measuring diffuse reflectance in the near infrared. J. Opt. Soc. Am. 45:781-785.
343. JANSANSKI, A., and BILANSKI, W. K. 1973. Spectral absorbance of soybeans and its importance for efficient heat processing. Can. Inst. Food Sci. Technol. J. 6:151-155.
344. JENSEN, S. A. 1982. Analytical application of NIR to determine moisture and protein in barley. Autoanalyst, June, pp. 2-4.
345. JOHNSON, E. 1981. Failure analysis of near IR equipment with comparison of hardware failure vs. operator errors. (Abstr.) Cereal Foods World 26:491.
346. JOHNSON, E., and STARK, E. 1981. Long term reliability studies of NIR instrumentation analysis of system failures and cockpit errors. (Abstr.) Cereal Foods World 26:494.
347. JOHNSON, P. D. 1964. Optical absorption and diffuse reflectance of powders. J. Appl. Phys. 35:334-336.
348. JONES, C. 1982a. Near-infrared analysis in the process industry. Anal. Instrum. 20:21-34.
349. JONES, C. 1982b. Using near-infrared analysis for on-stream composition measurements. InTech, August, pp. 51-54.
350. JONES, C., and REED, J. E. 1978. Near infrared analyzers refine process control. Chem. Eng. (N.Y.) 85:111-114.
351. JUDGE, J. F. X., GENOVESE, D., and GRAHAM, T. 1975. Expanding the infrared reflectance technology. (Abstr.) Cereal Foods World 20:450.
352. KAFFKA, K. 1981. Non-destructive methods of analysis in the food industry. (In Hungarian) Prum. Potravin 32:634-639.
353. KAFFKA, K., and CZABAFFY, A. 1981. The correlation between quality parameters and optical transmittance of some stone fruits determined with a near-infrared composition analyser. Acta Aliment. Acad. Sci. Hung. 10:75-85.
354. KAFFKA, K., and KULCSAR, F. 1982. Attempts to determine egg content in pastry products using the NIR technique. Acta Aliment. Acad. Sci. Hung. 11:47-64.
355. KAFFKA, K., NORRIS, K. H., KULCSAR, F., and DRASKOVITS, I. 1982. Attempts to determine fat, protein and carbohydrate content in cocoa powder by the NIR technique. Acta Aliment. Acad. Sci. Hung. 11:271-288.
356. KAFFKA, K. J., NORRIS, K. H., PEREDI, J., and BALOGH, A. 1982. Attempts to determine oil, protein, water and fibre content in sunflower seeds by the NIR technique. Acta Aliment. Acad. Sci. Hung. 11:253-269.
357. KAFFKA, K. J., NORRIS, K. H., and ROSZA-KISS, M. 1982. Determining fat, protein and water content of pastry products by the NIR technique. Acta Aliment. Acad. Sci. Hung. 11:199-217.
358. KAO, C. 1985. Application of NIR instruments for barley protein determination. (Abstr.) Cereal Foods World 30:566.
359. KARVALY, B. 1970. Some comments on theories concerning the connections between the true (bulk) absorption and diffuse reflectance spectra of powdered solids. I. Bodo's theory. Acta Phys. Chem. 16:107-117.
360. KARVALY, B., and PINTER, F. 1970. Investigations of the connections between the bulk absorption and diffuse reflectance spectra of powdered solids. Part II. Acta Phys. Acad. Sci. Hung. 29:203-220.
361. KATO, A., OGURO, M., and TSUTSUI, Y. 1964. Near-infrared spectra of oleamide and saturated fatty acid amides, C_8-C_{18}. (In Japanese with English summary) Yukagaku 13:28.
362. KAYE, W. 1954. Near-infrared spectroscopy. I. Spectral identification and analytical applications. Spectrochim. Acta 6:257-287.
363. KAYE, W. 1955. Near-infrared spectroscopy. II. Instrumentation and technique. Spectrochim. Acta 7:181-204.
364. KAYE, W. 1960a. Cellulose and its derivatives: Near infrared spectroscopy. Pages 409-414 in: The Encyclopedia of Spectroscopy. G. L. Clark, ed. Reinhold, New York.
365. KAYE, W. 1960b. Near infrared spectroscopy. Pages 494-505 in: The Encyclopedia of Spectroscopy. G. L. Clark, ed. Reinhold, New York.
366. KEMENY, G., POKORNY, T., FORIZS, K., and LEKO, L. 1983. Use of the near infrared measurement technique in the wine industry. (In Hungarian) Borgazdasag 31:127-132.
367. KENNEDY, J. F., WHITE, C. A., and BROWNE, A. J. 1985a. Application of infrared reflectance spectroscopy to the analysis of milk and dairy products. Food Chem. 16:115-131.
368. KENNEDY, J. F., WHITE, C. A., and BROWNE, A. J. 1985b. The use of infrared reflectance spectroscopy for the automatic continuous monitoring and control of whey ultrafiltration processes. Food Chem. 18:95-112.
369. KEYWORTH, D. A. 1961. Determination of water by near infrared spectrophotometry. Talanta 8:461-469.
370. KIRKBRIGHT, G. F., and CASTLEDEN, S. L. 1980. Spectroscopy by light and sound. Chem. Br. 16:661-666.
371. KLEPPER, L., and WILHELMI, K. 1979. Comparison between Kjeldahl and near infrared protein analysis on vegetative and head samples of wheat. Crop Sci. 19:923-925.
372. KNIGHT, R. A. 1979. Near-infrared reflectance analysis. Flour Milling Baking Res. Assoc. Bull. 5:200-201.
373. KORTÜM, G. 1969. Page 59 in: Reflectance Spectroscopy: Principles, Methods, Applications. Springer-Verlag, New York. 366 pp.
374. KORTÜM, G., BRAUN, W., and HERZOG, G. 1963. Principles and techniques of diffuse reflectance spectroscopy. Angew. Chem. Int. Ed. 2:333-341.
375. KRIKORIAN, E., and MAHPOUR, M. 1973. The identification and origin of N-H overtone and combination bands in the near-infrared spectra of simple primary and secondary amides. Spectrochim. Acta 29A:1233-1246.
376. KRISCHENKO, V. P. 1984. Research in the use of NIR techniques in the USSR with special reference to the determination of wheat gluten composition. Pages 110-118 in: Proc. World Cereal Bread Congr., 11th. Vienna.
377. KRISCHENKO, V. P., and BOLLING, H. 1972. Use of near reflectance spectroscopy for the determination of protein, gluten and ash in grain and its processed products. (In Russian) Agrokhimiya 82:124-129.
378. KRISCHENKO, V. P., PASHCHINSKY, V. P., LARCHENKO, V. I., KRUCHINIA, L. K., and ZHIGULEV, K. K. 1980. Determination of plant material humidity content by an infrared analyzer. Pages 82-85 in: Proc. Employment Infrared Spectrosc. Determining Quality Grain Other Products, Moscow.
379. KRISCHENKO, V. P., SAMOCHVALOV, S. G., GORPINCHENKO, T. V., MARTYANOVA, A. I., KHUDYAKOVA, F. K., ZHIGULEV, K. K., SOROCHINSKAYA, O. F., EVDOKIMOV, T. N., ZVYAGINTSEV, A. I., and GORSHKOVA, G. I. 1982. The employment of near-infrared spectrophotometry for determining the quality of characteristics of forage grasses, cereal grain and oil plant seeds. Agrokhimiya 6:112-124.
380. KRISCHENKO, V. P., SAZONOV, Yu. G., CHUIKOVA, L. A., VALITOVA, E. G., RUSAKOVA, M. P., and LITVINOVA, S. I. 1980. Gluten analysis by IR reflectance measurements. (In Russian) Agrokhimiya 7:103-107.
381. KRISCHENKO, V. P., SAZONOV, Yu. G., GORSHKOVA, G. I., and VESELITSKAYA, T. V. 1980. Determination of protein, starch, crude fibre, fat, oil and moisture in vegetable material and grain by infrared reflectance measurements. (In Russian) Agrokhimiya 6:128-133.
382. KRIVOSHIEV, G., and CHALUKOVA, R. 1981. Spectral transmittance characteristics of peaches, apricots, pears, plums, strawberries, tomatoes and peppers. Acta Aliment. Acad. Sci. Hung. 10:45-60.
383. KRONSTEIN, M., KRAUSHARR, R. J., and DEACLE, R. E. 1963. Sulfur as a standard of reflectance in the infrared. J. Opt. Soc. Am. 53:458-465.
384. KRUGGEL, W. G., FIELD, R. A., RILEY, M. L., RADLOFF, H. D., and HORTON, K. M. 1981. Near infrared reflectance determination of fat, protein, and moisture in fresh meat. J. Assoc. Off. Anal. Chem. 64:692-696.
385. KUBELKA, P. 1948. New contributions to the optics of intensely light scattering materials. J. Opt. Soc. Am. 38:448-457; errata 38:1067.
386. KUNZ, M. H., and JÄSCHKE, H. J. 1965. Analysis of superimposed infrared spectra using an electronic analogue computer. (In German) Z. Chem. 338-341.
387. KUSAMA, T., OCHIAI-YANAGI, S., and IWAMOTO, M. 1984. Comparison of rice (*Oryza sativa* L.) seed proteins from several varieties. Agric. Biol. Chem. 48:1649-1651.
388. LAI, F. S., POMERANZ, Y., TRAYLOR, D., and AFEWORK, S. 1984. Mathematical treatment of near-infrared reflectance data for the estimation of protein. Cereal Chem. 61:327-329.
389. LANDA, I. 1979a. Food constituents analysis using a monochromator with high speed scan and high energy

throughput. Pacific Scientific Tech. Pap. NIR 4005.

390. LANDA, I. 1979b. High-energy spectrophotometer for rapid constituent analysis in the range of 0.25–2.4 μm. Rev. Sci. Instrum. 50:34-40.
391. LANDA, I. 1979c. A rapid scanning spectrophotometer for NIR reflectance spectroscopy. Fed. Anal. Chem. Spectrosc. Soc. Meet., Philadelphia, September.
392. LANDA, I. 1980. Low cost near infrared instruments for rapid nondestructive measurements of grain constituents. (Abstr.) Cereal Foods World 25:518.
393. LANDA, I., and MOEN, R. 1979. Reduction of predictable errors in NIR measurements. (Abstr.) Cereal Foods World 24:460.
394. LANDA, I., and NORRIS, K. H. 1979. Rapid method for the measurement of monochromator parameters in the near infrared. Appl. Spectrosc. 33:105-107.
395. LANZA, E. 1983. Determination of moisture, protein, fat, and calories in raw pork and beef by near infrared spectroscopy. J. Food Sci. 48:471-474.
396. LANZA, E., and LI, B. W. 1984. Application for near infrared spectroscopy for predicting the sugar content of fruit juices. J. Food Sci. 49:995-998.
397. LARCHENKO, V. I., PASHCHINSKY, V. P., and KRUCHININA, L. K. 1980. Pages 99-102 in: Development of Methods, Instrumentation and Equipment for the Agrochemical Analyses of Soil, Plants, Fodder and Fertilizers. TSINAO, Moscow.
398. LARSEN, S. A. 1979a. Examination of wool top by infrared reflectance: A note on the influence of dyeing. J. Text. Inst. 70:526-528.
399. LARSEN, S. A. 1979b. Quantitative definition of wool-cashmere blends by infrared reflectance analysis. Am. Soc. Test. Mater. E13 Meet., Cleveland, October.
400. LARSEN, S. A. 1980. The effect of cashmere type on quantitative wool-cashmere blend analysis by infrared reflectance. Am. Soc. Test. Mater. E13 Meet., Cleveland, March.
401. LARSSON, A. 1980. Analytical work for milk quality asessment. (In Swedish) Nord. Mejeriindustri 7:598, 600.
402. LASCOMBE, J., SOURISSEAU, G., SAUMAGNE, P., CASTINEL, C., DIZABO, P., LEICKMAN, J. P., and JOSIEN, M. L. 1956. Application of the theory of Kirkwood-Bauer-Magat to the detection of molecules. Assoc. Cah. Phys. 73:44-46.
403. LASZTITY, R., and VARGA, J. 1984. Determination of some cereal constituents by NIR technique in Hungary—experience and future trends. Pages 106-109 in: Proc. World Cereal Bread Congr., 11th. Vienna.
404. LAUER, J. L., and ROSENBAUM, E. J. 1952. Near-infrared absorption spectrophotometry. Appl. Spectrosc. 6:29-46.
405. LAW, D. P. 1975. Near infrared reflectance studies: Spectra of wheat and wheat components. (Abstr.) Cereal Foods World 20:449.
406. LAW, D. P. 1985. Selection for malting quality in barley breeding using near infrared reflectance. Pages 84-92 in: NIR 84 Proc. Int. Symp. Near Infrared Reflectance Spectrosc., 1st. RACI, Cereal Chem. Div., Melbourne, Australia.
407. LAW, D. P., and TKACHUK, R. 1976a. Diffuse reflectance standards for the near infrared region. (Abstr.) Cereal Foods World 21:439.
408. LAW, D. P., and TKACHUK, R. 1976b. Moisture determination in wheat by near infrared reflectance spectroscopy. Allowance for different sample particle size. (Abstr.) Cereal Foods World 21:439.
409. LAW, D. P., and TKACHUK, R. 1977a. Near infrared diffuse reflectance spectra of wheat and wheat components. Cereal Chem. 54:256-265.
410. LAW, D. P., and TKACHUK, R. 1977b. Determination of moisture content in wheat by near-infrared diffuse reflectance spectrophotometry. Cereal Chem. 54:874-881.
411. LAW, S. E. 1973. Scatter of near infrared radiation by cherries as a means of pit detection. J. Food Sci. 38:102-107.
412. LAW, S. E., and NORRIS, K. H. 1973. Kubelka-Munk light scattering coefficients of model particulate systems. Trans. ASAE 16:914-917, 921.
413. LAZTITY, R., MERÉSZ, P., BÉKÉS, F., SMIED, I., and ZSIGMOND, A. 1981. Rapid methods for the determination of fat content in certain foods. (In Hungarian) Elelmiszervizsgalati Kozl. 27:113-122.
414. LEE, G. H. D. 1984. Rapid determination of selected meat components using near infrared reflectance spectroscopy. Diss. Abstr. Int. B. 46/01B:14.
415. LEMUR, R., and ROSENBERG, N. J. 1975. Reflectant induced modification of soybean canopy radiation balance. II. A quantitative and qualitative analysis of radiation reflected from a green soybean canopy. Agron. J. 67:301-306.
416. LEUCKX, M. 1984. Process monitoring with the Technicon InfraAlyzer 400D. Page 91 in: The 5th International NIRA-Symposium in Paris. O. Linsenmann, ed. (In German) Monatsschr. Brau. 37:88-91.
417. LIN, F. M., and POMERANZ, Y. 1965. Characterization of wheat components by infrared spectroscopy. 1. Infrared spectra of major wheat flour components. J. Assoc. Off. Anal. Chem. 48:885-891.
418. LINSENMANN, O., ed. 1984. The 5th International NIRA-Symposium in Paris. (In German) Monatsschr. Brau. 37:88-91.
419. LOCK, J. 1978. Infra-red analyser prevents product loss. Process 24:29.
420. LOFFLER, C. M., and BUSCH, R. H. 1983. Determination of protein percentage of wheat vegetation at two growth stages by near infrared reflectance. Crop Sci. 23:167-168.
421. LONG, T. M. 1983. Application of near infrared reflectance spectroscopy to tobacco analysis. Anal. Proc. (London) 20:69-72.
422. LONGERICH, H., and RAMALEY, L. 1974. Digital interface for a Cary 14 spectrophotometer. Anal. Chem. 46:2067-2071.
423. LOVE, R. 1981. The use of NIR reflectance instrumentation for improved fermentation research and control. (Abstr.) Cereal Foods World 26:494.
424. LOVE, R., STARK, E., and JUDGE, J. 1981. The InfraAlyzer 500, new fast scan computerized NIR spectrophotometer for grain research. (Abstr.) Cereal Foods World 26:494.
425. LUCHTER, K., and BELL, W. 1981. Universal NIR calibrations for soy meal. (Abstr.) Cereal Foods World 26:491.
426. LUCHTER, K., MADDOX, C., and HIRSCHFELD, T. 1980. NIRA of industrial samples. Fed. Anal. Chem. Spectrosc. Soc. Meet., Philadelphia, September.
427. LUCHTER, K., and MARK, H. 1982a. Development of near infrared dairy analysis system. Proc. NIRA Symp., 2nd. Tarrytown, NY, July.
428. LUCHTER, K., and MARK, H. 1982b. Regression and predictions—how to avoid pitfalls. Proc. NIRA Symp., 2nd. Tarrytown, NY, July.
429. MACK, A. R., BRACH, E. J., and RAO, V. R. 1978. Investigation of crop spectra and selection of optimal spectral channels. Can. J. Spectrosc. 23:42-51.
430. MADDIX, C. 1976. Survey of five different grinders in the United Kingdom for potential application to infrared reflectance analysis. Technicon Ind. Syst. News 2:7-9, 12.
431. MADDIX, C., LUCHTER, K., STARK, E., and MARCUS, H. 1980. The application of near infrared spectroscopic quantitative analysis to liquid samples. Pittsburgh Conf., Atlantic City, March.
432. MALCOLM, B. R. 1958. The mounting of fibres and organic crystals for spectroscopy in the near infrared region. J. Sci. Instrum. 35:423-424.
433. MANN, P. J. 1977. On-line moisture analyzer improves product quality. Food Eng. Int. 49:130, 132.
434. MARDEKIAN, J. 1982. The use of modeling in NIRA methods development. Proc. NIRA Symp., 2nd. Tarrytown, NY, July.
435. MARGOSHES, M. 1975. Infrared reflectance analysis. Technicon Bull. R-74.
436. MARIHART, J. 1984. Experiences with NIR-reflectance spectroscopy in routine analysis of starches, starch derivatives and starch saccharification products. (In German) Staerke 36:51-55.
437. MARK, H. 1986. Normalized distances for qualitative near-infrared reflectance analysis. Anal. Chem. 58:379-384.
438. MARK, H., and JACKSON, S. 1981. Operating a self-learning algorithm for quantitative analysis by NIRA. Fed. Anal. Chem. Spectrosc. Soc. Meet., Philadelphia, September.
439. MARK, H., and STARK, E. 1977. Theoretical considerations for accurate protein calibrations. (Abstr.) Cereal Foods World 22:461.
440. MARK, H. L., and TUNNELL, D. 1985. Quantitative near-infrared reflectance analysis using Mahalanobis distances. Anal. Chem. 57:1449-1456.
441. MARK, H., and WORKMAN, J. 1986a. Effect of repack on calibrations produced for near infrared reflectance analysis. Anal. Chem. 58:1454-1459.
442. MARK, H., and WORKMAN, J. 1986b. Critical values of several statistics at high probability levels for use with near-infrared

wavelength searches. Spectroscopy 1:39-46.

443. MARLER, J. W. 1980. Quantitative analysis of wool/synthetic blends by near infrared reflectance spectroscopy. Pacific Scientific Tech. Pap. NIR 5000.

444. MARLER, J. W. 1985a. Calibrating a spectrocomputer. Pages 31-41 in: NIR 84 Proc. Int. Symp. Near Infrared Reflectance Spectrosc., 1st. RACI, Cereal Chem. Div., Melbourne, Australia.

445. MARLER, J. W. 1985b. The application of NIR to measurements on washed wool. Pages 65-77 in: NIR 84 Proc. Int. Symp. Near Infrared Reflectance Spectrosc., 1st. RACI, Cereal Chem. Div., Melbourne, Australia.

446. MARTEN, G. C., HALGERSON, J. L., and CHERNEY, J. H. 1983. Quality prediction of small grain forages by near infrared reflectance spectroscopy. Crop Sci. 23:94-96.

447. MARTENS, H. 1978. Factor analysis of chemical mixtures. Non-negative solutions of cereal amino acid spectra. Int. Conf. Comput. Optim. Anal. Chem., Amsterdam, April.

448. MARTENS, H. 1980. On the calibration of a multivariate instrument for quantitative estimation of individual components in a mixture. Proc. Symp. Appl. Stat., Lyngby, Denmark.

449. MARTENS, H. 1981a. Qualitative and quantitative morphometry by multivariate image analysis. Chemometric theory, illustrated by a hypothetical example: Near infrared analysis of quality in bacon. Norw. Food Res. Inst. Rep. NR 4.

450. MARTENS, H. 1981b. Fat, water and protein determination in meat by near-infrared reflectance spectroscopy. Norw. Food Res. Inst.

451. MARTENS, H., BAKKER, E. A., and HILDRUM, K. I. 1981. Applications of near infrared reflectance spectrometry in the analysis of meat products. Text of lecture given at Eur. Meet. Meat Res. Work., 27th. Vienna, 24–28 August.

452. MARTENS, H., HILDRUM, K. Y., BAKKER, A., JENSEN, S. A., LEA, P., ESKELAND, B., VOLD, E., and RUSSWURM, H. 1980. Unscrambling multivariate data from mixtures. I: Fat, water and protein determination in meat by near-infrared reflectance spectroscopy. Proc. Eur. Meet. Meat Res. Work. 1:146-149.

453. MARTENS, H., and HRUSCHKA, W. 1982. Principal component analysis predicts protein and moisture from near infrared spectra of ground wheat. Pittsburgh Conf., Atlantic City, March.

454. MARTENS, H., and JENSEN, S. A. 1982. Partial least squares regression: A new two-stage NIR calibration method. Pages 607-647 in: Proc. World Cereal Bread Congr., 7th. Holas and Kratochvil, eds. Elsevier, Amsterdam.

455. MARTENS, H., and KNUDSEN, K. E. B. 1978. Fractioning barley proteins by computer factor analysis. Proc. Int. Congr. Food Sci. Technol., 5th. Kyoto, September.

456. MARTENS, H., and RUSSWURM, H., Jr., eds. 1983. Food Research and Data Analysis. Applied Science Publishers, New York.

457. MARTENSON, A. 1979. Applications of NIR reflection measurements. Proc. Int. Technicon Symp., 9th. Paris, November.

458. MASSIE, D. R. 1976. Fat measurement of ground beef with a gallium arsenide infrared emitter. Pages 24-26 in: Quality Detection in Foods. J. J. Gaffney, ed. Trans. ASAE Publ. 1-76.

459. MASSIE, D. R., and NORRIS, K. H. 1965. Spectral reflectance and transmittance properties of grain in the visible and near infrared. Trans. ASAE 8:598-600.

460. MASSIE, D. R., and NORRIS, K. H. 1971. A high intensity spectrophotometer interfaced with a computer for food quality measurements. Trans. ASAE 18:173-176.

461. MAURO, D. J., and OWEN, R. L. 1982. Rapid quantitation of ether and ester substituent groups in starch derivatives by NIR. (Abstr.) Cereal Foods World 27:469.

462. MAYFIELD, J. 1982. Measurement of germ value for flour milling monitoring. Baker 85:29-32, 34.

463. MAYFIELD, J. 1983. Calculating germ value to save wasted flour. Milling Feed Fert. 166:36-37.

464. McALLISTER, E. D., VILLA, J. J., and SALZBERG, C. D. 1956. Rapid and accurate measurement of refractive index in the infrared. J. Opt. Soc. Am. 46:485-487.

465. McCLURE, W. F. 1969. Design of a near-infrared moisture meter for tobacco. Internal Res. Rep. North Carolina State Univ., Raleigh.

466. McCLURE, W. F. 1975. A spectrophotometric technique for studying the browning reaction in tobacco. Trans. ASAE 18:380-383.

467. McCLURE, W. F. 1982. Fourier transform applied to NIR spectroscopy. Eur. Technicon NIRA Symp., London, October.

468. McCLURE, W. F. 1984. History of and future for near-infrared analysis. Anal. Proc. (London) 21:485-486.

469. McCLURE, W. F. 1985a. Fourier analysis of near infrared spectra. Pages 43-61 in: NIR 84 Proc. Int. Symp. Near Infrared Reflectance Spectrosc., 1st. RACI, Cereal Chem. Div., Melbourne, Australia.

470. McCLURE, W. F. 1985b. Status of NIR in the tobacco industry. Pages 127-133 in: NIR 84 Proc. Int. Symp. Near Infrared Reflectance Spectrosc., 1st. RACI, Cereal Chem. Div., Melbourne, Australia.

471. McCLURE, W. F., GIESBRECHT, F. G., and WEEKS, W. W. 1982. The use of Fourier transformations in NIRA of tobacco and other agricultural products. Proc. NIRA Symp., 2nd. Tarrytown, NY, July.

472. McCLURE, W. F., and HAMID, A. 1980. Rapid NIR measurement of the chemical composition of foods and food products. Part I: Hardware. Am. Lab. 12:57-69.

473. McCLURE, W. F., HAMID, A., GIESBRECHT, F. G., and WEEKS, W. W. 1984. Fourier analysis enhances NIR diffuse reflectance spectroscopy. Appl. Spectrosc. 38:322-329.

474. McCLURE, W. F., NORRIS, K. H., and WEEKS, W. W. 1977. Rapid spectrophotometric analysis of the chemical composition of tobacco. Part I. Total reducing sugars. Beitr. Tabakforsch. Int. 9:13-18.

475. McCLURE, W. F., and WILLIAMSON, R. E. 1982. Rapid spectrophotometric analysis of the chemical composition of tobacco. Part III. Polyphenols. Beitr. Tabakforsch. Int. 11:219-227.

476. McCUE, M., and MALINOWSKI, E. R. 1981. Target factor analysis of infrared spectra of multicomponent mixtures. Anal. Chim. Acta 133:125-136.

477. McDONALD, C. E. 1977. Methods of protein analysis and variation in protein results. Farm Res. 3, May/June.

478. McDONALD, R. S. 1982. Review: Infrared spectrometry. Anal. Chem. 54:1250-1275.

479. McFARLANE, I. D. 1981. Measurement of moisture content. U.K. Patent Application GB 2 078 938 A. Filed 13 April.

480. McGANN, T. C. A. 1978. Automated physicochemical methods for the analysis of milk—a review of major advances (1960-1978). Ir. J. Food Sci. Technol. 2:141-156.

481. McGUIRE, C. F. 1982. Near-infrared reflectance estimates of malt extract. Cereal Chem. 59:510-511.

482. McGUIRE, C. F. 1986. Quality evaluation of distillers' dried grain by near-infrared analysis. Cereal Chem. 63:155-159.

483. McKAY, D., and DORSHEIMER, W. 1982. Rapid analysis aids quality assurance. Food Eng. 54:78-79.

484. MEMELINK, A. A. 1966. Infra-red quantitative analysis of *trans* content in fatty acids and derivatives. Pages 388-391 in: Laboratory Handbook for Oil and Fat Analysis. L. V. Cocks and C. Van Rede, eds. Academic Press, London.

485. MEURENS, M. 1984. Analysis of aqueous solutions by NIR reflectance on glass fibre. Proc. Annu. Users Conf., 3rd. Pacific Scientific, Rockville, MD.

486. MILLER, B. S. 1979a. A feasibility study on use of NIR techniques. Cereal Foods World 24:88-89.

487. MILLER, B. S. 1979b. Studies into the use of reflectance analysis in the near-infrared for the investigation and assessment of grain character and grain production. (In German) Getreide Mehl Brot 33:249-251.

488. MILLER, B. S., AFEWORK, S., POMERANZ, Y., BRUINSMA, B. L., and BOOTH, G. D. 1982. Measuring the hardness of wheat. Cereal Foods World 27:61-64.

489. MILLER, B. S., and POMERANZ, Y. 1977. Protein determination in wheat Kj and IR reflectance methods compared. Presented at Hard Winter Wheat Quality Council, Kansas, City, MO.

490. MILLER, B. S., POMERANZ, Y., THOMPSON, W. O., NOLAN, T. W., HUGHES, J. W., DAVIS, G., JACKSON, N. G., and FULK, D. W. 1978. Interlaboratory and intralaboratoy reproducibility of protein determination in hard red winter wheat by Kjeldahl and near infrared procedures. Cereal Foods World 23:198-201.

491. MILLS, B. L., and VAN DE VOORT, F. R. 1982. Evaluation of CH stretch measurement for estimation of fat in aqueous fat emulsions using infrared spectroscopy. J. Assoc. Off. Anal. Chem. 65:1357-1361.

492. MINER, D. C., ZIOMEK, J. V., and LANDA, I. J. 1979. Near infrared instrumentation for controlling the quality of chocolate products. Pacific Scientific Tech. Pap. NIR 4008.
493. MINSON, D. J., BUTLER, K. L., GRUMMITT, N., and LAW, D. P. 1983. Bias when predicting crude protein, dry matter digestibility and voluntary intake of tropical grasses by near-infrared reflectance. Anim. Feed Sci. Technol. 9:221-237.
494. MINSON, D. J., TUCKETT, P. G., and LAW, D. P. 1985. A comparison of three methods of increasing the precision of regressions used for estimating nutritive value of forages by near infrared reflectance. Pages 138-148 in: NIR 84 Proc. Int. Symp. Near Infrared Reflectance Spectrosc., 1st. RACI, Cereal Chem. Div., Melbourne, Australia.
495. MISKELLY, D., LAW, D. P., and CLUCAS, T., eds. 1985. NIR 84 Proc. Int. Symp. Near Infrared Reflectance Spectrosc., 1st. RACI, Cereal Chem. Div., Melbourne, Australia.
496. MISKELLY, D. M., and MURRAY, I. C. 1985. The use of NIR in wheat protein testing in Australia. Pages 169-170 in: NIR 84 Proc. Int. Symp. Near Infrared Reflectance Spectrosc., 1st. RACI, Cereal Chem. Div., Melbourne, Australia.
497. MITCHELL, N. 1981. Near infrared absorption spectroscopy—a brief review. Varian Bull., September.
498. MOEN, R. 1979. Soybean processing and NIR measurement. Pacific Scientific Tech. Pap. 1015.
499. MOEN, R. D. 1980a. Statistical tools for evaluation of analyses of spectroscopic data. Fed. Anal. Chem. Spectrosc. Soc. Meet., Philadelphia, September.
500. MOEN, R. D. 1980b. N.I.R. measurement—past, present, and future. (Abstr.) Cereal Foods World 25:518.
501. MOEN, R. D. 1981. Easing the N.I.R. instrument calibration pains. (Abstr.) Cereal Foods World 26:494.
502. MOEN, R., and CARPENTER, F. M. 1981. Soybean processing and quality control program. Pacific Scientific Tech. Pap. NIR 2011.
503. MOLL, M., FLAYEUX, R., and LEHUEDÉ, J. M. 1976. Rapid determination of moisture and protein in malt by reflectance spectroscopy in the near infra-red. (In French) Bios (Nancy) 7:3-6.
504. MORGAN, A. G., and GOTHARD, P. G. 1979. Rapid prediction of malt hot water extract by near infrared reflectance spectroscopy studies on barley. J. Inst. Brew. London 85:339-341.
505. MORGAN, A. G., and GOTHARD, P. G. 1981. A rapid test for identifying potential malting quality in the early generations of a barley breeding program. J. Sci. Food Agric. 32:333-338.
506. MORGAN, A. G., and STARR, C. 1983. A comparison of the use of milled and whole rapeseeds for the measurement of nitrogen and oil content by near infrared spectroscopy. Pap. 6 in: NIR Analysis—How Near Infrared Reflects Composition. Chemistry Division, School of Agriculture, Aberdeen, Scotland.
507. MORGANTINI, M., D'ACUNZIO, G., PIOMBINO, O., and PERROTTA, F. 1976. Rapid automatic method for the simultaneous determination of some parameters of pasta foods. (In Italian) Boll. Chim. Prov. 27:278-282.
508. MURAKAMI, E. 1973. Reduced product loss with IR analyzer. Food Eng. Int. 45:105-106.
509. MURRAY, I. 1983a. Forage and feed assessment by near infrared reflectance (NIRR) spectrocomputer. Pap. 1983/1 Nutr. Chem. Train. Conf., Cambridge.
510. MURRAY, I. 1983b. Fundamentals of NIRR analysis of feedstuffs. Pap. 1 in: NIR Analysis—How Near Infrared Reflects Composition. Chemistry Division, School of Agriculture, Aberdeen, Scotland.
511. MURRAY, I. 1983c. Rapid silage evaluation by near infrared reflectance (NIRR) spectroscopy. Pages 83-87 in: Research Investigations and Field Trials 1982–83. School of Agriculture, Aberdeen, Scotland.
512. MURRAY, I. 1983d. Silage evaluation by near infrared reflectance (NIRR). Pages 24-30 in: Proc. Conf. Nutr. Chem. Scottish Agric. Colleges. Aberdeen, Scotland.
513. MURRAY, I. 1984. A correlation atlas of the NIR region—What NIRR sees in forage and feeds. Proc. Annu. Users Conf. NIR Res. Pacific Scientific, Rockville, MD.
514. MURRAY, I., GELMAN, A. L., and TOPPS, J. H. 1982. Evaluation of grass silage by near infrared reflectance or chemical analysis. Pages 391-394 in: Proc. IFIC Symp., Sydney, Australia.
515. MURRAY, I., and HALL, P. A. 1983. Animal feed evaluation by use of near infrared reflectance (NIR) spectrocomputer. Anal. Proc. (London) 20:75-79.
516. NAGAO, A., UOZUMI, J., IWAMOTO, M., and YAMAZAKI, M. 1985. Determination of fat content in meats by near-infrared reflectance spectroscopy. Yukagaku 34:257-261.
517. NAKAMURA, S., WATANABE, I., and KOYAMA, T. 1978. Some inquiries on usage of grain quality analyzer (GQA). (In Japanese) Rep. Tohoku Bur. Crop Sci. Jpn. 21:67.
518. NAKAZAWA, A. 1980. Application of partial regression coefficient ratio to infrared spectrophotometric analysis. (In Japanese) Bunseki Kagaku 29:617-621.
519. NIMTZ, M. 1982. Use of NIRA in the quality control laboratory. Proc. NIRA Symp., 2nd. Tarrytown, NY, July.
520. NIWA, E., NAKAYAMA, T., and HAMADA, I. 1980. State of water in fish flesh determined by near IR techniques. Nippon Suisan Gakkai-Shi 46:1147.
521. NOLAN, T. W. 1977. Selecting calibration samples for near infrared measurements. (Abstr.) Cereal Foods World 22:461.
522. NOLAN, T. W. 1982. Use of NIR in process control. Fed. Anal. Chem. Spectrosc. Soc. Meet., Philadelphia, September.
523. NORRIS, K. H. 1964a. Reports on design and development of a new moisture meter. Agric. Eng. 45:370-372.
524. NORRIS, K. H. 1964b. Simple spectroradiometer for 0.4 to 1.2 micron region. Trans. ASAE 7:240-242.
525. NORRIS, K. H. 1965. Measuring and using light transmittance properties of plant materials. Proc. Conf. Electromag. Radiation Agric., New York, October.
526. NORRIS, K. H. 1978. Near infrared reflectance spectroscopy—the present and future. Pages 245-251 in: Cereals '78: Better Nutrition for the World's Millions. Int. Cereal Bread Congr., 6th. Y. Pomeranz, ed. Am. Assoc. Cereal Chem., St. Paul, MN.
527. NORRIS, K. H. 1979. Near infrared reflectance spectrometry for compositional analysis. Fed. Anal. Chem. Spectrosc. Soc. Meet., Philadelphia, September.
528. NORRIS, K. H. 1980. Factors affecting quantitative analysis by near infrared reflectance. Fed. Anal. Chem. Spectrosc. Soc. Meet., Philadelphia, September.
529. NORRIS, K. H. 1983a. Extracting information from spectrophotometric curves. Predicting chemical composition from visible and near-infrared spectra. Pages 95-114 in: Food Research and Data Analysis. H. Martens and H. Russwurm, Jr., eds. Applied Science Publishers, New York.
530. NORRIS, K. H. 1983b. Multivariate analysis of raw materials. Pages 527-535 in: Chemistry and World Food Supplies: The New Frontiers. Chemrawn II (Chemical Research Applied to World Needs). L. W. Shemilt, ed. Pergamon Press, New York.
531. NORRIS, K. H. 1983c. Instrumental techniques for measuring quality of agricultural crops. Pages 471-483 in: Post-Harvest Physiology and Crop Preservation. M. Lieberman, ed. Plenum, New York.
532. NORRIS, K. H. 1984. Interpreting and using plots of correlation coefficients vs. wavelength. Proc. Annu. Users Conf. NIR Res., 3rd. Pacific Scientific, Rockville, MD.
533. NORRIS, K. H. 1985a. Art and science of near infrared spectroscopy. Page 27 in: NIR 84 Proc. Int. Symp. Near Infrared Reflectance Spectrosc., 1st. RACI, Cereal Chem. Div., Melbourne, Australia.
534. NORRIS, K. H. 1985b. Interpreting the diffuse reflectance spectra of two-component mixtures. Page 42 in: NIR 84 Proc. Int. Symp. Near Infrared Reflectance Spectrosc., 1st. RACI, Cereal Chem. Div., Melbourne, Australia.
535. NORRIS, K. H., and BARNES, R. F. 1976. Infrared reflectance analysis of nutritive value of feedstuffs. Pages 237-241 in: Proc. Int. Symp. Feed Composition Anim. Nutr. Require. Comput. Diets, 1st. P. V. Fonnesbeck, L. E. Harris, and L. C. Kearl, eds. Utah Agric. Exp. Stn, Logan.
536. NORRIS, K. H., BARNES, R. F., MOORE, J. E., and SHENK, J. S. 1976. Predicting forage quality by infrared reflectance spectroscopy. J. Anim. Sci. 43:889-897.
537. NORRIS, K. H., and BUTLER, W. L. 1961. Techniques for obtaining absorption spectra on intact biological samples. I.R.E. Trans. Bio-Med. Electron. 8:153-157.
538. NORRIS, K. H., and HART, J. R. 1965. Direct spectrophotometric determination of moisture content of grain and seeds. Pages 19-25 in: Principles and Methods of Measuring Moisture in Liquids and Solids. Vol. 4. A. Wexler, ed. Reinhold, New York.
539. NORRIS, K. H., and ROWAN, J. D. 1962. Automatic detection of blood in eggs. Agric. Eng. 43:154-159.
540. NORRIS, K. H., and WILLIAMS, P. C. 1977. Optimization of mathematical treatment of reflectance data for the estimation of protein in wheat. (Abstr.) Cereal Foods World 22:461.

541. NORRIS, K. H., and WILLIAMS, P. C. 1979. The determination of protein and moisture in HRS wheat by near-infrared reflectance spectroscopy. I. A comparative study of 12 instrumental methods. (Abstr.) Cereal Foods World 24:459.
542. NORRIS, K. H., and WILLIAMS, P. C. 1984. Optimization of mathematical treatments of raw near-infrared signal in the measurement of protein in hard red spring wheat. I. Influence of particle size. Cereal Chem. 61:158-165.
543. NORRIS, K. H., WILLIAMS, P. C., and MATTERN, P. 1977. Rapid estimation of amino acids in wheat by near infrared reflectance spectroscopy. (Abstr.) Cereal Foods World 22:461.
544. NYGAARD, B. 1985. Advances in analytical techniques: NIR—near-infrared reflectance. (In Swedish) Kem. Tidskr. 97:43-45.
545. O'CONNOR, R. T., DUPRE, E. F., and FEUGE, R. O. 1955. The infrared spectra of mono-, di- and triglycerides. J. Am. Oil Chem. Soc. 32:88-93.
546. O'HAVER, T. C., and GREEN, G. L. 1975. Derivative spectroscopy. Am. Lab. 7:15-21.
547. OLSON, D. G. 1982. Rapid composition analysis for processed meats. Proc. NIRA Symp., 2nd. Tarrytown, NY, July.
548. OPILA, R. L. 1982. Instrumentation needs for food process design. Instrum. Soc. Am. Trans. 20:49.
549. OSBORNE, B. G. 1980. Current and future applications of NIR analysis. Flour Milling Baking Res. Assoc. Bull. 5:212-221.
550. OSBORNE, B. G. 1981a. Application of near infrared reflectance spectroscopy to the analysis of food. Anal. Proc. (London) 18:488-489.
551. OSBORNE, B. G. 1981b. A near infrared research composition analyser. Eur. Spectrosc. News 35:33-34.
552. OSBORNE, B. G. 1981c. Principles and practice of near infra-red (NIR) reflectance analysis. J. Food Technol. 16:13-19.
553. OSBORNE, B. G. 1982. Application of NIR measurement of flour properties to predicting water absorption. Flour Milling Baking Res. Assoc. Bull. 4:152-155.
554. OSBORNE, B. G. 1983a. Applications of NIR in the baking industry. Anal. Proc. (London) 20:79-83.
555. OSBORNE, B. G. 1983b. Measurement of levels of bread improvers in concentrates by means of near-infrared reflectance spectroscopy. J. Sci. Food Agric. 34:1297-1301.
556. OSBORNE, B. G. 1983c. Investigation of the performance of an improved calibration for the determination of protein in U.K. home-grown wheat by near infrared reflectance analysis. J. Sci. Food Agric. 34:1441-1443.
557. OSBORNE, B. G. 1984a. Investigations into the use of near infrared reflectance spectroscopy for the quality assessment of wheat with respect to its potential for bread baking. J. Sci. Food Agric. 35:106-110.
558. OSBORNE, B. G. 1984b. Some application of NIRA in the UK to milling and baking industry. Proc. Int. Symp. NIRA, 7th. Technicon Ind. Syst., Tarrytown, NY.
559. OSBORNE, B. G. 1985a. Statistics of NIR calibration. Pages 1-12 in: NIR 84 Proc. Int. Symp. Near Infrared Reflectance Spectrosc., 1st. RACI, Cereal Chem. Div., Melbourne, Australia.
560. OSBORNE, B. G. 1985b. Progress towards standardization of NIR reflectance analysis. Pages 149-155 in: NIR 84 Proc. Int. Symp. Near Infrared Reflectance Spectrosc., 1st. RACI, Cereal Chem. Div., Melbourne, Australia.
561. OSBORNE, B. G., BARRETT, G. M., CAUVAIN, S. P., and FEARN, T. 1984. The determination of protein, fat and moisture in bread by near infrared reflectance spectroscopy. J. Sci. Food Agric. 35:940-945.
562. OSBORNE, B. G., and DOUGLAS, S. G. 1981a. Evaluation of grinders for NIR. Flour Milling Baking Res. Assoc. Bull. 2:64-67.
563. OSBORNE, B. G., and DOUGLAS, S. 1981b. Measurement of the degree of starch damage in flour by near infrared reflectance analysis. J. Sci. Food Agric. 32:328-332.
564. OSBORNE, B. G., DOUGLAS, S. G., and FEARN, T. 1980a. Evaluation of new NIR instruments. 1. InfraAlyzer 300B. Flour Milling Baking Res. Assoc. Bull. 4:165-173.
565. OSBORNE, B. G., DOUGLAS, S. G., and FEARN, T. 1980b. Evaluation of new NIR instruments. 2. Neotec 101. Flour Milling Baking Res. Assoc. Bull. 4:174-178.
566. OSBORNE, B. G., DOUGLAS, S., and FEARN, T. 1980c. Research into NIR analysis of flour with special reference to starch damage and baking quality. (Abstr.) Cereal Foods World 25:531.
567. OSBORNE, B. G., DOUGLAS, S. G., and FEARN, T. 1981. Assessment of wheat grain texture by near infrared reflectance measurements on Buhler-milled flour. J. Sci. Food Agric. 32:200-202.
568. OSBORNE, B. G., DOUGLAS, S., and FEARN, T. 1982a. The application of near infrared reflectance analysis to rapid flour testing. J. Food Technol. 17:355-363.
569. OSBORNE, B. G., DOUGLAS, S. G., and FEARN, T. 1982b. Recent progress in the application of NIR to the measurement of quality parameters in flour. NIR—analytical developments in cereal chemistry. Proc. World Cereal Bread Congr., 7th. 24 June–4 July, Prague.
570. OSBORNE, B. G., DOUGLAS, S., FEARN, T., and WILLIS, K. H. 1982. The development of universal calibrations for measurement of protein and moisture in UK home-grown wheat by near-infrared reflectance analysis. J. Sci. Food Agric. 33:736-740.
571. OSBORNE, B. G., and FEARN, T. 1981. Accuracy and precision of NIR protein measurement. Flour Milling Baking Res. Assoc. Bull. 4:185-188.
572. OSBORNE, B. G., and FEARN, T. 1983. Collaborative evaluation of universal calibrations for the measurement of protein and moisture in flour by near infrared reflectance. J. Food Technol. 18:453-460.
573. OSBORNE, B. G., and FEARN, T. 1984. Methods of wavelength selection in the development of NIR calibrations using a monochromator. Proc. Annu. Users Conf. NIR Res., 3rd. Pacific Scientific, Rockville, MD.
574. OSBORNE, B. G., and FEARN, T. 1983. Collaborative evaluation of near infrared reflectance analysis for the determination of protein, moisture and hardness in wheat. J. Sci. Food Agric. 34:1011-1017.
575. OSBORNE, B. G., FEARN, T., and BARRETT, G. M. 1982. FMBRA takes the grind out of measuring moisture. Milling Feed Fert. 165:34-35.
576. OSBORNE, B. G., FEARN, T., MILLER, A. R., and DOUGLAS, S. 1984. Application of near infrared reflectance spectroscopy to the compositional analysis of biscuits and biscuit doughs. J. Sci. Food Agric. 35:99-105.
577. OSBORNE, B. G., FEARN, T., and RANDALL, P. G. 1983. Measurement of fat and sucrose in dry cake mixes by near infrared reflectance spectroscopy. J. Food Technol. 18:651-656.
578. OSMER, D., and KETTENACKER, J. 1982. Practical application of Kubelka-Munk theory for the near infrared region. J. Coat. Technol. 54:35-41.
579. PANDEYA, R. S., ROSA, N., WHITE, F. H., ELLIOT, J. M., and DIRKS, V. A. 1978. Rapid estimation of some flue-cured tobacco chemical characteristics by infrared reflectance spectroscopy. Tob. Sci. 22:27-31.
580. PANFORD, J. A., and WILLIAMS, P. C. 1980. Monitoring systems for accuracy and precision in large scale NIR protein-testing of HRS wheat. (Abstr.) Cereal Foods World 25:531.
581. PANOZZO, J., and O'BRIEN, L. 1985. Application of near infrared reflectance spectroscopy to early generation selection for wheat quality. Pages 183-186 in: NIR 84 Proc. Int. Symp. Near Infrared Reflectance Spectrosc., 1st. RACI, Cereal Chem. Div., Melbourne, Australia.
582. PARK, Y. W. 1981. A study on predicting carotene, nitrate, soluble N and fibrous fractions in forages and vegetables with near infrared reflectance spectroscopy, and effects of processing methods on carotene contents. Ph.D. thesis, Utah State University (UM 81-21386).
583. PARK, Y. W., ANDERSON, M. J., ASAY, K. H., and MAHONEY, A. W. 1980. Predicting soluble nitrogen and fibrous fractions in crested wheatgrass with near-infrared-reflectance spectroscopy. Pacific Scientific Tech. Pap. NIR 2008.
584. PARK, Y. W., ANDERSON, M. J., and MAHONEY, A. W. 1982. Potential of near infrared reflectance (NIR) spectroscopy for estimating nutrient content of dehydrated vegetables. J. Food Sci. 47:1558-1561.
585. PELLETIER, M. F. 1984. Prediction of chromium and total nitrogen in chicken excreta with near infrared reflectance spectroscopy. Proc. Annu. Users Conf. NIR Res., 3rd. Pacific Scientific, Rockville, MD.
586. PETELIN, A. 1979. Theoretical and practical aspects of temperature effects on NIR measurements. (Abstr.) Cereal Foods World 24:460.
587. PETELIN, A. 1980. On the shapes of reflectance spectra in NIR. (Abstr.) Cereal Foods World 25:531.
588. PETERS, J. 1982. NIRA-possibilities in the continuous process

industry. Proc. NIRA Symp., 2nd. Tarrytown, NY, July.
589. PIRRITANO, R., and MELAND, P. 1983. A rapid method for soluble protein and percent extract analysis in malt extracts by NIRA. (Abstr.) Cereal Foods World 28:575-576.
590. PIRRITANO, R., MELAND, P., and GALAT, G. 1983. Rapid analysis of hops for alpha and beta acids using near infrared reflectance analysis. (Abstr.) Cereal Foods World 28:576.
591. PLYLER, E. K., BLAINE, L. R., and TIDWELL, E. D. 1955. Precise measurement of wavelengths in IR spectra. J. Res. Natl. Bur. Stand. 55:279-284.
592. POLESELLO, A., and GIANGIACOMO, R. 1983. Application of near infrared spectrophotometry to the nondestructive analysis of foods: A review of experimental results. CRC Crit. Rev. Food Sci. Nutr. 18:203-230.
593. POMERANZ, Y. 1976. Particle size of wheat in infrared analysis. Technicon Ind. Syst. News 2:4-6.
594. POMERANZ, Y., DAVIS, G. D., STOOPS, J. L., and HUBBARD, J. D. 1977. Estimation of protein content and quality in grain sorghum. (Abstr.) Cereal Foods World 22:472.
595. POMERANZ, Y., and MOORE, R. B. 1975. Reliability of several methods for protein determination in wheat. Baker's Dig. 49:44-48, 58.
596. POMERANZ, Y., MOORE, R. B., and LAI, F. S. 1977. Reliability of five methods for protein determination in barley and malt. Am. Soc. Brew. Chem. J. 35:86-93.
597. POOCK, S. J., and BELL, W. 1982. Multiple applications of NIRA in the corn wet milling industry. Proc. NIRA Symp., 2nd. Tarrytown, NY, July.
598. POOCK, S. J., KISER, D. L., and LAST, W. F. 1981. Near infrared reflectance analysis of several corn wet milling and distillery products. (Abstr.) Cereal Foods World 26:490.
599. POPINEAU, Y., and GODON, B. 1976a. The effect of sampling and grinding times on results from infrared reflectance analysis. Technicon Ind. Syst. News 2:9-10.
600. POPINEAU, Y., and GODON, B. 1976b. Testing of the Technicon InfraAlyzer. (In French) Tech. Ind. Cerealieres 152:7-12.
601. POSNER, E. S., and WETZEL, D. L. 1986. Control of flour mills by NIR on-line monitoring. Assoc. Oper. Millers Bull., April, pp. 4711-4722.
602. POWELL, L., and HIEFTJE, G. M. 1978. Computer identification of infrared spectra by correlation-based file searching. Anal. Chim. Acta 100:313-327.
603. PRUETT, R. E., and LAMBRIGHT, L. 1979. NIR analysis of cottonseed products: A feasibility study. (Abstr.) Cereal Foods World 24:455-456.
604. PRUETT, R., LAMBRIGHT, L., and CAVANAGH, G. C. 1978. Measurement of cottonseed constituents by use of a new near infrared reflectance instrument featuring an optical system of advanced design. (Abstr.) Cereal Foods World 23:462.
605. PSOTKA, J. 1982. Predicting the degree of gelatinization in processed food ingredients by NIRA. Proc. NIRA Symp., 2nd. Tarrytown, NY, July.
606. PSOTKA, J. J. 1983. Estimation of free fatty acid levels in crude vegetable oils using NIRA. (Abstr.) Cereal Foods World 28:576.
607. PSOTKA, J. J., and BAHL, J. 1982. Measurement of saccharide distribution in high fructose corn syrup by near-infrared spectroscopy. (Abstr.) Cereal Foods World 27:469.
608. PSOTKA, J. J., and CAREW, W. 1981. Use of near-infrared reflectance for moisture measurement in extruded ready-to-eat cereals. (Abstr.) Cereal Foods World 26:490.
609. PSOTKA, J. J., and RITTGERS, M. 1981. Prediction of degree of gelatinization in extruded ready-to-eat cereals by near-infrared reflectance. (Abstr.) Cereal Foods World 26:490.
610. RADER, B. R. 1966. Moisture in dried vegetables and spices. J. Assoc. Off. Anal. Chem. 49:726-730.
611. RAO, V. R., BRACH, E. J., and MACK, A. R. 1978. Crop discriminability in the visible and NIR regions. Photogramm. Eng. Remote Sensing 44:1179-1184.
612. REDMAN, D. G. 1980. Recent research on ash content. Flour Milling Baking Res. Assoc. Bull. 6:276-284.
613. REDSHAW, E. S., MATHISON, G. W., MILLIGAN, L. P., and WEISENBERGER, R. D. 1986. Near infrared reflectance spectroscopy for predicting forage composition and voluntary consumption and digestibility in cattle and sheep. Can. J. Anim. Sci. 66:103-115.
614. RESSLER, N., and ZIAUDDIN, C. 1973. An improved technique for the near infrared study of water binding by proteins. Illustration with gelatin. Biochim. Biophys. Acta 295:30-36.
615. RICE, G. G. 1978. An economical and practical evaluation of an infrared reflectance instrument for the segregation of wheat by protein content at two Kansas country elevators. (Abstr.) Cereal Foods World 23:461.
616. RINGLUND, K. 1980. Use of near-infrared reflectance spectroscopy for the determination of total nitrogen. (In Norwegian) Nord. Jordbrugsforsk. 62:273-275.
617. RINNE, R. W., GIBBONS, S., BRADLEY, J., SIEF, R., and BRIM, C. A. 1975. Soybean protein and oil percentages determined by infrared analysis. U.S. Dep. Agric. Agric. Res. Publ. ARC-NC-26.
618. ROBERTS, G. M. 1977. Infra-red analysis of milk powders and casein products. Pacific Scientific Tech. Pap. NIR 4007.
619. ROBERTS, G. 1980. Near infrared analysis in the confectionery industry. Neotec Ind. News Items 12.
620. ROBERTSON, J. A., and BARTON, F. E., II. 1984. Oil and water analysis of sunflower seed by near-infrared reflectance spectroscopy. J. Am. Oil Chem. Soc. 61:543-547.
621. ROBERTSON, J. A., and WINDHAM, W. R. 1981. Comparative study of methods of determining oil content of sunflower seed. J. Am. Oil Chem. Soc. 58:993-996.
622. ROBERTSON, N. H., DIXON, A., NOWERS, J. H., and BRINK, D. P. S. 1981. The influence of lipolysis, pH and homogenization on infrared readings for fat, protein, and lactose (in milk). S. Afr. J. Dairy Technol. 13:3-7.
623. ROSE, J. 1982. Quantitative and qualitative analysis with NIRA. Proc. NIRA Symp., 2nd. Tarrytown, NY, July.
624. ROSE, J. J., PRUSIK, T., and MARDEKIAN, J. 1982. Near infrared multi-component analysis of parenteral products using the InfraAlyzer 400. J. Parenter. Sci. Technol. 36:71-78.
625. ROSENTHAL, R. D. 1971. Introducing the Grain Quality Analyzer. Presentation at Annu. Meet. Kansas Wheat-growers' Assoc., Hutchinson.
626. ROSENTHAL, R. D. 1972. The Grain Quality Analyzer. A rapid and accurate means of determining the percent moisture, oil and protein in grain and grain products. Text of lecture given at Am. Assoc. Cereal Chem., Milling and Baking Division, Spring Meeting, April, Denver.
627. ROSENTHAL, R. D. 1973b. The Grain Quality Analyzer. Pac. Northwest Grain Inspect. Workshop, 3rd. Portland, April.
628. ROSENTHAL, R. D. 1979. Introduction to quantitative analysis in the near infrared. (In Hungarian) Elelmez. Ipar 33:371-382.
629. ROSENTHAL, R. D., HAYDEN, R., and ENGLER, S. 1976. A highly flexible 9.0 OD light transmittance/reflectance research spectrophotometer. Pages 16-19, 23 in: Quality Detection in Foods. J. J. Gaffney, ed. ASAE Publ. 1-76.
630. ROSENTHAL, R. D., and ROSENTHAL, S. B. 1981. Apparatus for near infrared quantitative analysis. U.S. patent 4,286,327. 25 August.
631. ROSENTHAL, R. D., and WEBSTER, D. R. 1973. On-line system sorts fruit on basis on internal quality. Food Technol. 27:52-60.
632. ROTH, R. 1984. Analysis by infrared reflection in the brewery. Cerevisin 9:59-62, 65-66.
633. ROTOLO, P. 1978. NIR reflectance applied to baking. Baker's Dig. 52:24-29, 36.
634. ROTOLO, P. 1979. Near infrared reflectance instrumentation. Cereal Foods World 24:94-98.
635. ROTOLO, P., and JUDGE, J. 1976. Application of infrared reflectance to flour. (Abstr.) Cereal Foods World 21:439-440.
636. ROTOLO, P., and PIRRITANO, R. 1977. Practical considerations for accurate protein calibrations. (Abstr.) Cereal Foods World 22:461.
637. ROY, E. B., and BAHL, J. 1982. Theoretical basis for the analysis of oils and fats by NIRA. Proc. NIRA Symp., 2nd. Tarrytown, NY, July.
638. RUBENTHALER, G. L., and BRUINSMA, B. L. 1978. Lysine estimation in cereals by near infrared reflectance. Crop Sci. 18:1039-1042.
639. RUDZIK, L. 1981. Remarks on determination of fat, protein and lactose in milk by methods of IR-spectrophotometry. (In German) Molk.-Ztg. (Berlin) 35:1321-1322, 1324.
640. RUDZIK, L., and WOBBECKE, R. 1982. Need for homogenization in infrared measurements on milk. (In German) Molk.-Ztg. (Berlin) 36:298, 307.
641. SABBAGH, M., and LARSEN, S. 1978. Wool yield by infrared reflectance analysis. Proc. Am. Soc. Anim. Sci. (West. Sect.)

29:108-109.

642. SAINT, K. J. 1985. Use of NIR equipment by the Australian Barley Board. Pages 171-172 in: NIR 84 Proc. Int. Symp. Near Infrared Reflectance Spectrosc., 1st. RACI, Cereal Chem. Div., Melbourne, Australia.
643. SALIMOV, M. A. 1981. Low-frequency infrared spectra of some disaccharides and determination of the type of glycoside bonds in them. (In Russian) Nauchn. Tr. Mosk. Lesotekh. Inst. 134:116-118.
644. SALMON, J., DORTON, B., and WETZEL, D. L. 1981. Reference data for moisture in dough by NIR. (Abstr.) Cereal Foods World 26:491.
645. SAMPSON, D. R., FLYNN, D. W., and JUI, P. 1983. Genetic studies on kernel hardness in wheat using grinding time and near infrared reflectance spectroscopy. Can. J. Plant Sci. 63:825-832.
646. SATO, T., IWAMOTO, M., HASHIZUME, K., YOSHINO, M., FURUKAWA, S., SOMEYA, Y., and YANO, N. 1985. Near infrared spectrophotometric analysis for measuring major constituents of raw milk. Nippon Chikusan Gakkai-Ho 56:878-882.
647. SAURER, W. 1978. Use of infra-red reflectance measurement for the determination of protein and moisture content and kernel hardness in wheat. (In German) Getreide Mehl Brot 32:272-276.
648. SAVITZKY, A., and GOLAY, M. J. E. 1964. Smoothing and differentiation of data by simplified least squares procedures. Anal. Chem. 36:1627-1638.
649. SAWAI, A., KONDO, T., and ARA, S. 1984. Application of near-infrared reflectance analysis to a forage quality test. Nippon Sochi Gakkai-Shi 30:204-206.
650. SCHAFER, K., and WEINARD, W. 1978. Infrared reflectance spectroscopy: A new method for simultaneous rapid determination of protein, fat and moisture. (In German) Muehle Mischfuttertechnik 115:240-243.
651. SCHMUETZ, W., and DO, Q. N. 1982. Improved fast protein determination in barley by IR reflectance spectroscopy. (In German) Muehle Mischfuttertechnik 119:258-261.
652. SCHORCH, K. H. 1983. Potential for production in wheat flour mills using infrared spectroscopy. (In German) Getreide Mehl Brot 37:23-25.
653. SCOTT, K. 1981. Measuring moisture: Infra-red method comes to the rescue. Control Instrum. 13:41-43.
654. SCOTT, R. E., and ROBERTS, E. M. 1978. Objective measurement of clean fleece weight by infrared reflectance spectroscopy. Wool Technol. Sheep Breed. 27 (6/7).
655. SHENK, J. S. 1977. Infrared grain and forage analyses promising. Forage Grassl. Prog. 18:1-2.
656. SHENK, J. S. 1978. Instant analysis of forage and grain quality. Crops Soils 30:708.
657. SHENK, J. S., and BARNES, R. F. 1977. Current status of infrared reflectance. Pages 57-62 in: Proc. South. Pasture Forage Crop Improve. Conf., 34th. Auburn, AL.
658. SHENK, J. S., and HOOVER, M. R. 1977. Infrared reflectance spectrocomputer design and application. Pages 122-125 in: Advances in Automated Analysis. Technicon International Congress (1976), Vol. 2. Industrial Symposia, Mediad.
659. SHENK, J. S., HOOVER, M. R., and WESTERHAUS, M. O. 1979. Infrared analysis of forages and feeds. Pacific Scientific Tech. Pap. NIR 2006.
660. SHENK, J. S., LANDA, I., HOOVER, M. R., and WESTERHAUS, M. O. 1981. Description and evaluation of a near infrared reflectance spectrocomputer for forage and grain analysis. Crop Sci. 21:355-358.
661. SHENK, J. S., MASON, W. N., RISIUS, M. L., NORRIS, K. H., and BARNES, R. F. 1976. Application of infrared reflectance analysis to feedstuff evaluation. Pages 242-248 in: Proc. Int. Symp. Feed Composition Anim. Nutr. Require. Comput. Diets, 1st. P. V. Fonnesbeck, L. E. Harris, and L. C. Kearl, eds. Utah Agric. Exp. Stn., Logan.
662. SHENK, J. S., NORRIS, K. H., BARNES, R. F., and FISSEL, G. W. 1977. Forage and feedstuff analysis with infrared reflectance spectro/computer system. Pages 454-463 in: Int. Grassl. Symp., 13th. Leipzig, GDR.
663. SHENK, J. S., and WESTERHAUS, M. O. 1984. Transfer of NIR spectra and equations. Proc. Annu. Users Conf. NIR Res., 3rd. Pacific Scientific, Rockville, MD.
664. SHENK, J. S., WESTERHAUS, M. O., and HOOVER, M. R. 1978. Infrared reflectance analysis of forages. Pages 242-244 in: Proc. Int. Grain Forage Harvest. Congr. ASAE, St. Joseph, MI.
665. SHENK, J. S., WESTERHAUS, M. O., and HOOVER, M. R. 1979. Analysis of forages by infrared reflectance. J. Dairy Sci. 62:807-812.
666. SHENK, J. S., WESTERHAUS, M. O., HOOVER, M. R., MAYBERRY, K. M., and GOERING, H. K. 1978. Predicting forage quality by infrared reflectance spectroscopy. Pacific Scientific Tech. Pap. NIR 2009.
667. SHENK, J. S., WESTERHAUS, M. O., and TEMPLETON, W. C., Jr. 1985. Calibration transfer between near infrared reflectance spectrophotometers. Crop Sci. 25:159-161.
668. SHERRY, N. B. 1982. Rapid, non-destructive analysis of hops for alpha acids and moisture. (Abstr.) Cereal Foods World 27:462.
669. SHERRY, N. B., and MELAND, P. 1982. NIRA—a creative alternative for soluble protein analysis in malt extracts. (Abstr.) Cereal Foods World 27:462.
670. SIMIC, V., and MARINKOVIC, Z. 1984. Optical ceramics for the near infrared. Naucno-Tekh. Pregled 34:38-46.
671. SIMMONS, C. T., and McDONALD, C. E. 1985. A near infrared reflectance method for the glycolipid fraction in wheat flour. (Abstr.) Cereal Foods World 30:566.
672. SIMMONS, J. G., FERNANDEZ, C. J., WADSWORTH, J. I., and BERARDI, L. C. 1976. Rapid measurement of oil, protein, and moisture in cotton seed and cottonseed meals. Oil Mill Gazet. 81:7-9.
673. SKLENSKY, A. F., ANDERSON, J. H., Jr., and WICKERSHEIM, K. A. 1967. A method for obtaining improved diffuse reflectance spectra in the near infrared. Appl. Spectrosc. 21:339-342.
674. SLACK-SMITH, T., FONG, D., and DOUGLAS, S. A. S. 1979. The potential application of near infrared reflectance to estimate the alcohol extractable matter content of scoured wool. J. Text. Inst. 70:33.
675. SMALL, D. M., and HARRY, S. P. 1985. Introduction of protein testing by NIR for the segregation of Australian hard wheat in South Australia. Page 126 in: NIR 84 Proc. Int. Symp. Near Infrared Reflectance Spectrosc., 1st. RACI, Cereal Chem. Div., Melbourne, Australia.
676. SMITH, D. B., and STARR, C. 1986. Near infrared reflectance analysis in plant breeding. Anal. Proc. (London) 23:125-126.
677. SMOCZKIEWICZOWA, A., WIENSKOWSKI, K., and ZALEWSKI, R. 1976. Application of IR spectroscopy to investigations of changes in vegetable oils during heating. (In Polish) Chem. Anal. (Warsaw) 21:891-896.
678. SPIES, R. D. 1982. Prediction of flour baking potential with NIR. (Abstr.) Cereal Foods World 27:462.
679. SPIES, R. D. 1983. The use of NIR spectroscopy of flour-water doughs to predict mixing time. (Abstr.) Cereal Foods World 28:575.
680. STALNAKER, J. B., LEVIN, J. A., and WETZEL, D. L. 1983. NIR analysis of Kansas hard white winter wheat. (Abstr.) Cereal Foods World 28:575.
681. STALNAKER, J. B., and WETZEL, D. L. 1980. NIR analysis of U.S. commercial bread flours. (Abstr.) Cereal Foods World 25:530.
682. STARK, E. 1979. Instrumentation requirements for near infrared reflectance composition analysis. Fed. Anal. Chem. Spectrosc. Soc. Meet., Philadelphia, September.
683. STARK, E. 1980. NIRA instrumentation state of the art. Proc. NIRA Symp., 2nd. Tarrytown, NY, July.
684. STARK, E. 1982a. Infrared without spectroscopists: Near infrared reflectance analysis (NIRA) by correlation transform spectroscopy. Fed. Anal. Chem. Spectrosc. Soc. Meet., Philadelphia, September.
685. STARK, E. 1982b. The design and development of a computerized NIR research scanning machine. Eur. Technicon NIRA Symp., London, October.
686. STARK, E. 1982c. How the InfraAlyzer 500 makes the math easy for the operator. Eur. Technicon NIRA Symp., London, October.
687. STARK, E. 1983. Correlation transform spectroscopy. Pittsburgh Conf., Atlantic City, March.
688. STARK, E., and HIRSCHFELD, T. 1980. Agrochemical analysis via NIRA. Pittsburgh Conf., Atlantic City, March.
689. STARK, E., HONIGS, D., and MARK, H. 1983a. Optimized wave-length search and calibration techniques for NIRA. (Abstr.) Cereal Foods World 28:571.
690. STARK, E., HONIGS, D., and MARK, H. 1983b. Minimizing reference sample analysis requirements for NIRA methods development. (Abstr.) Cereal Foods World 28:571.

691. STARK, E., JACKSON, S., MARK, H., and HONIGS, D. 1982. New data analysis techniques for near infrared reflectance analysis—(NIRA). (Abstr.) Cereal Foods World 27:463.
692. STARK, E., JUDGE, J., and HIRSCHFELD, T. 1980a. High performance NIRA spectrometer. Fed. Anal. Chem. Spectrosc. Soc. Meet., Philadelphia, September.
693. STARK, E., JUDGE, J., and HIRSCHFELD, T. 1980b. A new spectrophotometer for near infrared reflectance analysis (NIRA). Pittsburgh Conf., Atlantic City, March.
694. STARK, E., LIPSCHITZ, V., and RACHLIS, R. 1980. Performance of near infrared reflectance filter photometers. Pittsburgh Conf., Atlantic City, March.
695. STARK, E., LUCHTER, K., and HIRSCHFELD, T. 1981. Near infrared reflectance analysis of dairy products. Fed. Anal. Chem. Spectrosc. Soc. Meet., Philadelphia, September.
696. STARK, E., and MARK, H. 1978. Evaluation of a new approach to detection of near infrared reflected energy in commercially available instrumentation. (Abstr.) Cereal Foods World 23:470.
697. STARK, E., and MARK, H. 1982. A comparison of photometric data transformations for near infrared reflectance analysis (NIRA). Pittsburgh Conf., Atlantic City, March.
698. STARR, C., MORGAN, A. G., and SMITH, D. B. 1981. An evaluation of near-infrared reflectance analysis in some plant breeding programmes. J. Agric. Sci. 97:107-118.
699. STARR, C., SMITH, D. B., BLACKMAN, J. A., and GILL, A. A. 1983. Applications of near infrared reflectance analysis in breeding wheats for bread-making quality. Anal. Proc. (London) 20:72-74.
700. STARR, C., SUTTLE, J., MORGAN, A. G., and SMITH, D. B. 1985. A comparison of sample preparation and calibration techniques for the estimation of nitrogen, oil and glucosinolate content of rapeseed by near infrared spectroscopy. J. Agric. Sci. 104:317-323.
701. STEEL, P., and BELL, C. 1983. Multicomponent meat measurement. Food Flavourings Ingred. Process. 5:45-46.
702. STERMER, R. A., POMERANZ, Y., and McGINTY, R. J. 1977. Infrared reflectance spectroscopy for estimation of moisture of whole grain. Cereal Chem. 54:345-351.
703. STERMER, R. A., WATSON, C. A., and DIKEMAN, E. 1977. Infrared spectra of milled rice. Trans. ASAE Pap. 76-3030:547-549, 557.
704. STEVENS, D. J., and STEWART, B. A. 1982. Twenty-fifth edition of a revised classification of home-grown wheat varieties according to their value for milling and baking purposes. Flour Milling Baking Res. Assoc., Chorleywood, England.
705. STEVENSON, S. G., and WILLIAMS, P. C. 1978. The determination of protein, moisture and oil in diverse cereals, cereal products and oilseeds by NIRS. (Abstr.) Cereal Foods World 23:462.
706. STEWART, J. E. 1970. Infrared Spectroscopy. Marcel Dekker, New York.
707. STOPFORD, S., HILLIER, G., and CLUCAS, A. 1985. Scanning NIR-prediction of α-amylase in wheat samples. Pages 173-174 in: NIR 84 Proc. Int. Symp. Near Infrared Reflectance Spectrosc., 1st. RACI, Cereal Chem. Div., Melbourne, Australia.
708. STORDAHL, B. 1977. InfraAlyzer proves best apparatus for determining protein for small grain business. Pages 137-139 in: Advances in Automated Analysis. Technicon International Congress (1976), Vol. 2. Industrial Symposia, Mediad.
709. STORY, A. G. 1976. Spectral reflectance of light and infrared radiation by potatoes, stones and soil clods. Pages 83-85 in: Quality Detection in Foods. J. J. Gaffney, ed. ASAE Publ. 1-76.
710. STORY, A. G., and RAGHAVEN, G. S. V. 1976. Sorting potatoes from stones and soil clods by infrared reflectance. Pages 177-182 in: Quality Detection in Foods. J. J. Gaffney, ed. ASAE Publ. 1-76.
711. STREETER, G. 1982. The advantages of NIRA for food process control and quality assurance. Proc. NIRA Symp., 2nd. Tarrytown, NY, July.
712. SURLES, T., and ERICKSON, J. 1974. Near infrared spectroscopy: Instrumentation and applications. Varian Instrum. Appl. 8:5-7.
713. SURLES, T., ERICKSON, J. O., and PRIESNER, D. 1975. Practical aspects of diffuse reflectance spectroscopy. Am. Lab. 7:55-56, 58-64.
714. SURLES, T., and PREISNER, D. 1975. Diffuse reflectance spectroscopy. Varian Instrum. Appl. 9:4-7.
715. SUSI, H., and ARD, J. S. 1966. Near infrared investigation of interamide hydrogen bonds in hydrogen-bonding solvents. Arch. Biochem. Biophys. 117:147-153.
716. SUZUKI, K., and McDONALD, C. E. 1984. Near infrared reflectance (NIR) analysis on bread. (Abstr.) Cereal Foods World 29:511.
717. SUZUKI, T., IWAMOTO, M., CHO, R. K., UOZUMI, J., and IMAI, T. 1984. Applicability of near infrared reflectance method to moisture, protein and ash measurements of buckwheat flours. Nippon Shokuhin Kogyo Gakkai-Shi 31:200-202.
718. SVENSSON, G. 1981. Varietal and environmental effects on wheat milling quality. Agric. Hortic. Genet. 39:1-103.
719. SWIFT, J. R. 1971. Moisture measurement in the food industry: An infrared absorption technique. Food Trade Rev. 41:25-26.
720. TAKEUCHI, T., TSUGE, S., and SUGIYAMA, Y. 1969. Near-infrared spectrophotometric analysis of ethylene-propylene copolymers. Anal. Chem. 41:184-186.
721. TANDON, S. P., and GUPTA, J. P. 1970. Comparative assessment of the effectiveness of different diffuse reflectance parameters in absorption studies. Indian J. Pure Appl. Phys. 8:362-363.
722. TATTERSALL, A. K. 1985. Transferring near infrared reflectance calibrations: Evaluation of protein in wheat. Pages 28-29 in: NIR 84 Proc. Int. Symp. Near Infrared Reflectance Spectrosc., 1st. RACI, Cereal Chem. Div., Melbourne, Australia.
723. TEMPLETON, W. C., Jr., SHENK, J. S., NORRIS, K. H., FISSEL, G. W., MARTEN, G. C., ELGIN, J. H., Jr., and WESTERHAUS, M. O. 1978. Forage analysis with near-infrared spectroscopy—current status and outline of national research project. Pages 528-531 in: Proc. Int. Grassl. Congr., 14th. J. A. Smith and V. W. Hays, eds. Westview Press, Boulder, CO.
724. THOMASOW, J., and PASCHKE, M. 1981. Determination of fat, protein and lactose in milk by infrared absorption using the Multispec instrument. (In German) Milchwissenschaft 36:65-68.
725. THOMPSON, B. N., WILLIAMS, P. C., McLAY, G. W., KENT, W., and WETZEL, D. 1979. Flour quality control by near-infrared reflectance spectroscopy. (Abstr.) Cereal Foods World 24:460.
726. TINCHER, W. C., CASEY, P. G., and MAYFIELD, A. R. 1986. Characterization of heat set nylon carpet yarn. Text. Chem. Color. 18:25-30.
727. TINCHER, W. C., and LUK, A. 1985. NIRS analysis of cotton/polyester yarns. Text. Chem. Color. 17:200-204.
728. TKACHUK, R. 1979. Whole kernel analysis with a computerized double beam Cary 171 spectrophotometer. (Abstr.) Cereal Foods World 24:456.
729. TKACHUK, R. 1980. Nondestructive analysis of cereals and oilseeds by near infrared reflectance. (Abstr.) Cereal Foods World 25:527.
730. TKACHUK, R. 1981a. Oil and protein analysis of whole rapeseed kernels by near infrared reflectance spectroscopy. J. Am. Oil Chem. Soc. 58:819-822.
731. TKACHUK, R. 1981b. Protein analysis of whole wheat kernels by near infrared reflectance. Cereal Foods World 26:584-587.
732. TKACHUK, R. 1984. The effect of temperature on NIR analytical results. (Abstr.) Cereal Foods World 29:510.
733. TKACHUK, R., and ANTONISZYN, J. 1985. Protein, moisture and chlorophyll analysis with a six-filter VIS/NIR instrument. (Abstr.) Cereal Foods World 30:566.
734. TKACHUK, R., and KUZINA, F. D. 1978. Sulfur as a proposed near infrared reflectance standard. Appl. Opt. 17:2817-2820.
735. TKACHUK, R., and KUZINA, F. D. 1982. Chlorophyll analysis of whole rapeseed kernels by near infrared reflectance. Can. J. Plant Sci. 62:875-884.
736. TKACHUK, R., and LAW, D. P. 1978. Near infrared diffuse reflectance standards. Cereal Chem. 55:981-995.
737. TKACHUK, R., MELLISH, V. J., and LOWE, S. 1985. Prediction of amino acid levels in field peas by NIR diffuse reflectance spectroscopy. (Abstr.) Cereal Foods World 30:566.
738. TOOHEY, F. S. 1985. Testing of dried milk products by NIR. Pages 114-118 in: NIR 84 Proc. Int. Symp. Near Infrared Reflectance Spectrosc., 1st. RACI, Cereal Chem. Div., Melbourne, Australia.
739. TREVIS, J. E. 1974. Seven automated instruments. Cereal Sci. Today 19:182-190, 195.
740. TUNNELL, D. A. 1981. Near infrared bibliography. Technicon Ind. Div., Basingstoke, UK.
741. VALDES, E. V., and SUMMER, J. D. 1986. Determination of crude protein and fat in carcass and breast muscle samples of

poultry by near infrared reflectance spectroscopy. Poult. Sci. 65:485-490.

742. VAN DEN AKKER, J. A. 1968. Theory of some of the discrepancies observed in application of the Kubelka Munk equations to particulate systems. Pages 27-47 in: Reflectance Spectroscopy. W. W. Wendtland, ed. Plenum Press, New York.
743. VANDER MEULEN, D. L., and RESSLER, N. 1980. A near-infrared analysis of water-macromolecule interactions: Hydration and the spectra of aqueous solutions of intact proteins. Arch. Biochem. Biophys. 199:197-205.
744. VAN LONKHUYSEN, H. J. 1984. Some experience with the InfraAlyzer 500 on feeding stuffs. Pages 103-105 in: Proc. World Cereal Bread Congr., 11th. Vienna.
745. VERBIESE, Y. 1981. Continuous moisture determination by infrared absorption in the agricultural and food industries. (In French) Ind. Agric. Aliment. 98:759-768.
746. VOLIN, H. 1982. NIRA and its advances to date. Proc. NIRA Symp., 2nd. Tarrytown, NY, July.
747. VORNHEDER, P. F., and BRABBS, W. J. 1970. Moisture determination by near infrared spectrometry. Anal. Chem. 42:1454-1456.
748. WAKAI, Y., INOUE, Y., NISHIKAWA, Y., MURATA, J., and MIURA, T. 1984. Determination of various properties of sake and rice using the near-infrared reflectance analyzer. (In Japanese) J. Brew. Soc. Jpn. 79:445.
749. WALTL, K. 1982. The determination of protein content in grain and mill products. (In German) Allg. Muehlen-Markt 83:264-267.
750. WARD, R. G., SMITH, G. S., WALLACE, J. D., URQUHART, N. S., and SHENK, J. S. 1982. Estimates of intake and quality of grazed range forage by near infrared reflectance spectroscopy. J. Anim. Sci. 54:399-402.
751. WASHBURN, W. H., and MAHONEY, M. J. 1958. Near infrared spectra with sodium chloride optics. Appl. Spectrosc. 12:127-128.
752. WATADA, A. E., and NORRIS, K. H. 1978. Quality of fresh commodities estimated by spectrophotometric technique. Pages 648-653 in: Encyclopedia of Food Technology and Food Science Series. Vol. 3. Peterson and Johnson, eds. AVI Publishing Co., Westport, CT.
753. WATADA, A. E., NORRIS, K. H., WORTHINGTON, J. T., and MASSIE, D. R. 1976. Estimation of chlorophyll and carotenoid contents of whole tomato by light absorbance technique. Food Sci. 41:329-332.
754. WATSON, C. A. 1976. Factors that affect the particle size of samples prepared for near infrared instruments. Technicon Ind. Syst. News 2:6-7.
755. WATSON, C. A. 1977. Near infrared reflectance spectrophotometric analysis of agricultural products. Anal. Chem. 49:835a-840a.
756. WATSON, C. A., CARVILLE, D., DIKEMEN, E., DAIGGER, G., and BOOTH, G. D. 1976. Evaluation of two infrared instruments for determining protein content of hard red winter wheat. Cereal Chem. 53:214-222.
757. WATSON, C. A., ETCHEVERS, G., and SHUEY, W. C. 1976. Relation between ash and protein contents of flour mill streams determined with the InfraAlyzer and by standard approved methods. Cereal Chem. 53:803-804.
758. WATSON, C. A., and SHUEY, W. C. 1976. Observations of infrared reflectance measurements of wheat and flour. (Abstr.) Cereal Foods World 21:439.
759. WATSON, C. A., SHUEY, W. C., BANASIK, O. J., and DICK, J. W. 1977. Effect of wheat class on near infrared reflectance. Cereal Chem. 54:1264-1269.
760. WEBSTER, D. 1976. Theory of near-infrared quantitative measurements. Pacific Scientific Tech. Pap. NIR 6002.
761. WEBSTER, D. R. 1982. Developing NIR test methods for analyzing constituents in materials. Fed. Anal. Chem. Spectrosc. Soc. Meet., Philadelphia, September.
762. WEBSTER, D. R. 1985. Methods of calibration verification without existing prediction sample sets. Page 30 in: NIR 84 Proc. Int. Symp. Near Infrared Reflectance Spectrosc., 1st. RACI, Cereal Chem. Div., Melbourne, Australia.
763. WEINARD, W. 1981. New opportunities for protein determination in small and medium-sized mills. (In German) Muehle Mischfuttertechnik 118:482-483.
764. WETZEL, D. L. 1977. Particle size as a variable in near infrared reflectance analysis. (Abstr.) Cereal Foods World 22:461-462.
765. WETZEL, D. L. 1979. Contributions of constituents of components to practical NIR measurements of complex mixtures with a grain matrix. (Abstr.) Cereal Foods World 24:461.
766. WETZEL, D. L. 1981. Near infrared (diffuse) reflectance analysis: State of the art. Cereal Foods World 26:306.
767. WETZEL, D. L. 1982a. Near infrared (diffuse) reflectance analysis: Grating and interferometric instruments. Cereal Foods World 27:415.
768. WETZEL, D. L. 1982b. Quality measurement in the processing of agricultural products. Proc. NIRA Symp., 2nd. Tarrytown, NY, July.
769. WETZEL, D. L. 1983a. Near-infrared reflectance analysis. Anal. Chem. 55:1165A-1176A.
770. WETZEL, D. L. 1983b. Timely analysis and computation for control. Cereal Foods World 28:256.
771. WETZEL, D. L. 1983c. Lipid structures and NIRA. Proc. Int. Symp. NIRA, 4th. Technicon Ind. Syst., Tarrytown, NY.
772. WETZEL, D. L. 1984a. On-line monitoring for control of process. Proc. Int. Symp. NIRA, 7th. Technicon Ind. Syst., Tarrytown, NY.
773. WETZEL, D. L. 1984b. Physical sample characterization by granulation sorting from diffuse reflectance measurements in the near-infrared. Proc. Annu. Users Conf. NIR Res., 3rd. Pacific Scientific, Rockville, MD.
774. WETZEL, D. L., and ALEXANDER, R. J. 1979. Prospects of dry corn milling Q.C. using a later generation NIR analyzer. (Abstr.) Cereal Foods World 24:461.
775. WETZEL, D. L., and BRUINSMA, B. 1984. Relation of near-infrared analysis to baking qualities of dough. (Abstr.) Cereal Foods World 29:511.
776. WETZEL, D. L., EGEBERG, R., and MARTENSSEN, A. 1982. Computer post-run calculation and artificial intelligence in NIR analysis. Pap. 314 in: Proc. Cereal Bread Congr., Prague.
777. WETZEL, D. L., and LEVIN, J. A. 1982. Grinding optimization for NIR wheat hardness field testing. (Abstr.) Cereal Foods World 27:462.
778. WETZEL, D. L., and MARK, H. 1977. Spectroscopic cellulose determination as criterion of flour purity with respect to bran. (Abstr.) Cereal Foods World 22:481.
779. WETZEL, D. L., and MARK, H. 1978. Scanning NIR of grains, oil seeds, and their components. (Abstr.) Cereal Foods World 23:469.
780. WETZEL, D. L., MAURO, D., and GOTTNEID, D. 1980. NIR analysis of corn and corn by-products. (Abstr.) Cereal Foods World 25:531.
781. WETZEL, D. L., and WEHLING, R. L. 1980. Individual millstream analysis by NIR. (Abstr.) Cereal Foods World 25:531.
782. WETZEL, D. L., WILLIAMS, P. C., and MARK, H. 1978. NIR flour QC. (Abstr.) Cereal Foods World 23:462.
783. WHEELER, O. H. 1959. Near infrared spectra of organic compounds. Chem. Rev. 39:629-666.
784. WHEELER, O. H. 1960. Near infrared spectra. A neglected field of spectral study. J. Chem. Educ. 37:234-236.
785. WHETSEL, K. 1968. Near infrared spectrophotometry. Appl. Spectrosc. Rev. 2:1-67.
786. WHETSEL, K. 1966. Near infrared spectrophotometry. Pages 648-691 in: Encyclopedia of Industrial Chemical Analysis, Vol. 2. F. D. Snell and C. L. Hilton, eds. Interscience, New York.
787. WHITE, C. H., MUCK, G. A., BULTHAUS, M., and ROTOLO, P. 1978. The use of infrared reflectance analyzer for measurement of fat and total solids of dairy products. Annu. Meet. Am. Dairy Sci. Assoc., 73rd. East Lansing, MI, July.
788. WILKS, P. A., Jr. 1979. Quantitative infrared analysis. Int. Lab. 9:49-58.
789. WILLIAMS, P. C. 1973a. The application of the Neotec Grain Quality Analyzer to the analyses of cereal grains and oilseeds. I. Preliminary observations. (Abstr.) Cereal Sci. Today 18:284-285.
790. WILLIAMS, P. C. 1973b. The application of the Neotec Grain Quality Analyzer to the analyses of cereal grains and oilseeds. II. Technical aspects of operation. (Abstr.) Cereal Sci. Today 18:285.
791. WILLIAMS, P. C. 1974a. The application of infra-red reflectance spectroscopy to the analysis of cereal grains. III. The influence of grinding on the analysis of hard red spring wheat and other grains. (Abstr.) Cereal Sci. Today 19:410.
792. WILLIAMS, P. C. 1974b. The application of infra-red reflectance spectroscopy to the analysis of cereal grains. IV. The influence of season and growing environment upon the analysis of hard red spring wheat and other grains. (Abstr.) Cereal Sci. Today 19:410-411.

793. WILLIAMS, P. C. 1974c. The ADA—an automated digital system for the large scale analysis of cereal grains, using infra-red reflectance spectroscopy. Cereal Sci. Today 19:411.

794. WILLIAMS, P. C. 1975. Application of near infrared reflectance spectroscopy to analysis of cereal grains and oilseeds. Cereal Chem. 52:561-576.

795. WILLIAMS, P. C. 1976a. Sampling and sample preparation for infrared reflectance spectroscopy (IRS) analysis of grains. Technicon Ind. Syst. News 2:1, 3-4.

796. WILLIAMS, P. C. 1976b. Further observations on the factors affecting the accuracy of IRS (infrared reflectance spectroscopy) analysis of cereals. (Abstr.) Cereal Foods World 21:440.

797. WILLIAMS, P. C. 1977a. Technical aspects of instituting a protein grading system at a grain elevator, using near infrared reflectance spectroscopy (NIRS). Pages 118-121 in: Advances in Automated Analysis. Vol. 2, Industrial Symposia. Mediad Inc., Tarrytown, NY.

798. WILLIAMS, P. C. 1977b. The economics of near infrared reflectance spectroscopy in testing grain for protein and moisture. Pages 131-136 in: Advances in Automated Analysis. Vol. 2. Industrial Symposia. Mediad Inc., Tarrytown, NY.

799. WILLIAMS, P. C. 1978. Bulk-handling of wheat by protein content in Canada. Pacific Scientific Tech. Pap. NIR 1008.

800. WILLIAMS, P. C. 1979. Screening wheat for protein and hardness by near infrared reflectance spectroscopy. Cereal Chem. 56:169-172.

801. WILLIAMS, P. C. 1981. Choosing grinders for sample analysis. Milling Feed Fert. 164:30-31, 33.

802. WILLIAMS, P. C. 1984. The significance of outliers. Pages 90-92 in: Proc. World Cereal Bread Congr., 11th. Vienna.

803. WILLIAMS, P. C. 1985. New applications for near-infrared reflectance (NIR) spectroscopy. Pages 156-168 in: NIR 84 Proc. Int. Symp. Near Infrared Reflectance Spectrosc., 1st. RACI, Cereal Chem. Div., Melbourne, Australia.

804. WILLIAMS, P. C. 1987. A simple two-sample method for calibration of near-infrared computerized spectrophotometers. Appl. Spectrosc. In press.

805. WILLIAMS, P. C., and CORDEIRO, H. M. 1981. Determination of protein and moisture in hard red spring wheat by near-infrared reflectance spectroscopy. Influence of degrading factors, dockage, and wheat variety. Cereal Foods World 26:124-128.

806. WILLIAMS, P. C., and CORDEIRO, H. M. 1985. Effect of calibration practice on correction of errors induced in near-infrared protein testing of hard red spring wheat by growing location and season. J. Agric. Sci. 104:113-123.

807. WILLIAMS, P. C., JABI EL-HARAMEIN, F., ORTIZ FERERRA, G., and SRIVASTIVA, J. P. 1987. Preliminary studies of the determination of strength in wheat with near infrared reflectance spectroscopy. Unpublished.

808. WILLIAMS, P. C., and KRISCHENKO, V. P. 1986. A comparison of two near-infrared computerized spectrophotometers for the determination of protein in wheat. Appl. Spectrosc. 40:785-790.

809. WILLIAMS, P. C., MacKENZIE, S. L., and STARKEY, P. M. 1985. Determination of methionine in peas by near-infrared reflectance spectroscopy (NIRS). J. Agric. Food Chem. 33:811-815.

810. WILLIAMS, P. C., and NORRIS, K. H. 1979. Qualitative applications of near-infrared reflectance spectroscopy. I. Cartography of absorbing units in the area from 1–2.6 microns. (Abstr.) Cereal Foods World 24:460.

811. WILLIAMS, P. C., and NORRIS, K. H. 1983. Effect of mutual interactions on the estimation of protein and moisture in wheat. Cereal Chem. 60:202-207.

812. WILLIAMS, P. C., NORRIS, K. H., GEHRKE, C. W., and BERNSTEIN, K. 1983. Comparison of near-infrared methods for measuring protein and moisture in wheat. Cereal Foods World 28:149-152.

813. WILLIAMS, P. C., NORRIS, K. H., JOHNSEN, R. L., STANDING, K., FRICIONI, R., MacAFFREY, D., and MERCIER, R. 1978. Comparison of physicochemical methods for measuring total nitrogen in wheat. Cereal Foods World 23:544-547.

814. WILLIAMS, P. C., NORRIS, K. H., and SOBERING, D. C. 1985. Determination of protein and moisture in wheat and barley by near-infrared transmission. J. Agric. Food Chem. 33:239-244.

815. WILLIAMS, P. C., NORRIS, K. H., and ZAROWSKI, W. S. 1982. Influence of temperature on estimation of protein and moisture in wheat by near-infrared reflectance. Cereal Chem. 59:473-477.

816. WILLIAMS, P. C., and PANFORD, J. A. 1979. Use of the Neotec Model 6350 Research Composition Analyzer for optimization of wavelength and mathematical treatment for analysis of hard red spring wheat for protein. (Abstr.) Cereal Foods World 24:455.

817. WILLIAMS, P. C., and PANFORD, J. A. 1980. The development and use of universal calibrations for the measurement of protein and moisture in HRS wheat by near-infrared reflectance spectroscopy (NIR). (Abstr.) Cereal Foods World 25:530.

818. WILLIAMS, P. C., PRESTON, K. R., NORRIS, K. H., and STARKEY, P. M. 1984. Determination of amino acids in wheat and barley by near-infrared reflectance spectroscopy. J. Food Sci. 49:17-20.

819. WILLIAMS, P. C., and SOBERING, D. C. 1986. Attempts at standardization of hardness testing of wheat. II. The near-infrared reflectance method. Cereal Foods World 31:417-420.

820. WILLIAMS, P. C., and STARKEY, P. M. 1980. Influence of feed ingredients upon the prediction of protein in animal feed mixes by near-infrared reflectance spectroscopy. J. Sci. Food Agric. 31:1201-1213.

821. WILLIAMS, P. C., STEVENSON, S. G., and IRVINE, G. N. 1978. Testing wheat for protein and moisture with the Automated Digital Analyzer. Cereal Chem. 55:263-279.

822. WILLIAMS, P. C., STEVENSON, S. G., STARKEY, P. M., and HAWTIN, G. C. 1978. The application of near infrared reflectance spectroscopy to protein testing in pulse breeding programmes. J. Sci. Food Agric. 29:285-292.

823. WILLIAMS, P. C., and THOMPSON, B. N. 1978. Influence of whole meal granularity on analysis of HRS wheat for protein and moisture by near infrared reflectance spectroscopy (NRS). Cereal Chem. 55:1014-1037.

824. WILLIAMS, P. C., THOMPSON, B. N., WETZEL, D., McLAY, G. W., and LOEWEN, D. 1981. Near-infrared instruments in flour mill quality control. Cereal Foods World 26:234-237.

825. WILLIS, H. A., and MILLER, R. G. J. 1958. Diffusion spectroscopy in the near infrared. Conf. Mol. Spectrosc., London, February.

826. WILSON, J. M. 1970. Quantitative determination of the composition of soybean samples by the infrared attenuated total reflectance technique. Ph.D. thesis, University of Maryland (UM 71/2926).

827. WILSON, J. M., BEN-GERA, I., and KRAMER, A. 1971. Quantitative determination of composition of models of food systems by the infrared attenuated total reflectance technique. J. Food Sci. 36:162-165.

828. WILSON, J. M., KRAMER, A., and BEN-GERA, I. 1973. Quantitative determination of fat, protein, and carbohydrate of soy products with infrared attenuated total reflectance. J. Food Sci. 38:14-17.

829. WINCH, J. E., and MAJOR, H. 1981. Predicting nitrogen and digestibility of forages using near infrared reflectance photometry. Can. J. Plant Sci. 61:45-51.

830. WINGFIELD, J. 1979. N.I.R. determination of cellulose as a milling control parameter. Assoc. Oper. Millers Bull., June, pp. 3769-3770.

831. WOOLLARD, D. C. 1985. NIR technology within the New Zealand dairy industry—its dramatic impact. Pages 95-104 in: NIR 84 Proc. Int. Symp. Near Infrared Reflectance Spectrosc., 1st. RACI, Cereal Chem. Div., Melbourne, Australia.

832. WORTHINGTON, J. T., MASSIE, D. R., and NORRIS, K. H. 1976. Light transmission technique for predicting ripening time for intact green tomatoes. Pages 46-49 in: Quality Detection in Foods. J. J. Gaffney, ed. ASAE Publ. 1-76.

833. ZEHNDER, H. J. 1983. European launch of the InfraAlyzer 500. (In German) Alimenta 22:37-38.

834. ZHONG, S. 1982. Direct spectrophotometric determination of total nitrogen in soils. (In Chinese) Fenxi Huaxue 10:191-192.

835. ZWINGELBERG, H. 1983. Infrared spectroscopy—significance and measurement of grain hardness. (In German) Getreide Mehl Brot 37:25-28.

Supplementary References

CHEMISTRY

836. BAIN, O., and GIGUERRE, P. A. 1955. Hydrogen peroxide and its analogues. VI. Infrared spectra of H_2O_2, D_2O_2, and HDO_2. Can. J. Chem. 33:527-545.
837. BAKER, A. W. 1958. Hydroxyl stretching frequencies and chemical reactivities of substituted phenols. J. Phys. Chem. 26:744-747.
838. BAKER, A. W., and SHULGIN, A. T. 1958. Intramolecular hydrogen bonds to electrons and other weakly basic groups. J. Am. Chem. Soc. 80:5358-5363.
839. BARROW, G. M. 1955. The intensities of infrared hydroxyl bands. J. Phys. Chem. 59:1129-1132.
840. BAYLY, J. G., KARTHA, V. B., and STEVENS, W. H. 1963. The absorption spectra of liquid phase H_2O, HDO and D_2O from 0.7 μm to 10 μm. Infrared Phys. 3:211-223.
841. BECKER, E. D., and PIMENTEL, G. C. 1956. Spectroscopic studies of reactive molecules by the matrix isolation method. J. Chem. Phys. 25:225-228.
842. BENEDICT, W. S., GAILOR, N., and PLYLER, E. K. 1956. Rotation-vibration spectra of deuterated water vapor. J. Chem. Phys. 24:1139-1165.
843. BENSON, A. M., Jr., and DRICKAMER, H. G. 1957. Stretching vibrations in condensed systems. Especially bonds containing hydrogen. J. Chem. Phys. 27:1164-1174.
844. BONNER, O. D., and CHOI, Y. S. 1974a. Hydrogen bonding of water in organic solvents. I. J. Phys. Chem. 78:1723-1727.
845. BONNER, O. D., and CHOI, Y. S. 1974b. Hydrogen bonding of water in organic solvents. II. The change of water structure with composition. J. Phys. Chem. 78:1727.
846. BROWN, T. L. 1957a. The integrated intensity of the infrared OH absorption in phenols. J. Phys. Chem. 61:820-821.
847. BROWN, T. L. 1957b. The solvent effect in infrared intensities. Spectrochim. Acta 10:149-153.
848. BROWN, T. L., and ROGERS, M. T. 1957. The integrated intensity of the OH stretching band in aliphatic alcohols. J. Am. Chem. Soc. 79:577-578.
849. BUCKINGHAM, A. D. 1958. Solvent effects in infrared spectroscopy. Proc. R. Soc. London 248:169-182.
850. BULANIN, M. O. 1957. On the structure of absorption bands of liquid H_2O, D_2O and HDO. Determined by their molecular spectra. Opt. Spektrosk. 2:557-561.
851. BURGOVA, M. P., KUCHIERK, Y., and PROSKURINA, L. 1958. Anharmonicity as one of the characteristics of intermolecular interaction. Opt. Spektrosk. 5:141-146.
852. BYERS, F., and SPURR, R. A. 1955. Infrared intensity of the S-H stretching vibration. (Abstr.) Anal. Chem. 27:321.
853. CANNON, C. G. 1955. The interactions and structure of the -CONH- group in amides and polyamides. Mikrochim. Acta 555-588.
854. CANNON, C. G. 1956. Infrared frequency shifts and amide group interactions. J. Chem. Phys. 24:491-492.
855. CANNON, C. G. 1958. The nature of hydrogen bonding—a review of published work and discussions. Spectrochim. Acta 10:341-368.
856. CHAUDHRY, J. S., and MISHRA, R. K. 1970. D_2O equilibration measurement of total water content in biological tissues—a near infrared absorption study in vitro. Stud. Biophys. 20:117-130.
857. CHAUDHRY, J. S., SUBRAMANIAM, M., and MISHRA, R. K. 1971. Near infrared reflectance study of water in the monkey cornea in vitro. Stud. Biophys. 26:191-198.
858. CHIRGADZE, Yu. N., VENYAMINOV, S. Yu., and LABACHEV, V. M. 1971. Optical rotary dispersion of polypeptides in the near infrared region. Biopolymers 10:809-820.
859. COBLENTZ, W. K., and SHENK, J. S. 1979. Characterization of near infrared reflectance of protein and related compounds. Fed. Anal. Chem. Spectrosc. Soc. Meet., Philadelphia, September.
860. CORDES, H. F., and TAIT, C. W. 1957. Determination of water in several hydrazines. Anal. Chem. 29:485-487.
861. CORISH, P. J. 1959. Analysis of *cis*- and *trans*-1:4 contents of polyisoprenes by near infra-red spectroscopy. Spectrochim. Acta 15:598-604.
862. CURCIO, J. A., and PETTY, C. C. 1951. The near infrared absorption spectrum of liquid water. J. Opt. Soc. Am. 41:302-304.
863. DAVIES, M. M. 1940. The infra-red absorptions of organic hydroxyl compounds in the solid state in the 3μ region. J. Chem. Phys. 8:577-587.
864. DEBARRE, D., BOCCARA, A. C., FOURNIER, D., and LEUSCHER, E. 1984. Pages 147-153 in: Photoacoustic Effect: Principles and Applications. Proc. Int. Conf. on Photoacoustic Effect, 1st. Feb. 1981. Bad Honnef, Germany. E. Luscher, ed. F. Vieweg, Braunschweig.
865. DEPECKER, C., SOMBRET, B., and LEGRAND, P. 1985. Diffuse reflectance combined with Fourier transform infrared spectrometry: A new method of analysis. Analysis 13:349-355.
866. ELLIS, J. W., and BATH, J. 1938. Modifications in the near infrared absorption spectra of protein and of light and heavy water molecules when water is bound to gelatin. J. Chem. Phys. 6:723-729.
867. ENS, A., and MURRAY, F. E. 1957. A quantitative infrared spectrometric study of molecular association in alcohols. Can. J. Chem. 35:170-175.
868. FAHRENFORT, J. 1961. Attenuated total reflection. A new principle for the production of useful infra-red reflection spectra of organic compounds. Spectrochim. Acta 17:698-709.
869. FINCH, J. N., and LIPPINCOTT, E. R. 1956. Hydrogen bond systems: Temperature dependence of OH frequency shifts and OH band intensities. J. Chem. Phys. 24:908-909.
870. FINCH, J. N., and LIPPINCOTT, E. R. 1957. Hydrogen bond systems: Temperature dependence of OH frequency shifts and OH band intensities. J. Phys. Chem. 61:894-902.
871. FISHER, J. F. 1984. Reduction of the water interference in the near infrared measurement of alcohols. Pages 29-32 in: Proc. Joint Symp. Instrum. Control in the 80's. Instrument Soc. Am.
872. FISHMAN, E., and DRICKAMEN, H. G. 1956. Effect of pressure on the frequency of the OH band in butanol solutions. J. Chem. Phys. 24:548-553.
873. FLETT, M. St. C. 1957. Studies of the band near 3 μ in some hydroxyl compounds. Spectrochim. Acta 10:21-37.
874. FOSTER, G. N., ROW, S. B., and GRISKEY, R. G. 1964. Infrared spectrometry of polymers in the overtone and combination regions. J. Appl. Polym. Sci. 8:1357-1361.
875. FOX, J. J., and MARTIN, A. E. 1938. Investigations of infra-red spectra. Absorption of the CH_2 group in the region of 3μ. Proc. R. Soc. London A 167:257-281.
876. FOX, J. J., and MARTIN, A. E. 1940. Investigations of infra-red spectra (2.5–7.5μ). Absorption of water. Proc. R. Soc. London A. 174:234-262.
877. FREYMANN, M., and FREYMANN, R. 1941. The identification of homologous or isomeric organic compounds by their near-infrared absorption spectra. (In French) C. R. Hebd. Seances Acad. Sci. 213:174-177.
878. FREYMANN, R. 1938. On the question of the intra- and intermolecular "hydrogen bond" and the absorption spectra in the near-infrared. J. Chem. Phys. 6:497-498.

879. GAFFEY, S. J. 1985. Reflectance spectroscopy in the visible and near-infrared (0.35–2.55): Applications in carbonate petrology. Geology 13:270-273.

880. GAUNT, J. 1956. The analysis of heavy water by infrared spectrometry. Spectrochim. Acta 8:57-65.

881. GEPPERT, G., and KUNZ, M. 1962. Infrarotspektroskopische untersuchungen an kohlenwasserstoffen und ihren derivaten im bereich der kombinationsschwingungen (2.2 bis 2.4 mu). J. Prakt. Chem. 15:244.

882. GLATT, I., and ELLIS, J. W. 1951. Near infrared pleochroism. 2. The 0.8–25 mu region of some linear polymers. J. Chem. Phys. 19:449-457.

883. GOODWIN, S. F., and MITCHNER, M. 1986. Measurements of the near infrared optical properties of coal slags. Chem. Eng. Commun. 44:241-255.

884. GREINACHER, E., LUTTKE, W., and MECKE, R. 1955. Infrared spectroscopy of water in organic solvents. Z. Elektrochem. 59:23-31.

885. GROENWEGE, M. P., and VAN VUCHT, H. A. 1955. Analytical applications of the near-infrared using the Beckman U.V. spectrophotometer. (In German) Mikrochim. Acta 2-3:471-480.

886. HAAS, C., and KETELAAR, J. A. A. 1956. The effective field and the frequency of overtones and combination bands in the vibration spectra of crystals. Physica 22:1286-1290.

887. HAEUSLER, C. 1955. Calibration of the infrared ammonia band at 1.98 μ. C. R. Hebd. Seances Acad. Sci. 240:507-509.

888. HANLON, S., and KLOTZ, I. M. 1967. Near infrared spectroscopy in structural problems of biochemistry. Dev. Appl. Spectrosc. 6:219-235.

889. HEDIGER, H. J., and GUNTHARD, H. H. 1954. Optical rotatory power of organic substances in the infrared region. Helv. Chim. Acta 37:1125-1133.

890. HERMANS, J., Jr., and SCHERAGA, H. A. 1960. Structural studies of ribonuclease. IV. The near infrared absorption of the hydrogen-bonded peptide NH group. J. Am. Chem. Soc. 82:5156-5160.

891. HOFFMAN, K. H., and FISCHER, L. 1955. Determination of small amounts of water in films of organic liquids. Chem. Ing. Tech. 10:604-607.

892. HOLLENBERG, J. L., and IFFT, J. B. 1982. Hydration numbers by near-infrared spectrophotometry. 1. Amino acids. J. Phys. Chem. 86:1938-1941.

893. HONIGS, D. E., HIEFTJE, G. M., and HIRSCHFELD, T. 1982. Fast correlation transform in scanning NIRA. Proc. NIRA Symp., 2nd. Tarrytown, NY, July.

894. HONIGS, D. E., HIEFTJE, G. M., and HIRSCHFELD, T. 1983. Spectral reconstruction in NIRA. Pittsburgh Conf., Atlantic City, March.

895. HONIGS, D. E., HIEFTJE, G. M., and HIRSCHFELD, T. 1984. A new method for obtaining individual component spectra from those of complex mixtures. Appl. Spectrosc. 38:317-322.

896. HONIGS, D. E., HIRSCHFELD, T. B., and HIEFTJE, G. M. 1985. Near-infrared determination of several physical properties of hydrocarbons. Anal. Chem. 57:443-445.

897. HUGGINS, C. M., and PIMENTEL, G. C. 1956. Systematics of the infrared spectral properties of hydrogen bonding systems: Frequency shift half width and intensity. J. Phys. Chem. 60:1615-1619.

898. INGRAHAM, L. L., CORSE, J., BAILEY, G. P., and STITT, F. 1952. Correlation of OH stretching frequencies in phenols and catachols with chemical reactivities. J. Am. Chem. Soc. 74:2297-2299.

899. IWAOKA, T., WANG, S. H., and GRIFFITHS, P. R. 1985. Diffuse reflectance infrared spectrometry of organic materials. Spectrochim. Acta A 41A:37-41.

900. JAFFE, J. H., RANK, D. H., and WIGGINS, T. A. 1955. High resolution in the infrared by interferometry unambiguous display without overlapping orders. J. Opt. Soc. Am. 45:636-643.

901. JOSIEN, M. L., CASTINEL, C., and SAUMAGNE, P. 1957. Determination of infrared spectroscopy of hydrogen bonds formed by the SH vibrator. II. The monohalogen benzenethiols. Bull. Soc. Chim. Fr. 648-650.

902. KARYAKIN, A. V., and PETROV, A. V. 1964. Determination of the water content of tributyl phosphate on the basis of its absorption spectra in the near infrared. J. Anal. Chem. USSR (Engl. Transl.) 19:1146-1149.

903. KARYAKIN, A. V., PETROV, A. V., and GERLIT, Yu. B. 1964. Investigation of the state of water in some ketones by the near infrared absorption spectra. Vodorodnaya Svyaz 1964, p. 181.

904. KLOTZ, I. M., and FRANZEN, J. S. 1962. Hydrogen bonds between model peptide groups in solution. J. Am. Chem. Soc. 84:3461-3466.

905. KOGON, I. C. 1957. Evidence for intramolecular hydrogen bonding in aryl allophanates and alkyl or aryl-alkyl biurets. J. Am. Chem. Soc. 79:2253-2254.

906. KRIKORIAN, S. E., and MAHPOUR, M. 1973. The identification and origin of N-H overtone and combination bands in the near-infrared spectra of simple primary and secondary amides. Spectrochim. Acta 29A:1233-1246.

907. KUNZ, M. H. 1965. Spectroscopic determining of structure groups of hydrocarbons in the near infrared. (In German) Z. Chem. 5:333-338.

908. LAUER, J. L., LAU, K. K. P., and KING, V. W. 1982. Fast remote analysis of organic matter in minerals. Am. Lab. 14:40, 41-43. (December).

909. LIDDEL, U., and KASPER, C. 1933. Spectral differentiation of pure hydrocarbons: A near-infrared absorption study. J. Res. Natl. Bur. Stand. 11:599-618.

910. LIDDEL, U., and WULF, O. R. 1933. The character of the absorption of some amines in the near-infrared. J. Am. Chem. Soc. 55:3574-3583.

911. LOHMAN, F. R., and NORTEMAN, W. E., Jr. 1963. Determination of primary and secondary aliphatic amines by near infrared spectrophotometry. Anal. Chem. 35:707-711.

912. McIVOR, R. A., and HUBLEY, C. E. 1959. Infrared spectra or organophosphorus compounds. III. Phosphoramidates, phosphoramidothionates and related compounds. Can. J. Chem. 37:869-876.

913. McIVOR, R. A., HUBLEY, C. E., GRANT, G. A., and GREY, A. A. 1958. Infrared spectra of organophosphorus compounds. II. The 1–2.6 μ region. Can. J. Chem. 36:820-834.

914. MELDRUM, B. J., ORR, J. C., and ROCHESTER, C. H. 1985. Diffuse reflectance Fourier-transform infrared spectroscopy: In situ surface oxidation of carbon. J. Chem. Soc. Chem. Commun. 17:1176-1177.

915. MENEFEE, A., ALFORD, D., and SCOTT, C. B. 1956. Hydrogen bonding of the thiol group. J. Chem. Phys. 25:370-371.

916. MILLER, R. G., and WILLIS, H. A. 1956. Quantitative analysis in the 2 μ region applied to synthetic polymers. J. Appl. Chem. 6:385-391.

917. MITCHELL, J. A., BOCKMAN, C. D., Jr., and LEE, A. V. 1957. Determination of acetyl content of cellulose acetate by near infrared spectroscopy. Anal. Chem. 29:499-502.

918. MIZUSHIMA, S.-I., SHIMANOUCHI, T., TSUBOI, M., and ARAKAWA, T. 1957. Near infrared spectra of compounds with two peptide bonds and the confirmation of a polypeptide chain. VI. Further evidence of the internal hydrogen bonding and an estimation of its energy. J. Am. Chem. Soc. 79:5357-5361.

919. MIZUSHIMA, S.-I., TSUBOI, M., SHIMANOUCHI, T., and TSUDA, Y. 1955. Spectroscopic investigation of the strength of hydrogen bonds formed by amides. Spectrochim. Acta 7:100-107.

920. PINCHAS, S. 1955. Infrared absorption of the aldehydic C-H group. Anal. Chem. 27:2-6.

921. RAKSHIT, A. K., and SINGHAL, G. S. 1985. Near-infrared spectroscopic study of aquo-organic mixtures. J. Indian Chem. Soc. 62:330-331.

922. RUITER, E. DE. 1962. Eine schnellmethode zur bestimmung wasserstoffverteilung HAr/HA1 aus der absorption im nahen infrarot. Erdoel Kohle 15:616-619.

923. SAPPENFIELD, J. W. 1929. The absorption spectra of certain organic liquids in the near infrared. Phys. Rev. 33:37-47.

924. SAVAGE, C. M., and EDWARD, T. H. 1957. Infrared spectrum of hydrogen sulfide in the 2.6 μ region. J. Chem. Phys. 27:179-181.

925. SCHAUENSTEIN, E., and PUCHNER, H. 1962. The value of the intensity of the OH band at 1.4 μ for quantitative determinations. Monatsh. Chem. 93:253-262.

926. SCHLUETER, D. J., and PECK, E. R. 1958. Refractivity of air in the near infrared. J. Opt. Soc. Am. 48:313-315.

927. STONE, P. J., and THOMPSON, H. W. 1957. Vibrational and band intensity of the hydroxyl group in phenols. Spectrochim. Acta 10:17-20.

928. STUDER, F. J., and CUSANO, D. A. 1953. Titanium dioxide films as selective reflectors of the near-infrared. J. Am. Opt. Soc. 43:522-525.

929. SUHRMANN, R. 1950. Position and form of absorption bands

of liquids in the near infrared. (In German) Angew. Chem. 62:507-512.

930. TERENIN, A. N., YAROSLAVSKY, N. G., KARYAKIN, A. V., and SIDOROVA, A. I. 1955. Infrared spectroscopy of molecules absorbed on porous earth. Mickrochim. Acta 1955:467-470.
931. THIEL, M. W., BECKER, E. D., and PIMENTAL, G. 1957. Infrared studies of hydrogen bonding of water by the matrix isolation technique. J. Chem. Phys. 27:486-490.
932. TOMELLINI, S. A., SAPERSTEIN, D. D., STEVENSON, J. M., SMITH, G. M., WOODRUFF, H. B., and SEELIG, P. F. 1981. Automated interpretation of infrared spectra with an instrument based minicomputer. Anal. Chem. 53:2367-2369.
933. TROTT, G. F., WOODSIDE, E. E., TAYLOR, K. G., and DECK, J. C. 1973. Physico-chemical characterization of carbohydrate-solvent interactions by near-infrared spectroscopy. Carbohydr. Res. 27:415-435.
934. TSUBOMURA, H. 1955. Nature of the hydrogen bond. II. The effect of hydrogen bonding and other solute-solvent interactions upon the shape and intensity of the infrared OH band of phenol. J. Chem. Phys. 23:2130-2133.
935. TSUBOMURA, H. 1956. Nature of the hydrogen bond. III. The measurement of the infrared absorption intensities of free and hydrogen bonded OH bands. Theory of the increase of the intensity due to the hydrogen bond. J. Chem. Phys. 24:927-931.
936. WENDLANDT, W. W., and HECHT, H. G. 1966. Reflectance Spectroscopy. Chemical Analysis Monogr. Ser., Vol. 21. Interscience, New York.
937. WEYER, L. G. 1985. Near-infrared spectroscopy of organic substances. Appl. Spectrosc. Rev. 21:1-43.
938. WHEELER, O. H. 1959. Near infrared spectra of organic compounds. Chem. Rev. 59:629-666.
939. WHETSEL, K. B., ROBERSON, W. E., and KRELL, M. W. 1958a. Near infrared analysis of mixtures of primary and secondary aromatic amines. Anal. Chem. 30:1594-1597.
940. WHETSEL, K. B., ROBERSON, W. E., and KRELL, M. W. 1958b. Near infrared spectra of primary aromatic amines. Anal. Chem. 30:1598-1604.
941. WILKINSON, G. R., FORD, M. A., and PRICE, M. C. 1955. The emission spectra of molecules and radicals in the infrared. Pages 192-202 in: Proc. 1954 Rep. Conf. Mol. Spectrosc. Institute of Petroleum, London.
942. WILLIAMS, P. C. 1974. Errors in protein-testing and their consequences. Cereal Sci. Today 19:281-286.
943. WOOD, D. L. 1954. The near-infrared spectrum of the peptide group. (Abstr.) Anal. Chem. 26:428.
944. WORLEY, J. D., and KLOTZ, I. M. 1966. Near infrared spectra of H_2O and D_2O solutions. J. Chem. Phys. 45:2868-2871.
945. YAMANOUCHI, T., and TANAKA, M. 1985. Absorption properties of the near-infrared water vapor bands. J. Quantitative Spectrosc. Radiat. Transfer 34:463-472.

CHEMICAL INSTRUMENTATION

946. DMITRIEVSKII, O. D., NEPORENT, V. S., and NIKITEN, V. A. 1957. A high speed infrared spectrometer for the region 0.8–3.0 μ. Opt. Spectrosk. 3:180-181.
947. GIOVANELLI, R. G. 1955. Reflection by semi-infinite diffusers. Opt. Acta 2:153.
948. MORREN, L., VANDERMEERSCH, G., and ANTOINE, P. 1972. A study of the reflection factor of usual photometric standards in the near-infrared. Light Res. Technol. 4:243-249.
949. PARSONS, J. L., IRELAND, M. J., and BRYAN, F. R. 1956. Continuous flow infrared gas analyzer employing interference filters. J. Opt. Soc. Am. 46:164-166.
950. SAKAYANAGI, Y., and SATO, Y. 1955. A near-infrared spectrometer using a plane grating. Sci. Light (Tokyo) 3:84-89.
951. SKJOLDEBRAND, R. 1956. An infrared spectrophotometer and its use for the isotopic analysis of heavy water. Appl. Sci. Res. Sect. B 5:401-408.
952. VON PLANTA, P. C. 1957. Experimental study of the slit functions of an Ebert spectrometer in the visible and near infrared. J. Opt. Soc. Am. 47:629-631.

INDUSTRIAL CHEMISTRY

953. ALFREY, T., Jr., and PFEIFER, C. R. 1966. Rates of copolymerization of acrylonitrile and ethylenesulfonic acid. J. Polym. Sci. 4:2447-2460.
954. BAIN, B. H. 1956. A direct infrared method for determining 4-methyl (2,6 di-t-butyl) phenol in mineral oil. Appl. Spectrosc. 10:193-194.
955. BARCHEWITZ, P. 1943. Use of near infrared in chemical analysis (application to the analysis of fuels). J. Chem. Phys. 45:40-43.
956. BYRON, T. P. 1982. Fast OH number-determination for process control. Proc. NIRA Symp., 2nd. Tarrytown, NY, July.
957. CHEEVER, G. D. 1978. Nondestructive determination of zinc phosphate coating weights using specular reflectance infrared absorbance. Coat. Technol. 50:78-85.
958. DANNENBERG, H. 1963. Determination of functional groups in epoxy resins by near-infrared spectroscopy. Soc. Plastics Eng. Trans. 3:78-88.
959. DIAMOND, W. J. 1958. Water determination in oil-contaminated 'Freon 12' by infrared spectrophotometry. Appl. Spectrosc. 12:10-11.
960. ELLIOT, A. 1952. The alpha-beta transformation in stretched hair. Text. Res. J. 22:783-786.
961. FREYMANN, M., and FREYMANN, R. 1943. Identification of fuels by near-infrared spectrometry. Application to the analysis of ketones. Group. Fr. Dev. Rech. Aeronaut. Note Tech. 11:53.
962. OGILVIE, G. D., SLATER, C., and FERGUSON, A. 1980. Measuring oil levels in polyester yarn via their infrared reflectance. ICI Fibres Rep., Harrogate, N. Yorkshire, England.
963. SNELLER, J. 1979. New sensors tackle the tough jobs: Coextrusions, foam filled material. Mod. Plast., September, p. 88.

PHYSICS

964. BUTLER, W. L. 1962. Absorption of light by turbid materials. J. Opt. Soc. Am. 52:292-299.
965. BUTLER, W. L., and NORRIS, K. H. 1960. The spectrophotometry of dense light-scattering material. Arch. Biochem. Biophys. 87:31-40.
966. HECHT, H. G. 1976. Comparison of continuum models in quantitative diffuse reflectance spectrometry. Anal. Chem. 48:1775-1779.

PROTEIN CHEMISTRY

967. AMBROSE, E. J., and ELLIOT, A. 1951. IR spectroscopic studies of globular proteins. Proc. R. Soc. London 208A:75.
968. AMBROSE, E. J., ELLIOT, A., and TEMPLE, R. B. 1951. IR spectra and structure of fibrous proteins. Proc. R. Soc. London 206A:206.
969. ELLIOT, A., and AMBROSE, E. J. 1950. Evidence of chain folding in polypeptides and proteins. Discuss. Faraday Soc. 9:246-251.
970. FRANCE, W. L., and DICKEY, F. P. 1955. Fine structure of the 2.7-μ carbon dioxide rotation-vibration band. J. Chem. Phys. 23:471-474.
971. FRASER, R. D. B. 1956. Interpretation of infrared dichroism in fibrous proteins—the 2 μ region. J. Chem. Phys. 24:89-95.
972. HECHT, K. T., and WOOD, D. L. 1956. The near infrared spectrum of the peptide group. Proc. R. Soc. London 235:174-188.
973. MIZUSHIMA, S.-I., TSUBOI, M., SHIMANOUCHI, T., and ASAI, T. 1954. Near infrared spectra of compounds with two peptide bonds and the configuration of a polypeptide chain. J. Am. Chem. Soc. 76:6003-6006.
974. RESSLER, N., ZIAUDDIN, C., VYGANTAS, C., JANZEN, W., and KARACHORLU, K. 1976. Improved techniques for the near infrared study of water binding by globular proteins and intact tissues. Appl. Spectrosc. 30:295.
975. SUBRAMANIAN, S., and FISHER, H. F. 1972. Detection of polar side-chain hydration in polypeptides by near-infrared spectroscopy. Biopolymers 11:1305-1310.

MISCELLANEOUS FOOD AND AGRICULTURE

976. BOWERS, S. A., and SMITH, S. J. 1972. Spectrophotometric determination of soil water content. Soil Sci. Soc. Am. Proc. 36:978-980.
977. COLLINS, F. I., ALEXANDER, D. E., RODGERS, R. C., and L. SILVELA, S. 1967. Analysis of oil content of soybeans by wideline NMR. J. Am. Oil Chem. Soc. 44:708-710.
978. CONWAY, T. F. 1963. Nuclear magnetic resonance for determining oil content of seeds. J. Am. Oil Chem. Soc. 40:265-268.

979. KRISHNAN, P., ALEXANDER, J. D., BUTLER, B. J., and HUMMEL, J. W. 1981. Reflectance technique for predicting soil organic matter. Soil Sci. Soc. Am. J. 44:1282.

980. MARTEN, G. C., SHENK, J. S., and BARTON, F. E., II. 1985. Near infrared reflectance spectroscopy (NIRS): Analysis of forage quality. U.S. Dep. Agric. Agric. Handb. 643. 96 pp.

981. MOLNAR, P. 1982. The design and practical use of an overall quality index for food products. Pages 115-147 in: Food Research and Data Analysis. H. Martens and J. Russwurm, Jr., eds. Applied Science Publishers, London.

982. MORREY, J. R. 1968. On determining spectral peak positions from composite spectra using a digital computer. Anal. Chem. 40:905-914.

983. NIERNBERGER, F. F. 1978. Near-infrared reflectance instrument analysis of grain constituents: A cost study. ESCS-20, Economics, Statistics, and Cooperatives Services. U.S. Dep. Agric., June.

984. PARKER, P. E., and NIERNBERGER, F. F. 1980. Oil content analysis of sunflower by nuclear magnetic resonance, solvent extraction, and near-infrared reflectance: A cost study. U.S. Dep. Agric. Fed. Grain Inspection Serv. Market. Res. Rep. 1106.

985. ROBERTSON, A., and MORRISON, W. H., III. 1979. Analysis of oil content of sunflower seed by wide line NMR. J. Am. Oil Chem. Soc. 56:961-964.

986. WILLIAMS, P. C. 1984. A study of grinders used for sample preparation in laboratory analysis of grains. Cereal Foods World 29:770-775.

Index of Topics in the Bibliography

Numbers refer to the entries in the Bibliography.

Absorption, relation to diffuse reflectance spectra, Bodo theory of, 359
Acetyl content, determination of, in cellulose acetate, 917
Agriculture and agricultural products, analysis of, 70, 82, 83, 85, 225, 745, 755, 768
Agrochemicals, analysis of, 688
Air, refractivity of, in NIR, 926
Alcohol, determination of, in wine, 11
Alcohols, molecular association in, study of, 867
Aldehydic C-H groups, studies of, 920
Alfalfa root, determination of carbohydrates in, 104
Alkaloids, determination of, in tobacco, 244
Alpha-acids, determination of, in hops, 23
Alpha-amylase, prediction of, in wheat, 707
Amide bands, studies of, 361, 854, 906
Amides, interactions of, 853
Amine bands, studies of, 52, 910, 940
Amines, determination of, 911, 939
Amino acids, determination of
  in barley and wheat, 15, 106, 227, 543, 638, 818
  in field peas, 737
Ammonia band, measurement of, 887
Anharmonicity, studies of, 851
Animal products, lysine in, study of, 120
Aqueous fat emulsions, determination of fat in, 491
Aqueous solutions, analysis of, by NIR, on fiberglass, 485
Aquo-organic mixtures, NIR spectroscopy of, 921
Ash, determination of, general, 612
  in buckwheat flour, 717
  in grains, 377
  in wheat flour, 51, 334, 337, 757, 824
Attenuated total reflection, use of, in producing spectra, 868

Baking industry, use of NIR in, 171, 554, 557, 558, 633
Band overlap, study of, 226
Barley
  amino acids in, determination of, 818
  analysis of, 28, 29, 119, 167, 247, 264–267, 290, 642
  breeding of, NIR in, 227, 405, 505
  crude fiber in, determination of, 220
  malt quality of, 8, 264, 265, 503–505
  moisture in, determination of, 344, 814
  protein in, determination of, 167, 243, 344, 358, 596, 651, 794, 814
  soluble β-glucan in, 9, 167
Beer
  gravity in, determination of, 28
  pH in, determination of, 28
  quality assessment of, 145, 241, 242, 264, 317
  water in, determination of, 189
Beta-glucan, determination of, in barley, 167, 267
Biological samples, techniques for obtaining absorption spectra of, 537
Biological tissues, determination of water in, 856
Biscuit and biscuit dough, compositional analysis of, by NIR, 576
Blood, detection of, in eggs, 539
Blue cheese, determination of fat, protein, and water in, by NIR, 223
Bread
  fat in, determination of, 561
  improvers in, determination of, 555
  moisture in, determination of, 561
  protein in, determination of, 561
Bread baking quality, prediction of, 169, 557, 566, 678, 681, 699, 704, 775
Breakfast cereals
  fiber in, determination of, 30
  process control of, by NIR, 27
  sugar in, determination of, 32
Brewing industry, NIR in, 28, 118, 173, 174, 242, 243, 632
Buckwheat, quality assessment of, 717

Cake mixes, analysis of, 577
Calibration
  factors affecting, 146, 297, 299, 441
  methods of, 144, 198, 200, 264, 297, 299, 439, 442, 444, 448, 454, 501, 521, 560, 636, 689, 690, 691, 700, 804, 806
  verification of, 121, 151, 762, 776
Calibrations
  transferability of, 142, 271, 667, 722
  universal, 46, 570, 572, 806, 817
Calories, determination of, in meat, 395
Carbohydrates, determination of, general, 263
  in cocoa powder, 355
  in snack foods, 30
  in soy products, 828
Carbohydrate-solvent interactions, characterization of, 933
Carbon, surface oxidation of, in situ NIR studies of, 914
Carbonate petrology, applications of NIR in, 879
Carbonates, determination of, 299
Carotene, determination of
  in forages, 582
  in vegetables, 582
  in whole tomatoes, 753
Cary spectrophotometer, digital interface for, 422
Casein products, analysis of, 618
Cellulose, studies of, 364
  determination of, 830
  as indication of flour purity, 778
Cellulose acetate, determination of acetyl content in, 917
Cereal grains, quality assessment of, 379
Cereal processing, industrial use of NIR, 27
Cheese, analysis of, 66, 207, 208, 223
Chemometric method of sample selection, 36, 449
Chicken feces, analysis of, 585
Chlorophyll, determination of
  in whole rapeseed, 735
  in whole tomatoes, 753
Chocolate processing, use of NIR, 219, 492
Chromium, determination of, in chicken feces, 585
Coal, NIR spectroscopy of, 213
Coal slags, NIR spectroscopy of, 883
Cocoa powder, quality assessment of, 355
Coextrusions, determination of, 963
Collaborative studies, in wheat and flour, 65, 318–320, 322, 490, 572, 574
Color, measurement of, 132
Color graphics, use of, in interpretation of NIR data, 161
Commodity, review and outlook, FAO, 193
Confectionary industry, NIR in, 619
Copolymerization rates, determination of, 953
Corn
  analysis of, by NIR, general, 705, 780
  dry milling, NIR in quality control, 774
  moisture in, determination of, 203
  wet milling, NIR in control of, 50, 597, 598
Corn syrup, high fructose, determination of saccharides in, 607
Correlation transform, use of, 274, 684, 893
Cotton/polyester yarns, analysis of, by NIR, 727
Cottonseed
  gossypol in, determination of, 80
  oil, protein, and moisture in, determination of, 603, 604, 672
Cottonseed meal, determination of oil, protein, and moisture in, 672
Cream, analysis of, by NIR, 724
Crop, remote sensing of, 611
Crystals
  organic, mounting of, for spectroscopy, 432
  vibration spectra of, 886

Dairy products
  dried, analysis of, 25, 94, 170, 172, 191, 427, 618
  fluid, analysis of, 234–236, 367, 695, 787, 831
Data-processing, 225, 388, 428, 434, 438,

449, 452, 454–456, 493, 494, 513, 518, 532, 542, 573, 602, 637, 648, 697, 721, 770
Degrading factors, influence of, on NIR analysis, 805, 814
Derivative spectroscopy, use of, in NIR, 546
Deuterated water vapor, spectra of, 842
Deuterium oxide, spectra of, 840, 850, 944, 951
Dichroism in fibrous products, interpretation of, 971
Diffuse reflectance
  spectra, development of, in NIR, 673
  standards, use of, in NIR technology, 406
  theory, present status of, 259
Diffusers, semi-infinite, reflection by, 947
Diffusion spectroscopy, 825
Disaccharides, determination of glycosidic bonds in, 643
Distillers' dried grains, quality evaluation of, by NIR, 482
Doughs, prediction of mixing time in, 679
Dry matter digestibility, determination of, 493

Ebert spectrometer, use of, in NIR, 952
Economics of NIR testing, 798, 983, 984
Edible oils
  free fatty acids in, determination of, 13
  iodine value of, determination of, 13
  peroxide value of, determination of, 13
Eggs, detection of blood in, 539
Environmental factors, effects of, on NIR, 806
Epoxy resins, functional groups in, determination of, 958
Errors
  in NIR testing, methods for reduction, 393
  in protein testing, 942
Ethylene/propylene copolymers, spectrophotometry of, 720
Ewe milk
  crude protein in, determination of, 251
  fat in, determination of, 251
  lactose in, determination of, 251
Extruded cereal foods
  degree of gelatinization in, determination of, 609
  moisture in, determination of, 608

Factor analysis, use of, in NIR, 310
Fat, determination of, general, 412, 637, 650
  in aqueous fat emulsions, 491
  in bread, 561
  in cocoa powder, 355
  in dried milk, 4
  in dry cake mixes, 577
  in foods, 333
  in grain, 381
  in meat, 384, 395, 450, 452, 458, 516
  in meat products, 56
  in milk, 65, 67, 249, 622, 639, 724, 787
  in pastry products, 357
  in plant materials, 314, 381
  in soy products, 828
Fats, degree of unsaturation, determination of, 20
Fatty acids
  NIR spectra of, 291, 309
  trans-isomer content in, determination of, 484
Feed mix ingredients, interactions of, influence on NIR analysis, 820
Feed mixes, analysis of, 195, 744, 820
Fermentation control, use of NIR, 423
Fertilizers, analysis by NIR, instrumentation for, 397
Fiber, determination of
  crude, in barley, 220
  dietary
    in carrots, 307
    in wheat bran, 307
  in food, 18, 30, 31, 33
  in forages, 582, 583
  in silage, 261
  in sunflower seeds, 356
  in vegetables, 584
Fiber optics, use of, in NIR, 73
Fiberglass, use of, in NIR analysis, 485
Fibers
  binary mixtures of, analysis by NIR, 131
  mounting of, for spectroscopy, 432
Field monitoring, use of NIR, 325
Field peas
  amino acids in, determination of, 737
  fat in, determination of, 160
  protein in, determination of, 164
  starch in, determination of, 160
Filters, interference, use of, 258
Fish flesh, state of water in, 520
Flour
  ash in, determination of, 334, 337, 725, 824
  bread baking quality, prediction of, 123, 566
  glycolipids in, determination of, 671
  moisture in, determination of, 334, 572, 824
  prediction of water absorption of, 553
  protein in, determination of, 334, 572, 824
  quality assessment of, 568, 569, 635, 681, 725
  starch damage in, determination of, 563, 566
Fluoride, determination of, in plants, 217
Food components, analysis of, by NIR, 328, 330, 331
Food systems, composition of, models for, 827
Foods
  moisture in, determination of, 719, 747
  powdered, spectral analysis of, 22
  quality assessment of, 76, 93, 116, 156, 245, 283, 329, 332, 352, 389, 460, 472, 550, 745
  state of water in, study of, 341
  total solids in, determination of, 69
Forages
  digestibility of, determination of, 829
  nutritive value of, determination of, 57, 493, 494, 535
  quality assessment of, 34, 37, 38, 40, 41, 57, 109, 129, 152, 155, 204, 216, 218, 314, 379, 446, 493, 494, 509, 510, 513, 515, 536, 582, 584, 613, 649, 655–657, 659–662, 664–666, 723, 744, 750, 980
  voluntary intake of, prediction of, 750
Fourier transform, use of, in NIR, 163, 212, 311, 467, 469, 471, 473, 865
Free fatty acids, determination of, in vegetable oils, 606
Fructose, determination of, in corn syrup, 49
Fruit juices, sugar content of, determination of, 396
Fruits, quality assessment of, 87, 105, 110, 214, 353, 382, 631, 752
Fuels
  analysis of, by NIR, 955
  identification of, by NIR, 961
Functional groups, determination of, in epoxy resins, 958
Fungal pathogens, spore count of, determination of, 21

Gallium arsenide, use of, in NIR instrumentation, 458
Gas-analyzer, continuous flow, 949
Gauss-Jordan linear algebra, use of, in NIR, 295
Gelatinization, degree of, prediction of, 605
Globular proteins, water binding by, determination of, 974
Globulins, study of, in NIR, 127
Glucosinolate, determination of, in rapeseed, 700
Gluten content, determination of, in wheat, 376, 377, 380
Glycolipids, determination of, in flour, 671
Grain
  quality assessment of, 487, 626, 656, 749, 779, 794
  spectral reflectance and transmittance properties of, 459
  whole
    moisture in, determination of, 702, 814
    protein in, determination in, 728, 731, 814
Granularity, influence of, on wheat analysis, 599, 823
Grating spectrophotometer, high resolution, development of, 205
Greasy wool, yield of, determination of, 12, 138, 139
Grinders, in NIR analysis, 196, 321, 430, 562, 777, 801, 986
Ground beef
  fat in, determination of, 458
  soybean flour in, determination of, 90
Growing crops, spectra of, 429
Growing location, influence of, on wheat analysis by NIR, 806
Growing season, influence of, on wheat analysis by NIR, 806
Guayule
  moisture in, determination of, 91
  resin in, determination of, 91
  rubber in, determination of, 91

Hardness in wheat, determination of, 300, 488, 574, 645, 647, 800, 819, 835
Hat matrix, use of, in regression, 287
Heat set nylon carpet yarn, characterization of, by NIR, 726
Heat set temperature, determination of, 221
Heavy water, analysis of, 880, 951
High energy spectrophotometer, use of, in NIR, 389, 390, 460, 629, 685
High speed spectrometer, development of, 946
Homogenization, in milk analysis, by NIR, 640
Hops
  α-acids in, determination of, 23, 590, 668
  β-acids in, determination of, 24, 590
  moisture in, determination of, 23, 24, 668
Hydration numbers, determination of, in amino acids, 892
Hydrocarbon-producing plant species, analysis of, 130
Hydrocarbons
  spectral differentiation of, 909
  structure groups of, determination of, 907
  studies of, using NIR, 881, 896
Hydrogen bonds
  interamide, in solvents, 715

intermolecular, 878
intramolecular, 838, 878, 905, 918, 922
of peptides, 904
review of, 855
strength of, 919
in sulfhydryl compounds, 901, 915
systems
influence of temperature, 869, 870
study of, 897, 934, 935
of water, 844, 845, 849, 850
Hydrogen peroxide, infrared spectra of, 836
Hydrogen sulfide, spectrum of, 924
Hydroxyl bands
effect of pressure on, 872
effect of temperature on, 870
intensity of, 839, 925
in aliphatic alcohols, 848
in phenols, 846, 927
studies of, 873
Hydroxyl number, determination of, 956
Hydroxyl stretching frequencies, relation to reactivity, 898

Individual component spectra, method for obtaining, 296
Infrared spectroscopy, general, 706
of molecules, study of, 930
solvent effects in, 849
Instrumentation, NIR, general, 117, 133, 166, 168, 179, 248, 284, 285, 301, 303, 363, 389–392, 397, 418, 419, 422, 424, 472, 523, 524, 548, 551, 552, 564, 565, 600, 625–627, 630, 631, 634, 658, 660, 682, 683, 685, 686, 692–694, 708, 712, 739, 756, 767, 807, 833, 885
Integrating sphere, use of, in NIR, 342
Interferometry, use of, in resolution of band overlap, 900
Internal quality, determination of, general, 186
in cherries, 410
in fruit, 631
Isomeric organic compounds, identification of, 877

Kale, digestibility of, determination of, 7
Kirkwood-Bauer-Magat theory, use of, 402
Kjeldahl test, comparison with NIR, 308, 371, 489, 490, 812
Kubelka-Munk light-scattering coefficients, study of, 411
Kubelka-Munk theory, use of, in NIR, 578, 742

Lactose, determination of, in milk, 65, 67, 622, 639, 724
Laver (seaweed), quality assessment of, by NIR, 336, 339
Least squares curve-fitting, use of, in NIR, 312
Light
interaction with foods, 75
relation to spectroscopy, 370
spectral reflectance of, by potatoes, 709
Light scattering, measurement by diffuse reflectance, 78
Light scattering material
optics of, 385
spectrophotometry of, 965
Light transmittance properties, measurement of, 525
Lipid
in pea flour, determination of, 160
structures of, 771
Liquid baking ingredients, analysis of, by NIR, 48
Liquids
absorption bands of, 929
analysis of, by NIR, 431, 485
Lupinus seeds, analysis of, by NIR, 86
Lysine, determination of, 15
in barley breeding, 227
in cereals, 106, 818

Mahalanobis distances, use of, in NIR, 437, 440
Maize, determination of oil in, 6
Malt
determination of protein in, 596
quality assessment of, 119, 167, 173, 174, 243, 264, 265, 481, 503, 504, 589
Malt extracts, determination of soluble protein in, 589, 669
Malting industry, use of NIR in, 118, 173, 243
Malting quality, determination of, by NIR, 264–267, 288, 405, 504
Mathematical treatments
optimization of, 540, 816
use of, in mixture analysis, 62
Matrix isolation, use of, in hydrogen bonding studies, 931
Mayonnaise, determination of oil in, 159
Meat
fat in, determination of, 194, 516
general analysis of, by NIR, 256, 338, 384, 450, 451, 701
moisture in, determination of, 194
salt in, determination of, 44, 45
Meat products
fat in, determination of, 56, 88, 89, 194, 450, 452
general analysis of, 256, 414, 449, 451, 547
moisture in, determination, of 450, 452
protein in, determination of, 450, 452
Methionine, determination of, in peas, 809
Milk
absorption spectrum, influence of fat on, 54
fat in, determination of, 14, 65, 67, 68, 622, 639, 646, 724, 787
lactose in, determination of, 65, 67, 622, 639, 724
NIR analysis of, precision in, 64, 68
protein in, determination of, 65, 67, 622, 639, 724
raw, analysis of, by NIR, 92, 234, 367, 401, 480
total solids in, determination of, 14, 787
Milk powders, NIR analysis of, 208, 618
Milk products
dried, quality control of, 738
quality control of, 188
Milled rice, NIR spectra of, 703
Millet, determination of protein in, 705
Milling, production control by NIR, 98–100, 133, 171, 462, 463, 652, 824, 830
Milling quality
influence of environment on, 718
influence of variety on, 718
prediction of, 704
Mineral oil, quality control of, 954
Moisture
comparison of methods, 812
determination of, general, 206, 289, 479, 575, 650, 653, 747
in agricultural industry, 745
in barley, 167, 181, 344, 814
in bread, 561
in buckwheat flour, 717
in corn, 203, 702
in cottonseed, 672
in cottonseed meal, 672
in dairy products, 235
in dough, 644
in dried vegetables, 610
in flour, 334, 572, 824
in food, 719, 745
in malt, 167, 503
in meat, 384, 395, 452
in meat products, 56, 450
in pastry products, 357
in plant material, 314, 378, 381
in powdered coffee extracts, 126
in seeds, 254, 538
in single cell protein, 122
in soybeans, 55, 794
in spices, 610
in starch, 53
in sunflower seeds, 356, 620
in walnut seeds, 115
in wheat, 253, 312, 381, 407, 409, 453, 538, 570, 574, 647, 794, 798, 805, 811, 812, 814, 815, 817, 823
in winged bean seeds, 268
Molecules, emission spectra of, 941
Monochromator parameters, measurement of, 394

Natural fibers, NIR spectra of, 190
Near-infrared absorbers, cartography of, 810
Near-infrared analysis
of alcohols, reduction of water interference in, 871
comparison with classical IR, 282
Near-infrared instrumentation
Japanese patent, 286
reliability of, 345, 346
Near-infrared reflectance
applications of UV spectrophotometry, 885
bibliographies, 240, 279, 740
physics of, 77
Near-infrared reflectance analysis, automated, 101, 601, 772, 793, 821
Near-infrared reflectance data, interpretation of, 148, 149, 161, 255, 469, 532, 534, 932
Near-infrared reflectance spectroscopy, general, 2, 10, 19, 42, 43, 47, 59, 60, 102, 107, 108, 111, 128, 136, 147, 157, 158, 165, 175, 177, 197, 209, 211, 230, 239, 250, 272, 273, 275, 276, 283, 292–294, 302, 313, 316, 351, 362, 363, 365, 372–374, 403, 418, 419, 426, 435, 457, 468, 478, 486, 495, 497, 500, 517, 526, 527, 529–531, 533, 544, 549, 552, 592, 615, 623, 628, 657, 706, 712–714, 746, 758, 760, 761, 765, 766, 769, 785, 786, 788, 803, 936
factors affecting, 303, 528, 759, 790–792, 796, 805, 806, 811, 820, 823
Near-infrared spectra
studies of, 134, 135, 150, 190, 257, 260, 262, 291, 309, 375, 386, 409, 417, 534, 545, 587, 643, 663, 703, 751, 783, 784, 862, 873, 875, 876, 878, 937, 938, 941, 943, 944
transfer of, 663
use of, in factor analysis, 447
Nitrate, determination of
in forages, 582
in vegetables, 582
Nitrogen, total
comparison of methods, 813
determination of
in forages, 582, 584, 829
in ground rapeseed, 506
by NIR, 227, 324–327, 616, 813
in vegetables, 582, 584
in whole rapeseed, 506, 700
Normalization of NIR optical data, 142

Nuclear magnetic resonance, in oil determination
in soybeans, 977, 978
in sunflower, 984, 985
Number of samples for calibration, 299, 794
method for reducing, 299

Oat groats, determination of protein in, 84
Oil, determination of
in corn, 323
in cottonseed, 672
in cottonseed meals, 672
in grains, 269, 381
in oats, 323
in rapeseed, 506, 700, 730, 794
in soybeans, 269, 318, 319, 617, 794
in sunflower seeds, 356, 620, 621, 984, 985
in vegetable matter, 381
Oil content, determination of
in soybeans, by NMR, 977, 978
in sunflower, by NMR, 984, 985
Oils, vegetable
changes in, by heating, 677
NIR spectra of, 35
Oilseeds, quality assessment of, 140, 379, 779, 794
Oleamides, spectral studies of, 361
Olive oil, determination of hydroxyl number of, 237
Olive oil cake, determination of hydroxyl number of, 237
Onions, determination of soluble solids in, 81
On-line analysis of flour, 101, 601, 772
On-line applications of NIR, factors involved in, 143
Optical ceramics, use of, in NIR instrumentation, 670
Optical scattering coefficient, interpretation of, 74
Organic compounds, NIR spectra of, 868, 938
Organic hydroxyl compounds, NIR spectra of, 863
Organic liquids, NIR spectra of, 923
Organic materials, NIR spectrometry of, 899
Organic matter, analysis of, 908
Organic substances, optical rotating power of, 889
Organo-phosphorus compounds, spectra of, 912, 913
Outliers, significance of, 802

Packaging industry, use of NIR in, 248
Packaging laminates, analysis by NIR, 162
Parenteral products, analysis by NIR, 624
Partial least squares regression, use of, in NIR, 454
Particle size
effects on wheat analysis by NIR, 593, 764, 823
factors affecting, 754
influence of, on soybean spectra, 113
Pasta, quality assessment by NIR, 507
Pastry products, quality assessment by NIR, 354, 357
Pathlength error, method of correction, 281
Peat
bulk density of, determination of, 182
moisture in, determination of, 182
Peppers, spectral transmittance of, 382
Peptides
chain folding in, 969
hydrogen bonded, studies of, 904, 918, 975
NIR spectra of, 260, 943, 972, 973
polar side chain hydration of, 975
Phenolic hydroxyl, determination of, 229
Phenols
chemical reactivities of, 837
hydroxyl stretching in, 837
Photoacoustic spectra of grain, comparison with NIR, 262
Photoacoustic spectrophotometer, use of, in NIR region, 95
Photoacoustic technology, applications of, 370, 864
Photometric standards, reflection properties of, 948
Plane-grating, use of, in NIR spectrophotometry, 950
Plant-breeding programs, use of NIR in, 505, 676, 698
Plants, determination of fluoride in, 217
Polyester yarn, oil levels in, determination of, 962
Polyisoprenes, determination of *cis* and *trans* 1:4 contents of, 861
Polymers
NIR spectrometry of, 874
pleochroism of, 882
synthetic, quantitative analysis of, 916
Polypeptides, optical rotatory dispersion of, 858
Potatoes
chemical composition of, determination of, 338
internal discoloration in, detection of, 72
separation of, from stones and soil, 710
spectral reflectance of, 709
Poultry muscle
fat in, determination of, 741
protein in, determination of, 741
Powders
diffuse reflectance of, 347, 359, 360
optical absorption of, 347
Primary amides, NIR studies of, 375
Principal component analysis, use of, in NIR analysis, 149, 277, 453
Process control, use of NIR in, 27, 141, 163, 348–350, 368, 416, 427, 433, 508, 522, 588, 711, 770, 772, 830
Processed cereal foods, determination of fiber in, 30
Protein
determination of, general, 1, 180, 206, 477, 571, 651, 812, 813
in barley, 167, 344, 358, 596, 651, 794, 814
in bread, 561, 716
in buckwheat flour, 717
in cocoa powder, 355
in corn, 323
in cottonseed, 672
in cottonseed meal, 672
in forages, 39
in malt, 167, 503, 596
in meat, 384, 395, 450, 452
in milk, 65, 67, 622, 639, 724
in mill products, 749, 763
in mill streams, 757, 781, 824
in millet, 705
in oat groats, 84
in oats, 323
in pastry products, 357
in pea flour, 164
in plant material, 314, 327, 381
in pulses, 822
in silage, 261, 511, 512, 514
in sorghum, 705
in soy products, 828
in soybeans, 269, 318, 319, 617, 794, 826
in sunflower, 356
in tropical grasses, 493
in vegetative material, 381, 420
in wheat, 181, 183–185, 252, 253, 312, 318, 320, 379, 453, 489, 496, 541, 542, 556, 570, 574, 580, 595, 647, 675, 680, 708, 722, 749, 756, 789, 794, 797–800, 805–808, 811–815, 817, 821, 823
in wheat flour, 337, 572, 794, 824
in wheat kernels, 728, 731, 814
in whole rapeseed, 730
use of, in flour mills, 763, 824
Protein complexes, NIR spectra of, 257
Protein testing, errors in, 942
Proteins
chain-folding in, 969
characterization of, by NIR, 859
hydration of, study of, 743
infrared studies of, 967, 968, 970
structure of, 970

Quality control, use of NIR in, 316, 483, 502, 519, 711, 725, 782, 824
Quality index, use of, for food products, 981
Quantitative diffuse reflectance spectrometry, continuum models for, 966

Radicals, emission spectra of, 941
Radish, grading of, by light reflectance, 215
Rapeseed
chlorophyll in, determination of, 733
glucosinolates in, determination of, 700
moisture in, determination of, 733
nitrogen in, determination of, 506, 700, 733
oil in, determination of, 506, 700, 730
whole
chlorophyll in, determination of, 735
oil in, determination of, 728, 729
protein in, determination of, 728, 729
Reactive molecules, spectroscopic studies of, 841
Red peppers, determination of capsaicin in, 335
Reflectance spectroscopy, general, 936
Reflected energy, detection of, 696
Refractive index, measurement of, 464
Regression models, testing of, 151
Regression transforms, principles, 274, 278
Repack, influence of, on NIR calibrations, 441
Ribonuclease, structural studies of, using NIR, 890
Rice
amino acids in, determination of, 340
analysis of, by NIR, 304, 340, 705, 748
milled, determination of oil in, 304
protein in, determination of, 340
Rice seed proteins, comparison of, 387

Sake, analysis of, by NIR, 748
Sample absorbance, optimization of, 281
Sample area optimization, use of, in NIR, 280
Sample characterization, by diffuse reflectance, 773
Sample preparation
of crystals, 432
of fibers, 432
methods of, 238, 432, 537, 599, 700, 795, 986
Sample selection, use of spectral subtraction for, 298
Sampling, for NIR analysis, 599, 795
Secondary amides, NIR spectra of, 375
Seeds, determination of moisture in, 254
Segregation of wheat, by NIR, 615, 675, 797, 799
Short dough, analysis of, by NIR, 177, 576
Silage
fiber in, determination of, 261, 511, 512, 514
protein in, determination of, 261, 511, 512, 514

Smut, determination of, in wheat, by NIR, 71
Snack foods
  fiber in, determination of, 31
  starch in, determination of, 31
  total carbohydrate in, determination of, 31
Sodium chloride optics, use of, in NIR, 751
Soil
  analysis of, by NIR, instrumentation for, 397
  organic matter in, determination of, 154, 979
  total nitrogen in, determination of, 154, 834
  water in, determination of by spectrophotometry, 154, 976
Solvent, effects of, in infrared spectroscopy, 847, 849
Sorghum, determination of protein in, 594, 705
Soy meal, universal NIR calibrations, 425
Soy products, analysis of, by NIR, 828
Soybean oil, hydrogenated, analysis of, 26
Soybean vegetative canopy, reflectance studies of, 415
Soybeans
  moisture in, determination of, 55, 794
  oil in, determination of, 318, 617, 794
  protein in, determination of, 318, 617, 794
  quality assessment of, by NIR, 338, 425, 498, 502, 826
  spectral absorbance of, importance in processing, 343
Spectra
  composite, resolution of, 982
  individual, from complex mixtures, 895
  studies of, 16, 232
Spectral reconstruction in NIR, 894
Spectral reflectivity, measurement of, 225
Spectrophotometer
  computerized, comparative study of, 808
  development of, 112, 462, 526
Spectroscopy
  absorption, general, 42
  correlation transform, 684, 687
  derivative, 43, 546
  diffusion, 825
  internal reflection, 250
Spheroid mirror, use of, in NIR, 96
Spices, determination of moisture in, by NIR, 610
Standardization
  of NIR analysis, 560
  of NIR hardness testing, 819
Standards, reflectance, use of, in NIR, 383, 406, 670, 734, 736
Starch
  determination of, general, 436
    in grain, 381
    in pea flour, 160
    in snack foods, 30
    in vegetative material, 381
  moisture in, determination of, 53
  tapioca, determination of, in wheat starch, 270
Starch damage, determination of, in flour, 178, 192, 563
Starch derivatives, analysis of, 436, 461
Statistics, use of, in NIR, 199–201, 233, 499, 559, 802
Stepwise linear regression, use of, in NIR, 245
Stretched hair, alpha-beta transformation in, 960
Stretching vibrations, study of, 843
Structure, determination by NIR spectroscopy, 232, 888
Sucrose, determination of, in dry cake mixes, 577
Sugar, determination of, in breakfast cereals, 32
Sugars, determination of, in dry mixtures, 222
Sulfhydryl stretching, intensity of, 852
Sulfur, use of, as reflectance standard, 383, 734, 736
Sunflower seed
  fiber in, determination of, 356
  moisture in, determination of, 356, 620
  oil in, determination of, 356, 620, 621, 984
    by NMR, 984, 985
  quality assessment of, 620, 621
Sunflower seed meal, determination of residual oil in, 114
Superimposed NIR spectra, analysis of, 386

Target factor analysis, use of, in NIR, 476
Temperature, influence of, on NIR analysis, 315, 586, 732, 815
Terminal epoxides, determination of, by NIR, 231
Texture, determination of, in wheat, 567, 647, 819
Thiol group, hydrogen bonding of, 915
Titanium dioxide films, use of, as reflectors, in NIR, 928
Tobacco
  browning reaction, study of, 466
  moisture in, determination of, 421, 465
  nicotine in, determination of, 421, 470
  polyphenols in, determination of, 475
  quality assessment in, 233, 244, 421, 470, 474, 475, 579
  stem in, determination of, 421
  sugars in, determination of, 474
  total alkaloids in, determination of, 244
  total volatile bases in, determination of, 58
Tomato juice, quality assessment of, 305
Tomatoes
  carotenoids in, determination of, 753
  chlorophyll in, determination of, 753
  green, prediction of ripening time, 832
  internal color of, determination of, 79
  spectral transmittance characteristics of, 382
Transflectance, use of, in beer analysis, by NIR, 241
Turbid materials, absorption of light by, 964
Two-component systems, simultaneous analysis of, 210

Universal constants, development of, 46, 572, 817
Unsaturation
  cis-, determination of, in oils, 202
  determination of, general, 228
    in milk fat, 3
    in oils, 3, 20

Variable subsets, selection of, 5
Variety, effect of, on wheat analysis by NIR, 759, 805
Vegetable oils, determination of free fatty acids in, 606
Vegetables
  dehydrated
    moisture in, determination of, 610
    nutrient content of, determination of, 584
  quality assessment of, 72, 79, 81, 103, 215, 305–307, 335, 338, 382, 582, 709, 710, 752, 753, 832
Vibration spectra of crystals, frequencies of, 886
Vicia faba
  amino acids in, determination of, 86
  protein in, determination of, 86, 822
Voluntary intake, prediction of, 493, 613

Water
  absorption bands, structure of, 850
  absorption spectra of, 840, 862, 876, 944
  determination of, general, 369
    in glycerol, 125
    in hydrazines, 860
    in oil-contaminated freon, 959
    in organic liquids, 891
    in soil, 976
    in tributyl phosphate, 902
  heavy
    influence of gelatin on, 866
    spectra of, 850
  liquid, spectra of, 840, 850, 862, 866, 884
  in monkey cornea, study of, 857
  spectroscopy of, in organic solvents, 884
  study of state of, 341, 520, 903
Water binding, determination of, in proteins, 614
Water vapor bands, NIR studies of, 945
Water-macromolecule interactions, study of, 743
Wavelength
  measurement of, in infrared spectra, 591
  optimization of, 152, 297, 362, 429, 532, 540, 542, 804, 811, 816
Wavelength accuracy, method for checking, 187
Wavelength selection, methods of, 573, 816
Wheat
  classification of, and storage, 63, 290
  α-amylase in, determination of, 707
  amino acids in, determination of, 543, 818
  early generation, selection by NIR, 581, 699, 800
  hardness of, determination of, 300, 488, 567, 574, 647, 800, 819
  moisture in, determination of, 181, 253, 312, 408, 538, 570, 626, 647, 650, 794, 798, 805, 811, 812, 814, 815, 817, 823
  protein in, determination of, 180, 183–185, 252, 253, 312, 318, 320, 321, 371, 377, 381, 489, 490, 496, 540–542, 556, 557, 570, 574, 581, 595, 625–627, 647, 675, 680, 708, 722, 728, 731, 756, 759, 789, 790, 794, 797–800, 805, 806, 808, 811, 812, 814–817, 823
  strength in, determination of, 807
  quality, assessment of, by NIR, 17, 97, 153, 180, 320, 338, 377, 413, 557, 581, 626, 704, 807
  variety identification by NIR, 61, 169
  whole-grain analysis of, by NIR, 702, 728, 729, 731, 814
Wheat class, effect of, on NIR testing, 759
Wheat components, NIR spectra of, 404, 408
Wheat flour, determination of ash in, 334, 337, 824
Wheat gluten
  addition to bread flour, 124
  determination of, 376, 377, 380
Whey, quality control in, 368
Wine
  alcohol in, determination of, 11, 366
  quality assessment of, 366
Winged bean
  oil in, determination of, 268
  protein in, determination of, 268
Wool
  greasy, determination of yield of, by NIR, 12, 138, 139
  scoured, determination of alcohol extractables in, 176, 674

testing by NIR, general, 246, 654
washed, testing by NIR, 445
yield of, determination by NIR, 137–139, 641

Wool blends, analysis of, by NIR, 399, 400, 443
Wool top, examination of, by NIR, 398
Wort, analysis of, by NIR, 243

Yogurt, analysis of, by NIR, 724

Zinc phosphate coating, determination of, by NIR, 957

Index to Text

Abnormal samples, detection of, 71
Absorbance
  intensity of, and dipole, 21
  at specific wavelength points, use of, in generation of NIR calibration constants, 187–188
Absorption
  and transmittance, 21
  and wavelength, 12
Absorption bands
  changing of number and position of, caused by moisture, detection of, by NIR, 242
  of common constituents, illustration of, 241, 243
  of water
    assignment of, 27–29
    in high-moisture materials, study of, by NIR, 226
Absorptivity, definition of, 3
Acceptable error, definition of, 39
Accuracy
  of analysis by near-infrared transmittance in GT-90, factors affecting, 126
  definition of, 39, 136, 143
  determination of, 136, 143
  factors involved in, 145
  method of expression of, 143
  of NIR analysis of silage, factors affecting, 196
Acid detergent fiber, and digestibility, 170
Aeration, importance of, in sample storage, 164
Aging
  of components, as source of noise, 149
  of samples, 137
Agriculture, importance of, in Asia, 223, 224
Air-jet sieve size of wheat flours, NIR calibration for, 189
Alcohol content of beer and wine, determination of, by NIR, 194
Alcohol extractables in wool, determination of, by NIR, 227
Alcohols, homologous series of, spectra of, 22
Aleurone color in barley, determination of, by NIR, 231
α-Amylase in wheat, determination of, by NIR, 230
Alveograph W value, prediction of, by NIR, 221
Amino acids, determination of, by NIR
  in cereals, 195
  in rice, 224–225
  wavelength selection for, 130
Ammonia nitrogen, loss of, in drying, as source of error in NIR analysis of silage, 196
Amylose in rice, determination of, by NIR, 224
Anacon NIR Analyzer, 116
Analog computer circuitry, use of, in early near-infrared instruments, 114
Analytical (open) calibration, 131
Anharmonicity, 23–24
Anharmonicity constant, *x*, definition of, 24
Animal products, methods of sample presentation of, 160–161
Animal proteins, assignment of absorbing wavelengths in, 208
Anomalies in NIR spectroscopy, correction of, by computer, 94
Apples, determination of stage of maturity of, by NIR and visible transmittance, 206
Application of near-infrared instruments, steps in, 141
Arabinoxylan, in association with lignin complex, 174
Artificial calibration samples, use of, to illustrate introduction of errors in analysis, 73–75
Ash, determination of, by NIR
  in dairy products, 197
  in flour, 191, 202, 225
  in rice, 224
Assignment of absorption wavelengths, 31, 34
Australian Wheat Board, use of NIR by, for protein determination, 230
Automated sample presentation, use of, in Automated Digital Analyzer, 117
Automatic internal troubleshooting, use of in near-infrared instrumentation, 124
Automatic self-calibration, use of, in near-infrared instruments, 119
Automatic slope/bias correction, use of, in near-infrared instrumentation, 126
Auxiliary computers
  desktop, use of, in programming Technicon InfraAlyzer series, 124
  interfacing to simple bench instruments to facilitate wavelength selection, 128
  use of, in near-infrared instrumentation, 125
Avogadro's law, significance of, in NIR analysis, 157
Avogadro's number, implications of, in analytical chemistry, 21

Backward stepwise multiple linear regression, 138
Bacteria, influence of, on physical texture, 160
Baked goods, determination of moisture, protein, and sugar in, by NIR, 207
Band overlap, definition of, 109
Bandwidth
  definition of, 113
  dimensions of, 95
  influence of, on spectra, 241
  of tilting filters, variability in, 96
Barley. *See also* Malting barley
  moisture in, determination of, by NIR, 201–202, 239
  nitrogen content, viscosity, and glucose content of, determination of by NIR, 194
  protein in, determination of, by NIR, 201–202, 231
$BaSO_4$ as standard for visible reflectance, 233
Bathochromy, definition of, 22
Beer-Lambert law, 2
  and absorbance, 21
  definition of, 107
  and transmittance, 8
Beer's model, 67–69
Best pair of wavelengths, 139, 152
Best triple wavelength, selection of, features of, 139
β-Glucan in barley, determination of, by NIR, 194, 231
BIAS, definition of, 38
Bias, patterns of, 147
Biased errors, definition of, 143
Biscuit doughs, determination of fat, sucrose, and water in, by NIR, 192
Blazing of gratings, 93
Bleaching, influence of, on physical texture of wheat, 160
Blending after grinding, 164
Blind duplicates, definition of, 136, 143
Blue cheese, determination of fat, protein, and water in, by NIR, 207
Body reflection, definition of, 4
Bound starch in corn fiber residues, determination of, by NIR, 203
Boxcar smoothing
  definition of, 42
  procedure for, 155
Bran cleanliness, determination of, by NIR, 189
Bread, determination of protein, fat, and moisture in, by NIR, 193
Bread score, prediction of, by NIR, 221
Breakfast cereal, determination of sugar in, by NIR, 207
Breeding of plants, changes caused by, analysis of, by NIR, 222
Breeding programs for malting barley, use of NIR in, 194
Brewster's angle, 2
Brownouts, influence of, on NIR analysis, 156
Buckwheat, determination of ash and protein in, by NIR, 225
Building power supply, influence of, on NIR analysis, 156
Built-in burr mill, use of, in near-infrared instrumentation, 123
Built-in sample presentation, in near-infrared instrumentation, 123
Bulk density, of ground commodities, 159–160
Butter, analysis of, by NIR, 196

Buttermilk powder, use of NIR in quality control of, 229

C values, definition of, 116
Caffeine in green tea, determination of, by NIR, 225
Calibration
  definition of, 37
  development of
    by NIR difference spectrum, 242
    steps in, 39–40
  factors involved in, 57–58, 176–177
  methods of, classification of, 60
  prerequisites of, 130
  principles of, 37–38
  transfer of, 133, 196, 230, 231
  types of, for simple bench near-infrared instruments, 132
  use of NIR chemical data for, 220
  use of variance in selection of samples for, 188
Calibration coefficients, significance of, 58
Calibration model, development of, 37, 39–40
Calibration samples, selection of, using spectral data, 63
Calibration set, definition of, 37
Calories in raw pork and beef, determination of, by NIR, 207
Cam-driven holographic grating, use of, for rapid scanning, in near-infrared instrumentation, 126
Carbohydrate in meat pie ingredients, determination of, by NIR, 207
Carbon-13 nuclear magnetic resonance
  applications of
    to fibrous components analysis, 173
    to protein analysis, 179
  limitation of, 178–179
  use of, in calibration of NIR instruments, 181
Carbonyl compounds, presence of, detected in lignin complexes, 174
Carelessness, as error factor, sources of, 166
Casein and caseinates, use of NIR in quality control of, 229
Cell covers, influence of, on NIR analysis, 156–157
Centering of reflectance data, in NIR forage analysis, 229
Central wavelength, definition of, 113
Ceramic, as reflectance standard for NIR, 107, 111, 112, 233
Cereals
  breeding programs for, use of NIR in, 224
  fiber in, determination of, by NIR, 169
  production of, in Asia, 223
-CH stretch first overtone, use of, in NIR determination of metabolizable energy in silages, 196
Characteristic color curve, definition of, 10
Characteristic vibrations, definition of, 22
Check samples, use of, in detection of errors, 143
Chemical and physicochemical characteristics, use of, in sample selection for calibration, 134
Chemical constituents
  effects of, on comminution, 157
  imbalance of, examples of, 161
  influence of
    on physical nature of ground commodity, 157–158
    on wavelength selection, 157
  interactions between, 157
Chemical digestion, and digestion by rumen microorganisms, 171
Chemical loadings, definition of, 58
Chemical residual, 70, 72
Chemometric analysis, 180–181
Chickpea (garbanzo)
  breeding programs for, use of NIR in, 208
  variation in nitrogen-to-protein conversion factor in, 161
Chlorophyll in whole rapeseed, determination of, by NIR, 239
Chopper wheel, use of, in Grain Analysis Computer Model III, 120–121
Christy-Norris hammer mill, use of, in two-stage grinding, 160
Christy-Norris laboratory mill, use of, in sample preparation of dried grasses for NIR analysis, 196
Circular variable interference filters, 95
cis- Unsaturation, determination of, by NIR, 204
Clean wool, NIR analysis of, 227–228, 230
Cocoa beans, liquor, and powder, determination of oil and moisture in, by NIR, 208
Coconut and copra industries, use of NIR in, 224
Coefficient of correlation, 38, 43
  significance of, in slope/bias correction, 147, 148
  use of, in interpretation of standard error of performance, 147, 148
Coefficient of determination
  definition of, 38, 147
  significance of, 148
Coefficient of variability, use of, in interpretation of standard error of performance, 146
Coffee, quality control of, use of NIR in, 224
Collimating lens, use of, in near-infrared instruments, 121
Collimation, and light source, 90
Color, origin of, 1
Combination area, definition of, 107
Combination bands, definition of, 20
Commercial near-infrared instruments
  bench-type, operational principles of, 130–136
  definition of, 107
  list of, 108
Comminution of wheat, 213
Commodities of animal origin
  sources of variance in, 162
  storage of, 162
COMP/SPEC computerized spectrophotometer, 89
  general design of, 97–101
  scanning range of, 98
Components of NIR spectrophotometers, nonlinearity in, 89
Composition of indigestible residue of warm- and cool-season grasses, 169
Compound feeds, use of NIR in analysis of, 195, 196
Compressed sulfur, as reflectance standard for NIR, 112, 233
Computer. *See* Auxiliary computers; COMP/SPEC computerized spectrophotometer; Computerized spectrophotometer
Computerized spectrophotometer. *See also* COMP/SPEC computerized spectrophotometer
  applications of, 130
  calibration of, factors involved in, 230
  sample presentation to, methods of, 235
  software for, 234, 235
  specifications of, 233–234
  steps in operation of, 136–141
  use of, for NIR analysis of grass silages and compound feeds, 196
Constituent, definition of, 107
Constraining of regression, use of, 46
Contaminants in wool, determination of, by NIR, 227–228
Conversion factors, influence of, on NIR analysis, 161
Cooking methods, as variable in NIR analysis, 161
Cooling of lead-sulfide detectors, influence of, on efficiency, 233
Copra industry, use of NIR in, 224
Corn
  high-moisture, moisture in whole kernels of, determination of, by NIR, 203, 239
  moisture in individual whole kernels of, determination of, by near-infrared transmittance, 203
  oil in, determination of, by NIR, 203
  protein in, determination of, by NIR, 203
Corn feed, determination of protein, moisture, and oil in, by NIR, 203
Corn gluten, determination of protein, moisture, and oil in, by NIR, 203
Correlation peak, influence of, on accuracy and stability of NIR calibrations, 127
Correlation plot
  influence of quotient treatment on, 127
  interpretation of, 127, 152
  use of
    in calibration, 137
    in evaluation of regression in NIR calibration, 43–45
    in evaluation of reflectance data, 236, 238
Correlograms, 236
Cosmic rays, occurrence of, 17
Cream, analysis of, by NIR, 196
Cross-polarization/magic-angle spinning carbon-13 nuclear magnetic resonance, application of, to spectral analysis of grasses, 179
Cross-talk noise, definition of, 98
Crude fiber, determination of, by NIR
  in barley, 195
  in wheat and rice plants, 218
Curve fit coefficients, definition of, 53
Curve fitting, use of, in NIR technology, 113

Dark-surface reflectance, as source of nonlinearity, 151
Data compression
  benefits of, 58
  definition of, 57–58
  methods of, 50, 52–53
Data treatment, 37
Degradation of starches, determination of degree of, by NIR, 197
Degrees of freedom, and fundamental absorbances, 19
$\Delta \log (1/R)$, use of, in NIR analysis, 113
$\Delta R/R$ algorithm, use of, in Automated Digital Analyzer, 117
Demodulation of preamplifier output, method of, 99
Dependent variables, definition of, 144
Derivative
  formulae for, differences between methods of expression of, 48
  limitations of, 48
  methods of calculation of, 48
  objectives of, 47
  size of, 154

Derivative mathematics, use of, in log ($1/R$) instruments by interfacing to auxiliary computer, 128
Derivatization, influence of, on reducing interference, 61
Detection, definition of, 37
Detectivity, 97
Detectors. *See also* Lead-sulfide detectors; Silicon detector
  characteristics of, 96–97
  operating ranges of, in COMP/SPEC, 98
  positioning of, importance of, 12
  preamplifier output of, format of, 99
Determination, definition of, 36, 151
Dextroses, analysis of, by NIR, 197
Diatomaceous earth, use of, in grinding oilseeds, 163
DICKEY-john Grain Analysis Computer
  Model I, 107, 114–115
  Model II, 115
  Model III, 120–122
Difference curves, use of, in wavelength selection, 112, 242–243
Differential calibration, 134
Differential hydration of constituents, definition of, 159
Differential light-to-dark cycles, use of, in near-infrared instrumentation, 128
Differential moisture absorption, as source of error in chemical analysis, 181
Differential response of plant species to chemical reagents, 171, 180
Diffuse reflectance
  definition of, 1, 4, 108
  ideal, 11
  measurement of, 10–13
  standards for, 11
Diffusion of light, illustration of, 4–7
Digestibility of feeds and forages, determination of, by NIR, 195, 196, 205
Digital computation of NIR data, use of, in Automated Digital Analyzer, 117
Digital interface for COMP/SPEC, 99–100
Digital microprocessors, use of, in near-infrared instrumentation, 119
Dipoles, definition of, 21
Direct calibration, definition of, 59
Discrete filter, use of, in wavelength generation, 112, 113
Discriminant analysis
  of forage and feed components, to detect defective spectra, 196
  principles of, 244
  use of, 52, 140, 208
  of wheat
    for baking quality, using NIR, 221
    for identification of varieties, 189, 244
Disease resistance in wheat, determination of, by NIR, 194
Double-beam irradiation, 11–12, 111
Drawer-type keyboard, use of, in near-infrared instrumentation, 119
Dried apples, determination of sugars in, by NIR, 203
Dried gluten, analysis of, by NIR, 231
Dried molasses, sugar bands in, 241
Dry cake mixes, determination of fat and sucrose in, by NIR, 193
Dry matter digestibility of forages, determination of, by NIR, 229, 230
Dryer type, influence of, on NIR analysis of milk powder, 229
Dunne effect, as source of error in calibration, 131–132
Dust, accumulation of, in near-infrared instruments, 155, 166

Echelette-grating dispersing elements, use of, in computerized spectrophotometer, 233
Edible-oil industry, use of NIR in quality control in, 226
Efficiency of dialysis, determination of, by NIR, 242
Eggs in pasta, determination of, 197
Electromagnetic spectrum, 17
Electron microscopy, use of, in digestibility studies, 171
Electron spin resonance, 17
Elementary spectra, definition of, 53
Ellipsoidal mirror, use of, in focusing light source in COMP/SPEC, 98
Empirical reference methods, use of, in NIR calibrations for fibrous materials, 169
Encoder, use of, in near-infrared instruments, 114, 115
Endosperm modification in barley, determination of, by NIR, 231
Energy
  available to monochromator, determination of, 92
  interchange of, factors affecting, 109
  levels of, definition of, 18
  output, of monochromator, factors involved in, 91
Enumerative (closed) calibration, 131
EPROM (erasable programable read-only memory) modification of tilting-filter NIR instrument, to increase derivative segment width, 177
Errors
  in composition determination, influence of moisture on, 242
  detection of, methods for, 83
  in laboratory analysis, 143
  in oil determination in wheat, by NIR, reasons for, 217–218
  relative significance of, 143
  sources of, in NIR testing, 42, 148
Ether and ester substituents in starch derivatives, determination of, by NIR, 203
Extended Beer's model, use of, in calibration, 68
External validation, method of, 69–70
Extraction of fiber, definition of, 178

f/Number, 89–90
F-value exclusion, use of, in NIR calibration, 134–135
Faba bean
  breeding programs for, use of NIR in, 208
  difficulty in grinding of, 160
Factor analysis, definition of, 53
Factory calibration, use of, of near-infrared instrumentation, 124
Fail-safe microswitch, use of, in Automated Digital Analyzer, 118
Falling Number KT 3100 mill, use of, for sample preparation for NIR analysis, 186
Fast correlation transform, 51
Fat
  determination of
    in meats and meat products, by NIR, 197, 207, 226
    in milk powder, by NIR, 228, 229
    in raw milk, by near-infrared transmittance, 207
  influence of, on water absorption band, 207
  unsaturation of, determination of degree of, by infrared spectroscopy, 203–205
Fatty acids, spectra of, 22
Feathermeal, protein, water, and oil bands in, 241
Fecal analysis for protein and lignin, by NIR, 195
Feeds. *See also* Forage
  analysis of, use of NIR technology in, 223, 224, 225
  interactions between ingredients of, effects of, on NIR analysis, 195
Fermi resonance, definition of, 20
Fiber, crude, in breakfast cereals, determination of, 207
Fiber optics extension, use of, in remote scanning by near-infrared instrumentation, 128, 130
Fiberglass disk NIR technology, 139
Fibrous components
  of forages, determination of, by NIR, 205
  influence of
    on grinding and bulk density, 158
    on precision of NIR analysis, 158
  interaction of, with static electricity, 158
Filter crossing, steps to avoid, 141
Filters, narrow-band interference, characteristics of, 94–95. *See also* Discrete filter; Interference filters; Tilting filter
Finger millet, determination of protein and moisture in, by NIR, 204
Finite-difference derivative, 46, 48
First derivative
  of log ($1/R$)
    algorithm for, use of, in NIR technology, 113
    use of, in analysis of wool by NIR, 230
  and second derivative, comparison of 47
First-derivative spectra, and second-derivative spectra, comparison of, 241
First-generation near-infrared instruments, 114–119
Fishmeal, protein and oil bands in, 241
Fixed-filter instruments, limitations of, 95
Flaxseed, errors in testing for protein in, 214
Flexible disks, use of, for data storage in near-infrared instrumentation, 124
Flour
  ash in, determination of, by NIR, 202, 207, 225
  grade color of, determination of, by NIR, 190
  improvers in, analysis of, using NIR, 192
  water absorption by, prediction of, using NIR, 192
Flour milling
  quality control in, use of NIR in, 191–192, 223, 224
  stream blending in, use of NIR in, 191
  yield of, determination of, by NIR, 230
Flue-cured tobacco, analysis of, by NIR, 206
Fluid and dry milk powder, analysis of, by NIR, 196
Foam rubber compression lid, for NIR sample cell, design of, 118
Focal ratio, definition of, 89
Food substances, spectra of, 22
Forage
  analysis of, use of NIR in, 195–196, 215–216, 223, 224, 225
  chemical composition and nutritional value of, prediction of, by NIR, 205
  fractions of, classification of, 170
  preservatives added for ensiling of, prediction of need for, by NIR, 195
  quality of, definition of, 205
  quality control of, use of NIR in, 224, 225

Force constant, definition of, 21
Forward stepwise multiple linear regression, 45, 138
Fourier analysis, use of, in monitoring instrument performance, 102–103
Fourier coefficients, 49, 50
Fourier data compression, use of, in conserving computer space, 50
Fourier spectrum, definition of, 49–50
Fourier transform
  applications of, 48, 50–51
  definition of, 49
  notation for, 54
  power spectrum of, definition of, 49–50
  regression by, as calibration method, 63–64
Fourth derivative of log $(1/R)$, use of, in NIR technology, 113
Foxtail millet, determination of protein and moisture in, by NIR, 204
Free amino acids
  differential response of, to NIR and chemical reagents, 180
  in green tea, determination of, by NIR, 225
Free starch in corn fiber residues, determination of, by NIR, 203
Freeze-dried material, determination of purity of, by NIR, 241–242
Frequency, definition of, 17
Fresnel equations, 2
Fresnel reflectance, 2
Frost, influence of, on physical texture of wheat, 160
Frozen green forages, analysis of, by NIR, 195
Fructose syrups, quality control of, by NIR, 226
Fruits and vegetables, analysis of, by NIR, 197
Full-spectrum calibration, definition of, 57
Functional groupings, influence of, on spectra, 22
Fundamental absorption, definition of, 18
  number of bands of, 20

Gain, definition of, 150
Gain setting of instrument, and nonlinearity, 150
Gap size, definition of, 113
Gaussian assembly of samples for calibration, advantage of, 133
Gelatin in meat, determination of, by NIR, 206
Generalized least-squares method, use of, in calibration, 69
Glucoarabinoxylans, in hemicellulose, detected by carbon-13 nuclear magnetic resonance, 173–174
Glucose content of barley, determination of, by NIR, 194
Glucosinolates in whole rapeseed, determination of, by NIR, 239
Gluten protein and fat in maize, determination of, by NIR, 196
Gold, as reflectance standard
  for diffuse reflectance, 107, 111
  for NIR, 233
Grain
  fiber in, determination of, by NIR, 169
  texture of, influence of, on NIR analysis, 185–186
Grass/legume forage mixture, NIR analysis of, 195
Gratings, types of, in near-infrared instruments, 92, 126
Gravity-fed sample cell, use of, in near-infrared instrumentation, 123
Grazed forages, prediction of intake of, by NIR, 195
Grazed pasture grass analysis, use of NIR in, 196
Greasy wool, determination of wool base of, by NIR, 228
Green tea, determination of total nitrogen in, by NIR, 225
Grinders, 160, 163
  condition of, influence of, on NIR analysis, 163–164
  maintenance of, frequency of, 163–164
  for wheat, comparison of, 213
Grinding
  fine, influence of, on analysis of forage by NIR, 229
  methods of, 163
  normal, definition of, 163
  resistance to, in wheat, calibration for, 189
  two-stage, 160, 163
  as variable in NIR analysis of oilseeds, 204
Gross composition, determination of, by NIR, 241
Ground beef, determination of fat, protein, and moisture in, by NIR, 207
Ground loops, as cause of spikes in power supply, 156
Ground state of molecules, definition of, 17, 18
Ground wheat, as reflectance standard in NIR analysis, 112
Growing environment, influence of, on physical texture, 160
Growing locality and growth stage, as variables in calibration for grass and legume forages, 195
Growing season, effect of, on wheat analysis, 239
Growth regulators, prediction of effects of, on vegetables, by NIR, 221

Hard and soft wheats, separate calibrations needed for, 186
Hard disk, use of, for data manipulation in NIR technology, 126
Hardness of wheat
  determination of, by NIR, 202, 230
  particle size index in testing for, 202
  and pearling resistance, calibration for, 189
Heat-treating unit, for soybeans, use of NIR spectra in development of, 206
Hemicellulose, and lignin, 173
High-fructose corn syrup, analysis of, by NIR, 203
High-pass cutoff filter, use of, in elimination of overlapping orders, 92
High-pressure liquid chromatography, application of, in study of fibrous constituents, 172, 175
Higher-order derivatives, characteristics of, 47–48
Holographic grating, 92, 126
Homogeneous and heterogeneous feed components, definition of, 195
Hops, NIR analysis of, 226
Horizontal paddle-wheel filter assembly, use of, in near-infrared instrumentation, 122
Hot water extract in barley malt, determination of, by NIR, 194, 231
Hruschka regression, 63
Hydration of complex molecules, degree of, effect of
  on NIR analysis, 159
  on vibration pattern, 19
Hydrocarbons, spectra of, 22
Hydrogen atom and absorption, 22
Hydrogen bond, definition of, 28
Hydrogen combined atoms, use of, in definition of structural forms, 29
Hydroperoxides in fatty acids, determination of, by NIR, 204
Hypsochromy, definition of, 22

Impurities, detection of, by NIR, 241–242
In vivo digestibility of forages, determination of, by NIR, 195–196
Independent variables, definition of, 144
Index of refraction
  definition of, 2
  effect of
    on reflectance, 5, 7–8
    on scatter coefficient, 13
  relative, of water and ground wheat, 151
  of tilting filters, 114
Indirect calibration, definition of, 59
Inductive noise, as cause of spikes in power supply, 156
Industrial applications of near-infrared technology, 141
Infrared, use of, in milk analysis, 207
Infrared light-emitting diodes, 12
Insects, influence of, on physical texture, 160
Instrument drift, 216
  and single-beam irradiation, 111
Instrument geometry, 110, 156
Instrument noise, 148, 149
Instrument normalization, 134
Instrument/sample interaction, example of, 153
Integrating sphere
  coatings of, methods of preparation of, 233
  in measurement of diffuse reflectance, 11
  in near-infrared instrumentation, 119, 123
Interamide hydrogen bonds in hydrogen-bonding solvents, study of, by NIR, 208
Intercept, adjustment of, prerequisites for, 146–148
Intercorrelation
  between constituents, 131, 134, 135
  of data, 60
Interfacing of spectrophotometers to computers, 97
Interference by chemical constituents, 109
Interference filters, use of
  in controlling stray light, 150
  with near-infrared light-emitting diodes, in GT-90, 125
  in NIR analysis, 109
  in transmittance measurement, 12
Interlaboratory NIR test errors, determination of, 222
  for fiber analysis, 178
Internal degrees of freedom of vibrations of polyatomic molecules, 19
Internal reflection, loss of measured reflectance caused by, 73
Internal validation, method of, 70
Intralaboratory NIR test errors, determination of, 222
Inverted optical system, in Automated Digital Analyzer, 117
Iodine value of milk fat, determination of, by infrared absorbance, 207
Irradiation, effect of, on molecules, 18
Iterative procedure, definition of, 44

K values, definition of, 116
Kernel size, effect of, on barley analysis, 239

Kjeldahl test for total nitrogen, main sources of error in, 145
Krasnozernaya wheat types, use of, in NIR calibrations, 213
Kubelka-Munk constant, and concentration, 10
Kubelka-Munk theory, 9–10
  treatment of near-infrared data by, 78–79, 113

Laboratory error, 39, 41–42
Labormim Infrapid
  model 31, 123
  model 61, 128
Lack of fit, detection of causes of, 70–71
Lactose in milk powders and casein, determination of, by NIR, 229
Lamb, determination of fat, protein, and moisture in, by NIR, 207
Latent (loading) variables, definition of, 53
Latent regression factors, definition of, 61
Latticed monochromator, use of, in near-infrared instrumentation, 128
Laver (seaweed), determination of protein in, by NIR, 225
Lead-sulfide detectors
  effective range of, 19
  location of, in near-infrared instruments, 110, 115, 121
  operating range of, 97
  temperature sensitivity of, 97, 149, 233
Least interference, concept of, 112
Legume forages, NIR analysis of, for protein and digestibility, 195
Lenses, properties of, 89, 90
Lentils, breeding programs for, use of NIR in, 208
Lettuce, determination of stage of maturity of, by NIR in greenhouses and field, 206
Leverage
  definition of, 70, 71–72
  use of, in outlier detection, 71, 77
Light scattering, 1, 7
Lignified plant cell walls, influence of, on potential energy of forages, 173
Lignin
  and digestibility of organic matter in dry matter, prediction of, by NIR, 196
  influence of, on comminution, 160
  mode of association with cell wall, influence of, on digestibility, 172
  structural differences in, influence of, on reaction to chemical reagents, 180
Limiting amino acids in barley and wheat, determination of, by NIR, 202–203
Line filters, use of, in power supply regulation, 156
Linear regression, examples of, 37
Linearity of Beer-Lambert law, factors affecting, 22
Lipid, starch, and protein in peas, determination of, by NIR, 194
Liquid nitrogen, use of
  in grinding of oilseeds, 163
  in sample preparation for NIR analysis, 194
Loading error, definition of, 40
Loaf volume, prediction of, by NIR, 192, 221
log $(1/R)$
  algorithm for, advantage of, 112
  explanation of, 35
  of plant constituents, illustration of, 115
Log reference values, definition of, 118
LT Analyzer Model 7000, 128
LT Quantum 1200, 129–130
Lysine, determination of, by NIR
  in barley, 194, 202
  in maize, 216

Macbeth Color Eye, use of, in wool analysis, 227
Mahalanobis distance, use of, in calibration, 131
Maize. *See also* Corn
  NIR analysis of, 196
Major functional groups, identification of, in NIR, 208
Malt, determination of protein in, by NIR, 202
Malt extract, determination of, by NIR, 203, 231
Malting and brewing industry, use of NIR in quality control in, 194
Malting barley, analysis of, by NIR, 193, 225–226
Malting quality of barley, prediction of, using NIR, 222
Master instrument, definition of, 134
Mathematical ball-milling, definition of, 52
Mathematical symbols, 53, 54
Mathematical treatments of log $(1/R)$ data, 44, 112–113, 153, 208
Matrix inversion, use of, in optimizing wavelength selection, 121
Mean particle size, factors affecting, 164
Measurement, definition of, 36, 151
Meat and meat products, analysis of, by NIR, 197, 206–207
Mechanical noise, 149
Metabolizable energy in grass and forage crops, determination of, by NIR, 195–196
Methionine in peas, determination of, by NIR, 208
Methyl groups, predominance of, in short-chain fatty acid NIR spectra, 204
Methylene groups, predominance of, in long-chain fatty acid NIR spectra, 204
MgO, as standard for visible reflectance, 233
Micrometer, definition of, 17
Micron, definition of, 17
Milk powder, NIR analysis of, 228–229
Millet, influence of species of, on NIR analysis, 204
Minced ham, determination of protein and moisture in, by NIR, 226
Mineral (ash) content of wool, determination of, by NIR, 227
Minimal adequate subset, in calibration, 188
Mironovskaya wheat, use of, in calibrations, 217
Mirrors, properties of, 89
Mixed feed, NIR analysis of, 224, 225
Modem telephone system, for monitoring near-infrared instruments, 124
Moisture
  in corn, economic implications of, 203
  detection of trace amounts of, by NIR, 242
  determination of
    in fertilizers, 221
    in flour, by NIR, 202, 225
    in lima beans, by near-infrared transmittance, 201
    in meat, by NIR, 197
    in meat pie ingredients, by NIR, 207
    in milk powder, by NIR, 228, 229
    by near-infrared transmittance, wavelengths for, 201
    by NIR, using rapid meter, factors affecting, 218
    in peanuts, by near-infrared transmittance, 201
    in wheat, by near-infrared transmittance, 201
    in wool, by NIR, 230
  increase of, during refrigerated storage, 164
  influence of
    on grinding of wheat, 159
    on light scattering, 6, 8
    on NIR calibration constants, 221
    on other constituents, 109
    on sample presentation, 109
    on wavelength shift caused by hydrogen bonding, 208
  loss of
    during grinding, NIR calibration for, 186
    in storage in metal containers, 164, 165
  range of, in foods, 109
Moisture meter, type EVI, NIR, 213, 218
Molds, influence of, on physical texture, 160
Molecular vibrations, in NIR analysis, 17
Molecular weight, and rotational frequencies, 23
Monosodium glutamate in green tea, determination of, by NIR, 225
Monochromatic light, stray light in, 150
Monochromator
  function of, 92–94
  power supply to, necessary features of, 91
  use of, in wavelength generation, 112
Multicollinearity, 57, 60
Multiple interference filter NIR instruments, application of, to forage analysis, 205
Multiple linear regression. *See also* Multiterm linear regression
  limitations of, 230
  stepwise, 61, 138
Multiple-range wavelength selection, 139, 140
Multiple reflectance, definition of, 109
Multiplicative scatter correction, method for, 73
Multiterm linear regression. *See also* Multiple linear regression
  calculation of, 46–47
  example of, 37
  properties of, 45–46
Multivariate equation, definition of, 57
Mutual irradiation, 1

Nanometer, definition of, 17
Narrow band-pass interference filters, use of, in near-infrared instruments, 113–114. *See also* Discrete filter; Tilting filter
Near-infrared analytical system, 114
Near-infrared index, as test for wheat hardness, 202
Near-infrared instruments, commercial, 107, 108, 114–130
  bench-type, operation of, 130–136
  comparison of, 214, 216
  distribution of, in Asia, 223
Near-infrared light-emitting diodes
  intermittent energizing of, 125
  use of
    in controlling stray light, 150
    in wavelength generation, 112, 125
Near-infrared loading weights, and established wavelengths, 79
Near-infrared loadings, definition of, 58, 60
Near-infrared region

advantages of, for quantitative analysis, 29
characteristics of, 177
Near-infrared residual, definition of, 70
Near-infrared technology
development of, 89
principles of, 107
use of, in breeding programs for wheat, 189–190
Near-infrared transmittance analysis, use of
for determination of protein and moisture in whole barley and whole wheat kernels, 238
in GT-90, 109, 125
Neotec Automated Digital Analyzer, 116–118
Neotec Feed Quality Analyzer. *See* Pacific Scientific/Neotec Feed Quality Analyzer
Neotec Grain Quality Analyzer
Model I, 107, 115–116
Model 21, 119
Model 31, 119, 228
Model 31EL, use of, in wool analysis, 228
Model 41, 119
Model 101, 122
Model 102, 122–123
Model 3000 on-line NIR analyzer, 123
Ultraspeed, 123
Neutral detergent fiber
in breakfast cereals, determination of, 207
in forage grass analysis, 170
NIR, applications of
in analysis
of cereal grains and oilseeds, 201
of corn and sorghum, whole kernels of, 203
of feeds, standardized, 196
of forages, 175–180, 229–230
of high-moisture whole wheat, barley, corn, and sorghum, 239
of milk powders, 228–229
of pea starch and lipids, bimodal effect in, 194
of rapeseed, whole, 239
of wheat, 186, 201, 202
in crop protein translocation studies, 209
in error detection in reference methods, 41–42
NIR analytical data, conversion of, to constant moisture basis by near-infrared instruments, 117, 121
NIR spectroradiometry, use of, in discrimination between field crops, 209
NIR spectroscopy
of fatty acids, 204
principles of, 107–114
Nitrogen, determination of
in barley, by NIR, 194
in forages, 216
in soils, by NIR, 224
Nitrogen status of cereal plants, prediction of, by NIR, 222
Nitrogen-to-protein conversion factor, as source of error in NIR analysis, 41, 161
Nitrogenous substances, assignment of absorbing wavelengths in, 208
Noise
reduction of, in computerized spectrophotometer, methods of, 233–234
removal of, from reflectance spectra, 236, 238
sources of, 149, 156
in spectra, 42, 149, 235–236
Noise equivalent power, 96–97
Nonfatty substance of cheese,
determination of water in, by NIR, 207
Nonlinearity, sources of, 150–151
Nonprotein nitrogen, influence of, on NIR analysis, 161
Normalization
of instruments, 134
as method for controlling particle-size differences, 236
of spectra, 52
Nuclear magnetic resonance, 17. *See also* Carbon-13 nuclear magnetic resonance
use of, in oilseed analysis, 204
Number of samples for calibration, 162, 217
Nylon/wool blends, determination of wool in, by NIR, 228

Oats, analysis of, by NIR, 195, 201, 203, 204
Objective analysis, definition of, 140
Offset lead-sulfide detector, use of, in near-infrared instruments, 121
Oil
determination of, by NIR
in cereals, 193
in oilseeds and oilseed meals, 193
in pea flour, 194
in rapeseed, 239
in soybean, 214
in wheat, 217
extrusion of, in grinding and cell loading, influence of, on NIR analysis, 158, 194
as pasting agent in ground commodities, 158
in soybean, dependence of error on range of, 214
and storage of ground samples, influence on, 158
structure of
determination of, by midinfrared spectroscopy, 204
and NIR response, 205
unsaturation of, determination of degree of, by infrared spectroscopy, 204–205
On-line application of NIR in milling and baking industries, 193, 202
Onions, direct determination of soluble solids in, by near-infrared transmittance, 206
Open sample cup, use of, in DICKEY-john Grain Analysis Computer Model I, 115
Optical correction
definition of, 57
effect of, in improving NIR analysis, 79
Optical density, 110
and absorbance, 21
Optical glass windows, use of, in near-infrared instrument sample cells, 115
Optical interface, definition of, 2
Optical systems, in commercial near-infrared instruments, 113–114
Optimum harvest time for forages, prediction of, using NIR, 195
Organic matter digestibility in forages preserved by different methods, determination of, by NIR, 195–196
Orthogonality, definition of, 60
Oscillating mirror, use of, in near-infrared instrumentation, 123
Outliers
detection of, 63, 70, 71–72, 76–77, 83
effects of, 133
Overfitting of data
definition of, 39, 59, 138
effect of, 77
in NIR analysis of whole wheat kernels, 238
symptoms of, 39
Overlapping spectral orders, definition of, 92
Overtone area, definition of, 107
Overtones, 18, 31

Pacific Scientific/Neotec Dynascan on-line NIR analyzer, 124
Pacific Scientific/Neotec Feed Quality Analyzer
Model 51, 119–120
Model 51A, 124
Model 7000, 124
Pacific Scientific/Neotec Grain Quality Analyzer. *See* Neotec Grain Quality Analyzer
Pacific Scientific/Neotec Model 3000 on-line NIR analyzer, 123
Pacific Scientific/Neotec Research Composition Analyzer, 126–128
Pacific Scientific/Neotec Ultraspeed, 123
Packaged desserts, determination of moisture, protein, and sugar in, by NIR, 207
Packing density, influence of, on NIR determination of moisture in whole wheat and barley, 239
Paddle-wheel filter arrangement, use of, in near-infrared instruments, 115, 122
Paper envelopes, changes in moisture content of samples stored in, 164
Paprika, analysis of, by NIR, 197
Parameter, definition of, 107
Parenteral products, analysis of, by NIR, 208
Partial least-squares regression, use of, in calibration, 65–67, 75–76, 113
truncated, 80–82
Particle shape, influence of, on NIR analysis, 164
Particle size
distribution of, factors affecting, 164
effect of, on wheat spectra, 35
of ground commodities, 158–159, 160
mean, factors affecting, 164
and nonlinearity, 151
as reference in log $(1/R)$ algorithm, 138
Pasta, analysis of, by NIR, 197
Pasture hay, analysis of, by NIR, 229–230
Path length of radiation, difference between absorption and reflectance spectroscopy, 2
PDP 11 computer, use of, 234–235, 236
Peaches, determination of stage of maturity of, by NIR and visible transmittance, 206
Peak energy emission of lamps, 150
Peak-to-peak noise, in COMP/SPEC, 101–102
Pearl millet, determination of protein and moisture in, by NIR, 204
Pears, determination of stage of maturity of, by NIR and visible transmittance, 206
Peat, NIR analysis of, 197
Penetration of radiation into sample, factors affecting, 9
Percon Inframatic
model 8100, 124
model 8120, 123
Permacell
definition of, 145
use of, in NIR analysis, 112
in determination of repeatability, 121, 145
Phenolic antioxidants, determination of, by NIR, 205
Phenylpropanoid composition of lignins,

determination of, by carbon-13 nuclear magnetic resonance, 174
Photometric noise, determination of, in COMP/SPEC, 101
Photon
 definition of, 17
 energy of
  absorption of, by molecules, 18
  and wavelength, 2
Photon detectors, 96
Photoswitch, use of, in Automated Digital Analyzer, 118
Physical texture of samples, 160
Physics of NIR, 1
Planck's constant, 17
Plastic bags for sample storage, limitations of, 165
Polycarbonate cell, use of, in near-infrared instrumentation, 126
Polyperfluoroethylene, as reflectance standard for NIR, 233
Polystyrene, as standard for checking wavelength accuracy and precision, 137
Poppyseed, errors in testing for protein in, 214
Pork, determination of protein, fat, and moisture in, by NIR, 207
Power supply
 to monochromator, necessary features of, 91
 variations in, influence of, on NIR analysis, 156
Precision
 of bench instruments, example of, 149
 and constituent, influence of, 145, 158
 definition of, 39, 136, 143
 determination of, 143
 establishment of, steps in, 144–145
 guidelines for, 136
 importance of, in accuracy evaluation, 144
 improvement of, 121, 146
 of NIR analysis of wheat, 214–215, 216
 relative, of NIR and standard analysis for protein, 121
Prediction, definition of, 37
Preservation of ground samples, 164–165
Preservatives added for ensiling of forages, prediction of need for, by NIR, 195
Principal components, definition of, 53
Principal components regression, in calibration, 64, 75
Prism, use of
 in controlling stray light, 150
 with grating monochromator, 93
Processed food industry, importance of, in Japan, 223
Propylene oxide and ethylene oxide in starch products, determination of, by NIR, 203
Proso millet, determination of protein and moisture in, by NIR, 204
Protein
 and comminution of wheat, influence on, 157–158
 determination of, by NIR
  in barley, 231
  in dry gluten, 220
  in flour, 202, 225
  in forages, 205, 29, 230
  in milk, precision of, 229
  in meat, 197
  in meat pie ingredients, 207
  in peas, 194
  in rapeseed, 239
  in wheat, 214
 effect of, on wheat spectra, 35
 prediction of, by NIR analysis of vegetable-stage plant material, in corn and wheat, 209
 in wheat, dependence of error on range of, 214
Protein segregation of wheat, use of NIR in, 230
Pulse point, definition of, 115
Pulses
 breeding programs for, use of NIR in, 224
 use of, in livestock feeds, 206
Push-button operation, of near-infrared instruments, 119, 121, 122, 131
Pyrex, optical properties of, 89

Quadratic equations, use of, in computing calibration constant, 228
Quality control of liquid milk, on-line use of NIR in, 197
Quark, analysis of, by NIR, 196
Quartz cell windows, static electricity buildup in, 155, 165, 166
Quotient mathematics
 in calibration, 133
 in near-infrared instrumentation, 120, 122
 principle of, 121, 133
 use of, 47

Radiation in solid materials, illustration of, 5–7
Radiation source, in near-infrared spectroscopy, 91
Radio frequency noise, as cause of spikes in power supply, 156
Random errors, definition of, 143
Rapeseed
 chlorophyll in, determination of, by NIR, 239
 moisture in, determination of, by NIR, 201
 oil in, determination of, by NIR, 194, 201, 204, 239
 protein in, determination of, by NIR, 201, 204
Raw milk, determination of lactose, protein, fat, and total solids in, by NIR, 226
Ray tracing, use of, in study of reflectance, 1–2
Reciprocal dispersion, 93
Red pepper, determination of capsaicin in, by NIR, 225
Reduced mass, definition of, 21
Reference method
 data from, and NIR data, correlation between, 38
 definition of, 37
Reference wavelength, 45
 at 1,680 nm, explanation of, 115
Reflectance
 decreased, of whole kernels, methods for compensation for, 237
 definition of, 109
 diffuse. *See* Diffuse reflectance
 Fresnel, 2, 107–108
 illustration of, 2
 regular, definition of, 107–108
 relative, definition of, 233
 specular, definition of, 107–108
 and transmittance, 9, 21, 109
 for whole wheat NIR analysis, advantages of, 237–238
Reflectance spectroscopy, principles of, 107–114
Reflectance standards, 111–112
Reflected energy, of ground and whole wheat kernels, difference between, 237
Refractive index. *See* Index of refraction
Regression coefficient
 definition of, 43
 large, effects of, 138, 151
Regression constant, definition of, 43
Regression factors, number of, methods for selection of, 65, 236
Relative aperture, definition of, 89
Relative refractive index, of water and ground wheat, 151
Removable cartridge, for programming InfraAlyzer Model 200, 124
Repeatability, 121, 143
RER (ratio of standard error of performance to range of standard data), definition of, 147
Residual error, definition of, 43
Residual grease of wool, determination of, by NIR, 228
Response linearization, definition of, 57, 73
Reverse scanning, control of, in COMP/SPEC, 100
Ribonuclease, structural studies of, in near-infrared region, 208
Rice
 breeding of, use of NIR in, 224
 genotypes of, with improved lysine content, detection of, with NIR, 224
 milling of, prediction of degree of, by NIR lipid determination, 203–204
Rice starch, determination of gelatinization of, by NIR, 224
Ridge regression, 60
Rigid-plate sample presentation, use of, in near-infrared instruments, 122
RMS noise, definition of, 102
Rollaway shelves, use of, for sample storage, 164
Root mean square
 of cross-validation, in validation, 70
 of estimation, 70
 of prediction, definition of, 69–70
Root mean square deviation, 39
Rotating sample cell, use of, in Grain Analysis Computer Model III, 120
Rotational energy, 18
Rotational frequencies, and molecular weight, 23
Rotational motion, origin of, 19
RPD (ratio of standard error of performance to standard deviation), 135, 147
Ruled plane grating, 92
Rumen digestible organic matter, compared with NIR, for estimation of in vivo metabolizable energy in forages, 195–196
Rumen microorganisms, in digestibility studies, 171
Running mean, definition of, 42

Saccharification products, NIR analysis of, 197
Sake, determination of alcohol, acid, sugar, and amino acids in, by NIR, 226
Salad cream, determination of oil and egg in, by NIR, 197
Salinity, influence of, on reflectance from leaves, 2
Sample-and-hold modules, in COMP/SPEC, 99
Sample cell
 loading (packing) of
  error in, determination of, 217
  factors affecting, 159, 165
  influence of, on NIR analysis, 165
 overfilling of, steps to avoid, 165
Sample sensitivity of regression

coefficients, 151
Sample-to-sample contamination, steps to avoid, 165–166
Samples. *See also* Sampling
abnormal, detection of, 71
accumulation of, difficulties in, 131
aging of, 137
constituents of, intercorrelation between, 188
containers for, 164, 165
contamination of, 165–166
Gaussian assembly of, for calibration, 133
and instrument, interaction of, 151, 153
loading of. *See* Sample cell, loading (packing) of
mixing of, before loading, 165
number of, for calibration, 162, 217
for homogeneous and heterogeneous feed components, 195
presentation of
to computerized spectrophotometer, methods for, 235
of oilseeds, methods of, to minimize moisture contamination, 242
specialized systems of, for powders, slurries, and liquids, 124
reflectance of, and nonlinearity, 150–151
selection of
by chemical and physicochemical characteristics, for calibration, 134
methods of, for forage analysis, 175
principles of, 162
by spectral characteristics, 133–134, 152–153
and variance in commodity, 162
size of
examples of, for different operations, 46
and sample error, 40
small, qualitative examination of, 241
storage of, 158, 164–165
surface area of, influence of, on NIR analysis, 158
temperature of, effect of, on analytical results, 160–161, 233
texture of, factors affecting, 157–161
thickness of, influence of, on determination of moisture by near-infrared transmittance, 201
unknown, 137
variance in, sources of, 131
Sampling, 162–163
error in, 40–41
systems for, 163
Sand-blasted gold, as reflectance standard for diffuse reflectance, 107, 111
for NIR, 233
Saturation of detector/amplifier signal, and nonlinearity, 150–151
Saunderson theory, 73
Savitzky-Golay smoothing, 42–43, 48
Scanning electron microscopy, in chemical digestion studies, 171
Scanning NIR spectrophotometers, use of, in compositional analysis of mixed feeds and forages, 195
Scanning rate, control of, in COMP/SPEC, 100
Scatter, 3, 6–7
Scatter coefficient
of grains, ground and whole, 14
and index of refraction, 13
wavelength dependence of, 10, 13
Screw potentiometers, use of, in near-infrared instruments, 115
SDS sedimentation, determination of, by NIR, 190
Second derivative
of log (1/*R*)
in forage grass analysis, by NIR, 196
in NIR technology, 113
of plant constituents, illustration of, 115
in wool analysis, by NIR, 230
of spectra, 47–48
Second-generation near-infrared instruments, 119–124
Segment size, definition of, 153–154
Semihard cheeses, determination of fat, protein, and moisture in, by NIR, 207
Signal-to-noise ratio, definition of, 111, 148
Silages, detection of changes in spectral patterns of, by NIR, 196
Silicon detector
in diffuse reflectance and transmittance measurements, 12
effective range of, 19
in GT-90, 125
in very-near-infrared technology, 126
Simple calibration, 133
Simulation of bench instruments in wavelength selection, 139
Single-beam irradiation, 111
advantages of, 98
Single-term calibration
advantages of, 138
use of, in near-infrared instrumentation, 122
Single-term linear regression, 37, 43–44
Skim milk powder, NIR analysis of, 228–229
Slave instrument, definition of, 134
Slope, definition of, 43
Slope/bias adjustment, 135, 147
automatic, in near-infrared instruments, 126
Small samples, qualitative examination of, 241
Smoothing, 67
and analytical efficiency, 80
by Fourier transform, 50
instrumental method for, 154–155
of NIR spectral data, effects of, 42
of reflectance data, method for, 236
Snack foods, determination of fiber, starch, and total carbohydrate of, by NIR, 207
Snell's law, definition of, 2
Sodium chloride in cured and salted fresh ham and ground beef, determination of, by NIR, 207
Software for COMP/SPEC, 103–104
Soil, determination of water content of, by NIR, 209
Solids in dairy products, determination of, by NIR, 197
Solids-not-fat, determination of
in dairy products, by NIR, 197
in raw milk, by near-infrared transmittance, 207
Soluble β-glucan in barley, determination of, by NIR, 231
Solvent-extractable material in wool, determination of, by NIR, 227, 230
Sophistication of computerized spectrophotometers, definition of, 113
Sorghum
ash in, determination of, by NIR, 204
growing location of, influence of, on NIR analysis, 204
high-moisture, moisture in whole kernels of, determination of, by NIR, 203
moisture in, determination of, by NIR, 204, 239
protein in, determination of, by NIR, 204
Soups and soup bases, determination of moisture and protein in, by NIR, 207
Sources of variance in samples, 131
Soybean
breeding of, use of NIR in, 223
moisture in
determination of, by near-infrared transmittance, 206
NIR analysis for, 201
oil in, determination of, by NIR, 201, 203, 204, 206, 225
protein in
determination of, by NIR, 201, 203, 204, 206, 225
errors in testing for, 214
denaturation of, determination of, by NIR, 225
wet processing of, use of NIR in quality control of, 206
Species specificity of NIR filters for forage analysis, 195
Specificity of wavelengths for constituents, verification of, 158
Spectra
of alcohols, homologous series of, 22
and bandwidth, 241
elementary, definition of, 53
factors affecting, 22
of fatty acids, 22
files for, organization of, in COMP/SPEC, 103–104
first-derivative and second-derivative, comparison of, 241
of food substances, and composition, 22
of hydrocarbons, 22
interpretation of, 26
noise in, 42, 149, 236, 238
normalization of, 52
origin of, 21–22
smoothed and unsmoothed, comparison of, 42
of water, 35
of wheat, 35
Spectral characteristics, in sample selection for calibration, 133–134
Spectroscopic symbols, 53
Spectrum reconstruction, 51
Specular reflectance, 2, 107–108
loss of measured reflectance caused by, 73
Spelt wheat, analysis of, by NIR, 195
Spex mill, use of, in NIR analysis of forage, 229
Spikes in power supply, influence of, on NIR analysis, 156
Spring and winter barleys, NIR analysis of, 194
Stability of calibrations
definition of, 130
methods for improving, 152
Stage of maturity, influence of, on physical texture, 160
Standard and selected filters, comparison of, for NIR determination of in vivo and in vitro digestibility, 195
Standard deviation of original data, significance of, in interpretation of standard error of performance, 148
Standard error of calibration, definition of, 38
Standard error of estimate, definition of, 43
Standard error of performance
definition of, 38, 143
interpretation of, 146–147, 148
theoretical limit of, for determination of

protein in wheat by NIR, 146
in validation, 70
Standard error of prediction, definition of, 38, 143. *See also* Standard error of performance
Standard error of single test, definition of, 41
Standard laboratory analysis, accuracy of, in NIR analysis, 162
Standard methods of analysis, 37
recommended procedures for, 152, 153
Starch
damage of, in flour, determination of, by NIR, 191–192
determination of, by NIR
in barley, 214
in peas, 194
in wheat, 217
NIR analysis of, 197
Starting wavelength, definition of, 113
Static electricity, influence of, on NIR analysis, 155, 158, 165
Statistical abbreviations, 54
Statistical calibration methods, 60–69, 82–83
Statistical formulas, 54
Statistical symbols and abbreviations, 53–54, 58–59
Statistical terms in evaluation of NIR accuracy, 143–144
Step-down (backward) multiterm linear regression, definition of, 45
Step-up multiterm linear regression, definition of, 45
Stepping filter wheel, use of, in Grain Analysis Computer Model III, 120, 121
Stepping motor, use of, in computerized spectrophotometer, 98
Stepwise derivative linear regression, illustration of, 61
Stepwise multiple linear regression, 61, 138
Stone-fruit pulp, analysis of, by NIR, 197
Storage, of samples, 164–165
Stray light, 149–150
Studentized residuals, definition of, 71
Substitution in starches, determination of degree of, by NIR, 197
Succinic anhydride and acetic anhydride in starch products, determination of, by NIR, 203
Sugar
interference by, in determination of fiber in cereal foods, by NIR, 207
and protein in forages, ratio of, in prediction of need for preservatives for ensiling, use of NIR values in, 195
in wine, determination of, by NIR, 194
Suint in wool, determination of, by NIR, 228
Sulfur, as reflectance standard in NIR technology, 112
Sunflower seed, determination of oil in, by NIR, 204
Supporting wavelengths, 151
Suprasil-W fused silica, use of, in spectrophotometry, 233
Surface area of sample, influence of, on NIR analysis, 158
Surface reflectance, 2

t-Value exclusion, use of, in NIR calibration, 134–135
Tapioca starch, contamination of, with wheat starch, use of NIR in determination of, 230–231
Tecator Fibertec system, in fiber analysis, 178
Technicon/DICKEY-john InfraAlyzer
Model 2.5, 118–119
Technicon InfraAlyzer
Model 200, 124
Model 250, 124
Model 300, 124
Model 400, 123
Model 500, 128
Teff, determination of protein and moisture in, by NIR, 204
Teflon, as reflectance standard for NIR, 112, 115
Temperature
ambient
and NIR analysis, 161
and NIR instruments, 155–156
and comminution, 160
errors induced by, correction for, by calibration in GT-90, 125
and hydrogen bonding and hydration status, 161
and protein determination in whole wheat by near-infrared transmittance, 238
and wavelength, 161
and wheat flour testing by NIR, 186–187
Terminal and *cis*- unsaturation in aliphatic hydrocarbons, determination of, by NIR, 204–205
Terminal epoxides, determination of, by NIR, 205
Terylene/wool blends, determination of wool in, by NIR, 228
Thermal detectors, principles of, 96
Thermoelectric elements, for temperature control of near-infrared instruments, 97, 155
Thickness of material, influence of, on light scattering, 6–7
Third derivative of log $(1/R)$, in NIR technology, 113
Tilting filter
effective range of, 114
principles of, 95–96
and refractive index, 114
in wavelength generation 96, 112, 113–114
Tobacco
industrial applications of NIR in testing of, 206, 224
NIR analysis of, 197, 206
Tomatoes and tomato products, analysis of, by NIR, 197
Total β-glucan in barley, determination of, by NIR, 231
Total nitrogen, determination of, by NIR
in forages, 216
in soils, 224
Transferability of calibration constants, 133
Transflectance
definition of, 111
in liquid analysis by near-infrared instruments, 111, 128
and wavelength area, effect of, 111
through wheat, effect of layer thickness on, 237
Transmittance
and absorption, 8, 21
analysis by, in near-infrared instruments, 109, 125
definition of 8, 109
measurement of, 12
and reflectance, 9, 21, 109
Transmitted radiation, use of, in diffuse reflectance measurement, 12
Treatment specificity of NIR filters for forage analysis, 195
Trebor instruments
GT-70 and GT-7700, 129
GT-80 small-sample whole-grain analyzer, 126
GT-90 whole-grain analyzer, 125–126
GT-99, 126
Tropical and temperate grasses, separate NIR calibrations needed for, 195
True test error, 146
Truncation of NIR loading vectors, 80–82
Tungsten-halogen lamps, advantages of, 91
Two-piece sample cell, use of, in NIR chocolate analysis, 231
Two-point calibration, 122, 133–134
in determination of protein in wheat, 208
Two-stage grinding, 160, 163

Udy cyclone grinder, use of, for sample preparation for NIR analysis, 186
Underpowered lamp, for prolonged lamp life, 118–119
Universal calibration constants, 132–133
development of, 152
in USSR, 216
for forages, factors affecting development of, 195
Unknown samples, 137
Unsaturation of fat, determination of degree of, by NIR, 207
Updating of calibrations, 83–84
User-written software for computerized spectrophotometer, advantages and disadvantages of, 234

Validation, 38, 69–70
Variable (wedge) interference filters, construction of, 95
Variety and genotype, influence of, on physical texture, 160
Vegetable matter content in wool, determination of, by NIR, 227
Vegetables, analysis of, by NIR, 197
Verification, definition of, 38
Vertical sample-cell drawer, use of, in near-infrared instruments, 122
Very large regression coefficients, 138
Vibration, molecular
bending, definition of, 19
characteristic, 22
frequencies of, 17
origin of, 19
stretching, definition of, 19
types of, 19
Vitreousness of wheat, use of, as grading factor, 202
Volatile constituents, loss of, from fertilizers, effects of, on NIR moisture determination, 221
Voluntary intake of forages, determination of, by NIR, 229

Walnut, NIR analysis of, 207
Warm- and cool-season grasses, comparison of, 169
Washed wool, determination of wool content of, by NIR, 227–228
Water
absorption bands of, 27–29, 207, 226
absorption by, in near-infrared region, 94
content, of samples, and scatter properties and nonlinearity, 151
effect of, on wheat spectra, 35
fundamental absorbances of, 27
interference by, in nuclear magnetic resonance analysis of oilseeds, 204
spectrum of, 35
state of, in fish flesh, fresh fruit, milk, and soybean curd, determination of,

by NIR, 226
Wave number, 17
Wavelength point, definition of, 117, 121
Wavelength range display, expansion of, in near-infrared instrumentation, 112
Wavelength selection, for near-infrared analysis, 112
  accuracy level required for, 112
  and composition, influence of, 128
  errors in, 151–153
  methods of, 43–45
  operator-controlled, use of, in Automated Digital Analyzer, 117
  order of, and variance, 151
Wavelength term, definition of, 117
Wavelengths
  and absorption, 12
  of absorption bands, and chemical bonds, 22
  accuracy of, importance of, 102
  assignment of, 31, 34
  best pair of, 139, 152
  best triple, 139
  central, definition of, 113
  definition of, 17
  generation of, methods of, 112
  in InfraAlyzer 400, 123
  large numbers of, use of, in NIR forage analysis, 229
  for NIR analysis, commonly used, 157, 158
  and nonlinearity, 150
  and photon energy, 2, 17
  precision of, 102
  reference, definition of, 45
  and regression coefficient, 151
  and scatter coefficient, 10, 13
  selection of. *See* Wavelength selection
  starting, definition of, 113
  supporting, 151
  and temperature, effect of, 161
  of tilting filters, variability in, 96
  and transflectance, 111
  types of, for use in near-infrared technology, 112
  for water determination by NIR, influence of moisture content on, 226
  for wool analysis by NIR, 227, 228
Wedge interference filter
  construction of, 95
  use of, in near-infrared monochromator, 201
Weed seed, influence of, in determination of protein in whole wheat by near-infrared transmittance, 238
Weend procedure, for proximate analysis of fibrous foods, 170
Weighted least-squares estimator, use of, in calibration, 63
Weighting of NIR data, 60
Wet and dry gluten in wheat, determination of, by NIR, 219–221
Wheat
  constituents of, assignment of absorbing wavelengths in, 208
  grade of, influence of, in determination of protein in whole wheat, by near-infrared transmittance, 239
  hardness of, influence of, on log $(1/R)$ values, 186
  high-moisture, determination of moisture in, by NIR, 239
  middlings of, determination of protein, fat, and fiber in, by NIR, 196
  moisture in, determination of, by NIR, 201–202
  nitrogen-to-protein conversion factor for, variation in, 161
  protein in
    determination of, by NIR, 190–191, 201–202, 230
    errors in testing for, in cargoes, 41
    influence of season on NIR testing for, 187
  spectrum of, 35
  strength of, NIR in testing for, 202
  varieties of, identification of, by NIR discriminant analysis, 189, 244
Whey protein concentrates, NIR in quality control of, 229
Whole milk powder, NIR analysis of, 228–229
Willstätter-Stoll theory of critical reflection, 1
Winged bean, determination of protein and oil in, by NIR, 206
Wood's anomalies, 94
Wool
  analysis of, by NIR, 227–228
  dye on fibers of, determination of, by NIR, 208

Y-intercept, definition of, 43
Yogurt, analysis of, by NIR, 196

Zeleny sedimentation of wheat, determination of, by NIR, 190